Ecology of Soil Borne Plant Pathogens

NIPA® GENX ELECTRONIC RESOURCES & SOLUTIONS P. LTD.
New Delhi-110 034

About the Author

Dr. N.G. Ravichandra, Former Professor of Plant Pathology and Scheme Head, All India Co-ordinated Research Project (Nematodes), Department of Plant Pathology, University of Agricultural Sciences, GKVK, Bangalore, Karnataka, India, was involved in teaching, research and extension activities. He has guided the Post-Graduate students specialized in Plant Nematology for M.Sc., (Agri.) and Ph.D. Degree Programmes.

Dr. N.G. Ravichandra has authored 14 text books in Plant Pathology & Nematology, published 276 scientific research papers in National and International journals, contributed chapters to 15 text books, edited 6 books and also authored 10 books in Kannada language.

Dr. N.G. Ravichandra has identified and reported a new species of root-knot nematode (*Meloidogyne enterolobii*) on Guava and a new host, Lac plant (*Flemingia semialalta* L.) for root-knot nematode (*Meloidogyne incognita*) from Karnataka. These are the first reports from India.

Dr. N.G. Ravichandra has been bestowed with three National awards for his research accomplishments and contributions. He has been recognized as the International Editor and Reviewer for three prestigious International Journals and National Editor for the eminent 'Indian Phytopathology' Journal. He has developed nine technologies for the management of root-knot and burrowing nematodes infecting tomato, brinjal, rice and banana, which have been included in the Package of Practices of University of Agricultural Sciences, Bangalore, Karnataka, India.

Dr. N.G. Ravichandra has operated research projects funded by the Department of Bio-Technology & Indian Council of Agricultural Research, New Delhi, India as Principal Investigator and Co- Principal Investigator ; participated and presented original research papers on various aspects of plant diseases in National and International conferences; undergone advanced training programs sponsored by the Department of Biotechnology & Indian Council of Agricultural Research, New Delhi, India.

Dr. N.G. Ravichandra is an active life member of seven professional societies including Nematological Society of India, Indian Phytopathological Society, Society of Mycology & Plant Pathology, National Environmental Science Academy, Institute of Agricultural Technology, Glacier Journal Research Foundation and Institute of Scholars.

Scan for other Books by Same Author

Ecology of Soil Borne Plant Pathogens

Dr. N.G. Ravichandra
Former Professor of Plant Pathology
College of Agriculture
University of Agricultural Sciences
GKVK, Bangalore- 560 065
Karnataka, India

NIPA® GENX ELECTRONIC RESOURCES & SOLUTIONS P. LTD.
New Delhi-110 034

NIPA® GENX ELECTRONIC
RESOURCES & SOLUTIONS P. LTD.

101,103, Vikas Surya Plaza, CU Block
L.S.C.Market, Pitam Pura, New Delhi-110 034
Ph : +91 11 4386 0225, 9717133558, 9540816132
E-mail: newindiapublishingagency@gmail.com
Website: www.niparesources.com

Print ISBN: 978-93-58875-25-6

ebook ISBN: 978-93-58876-77-2

Composed and Designed by NIPA®.

To
Anusha Vijayakumar

Acknowledgements

I would like to express my deep sense of gratitude to my beloved parents late Sri. N. Gurushankar and Smt. Parvathamma for their divine blessings.

I am highly grateful to my in-laws late Sri. M.S. Chandrashekhar and Smt. Kamala for their continued affection and encouragement.

I wish to record my appreciation, affection and support to my beloved wife Ms. Deepthi and dearest daughter, Ms. Anusha Vijaykumar for their care, patience and keen interest throughout the preparation of the manuscript. I fondly acknowledge with affection, their constant encouragement.

My heartfelt thanks are due to my dear brother-in-law Mr. B. Sudesh Kumar and my beloved sister Ms. Tara, for their sustained moral support, affection and special care. I fondly acknowledge their keen interest and blessings being showered on me.

My special appreciation and thanks to M/s. NIPA® Genx Electronic Resources & Solutions P. Ltd., New Delhi, India, who deserves commendation for their professional contribution.

Preface

Soil is a complex, heterogeneous and dynamic natural system. Soil environment is important in ecology as it influences several processes in plants. Plant growth, abundance and distribution are naturally influenced by soil environment. The soil environment can have a significant impact on microbial habitat through its influence on water and gaseous movement in the soil system, which affects microbial diversity, activities and functions. It directly affects the microbes including pathogenic forms, as well as the rates of processes they perform. The degree of pathogenicity of soil-borne micro-organisms observed under laboratory conditions is not usually repeated in the field, where biological and non-biological influences are less controlled. Ecological study of soil borne plant pathogens, thus, includes the detailed information on major aspects of the soil, which helps in a proper understanding of the damage and yield loss caused by soil borne pathogens.

'Ecology of Soil-borne Plant Pathogens" is being taught as a core course in PG degree programmes at various universities, with a common syllabus. Not a single text book covering the prescribed syllabus is available. Thus, a dire need of such a textbook has been felt for quite some time for the benefit of students specializing in the subject. This book, therefore, has been designed to cover the course outline of "Ecology of Soil-borne Plant Pathogens", prescribed by the Indian Council of Agricultural Research, New Delhi, India.

The book comprises eight chapters covering the latest information on "Soil as an environment for plant pathogens, Nature and importance of rhizosphere and rhizoplane, Soil and root Inhabiting fungi, Types of biocontrol agents, Inoculum potential and density in relation to host and soil variables, Competition, Predation, Antibiosis and Fungistasis, Conducive and Suppressive soils and finally, Biological control-Concepts & potentialities of bioagents for managing soil borne pathogens". Appropriate diagrams, convincing tables and suitable graphs / illustrations have been furnished at the right places. A bibliography providing the list of references cited has also been included at the end of each chapter.

The contents of this book, reflecting an extensive literature search, will also be useful for the teaching, research and extension faculty in Agricultural and Horticultural Universities, the State Departments of Agriculture, Horticulture, Forestry, Sericulture & Fisheries, Plant Protection Organizations, Plant Quarantine Units, Administrators & Policy makers and all those who are interested and concerned with plant protection.

The author appreciates receiving suggestions and constructive criticism that would improve the quality of the book

Bangalore **N.G. Ravichandra**

Contents

Acknowledgements .... *ix*

Preface .... *xi*

1. Soil as an Environment for Plant Pathogens 1

1.1 The Soil Environment 2
1.2 Soil-microbes-plant Relationship 6
1.3 Survival of soil-borne inoculum 8
1.4 Physiological Adaptations and Soil Fungistasis 9
1.5 Dispersal of Soil-borne Inoculum 11
1.6 Classification of Soil Biota 13
1.7 Bacterial and Fungal Population in Indian Soils 17
1.8 Soil Microbes and Soil Structure 19
1.9 Soil Biota Regulates Detrimental Organisms and Processes 20
1.10 Soil Biota and Plant Pathogens 22
1.11 Effect of Agricultural Practices on Role of Soil and Soil Biota 24
1.12 Characteristics of Soil Microbial Communities 29
1.13 Soil Biodiversity and Root Pathogens in Agroecosystems 33
1.14 Trophic Webs and Complementary or Interference Interactions in Soil 34
1.15 Root pathogens- Soil borne Plant Pathogens 36
1.16 Soil Food Web and Resilience in Soils 37
1.17 Interactions of Soil Nutrient Environment, Pathogenesis and Host Resistance 39
1.18 Role of Soil Health in Plant Disease Management 40
1.19 The Soil Food Web and Nutrient Cycling 44
1.20 Soil as a Huge Laboratory for Microorganisms 47
1.21 Biochar –Its Impacts and Mechanisms on soil microorganisms 48
1.22 Grassroots Ecology- Plant Microbe Soil Interactions 51
1.23 Soil Conditions and Soil-borne Pathogens 53
Bibliography *81*

2. Nature and Importance of Rhizosphere and Rhizoplane 86

2.1 Roots and the Rhizosphere 86
2.2 Historical Background 89
2.3 Composition and Importance of the Rhizosphere 91
2.4 Rhizosphere Soil Environment 96

2.5 Rhizoplane ... 98
2.6 Host Exudates ... 111
2.7 Plant Exudates and Root Penetration ... 120
2.8 Plant Exudates and Mucilage Diversity ... 125
2.9 Haustorial mucilage ... 127
2.10 Plant Exudates and Soil ... 131
2.11 Complex Interactions in the Rhizobiome that Improve Plant Health ... 132
2.12 Plant-microbe Cross-talk and Rhizosphere Effect ... 134
2.13 Rhizosphere Microbiome ... 136
2.14 Rhizomicrobiome Signalling ... 145
2.15 The Rhizosphere Microbial Complex in Plant Health ... 151
2.16 Root Exudates ... 155
2.17 Dynamic Root Exudate Chemistry and Microbial Substrate ... 164
Bibliography ... *174*

3. Soil and Root Inhabiting Fungi ... 180
3.1 Soil Inhabiting Fungi ... 180
3.2 Ecological Plant-Microbe Interactions ... 188
3.3 Global Diversity and Geography of Soil Fungi ... 190
3.4 Global patterns in Endemicity and Vulnerability of Soil fungi ... 194
3.5 Soil Fungal Diversity and Assembly along a Xeric Stress Gradient ... 202
3.6 Global Diversity and Biogeography of Phytopathogenic Fungi in a Changing World ... 203
3.7 Fungal Communities in Soils under Global Change ... 206
3.8 Impact of Abiotic Conditions on Fungal Diversity ... 209
3.9 Potential Links between Wood-inhabiting and Soil Fungal Communities ... 211
3.10 Biodiversity of Soil-Inhabiting Fungi ... 215
3.11 Methods and Recent Achievements in Studies of Soil-Borne Fungi ... 216
3.12 Fungal Community Composition, Diversity, Soil Depth and Landscape Position ... 222
3.13 Phylotype Diversity Within Soil Fungal Functional Groups ... 224
3.14 Novel Clades of Fungi ... 225
3.15 Diversity of Soil Borne Plant Pathogens ... 228
3.16 Models for Studies of Soil Biodiversity and Its Impact on Soil Borne Fungal Pathogens ... 230
3.17 Root Inhabiting Fungi ... 236
3.18 Root-Inhabiting and Rhizosphere Mycobiomes ... 244
Bibliogrphy ... *258*

4. Types of Biocontrol Agents ... 263
4.1 Need for Biological Control, Merits and Demerits ... 264
4.2 Bacterial and Fungal Biocontrol Agents Against Plant Diseases ... 266
4.3 Factors Affecting Biocontrol of Plant Diseases and Selection of Potential BCAs ... 273

4.4 Challenges in Establishing Beneficial Microbes as BCAs.................275
4.5 Biotechnological and Omics Techniques for Biocontrol Strategy Improvement 277
4.6 Research Gaps and Future Direction...281
4.7 Present Status, Historical Perspective and Future Prospects of Bio-Agents in Agriculture...282
4.8 Research and Development in Biopesticide Based Pest Management285
4.9 Biocontrol Strategy- An Eco-smart Tool 288
4.10 Factors Affecting the Success of Bio-agents against Phytonematodes .293
4.11 Global Agriculture Biological Control Agents Market- Analysis and Insights 296
4.12 Modes of action of Biocontrol Agents...299
4.13 Identification and Development of New Microbial Strains 315
4.14 Types of Microbes used as Biocontrol Agents 316
4.15 Identification of New Potent Microbial Traits 329
4.16 Potential Postharvest Biological Control Agents...............................335
4.17 Virus-based Bioagents and Phage against Plant Diseases 338
4.18 Novel Biocontrol Approaches and Mode of Action 342
Bibliography...*345*

5. Inoculum Potential and Density in Relation to Host and Soil Variables 351
5.1 Important terms...351
5.2 Soil-borne Inoculum...353
5.3 The Biological Soil Environment and Inoculum Survival.................355
5.4 Impact of Inoculum Density of *Fusarium oxysporum* f. sp. *zingiberi* on yield...359
5.5 Influence of Soil-borne Inoculum of *Plasmodiophora brassicae* disease severity...360
5.6 Rapid Quantification of Soil Borne Pathogen Communities 362
5.7 Disease Potential of Soil- and Tuber Borne Inocula of Colletotrichum Coccodes and Black Dot Severity on Potato 363
5.8 Synergistic Effect of Organic and Inorganic Fertilization on the Soil Inoculum Density of the Soil Borne Fungal Pathogens.........364
5.9 Influence of Inoculum Density on Disease Development...................366
5.10 A Conceptual Framework to Reduce Inoculum Potential and the Effect of Bulk Density 368
5.11 Role of Inoculum Density and Incubation Period on Saprophytic Behaviour of *Pythium ultimum* and *Sclerotium rolfsii* and *Rhizoctonia* spp. 370
5.12 The Importance of Primary Inoculum to Crop Health, Food Security and Rice Sheath Blight Disease Development 373
5.13 Inoculum Levels on the Aggressiveness of *Fusarium oxysporum* f.sp. ciceris 375

5.14 Inoculum dose–Disease Response Relationships ... 377

5.15 Inoculum Density and Spatial Patterns of *Plasmodiophora brassicae* at the Field Level ... 380

5.16 Effects of Inoculum Density and Cultivar Susceptibility on Rhizoctonia Damping-off and Crown & Root rot ... 382

5.17 The effect of Aggregation of Pathogen and Biocontrol Microbe Inocula on Biocontrol Potential ... 384

5.18 Inoculum Potential of *Sclerotinia sclerotiorum* sclerotia ... 385

5.19 Inoculum Potential of *Fusarium* species ... 387

5.20 Evaluation of Inoculum Potential of Pathogens in Seeds ... 393

5.21 Effect of Inoculum Levels on Disease Development and Survival potential ... 394

5.22 Relationship between Infective Inoculum and Pathogenic Potential .. 395

5.23 Effect of Inoculum Load on Disease Initiation, Threshold Point and Population Build up ... 397

5.24 Effects of Inoculum Density and Substrate Type on Saprophytic Survival Of Fungi ... 399

5.25 Inoculum Density and Virulence-dependent Interactions ... 400

5.26 Effects of Inoculum Density on Disease Severity Caused by Pythiaceous fungi ... 402

5.27 Inoculum Density Relationships for Infection by Phytophthora ramorum, Rhizoctonia solani and Xanthomonas cucurbitae ... 403

5.28 Nematode Inoculum Density on Host Growth and Nematode Reproduction ... 407

5.29 Effect of Inoculum Densities on Seed Germination, Seedling Growth Parameters and Disease Incidence ... 411

5.30 Effect of inoculum density on the early reproduction and damage potential of the burrowing nematode, Radopholus similis ... 413

5.31 The effect of different initial densities of Meloidogyne javanica on the build-up of Pasteuria penetrans population ... 415

5.32 Synergies between inoculum size & heat disturbance and Primer inoculum sources ... 416

5.33 Bacterial Inoculum and Fungal Disease Development ... 418

5.34 Effect of Temperature on Wheat Streak Mosaic Disease Development in Winter Wheat ... 420

5.35 Inoculum potential of Polymyxa Betae, Beet Necrotic Yellow Vein Furovirus and Narcissus Mosaic Virus ... 421

Bibliography ... *422*

6. Competition, Predation, Antibiosis And Fungistasis ... 428

6.1 Competition ... 428

6.2 Bacterial Competition ... 434

6.3 Native Soil Borne Pathogens Equalize Differences in Competitive Ability ... 438

6.4 Microbial Interactions Promote Rhizosphere Colonization 441
6.5 Predation 450
6.6 Key Predatory Groups and their Direct Effects on the Soil Microbiome 463
6.7 Antibiosis 477
6.8 Fungistasis 492
Bibliography *509*

7. Conducive and Suppressive Soils 515
7.1 Conducive Soils 515
7.2 Suppressive Soils 521
7.3 Agricultural Management Regime and Disease Suppression 535
7.4 Antibiotics and Drugs from Soil Microbiome 538
7.5 Complexities & Climate Change and Disease Suppressing Soils 539
7.6 Current Insights into the Role of Rhizosphere Bacteria in Disease Suppressive Soils 542
7.7 Disease Suppressive Soils- New insights from the Soil Microbiome. 446
7.8 New Approaches for understanding Disease-Suppressive Microbiomes 549
7.9 Disease Suppression as a Tool for Disease Management 551
7.10 General Suppressive Soil vs. Specific Suppressive Soil and Characteristics 554
7.11 Examples and Case History Studies of Suppressive Soils 556
7.12 Characterizing the Microbial Components Responsible for Soil Suppressiveness 564
7.13 Functional Analysis of Key Populations Involved in Soil Suppressiveness 562
7.14 Microbial Community Composition in Take-all Suppressive Soils ... 564
7.15 Rhizosphere Suppression Hinders Antibiotic Resistance Gene (ARG) spread under Ralstonia solanacearum invasion 566
7.16 Microbial Diversity in Soils Suppressive to *Fusarium* Diseases 568
7.17 Suppression of Banana *Fusarium* wilt Disease with Soil Microbial Mechanisms 570
7.18 Functional Amplicon Sequencing and 10× Metagenomics to Dissect Disease-suppressive Rhizosphere Microbiomes 572
7.19 28S LSU Gene Sequencing in assessing fungal community structure in disease suppressive soils 574
7.20 Involvement of Burkholderiaceae and Sulfurous volatiles in Disease-suppressive Soils 576
7.21 The Microbiome of Suppressive Soils Against Vascular wilt 578
7.22 Multiple Parallel Mineralization Technique in Feneration of *Fusarium oxysporum*-suppressive Soil with Non-soil Carriers 581
7.23 Suppressive Soils in Plant Disease Management-Examples
7.24 Managing Soil Suppressiveness in Agro-Ecosystems 585

7.25 Chemical Characters of Disease Suppressive and Conducive Soils: .. 588
7.26 Responses of Microbial communities to Banana *Fusarium* wilt in Suppressive and Conducive Soils 590
7.27 The Metagenomics of Disease-suppressive Soils 592
Bibliography *596*

8. Biological Control- Concepts and Potentialities of Bioagents for Managing Soil Borne Pathogens 602
8.1 Antagonistic Microorganisms 604
8.2 Strategy and Principles of Biological Control 607
8.3 Mechanisms Employed by Bio-Control Agents for Plant Disease Management 609
8.4 Soil and Phytomicrobiome for Plant Disease Suppression and Management 614
8.5 Major Bioagents 616
8.6 Biocontrol 671
Bibliography *688*

List of Figures

Fig. 1: An illustration describing the interaction of some microbial processes within the soil, plant, and atmosphere ... 7
Fig. 2: Soil environment factors influence MOs population and diversity in soil ... 17
Fig. 3: The entry points for beneficial and pathogenic soil microbes, and impacts of cropping practices, on plant production. ... 25
Fig. 4: Topological Soil Food web ... 37
Fig. 5: Trophic levels in the soil food web ... 44
Fig. 6: Soil organisms in soil stabilization ... 46
Fig. 7: Relative abundance of the top-10 ... 58
Fig. 8: Rhizosphere structure ... 88
Fig. 9: Some rhizosphere processes in the soil ... 93
Fig. 10: Schematic of a root section showing the rhizosphere structure ... 96
Fig. 11: Overview of (micro)organisms present in the rhizosphere zoo. ... 100
Fig. 12: Rosin ... 114
Fig. 13: Capsaicin, Capsanthin and Capsorubin ... 115
Fig. 14: Guggulsterone ... 115
Fig. 15: Tolu Balsam ... 115
Fig. 16: Chemical structure of Guar gum ... 117
Fig. 17: Overview of plant mucilage types. ... 126
Fig. 18: Schematic representation of interactions between rhizo-microorganisms and plants ... 146
Fig. 19: Schematic representation of quorum sensing (QS) and quorum quenching (QQ) inhibition of signal perception pathways. The binding of QQ agents to the LuxR receptors either inactivates quorum sensing receptors or reduces the quantity of receptors in the QS molecules for targeted gene expression ... 149
Fig. 20: Overview of rhizosphere communication in intra- or interspecies signaling among microorganisms and interkingdom signaling between microbes and plants ... 151
Fig. 21: Root exudates link soil nutrients to plant nutrient strategy ... 164
Fig. 22: Average endemicity of soil fungi across ecoregions ... 196
Fig. 23: Aspects of soil fungal biodiversity ... 215
Fig. 24: Soil fungal biodiversity as an interdisciplinary research ... 217
Fig. 25: Outlook on the beneficial characteristics of rhizosphere-resident plant growth promoting fungi (PGPF) for plant growth ... 238
Fig. 26: Mechanism of resistance offered by PGPF against the invading pathogens ... 239
Fig. 27: Application of PGPF and its effect on the induced resistance in plants ... 240
Fig. 28: Different antimicrobial products produced by biological control agents, i.e., bacteria and fungi, and their impact on plant pathogens and plant growth parameters. ... 272

Fig. 29: Multi-omics approaches such as genomics, metagenomics, proteomics, transcriptomics, and metabolomics need to be used to identify potential microbial strains, a source of antimicrobial products. 280
Fig. 30: Global Agriculture Biological Control Agents Market 296
Fig. 31: Mechanism of biocontrol agent against plant pathogens 315
Fig. 32: Microbiological tools, available in 2021 for agriculture with published field test reports by (A) application, (B) organism, and (C) type 329
Fig. 33: Effect of *M. incognita* inoculum densities on reduction in shoot length. Data are means of ten replicates. Vertical error bars represent standard errors of differences of means 408
Fig. 34: Effect of *M. incognita* inoculum densities on reduction in root length. Data are means of ten replicates. Vertical error bars represent standard errors of differences of means 409
Fig. 35: Effect of *M. incognita* inoculum densities on reduction in fruit yield. Data are means of ten replicates. Vertical error bars represent standard errors of differences of means 409
Fig. 36: Effect of *M. incognita* inoculum densities on number of galls. Data are means of ten replicates. Vertical error bars represent standard errors of differences of means. 410
Fig. 37: Effect of *M. incognita* inoculum densities on reproductive factor. Data are means of ten replicates. Vertical error bars represent standard errors of differences of means. 410
Fig. 38: Competition among biocontrol agent and target pathogen. It occurs for suitable niches and available resources in the rhizosphere. Possession of traits listed in the figure help the biocontrol agent to out-compete the pathogen thereby controlling the corresponding disease 430
Fig. 39: Conceptual representation of the intimate link between soil microbiome predators and the microbiome leading to changes in ecosystem functions. 458
Fig. 40: The road to the application of soil microbiome predators. 460
Fig. 41: Antibiotics produced by *Trichoderma* species 479
Fig. 42: The interplay of the soil inherent factors and environmental changes in the development of disease-suppressive soils 526
Fig. 43: Effects of soil abiotic factors on disease suppressiveness 529
Fig. 44: Flowchart depicting the general process to study specific disease-suppressive soils 530
Fig. 45: Possible mechanisms of disease-suppressive soils—a close view 531
Fig. 46: Negative feedbacks illustrated by take-all specific suppression 548
Fig. 47: Hypothesized positive feedbacks in the development and maintenance of general pathogen suppression 548
Fig. 48: Trichoderma modes of action as bioagent 617
Fig. 49: Mechanisms of *Trichoderma* induced resistance and signaling pathways in plants against plant diseases. 618
Fig. 50: Regional contribution to the registered *Trichoderma*-based biological control formulations 621
Fig. 51: Mechanisms of *Bacillus subtilis* in biological control of biotic stress 634
Fig. 52: Benefits of *Streptomyces* in Agriculture 649
Fig. 53: Schematic representation of the most commonly studied phages 660

1

Soil as An Environment for Plant Pathogens

Soil is a strange full of wisdom particularly in terms of biodiversity. Soil biota consists of the microbes (bacteria, fungi, archaea and algae), soil animals (protozoa, nematodes, mites, springtails, spiders, insects and earthworms) and plants, living all or part of their lives in or on the soil or pedosphere. Millions of species of soil organisms exist but only a fraction of them have been cultured and identified. Micro organisms (MOs) (fungi, archaea, bacteria, algae and cyanobacteria) are members of the soil biota but are not members of the soil fauna (Sachidanand et al., 2019). The soil fauna is the collection of all the microscopic and macroscopic animals in a given soil. The size of a soil organism can restrict its location in the soil habitat. Smaller members of the microfauna like nematodes are basically aquatic organisms that live in the thin water films or capillary pores of aggregates preying or grazing on other aquatic microfauna such as amoebas. Soil has a direct effect on the environmental conditions, habitat and nutrient sources available to the soil biota. The term pedosphere is often used interchangeably with soil and captures the concept that the soil is a habitat where the integration of spheres occurs. These spheres include the lithosphere, atmosphere, hydrosphere and the biosphere. Numerous biogeochemical processes regulated by soil biota occur in the pedosphere.

The physical structure, aeration, water holding capacity and availability of nutrients are determined by the mineral constituents of soil, which are formed by the weathering of rock and the degradative metabolic activities of the soil MOs. Soil microbiology is the study of organisms in soil, their functions, and how they affect soil properties. It is believed that between two and four billion years ago, the first ancient bacteria and MOs came about in earth's oceans. These bacteria could fix N_2, in time multiplied and as a result released oxygen into the atmosphere. This led to more advanced MOs. Soil biology plays a vital role in determining many soil characteristics. The decomposition of organic matter by soil organisms has an immense influence on soil fertility, plant growth, soil structure, and C storage. As a relatively new science, much remains unknown about soil biology and their effects on soil ecosystems. The soil is home to a large proportion of the world's biodiversity

Nutrients are extremely important and directly influence growth, yield and quality of crops. Soil MOs can provide nutrients to plants through the fixation of atmospheric N_2 and or by enhancing nutrient mobilization/uptake through their biological activities such as mineralization, and through the production of siderophores, organic acids and phosphatases. As microbial number and activity are more intense in the rhizosphere than in soil it hardly seems conceivable that the development of rhizosphere microflora would not directly affect the development of roots. However, particularly attention has been paid to the beneficial effect exerted by rhizosphere inhabitants. Typical rhizosphere bacteria such as *Arthrobaeter, Seudoponas* and *Azrobacterium* were found long ago to be able to produce substances promoting plant growth. Ectomycorrhizal fungi also provide the host plant with phytohormones and growth-regulating vitamins B. Understanding soil ecology is considered to be important for the sustainability of ecosystems as well as for the restoration of disturbed habitats. Soil MOs contribute to soil quality and play an important role in soil ecosystem processes, such as nutrient cycling, organic matter decomposition and bioremediation. Both changes in soil microbial communities resulting from ecosystem management and global change can have significant impacts on ecosystem dynamics and microorganisms respond sensitively to changes and environmental stress because they have intimate relations with their surroundings , therefore, interest in quantifying impacts on the biotic and abiotic component has increased with concern for the sustainability of agricultural ecosystem.

1.1 The Soil Environment

The soil environment consists of a variety of physical, biological and chemical factors that affect the abundance and diversity of microbes found in the soil. At its basic level, the soil environment consists of a solid and porous fraction. Within these fractions, a variety of chemical and physical factors are affected by and and affect microbes. These include, but are not limited to texture, temperature, pH, oxygen, cation exchange capacity and redox reactions. The soil environment directly affects the types of microbes, as well as the rates of processes they perform. For example, microbial activity increases with temperature, which in turn affects rates of decomposition. On the other hand, microbial processes directly affect their environments as well, contributing to the carbon and nitrogen cycles, which are important for microbial and plant health (Brady et al., 2010). At the micro scale, bacteria and other microbes participate in a variety of reactions that affect nutrient cycling, pH, as well as oxygen and CO_2 content. At the macro scale, these processes can change the landscape in drastic ways, assisting in weathering of the soil and development of soil layers.

Soil accommodates the ‘hidden’ half of terrestrial plants. Although diseases caused by pathogens in soil are as important as leaf diseases, microbial communities, biological processes and the potential for interactions in the soil are poorly understood. The importance of such interactions is sometimes overlooked. For example, the effects of arbuscular mycorrhizal fungi on root morphology and function, on microbial populations in the rhizosphere and on soil-borne disease have only received broad scientific attention since the l98Os (Nehl David et al., 2003). Frequently the degree of pathogenicity of soil-borne micro-organisms observed under laboratory conditions is not repeated in the field where biological and nonbiological influences are less controlled.

The physical soil environment : Soil consists of particles of various size and pores of varying diameter. Particles may be sand (>50 um diameter), silt (2-5O um) or clay (less than 20 um). The total porosity of soils varies, but even more important is the variation in pore size distribution. The smaller the pores the harder it is to remove water from the soil. A soil is said to be at field capacity when it is at the moisture content to which it drains under gravity. The force required to remove water from soil is known as the matric potential and is measured in MPa or cm H_2O. The availability of water is the major physical factor limiting the activity of soilborne fungi, bacteria and nematodes (Nehl David et al., 2003). Water availability varies with soil pore size (matric potential) and solutes (osmotic potential) which are primarily derived from clays and organic materials. Temperature changes the availability of water through effects on osmotic potential, freezing and evaporation, especially near the soil surface. When water potential falls below -0.1 MPa, the rate of utilisation of organic substrates declines rapidly. Nevertheless, many fungi remain active when soil water potential falls to -30 to -50 MPa, or lower. Most bacteria are much less tolerant of dry conditions and their activity ceases below the range -1 to -15 MPa. Movement of nematodes is prevented at higher matric potentials than those required to cause desiccation. At the opposite end of the moisture scale, soil borne inoculum faces lack of oxygen when soil pores become saturated during flooding.

Most plant parasitic nematodes are between 10 and 40 pm in diameter and therefore optimum movement occurs when soil pores are in the range 30-l00 pm. Recently drained pores (at field water holding capacity) favour nematode movement because nematodes are not efficient swimmers and move with a wavelike motion, pushing against soil particles. Further drying increases the surface tension of water and this prevents migration. Zoospores such as those produced by *Phytophthora* spp. are generally about 4 pm in diameter and pores appreciably larger than this are required to allow dispersal. Bacteria are motile within films of water and their dispersal is restricted when pore necks are less than 1.5 pm. The growth of fungal mycelium through soil is less restricted than that of zoospores or bacteria because cytoplasmic streaming in hyphae negates the need for free water in soil pores. Within specific ranges, temperature is directly related to microbial metabolism and growth. Temperature extremes lead to death of micro-organisms. This effect is exacerbated near the soil surface. *Ratstonia* (*Pseudomonas*) *solanacearum* is eliminated from soil when temperatures rise above 43°C. In naturally infested potato fields in northern New South Wales, *R. solanacearum* is absent from the upper layers of soil. Apart from the effects of water availability, pore size and temperature, pathogens may also face shearing or compression from cultivation or as a drying soil shrinks. These effects will be greater for large propagules or hyphae.

The chemical soil environment: The nutritional status of a soil has both direct and indirect effects on the activities of the soil micro flora. All the essential mineral elements are reported to influence disease incidence or severity but it is difficult to distinguish whether this is due to effects on the pathogens or the host (Nehl David et al., 2003). A particular element may inhibit some diseases but stimulate others and have an opposite effect with modification of the environment or rate. For example, inhibiting the nitrification of ammonium reduces Verticillium wilt but may increase rhizoctonia canker in potato. In acid soils the availability of manganese and aluminium can reach todc levels. Increased manganese availability is a factor in soils that are suppressive to take-all in wheat and *Verticillium* wilt and *Rhizoctonia* root rot in cotton. At least part of this effect is a direct interaction between manganese and the pathogen. Many bacteria and fungi oxidise and/or reduce inorganic compounds and affect their availability, the soil pH and redox potential. Iron is an essential micro-element. Some fungi and bacteria produce chelating agents (siderophores) which bind available iron and make it inaccessible to other micro-organisms. The suppressiveness of certain soils to *Gaeumannomaces graminis*, the cause of takeall in wheat, is associated with siderophore production by pseudomonad bacteria. Plant residues are a significant component of soil, affecting its structure, chemistry and biology. Soil organic matter is derived directly from breakdown of plant residues and indirectly as the remains and by-products of successive generations of saprobic and parasitic micro-organisms. Toxic compounds produced during the breakdown of organic matter play an important role in soil microbial ecology. Ammonia is produced in the soil when legumes are slashed and incorporated as 'green manure' or 'cover' crops. Fungicidal concentrations of ammonia ranging from 0.4 to 3O ppm have been detected in soil seven days after the incorporation of hairy vetch residues. There is interest in 'biofumigation' using residues of some brassica crops. The glucosinolates that are present in most brassicas break down to produce isothiocyanates which are active against several soil-borne pathogens.

The biological soil environment: The biological component of the soil includes an extensive flora and fauna ranging in size from bacteria to earthworms. It has been estimated that there are up to 10^9 micro-organisms representing up to 10^4 species in a gram of soil. The types of relationship that exist between the various components of this dynamic microflora vary from synergism and mutualism through no interaction to inhibition, antagonism, competition and parasitism (Nehl David et al., 2003). The biology and consequently Bulk soil differs greatly from that influenced by roots). Rhizosphere the chemistry of the soil, changes near roots. in the rhizosphere (zone of soil adjacent to and pH may differ from the surrounding soil by up to 2 pH units due to root exudation, which occurs by both passive leakage and active release. Root exudates support a dense and highly competitive community of micro-organisms, usually within 1 to 3 mm of the root, which is very different to that in non-rhizosphere soil. While root exudates play an important role in attracting root pathogens the microbial flora of the rhizosphere is potentially antagonistic to them. All root-infecting pathogens must pass through the rhizosphere, where microbial competition is intense. Although such competition is difficult to quantiff, the carbon and nitrogen supplied by root exudates is important to successful infection of the host by many pathogens. Soil micro-organisms increase their competitive saprophytic ability by producing substances which antagonise others or by sequestering substrates. Toxins include volatile and non-volatile by-products, lytic enzymes and synthesised antibiotics. Rhizosphere pseudomonads produce phenazine antibiotics and siderophores which increase their competitiveness and are involved in the suppression of take-all in wheat.

Certain siderophores can also act as powerful antibiotics, thus increasing their function in competition. Most, if not all, soil-borne pathogens are subject to predation or parasitism by other organisms. Bacteria are infected by bacteriophages and consumed by protozoans and nematodes. Fungi are lysed by bacteria, grazed by nematodes and arthropods and colonised by hyperparasitic fungi. Nematodes are infected by bacteria, trapped and parasitised by fungi and consumed by predacious nematodes. Life in the soil is tough.

Soil biology is important for soil health and the soil biological community encompasses all living things including earthworms, insects, nematodes, plant roots, animals, and microbes. Beneficial soil microbes perform fundamental functions such as nutrient cycling, breaking down crop residues, and stimulating plant growth (Terrence Bell et al., 2021). While the role of microbes to maintain soil health and contribute to crop performance is clear, the soil biological component is extremely difficult to observe and manage. Soil microbes, biological communities, and the functions they perform are dynamic, complex, and not easily interpreted for field practices. Yet, they are getting more attention as farmers are encouraged to practice more holistic management approaches on their farms.

It has been estimated that a single gram of soil can contain up to several billion bacteria alone. In agricultural systems, soil microbes are directly associated with soil carbon because of the essential role they play in residue decomposition and nutrient cycling. In soils, carbon-based materials include plant litter, compounds released from plant roots, living or dead soil organisms (including microbes), and larger soil fauna like worms and insects, as well as their waste. Collectively, all carbon-based materials, including living things, are known as soil organic matter. Like all living things, soil microbes need access to nutrition and water and therefore tend to concentrate near sources of soil organic matter. Microbial biomass and diversity tend to be much higher in soil immediately surrounding plant roots, an area known as the rhizosphere, and other organic matter deposits. Due to their close proximity to plant roots, soil microbes significantly affect soil and crop health. Some of the activities they perform include nitrogen-fixation, phosphorus solubilization, suppression of pests and pathogens, improvement of plant stress, and decomposition that leads to soil aggregation. However, soil microbes can also be harmful to crops if they cause disease or compete for nutrients.

Although microbial diversity is necessary for multiple microbial functions to occur in the soil, including disease suppression, clear relationships between microbial diversity and soil or crop health continues to be difficult to decipher. In general, soils with higher microbial abundance and diversity have a greater chance of containing individuals that perform a particular function under a range of conditions, but there is often a saturation point where more diversity does not result in noticeably improved soil function. Despite understanding their importance on farms, the diversity and complexity of soil microbial communities make them incredibly difficult to manage (Terrence Bell et al., 2021). Many strategies do exist for managing soil microbes, however, and range from adding beneficial microbes (inoculants or commercial products) to suppressing harmful microbes (soil fumigation, soil steaming, anaerobic disinfestation, and solarization). These approaches range in cost, labor and equipment needs, scale of application, and measurable efficacy. Additionally, common crop management practices often target other agronomic needs and simultaneously influence soil microbial communities. Examples include tillage, crop rotation, cover cropping, and adding compost, manure, or mulch.

1.2 Soil-Microbes-plant Relationship

Organisms in the soil need food for two distinct purposes: to supply energy for their essential and vital processes and to build their body tissues. However, the sources of food for various organisms can be different for these two purposes. Fully autotrophic organisms require much more energy for growth than many heterotrophs, since greater energy is needed to convert the carbon (C) of inorganic sources than that of organic substances such as sugar (Mahdi M. Al-Kaisi et al., 2017). Also, autotrophic organisms can meet their nitrogen (N) need from ammonium or nitrate salts as preferred sources. However, heterotrophic organisms are classified by their nutritional needs or by the biochemical changes they bring. The majority of heterotrophs can use glucose as their primary source and produce an enzyme, if necessary, to convert a wide range of carbohydrates into sugar. Plant and soil interaction is characterized by a defined-zone called the "rhizosphere," where the root system influences microbial activity. This zone is distinguished from the rest of the soil mass by the active interaction between soil–plant–microbial communities. The interaction between soil, plant, and microbes is essential to nutrient cycling processes, such as soil organic nitrogen (SON) mineralization and phosphorus (P) transformation for plant nutrient supply. In the root zone within the soil environment the contact between microorganisms and the root system involves a range of activities that contribute to the plant growth. Such soil–microbial interaction plays a significant role in increasing bioavailability and uptake of mineral nutrients by plants. These biofertilizers such as, plant growth-promoting bacteria or fungi, as Mycorrhizae and *Penicillium bilaii*, can increase nutrient bioavailability. Plant growth-promoting bacteria can contribute to the development of the plant that is more positioned to tolerate adverse growing conditions such as disease, pathogens, and. Some of the mechanisms that are associated with the promotion of healthy plant growth may include Associative N fixation; The lowering of ethylene levels that are otherwise an impediment to plant growth; The sequestration of iron by siderophores; The production of photohormones such as auxin and cytokinins; The introduction of pathogen resistance in the plant; The solubilization of nutrients such P; Promotion of mycorrhizal functioning; Modification of root morphology; Enhancement of legume-rhizobia symbioses; and Decreasing (organic or heavy metal) pollutant toxicity.

The soil is home to a variety of microorganisms involved with different processes that enable it to support crop production sustainably. These biological processes affect nutrient cycling and availability, soil aggregation, organic matter decomposition, and crop performance (John Idowu et al., 2023). Some of these important microbially mediated processes are described in this guide: mineralization of the soil organic matter (breakdown of organic materials to release nutrients), immobilization (temporary tying up of nutrients by microbes, thus limiting their availability to plants), nitrogen fixation (conversion of atmospheric nitrogen to forms that crops can use), nitrification (conversion of the ammonium form of nitrogen to nitrate-nitrogen), denitrification (loss of nitrate-nitrogen in the soil through microbial conversion to nitrogen gas), and mycorrhizal associations (the interaction of beneficial soil fungi with plants) (Fig.1).

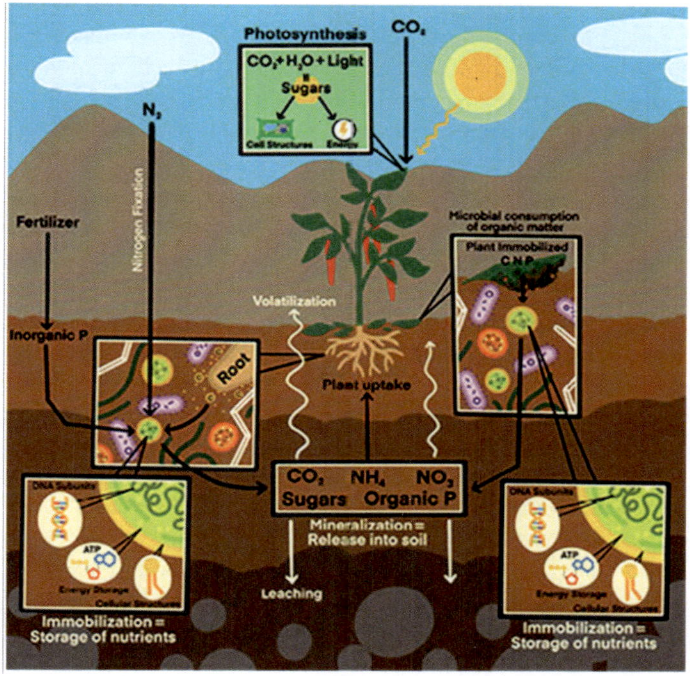

Fig. 1: An illustration describing the interaction of some microbial processes within the soil, plant, and atmosphere

The soil ecosystem: While the physical, chemical and biological attributes of soils can be measured independently. the soil is an ecosystem in which no factor functions in isolation. A beautiful example of interacting factors is shown by *Fusanum roseum* infection of wheat. High soil moisture leads to a high rate of diffusion of root exudates which leads to a high rate of germination of chlamydospores of *F. roseum* (Nehl David et al., 2003). However, this does not lead to high infection because the high moisture level also stimulates bacterial activity. The bacteria attack the young germ tubes of *F. roseum* causing them to lyse. At much lower soil moistures, there is less diffusion of exudates and lower spore germination but at low moisture levels, bacterial activity is restricted and the germ tubes are not lysed. The result is a high level of infection. Development of club root of brassicas, caused by *Plasmodiophora brassicae,* provides another example of interaction between soil properties and pathogen activity. Club root occurs under a wide range of conditions in acid soil but is restricted to a narrow range in alkaline soil, where high temperature, high moisture content and a high spore load are required. Furthermore, disease severity is proportional to inoculum density at low, but not high, light intensities.

1.3 Survival of Soil-borne Inoculum

Survival effectively represents 'dispersal in time'. Inoculum of a pathogen must be able to survive the period between inoculum production on one host and either re-infection of the same host or infection of another susceptible host. This interhost or inter-crop period may vary from a few days to a few months or several years. The soil environment produces several obstacles to successful survival that must be overcome by soil-borne inoculum in the absence of susceptible hosts. The pathogen or its inoculum must be able to either avoid, compete with, or withstand antagonism or parasitism from other soil micro-organisms. The inoculum must also be able to tolerate extremes and fluctuations of temperature and moisture in the soil.

Alternative hosts: If a pathogen has a large host range then the inter-host period can be significantly reduced and in most cases the pathogen not only survives the intercrop period but multiplies as well. *Verticillium dahliae* is a pathogen of cotton and has numerous alternative weed hosts which include *Xanthrum* spp. (Noogoora and Bathurst burrs), *Hibrscus trtonum* (bladder ketmia) and *Rapustrum rllgosum* (turnip weed). Many of the alternative weed hosts are common in cereal fallow rotations used in cotton production systems (Nehl David et al., 2003). Similarly, *Thielaviopsis basicola*, which causes black root rot of cotton, can survive and multiply on legume crops grown in rotation with cotton. Colonisation of an alternative weed or crop host allows the pathogen to avoid direct interaction with the soil microflora while surviving the intercrop period. It is possible for some pathogens to survive on an alternative host without causing any apparent symptoms. *Fusarium oxysporum* f.sp *vasinfectum*, which causes *Fusarium* wilt of cotton in Australia, has been isolated from several symptomless weed hosts including *Sesbanta cannabina* (sesbania pea) and *H. trtonunt.*

Within crop residues: A second strategy for avoiding direct interaction with the soil microflora is survival within crop residues. The inoculum of a pathogen which is embedded within stem or root residues is buffered from environmental extremes and partially protected from exposure to other soil micro-organisms. *Armillarta mellea* survives in infected stumps and roots while *Vertictllium dahliae* and *Macrophomina phaseoli* survive in the roots and stems of their hosts. Even foliar pathogens such as *Alternaria heltanthi* and *Alternarta macrospora* can survive in infected crop debris (Nehl David et al., 2003). Bacterial pathogens such as *Ralstonia solanacearum* and *Xanthomonas campestris* pv. *malvacearum*, root lesion nematodes of the genus *Pratylenchus* and several viruses, including tobacco and sugarcane mosaic virus and cymbidium ring spot virus, may also survive in association with crop residues.

As a saprophyte: Some plant pathogenic fungi have been described by S. D. Garrett as 'soil inhabitants'. They have a high competitive saprophytic ability and as a consequence can survive indefinitely as saprophytes. Such fungal pathogens often have a wide host range and are widely distributed in most soils. *Pythium* spp. and *Rhizoctonta solani* are typical soil inhabitants which surrive readily as saprophytes on plant residues. Garrett also described some pathogenic fungi as 'root inhabitants' (Nehl David et al., 2003). Root inhabitants generally have a low competitive saprophytic ability and do not survive very long as saprophytes. These fungi are frequently specialised for parasitism on only one or a few hosts. Some of the factors which are known to increase the competitive saprophytic ability of fungi are high growth rates, proficient enzyme production, antibiotic production and antibiotic tolerance. The saprophytic competence of soil-borne bacteria is based upon their small size, their metabolic versatility and their ability to rapidly respond to changing environmental conditions. The small size of bacterial cells enables them to absorb substrates quickly, multiply prolifically and escape predation. Metabolic tricks, including the sequestering of inorganic nutrients and carbon substrates, contribute to the competitiveness of soil borne bacteria. Several species of *Pseudomonas* have membrane bound enzymes which enable the extracellular conversion of glucose, when in plentiful supply, to gluconate and thence to 2-ketogluconate, compounds which many microorganisms cannot utilise. Plant pathogenic nematodes are obligate parasites and cannot survive saprophytically.

Resistant resting structures: Many soil-borne plant pathogens produce resistant resting structures such as the large sclerotia formed by *Sclerotinia sclerotiorum* and *Sclerotium rolfsii*, microsclerotia formed by *V. dahliae*, chlamydospores produced by *Fusarium oxysporum* and *T. basicola,* oospores produced by *Phytophthora cinnamomi* and cysts produced by nematodes such as *Globodera* spp. and *Heterodera* spp. These resistant structures allow the pathogen to survive in the soil from season to season or sometimes over several seasons (Nehl David et al., 2003). If they are embedded in crop residues then the life of these resistant structures may be extended even further. While thickened walls serve as a physical barrier to desiccation and attack by antagonists, most resting structures also have chemical defences. The insoluble phenolic compounds known as melanins have a role in the defence systems of most organisms, including fungi. Melanins and their precursors can exhi.bit antibiotic activity but the major role for melanins is protection against the hydrolytic enzymes of antagonistic organisms. Hyaline spores or hyphae in soil are attacked and lysed quickly while melanised ones may survive for years. Consequently, most soil-borne fungi produce melanins. The rind of fungal sclerotia is pigmented and in most cases contains melanins. Plant pathogenic bacteria do not produce spores but many are tolerant of desiccation and survive for long periods under dry conditions. Survival of soil borne bacteria is enhanced by exudates (polysaccharides) which form a protective coating around micro-colonies.

1.4 Physiological Adaptations and Soil Fungistasis

Although host recognition events in soils are not yet well understood, it is clear that the greater the degree of host specificity, the more precise is the recognition factors. Sclerotia of *Sclerotium ceptuorum* are stimulated to germinate by volatile sulphur compounds released by the host *Allium* spp. Zoospores of a grass infecting *Pythium* sp. are preferentially chemotactic for, and encyst in greater numbers on, grass roots. When spores are produced it is important that they do not germinate immediately (Nehl David et al., 2003). They are still attached to a dead or dying plant and it would be suicidal to germinate in situ because there would be no susceptible healthy plant to infect. Fungi overcome this problem by self-inhibition. When a substrate is very heavily colonised by a particular fungus, the spores produced by that fungus do not germinate in situ. This can be readily seen on an agar plate colonised by a sporulating fungus. Infested plant residues will eventually disintegrate. If the soil is cultivated, the spores will be distributed through the soil. Self inhibition no longer operates because the spores are no longer located on a heavily colonised substrate. Again, it would be suicidal to germinate in this situation because, after cultivation, no susceptible plant would be available for infection. This problem is overcome because the spores of soil fungi are subject to soil fungistasis. Fungistasis prevents the germination of fungal spores in the soil in the absence of a potentially colonisable substrate or in the presence of high populations of microorganisms competing for nutrients. Bacteria and actinomycetes are similarly restricted, apparently by the same mechanisms as fungi. Several theories have been advanced to explain the phenomenon.

The 'nutrient deprivation' hypothesis suggests that fungistasis results from insufficient nutrients to support spore germination. Fungal spores can 'test' the environment to ensure it is suitable for germination. Once spores are imbibed, they release a small pulse of sugars and amino acids. Unless the equivalent amounts of sugar and amino acids are then absorbed by the spore from its environment, the metabolism of the spore closes down and it becomes inactive again. In some soils, volatile or non-volatile inhibitors of biotic or abiotic origin may play a role in fungistasis. Possible fungistatic agents include ammonia, which is detectable in some soils, and siderophores of microbial or plant origin. Simple nutrients like sugars and amino acids can override inhibition due to fungistasis. Plant roots exude amino acids and sugars as they grow through the soil. When a root passes close to an inactive spore, the exuded nutrients may stimulate spore germination. Exudates from the roots of resistant plants may also break fungistasis and stimulate germination. If this occurs then the pathogen can often produce more spores as the food supply again becomes limiting. Genetic heterogeneity within a dormant pathogen population provides a selective advantage by preventing all propagules from germinating in response to a single stimulus. Thus the chances of mass suicide in the event of rapidly changing conditions are reduced.

Nematodes are motile and are attracted to roots. At least one nematode, *Meloidogyne javanica,* is subject to a hatching inhibitor that is present in the soil. This effect is similar to fungistasis. Some nematodes survive from one crop to the next by producing resistant structures such as cysts but many plant parasitic nematodes are subject to a phenomenon very like diapause in insects which interrupts maturation. It is like self-inhibition in fungi but the effect lasts for many months and is not influenced by the substrate. Hatching and further development occur when the next crop is planted. With some nematodes hatching will not occur, even when the eggs are mature, unless a specific chemical is present. In *Globodera rostochiensis*, hatching is induced by a chemical produced by the roots of potatoes and other closely related species. Thus hatching only occurs when a susceptible host is present. The greater the supply of energy and / or nutrients within inoculum, the greater is its competitiveness with other organisms. In the rhizosphere, inhibition of germination of two formae speciales of *F. oxysporum* by siderophore-producing pseudomonads was effective when chlamydospores contained I ng of iron, but ineffective when they contained 3 ng of iron. Clearly, pathogens with small propagules will rely more heavily on exogenous nutrients for infection.

1.5 Dispersal of Soil-borne Inoculum

Dispersal enables a pathogen to access a host under favourable environmental conditions. Furthermore, dispersal increases the chances of at least part of a pathogen population escaping adverse conditions. In comparison to dispersal of airborne inoculum, the soil is physically more limiting (Nehl David et al., 2003). Consequently gradients of dispersal for soil-borne pathogens are much steeper. The dispersal of soil-borne inoculum is directly affected by factors such as soil particle size, soil structure, moisture content and location in the soil profile. If the energy for dispersal is derived from within the inoculum itself and the pathogen controls the direction of movement then this may be described as active dispersal. In passive dispersal the energy for dispersal is derived from an external source such as human activity, wind or water movement and the pathogen has no control of direction. The production of large amounts of inoculum can partly compensate for the vagaries of passive dispersal. Some fungi, though generally considered to be soil-borne pathogens, produce inoculum which is dispersed above ground in air or with seed. This enables dispersal over large distances and overcomes the limitations associated with dispersal within soil. For many soil-borne pathogens there is little dispersal at all and the inoculum remains in situ until intercepted by host roots. In permanent bed systems, where subsequent crops are sown in exactly the same position season after season, this results in a more rapid build-up of disease after the initial introduction. Inoculum of *Thielaviopsis basicola*, which causes black root rot of cotton in northern New South Wales, becomes concentrated in a band in the centre of the bed directly beneath the planting line for the following season.

Active dispersal: Most soil-borne plant pathogens spread very slowly through soil, either as motile organisms or by growth of hyphae. Most diseases caused by these pathogens occur in well defined patches which expand slowly from season to season. The production of rhizomorphs by *Armillaria meltea* enables the pathogen to grow from an infected tree to an adjacent healthy tree. The mycelial strands produced by *Phymatotrichopsis omnivora* (Nehl David et al., 2003), which causes root rot of more than 2,000 species, also function in the same way. As discussed earlier the movement of nematodes, zoospores and bacteria is significantly affected by soil texture and soil moisture content. Nematodes are not generally associated with heavy clay soils where the smaller pore sizes restrict movement. Under ideal conditions nematodes can move 20-30 cm per year. Zoospores are only motile in soil for periods of minutes to hours, depending on their energy reserves and factors that induce encystment. Individual zoospores of *Phytophthora* spp. migrate 2-35 cm in soil depending on the species and the soil texture. In Victorian eucalypt forests, *P. cinnamomi* dieback can extend downhill at rates of 400 metres per annum, mainly by zoospore dispersal through the soil water system. Although this movement involves passive dispersal in flowing water, it is motility that enables the zoospores to access and remain suspended in this water until a potential host is detected. Motile bacteria can migrate through soil over horizontal distances of 1-30 cm within one to a few days, depending on soil moisture, soil texture and the presence of a host. Over time, populations of bacteria can migrate along entire root systems. Root exudates act as attractants for motile inoculum. The zoospores of *Phytophthora* spp. are chemically attracted to wounds or to the region of elongation just behind the root tip.

Passive dispersal by carriers or vectors: Animals of all kinds are continuously disturbing and transporting soil and plant material from place to place. The spores of *Phytophthora* spp. can survive passage through the gut of the giant African snail by this means the pathogen has been dispersed from one garden to the next. Similarly feral pigs on the island of Hawaii transport *P. cinnamomi* in soil attached to their hoofs from one area to another (Nehl David et al., 2003). The transmission of viruses by several hyphal and zoosporic fungi, nematodes, and eriophyid mites within the soil has been demonstrated. Conditions which favour spread of the vector favour spread of the virus.

Passive dispersal in or with soil moved by wind and floodwater: Strong wind gusts frequently move soil. Organic components of the soil are preferentially airborne. Plant pathogenic nematodes have been recovered from dust traps set 2 metres above the ground during high winds. Flood waters wash across agricultural land and carry soil and crop residues downstream. The distribution of Fusarium wilt of cotton in the Darling Downs area of Queensland closely matches the areas inundated by floodwater in previous years.

Passive dispersal in or with soil and water moved by human activities: There can be no doubt that humans are the most important factor in the long range dispersal of soil-borne plant pathogens. Humans are also an important factor in short-range dispersal through soil cultivation and the movement of vehicles and earth-moving equipment. Any activity that moves soil or plant debris from one place to another can disperse pathogens. These activities include the movement of second-hand or dirty farm machinery, the trade in potted plants and rootstocks between nurseries and transport of plant produce or residues (Nehl David et al., 2003). The introduction of the potato cyst nematode (*Globodera rostochiensis*) from South America to Europe in soil attached to potato tubers is well documented. Jarrah dieback, caused by *Phytophthora cinnamomi*, was first noticed in a very restricted area of forest in Western Australia in 1921. It has now destroyed large areas of forest. Earth-moving equipment used in the making of roads spread the pathogen extensively and run-off water carried it from the roads into the gullies. The pathogen which causes Verticillium wilt in a large range of crops can be introduced when cotton gin trash is used as garden mulch. There is ample evidence of pathogens being effectively dispersed in irrigation water. Long term studies of the spread of Fusarium wilt of cotton in Israel showed that the pathogen was dispersed along the rows by furrow irrigation. Cultivation across the rows between crops had only a small impact on the spread of the disease. An outbreak of Phytophthora boll rot of cotton in California was attributed to 'inoculation' of the pathogen in water applied to the crop through the overhead sprinkler system from a storage dam. In Queensland, rainwater run-off from soil beneath a single plant infected with *Fusanum oxgsporum* f.sp. *cubense*, the cause of Panama disease, is enough to contaminate irrigation dams with the pathogen.

The vast biodiversity of soil is crucial in regulating the impacts of pathogens and pests on microorganisms, plants and animals, including humans (Samaddar et al., 2021). The complexity of soils' diverse communities and micro-environments provide a myriad of mechanisms, and potential solutions, for regulating detrimental organisms. Many of these biological processes are indirect, not visible, involve the actions of complex consortia of organisms, and are intimately linked to their physical environment and its management. Nonetheless, over the past few decades, the growing availability of new tools, both molecular and imaging, has expanded our ability to investigate and elucidate these complex soil phenomena.

Soil hosts arguably the most diverse biological communities on Earth. Soil's heterogeneity, created by complex arrangements of minerals and organic matter of different structure and composition, organized into aggregates of different sizes, provides a vast variety of ecological niches (Barrios, 2007). Many of soil's inhabitants are either directly or indirectly involved in regulating organisms that may have detrimental effects on humans. At the same time, soil is also a reservoir of a variety of bacterial, fungal and viral pathogens capable of causing diseases in plants, animals or humans. Pathogens are common members of soil communities who may be transient inhabitants, consistently present for parts of their life cycle, or permanent fixtures within the community. Whether or not potential pathogens have negative impacts on humans and other biota depends on if they successfully colonize and establish themselves in soil (if they are not resident species) and come in contact with a susceptible host. Some pathogens require transfer via vectors or carriers who live in the soil. Other important factors include whether pathogen populations reach high enough levels to cause infections (inoculum potential) and whether they are metabolically capable of causing infection.

Understanding the complex interactions and ecological phenomena that govern suppressiveness in a particular place and time will provide a foundation for new tools and indicators to predict potential outbreaks and develop preventative solutions. Better knowledge of which tools to use, and when to stay out of the way of soil's native ability to regulate detrimental organisms, will come from ecological insights gained from studies of soil biodiversity (Fierer et al., 2021). Soil health has proven to be a powerful conceptual framework that helps integrate across multiple functions, processes and solutions at play and could increasingly play a role in how we think about regulating deleterious organisms in soil. When one considers, however, the vast set of interactions among pathogens, vectors, hosts, antagonists, abiotic environment and other players, it is evident our perspective needs to be broader than soil to transcend the silos of specific disciplines and taxa.

1.6 Classification of Soil Biota

Soil organisms constitute the living organic component of soil and range in size across five orders of magnitude (viz. 10-7 to10-2m). Organisms present in soil are categorized into two main groups: Soil flora and fauna. These again divided into subgroups depending upon their size, soil macro-flora and micro-flora andsoil macro-fauna and micro-fauna. Soil organisms can be classified based on their size (Table

1) (Sachidanand et al., 2019). Sizes for soil organisms range from 1µm to 20 mm. All organisms need organic C to build their cells. Chemoheterotrophs obtain the organic C from organisms which convert carbon dioxide (CO_2) to organic compounds with either light energy (photoautotrophs) or chemical energy(chemoautotrophs). Photoheterotrophs utilize light energy and also utilize organic C. Organisms can be classified based on theenergy and C sources that they use (Table 2). Photoautotrophs conduct photosynthesis, a process in which they utilize atmospheric CO_2 and water to produce organic C and oxygen in the presence of light. Photosynthesis is conducted by bacteria, algae and the higher plants, and is the foundation of food webs in diverse ecosystems.

Table 1: Classification of soil organisms

Classification	Body width	Examples
Microflora	<10 µm	Bacteria and Fungi
Microfauna	<100 µm	Protozoa and Nematodes
Mesofauna	100 µm to 2 mm	Enchytraeids, Acari and Collembola
Macrofauna	2 mm to 20 mm	Earthworms,Millipedes,Wood lice,Snails and Slugs

Note: Clay size particles are less than 2um

Photoheterotrophs utilize organic C compounds can alsoconduct photosynthesis. Therefore, it is possible for organisms,such as green and purple bacteria, to utilize organic and inorganic forms of C. Chemoautotrophs generate energy through theoxidation of inorganic compounds. Nitrifying bacteria can oxidize ammonium to nitrate under aerobic conditions and use the energy to convert CO_2 to organic C. Aerobic chemoheterotrophs, most bacteria, fungi, protozoaand soil fauna, utilize organic C compounds for biomass and energy production.

Table 2: Grouping of organisms based on energy and C sources

Energy Source	Carbon Source	
	Autotrophs (fix CO_2)	Heterotrophs (Utilize organic C)
Light (Photo-)	Photoautotrophs	Photoheterotrophs
Chemical (Chemo)	Chemoautotrophs	Chemoheterotrophs

A diverse biological community in soils is essential to sustaining a healthy environment for plant roots. There may be over 100,000 different types of organisms living in soils. Most are providing numerous functions that assist plants, such as making nutrients more available, producing growth-stimulating chemicals and helping form soil aggregates. In a teaspoon of agricultural soils it is estimated that there are from 100million to 1billion bacteria, several yards of fungi, and several thousand protozoa. It may hold 10 to 20 bacterial-feeding nematodes and a few fungal-feeding and plant parasitic nematodes. Arthropods can number up to 100 square-1 foot, and earth worms from 5 to 30 square-1 foot.

Of all the organisms in soils, only a small number of bacteria, fungi, insects and nematodes might harm plants in any given year. Diverse populations of soil organisms maintain a system of checks and balances that can keep disease organisms or parasites from becoming major plant problems. Some fungi kill nematodes and others kill insects. Still others produce antibiotics that kill bacteria. Protozoa feed on bacteria and may attack fungi. Some bacteria kill harmful insects. Many protozoa, springtails and mites feed on disease-causing fungi and bacteria. Beneficial MOs, such as fungus *Trichoderma* and bacteria *Pseudomonas fluorescens*, colonize plant roots and protect them from attack by harmful organisms. Some of these organisms, isolated from soils, are now sold commercially as biological control agents. The effects of bacteria and fungi that suppress plant disease organisms are thought to arise from competition for nutrients, production of antagonistic substances, and /or direct parasitism. In addition, a number of beneficial soil organisms induce the immune systems of plants to defend the plants. Also, roots of field crops usually have their own characteristic microbial communities with numerous interactions (Table 3) (Sachidanand et al., 2019).

Soil organisms contribute to the maintenance of soil quality because they control many key processes. Soil MOs and their communities are continually changing and adapting to changes in their environment. A high quality soil is biologically active and contains a balanced population of MOs. In ecological theory, succession indicates the somewhat orderly and predictable species changes in space and time following the colonization of a new environment and it has been one of the central ecological topics for over a century (Fierer et al., 2010). Several studies, the use of space-fortime substitution (chronosequence) succession, despite recent criticisms, has offered a unique opportunity to examine soil macro-community development in terms of biodiversity nutrient cycling, natural and anthropogenic disturbances and land-use changes across multiple time scales. The physico-chemical and structural characteristics of soil provide many microenvironments in which complex bacterial populations can evolve. As well as the physical properties of the soil, bacteria are also influenced by nutrient availability resulting from rhizodeposition and decomposition. The diversity, abundance and activity of bacterial communities are therefore structured in relation to depth, since the primary source of nutrient input in grasslands is above ground. Several authors have attempted to examine the vertical distribution of bacteria in soils. Microbial biomass and activity (CO_2 evolution) was highest in the top 5cm of soil and decreased down to a depth of 15 cm. There are many soil factors (Figure 2) which affect the activity of the soil MOs (Sachidanand et al., 2019).

Table 3: Relationships between soil functions and ecological services

Soil Functions	Ecological Services	Examples of Related Soil Biota
Food and fibre production	C-cycling	Microbial biomass, methanogens
	Decomposition of organic matter	Microarthropods, saprotrophic fungi
	N-cyding	Nitrifiers, denitrifiers
	P-cyding	Phosphatase, mycorrhiza
	S-cyding	S-reducing bacteria
	N fixation	Rhizobia
	Primary (microbial) activity	Microbial community structure and activity
	Soil food web transfers	Microbial community and food web structure
	Disease and pest transmission/ suppression	Predators, pathogens
	Nutrient supply from symbioses	Mycorrhiza, N-fixers
	Redistribution by bioturbation	Earthworms, ants
	Bio-aggregation of soil	Fungi, worms
Environmental interactions	Degradation/immobilization of pollutants	Fungi, worms
	C retention/release	Microbial biomass, methanogens
	N retention/release	Nitrifiers, denitrifiers
	P retention/release	Microbial activity, mycorrhiza
	S retention/release	S-reducing bacteria
	Tolerance/resistance (toxins)	Soil community structure and activity
	Redistribution by bioturbation	Earthworms, ants
	Bio aggregation of soil	Fungi, worms
Supporting ecological habitats and biodiversity	Habitat for rare soil species	Wax cap fungi, Southern Wood Ant
	Germination zone of plants	Plant roots, mycorrhiza
	Nutrient supply from symbioses	Mycorrhiza
	Food source (above ground)	Fungi, insects
	Reservoir for soil biodiversity (taxonomic)	Soil species and diversity
	Reservoir for soil biodiversity (genetic)	Community DNA and RNA
	Reservoir for soil biodiversity (functional)	Nitrifiers, trophic structure, worms

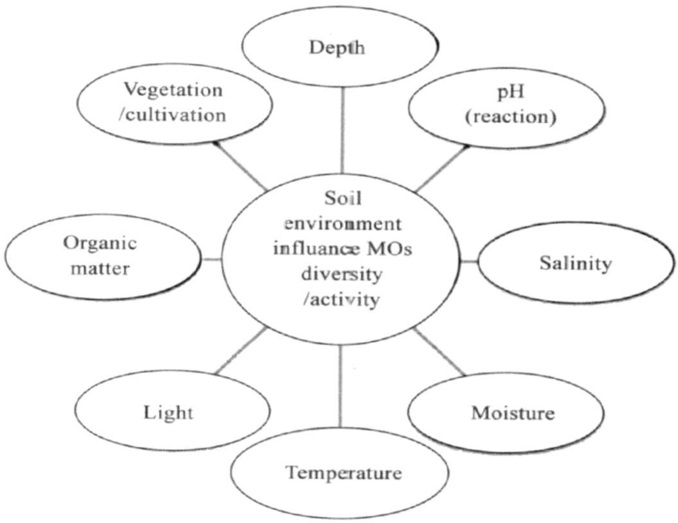

Fig. 2: Soil environment factors influence MOs population and diversity in soil

1.7 Bacterial and Fungal Population in Indian Soils

Soil microbes require what all other living organisms require a source of C and energy, adequate moisture and aeration, essential inorganic nutrients and trace elements necessary to build cells. Ideally, the soil should contain about 50% water filled pore space, allowing for a good balance of air and water filled pores for growth and activity of aerobic microbes (Sachidanand et al., 2019). The basic plant nutrients cycle highlights the central role of soil organic matter. Cycling of many plant nutrients, especially N, P, S and B, closely follows parts of the C cycle. Plant residues and manure from animals fed forage, grain and other plant-derived foods are returned to the soil. This organic matter pool of C compounds becomes food for bacteria, fungi and other decomposers. Abiotic soil factors, including availability of nutrient, reaction, topsoil depth, water content, aeration, temperature and management practices, may affect the structure and activity of soil microbial communities. Contrarily, low microbial population in forest hill and acid soil may be because addition of plant remains and wide C:N ratio and low cation (particularly Ca2+) content in the soil.

It is evident from the data that there is a wide range of soil type variation in bacterial and fungal population were studied by various researchers are given in (Table 4). Alluvial soil showed maximum abundance of bacterial population counts (36.0×107cfu g-1 soil) followed by red soil (20.8×106cfu g-1 soil), while minimum under forest / hill and acid soils (18.1×103 and 41.8×103cfu g-1 soil, respectively) in rhizosphere soils. Whereas, red and black soils showed higher fungal population (24.8×104 and 19.8×104cfu g-1 soil, respectively) as compared to acid and forest /hill soils.

Table 4: Bacterial and fungal population in the rhizosphere soils under different types in Indian soil

Soil type	Population (cfu/g soil)	
	Bacteria	**Fungi**
Alluvial	36.0 x10^7	90.0 x10^3
Red	20.8 x10^6	24.8 x10^4
Black	29.8 x10^5	19.8 x10^4
Acid	41.8 x10^3	53.2 x10^2
Forest / hill	18.1 x10^3	14.1 x10^2

It is evident from the data that there is increase in population counts with improving soil aeration and optimum soil pH. Biological nitrogen fixation (BNF) is the major way to introduce N into arable land ecosystems and balance the C:N ration in the soil. BNF in deserts is mediated mainly by some heterotrophic bacteria, associative bacteria, cyanobacteria, actinorhizal plants and legumes. Several symbiotic systems of legumes which are tolerant to extreme conditions of salinity, alkalinity, acidity, drought, fertilizer, metal toxicity, etc., were identified (Sachidanand et al., 2019). In fact, the existence of Rhizobium-tree legume symbioses, which are able to fix appreciable amount of N2 under arid conditions, is fascinating. Soil acidity is a significant problem facing agricultural production in many areas of the world and limits legume productivity. Most leguminous plants require a neutral or slightly acidic soil for growth, especially when they depend on symbiotic N_2 fixation. It has been recently reported that pasture and grain legumes acidify soil to a greater extent and that the legume species differ in their capacity to produce acids. Legumes and their rhizobia exhibit varied responses to acidity. Soil salinity and acidity are usually accompanied by mineral toxicity (specific ion toxicity), nutrient deficiency and nutrient disorder. Salt damage to non-halophytic plants grown in nutrient solution is often due to the effect of ion imbalance (disorder) rather than the osmotic potential. In India, alluvial and black soils are commonly noticed normal soil pH and suitable for legume/oil seed production and these soils maintain a narrow C:N ratio as compared to acid/ forest soil. In this might due to alluvial and red soil more support bacterial and fungal population.

Microbisols: Mishra (2015) proposed a soil group called Microbisols, which represents a soil type usually characterized by effective group of microorganisms. Regulating ecosystem services control the processes of water flow, energy transfer, nutrient uptake and release, carbon transfer, and chemical processing, as well as services provided by creating and maintaining environments for diverse plant, animal, and microbial communities. The role of MOs in the genesis and maintenance of soil structure has recently been reviewed (Sachidanand et al., 2019). The aim was to emphasize the importance of this process in the rhizosphere. It has been demonstrated that there are more water-stable aggregates in the rhizosphere than in the non-rhizosphere soil. Since the number of polysaccharide producing MOs is characteristically higher in the rhizosphere, it can be assumed that soil stabilization around the root can, at least to some extent, be due to 'the rhizosphere microflora. In tropical soils, where most of the microbial population is concentrated in the root zone, it would be worthwhile to elucidate the relative importance of the root itself and that of associated MOs in soil structure stabilization. Such investigations should not be restricted to free-living MOs, but should be extended to corrhizae which were reported to be involved in sand aggregation and' dune stabilization in colder climates. In contrast to this beneficial activity, MOs can be harmful in two ways: by decomposing the aggregating compounds originating from plants or microorganisms; and by coating soil particles with water-repellent films.

Soil health is worth quantifying because soils and their biota provide ecosystems functions that benefit humans. These ecosystem services can be of considerable value and include soil functions of storing and releasing water, decomposing plant and animal residues, transforming and recycling nutrients, sequestering and detoxifying organic toxicants, and promoting plant health by suppressing plant-pathogenic microbes and phytophagous fauna. It is often possible and desirable to measure soil function directly (Sachidanand et al., 2019). Direct measurements of some soil functions may be too expensive (for example, direct measurements of nutrient transformations) or require observations across too much time (for example, the capacity of a soil to supply water for plant growth during a drought maybe observable only during rare drought years). In such cases, rather than measuring the soil function directly, it may be preferable to measure surrogates or proxies that are well correlated with the soil function. Soil organisms meet this criterion, because the abundance and diversity of soil organisms often are well correlated with many beneficial soil functions. Indicators Relationship to soil condition and Rationale as a priority measurement Microbial biomass C and N Microbial catalytic potential and respiratory for C and N Potentially mineralizable N (anaerobic incubation) Soil productivity and N supplying potential Dehydrogenase activity, Soil respiration, water content and temperature Microbial activity measure At the beginning of any soil-monitoring program it may be possible to identify a suitable reference site where the actual values of SMB approach the attainable SMB value for that ecosystem (i.e. determined by localised climate, soil type, and land use). This is often the case for land restoration projects where either the characteristics of the site before disturbance have been well documented and/or a parallel undisturbed system can be identified. In the case of forest rehabilitation (e.g. after mining), the natural forest (under equilibrium for current climatic conditions) surrounding the rehabilitated areas sets the attainable value and, hence, the target for the recovery of the mined and rehabilitated soils.

1.8 Soil Microbes and Soil Structure

From an agricultural point of view, a well-structured soil has low bulk density, is well aerated, absorbs rainfall and irrigation water rapidly, and is easily penetrated by plant roots and soil animals. The growth of fine plant roots and fungal hyphae knits soil particles and small aggregates together into larger units called macroaggregates (Sachidanand et al., 2019). Soil organisms play important role in soil aggregation. Constituents of soil are viz. organic matter, polysaccharides, lignins and gums, synthesized by soil microbes play important role in cementing/ binding of soil particles. Further, cells and mycelial strands of fungi and actinomycetes, Vormicasts from earthworm is also found to play important role in soil aggregation. Different soil MOs, having soil aggregation/soil binding properties are graded in the order as fungi > actinomycetes > gum producing bacteria>yeasts. examples are: fungi like *Rhizopus, Mucor, Chaetomium, Fusarium, Cladasporium, Rhizoctonia, Aspergillus, Trichoderma* and bacteria like *Azotobacter, Rhizobium, Bacillus* and *Xanthomonas*. Grasses have a greater effect than other plants on stable aggregate production because a large proportion of plant biomass is maintained underground in the root system. Production of polysaccharide gums by rhizosphere MOs and plant roots enhances the formation and stabilization of macroaggregates because the materials act as glues to bind particles together. Improved aggregate stability was found in cover cropped soils as compared to conventional soils. Cover crops with a higher C:N ratio (e.g., mixtures containing grasses) may promote greater polysaccharide production and thus greater aggregate stability than cover crops with lower C:N ratio (e.g., legumes).

Soil microbiomes drive key functions in agroecosystems, determining soil fertility, crop productivity and stress tolerance. The microbiome is intricately linked with soil structure, such as aggregation and pore connectivity, because this structure regulates the flow of water, oxygen and nutrients through the system (Hartmann and Six, 2023). The key functions of soil microbiomes in agroecosystems were summarized, highlighted the dependence of these functions on the structural integrity of the soil, and discussed how agricultural practices influence the link between soil structure and microbiome functioning. System-level agricultural management practices can induce structural alterations to the soil, thereby changing the microbial processes occurring at the microscale. These changes have large-scale consequences, such as soil erosion, reduced soil fertility and increased greenhouse gas emissions. Sustainable approaches such as integrated soil fertility management and integrated pest management seek to improve soil structure and enhance microbial biodiversity, but we lack a mechanistic understanding of how multifaceted decisions at the farm level shape these context-dependent small-scale processes in the long term. Future research needs to bridge the microscale and field scale to inform agricultural management decisions for building climate-smart, resource-efficient and stress-resilient agroecosystems, and to harness the soil microbiome as a nature-based solution for sustainable agriculture.

1.9 Soil Biota Regulates Detrimental Organisms and Processes

The survival of any organism in soil, pathogenic or not, depends on its interactions with other members of their biological communities (biotic factors). In soil, a diversity of interspecific interactions-mutualism, antagonism (including parasitism and predation), competition and commensalism-determine the presence and impacts of soil organisms, including those who become pathogenic (Samaddar et al., 2021). Some relationships are indirect, with pathogen invasion, establishment and survival depending on relatively general changes in soil communities, such as an increased population density or biomass. For example, diffuse competition with a broad swath of the microbial community could regulate pathogens in soils by limiting their access to resources. However, in other cases, more direct antagonistic or competitive interactions can prove critical. Kinkel *et al.* (2011), for example, present a coevolutionary framework for investigating and managing soil communities capable of suppressing plant diseases. Drawing on the importance of antagonistic coevolutionary relationships in disease suppression, they demonstrate the importance of direct interactions between pathogens and non-pathogenic indigenous soil microbes for promoting and sustaining disease suppression in soils.

The remarkable diversity of soil microbial communities is crucial in deterring pathogen establishment and growth. Soil communities with greater diversity may be more likely to include antagonists or competitors that are particularly effective at suppressing invasive pathogens (i.e., the 'sampling effect'). They may also contain more competitors or antagonists that act in concert to suppress pathogens (i.e. niche complementarity). Strong relationships between diversity and invasibility are observed in many ecological systems; for example, non-native plants are more likely to invade grasslands as native plant diversity is lost. Similarly, reduction of bacterial or fungal diversity in soil, demonstrated experimentally, increases vulnerability to invasion by pathogens. Another consequence of high levels of biodiversity is greater ecological resilience; that is, 'the capacity of a system to absorb disturbance and reorganize while undergoing change so as to still retain essentially the same function, structure, identity, and feedbacks'. Soil microbiomes have a high degree of functional redundancy whereby multiple actors play similar roles in the soil ecosystem. The resilience of complex biological systems is defined not just by the number of different species that respond to a specific disturbance, but more importantly by response diversity, i.e. the range of responses to the disturbance (Ludwig et al., 2018). Diverse soil communities also have the capacity to bounce back more quickly from perturbations (such as invasion of a pathogen) and not bounce as far in subsequent exposures. However, repeated disturbances that reduce the diversity of soil microbial communities, as is common in large-scale, intensive agricultural management, can reduce the response diversity, and, therefore, reduce the resistance and resilience of agroecosystems, creating opportunities for invasion by non-native organisms. Fortunately, targeted management practices that increase soil microbial diversity and biomass can strengthen direct and indirect antagonistic interactions and consequently help regulate deleterious organisms.

Changes in soil abiotic factors are interdependent and exert substantial impacts on soil function and biodiversity (Samaddar et al., 2021). For example, disturbing soil aggregates, key components of soil structure, homogenize the soil environment, expose protected organic matter to degradation, reduce niche complexity and lead to biodiversity losses. Intensive agricultural practices, land-use changes and global climate change are recognized as major threats that modify the soil environment and reduce soil biodiversity. Even so, none of these factors can be considered in isolation to test their effects on biodiversity as they interact and influence one another. Better understanding of mechanisms underlying management practices that focus on soil environment can facilitate the conservation of soil biomass and biodiversity. Relationship between the concept of soil health and the regulation of detrimental organisms: The concept of soil health has re-emerged over the last few decades and grown in importance in sustainable agriculture and climate resilience. Healthy soils are broadly defined as those being 'capable of supporting the production of food and fiber, to a level and with a quality sufficient to meet human requirements, together with continued delivery of other ecosystem services that are essential for maintenance of the quality of life for humans and the conservation of biodiversity'. Much of the emphasis in soil health assessments has been placed on soil organic carbon, nutrient cycling, soil structure and water relations, and general microbial activities. Most existing conceptual frameworks or tools for soil health do not directly consider suppression of detrimental organisms to plants and humans, despite the fact that suppressiveness is often an additional benefit of practices targeting other soil health indicators.

Two decades ago, van Bruggen and Semenov (2001) proposed considering incidences of plant and animal disease as symptoms of an ecosystem in poor health. They argued that the ability of the biological community to suppress or reduce populations of pathogens (e.g. 'disease suppression') is an important indicator of a stable and healthy soil ecosystem. Several years later, Janvier *et al.* ((2007) proposed including suppressiveness as an indicator of soil health. More recently, Lehmann *et al.*(2020) proposed a 'new generation' of soil health indicators and recommended the inclusion of biological assessments of ability of soil to suppress disease. Thus, an emerging consensus is disease suppressiveness is an integral component of soil health. A challenge is the difficulty in assessing disease potential in soils and thus in identifying appropriate and feasible indicators, as well as methodologies for measuring them.

1.10 Soil biota and Plant Pathogens

i. Pathogens of plants in soil: Plants face different challenges than do animals from the perspective of soil-borne diseases. Humans and most terrestrial animals are primarily in contact with the soil's surface and exposed to small amounts via ingestion, inhalation and dermal contact. Plants, on the other hand, are literally rooted in soils (Samaddar et al., 2021). To access the diffusely distributed nutrients and water in soil, their roots are in intimate contact with soil throughout their life span. A draw-back is that the plant has no 'down time' nor can it escape from the soil, and thus strategies are needed to deal with continuous assaults from soil pathogens. A benefit of this intimacy, however, is there is time (including evolutionary time) for the plant to develop collaborations with rhizosphere organisms to regulate impact of pathogens.

What are termed 'disease suppressive soils', soils whose indigenous microorganisms reduce establishment, persistence or impacts of pathogen are excellent examples of soil's potential to regulate detrimental organisms (Schlatter et al., 2017). Research differentiates between general and specific suppression of pathogens, yet recognizes there is a continuum from general to specific, 'with the former underlying and potentially give rise to the latter over time'. For a disease to emerge in a crop requires a susceptible host plant, a pathogen that is virulent and an environment conducive to infection. Both abiotic and biotic processes in soil regulate the potential for and severity of a pathogen's impact on plants. Soil-borne plant pathogens are taxonomically quite diverse and include bacteria, archaea, fungi, viruses, nematodes and protozoa. A number of classification schemes have been employed for plant pathogens, ranging from plan nutritional habit, plant physiology, parasitism to ecological perspectives.

Whether pathogens are (i) part of the soil microbial community, (ii) survive in soil for an extended period, or (iii) are transient entities that can only briefly exist outside of a suitable plant host determines the dynamics of their interaction with soil communities and abiotic factors and, therefore, determines the mechanisms by which soils may suppress them. More extensive overviews of relationships between the plant rhizosphere, pathogens and beneficial organisms can be found elsewhere.

ii. Mechanisms of suppressiveness of plant pathogens: What is termed 'general suppressiveness' results from the activities of consortia or communities of microorganisms, via competition for resources or antagonistic activities, and is often effective against a variety of plant diseases (Samaddar et al., 2021). Antagonistic interactions are influenced by the nutrient and energy supply available for growth in soil of both the pathogen and to its host. Properties of general suppression are (i) not transferable from a suppressive soil to another soil, (ii) can be reduced or removed by sterilization of soil, (iii) often enhanced by inputs of organic amendments and tied to increases in abundance and activity/diversity of the microbial community. Stimulation of indigenous microbial communities is thought to deplete limiting resources for pathogen growth and infection; suppression of *Phytophthora* root rot is an example. Sometimes the impact is linked to major subgroups of the community. Specific amendments, like debris of wild rocket or rice bran, can enrich *Streptomycetes* spp. and help suppress the pathogen *Fusarium oxysporum* and potato scab diseases. Secondary metabolites, both volatile and soluble, produced by many bacterial species in response to interspecies interactions and competition is also considered to play a role in general suppression.

Specific suppressiveness results from individual taxa of microorganisms and, unlike for general suppression, the benefit is often transferable from one soil to another (e.g. via inoculating with suppressive soil). Antibiosis is involved in specific suppression of take-all disease of cereals and often emerges after long periods of monocropping. The specific suppression of the disease take-all, caused by *Gaeumannomyces graminis*, has been directly linked to antibiotic production by fluorescent *Pseudomonas* spp. An example of a parasitic antagonistic interaction is the suppression of the nematode, *Heterodera schachtii* via the fungi *Dactylella oviparasitica* and *F. oxysporum* infecting nematode cysts and eggs. Many plant growth-promoting rhizobacteria (PGPR) release antimicrobial or antifungal compounds that deter plant pathogens. For example, fluorescent pseudomonads produce the antibiotic 2,4-DAPG which has been extensively studied as a protectant against soil-borne diseases.

Far from the view of plants as passive factories that transform sunlight into organic molecules, it is now recognized that plants actively respond to a complex assortment of physical and chemical environmental cues and precisely regulate their immediate soil environment. Recent advances in omics and visualization techniques have revealed complex and dynamic (i.e. responsive) relationships between plants and their rhizosphere microbial communities, and that plants recruit specific taxa and functional groups to help them uptake nutrients, promote growth, increase stress tolerance and avoid or fight off disease (Trivedi et al., 2020). Responses include the recruitment of beneficial rhizosphere microbes. When the plant cell membrane perceives a stressor, it releases a 'cry for help' transmitted by downstream signalling networks which trigger an immune response in the plant. For example, kinases are produced leading to accumulation of plant hormones like abscisic acid, salicylic acid (SA), jasmonic acid (JA) and ethylene. They stimulate changes in the plant root's exudates which target and recruit selective groups of microbes to colonize its rhizosphere. These recruitments essentially create a 'suppressive soil memory' which protects successive plants of the same type growing in the same location from infection by those pathogens.

Elucidating who, and what mechanisms, are involved in the consortia of microbes conferring general soil suppressiveness is increasingly possible through coupling culture-dependent and culture-independent methods. In studies of soils suppressive of *Rhizoctonia solani*, Mendes *et al*. (2001) could link higher relative abundance of key bacterial groups like Proteobacteria, Firmicutes and Actinobacteria, and genes coding for non-ribosomal peptide synthetases, to pathogen suppression. Similarly, in a study of soils that suppress *Ralstonia solanacearum*, specific rhizosphere taxa associated with whether or not plants developed disease symptoms included members of *Pseudomonas* and *Bacillus*, along with high abundance of genes encoding antimicrobial compounds. This growing knowledge provides the foundation to design strategies to better select for beneficial native organisms, identify viable microbial inoculants or engineer plant rhizosphere microbiomes to reduce disease incidence in plants. Abiotic properties of soil also play many roles in regulating plant disease, both directly and through their interactions with soil biota. Higher available nitrogen can increase disease susceptibility by favouring pathogen growth; for example, increasing colonization by *Pseudomonas syringae* on winter wheat and higher soil moisture content can increase impacts of pathogenic *Ra. solanacearum* on ginger. These interactions are reviewed in greater detail elsewhere.

1.11 Effect of Agricultural Practices on Role of Soil and Soil Biota

It is long known that environmental conditions, and consequently how we manipulate the environment to manage soil, profoundly affect soil biota and soil health (Samaddar et al., 2021). Traditional and indigenous farming typically include practices of using organic inputs and diversified rotations, at least in part to reduce crop losses from detrimental organisms. Many of these same practices are used today in organic farming, in some cases being re-integrated into conventional farming systems to both sustain and increase resiliency of agroecosystems. There are co-benefits and trade-offs associated with many management practices. Some implemented with the goal of improving fertility or soil structure also impact, either negatively or positively, soil's ability to regulate detrimental organisms.Figure 3 depicts the major components of agricultural systems (boxes) and how biological processes mediate their interactions (arrows). Also shown are the impacts of different management practices on the biological underpinnings of the agroecosystem. Whether intentional or not, farm managers choose practices that work with (or against) the ecological principles underpinning soil suppressiveness discussed earlier. Below, we briefly consider how selected management practices affect the ability of soil to regulate detrimental organisms and processes.

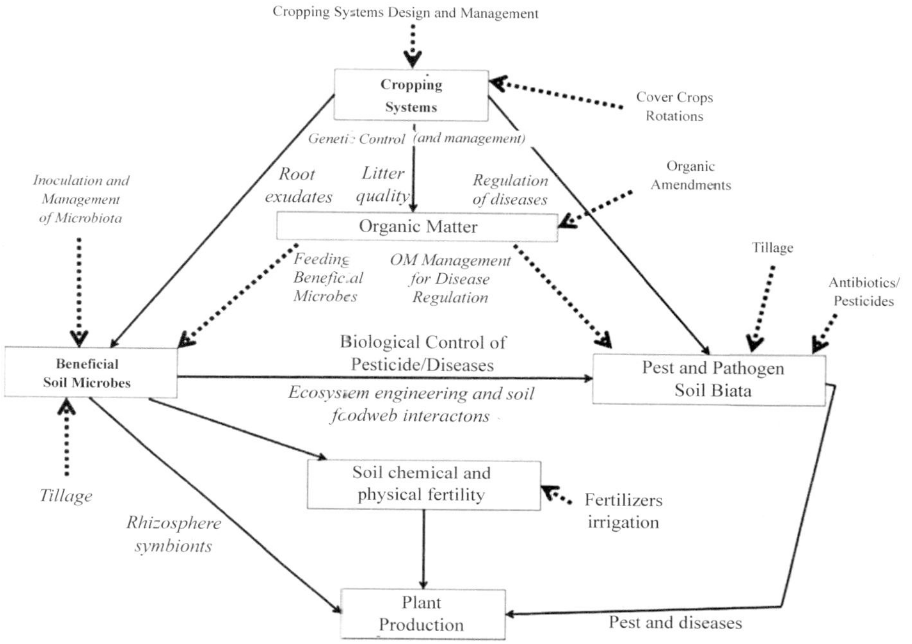

Fig. 3: The entry points for beneficial and pathogenic soil microbes, and impacts of cropping practices, on plant production. Modified from Brussaard et al.,(2007). The boxes refer to major relevant components of the agroecosystem. The text in red and the dotted arrows refer to management practices

i. Organic amendments: Soil organic amendments are among the most frequently used practices to reduce deleterious impacts of soil pathogens. Amendments include composts made from a variety of sources-animal manure, food waste, green wastes- and non-composted materials such as green waste, biodigestates, biochar and biosolids from sewage treatment plants (Samaddar et al., 2021). Organic amendments can have positive or negative effects on soil's ability to regulate detrimental organisms resulting from the introduction of nutrients or toxins, and beneficial organisms or pathogens. Of all soil amendments, compost has received the most attention because it is widely available and less associated with negative impacts (e.g. risk of adding food-borne pathogens) than, for example, untreated animal manures. Compost increases soil micro- and macronutrients, alters physical properties like soil structure and increases microbial biomass, activity and diversity. Mechanisms of disease suppression include competition among microbial populations, antagonism via antibiosis, hyperparasitism involving direct attacks on pathogens and systemic-induced resistance (SAR) of the plant. The provision of a variety of carbon and nutrient inputs also modifies the interactions and equilibrium among members of the soil community and may help the plant in recruiting beneficial organisms.

Nonetheless, improperly processed composts, as well as untreated animal manures, can introduce pathogens to soil. For example, greater concentrations of *Salmonella* spp. were detected in soils amended with poultry-based organic amendments than synthetic fertilizer (Gu et al., 2018). Factors determining the efficacy of compost inputs are type of manure, method of application, type of soil, storage, composting protocols, nutrient ratios, microbial diversity of the amended soils, as well as geographical and environmental factors. Concerns that animal-based composts may contribute to soil pathogens in leafy greens and other vegetables have led to changes in compost processing and application to minimize pathogen risks. Compost prepared following established guidelines greatly reduces the potential for introducing pathogens. Recent meta-analyses showed soil-borne fungal plant pathogen suppression in more than half of studies that used traditional organic amendments; however, in 20% of cases disease incidence increased. Specific inputs can be designed based on understanding of soil microbial interactions. For example, adding biopolymers such as chitin and chitosan can stimulate the prevalence and activity of taxa that specifically break down cell walls of pathogenic fungi. Organic inputs can also be combined with flooding of soils in anaerobic soil disinfestation (ASD) to select for anaerobic communities that generate pathogen-suppressing fermentation products (Samaddar et al., 2021).

Recently, biochar has received increasing attention for its potential benefits for soil health, water relations and disease suppression. Biochar amendments have been shown to be effective in suppressing a wide range of pathogens including bacteria, fungi, virus and nematodes (Bonanomi et al., 2020). Some of the mechanisms involved may be similar to those of compost as discussed above; however, biochar's strong sorptive properties may also play a role. Biochar sorption and inactivation of pathogen-derived toxins and enzymes are potential mechanisms in disease suppression. Broad generalizations about the efficacy of biochar in regulating deleterious organisms are difficult to make at this time however, given how dependent results are on biochar feedstock, synthesis conditions, application rate, as well as soil and crop properties.

ii. Cover crops: Cover crops are typically included in rotations to improve soil parameters such as aggregation, water infiltration and water-holding capacity, and provide nitrogen if legumes are used. However, general suppressive activity can also result from using cover crops due to their stimulation of microbial biomass, diversity and activity (Samaddar et al., 2021). Incorporated grasses (sorghum, millet, oats, rye), brassicas (rapeseed, mustard), legumes (alfalfa) and vines (kudzu) as cover crops have been shown to decrease specific plant pathogen populations. Mixtures of cover crops, especially increasing plant functional group richness, may increase the resistance to pathogens in the following crop; however, single cover crop species can also be effective. There are instances, however, where cover crops increase enteric pathogen survival so choosing the right cover crops is vitally important.

iii. Diversifying crop rotations: Recognition of host–pathogen cycles have influenced development of diversified cropping systems because they reduce the build up of soil-borne plant pathogens that results when a crop is grown continuously (Samaddar et al., 2021). A recent meta-analysis showed crop diversification can boost soil biodiversity and fertility while also regulating disease incidence. Rotations regulate pathogen populations by disrupting the host–pathogen cycles; altering the soil's physico-chemical traits, or biological communities that make the environment less conducive to pathogen development or survival (general suppression); or through direct inhibition of pathogens via production of toxic chemicals or stimulating specific antagonists (specific suppression). The specific choice of crops in rotation is important to ensure planted crops are not compatible with pathogens selected for by the previous crop. Examples of successful rotations include red clover mitigating tuber disease in a potato and using brassicas to combat *Verticillium dahliae* infestation in strawberry production. The integration of animal and cropping systems is another strategy. For example, to reduce parasitic loads in cattle, grazing management using pasture diversification or rotation are recognized approaches to improve livestock husbandry (Grace et al., 2019).

iv. Tillage: Reducing tillage of soils decreases erosion and loss of soil organic matter; however, its impacts on detrimental soil organisms are more complex and sometimes contradictory (Samaddar et al., 2021). Reduced or conservation tillage which leaves crop residues on the soil surface, or partially buried in soil, can create a competitive environment which facilitates competition and antagonistic interactions between microbes. This can result in disease suppression and lead to niche differentiation and increased microbial diversity deeper in the soil profile. In other cases, however, retention of crop residues can facilitate survival of pathogens by protecting them from microbial attack. Some pathogens, even in the absence of their host plants, can survive as saprophytes or as spores until the host returns (Wang et al., 2020). Conventional deep tillage may translocate the pathogens deeper into soil where the environment is less conducive to survival. Other studies found no impact of tillage management on *E. coli* and *Salmonella* numbers.

v. Use of antibiotics and agrichemicals: Numerous agrichemicals including herbicides, insecticides and fungicides are routinely used in intensive, large-scale agricultural management, often with negative consequences for soil communities, particularly fungi and soil fauna (Samaddar et al., 2021). Pesticides use has increased over the past decades. Of what is applied to soil, only a small portion of the pesticide reaches the target organism and the remainder stays in the soil where it may be biodegraded but may also expose soil communities (Jacobsen and Hjelmsø, 2014). Seed dressing with insecticides and fungicides can reduce activities of earthworms, mycorrhizal fungi and other microbial populations and processes. Many processes governing nutrient cycling, e.g. nitrification, are sensitive to pesticides and other organic chemicals. The available concentrations and bioavailability of pesticide on soil microbes depend on soil properties as well, but a common impact is the disruption of specific soil functions and reduced diversity. Although the impacts of agrichemicals on soil communities have been well documented, there is little direct information about how these chemicals impact the soil's ability to regulate pathogens. However, in aquatic ecosystems, the use of herbicides and fungicides was associated with increases in populations of pathogens by reducing densities of protozoan predators and by altering competition with indigenous microbes, highlighting need for performing similar studies in soil.

Similarly, antibiotics have been widely used in the suppression or management of diseases in livestock and some crops (Leigh Archer, 2020). However, overuse of such antibiotics has increased potential threats to the soil microbial community and also elevated antibiotic resistance in the soil environment. Impacts of high concentrations of antibiotics can, but not always, cause changes in enzyme activity and carbon use, reduction in microbial biomass and shifts in the community composition. Low, sub lethal concentrations of antibiotics are problematic because they exert a selective pressure for antibiotic resistance in microbial populations and may promote persistence of certain human and other animal pathogens in soil. Although much effort has gone into reducing antibiotic use in livestock production worldwide, the spread and dissemination of antimicrobial resistant bacteria and genes are still a global health concern. Efforts to improve antimicrobial stewardship worldwide need to continue and expand to reduce impacts on human and environmental health.

Soils contain very substantial numbers of phytopathogenic microbes capable of devastating crops worldwide, viz., the, fungi, viruses, bacteria, nematodes, mycoplasma, protista etc., Some pathogens have very limited and highly specific host ranges while others are generalists, to a lesser or greater degree, causing diseases across many host taxa (Dixon and Tilston, 2010). Similarly, their geographical ranges may be restricted or alternatively spread around the entire globe. The intensity of pathogenesis varies also from those which devastate crops by ultimately killing their hosts in the processes of colonisation and reproduction to those which are only mildly aggressive and possibly almost commensalists. The forms of disease syndrome generated in crop hosts vary from simple root invasion and rotting through to altered root growth, e.g. root galling and clubbing, to collar rotting and damping-off into vascular wilting and colonisation throughout all the aerial organs of the plant. Some pathogens invade early in the host life-cycle and may then enter a quiescent or dormant phase from which they re-emerge devastating maturing flowers and fruit or destroying products stored after harvesting from the crop. Indeed some pathogens are only apparent from the toxins which they elicit in stored products.

The extensive range of pathogenic lifestyles available to soil-borne microbes increases the problems facing plant pathologists striving to unravel their biology. These difficulties are compounded by the sheer physical obscurity resulting from dwelling either wholly or at least partially in soil. Attempting to ascribe taxonomic identity and to understand the life-cycle biology of organisms that conduct all or part of their life-cycles in soil and subsequently are enclosed in the bodies of their host plants is a complex and at times confusing task. Adding to these problems are barriers in understanding the physiology and metabolism of host roots which are the primary targets for invasion by soil-borne pathogens. Roots possess a modular structure permitting responses to their soil environment and adaptation to changes, some of which are a result of the presence of pathogenic microbes. In achieving adaptation the roots differentiate between adopting partnership modes with benign microbes which enhance the potential efficiency for resource capture and adopting defensive modes as a reaction to the presence of pathogens. The root-cap region appears to be the main environmental sensing and response control centre. Recent research indicates apparent root-to-root interactions and the capacity for recognising 'self' and non-self' roots. Root exudates possibly form one means for positive communication with benign microbes and negative responses to pathogens.

1.12 Characteristics of Soil Microbial Communities

i. Inoculum Potential: Balanced in the equation of soil factors and husbandry techniques which support sustainable crop growth are the general traits of microbes which give them advantages as pathogens. Pathogens can only dominate the soil environment and cause disease when they are present in sufficiently large quantities that they are able to breakdown the genetic barriers possessed by crop hosts and overcome edaphic biological, chemical and physical soil environments which are adverse to them. Raising the microbial diversity of soils is associated for example, with increased suppressiveness towards soil-borne pathogens. Grassland soils for example, possess higher microbial diversity compared with arable ones and are able to diminish the effects of R. solani (van Elsas et al. 2002). A cardinal characteristic for successful pathogenesis by soil-borne organisms is the inoculum potential of a pathogen. This concept embraces for the pathogen all of those factors which cumulatively lead to successful invasion, colonisation and pathogenesis. Where inoculum potential is high one or possibly a few disease-causing microbes form a substantial proportion of the total soil microflora population and incite ill-health in the host. Despite this concept being conceived by Garrett in 1956 it remains one of the corner stones of our understanding of host and pathogen biology in the soil. The concept of inoculum potential applies to both pathogens and to the benign microbes which antagonise them and hence may have value in evaluating biocontrol agents.

The biodiversity of benign soil microbes which are antagonisitic to pathogenic forms is critical in order to maintain soil health and quality (Dixon and Tilston, 2010). Diversity is affected by crop type, soil type and their management as discussed earlier in this chapter. Characterising the diversity of benign microbes is frequently defeated by sheer scale of soil inhabitants. In the top layers of soil even with only moderately fertility there are 109 bacterial cells per gram most of which resist culturing. Less than 5% of soil microbial biomass has been cultured and taxonomically characterised. The relationship between the size of microbial biomass and the capacity of a soil to suppress pathogens was demonstrated. A *Pythium* sp. was suppressed by high microbial biomass which reduced the impact on bulbous Iris. Mechanisms suggested which underpined suppression included nutrient competition, commensalism, microbial antagonism, parasitism and systemic induced resistance. Host effects which lead to suppressivity have been ascribed to continuous monocropping leading to: disease decline, cover cropping, compost application, tillage and rotation. Adding organic amendments such as manure, compost and cover crops especially when combined with biocontrol agents such as *Trichoderma* or *Gliocladium*, limit pathogenesis as seen with reductions of sheath blight in rice. Pathogens emerge as a result of cultivation where a homogeneous host population and a conducive environment coincide. If this is accepted then it may be reasonable to assert that the contrary state of suppressiveness also results from the appropriate combination of microbial populations and soil environmental factors. Understanding the development of suppressive soils may be obtained from firstly following the development of suppressiveness over time, secondly examining naturally occurring suppressive soils and thirdly using soil treatments which encourage suppressiveness.

The horizontal and vertical distribution of *Aphanomyces* spp. causing root rot disease of field peas is reported by Moussart et al. (2009) and provides clues in understanding the distribution patterns of pathogen propagules and the onset of disease. Measurements of inoculum potential indicated a horizontal distribution of propagules among several foci in the field. Total soil nitrogen correlated with host root mass and negatively with the root colonisation. Stimulating vigorous host root growth apparently discouraged the pathogen. In another example intercropping apparently aided the inoculum potential of beneficial microbes such arbuscular mycorrhizas in tropical agroforestry with trees such as *Albizzia procera* (safed siris), *Ambilica officinalis* (aonla) and *Eucalyptus teritciornis* using black gram and wheat or green gram and mustard. Knowledge of inoculum concentration is equally important as that for the environment. Thus studies of soil inoculum densities and Verticillium wilt (*V. dahliae*) in *Acer platanoides* and *Catalpha bignonioides* indicated that 5% diseased plants occurred at 1–2 microsclerotia per gram of soil. Bio-assay techniques used to establish concentrations of propagules of pathogens in soil have tended to remain empirical and lack analytical precision. Recently, however, improvements have been achieved using enzyme-based and PCR-based systems as reviewed for P. brassicae, the cause of club root in brassicas.

The colonisation profile is affected by soil physical properties and this character is analogous to the pathozone profile for the growth or movement of fungal parasites during primary and secondary infection. In primary infection the fungus grows saprophytically from a source of previously colonised host material towards a susceptible host. The analogies between parasitic and saprotrophic invasion can be extended by considering saprotrophy by soil microbes as a special case in the general epidemic processes in which colonisation replaces infection. From this there can be a general interpretation of the dynamics and spread of pathogens and saprophytes, parasites spread horizontally through a host population (such as roots) and saprophytes move through a discrete series of nutrient sites. Unlike saprophytes the parasitic fungus could experience a switch from finite to invasive growth. At low root densities the pathogen is constrained by the availability of host tissue as this changes there is a switch to rapid invasive spread. This could explain phenomena such as sudden increases in the rate of infection during the spread of take-all (*G. graminis* var. *tritici)* on wheat as the epidemic switches from primary to secondary infection in response to changes in the density of roots.

ii. Entry and Colonisation: Root colonisation is an important first stage in infection by pathogens. Early research by Hiltner in 1904 (Hiltner 1904) suggested the "rhizosphere effect" whereby microbes are attracted to nutrient supplies exuded by roots. This has been reviewed by Bais et al., (2006) who contend that in addition to providing a carbon rich environment the roots initiate cross-talk signalling with soil microbes which in turn produce messages initiating colonisation. Motility is an important trait of both pathogens and beneficial microbes enabling participation in this cross-talk. Chemical attraction (chemotaxis) is involved in cross-talk between microbes and root. Electrical potentials resulting from electrogenic ion transport at the root surface, attracts swimming zoospores of oomycete organisms to the root surface. Some rhizobacteria create suppressive soils which inhibit the activities of plant pathogenic microbes. Mechanisms for this form of biocontrol include competition for nutrients as supplied in root exudates, niche exclusion, induced host systemic resistance (ISR) and the production of antifungal metabolites.

Root exudates substantially increase microbial activity in the rhizosphere. Their role in pathogenesis caused by root invading bacteria and fungi has yet to be fully appreciated. This deficiency is the result of inadequate technology in the methods of analysis (Dixon and Tilston, 2010). It is however, recognised that there is constant "underground warfare" alleviated for host plants by production of phytoalexins and defense proteins. Sequencing the genomes of crop plants reveals that they are rich sources of antimicrobial indole, terpenoid, benzioxazinone and flavanoid / isoflavanoid natural products. This chemical diversity could yield direct forms of sustainable control of pathogenic microbes and/or indicate indirect means via the stimulation of their secretion. The rhizosheath is the soil adhering directly to the roots and the rhizosphere is the zone of soil directly under the influence of the roots, the pathozone is the region of soil surrounding a root, seed or hypocotyls within which fungal propagules must exist if they are to have any chance of infecting the host. The ability of a fungus to invade was not determined by the furthest extent of fungal growth but instead by the distribution and density of fungal biomass as captured by the pathogens.

Invasion and persistence by pathogen in host populations are defined by Gilligan and van den Bosch (2008) in straightforward terms. Invasion they identify as "the introduction and subsequent increase of the pathogen in a host population" while persistence requires the host population to contain sufficient susceptible individuals such that pathogenesis continues for a prolonged period. Modifying these characteristics by forms of control limits the scale and extent of pathogen epidemics. The form of an epidemic results from the integration of the factors: inoculum, host, environment and time, environment includes aerial and edaphic meteorological factors and those of the soil itself such as pH, texture, bulk density and chemistry. The transition from inoculum to epidemic is driven by fungal (pathogen) growth, the infection dynamics of a single host and invasive spread of the disease-causing microbe through a population of hosts. Like soil, the rhizosphere and more specifically the rhizoplane, has microsites where the microbial residents are concentrated. These microsites are often a function of the architectural complexity of roots and provide an array of potential niches for antagonists.

iii. Pathogenesis: The invasion and disruption of root tissues following infection of a root system by a pathogen has a profound effect on the efficiency of the supply of water, ions and growth regulatory substances and on the rate of root growth (Dixon and Tilston, 2010). The position and depth of lesions on wheat roots dictate the emergent physiological effects. More severe take-all (*G. graminis* var. *tritici*) for example, may develop following infection of newly formed, undifferentiated tissues than after infection of differentiated tissues where subsequent vascular colonization can by limited by the presence of an endodermis, particularly if the lesions are close to the shoot. Penetration to the stele effectively excises the distal portions of roots, though this may not necessarily translate into reduced growth rates as there is a state of functional equilibrium in the partition of resources between the roots and the shoot. Low levels of infection can stimulate production of nodal roots (and account for increases in root-to-shoot ratios), with the result of at least off-setting and possibly over-compensating for those lost due to penetrating infections, though the reverse may also be true. The regenerative capacity depends on the growth stage and of agricultural cereals, barley has greater regenerative capacity than wheat and this is the reason why wheat is affected more severely by take-all (*G. graminis* var. *tritici*) than barley.

Initially crop growth is vigorous so there is increased rate of water uptake and its rapid depletion consequently waterstress takes effect earlier, predisposing the plants to attack by *G. graminis* var. *tritici.* The low water content restricts the activities of other microflora, especially bacteria so levels of antagonism by *Pseudomonas* spp., for example, are often low. In addition, the area of the root–soil interface will decrease as the dehydrating roots shrinks away from the soil particles as its volume decreases. Disease incidences arising from primary infection (soil-borne inoculum, auto-infection) and secondary infection (root-to-root passage, allo-infection) are distinguished in modelling *Pythium violae* (a cause of cavity spot in carrot) resulting in the confirmation that this disease syndrome is polycyclic.

Necrotic host tissues, such as diseased roots, are known to support a diverse array of saprophytic organisms and fluorescent pseudomonads often colonize and dominate in the community arising in take-all lesions. Within colonized *G. graminis* lesions, it is suggested that antagonists prevent pathogen outgrowth from the lesion and limit subsequent autoinfection and infection by other pathogens resulting in further tissue destruction. The bacteria present within the rhizosphere are less fastidious with their requirements for pre-formed organic compounds, vitamins and growth factors than those present in non-rhizosphere soil. The quantitative and qualitative composition of sugars, organic acid, amino acids, glycosides, nucleotides and their bases, enzymes, vitamins, indole-derivatives, phenolic substances and other substances within root exudates changes during plant growth and root development, and in response to foliar applications of chemicals and environmental and climatic factors. With increasing root age cell debris augments the diversity of nutrients within the rhizosphere. The amount of organic material released by roots can be considerable with 1–2% of the carbon translocated to the roots being released into the soil as water-soluble exudate (0.2–0.4%) and 0.8–1.6% as insoluble mucilaginous material.

If the substances actively secreted by roots, the lysates of root material and gases released from the roots, referred to collectively as rhizodeposits are included in such calculations then 12–40% of photosynthate enters the soil in this manner (Dixon and Tilston, 2010). As well as stimulating antagonistic organisms and other metabolic types of bacteria such as carbohydrate fermenters, cellulose decomposers, ammonifiers and denitrifiers; pathogens are attracted towards and stimulated by root exudates. The rhizosphere therefore, presents a suitable microhabitat in which events relating to disease suppression could occur. In suppressive soils, the zone of suppression around roots extends for up to 1.5 mm from the rhizoplane, but the rhizosphere can influence soil up to 2 cm from the rhizcplane. For rhizosphere organisms to effect disease suppression, they must be in place before arrival of the pathogen within the infection court.

In field-grown wheat the seminal roots harbour small, sparse populations of bacteria located mainly at the junctions of cortical cells. Whereas in old nodal roots large aggregates of bacteria colonize the troughs created by the extensive longitudinal corrugations and intercellular spaces in the cortex and may even be found in the xylem. To suppress, displace or pre-empt a pathogen in the invasion of an host the antagonist must be a specialized and aggressive colonist of host tissue. Furthermore an antagonist must be able to use most of the nutrients needed by the pathogen and preferably the antagonist should have some means of inhibiting the growth of the pathogen. Pathogens and their antagonists inhabit ecological niches with a high degree of equivalence and the only way in which they differ is that antagonists lack the ability to initiate the flow and transport of food out from the host.

1.13 Soil Biodiversity and Root Pathogens in Agroecosystems

Soil ecosystem is a living and dynamic environment, habitat of thousands of microbial species, animal organisms and plant roots, integrated all of them in the food webs, and performing vital functions like organic matter decomposition and nutrient cycling; soil is also where plant roots productivity represent the main and first trophic level (producers), the beginning of the soil food web and of thousands of biological interactions (del Pilar Rodríguez Guzmán, 2022). Agroecosystems are modified ecosystems by man in which plant, animal and microorganisms biodiversity has been altered, and sometimes decreased to a minimum number of species. Plant diseases, including root diseases caused by soil-borne plant pathogens are important threats to crop yield and they causes relevant economic losses. Soil-borne plant pathogens and the diseases they produce can cause huge losses and even social and environmental changes, for instance the Irish famine caused by *Phytophthora infestans* (1845–1853), or the harmful ecological alterations in the jarrah forests of Western Australia affected by *Phytophthora cinnamomi* in the last 100 years.

How can a root pathogen species increase its populations densities at epidemic levels? In wild ecosystems usually we expect the soil biodiversity (microbiome, nematodes, mycorrhiza, protozoa, worms, etc.) through the trophic webs and different interactions between soil species, are going to regulate each other and the pathogens populations, avoiding disease outbreaks. In agroecosystems where plant diseases and epidemics are frequent and destructive, soil-borne plant pathogens has been managed applying different strategies: chemical, cultural, biological agents and others; however so far, there is not enough knowledge about how important is soil biodiversity, mainly microbiome diversity and soil food webs structure and function in the management of root pathogens, in root and plant health, in healthy food production, and maybe more relevant in the conservation of soil as a natural resource and derived from it, the ecosystem services important for life in our planet.

Soil biodiversity: The soil is an ecosystem, a living system where interplay mineral and organic materials. Soils are built up through millions of years, from the parent rock layer to the small sand, lime and clay particles derived from physical and chemical intemperization processes, and from biological activities which perhaps be the most important factor in the soils formation (del Pilar Rodríguez Guzmán, 2022). Of the total soil components, organic material represents 5 to 10%, from this percentage 10 to 20% is the active fraction, from the active organic fraction only 20 to 40% are living organisms, and from these 50% are fungi, 30% are bacteria and actinomycetes, 10% are yeasts, algae, protozoa, and nematodes, and 10% are fauna. This means, the active microbial biomass represents 90% of the soil living organisms. Another essential component in soil biodiversity is the soil fauna which is divided in function of their size (mm) into micro-, meso- and macrofauna. Microfauna includes Protozoa, Rotifers from 0.005 to 0.2 mm; mesofauna is composed by nematodes, arthropods, enchytraeids, mites, springtails from 0.2 to 10 mm; macrofauna include animals like insects from 10 to 20 mm; and mega fauna includes earthworm (≥ 20 mm), macroarthropods and small mammals (cm).

Plant roots are another fundamental component of the soil biodiversity, they represent the primary source of organic matter in soil, and the amount of root production is relevant in the process of decomposition and cycling into organic matter in soils. Roots are the subterranean organ of the plant, and they fix them to the soil. Plants take water and nutrients from soil through the roots and the root vascular system transport them to the upper parts of the plant; in some cases, roots are a storage organ. Root exudates and rhizosphere are relevant components for root functioning, and they are essential for all the biological and soil microbial activities like attraction of mutualist symbionts and pathogenic microbes, release and cycling of nutrients, allelopathic processes and also for physical and chemical soil characteristics such as soil aggregation and structure, cation exchange capacity and pH (Gregory et al., 2013). Plant roots represent the first trophic level (autotrophs entities) in the soil trophic web, from which microbes and small fauna obtain their nutrients and energy. Soil-borne plant pathogens (SBPP) are a component of the soil ecosystem, and also members of the soil biodiversity; these microorganisms live part of their life in soil and in the plant rhizosphere, but they also infect and damage the plant roots from which they feed; in some way plant roots are their habitat. SBPP include organisms from bacteria, fungi, oomycetes, nematode, protozoa groups and mesobiotic entities like virus and viroids.

1.14 Trophic Bebs and Complementary or Interference Interactions in Soil

Since the beginning of soil formation from the parental rock, biological activity is fundamental, many biological interactions are established, and they initiate the soil trophic web (del Pilar Rodríguez Guzmán, 2022). At the beginning, trophic webs may be simple and with few components, but as root biomass and their exudates increase in amount and different types, the soil trophic webs are more complex in their biological diversity, structure and functions. Plant roots are key components in soil function, they provide most of the organic matter to the soils, they are the first trophic level in the soil food web and represent the autotroph organisms (photosynthesizers) from which heterotrophs organisms in the next trophic levels obtain their food and energy. Plant roots and specifically the rhizosphere region have a transcendental role in the dynamic of soil microbial activities through the development and release of rhizodeposits. Rhizodeposits include sloughed-off root cap and border cells, mucilage, and exudates. The exudates are made up of organic acids, amino acids, proteins, fatty acids, enzymes, sugars, phenolic metabolites and other metabolites which are used by microorganisms. Most of the root exudates are released at the root cap and the meristematic zone behind the root cap; therefore, these regions are considered important for determining the temporal and spatial activity and distribution of the microbial communities.

Bacteria and fungi together with protozoa, plant pathogenic nematodes and fungi, are in the second trophic level of the soil food web, they are heterotrophs and they may function as decomposers, mutualists, pathogens, parasites and root-feeders, and they are food for the third trophic level (heterotrophs) that includes also nematodes and protozoa (bacterivorous and fungivorous), and microarthropods which work as shredders, predators and grazers; organisms from this level are food for the fourth and fifth trophic levels (heterotrophs) which include higher level predators like arthropods, small mammals, birds (Garbeva et al., 2006).

Soil protozoa is a group of microorganisms which has not been studied so deep and frequently like bacteria or fungi, and their role in the soil food web is sometimes considered only like predators and grazers of microbes; however, they interact with the root systems and bacteria in several and particularly important modes. *Acanthamoebae* sp., an amoeba is considered the most common soil free-living protozoon, this amoebae have a positive effect on root elongation and branching in interaction with rhizobacteria, and mention that "Protozoa function as bacteria-mediated mutualists promoting plant growth by hormonal feed-back mechanisms and nutrient effects based on nutrient release from grazed bacterial biomass"; these bacteria may also be involved in the different phases of the soil nitrogen cycle. All these activities occur in the soil microbial loop, as a relevant component of the soil food web (del Pilar Rodríguez Guzmán, 2022). Soil nematodes are considered important component in the soil food web; their soil communities are usually large and species-rich, with different functional groups (bacterivores, fungivores, herbivores, omnivores, predators, parasites and pathogens), located in different trophic levels: root feeders nematodes in second trophic level (decomposers), fungi and bacterial feeder nematodes in third trophic level (grazers), nematode and protozoa predators in fourth trophic level (higher level predators), in this way nematodes participate in the regulation of soil microbial communities and indirectly in the flux of plant nutrients. Because of their biological characteristics (they are ubiquitous, abundant and diverse), they respond soon to changes in the soil environment, and for this reason they have been considered as important indicators of the soil health. Plant pathogens and therefore soilborne plant pathogens are also an important component inside the soil food web and the ecosystems (Dobson et al., 2005). They can be considered as microherbivores because they feed over root systems, and later they are food for the next trophic level, protozoa, nematode, rotifers, mites, and microarthropods.

1.15 Root Pathogens- Soil borne Plant Pathogens

Soil borne plant pathogens include organisms from bacteria, fungi, oomycetes, nematode, protozoa groups and mesobiotic entities like virus and viroids. SBPP penetrate, infect and invade the roots using different biochemical and physical mechanisms, causing cell and tissue damages; they feed and establish at different regions in the roots including xylem and phloem, but they also obtain their food and energy from rhizosphere root exudates (del Pilar Rodríguez Guzmán, 2022). SBPP are endo-, ecto-, or semi-endo-pathogens what means they can enter and live their whole life cycle inside roots, or they live some life stages in the soil. In plant pathogenic nematodes, there are species in which young and immature females penetrate only half of their body into the epidermal and cortex cells in the roots (*Tylenchulus semipenetrans*), until they mature and transform into a swollen body containing eggs that are released into the soil (*Meloidogyne* sp.); some plant nematodes are sedentary (*Xiphinema* sp.) while others migrate inside the roots (*Pratylenchus* sp., *Radophulus* sp.) or go up to the stem (*Ditylenchus* sp.). Soil borne plant pathogens produce localized or systemic damages; they damage the roots producing root rots, wilts, necrosis and death which impair nutrients and water uptake to the upper plant organs, where damage is manifested as seedlings damping-off, stunt, chlorosis, wilts, bark cracking, twigs and branch diebacks, drop of flowers and fruits, and in consequence biological and economical yield losses. SBPP are obligate or facultative pathogens; some of these pathogens may live also as soil saprophytic organisms depending on substrate availability and soil environmental conditions. When soil environment conditions are adverse, many SBPP develop resistance structures (e.g., sclerotia, cysts, oospores, chlamydospores), and they enter in a dormancy stage for until 20 (*Sclerotium cepivorum*) or 30 years (*Rhizoctonia solani*).

SBPP produce different kinds of propagules, which refers to any entity or unit able to multiple, disperse and conserve the pathogen population, e.g., in fungi: spores, conidia, sporangia, sclerotia; in nematode: eggs, cysts, immature stages, adult male and female; in virus: they are vectored by fungi, nematodes and mites (Agrios, 2005). Propagules and resistance structures of SBPP remain in soil as inoculum that is dispersed by water, microfauna, and cultural practices. There are important and unique pathogen traits which contribute to their successful establishment, increase and eventual epidemic expression, and these are inoculum density, pathogenicity, virulence, dispersal ability, reproductive mode (sexual/asexual), secondary hosts, and long-term resistance and survival structures in soils. SBPP are another natural component of the soil communities and part of the soil food web, they interact with plant roots, but they also interact with other soil organisms in mutualist and antagonist relationships and all these interactions shape the SBPP population density and dynamics in time and space. Some examples of important SBPP affecting plant communities and crops are, fungi: *Fusarium oxysporum*, *Rhizoctonia solani*, *Verticillium dahliae*, *Armillaria mellea*, *Gaeumannomyces graminis*, *Sclerotium ceviporum*; Oomycetes: *Phytophthora cinnamomi*, *P. capsici*, *Pythium aphanidermatum*; nematodes: *Meloidogyne incognita*, *Nacobbus aberrans*; bacteria: *Ralstonia solanacearum*, *Agrobacterium tumefaciens*; mesobiotic entities: lettuce necrotic stunt virus (LNSV), spindle tuber of potato viroid.

Economic and yield crop losses caused by SBPP diseases are significant and may provoke loss of the total crop yield as in the white root rot of onion and garlic caused by the fungus *Sclerotium cepivorum* when inoculum density is high and persists in soil for long time.

1.16 Soil Food Web and Resilience in Soils

The soil food web is the community of organisms living all or part of their lives in the soil. It describes a complex living system in the soil and how it interacts with the environment, plants, and animals. Food webs describe the transfer of energy between species in an ecosystem. While a food chain examines one, linear, energy pathway through an ecosystem, a food web is more complex and illustrates all of the potential pathways. Much of this transferred energy comes from the sun. Plants use the sun's energy to convert inorganic compounds into energy-rich, organic compounds, turning carbon dioxide and minerals into plant material by photosynthesis. Plant flowers exude energy-rich nectar above ground and plant roots exude acids, sugars, and ectoenzymes into the rhizosphere, adjusting the pH and feeding the food web underground. Plants are called autotrophs because they make their own energy; they are also called producers because they produce energy available for other organisms to eat. Heterotrophs are consumers that cannot make their own food. In order to obtain energy they eat plants or other heterotrophs. An example of a topological food web (image courtesy of USDA) is illustrated in fig. 4.

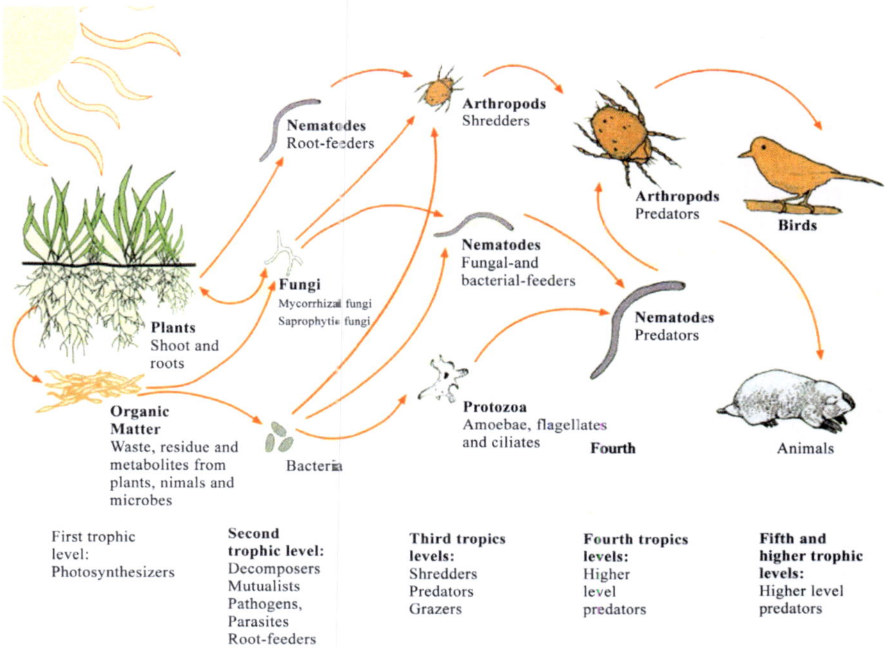

Fig. 4: Topological Soil Food web

Resilience in soils: Soil resilience refers to the ability of a soil to resist or recover their healthy state in response to destabilising influences. This is a subset of a notion of *environmental resilience*. Soil resistance, a related term refers to the ability of soil to resist changes or the extent to which a soil will recover from any cropping or management change. The term is distinct from Soil resilience as resistance is the inherent capacity to withstand disturbance, while resilience is the capacity to recover after disturbance (Arthur, et al., 2012).

Belowground roots, exudates, microbial and microorganisms diversity conform a complex and diverse soil food web with multiple trophic and complementary relationships, which can be classified as mutualists (+) and antagonists (−) relationships that in theory must result in a well-balanced soil system where is expressed the suppressiveness to SBPP; these are call "suppressive soils" (del Pilar Rodríguez Guzmán, 2022). In the development of the biological control of SBPP the concept of suppressive soils has been a key one, because take in consideration that in the soil ecosystem there are a whole microbial (fungi and bacteria) community with the potential to interact with root pathogens and regulate their population densities and dynamic under certain physical and chemical soil environmental conditions. Baker and Cook (1974) originally defined suppressive soils as "*soils in which the pathogen is not able to establish or persist, the pathogen establishes but causes no damage, or the pathogen causes some damage, but the disease becomes progressively less severe, even though the pathogen persists in soil*".

Development of molecular biology methodologies has allowed to better understand that in the soil microbial community, certain groups of bacteria and fungi are involved in soil suppressiveness. Mendes et al. (2011) found thousands of bacteria and archaeal species in the groups of Proteobacteria, Firmicutes and Actinobacteria constantly associated with suppression to disease caused by *Rhizoctonia solani* in beet. In other research Penton et al. (2014), resolving soil disease suppression to *R. solani* strain AG8 and *Fusarium psudograminearum* in cereal fields in Australia, found that suppressive soils were attributed to less than 40 genera of fungi, including certain endophytic species and mycoparasites in the groups Ascomycota, Basidiomycota and Chytridiomycota; the fungi genera most associated with the suppressive fields were *Xyllaria* (endophyte), *Bionectria* (mycoparasite), *Anthostomella* (saprotrophic), and also with antifungal activity *Chaetonium*, *Corynascus* and *Microdiplodia*; these authors indicate the importance of analyzing soil suppressiveness including both fungi and bacteria, and their interactions with fungi and plants. It has been mentioned that ecosystems as open systems, have the ability of buffering negative stresses throughout their resilience and resistance properties expressed at the population, community, and ecosystem levels (Shade et al., 2012), but this ability depends on the strength and time of duration that the stress factor persists. Suppressiveness of soils to SBPP and diseases may be considered as an expression of the soil resilience capacity where soil microorganisms multifunctionality must play an important role. Soil resilience is defined as the ability of a soil to recover to its initial state after a stress event. Suppressiveness/resilience may not be necessarily manifested in every soil in natural environments and even less in agroecosystems.

1.17 Interactions of Soil Nutrient Environment, Pathogenesis and Host Resistance

Host plants and soil borne pathogens that attack them exist within an ecological matrix populated by numerous microbial species that may influence the access of pathogenesis. These events are moderated by physical and chemical components of the soil (Dixon, 2002). The impact of inorganic and organic nutrients on pathogenesis and the development of host resistance were discussed in this review using two host-pathogen combinations as examples. Calcium, boron, nitrogen and pH have been demonstrated to affect the processes of resting spore germination, host invasion and colonisation in the *Plasmodiophora brassicae-Brassica* combination that results in club root disease. Organic nutrients that have associated biostimulant properties have been demonstrated to influence the development of *Pythium ultimum-Brassica* combination that results in damping-off disease. This latter combination is affected by the presence of antagonistic microbial flora as demonstrated by increased ATP, extra-cellular enzyme and siderophore production. In both examples there are indications of the manner by which host resistance to pathogenesis may be enhanced by changes to the nutrient status surrounding host plants. These effects were discussed in relation to the development of integrated control strategies that permit disease control with minimal environmental impact.

The generally adequate availability of most nutrients for plant growth has induced an apathetic recognition of the vital role of minerals in defense (sic) against plant disease (Huber, 1980). It is doubtful that a satisfactory understanding of most host-pathogen interactions will be obtained until associated nutrient relationships are elucidated. And: The nutrition of a plant determines in large measure its resistance or susceptibility to disease, its histological or morphological structure or properties, the functions of tissues to hasten or slow pathogenesis, and the virulence and ability of pathogens to survive. Studies in this area require understandings not only of hosts and pathogens but also of their nutrient metabolism. Huber (1980) went so far as to identify that nutrients are associated with host defence against pathogens on at least three fronts, viz., Maximising the inherent resistance of plants ; Facilitating disease escape through increased nutrient availability; Stimulating plant growth and altering the external environment to influence survival, germination and penetration of pathogens. This list could be supplemented by considerations of the impact of nutrients on populations of benign soil flora and these are of at least as great significance as each of the others. Such holistic ecological studies have lapsed for almost two decades being overshadowed by molecular research.

Biostimulants provide an organic stimulus that encourages soil fungistasis and antibiosis. In each case this can result in the indirect stimulation of host plant growth and reproduction which is additional to the direct effects derived from adding extra nutrients for the host plants. Adding nutrients that encourage plant growth and reproduction results in more robust growth which is resistant (or tolerant) to invasion by plant pathogens (Dixon, 2002). Some fragments of scientific knowledge are beginning to emerge that demonstrate that forms of generalised host resistance (or tolerance) are increased when inorganic and organic nutrients are available. Distinguishing whether this is a direct result of the presence of nutrients acting on host metabolism or indirect effect derived from the enhanced presence of benign microbes will require substantial research effort especially where soil-borne pathogens and microbes are concerned. Nonetheless such studies are likely to yield improvements to the understanding of the manner by which hosts, pathogens and benign microbes interact within the soil environment. In turn this offers more ecologically sound philosophies for the control of plant pathogens that avoid damage to the environment since they do not require the use of site specific, artificial, toxic molecules for pathogen control. Integration of the manipulation of crop nutrition with plant pathology adds substantial dynamism to minimising losses caused by diseases. Seeking to understand the basic mechanisms is of major importance since this will make the processes of pathogen control more focused and enable them to be refined and improved while still sustaining the environment, increasing world food security, diminishing pollution and encouraging the preservation of biodiversity.

1.18 Role of Soil Health in Plant Disease Management

Soil health sustains the biological productivity, maintains environmental quality and promotes plant health. Soil borne diseases are most damaging when soil conditions are poor. Major factors influencing the disease in plants are soil moisture, soil temperature, soil pH and soil nutrients (Thakur Rajneesh et al., 2022). To manage disease, different methods are used like crop rotation, biological control, cover crops, suppressive soils, organic amendment, plant growth promoting rhizobacterium, vascular arbuscular mycorrhizal fungi, mulching, good compost, good aeration, *etc*. Implementation of these practices improves the soil health and reduces disease incidence in a sustainable manner. Cover crops, crop rotations and healthy sanitary practices keep the pathogenic populations at low levels and also add beneficial nutrients like nitrogen, phosphorus and potassium to the soil. These all manage the disease by creating physical barrier, releasing antagonistic chemicals, competing with pathogen, increasing nutrient uptake, *etc*. Biological control agents have different mode of action *viz*., parasitism, predator, antibiosis, competition for site and nutrition, as well as by inducing the resistance in plants against pathogen [Induced Systemic Resistance (ISR)]. However, all these management strategies are helpful in improving the soil health, decreases the disease incidence and subsequently increases the yield and productivity of the crop.

Soil borne plant pathogens such as *Rhizoctonia* spp., *Fusarium* spp., *Verticillium* spp., *Sclerotinia* spp., *Pythium* spp. and *Phytophthora* spp. can cause 50 to 75 per cent yield loss for many crops such as wheat, cotton, maize, vegetables, fruits and ornamentals (Baysal-Gurel and Kabir, 2018). The use of fungicides against soilborne plant pathogens can help to manage some diseases, in contrast, frequent and indiscriminate use can increase environmental and health concerns, and lead to development of fungicide resistance (Christopher et al., 2010). Some environment-friendly approaches have been developed to manage the soil borne diseases while maintaining the soil health. In soil, microorganisms reside in rhizosphere and constitute a complex organization of endophytes, saprophytes and actinomycetes, both harmful and beneficial ones (Baysal-Gurel et al., 2019). Various strains of Bacillus species are known as potential elicitors of ISR and exhibit significant reduction in the incidence or severity of various diseases on diverse hosts (Thakur Rajneesh et al., 2022). It is believed that plants have the ability to acquire enhanced level of resistance against pathogens after getting exposed to biotic stimulation provided by many plant growth-promoting rhizobacteria (PGPR's) and this is known as rhizobacteria mediated induced systemic resistance (ISR). Arbuscular mycorrhizal fungi (AMF) can suppress plant pests and diseases through induction of systemic resistance. Mycorrhiza-induced resistance (MIR) involves an ISR component elicited by bacteria in the mycorrhizosphere, as AMF, rhizobacteria possess MAMPs, which can trigger MAMP-induced immune responses. Green manuring can play an important role in this regard as it showed versatile impacts like improvement of soil physio-chemical and biological properties, fertility, nutrient supply to succeeding crops, checking erosion and plant protection. This review paper seeks to summarize the different methods used for management of soil borne diseases by improving soil health including mycorrhiza, cover crops, bioagents, mulching, good compost, good aeration, suppressive soils, nutrients, intercropping and crop rotation (Thakur Rajneesh et al., 2022).

Soil health: Soil quality (health) is defined as "The capacity of a soil to function, within ecosystem and land use boundaries, to sustain biological productivity, maintain environmental quality and promote plant and animal health." It is worth noting here that "soil health" and "soil quality" are essentially synonymous terms. Soil moisture, soil temperature, soil pH, organic matter, nutrients, beneficial microorganisms, organic amendments, crop rotation, cover crops and green manures are the various factors which influence soil health.

Soil moisture: Water contained in soil is called soil moisture. Optimum soil moisture is important for plant growth. There is a relation of moisture with diseases. Late blight of potato, apple scab, downy mildew of grapes and fire blight are favoured by high rainfall and high relative humidity. Most severe symptoms of *Phytophthora, Rhizoctonia, Sclerotinia* and *Sclerotium* can be observed on plants when the soil is wet but not flooded. In case of bacterial (*e.g., Erwinia* and *Pseudomonas*) and nematode infection, more severe symptoms are produced on plants when the soil is wet. *Streptomyces scabies* which causes the common scab of potato becomes more severe in soils drying out after wetting. *Fusarium roseum,* the cause of seedling blights, *Macrophomina phaseolina*, the cause of charcoal rot of sorghum and root rot of cotton and *Fusarium solani,* the cause of dry root rot of beans occur in dry environments. Verticillium and canker diseases of forest trees are more severe in water stress conditions. Soil moisture and temperature have direct relation with seedling blight of wheat and corn. All these diseases can be managed by maintaining the soil moisture level below or above optimal soil moisture required by the pathogen.

Soil temperature: Warm, moist soils with high levels of carbon to nitrogen ratio will have higher level of microbial activity and a relatively higher level of suppression. Soil temperature can greatly affect the activity of locomotion, infection and reproduction of nematodes. Some species of the fungi *Typhula* and *Fusarium,* which causes snow mold of cereals and turf grasses, thrive only in cool season or cold regions. Also, the late blight pathogen *Phytophthora infestans* is more serious in the Northern latitudes whereas in the subtropics it is serious only during the winter. Many diseases, such as brown rot of stone fruits caused by *Monilinia fructicola*, are favored by relatively high temperatures and are limited in range to areas and seasons in which such temperatures are prevalent. Several diseases, such as the Fusarial wilts, anthracnoses caused by *Colletotrichum* spp. and the bacterial wilts of solanaceous plants caused by *Ralstonia solanacearum*, are favored by high temperatures and are limited to hot areas, being particularly severe in the subtropics and tropics. At temperature much below or above the optimum for the pathogen, or near the optimum for the host, disease development is slower.

Soil pH and soil structure: Soil pH is an estimate of the activity of hydrogen ions in the soil solution. It is also an indicator of plant available nutrients. High activity is not desirable and the soil may require liming with base cations Ca or Mg in order to bring the solution back to neutral. Soil pH increases by adding lime while it decreases by adding sulphur and ammonium. The pH of the soil is important in the occurrence and severity of plant diseases caused by certain soil borne pathogens. For example, the club root of crucifers caused by *Plasmodiophora brassicae* is most prevalent and severe at about pH 5.7, whereas its development drops sharply between pH 5.7 and 6.2 and is completely checked at pH 7.8. Common scab of potato caused by *Streptomyces scabies* can be severe from pH 5.2 to 8.0 or above, but its development drops sharply below pH 5.2. Soil texture and structure could have effects on plant diseases because they affect water holding capacity, nutrient status and gas exchange as well as root growth. Poor soil aeration caused by poor soil structure, soil type or water logging is associated with the development of cavity spot (*Pythium* spp.) disease in carrot. The pea root rot complex *(Fusarium* spp.) is known to be affected by compaction, temperature and moisture of the soils. All these diseases can be managed by maintaining good aeration or soil pH and avoiding waterlogged conditions.

Soil microorganisms: Microflora (bacteria, fungi, actinomycetes, *etc.*) constitute up to 75 to 90 per cent of the soil-living biomass and are the primary decomposers of organic matter (Thakur Rajneesh et al., 2022). Two major components of microbial biota in soil are: a). Disease-inducing microbes: Fungi and bacteria can cause diseases to plants or degrade the soil quality by interfering with beneficial microorganism(s) thereby affecting plant health. Fungal pathogens like *Plasmodiophora brassicae* (club root of crucifers), *Spongospora subterranea* (powdery scab of potato), *Fusarium, Rhizoctonia, Verticillium, Phytophthora, Pythium* and *Sclerotinia* and bacterial pathogens like *Ralstonia solanacearum* (wilt), *Erwinia* sp. (soft rot) and *Streptomyces scabies* (potato scab) causes a great extent of damage to many crops. Soil-borne viruses also cause diseases such as *Lettuce big vein virus* (LBVV) and *Lettuce stunt necrotic virus* (LSNV). These soil-borne pathogens can survive in soil for many years. Biocontrol agents inhabiting in soil (Resident biocontrol agents): Soil residing microorganisms (bacteria/fungi) are used successfully for controlling diseases. Disease control and better crop health can be achieved through various activities performed by these soil microbes like hydrocyanic acid (HCN) production, siderophore production, nitrogen assimilation, antibiotic production, hydrolytic enzyme (lipase, chitinase, *etc.*) production, induced systemic resistance and systemic acquired resistance. *Bacillus* spp. and *Pseudomonas* spp. serve as excellent examples of biocontrol agents having prominent PGPR *(*plant growth-promoting rhizobacteria*)* activities and disease reduction ability (Saha *et al.*, 2016).

On the basis of microbial activity of soil, two types are as follows.

i. **Disease-inducing soils:** are disease causing soil which means that pathogenic microbes comprise of 5 to 20 per cent of total soil micro flora. When fresh organic matter is applied in this type of soil, incomplete oxidized products are released, which are hazardous to plants and in turn, are easily attacked by pathogens or insects. Such soils can be amended into disease suppressive soils by addition of inoculum of effective microorganisms

ii. **Plant growth promoting microorganisms (PGPM):** exert direct effect on plant growth promotion *e.g. Rhizobium* and *Glomus* spp. Bacteria which possess the tendency to colonize roots actively are called as PGPR. For soil-borne pathogens and disease management, rhizospheric microbes emerged as a biological weapon that triggers the mechanism of disease reduction through Systemic Acquired Resistance (SAR) and Induced Systemic Resistance (ISR). Soil micro flora facilitate uptake of nutrients from soil which results in enhanced yield as well as disease reduction or suppression. *Bacillus subtilis* has the potential in disease reduction and more than twenty antibiotics are produced by them. Efficacy of *Bacillus* spp. has been reported in different crop plants like tomato, chilli, brinjal *etc.* to control different pathogens like *Colletotrichum acutatum, C. capsici, C. gloeosporioides, Pythium aphanidermatum* and *R. solani.*

Antibiotic production: Rhizobacteria contributes in disease suppression with antibiotic production. There are six classes of antibiotic compounds (for which their modes of action are partly understood) that are related to the biocontrol of root diseases *viz.* phenazines, phloroglucinols, pyrrolnitrin, pyoluteorin, cyclic lipopeptides (all of which are diffusible) and HCN. *Pseudomonas fluorescens*-5 (Pf-5), source of pyoluteorin and pyrrolnitrin, was found to be effective in protecting damping off in cotton (*Pythium ultimium /Rhizoctonia solani*). Majority of *Bacillus* spp. for the production of antibiotics such as polymyxin, circulin and colistin, were found active against Gram-positive and Gram-negative bacteria and many pathogenic fungi. An antibiotic, Geldanamycin produced by *Streptomyces hygroscopicus* var. *Geldanus*, reduced the *Rhizoctonia* root rot of pea. Similarly, HCN is a volatile compound and is useful in the biocontrol of diseases.

Root colonization: The main aspect for biocontrol is the colonization of rhizosphere soil or external / internal root zone by microbes, especially bacteria. In biocontrol method, root colonization is completed in two phases: Firstly, bacteria get attached to rhizosphere and are then transported on the elongating root tip. Secondly, bacteria spread locally, proliferate to the boundaries of niche by competing with other indigenous microorganisms and survive. Root colonization involves recognition of pathogens by potential antagonists and it is important for base line of effective biocontrol strategies.

Induced systemic resistance: Induced systemic resistance (ISR) was explained and reported in carnation plant in which *Pseudomonas* strain (Pf WCS 417r) was found effective against *F. oxysporum* f. sp. *dianthi.* Induced resistance is the ability of plants to develop an enhanced defensive ability when appropriately stimulated. Some pathogenesis-related proteins (PRs) like 1, 3-glucanases and chitinases are capable of hydrolyzing fungal cell walls and insects. *Pseudomonas* and *Bacillus* spp. are the most popular rhizobacteria encompassing ISR.

1.19 The Soil Food Web and Nutrient Cycling

The soil food web is a common grouping of soil organisms according to what they eat (i.e. trophic groups). Trophic levels also reflect the flow of energy through the soil food web, as all organisms consume food to get the energy they need to grow and reproduce (Amanda Bennett, 2023).

Nutrient cycling: Every time an organism feeds, waste nutrients are excreted in a form that is more available to plants. Plant roots are hotspots for such feeding activity so plants can readily take up these nutrients leading to the production of more plant material, supplying new organic matter to the soil and completing recycling of nutrients. Predators (e.g. protists, nematodes, mites and springtails) are important to drive the release of nutrients and make them available for plant uptake. Soil fauna can account overall for 30–40% of net nitrogen released into plant-available forms. The remaining nutrient is released by microbes or the enzymes they produce.

Trophic levels in the soil food web: Figure 5 illustrates the various trophic levels in the soil food web (Amanda Bennett, 2023).

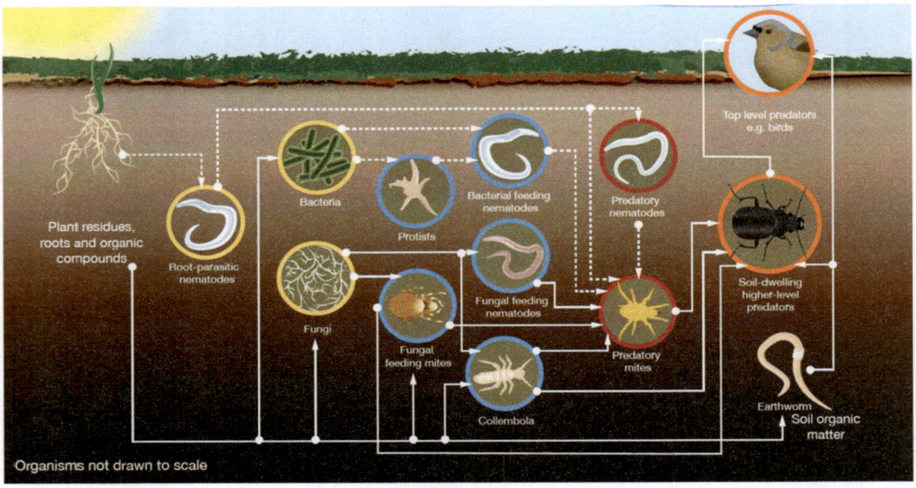

Fig. 5: Trophic levels in the soil food web

Basal trophic level: Food sources that are either living plant tissues or dead organic matter such as plant litter or the dung and dead remains of soil or above-ground organisms.

Second trophic level (yellow circles): Microorganisms feed on food sources in the basal trophic level and comprise decomposers feeding on dead organic matter, herbivores and plant parasites that feed on living plants. and microbes that feed on compounds exuded from roots.

Third trophic level (blue circles): Nematodes and microarthropods that graze on bacteria or fungi but may also eat dead organic matter.

Fourth trophic level (red circles): Larger microarthropods and nematodes that eat their smaller cousins.

Fifth and higher trophic level (orange circle): Higher level predators in the soil food web include predatory mites, centipedes, and beetles.

Root zone interactions: The high level of feeding activity around plant roots, driven by the compounds that plants exude, releases nutrients where plants can access them and provides a diverse range of predators that can attack root pests. These root exudates also support beneficial microbes that help produce plant growth compounds and fight off disease.

Soil organisms stabilise soil structure: Good soil structure is important for aeration, root proliferation (to access water and nutrients) and better drainage. Plant roots, bacteria, fungi, earthworms, and other soil biology help stabilise soil structure by binding soil into aggregates (Fig. 6) (Amanda Bennett, 2023).

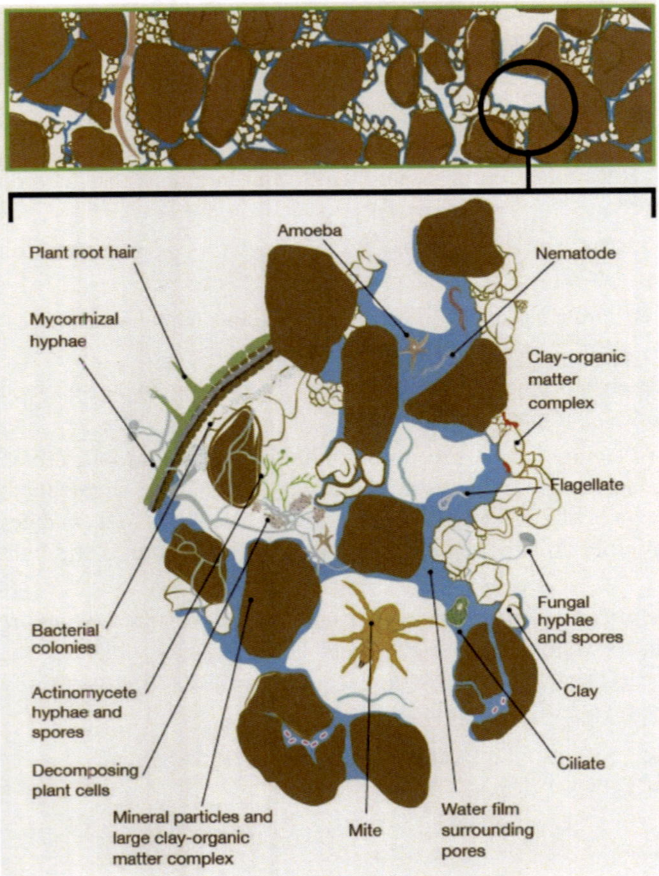

Fig. 6: Soil organisms in soil stabilization

Soil organisms help to develop stable aggregates

- Soil organisms produce a range of sticky compounds (glycoproteins and polysaccharide gums) that help bind clay, silt and sand particles into microaggregates
- Roots and fungi help to enmesh and bind these into larger aggregates or push tightly bound clay particles apart
- Bacteria can form protective biofilms over aggregate surfaces
- Plant roots also have a central role in structure development through drying and compression as they grow
- Earthworms change the structure of soil by burrowing and generating new stable aggregates as worm casts. This creates new microhabitats for other soil organisms and plant roots

The soil pore network defines habitat space: The soil environment shows extreme variation in space (vertical and horizontal) and time. A wide range of surface types, pore sizes, microclimates and resources form the landscape in which soil organisms live and interact (Amanda Bennett, 2023). Roots use pores of >100 µm as entry points, while root hairs, protists and fungi use pores of >10 µm. Bacteria can move in water films of only 1 µm depth.

Pore size distribution: The amount and nature of the pore space in soil is dependent on soil texture and the way in which mineral particles and soil organic matter are aggregated together. Pore size distribution controls the balance of oxygen and water available to organisms. It also regulates the access of soil organisms to one another and their resources. Larger pores can drain freely, but water is held on particle surfaces and in smaller pores. The dung and organic compounds excreted by soil organisms are efficient at storing water in soil. It also acts as a sponge for many plant nutrients, retaining them in the soil in a form that plant roots can release.

Soil borne plant pathogens and deleterious microorganisms: Soil borne pathogens may not cause obvious above-ground symptoms, although there may be stunting or areas of the field showing poor plant performance. Rotting roots will impair root functioning (Amanda Bennett, 2023). Damage occurring below ground can have a significant impact, particularly when the harvested product is a root crop, as symptoms may not be noticed until harvest. Disease in the field can be patchy rather than uniform. Control of soil borne pathogens is not simple, and chemical options are not really viable. Spores of most soil borne pathogens can survive for many years. Microbial interactions in the root zone (rhizosphere) that have a detrimental effect on the plant without visible symptoms are not well understood. The impact of deleterious microorganisms on plant growth and productivity has been demonstrated by soil sterilisation or soil transfer experiments, and various modes of action have been implicated.

1.20 Soil as a Huge Laboratory for Microorganisms

Biodiversity consisting of living organisms both plants and animals, constitute an important component of soil. Soil organisms are important elements for preserved ecosystem biodiversity and services thus assess functional and structural biodiversity in arable soils is interest (Sachidanand et al., 2019). One of the main threats to soil biodiversity occurred by soil environmental impacts and agricultural management. This review focuses on interactions relating how soil ecology (soil physical, chemical and biological properties) and soil management regime affect the microbial diversity in soil. It was proposed that the fact that in some situations the soil is the key factor determining soil microbial diversity is related to the complexity of the microbial interactions in soil, including interactions between microorganisms (MOs) and soil. A conceptual framework, based on the relative strengths of the shaping forces exerted by soil versus the ecological behavior of MOs, is proposed. Plant-bacterial interactions in the rhizosphere are the determinants of plant health and soil fertility. Symbiotic nitrogen (N2)-fixing bacteria include the cyanobacteria of the genera *Rhizobium, Bradyrhizobium, Azorhizobium, Allorhizobium, Sinorhizobium* and *Mesorhizobium.*

Free-living or associative N2-fixing bacteria belonging to the species *Azospirillum, Enterobacter, Klebsiella* and *Pseudomonas* have been shown to attach to the root and efficiently colonize root surfaces. Free-living soil bacteria play a vital role in plant growth, usually referred to as plant growth promoting rhizobacteria (PGPR) are capable of promoting plant growth by colonizing in plant root zone. These are associated with the rhizosphere, which is an important soil ecological environment for plant-microbe interactions.

Plant growth promotion and development can be facilitated both directly and indirectly. Indirect plant growth promotion includes the prevention of the deleterious effects of phytopathogenic organisms. This can be achieved by the production of siderophores, i.e. small metal-binding molecules. Biological management of soil-borne plant pathogens and the synthesis of antibiotics have also been reported in several bacterial species. Another mechanism by which PGPR can inhibit phytopathogens is the production of hydrogen cyanide (HCN) and /or fungal cell wall degrading enzymes, e.g., chitinase and ß-1, 3-glucanase. Direct plant growth promotion includes symbiotic and non-symbiotic PGPR which function through production of plant hormones such as auxins, cytokinins, gibberellins, ethylene and abscisic acid. PGPR also help in solubilization of mineral phosphates and other nutrients, enhance resistance to stress, stabilize soil aggregates, and improve soil structure and organic matter content. The review by Sachidanand et al., (2019) indicates that soil environment drivers and their interactions may cause changes in soil microbial community and activity overall abundance; however, for this addition of organic matter, loosening the soil mass, providing the optimum moisture in soil, reducing the heavy agricultural equipment, replacing chemicals with alternative source of manure, could be few steps towards ecofriendly soil with high microbial population. These results illustrated the potential for complex community changes in terrestrial ecosystems under soil management scenarios that alter numerous factors simultaneously.

1.21 Biochar –Its Impacts and Mechanisms on Soil Microorganisms

Biochar is an environmentally friendly material with excellent adsorption properties and high stability that is widely used in the field of water and soil pollution control. In recent years, biochar has attracted considerable attention because of its ability to improve the soil environment and its positive impact on the soil microbial community. Soil is a highly complex habitat that contains many soil microorganisms. It is important to investigate biochar's effects on these microorganisms' activities and the subsequent impacts on the soil environment and the sustainable development of agricultural production (Huang et al., 2023). Biochar promotes the growth of microorganisms, thus increasing their numbers and abundance. Moreover, the nutrients contained in biochar, such as carbon (C) and nitrogen (N), are important for increasing the diversity and abundance of most soil microbial communities. The increased soil nutrient content following biochar application is also a major factor in enhancing microbial metabolic activities. The water-holding capacity and the pore structure of biochar provide a good habitat for soil microorganisms and are conducive to their growth and reproduction. In addition, the high adsorption capacity of biochar, which is due to its high specific surface area, can immobilise pollutants in the soil, reduce the migration and transformation rates of pollutants in the soil, and reduce the toxic effect of pollutants on microorganisms. The raw materials used in the preparation of biochar and the application dose have significant effects on soil microorganisms. Biochar effectively increased the biomass of microorganisms in the soil, and the biomass of microorganisms in the experimental group increased with increasing doses of biochar compared to the control. The pH and pyrolysis temperature of biochar also had significant effects on the soil microbial community.

Biochar is a carbonaceous material derived from the pyrolysis of carbon-rich biomass that has attracted increasing research and attention because of its ability to enhance soil carbon storage, increase soil fertility, fix and transform pollutants in soil, and improve the soil environment. These enhancements directly or indirectly affect soil microorganisms' metabolic activities and community structure. Huang et al.,(2023) reviewed the effects of biochar on soil physicochemical properties, enzyme activities, nutrients, contaminants, and related microbial activities. They also summarised the possible mechanisms involved in the interaction between biochar and microorganisms and the potential hazards associated with biochar use.

Biochar as a habitat for soil microorganisms: Biochar is a pyrogenous organic material synthesised through the pyrolysis of different types of biomass. Biochar has a variety of effects on the soil environment (Huang et al., 2023). The porous structure of biochar is an important feature that can provide a habitat for microorganisms and can protect soil microorganisms from predation. The pore size of biochar (> 50 nm) is usually larger than most soil bacteria, fungi, and protozoa, allowing biochar to accommodate these soil microorganisms, but its pore size excludes large arthropods, which provides conditions for microorganisms to inhabit and colonise without the threat of predation. The colonisation of soil bacteria on biochar reduced the negative impacts of soil nutrient loss due to leaching and increased soil bacteria abundance, but also that biochar had no significant impact on the abundance of soil fungi, which may have been due to the weakened mobility of fungal mycelia caused by its reticular structure. The pore structure of biochar provided some protection to endomycorrhizal (AM) and ectomycorrhizal (EM) fungi from external environmental stress, and biochar also stimulated the germination of AM and EM fungal spores, which facilitated the growth and reproduction of AM and EM fungi. The rate of microbial colonisation of biochar is related to the degree of aging of the biochar. Freshly applied biochar may release some endogenous contaminants, such as Cd and Pb, into the soil, which can weaken the metabolic activity of bacteria and slow their colonisation rate on the biochar (Xiang et al. 2021). Unlike fresh biochar, aged biochar helps bacteria and fungi to grow and multiply because it releases fewer contaminants, and the environment created makes it easier for bacteria and fungi to colonise the biochar. The co-interaction between microorganisms also affects the time and rate of colonisation; for example, bacterial cells co-cultured with *Geobacter metallireducens* and *Methanosarcina barkeri* colonise the biochar surface within a very short time.

Effects of biochar on soil enzyme activity and microorganisms: The response of soil enzymes to biochar is highly variable, and their activity is influenced by the amount of biochar applied, the type of biochar, and the nature of the soil to which it is applied (Huang et al., 2023). Researchers found that the activity of urease and sucrase in the soil increased significantly with increasing doses and concentrations of biochar because biochar application increased the organic matter and nutrient contents in the soil, which created the conditions necessary for the growth of soil microorganisms and increased the activity of soil microorganisms, as well as the activity of enzymes. Biochar application increased the activity of redox enzymes, with significant increases in the activities of catalase and dehydrogenase but no significant changes in the activity of hydrolytic enzymes. The main reason for the increase in dehydrogenase activity is that the application of biochar increases the content of active organic matter in the soil, which in turn enhances the microbial activity in the soil and increases the activity of dehydrogenase. There is a close relationship between soil microorganisms and soil enzymes. The overall enzyme activity in soil consists of various intracellular and extracellular enzymes that originate from microorganisms (e.g., bacteria, fungi) or from plants and animals. Soil enzymes are also key factors affecting bacterial community structure. Biochar application increased the activity of soil urease and the abundance of soil microorganisms, and they identified a significant correlation between the activity of urease and the abundance of soil microorganisms.

The indirect effect of biochar on soil enzymes refers to the interaction between specific microorganisms and plant roots that is stimulated by biochar, thus affecting enzyme activity, expression, and secretion level. The mechanisms through which biochar affects soil enzymes and microorganisms need to be clarified for specific microorganisms and enzymes, as biochar stimulates specific microbial communities and thus affects the activity of their related enzymes. For example, biochar affects the fungal community, thus affecting the activity of ligninolytic enzymes. The research on the influence of biochar on soil enzymes and microorganisms is not sufficient, and research in this area should be further deepened.

Effect of biochar on microbial signalling molecules: N-Acyl homoserine lactone (AHL) is a signalling molecule that regulates inter-microbial behaviour. Biochar adsorbs AHL, thereby affecting gene expression and intraspecific communication in soil microorganisms, most commonly in gram-negative bacteria (Sun et al. 2021). The adsorption of signal molecules by biochar is related to its pyrolysis temperature, with biochar produced via high-temperature cracking having a higher capacity to adsorb signal molecules, as well as interfering more strongly with bacterial inter- and intra-specific signal communication. Biochar also affects the communication of microbial signalling molecules by adjusting the pH of the soil. The degree of hydrolysis of AHL, a signalling molecule that is more sensitive to changes in pH, increases with increasing pH, which reduces intercellular signalling by bacteria. However, farnesol, a signalling molecule that is not sensitive to changes in pH, has no significant effect on inter- and intra-species signalling by fungi. The effect of biochar on signalling molecules causes plants to respond to environmental stresses. Biochar application increased resistance to *Leveillula taurica* and *Botrytis cinerea* in tomatoes and pepper. The effects of different types of biochar on arbuscular mycorrhizal fungi (AMF) and *Fusarium oxysporum* in the tomato root system were found to differ, with waste biochar not enhancing AMF colonisation in the plant root system and wood-prepared biochar affecting the root secretions of the plant, which in turn inhibited the growth and development of *Fusarium oxysporum* mycelium. There are also differences in the contaminants contained in different types of biochar. The volatile organic compounds (VOCs) contained in biochar can inhibit microbial metabolism and may also be an uncertain factor influencing plant resistance. To better understand the effect of biochar on signalling molecules, more research is needed on the quantification of biochar, pyrolysis temperature, and endogenous contaminants.

Mechanisms: The key mechanisms through which biochar affects soil microorganisms after application to soil include, the pore surface structure of the biochar can provide a favourable habitat for microorganisms and can promote their growth; biochar promotes the growth and reproduction of soil microorganisms via improving soil pH, water content, and aggregate content; biochar provides sufficient nutrients for soil microorganisms; the adsorption effect of biochar reduces the concentration and content of toxic and harmful substances in the soil, reducing the toxic effects of harmful substances on microorganisms; and biochar affects microorganisms by influencing microbial signalling molecules (Huang et al., 2023).

Biochar shows good potential for tackling soil pollution and promoting agricultural development. Future research should explore the following, viz., the potential mechanisms of biochar-soil microbial interactions to gain further insight into the direct effects of biochar. application on soil microbes; the impacts of biochar on specific soil microbial communities, such as archaea, yeasts, protozoa, and other microorganisms that have a significant impact on biogeochemical soil cycles; criteria for evaluating the impact of biochar on soil microbial diversity and function to overcome the difficulties in judging this impact; and the response of soil microorganisms to biochar functional groups and the identification and application of specific biochar functional groups, which can clarify the deeper mechanisms of biochar effects on soil microorganisms.

1.22 Grassroots Ecology- Plant Microbe Soil Interactions

Ecological theory recognizes a variety of abiotic and biotic factors that shape the structure and dynamics of terrestrial plant communities. Climate, age, environmental harshness, area, isolation, disturbance, environmental heterogeneity, and biotic interactions may all play key roles in determining plant community composition and diversity in space and time (Heather L. Reynolds et al., 2003). Much of plant ecological theory has been dominated by debate over the relative importance of these factors, manifest in controversy over plant strategy theory and the relationships between competition intensity, species diversity, and habitat productivity. Among the biological forces structuring plant communities, plant-microbe interactions have received little attention. This is not surprising, given the overwhelming diversity of soil microbes, the technical challenges involved in their study, the tendency to view microbial associates as simply extensions of the plant and their outright exclusion from experimental designs through use of artificial soil mixes or soil amendments that result in depleted microbial communities.

A growing body of research on plant-microbe interactions in soil is contributing to the development of a new, microbially based perspective on plant community ecology. Soil-dwelling microorganisms are diverse, and interactions with plants vary with respect to specificity, environmental heterogeneity, and fitness impact. Two microbial processes that may exert key influences on plant community structure and dynamics are microbial mediation of niche differentiation in resource use and feedback dynamics between the plant and soil community. The niche differentiation hypothesis is based on observations that soil nutrients occur in different chemical forms, that different enzymes are required for plant access to these nutrients, and that soil microorganisms are a major source of these enzymes. It was predicted that plant nutrient partitioning arises from differential associations of plant species with microbes able to access different nutrient pools. Feedback dynamics result from changes in the soil community generated by the specificity of response in plantmicrobe interactions. It is suggested that positive feedback between plants and soil microbes plays a central role in early successional communities, while negative feedback contributes both to species replacements and to diversification in later successional communities. It was suggested that plant-microbe interactions in the soil are an important organizing force for large-scale spatial gradients in species richness. The relative balance of positive feedback (a homogenizing force) and negative feedback (a diversifying force) may contribute to observed latitudinal (and altitudinal) diversity patterns. Empirical tests of these ideas are needed, but a microbially based perspective for plant ecology promises to contribute to our understanding of long-standing issues in ecology, and to reveal new areas of future research.

In recent years, a growing body of ecological research on plant-microbe interactions in soil has developed. A new, microbially based perspective on plant community structure and dynamics is emerging from this work. Heather L. Reynolds et al., (2003) reviewed the diversity of plant-microbe interactions in soil and discuss the costs and benefits to each partner. The role of environmental heterogeneity and host specificity in generating a continuum of positive and negative effects, both direct and indirect was discussed. Two processes, mediation of niche differentiation in plant resource use and feedback dynamics between the plant and soil community, emerge from these fundamentals as key microbial influences on plant community structure and dynamics. The evidence for these processes and their applications to understanding temporal and spatial vegetation patterns was discussed and it was concluded with a look toward how a microbial perspective can inform future research directions in plant ecology.

Plant interactions with soil microorganisms have become more central to a variety of ecological research efforts, in spite of the fact that most ecologists lack training in microbiology, the techniques of microbial ecology are challenging to apply in the field, and the diversity of microbes and their interactions with plants are to a large extent still unknown. This paper has focused on microbial mediation of niche differentiation in resource use and feedback dynamics between the plant and soil community, but soil microbes may influence many other important ecological processes. Useful approaches range from observational studies of plant nutrient use and microbial associates in greenhouse or field to highly controlled greenhouse and field experiments, where the identity of nutrient resources and the composition of the microbial community are manipulated, and host plant responses are monitored. Advances in molecular and stable isotope techniques make these approaches feasible. Subsequent research might then focus on comparison of the phenomenon in early vs late-successional communities or in communities along latitudinal or altitudinal gradients. Such comparisons, using techniques of measuring plant growth in sterilized and "cultured" soil, are also needed to evaluate the relative balance of positive and negative feedback over temporal and spatial gradients. These relatively simple techniques have much to offer, although a disadvantage is that the microbial community is treated as a black box and only their net effect on plant growth is revealed. The growing array of molecular techniques will enhance our perspective on plant-soil feedback by allowing the manipulation of individual microbial groups, as well as quantification of microbial populations and their individual effects on plants in diverse natural communities.

1.23 Soil Conditions and Soil-borne Pathogens

1.23.1 Fungal pathogens: Soil-borne fungal plant pathogens are important determinants in the dynamics of plant populations in natural environments and in epidemics in agricultural environments. Examples of economically important soil-borne fungal plant pathogens include *Fusarium* spp., *Gaeumannomyces graminis*, *Verticilium* spp., *Phytophthora* spp., *Pythium* spp. and *Rhizoctonia solani* (Otten and Gilligan, 2006). Despite low initial densities of inoculum in soil, these pathogens can cause complete destruction of plants and, occasionally, total loss of yield. Infections occur initially at microscopic scales (approximately 10^{-3} m) following growth and movement of fungal plant pathogens through soil in a region surrounding a root or other subterranean organ. The region surrounding susceptible hosts from which a pathogen can infect has been defined as the pathozone. Epidemics, however, occur at macroscopic scales of metres and larger with cascading spread through populations of roots or plants. To date, most knowledge of the role of soil physical conditions in epidemics is obtained from data collected at the field scale. Here correlations are sought between particular environmental conditions or agronomic practices and the outcome of epidemics. However, different soil physical conditions operate at macro- and microscopic scales, and at each scale small changes in soil physical conditions can influence the outcome of soil-borne epidemics.

Many epidemics of root diseases involving soil fungi depend on the interplay between fungal growth and the spatial and temporal heterogeneity of the soil environment. Colonization or infection of a root occurs at fine scales with growth and movement of fungal mycelia through soil. However, epidemics are observed at coarser scales, and depend on a cascading spread through populations of roots. Otten and Gilligan (2006) briefly reviewed conventional analyses of soil-borne epidemics and argue that these treat soil physical conditions at scales too coarse to be meaningful for interactions between soil, plants and fungi, and fail to consider the effect of soil physical conditions on the underlying epidemiological processes. Instead, a conceptual epidemiological framework was proposed that integrates spatial scales and used this to review the effect of soil structure on the dynamics of soil-borne pathogenic fungi. Using the soil-borne fungal plant pathogen *Rhizoctonia solani* as an example, it was demonstrate that invasion of fungi into host populations is critically affected by environmental conditions operating at each of two scales, viz., at the microscopic scale (µm- cm) the fungus preferentially explores certain pathways in soil, and small changes in soil physical conditions make the fungus switch from small, dense colonies to large, sparse and rapidly expanding ones; at the larger scale (µm- cm) a critical density of susceptible hosts is required, in excess of which fungi switch from non-invasive to invasive spread. It was suggested that the approach will increase the applicability of research dealing with microscopic soil–plant–microbe interactions towards the solution of large-scale epidemiological problems.

Soil physics and fungal spread: The soil provides the physical habitat in which biological interactions take place. The spatial heterogeneity of the structure of soil is manifest in various ways and at various scales: in tilled soil it can manifest itself in the form of beds of aggregates; in non-tilled soils it may appear as a system with planar pores or cracks between soil peds (Otten and Gilligan, 2006). Biopores, together with variation in soil physical properties associated with structural heterogeneity, such as the availability of water and oxygen, contribute further to the variability in the fungal environment. Despite the overwhelming evidence that disruptions of the soil structure have profound effects on microorganisms, it is yet know too little of the changes they bring about in the microenvironment to pinpoint controlling mechanisms. Microcosm studies in which some control can be exerted over the soil environment offer a way forward by enabling specific hypotheses to be tested. It is, however, essential that they include heterogeneity at scales that are relevant to fungal growth. In this section we introduce the spatial scales that we consider relevant for fungal colonies and discuss the effect of heterogeneity at various scales on fungal spread.

i. Fungal spread through a homogeneous single grain structure: Perhaps the simplest physical environment we can construct for microbiological studies comprises sand with a narrow particle-size distribution, packed at a specific density. The geometry of the pore space through which fungal spread occurs is determined largely by the particle-size distribution of the sand, with pores containing either water or air. The simplicity of the structure enables us to address specific questions related to the effect of the physical environment upon fungal spread in replicated experimental systems. The following broad conclusions were drawn when we observed the spread of *R. solani* through simple homogeneous microcosms (Otten *al* *et*., 1999).

- Fungi spread predominantly, though not exclusively, through the air-filled pore volume, with further and faster spread with increased air-filled pore volume for otherwise identical media.
- Fungi spread preferentially along permeable soil surfaces when placed on such a surface, with faster expanding and up to four times larger fungal colonies, even though the same amount of biomass is formed.

- Changes in pore-size distribution at identical air-filled pore volume affect the rate and extent of colony growth, with more rapid and extensive spread in coarse media at identical air-filled pore volumes; and
- Radial spread of fungal colonies is more variable through soil than on surfaces.

ii. Fungal spread through homogeneous, aggregated soil: Narrow pore-size distributions are not common in natural soils. Pore spaces vary greatly both within and between aggregates, due mainly to varying clay content, organic matter content, and the degree of compaction caused by machinery and cultivation. Even within controlled environments one cannot prescribe soil structure precisely (Otten and Gilligan, 2006). One can, however, use bulk density and aggregate sizes as experimental variables to manipulate soil structure. Controlling these variables within specified ranges that are expected to occur in natural soils may subsequently identify the extent to which fungal spread can be affected. Using biological thin sections, which enable *in situ* visualization of fungal mycelium and soil structure, it was demonstrated the huge effect soil structure has on the spatial organisation of *R. solani* in soil. Colonization potential, defined as the volume of soil surrounding a source of inoculum within which a susceptible host has a finite probability of becoming colonized, was more than sixfold larger for soil packed at bulk densities >g cm 1.4 g cm^{-3} compared with less dense soil (1.2 g cm^{-2}) (Otten *al et.*, 2001). These results accord with greater fungal densities in compacted field soil, suggesting that the findings have generality beyond the isolate used in this study. Not surprisingly therefore, the spatial distribution of fungal hyphae in soil is patchy, in particular in the more porous soil. In such soil, larger voids within a soil are preferentially exploited by fungal mycelium; the denser volumes within the soil are partly avoided, even though these volumes are potentially accessible to fungal growth in denser soil containing no large voids.

iii. Fungal spread through heterogeneous, structured soil: Microcosms for biological studies mostly comprise repacked sieved soils. Additional large-scale heterogeneity in the physical environment is rarely introduced. Yet cracks, old root channels and bio-pores are common features in most natural soils, in particular in reduced tillage systems where they can prevail for a long time. Filamentous fungi are well adapted to explore macro-pores via hyphal extension, and they can readily span air-gaps of several mm *in vitro* and in experimentally controlled arrays of aggregates. The effect of geometry of larger pores on fungal spread through soil by introducing them artificially into replicated microcosms was reported (Otten *al* *et.*, 2004). Using biological thin sectioning techniques we showed that at the advancing edge of a fungal colony, mycelium occurs predominantly within or directly adjacent to cracks. Behind the advancing edge of the colony the fungal hyphae penetrated the surrounding bulk soil, but the results clearly showed that initial colonization occurs through rapid exploration of macro-porosity within soil. The preference for the crack was further enhanced if the conditions in the surrounding bulk soil were made less favourable. The precise mechanisms that make fungi follow these pathways preferentially in soil, however, remain a topic for further investigation. The preferential exploration of these soil features may prove decisive in the interactions between the soil, plants and fungi.

The spatial and temporal dynamics of microorganisms in a complex heterogeneous and competitive environment dictate the outcome of many ecologically and economically important processes in soil, such as organic matter dynamics, successful remediation of contaminated soil, invasion of beneficial organisms or the spread of an epidemic. The pore space provides the environment in which biological, physical and chemical interactions occur, and through which microorganisms and roots spread to interact and exploit resources in soil. It is known that soil structure affects microbiological processes, but this knowledge has not yet led to a level of understanding that would enable successful and predictable management of these processes in soil. Research should now focus more on how small-scale interactions relate to large-scale phenomena observed in fields, and how large-scale manipulation of soil structure is experienced by microorganisms in the rhizosphere. With recent progress in precision agriculture, together with advances in molecular diagnostics that focus on small-scale behaviour, understanding how intervention at smaller scales affects ecosystems is growing.

The effect of soil physical properties, and soil structure in particular, on the dynamics of soil-borne pathogenic fungi and the outcome of epidemics was reviewed. However, other properties can be considered in a similar way. For example, pathozone dynamics have been used to analyse the effects of antagonists, host exudates and genetic differences within pathogen species (Golldack *al et.*, 2004). The epidemiological approach was used to identify two processes operating at different spatial scales that drive soil-borne epidemics, namely the spread of fungal colonies leading to infection of a single host, and invasive spread through a population of hosts. Soil physical conditions affect infection at both scales. There is great potential for soil physical conditions to mediate fungal growth, with small changes in the soil physical environment provoking large and abrupt changes in fungal morphology. At the coarser scale, soil conditions can mediate the spatial distribution of roots, and it was shown how small changes to a population structure could make a fungus switch from non-invasive to invasive spread. This leads to the important conclusion that whilst small-scale heterogeneity in soil physical conditions regulates fungal growth dynamics and infection of a single host, the outcome of an epidemic depends on the non-linear relationship between distributions of hosts, such as root geometry, and the probability for a fungus to spread from an infected to a susceptible host. Whereas it is important that research remains focused on the physical habitat in which soils, plants and microbes interact, one must not isolate such microscopic interactions from the ecological processes under investigation. It was suggested that the epidemiological framework introduced here will increase the applicability of research dealing with microscopic interactions to the solution of large-scale epidemiological problems.

1.23.2 Bacterial pathogens: Indigenous microbial communities in the soil forms complex networks and manipulates plant health. Based on the composition of the resident microbial community, a biological barrier is formed, with microbes interacting with pathogens and defending against invasion by pathogens near the root surface (Liu et al., 2022). Therefore, assessing the relationship between the soil community and crop morbidity is a critical step toward understanding potential impacts of these communities on plant health. It has been revealed that species-rich biomes are more resistant than species-poor biomes to pathogen invasions and high incidence of soil-borne diseases could be due to the deterioration of the soil microecological environment. Additionally, tobacco farmlands with high biodiversity were more resistant to pathogen infection.

Bacterial wilt, caused by *Ralstonia solanacearum*, is one typical soil-borne disease and can bring severe losses to agricultural crops. In real communities, the complexity of the soil environment may largely decrease their inefficiency of pathogen suppression (Liu et al., 2022). However, the phenomenon of 'disease-suppressive soils' shows an ideal model by which plant protection can be triggered by soil microbes. Hence, a better knowledge of understanding the positive functions from indigenous microbial communities in the soil is essential for a sustainable and effective bacterial wilt management strategy.

Natural soil has the ability to suppress the soil-borne pathogen to a certain extent, and the assemblage of soil microbiome plays a crucial role in maintaining such ability. Long-term monoculture accelerates the forms of soil microbiome and leads to either disease conducive or suppressive soils. Liu et al., (2022) explored the impact of soil conditions on bacterial wilt disease (healthy or diseased) under long-term tobacco monoculture on the assemblage of bacterial and fungal communities in bulk and rhizosphere soils during the growth periods. With Illumina sequencing, the bacterial and fungal composition of soil samples from tobacco bacterial wilt diseased fields and healthy fields were compared in three growth periods. It was found that *Proteobacteria* and *Ascomycota* were the most abundant phylum for bacteria and fungi, respectively. Factors of soil conditions and tobacco growth periods can significantly influence the microbial composition in bulk soil samples, while the factor of soil conditions mainly determined the microbial composition in rhizosphere soil samples. Next, rhizosphere samples were further analyzed with LEfSe to determine the discriminative taxa affected by the factor of soil conditions. For bacteria, the genus *Ralstonia* was found in the diseased soils, whereas the genus *Flavobacterium* was the only shared taxon in healthy soils; for fungi, the genus *Chaetomium* was the most significant taxon in healthy soils. Besides, network analysis confirmed that the topologies of networks of healthy soils were higher than that of diseased soils. The results suggested that microbial assemblage in the rhizosphere will be largely affected by soil conditions especially after long-term monoculture.

Soil bacterial community diversity: *Proteobacteria* (36.94±0.64%; mean±SE) was the most dominated bacterial phylum in soil samples, followed by *Actinobacteria* (19.42±0.92%). Relative abundance of bacterial classes changed during different tobacco growth stages in both healthy and diseased soils (Fig. 7a) (Liu et al., 2022). For healthy soils, a clear shift was seen from dominance of *Alphaproteobacteria* (18.89%) in the bulk soil samples collected in March to that of *Betaproteobacteria* (13.65%) in the rhizosphere samples collected in both July and September. For diseased soils, the decrease in relative abundance of *Alphaproteobacteria* (17.94%) were also seen in rhizosphere samples; however, this compensation of *Betaproteobacteria* (13.21%) was only in July samples.

Ascomycota (60.35±2.81%) was the most abundant fungal phylum in soil samples, including *Sordariomycetes* (35.42±4.47%), *Eurotiomycetes* (9.85±3.17%), *Dothideomycetes* (7.09±2.50%), *Leotiomycetes* (2.66±1.22%) and *Pezizomycetes* (2.16±1.31%) listed as top-10 classes (Fig. 7b). *Sordariomycetes* were higher in healthy soils (47.08±0.50%) than in diseased ones (36.12±1.06%). For the rhizosphere samples, there was a consistent enrichment of *Eurotiomycetes* in diseased soils (10.37±3.92%), while *Pezizomycetes* was largely seen in healthy soils (6.27±0.58%). Besides, effects of soil conditions on both bacterial and fungal composition were clearly seen from the PCoA analysis. PERMANOVA analyses were further confirmed with significance of the factor of soil conditions in samples with and without bulk soil samples ($p<0.001$). Specifically, PCoA explained 22% and 12% of total variation in compositions of bacterial and fungal bacterial communities in rhizosphere soils. Additionally, there were few effects of tobacco growth periods on both bacterial and fungal community compositions in rhizosphere samples.

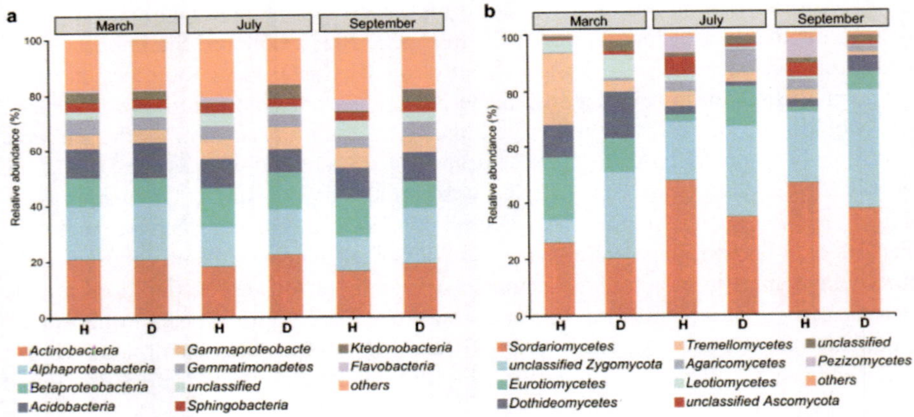

Fig. 7: Relative abundance of the top-10 a) bacterial and b) fungal classes for samples collected form healthy (H) and diseased (D) fields in March, July and September

Rhizosphere bacterial taxa discriminated by soil conditions: Rhizosphere soil samples collected in July and September were further compared with LEfSe to determine the discriminative taxa affected by the factor of soil conditions. In July samples, there were 10 taxa found in the heathy soils and 9 in the diseased soils (Liu et al., 2022). The genus *Ralstonia*, to which the pathogen *R. solanacearum* belongs, was found in the diseased soils (logarithmic LDA=4.25, $p=0.02$). The most significant taxon in the healthy soils was the phylum *Bacteroidetes* (logarithmic LDA=4.45, $p<0.001$). In September samples, a total of 35 taxa were found: 13 taxa in the healthy soils and 14 taxa in the diseased soils. The class *Alphaproteobacteria* (logarithmic LDA=4.78, $p=0.02$) was the most significant taxon, followed by the genus *Sphingomonas* (logarithmic LDA=4.62, $p<0.001$) in diseased soils, while the order *Flavobacteriales* (logarithmic LDA=4.78, $p<0.001$) had the largest effects in healthy soils. Among those discriminative taxa in healthy soils, there were 6 taxa shown in both July and September samples and *Flavobacterium* (in July samples, 1.64±0.11%; in September samples, 3.29±0.42%) was the only shared genera.

For bacterial networks, both the network size and the degree of connectivity showed large differences between healthy and diseased soils. The topologies of the healthy networks, namely, the number of nodes and the number of links, were higher than that of the diseased networks. In total, there were 231 nodes and 433 links in healthy soils, whereas 156 nodes and 216 links in diseased soils. Moreover, using betweenness centrality, *Flavobacterium* (OTU3316, OTU4570) and *Pseudomonas* (OTU1200, OTU3398, OTU9417) were found as key genera in the networks of healthy soils. The greater variety of potential key taxa observed in the rhizosphere samples might be beneficial to maintain plant health. Analyses of LEfSe and network analysis showed that *Flavobacterium* and *Pseudomonas* may be the most active microbial species in healthy soil. *Flavobacterium* can play a role in biological control by producing antibacterial effect factors, antibacterial substances, extracellular macromolecular degrading enzymes, etc. (Carrion et al., 2019). *Pseudomonas* which can produce antifungal/inhibitory compounds and siderophores that can control against bacterial wilt disease. High *Pseudomonas* diversity can reduce *R. solanacearum* density in the rhizosphere and decrease the disease incidence due to both intensified resource competition and interference with the pathogen.

Effect of Soil nutrients: *R. solanacearum* can survive in soil and shift between dormant and active states to persist, attack, and colonize plants, impairing their growth. Therefore, soil suppressiveness to *R. solanacearum* is crucial to supporting crop production. It is widely accepted that soil suppression, especially general suppression, is a function of soil nutrient characteristics (Yifan Cao et al., 2022). In recent decades, the effects of carbon, nitrogen, phosphorus and potassium supply on disease suppression have received growing attentions. For example, a higher content of soil organic matter, generally linked to a support of high microbial diversity, was found to be responsible for suppressing *R. solanacearum* wilt disease. Higher contents of soil available nutrients were also recognized as key factors in determining soil suppression to wilt disease through their direct effects on pathogens and indirect effects on soil microbiomes. Until now, however, there has been no general explanation of whether a particular nutrient can decrease the severity of a disease. Ecological stoichiometric theory uses elemental ratios and the concept of stoichiometric invariance to predict nutrient retention and biomass production at subcellular to ecosystem scales, which helps to understand the elemental balance in ecological interactions and processes. For example, C:N:P ratio was reported to influence organic matter decomposition and nutrient cycling in soil through its effect on microbiomes, with high C/N (C/P) leading to N (P) limitation while low C/N (C/P) leading to C limitation. Similarly, the excesses of N/P, N/K and P/K ratios may also result in relative nutrient deficiency, thus affecting resistance to soil-borne disease (Yang et al., 2018). However, fewer efforts have been devoted to exploring this relationship between nutrient supply and resistance to pathogen, despite they are both important soil determinants of crop production.

Soil nutrient supply and resistance to pathogen invasion are critical to crop yield during agricultural production. Crop production is often hampered by co-occurring nutrient and pathogen stresses, but the relationship between soil nutrient status and disease suppression remains unclear. Yifan Cao et al., (2022) conducted a meta-analysis based on 406 observations from 55 studies to decipher the relationships between soil nutrient supply and disease suppression against bacterial wilt disease caused by *R. solanacearum*. It was found that the contents of soil organic carbon (SOC), total nitrogen (TN), available potassium (AK) and available phosphorus (AP) were significantly higher in relatively disease suppressive soils than in conducive soils. Soil AP showed a significantly negative correlation with both crop disease incidence (DI) and disease severity (DSI). SOC was significantly and negatively correlated with DI and DSI in general, but a positive correlation with DI was found when SOC content was higher than 1.3%. Then for nutrient stoichiometry in soil, the ratio of carbon to nitrogen (C/N) was maximum to 22, was significantly and negatively correlated to the DI, but was positively correlated to DSI when C/N was higher than about 30. The ratio of TN to AP (TN/AP) showed significantly positive correlations with both DI and DSI in soil. Further in applied organic fertilizer, the ratio of nitrogen to phosphorus (N/P) was significantly and negatively correlated with DI when N/P was lower than about 5, but was positively correlated with DI when higher than that. These results demonstrated that the quantity and stoichiometry of phosphorus, organic carbon and nitrogen in soil and applied fertilizer, notably the prevention of nutrient deficiency and the balance of different nutrients, play an important role in supporting soil suppressiveness to *R. solanacearum* wilt disease, and this implied a possible trade-off between nutrient acquisition and pathogen defense in soil ecosystems.

Effect of Soil amendment: The effect of a soil amendment (SA) composed of urea (200 kg of N per ha) and CaO (5,000 kg/ha) on the survival of *Ralstonia solanacearum* in four Philippine soils was investigated in a series of laboratory experiments. Within 3 weeks after application, the SA either caused an initial decrease, a final decline, or no change in the pathogen population, depending on the particular soil type (Michel and Mew, 1998). An initial decrease occurred in a soil with a basic pH and resulted in a significantly ($P < 0.001$) lower pathogen population immediately and at 1 week after amending the soil. This decrease was probably due to the high pH in the soil during urea hydrolysis. A final decline in the *R. solanacearum* population after 3 weeks occurred in two soils in which nitrite accumulated after 1 week. In these soils, no decline in bacterial levels occurred when nitrite formation was inhibited by 2-chloro-6-trichloromethylpyridine. In the soil with low pH, no nitrite accumulated and the *R. solanacearum* population did not decline. The suppressive effects of pH and nitrite on *R. solanacearum* growth were confirmed by in vitro experiments. Ammonium reduced the growth of *R. solanacearum*, but was not suppressive. Interactions of pH with ammonium and nitrite also occurred, whereby ammonium reduced growth of *R. solanacearum* only at pH 9 and nitrite was suppressive only at pH 5. Nitrate had no effect on *R. solanacearum* growth in vitro.

Microbial Network and Soil Properties: It is hypothesized that the occurrence of bacterial wilt may be related to changes in soil chemical properties and microbial interactions. Qi Gaofu, et al., (2019) compared the soil chemical properties and microbial network structures of a healthy soil (HS) and a bacterial wilt-susceptible soil (BWS). The contents of available nitrogen, potassium, and phosphorus and the soil pH in the BWS were significantly lower than those in the HS. BWS showed nutrient deficiency and acidification in comparison with the HS. The structure and composition of the BWS network were quite different from those of the HS network. The BWS network had fewer modules and edges and lower connectivity than the HS network. The HS network contained more interacting species, more key microorganisms, and better high-order organization and thus was more complex and stable than the BWS network. Most nodes and module memberships were unshared by the two networks, while the ones that were shared showed different topological roles. Some generalists in the HS network became specialists in the BWS network, indicating that the topological roles of microbes were changed and key microorganisms were shifted in the BWS. To conclude, the composition and structure of the microbial network of the BWS were different from that of the HS. Many microbial network connections were missing in the BWS, which most likely provided conditions leading to higher rates of bacterial wilt disease.

Composition of soil microbiome: The data set was rarefied by randomly selecting reads from a sample to count the number of operational taxonomic units (OTUs) represented using mothur software to an even sequence depth of 20,000 reads per sample (Qi Gaofu, et al., 2019). The total number of OTUs binning at 97% similarity found in all the samples was 4,461. Rarefaction curves showed that the sequence depth was sufficient to access the most abundant taxa in each soil. Thus, we focused on interaction networks of the more abundant taxa. The analysis of the estimated richness indices (e.g., observed species, Chao1, and abundance-based coverage [ACE]) revealed that the HSs harbored higher bacterial richness. The Shannon diversity index indicated that the bacterial diversity of the HS was significantly higher ($P < 0.05$) than that of the BWS.

The plant-beneficial bacterial species viz., *Actinoallomurus*, *Actinocorallia*, *Actinospica*, *Bacillus*, *Blastococcus*, *Burkholderia*, *Conexibacter*, *Lysinibacillus*, *Lysobacter*, *Rhodanobacter*, *Spirillospora*) known to inhibit the growth of pathogens are more abundant in the HSs than in the BWSs. But the abundance of the pathogen *Ralstonia* in BWSs was higher than in HSs. Network analysis found that these beneficial bacterial species have more negative interactions in the HSs than in the BWSs. The abundances of 23 bacterial genera (*Acetivibrio*, *Acidisoma*, *Anaerolinea*, *Azohydromonas*, *Bacteroides*, *Chondromyces*, *Clostridium*, *Cryptanaerobacter*, *Denitratisoma*, *Hyalangium*, *Methylocystis*, *Nakamurella*, *Neochlamydia*, *Noviherbaspirillum*, *Oryzihumus*, *Parachlamydia*, *Pelomonas*, *Rhizobacter*, *Sideroxydans*, *Sinosporangium*, *Streptococcus*, *Terrabacter*, and *Variovorax*) were significantly different among BWSs and HSs.

The majority of the nodes in the HS and BWS networks belonged to 10 different bacterial phyla. Among them, *Acidobacteria*, *Actinobacteria*, and *Proteobacteria* were more dominant. The proportions of *Actinobacteria*, *Gemmatimonadetes*,

Proteobacteria, and *Verrucomicrobia* in the HS network were higher than those in the BWS network. The proportions of *Acidobacteria*, *Bacteroidetes*, *Firmicutes*, and *Planctomycetes* in the BWS network were higher than those in the HS network. Only 90 nodes were shared by the HS and BWS networks. Most of the nodes (290 nodes) were unshared by the two networks, suggesting that most microbial species had their special niches under the HS and BWS soils. The HS network possessed more specific nodes (176 nodes) than the BWS network (114 nodes). These results suggested that the compositions of nodes of the HS network were different from those of the BWS network.

1.23.3 Viruses in Soil Ecosystems: Viral abundance in soils can range from below detection limits in hot deserts to over 1 billion per gram in wetlands. Abundance appears to be strongly influenced by water availability and temperature, but a lack of informational standards creates difficulties for cross-study analysis (Kurt E. Williamson et al., 2017). Soil viral diversity is severely underestimated and undersampled, although current measures of viral richness are higher for soils than for aquatic ecosystems. Both morphometric and metagenomic analyses have raised questions about the prevalence of nontailed, ssDNA viruses in soils. Soil is complex and critically important to terrestrial biodiversity and human civilization, but impacts of viral activities on soil ecosystem services are poorly understood. While information from aquatic systems and medical microbiology suggests the potential for viral influences on nutrient cycles, food web interactions, gene transfer, and other key processes in soils, very few empirical data are available. To understand the soil virome, much work remains.

Relatively few plant viruses are transmitted through the soil, but these viruses have special characteristics that distinguish them from those with aerial vectors. Typically, they cannot rely on their vectors to transport them over long distances and, although they can have other mechanisms by which they move from site to site, their distribution is often locally limited (Roberts Alison G , 2014). However, once established at a site they may be very persistent and difficult to eradicate; the soil itself acts as a reservoir of infection between crops. Some soil-borne viruses are transmitted through the soil without the aid of a vector, but the majority have soil-living organisms, either fungi, plasmodiophorids or nematodes, as vectors. However, these vector organisms typically move only short distances, generally millimetres to centimeters, through the soil, and hence the distributions of the viruses they transmit tend to be patchy within a particular field. Movement of soil during cultivation is a significant factor in local spread, and the patches of infection are often elongated in the direction of cultivation or water flow. Soil movement may also be an important means of spread to new sites. This may be either natural soil movement, for example, by wind blow, or the result of human activities, for example on the wheels of agricultural machinery or together with planting material. Long distance movement in soil is probably most likely for those viruses whose vectors are fungi or plasmodiophorids, and which can be carried in the long-lived resting spores of their vector. A striking example is Beet necrotic yellow vein virus (BNYVV), the causal agent of rhizomania disease of sugar beet, which is carried in the resting spores of its vector, *Polymyxa betae*. Soil, spores and virus can be transported on sugar beet roots harvested from infected sites to processing factories, where they are washed off and discharged into waste water outlets. The viruliferous spores can travel long distances downstream and may contaminate water that is used to irrigate previously virus-free sites.

A number of plant viruses in different genera are harboured in and transmitted through the soil. They are found throughout the world and several of them cause significant economic losses to major crops. Vectors for spread of these viruses are fungal-like organisms or plantparasitic nematodes, but some viruses can also be exuded from infected roots and transmitted abiotically, without the aid of a vector. The long-term persistence of these viruses in soil, often for decades, and a lack of efficient control strategies mean that diseases caused by these viruses are very difficult to control. Eradication of inoculum from infected soil is almost impossible, particularly in those parts of the world where highly toxic soil fumigant chemicals have been banned in recent years. Soil-borne viruses are difficult to study and their biology is relatively poorly understood at present.

The ability of some viruses to be carried in the resting spores of their vectors is well known. These spores can survive in soil for many years until favourable conditions for growth recur, and this enables their associated virus to also persist for long periods of time. The nematode-transmitted viruses persist by different mechanisms. In general, either they have very wide plant host ranges, including wild plants and weed species, or their hosts are long-lived perennial species, such as fruit trees (Roberts Alison G , 2014). Thus, although they are not retained by the nematodes when they moult and are not transmitted through nematode eggs, the cycle of transmission from vector to plant and back to vector is unlikely to be broken by the absence of susceptible plant hosts. Moreover, non-feeding adult nematodes can survive and retain virus for months or years. Soil-borne viruses always infect their plant hosts through roots or other underground structures. In some instances, infection is confined to the roots. Pathological changes to root function may result in visible symptoms in the above-ground parts of the plants, even when these parts are not infected by the virus. Nevertheless, early detection of infection may be difficult. This, together with the persistence of the viruses, poses special difficulties for their control.

Trends in viral abundance in soils: Despite the critical importance of agricultural soils in feeding humanity, hot deserts are the most highly represented soil type for which viral abundance data are available (n = 24), followed by agricultural soils (n = 13). Cold deserts (n = 6), forest soils (n = 5), and wetlands (n = 5) are not as well represented in the literature, and fallow or turf systems (n = 3) and dune sands (n = 1) are poorly represented. Future studies should aim to examine viral abundance trends in more diverse and underrepresented soil biomes such as boreal forest, unmanaged grasslands, rain forest, coastal dune systems, and, importantly, agricultural soils (Kurt E. Williamson et al., 2017).

Reported viral abundance in soils ranges from 2.2×10^3 gdw^{-1} (per gram dry weight) in desert sands collected from Saudi Arabia to 5.8×10^9 gdw^{-1} in forest soil collected from eastern Virginia. Part of this range can be accounted for by differences in extraction-enumeration methods. Bearing in mind these methodological differences, a meta-analysis of the published data indicated that soil type was significantly correlated with viral abundance (Spearman $r = 0.803, p < 0.001$). In general, viral abundance appears to be lowest in hot deserts; intermediate in agricultural soils, fields, and cold deserts; and highest in forested and wetland soils. Soil viral abundance was significantly positively correlated with bacterial abundance (Spearman $r = 0.647, p < 0.001$), which supports the hypotheses that the majority of soil viruses infect bacterial hosts and that host abundance may be a key factor controlling viral abundance in soils. Meta-analysis of data available in the literature indicated that viral abundance was negatively correlated with soil pH (Spearman $r = -0.352$, $p = 0.009$). Soil pH is an important environmental factor influencing viral attachment to soil surfaces, and therefore viral persistence and extraction efficiency. Temperature is also known to influence viral persistence and abundance in soils (Wen et al., 2004), with lower temperatures likely fostering the persistence of viruses and higher temperatures likely promoting thermal decay of viral particles. Temperature appears to be a key difference that could explain the stark contrast in viral abundance between hot and cold deserts, but because soil temperatures are rarely reported in the available studies, meta-analysis could not be conducted to test for a relationship. It was strongly suspected that soil water content is a key environmental parameter driving both bacterial and viral abundance in soils, but again, lack of continuity in the existing data set precluded meta-analysis.

Indeed, a major problem we encountered in attempting cross-comparisons between studies was the inconsistent way in which the physical and chemical characteristics of soil samples are reported. A few studies included quantitative details on soil texture, pH, organic matter content, water content at the time of sampling, and other measures, such as cation exchange capacity. Some studies included information on C and N content and others performed analysis of multiple elements, but almost no studies reported all of these details together. Minimum information standards for reporting environmental metadata that accompany soil viral abundance or, indeed, other aspects of soil viral ecology are needed to enable more direct cross-study comparisons in the future. Guidance for such standards could be taken from the genomics research community, where community metadata standards have been adopted for some time (Yilmaz et al., 2011). A robust set of minimum information standards for soil viral ecology studies would include geography (latitude and longitude); edaphic factors, including soil texture (percentages of sand, silt, and clay), soil class, land use, and vegetation type; and factors most likely to influence biological activity, including pH, soil organic matter content, soil nitrogen, and soil water content at the time of sampling. All of the viral abundances are single-time-point measurements. Currently, few data exist regarding change in viral abundance over different timescales, nor have the process rates that govern changes in total viral abundance been assessed.

Viral diversity in soils: The diversity of the soil virome is severely underestimated. Although several studies support the idea that the majority of viruses in soils are phages that infect bacteria, many other organisms, including archaea, protists, fungi, nematodes, annelids, arthropods, plants, and burrowing animals, inhabit soil ecosystems and are likely hosts for soil viruses (Kurt E. Williamson et al., 2017). Bacterial viruses have received the most attention, but with an estimated 10,000 bacterial species g^{-1} soil, our present knowledge regarding phage diversity alone still scarcely scratches the surface. The ~2,000 reported bacterial viruses infect hosts within only 11 bacterial and 2 archaeal phyla, with ~85% of these viruses infecting hosts within the 3 phyla *Gammaproteobacteria*, *Firmicutes*, and *Actinobacteria*. Using the conservative numbers from the SILVA database, there exist 75 bacterial and 28 archaeal phyla, with rarefaction estimates indicating there may be as many as 1,100 bacterial and archaeal phyla combined. In short, the diversity of known viruses is highly limited. Considering other host groups within soils, viruses that infect insects and plants have been described, but this knowledge is incomplete and tends to focus on crop pests or viral biological control agents for pests. Almost nothing is known about viruses that infect nonbacterial soil microbes, such as the archaea, fungi, and soil protozoa. In all, there is yet a tremendous amount of work to be done in characterizing the diversity of the soil virome.

A more holistic understanding of soil viral populations and communities is necessary to characterize and quantify viral impacts on the critically important ecosystem services that soils provide. Viral infection of hosts can influence host community capacity to decompose soil organic matter, convert organic phosphorus into plant-available forms, and respire carbon dioxide, to name just a few examples of the myriad ways viruses may influence soil processes. Improving our understanding of the soil virome will also provide critical insights into the roles viruses have played, and continue to play, in shaping host fitness through selection against virus-sensitive genotypes, lysogenic conversion and host metabolic manipulation, and horizontal gene transfer. Soils remain the most poorly understood ecosystems on Earth. At the same time, viruses represent the largest pool of untapped genetic diversity and unexplored sequence space on the planet. In this regard, the soil virome comprises an unknown quantity within an unexplored territory: a vast new frontier, ripe with opportunities for discovery.

The Third Age of Virus Ecology: Viruses play important roles in biogeochemical nutrient cycles and act as genomic reservoirs in marine and freshwater environments, the understanding of which brought about the so-called 'third age' of virus ecology in aquatic environments. Unfortunately, the third age is in oceanography and limnology and outside the soil world. The main reason why virus ecology in soils has shown less progress is that agronomical and epidemiological interests were the primary motivation of viral studies by soil microbiologists (Makoto Kimura et al., 2008). In this review, past research on viruses in soils was summarized after the introduction of the ecological traits of viruses, which are the effects of viruses on beneficial bacteria and soil-borne plant pathogens, adsorption of viruses to soils, and soil factors affecting viral inactivation and survival in soils. Horizontal gene transfer (transduction) in soils is also reviewed. The abundance of viruses and their roles in biogeochemical nutrient cycles were summarized in aquatic environments. Five to 25% of the carbon fixed by primary producers is estimated to enter into the microbial loop via virus-induced lysis at different trophic levels in aquatic environments. The diversity of virus communities in aquatic environments estimated from analyses of the frequency distribution of capsid sizes and the morphology of virus populations were reviewed and recent findings on the genomic diversity of viruses and their roles as the greatest genomic reservoirs in aquatic environments follow in the subsequent section. Viral genomics is elucidating the viral diversity and phylogenetic relationships among viruses in different environments. As the soil environment is a more diverse habitat for viruses than aquatic environments, viruses in soils have great potential to play roles comparable in quantity, which are unique in quality, to those in aquatic environments. Therefore, the potentiality and characteristics of viruses in soils are discussed in the final section for future research on virus ecology in soils from the viewpoints of biogeochemistry and genomic diversity. Synecological approaches to viruses in soils may open up a new era of soil virus ecology.

Virus adsorption to soils - A key factor in virus ecology in soils: Adsorption of viruses to soils is a specific phenomenon profoundly influencing their infectivity to host organisms, and clay minerals play a central role in this phenomenon (Makoto Kimura et al., 2008). The adsorption follows either the Freundlich or Langmuir isotherm: the former under conditions of low surface coverage by viruses, and the latter under conditions of nearly saturated surface coverage. The amount of viral adsorption was found to be linearly related to the square root of time. In addition, it was suggested that viruses adsorb to colloidal clays using van der Waals forces. Phages were observed using TEM to adsorb to kaolinite particles by their tails.

The degree of adsorption of viruses to soils is generally more than 90% for all viruses. A variety of soil, environmental and viral factors affect the adsorption of viruses to soils include type of clay minerals, cation exchange capacity, soil-associated and dissolved organic matter, soil pH, ionic strength and its constituent, type of virus etc. These parameters interact in viral adsorption on soils because viruses are, in general, negatively charged above the isoelectric point, whereas soils have negatively and positively charged sites as well as hydrophobic sites. The pH influences the charge on both viruses and soils, and ionic strength and its constituents determine the binding force between them. In contrast, organic matter (both soil-associated and dissolved) weakens the electrostatic binding of viruses to soils. The adsorption of the T2 phage of *E. coli* and Type 1 poliovirus on clay particles (kaolinite, montmorillonite and illite) was first demonstrated with an increase in viral adsorption associated with an increase in Na and Ca concentrations. The mechanism of viral adsorption to clays was via the clay–cation–virus bridge. Among cations, divalent cations were more effective as bridging agents than monovalent cations. And naturally occurring river clays (the Missouri River) showed very similar responses of viral adsorption to those of pure clay minerals.

Environmental factors affecting the inactivation and survival of viruses in soils: A variety of soil and environmental factors affect viral inactivation and infectivity to host organisms, among which are temperature, pH, clay type and its content, organic matter content (nutrient status), heavy metals and acid pollutants, aerobicity (anaerobiosis), ionic strength (salinity), and moisture content (Makoto Kimura et al., 2008). The soil environment is very different to aquatic environments as a habitat for microorganisms and viruses, and the characteristics of soils may modify the roles of viruses in biogeochemical nutrient cycles and as genomic reservoirs. Soil consists of solid, liquid and gas phases, and the solid phase of sands, silts, clays and organic materials makes the pore space into countless, small compartments with sizes of nano- to millimeter scales (soil structure). Soil structure results in the preparation of aerobic and anaerobic habitats for microorganisms, and these habitats are potentially stable enough for microorganisms to prevail there over several generations. The neighboring pore spaces in soil are generally isolated from each other by solid barriers, forming isolated independent habitats for soil microorganisms. Therefore, heterogeneity is far greater in soils than in aquatic environments as a habitat for microorganisms, indicating that soil may have more potential as a genomic reservoir. Meanwhile, the degree of viral adsorption on soil particles is generally more than 90% for all viruses, which strongly modulates their transmigration from one zone to another. The horizontal gene transfer mediated by viruses may be less common in soils than in aquatic environments.

The dominant lifestyle of viruses in soils needs to be elucidated for evaluating viral roles in biogeochemical nutrient cycles. Phages occurring in oligotrophic pelagic waters with low bacterioplankton abundance have a tendency toward polyvalency, whereas phages found in nutrient-rich and host-rich coastal waters are generally host specific. The host range of indigenous soil viruses is an interesting subject for future research. The advent of molecular phylogenetic analysis enables a diversity analysis of the unculturable majority in the environment, resulting in elucidation of the high divergence of microbial communities in soil and marine environments and of the marked differences in community structure between them. These findings expect the existence of novel viruses with novel genes characteristic to soil environments. The finding of novel clades of *g23* genes from a paddy field may support this inference (Jia et al., 2007).

As a first step in the third age of virus ecology in soils, the development of reliable extraction methods of viruses is an immediate necessity. It is a prerequisite for evaluating the contribution of soil viruses to global biogeochemical nutrient cycles. At the same time, soil microbiologists need to redirect their interest to synecology of viral communities from autecology of specific viruses. The methodologies for the community analysis are common and familiar to soil microbiologists, for example, molecular approaches to conserved structure/functional genes of specific viral groups. Phylogenetic study on viral communities in soils is important to assess horizontal gene transfer on a global scale. Detection of comparable genes in soils to those found in aquatic environments expands the gene transfer from the scale within aquatic environments to the whole planet (Makoto Kimura et al., 2008). Thus, the third age of virus ecology in oceanography and limnology invites soil microbiologists to cooperate together to elucidate pertinent global issues in biogeochemistry and evolution biology.

Soil Viruses- A New Hope: As abundant members of microbial communities, viruses impact microbial mortality, carbon and nutrient cycling, and food web dynamics. Although most of our information about viral communities comes from marine systems, evidence is mounting to suggest that viruses are similarly important in soil. As abundant members of microbial communities, viruses impact microbial mortality, carbon and nutrient cycling, and food web dynamics (Joanne B. Emerson,2019).

Although most of our information about viral communities comes from marine systems, evidence is mounting to suggest that viruses are similarly important in soil. Joanne B. Emerson (2019) outlined soil viral metagenomic approaches and the current state of soil viral ecology as a field, and then highlighted existing knowledge gaps that we can begin to fill. We are poised to elucidate soil viral contributions to terrestrial ecosystem processes, considering: the full suite of potential hosts across trophic scales, the ecological impacts of different viral replication strategies, links to economically relevant outcomes like crop productivity, and measurable in situ virus-host population dynamics across spatiotemporal scales and environmental conditions. Soon, it is learnt how soil viruses contribute to food webs linked to organic matter decomposition, carbon and nutrient cycling, greenhouse gas emissions, and agricultural productivity.

Soil viruses are ubiquitous in the environment and important for their influence on host community composition and function. Viral infection influences host evolution, metabolism, function, diversity, community evenness, and more. As host bacteria play critical roles in agricultural systems, including nutrient cycling and soil aggregate formation, the influence which soil viruses have on their hosts makes them key players in these systems (McDearis Regan, 2022). However, many important questions remain regarding virus abundance and distribution under human imposed and natural conditions in agricultural systems. Understanding how these conditions impact virus abundance, transport, and community structure is critical for a broader understanding of soil microbial functional diversity.

i. Virus elution as predicted by structure: The observed shifts in viral community composition across pore volumes eluted could have been caused by many physiochemical factors, such as soil and virus surface properties, or shape and/or size of virus particles. Each family of viruses taxonomically identified has characteristic shapes and surface properties (McDearis Regan, 2022). Almost all of the eluted Mimiviridae family was eluted within the first pore volume. Mimiviridae was physically the largest (e.g., capsid diameter) viral family eluted, belonging to the nucleocytoplasmic large DNA virus clade (NCLDV) (Walker et al., 2020). In general, Mimiviridae have icosahedral geometries, and are enveloped viruses. The extent to which Mimiviridae's size and surface properties accounted for their early elution from the columns is unclear. It is also possible that new viral progeny of the Mimiviridae family were produced during the first flow interruption between column saturation with pore volume 1 and the subsequent elution of pore volume 1 after the 16 hour flow interruption, and that these new viruses were readily eluted when the flow resumed. In the second pore volume, the relative abundance of Myoviridae was greatest compared to all other pore volumes. Myoviridae are bacterial phages with non-enveloped head-tail shapes. The tails are intermediate in length and can be either contractile or non-contractile. Podoviridae and Siphoviridae are both non-enveloped head-tail viral families. Interestingly, while Podoviridae have a short tail and Siphoviridae have a long tail, both of these head-tail viruses with similar capsid diameters were eluted in similar patterns, with relative abundance increasing through the second to fifth pore volumes.

Without knowing the exact surface properties of each individual virus, it is impossible to clarify the role of virus surface properties in the differential elution of viruses from the soils. Mimiviridae was the largest virus eluted, which could have been an important factor in influencing their mobilization in the first pore volume; their large size may have preclude them from entering micro-pore spaces, making them more likely to have existed in the mobile liquid fraction. Interestingly, the predominant viral family in the second pore volume was the second largest viral family, Myoviridae, and the predominant viral families in the third to fifth pore volumes were smaller, Podoviridae and Siphoviridae. These results suggest that virus size might play a role in determining virus mobilization from soil under saturated flow.

ii. Viral production across pore volumes: Production of new viral progeny may have been an important contribution to the changes in viral community structure with increasing pore volumes (McDearis Regan, 2022). Enumeration of the total number of viruses eluted over time revealed an increase in overall abundance of viruses in the second pore volume, which could have potentially been due to production of new viruses, rather than only from viral release from the soil column. After the overnight (16 hour) flow interruption, Myoviridae experienced the largest increase in relative abundance in the effluent of the second pore volume, which contains some temperate and virulent species (Walker et al., 2020). Potentially, temperate Myoviridae could have been induced to lytic reproduction from a lysogenic cycle, causing the increase in Myoviridae identified. Prophage induction can occur in response to host stress, as shown by studies which have subjected hosts to the antibiotic Mitomycin-C, causing prophage to switch to lytic reproduction to escape the stressed/DNA-damaged host cell through cell lysis and release of viral progeny. As shown in previous examinations of these columns, measured ions in the effluent decreased significantly between the first and second pore volumes. Based on this data, if we assume that the nutrient concentration decreased significantly between the first and second pore volumes, this lack of nutrients in the soil solution may have potentially stressed bacterial hosts, thereby inducing viral induction.

iii. Uniformity of Residual Soil Virus Community Composition: The community composition of viromes extracted from residual column soil after the flow experiment were significantly different ($p < 0.001$) than that of the eluted viruses (McDearis Regan, 2022). Much of the difference was associated with the presence of Phycodnaviridae, which were present at significantly higher relative abundances in the residual soil viromes and nearly absent in the eluted viromes. Phycodnaviridae are icosahedral enveloped viruses (Walker et al., 2020). The release of Phycodnaviridae in the residual soil viromes could be due to interactions with the extraction buffer chemistry or the physical dispersion caused by blending soil samples. There were also a greater number of unassigned viruses in the extracted viromes than in the final pore volume of the eluted viromes. It is very likely that these viruses have some surface property which enables them to attach more strongly to soil surfaces but detach in the presence of chemical conditions that disrupt the adsorptive forces binding them to solid surfaces and/or physical dispersion of aggregates that may have entrapped the virus in small pores. The community composition of the viruses remaining in the soil was similar across the three sampled columns depths. This suggests that there was no redistribution of viruses along the flow path as a result of reattachment and/or entrapment deeper in the soil column. In general, the difference in viral community composition between the eluted and residual soil viromes shows that residual viral families, which were immobile during elution but subsequently removed during destructive extraction, likely had varying characteristics of surface charge and/or shape, allowing them to remain in the soil column until physiochemical conditions drastically changed upon column deconstruction. An alternative explanation is that the viruses eluted during column flow were new viruses produced during the flow-interruption intervals between pore volumes. Further experimentation is needed to determine if either or both mechanisms account for the virome composition patterns observed in this study.

This study showed that the transport of native soil viruses is highly affected by hydrological processes, and that viral families respond differently to interrupted flow conditions. McDearis Regan (2022) concluded that the most likely explanations for the observed variable mobilization of viral families involves the physiochemical properties of the viral families, including size, capsid envelopment, and shape and chemical surface properties. Levels of ammonium equal to nitrogen deposition did not have a meaningful impact on viral or bacterial community composition. In the future, the biological response to nitrogen additions would need to be measured by adding much greater quantities of ammonium for host stimulation to examine the specific responses of viruses infecting bacteria stimulated by N-addition. The results are very important for our understanding of the transport and retention of natural viral communities under rainfall or irrigation events. It was showed that virus transport may be size dependent, which is very important for predicting the evolution and spatial distribution of the diversity and abundance of native soil viruses under field conditions. Overall, this study provides important insights into the impacts of natural flow conditions on the mobilization of soil viral communities.

Soil-borne viruses of wheat: Temperature is an important environmental factor that affects virus infection because intracellular viral replication requires a specific temperature range and also because virus movement, virus transmission and plant defense systems are also affected by temperature (Yang et al., 2022). In infected wheat plants, bright yellow mosaic symptoms typically occur in early spring but disappear in new leaves during early summer. These findings suggest that low temperatures may be favorable for virus infection. For SBWMV, the optimal infection temperature was 17 °C. Substantial work has been conducted to determine the optimal infection temperatures of CWMV and WYMV. Following mechanical inoculation of the sap of virus-infected leaves onto wheat and *Nicotiana benthamiana* (*N. benthamiana*), CWMV completes its infection at 16 °C but not at 24 °C. Inoculation studies of wheat and *N. benthamiana*, using a full-length complementary infectious DNA clone of CWMV that was constructed using reverse genetics showed that the optimum temperature for replication was 15–17 °C. The optimum temperature for WYMV multiplication and systemic infection was 8 °C, but the optimal temperature for viral movement and silencing is not known (Zhang et al.,2021).

The underlying molecular mechanisms whereby low temperatures contribute to viral infections are not well understood. A previous study revealed that low temperatures can inhibit host antiviral defenses mediated by RNA silencing, including virus-derived small interfering RNAs (vsiRNAs). Recently, numerous vsiRNAs derived from CWMV-infected wheat were identified by next-generation or deep sequencing, and these vsiRNAs exhibited a strong bias in their 5′-terminal nucleotides. The significance of this bias needs to be investigated and it is unclear whether these vsiRNAs are temperature-dependent, but there is no denying that this phenomenon may be the result of the evolution of the virus during the long-term competition with the host.

1.23.4 Nematodes in Soil Ecosystem: Soil nematodes have been thriving in habitats around the world for at least 400 million years. Soil is an excellent habitat for nematodes and 100 cc of soil may contain several thousand of them. Because of their importance to agriculture, much more is known about plant-parasitic nematodes than about the other kinds of nematodes which are present in soil. Most kinds of soil nematodes do not parasitize plants, but are beneficial in the decomposition of organic matter (Robert McSorley, 2019). These nematodes are often referred to as free-living nematodes. Juvenile or other stages of animal and insect parasites may also be found in soil. Although some plant parasites may live within plant roots, most nematodes inhabit the thin film of moisture around soil particles. The rhizosphere soil around small plant roots and root hairs is a particularly rich habitat for many kinds of nematodes. As nematodes are dependent on the continuity of soil water films for movement, their activities are largely controlled by soil biological and physical conditions. Soil-dwelling species live in the water film of the interstices of soil particles. Nematodes must live and carry out all functions of life in these spaces, thus free-living or plant-parasitic soil-dwelling forms are usually extremely small. Morphological diversification of nematodes can operate only within the constraints of their life history parameters, and evolutionary pathways for nematodes living in an interstitial–soil or sediment/sand environment are limited. Thus these forms have limited abilities to diversify into the nonaquatic regions of the Earth, and their size is constrained by this fact.

Relationship between nematodes and some soil properties: Physical and chemical soil parameters affect the activity of living organisms within the soil. Five major properties, including temperature, moisture, and water holding capacity of the soil, as well as soil pH and electrical conductivity, play an important role in soil functioning and diversity, and various soil factors are reportedly responsible for the distribution of soil nematodes (Amal Ahmed Mohammed Al-Ghamdi, 2021). The chemical characteristics of soil affect the density and diversity of soil organisms, including the composition and structure of nematode communities. Additionally, soil type affects both the composition of nematode species in soil and how well the host plants can grow. Of note is the fact that soil acidity has been found to be an important physicochemical factor in soil that affects soil physicochemical and biological properties and, consequently, plant growth and soil biota. Soil salinity and salt accumulation are other important factors that are both affected by weather changes, improper watering, and plant fertilization.

Organic matter (OM) constitutes a fundamental soil component that serves as a source of carbon and energy for plants and soil organisms; however, OM also affects root infestation by nematodes. Field practices, such as crop irrigation and mechanical, chemical, or biological methods, may alter the soil environment and, consequently, affect root growth in soil. These alterations may include soil temperature and moisture and the activities of soil biota, including those of plant-parasitic nematodes. For example, investigations have shown that when compared to bare soil, mulched soil has a lower temperature. Similarly, when crops are grown too close together, the soil temperature in the area may be lower than normal, and the soil may be more humid than that in fields where fewer plants are grown. These findings are corroborated by research that showed a significant and positive correlation between mulch and plant density root necrosis in *Pratylenchus goodeyi.* At higher temperatures, reproduction is negatively affected in *P. goodeyi,* explaining why this species thrives in soils that are more heavily mulched.

A study by Amal Ahmed Mohammed Al-Ghamdi (2021) investigated the relationship between nematode occurrence and some soil parameters. The results showed that the proportion of nematodes in soil was negatively correlated with soil pH; however, it increased with increasing electrical conductivity (EC), soil moisture (SM) and organic matter (OM) content. A positive correlation was found between the proportion of nematodes in the roots and the amount of coarse sand, while a negative correlation existed between the proportion of nematodes and the amount of fine sand, clay, and silt. The degree of colonization by nematodes increased with increasing coarse sand content. Conversely, the soil nematode population decreased with increasing fine sand content, clay and silt content. The nematode population in banana roots correlated with soil phosphorus, potassium, calcium and magnesium levels. Conversely, a negative correlation was found between nitrogen concentration and the nematode population in banana roots. Overall, the population of nematodes in soil affects its pH, EC, SM, and OM content.

The microenvironment has a strong effect on nematodes, which help indicate soil health status and many soil functions (Gebremikael et al., 2016). Of note is that nematodes are functionally disparate organisms that are widespread and quickly respond to alterations in the environment. The interwoven relationships that have been described between soil characteristics and the proportion of nematodes in diverse functional guilds have been used to create criteria to assess soil. In agriculture, the proportion and diversity of nematodes are utilized to determine soil process rates and functions. A variety of abiotic and biotic factors affect the establishment of nematode communities in soil. Plant-parasitic nematodes are, in general, aerobic and originally aquatic organisms that require adequate moisture and aeration from adsorbed soil water films (moisture levels ranging between 40% and 60% of field capacity). Soil acidity has also been reported to affect nematode populations. For example, the proportion of *Meloidogyne incognita* and *R. similis* were generally decreased in either extremely acidic or alkaline soils; however, *R. similis* was more tolerant to soil with a very low pH than *M. incognita.*

Soil pH level was not associated with the proportion of nematodes in many studies. Additionally, no correlation was found between soil moisture and nematode diversity. Furthermore, a positive correlation was found between temperature and electrical conductivity and nematode diversity. Finally, a negative correlation was found between soil moisture and pH and the proportion of nematodes, but no correlation was found between soil moisture content and nematode diversity (Gade and Hiware, 2017). In a study that investigated the infection of pineapple by *Pratylenchus brachyurus*, a negative relationship was found between nematode population densities and soil pH.

Previously, it was reported that soil organic matter contents correlated with a lower proportion of plant-parasitic nematodes, suggesting that the addition of organic compounds is a way to suppress these parasites. However, soil OM contents were also positively correlated with free-living nematodes, probably by favoring the growth of microbes (bacteria and fungi), which increased the nematode population. Therefore, soil OM affects the spatial soil distribution of nematodes and has a significant effect on mycophagous nematode communities. Furthermore, soil OM affects the density of bacteriophage nematodes when the added organic material is rich in nutrients and has a low carbon: nitrogen ratio (Amal Ahmed Mohammed Al-Ghamdi, 2021).

Soil compactness and porosity are determined by several conditions such as soil texture, soil type (sand, silt, clay particles), and OM content. Nematodes rely on the availability of moisture and aeration to thrive in crop fields. In general, nematodes thrive more in light sandy soils than in heavy clay soils due to the better aeration in soils containing coarse particles; however, these parasites have a more nefarious effect on plants growing in conditions of water stress because water drainage is facilitated in coarse particulate sandy soils. *M. incognita* and *Hoplolaimus columbus* thrive in soils with a high sand content. In this study, the nematode density was increased when the proportions of sand, silt, and clay were 48%, 35%, and 17%, respectively. It was reported that *Tylenchulus semipenetrans* reproduced well in soils with a low proportion of clay (10% to 15%). *T. semipenetrans* can thrive in any soil texture, but they cause extensive damage to plants grown in shallow, poorly drained soils that have an organic matter content of approximately 2% to 3%. However, these findings showed that the population of nematodes in soils was highest in those with an organic carbon content of approximately 1.5%.

Previous reports have shown that soil texture had a significant effect on the occurrence of some *Pratylenchus* species (Amal Ahmed Mohammed Al-Ghamdi, 2021). Of note is that the amount of sand in the soil is an important factor that affects the density of root-lesion nematodes. Previous research showed that a higher proportion of root-lesion nematodes, especially *Pratylenchus penetrans*, was found in loamy sand, and these species were reportedly less prevalent in heavy soils. *P. penetrans* was more prevalent in light (0.0% and 7.5% of clay) than in heavy soils (45.0% and 60.0% of clay). The results showed that lesion nematodes were prevalent in sandy clay loam. *Pratylenchus crenatus* was mostly prevalent in loam and clay soils, including sandy clay loam. Additionally, they identified *P.neglectus* in various soil types; however, these nematodes were less frequent in heavy soil.

In citrus plants, standard critical levels of nitrogen, phosphorus, and potassium that were necessary for their growth were also adequate for nematode activities. Increased levels of these elements in soil caused an increase in the population density of nematodes, except for the case of nitrogen, where an increase in its level resulted in a decrease in the nematode population. The density of nematodes in the soil where citrus plants were thriving was negatively correlated with the nitrogen concentration but positively correlated with soil potassium concentration (Amal Ahmed Mohammed Al-Ghamdi, 2021). Other investigators have also reported that ammonia had a nematocidal effect, which was improved when ammonia-releasing fertilizers were combined with alkaline amendments. In a pot experiment, it was shown that when urea was applied at a rate of 160 kg ha-1, *Tylenchulus semipenetrans* could not infect or develop on rough lemon plants. Additionally, it has been reported that soil potassium content was positively correlated with the densities of J2 and males in soil.

Previous studies conducted on tobacco plants showed a positive relationship between nematode density and alkaline earth metals (magnesium and potassium) for root-knot nematodes. Similar results were reported about root-knot nematodes in tomato plants and *Scutellonema* spp. in maize and bean plants. The population density of *P. goodeyi* was significantly and positively correlated with soil potassium in Butare-Gitarama, which is an area known to have the lowest soil potassium concentration. In other ecoregions that did not have low potassium concentrations, no such relationships were found. Soil moisture impacts the speed of Nematodes**.** Nematodes in the driest soil were the slowest, showing little to no movement. This indicates that nematodes can't travel as quickly in soils with less moisture, which is useful information for farmers.

The effect of Soil type and Ecosystems: Soil nematode communities and microbial diversity and the properties of three soil types were analysed to assess the effect of these environmental variables on biological diversity in natural (forest), semi-natural (meadow), and managed (agriculture) habitats of the Slovak Republic (Renčo et al., 2020). The type of ecosystem and soil and the interaction of both factors had considerable effects on most monitored abiotic and biotic soil properties. The forest with a Chernozem soil had the most nematode species, highest nematode diversity, highest abundance of nematode within functional guilds, best values of ecological and functional indices, highest microbial biomass, highest microbial richness and diversity, and the highest values of various soil properties, followed by meadows with a Cambisol soil. The agricultural ecosystem with a Stagnosol soil had the lowest biological diversity and values of the soil properties. Several nematode species were new for Slovak nematode fauna. Sampling date and the interaction of all three factors (ecosystem × soil × date) had minor or no effect on most of the parameters, except soil moisture content, microbial richness, nematode channel ratio, nematode maturity index, and plant parasitic index. Both the biological indicators and basic soil properties indicated that the natural forest with a Chernozem soil was the best habitat from an ecological point of view. This ecosystem is thus the most appropriate for ecological studies.

Several particular studies have revealed that environmental conditions determines the degree of species diversity of soil nematodes or nematode abundance e.g. ecosystem type and its properties, soil type and its properties; vegetation and its species diversity. Similar, microbial activity and biomass have been evaluated in arable soils due to crop production as affected by tillage; fertilizers or management system; in forest soil as affected by forest type or in grasslands affected by plant diversity. The structure of soil nematode communities and microbial diversity, however, have not been investigated or compared amongst various land use (ecosystems) and main soil types in the territory of Slovak Republic in collaborative study.

Soil properties: The factorial ANOVA found that ecosystem type viz., forest, meadow and agricultural field (FOR, MEA and AGR) and soil type (SS, CS, and CM) significantly affected all soil properties (except SM content vs. soil type) and that sampling date (M, J, and S) affected only SM content ($P<0.01$) (Renčo et al., 2020). The bi-factorial interaction ecosystem × soil significantly affected all soil properties, ecosystem × date affected half of the properties, and soil × date and the interaction of all three factors (ecosystem × soil × date) had minor or no effects on the soil properties. The values of the soil properties were generally higher in the FOR soils (except pH) than the MEA and AGR soils and higher in CM (including pH) than CS and SS. pH and the C/N ratio were correlated negatively in FOR but positively in AGR and MEA.

The three ecosystems and soil types contained 133 nematode species (32 bacterivores, 26 fungivores, 9 carnivores, 24 omnivores, and 42 plant parasites). *Heterocephalobus eurystoma*, *Stegetellina leopolitensis*, *Ditylenchus parvus*, *Ditylenchus tenuides*, *Paraphelenchus obscurus*, *Boleodorus volutus*, *Cephalenchus intermedius*, and *Ecphyadophora tenusissima* were new to the list of Slovak nematode fauna, increasing the total number of soil nematode species to 732. The number of species (99) and diversity were highest in the FOR soils, followed by the MEA (90) and AGR (53) soils. Nematode species number and diversity were higher in CM than CS and SS (102, 81, and 60, respectively). The most abundant nematode species by trophic group were *Acrobeloides nanus* and *Chiloplacus propinquus* (bacterivores), *Aphelenchus avenae* and *Filenchus vulgaris* (fungivores), *Clarkus papillatus* and *Mylonchulus brachyuris* (carnivores), *Eudorylaimus carteri* (omnivores), and *Agienchus agricola*, *Boleodorus thylactus*, and *Bitylenchus dubius* (facultative and obligate plant parasites).

The ANOVA found that ecosystem and soil types significantly affected all descriptors (except PP vs. ecosystem type). Sampling date had a significant effect on MI, ΣMI, PPI, and NCR ($P<0.01$, 0.05). The interaction ecosystem × soil significantly affected all descriptors ($P<0.01$), ecosystem × date and soil × date significantly affected half of the descriptors, and the interaction of all three factors (ecosystem × soil × date) affected the majority of the descriptors. MI, ΣMI, PPI, H′, SI, and TD were generally higher in FOR than MEA and AGR soils and in CM than CS and SS. EI was highest in AGR and CS, CI was highest in MEA and SS, and NCR was highest in FOR and CS.

Relationships of ecosystem type with soil properties and nematode and microbial communities: Ecosystem type was an important factor shaping soil nematode and microbial communities and affecting soil properties. The abiotic and biotic soil properties and interactions amongst them were best for the FOR ecosystem. FOR had the highest SM, C, and N contents and C/N ratio but the lowest pH (Renčo et al., 2020). C and N contents were twice as high in FOR than AGR but were similar to those in MEA. The supposed benefits of management of agricultural land (e.g. tillage, fertilisation, and crop rotation) include increased soil C and N contents, fertility, water retention, and overall provision of ecosystem services. The low C and N contents in our AGR soils, however, suggested differences in the quantity and quality of inputs to the soil, nutrient inputs and losses, low plant diversity and stimulation of decomposition by soil disturbance compared to the semi-natural (MEA) and natural (FOR) ecosystems. These results are in agreement with many studies of differences in soil C and N contents and changes following conversion of forest to managed agricultural land. This review revealed that large amounts of C and N could be lost when forest is converted into cultivated land but that no changes in soil C and N contents were recorded when forests were converted to uncultivated pasture. In contrast, the abandonment and reforestation of agricultural land can substantially increase C and N storage, due to increase in plant diversity. Additionally, cultivated soils usually have lower C/N ratios than forest soils.

Plants and their root systems serve as food for plant parasitic nematodes before they serve as a food source for microbivorous nematodes during decomposition. Root systems are more diverse in natural ecosystems with rich communities of plant species than for monocultured crops. Root growth is also more extensive and less ephemeral in perennial plants than annual crops and supports a soil community with many species of plant parasites, omnivores, and predators. Plant parasites are common in natural grasslands. The abundance of plant parasites, such as *Boleodorus thylactus*, *Malenchus exiguus*, and *Paratylenchus microdorus* (Pp_2) or *Amplimerlinius macrurus*, *Geocenamus microdorus*, and *Bitylenchus dubius* (Pp_3) was highest in MEA. The importance and high population densities of plant parasitic nematodes in agriculture are mainly associated with specific crop pests, e.g. root-knot and cyst-forming endoparasites (e.g. *Meloidogyne*, *Heterodera*, and *Globodera*). The high overall abundance of Pp nematodes in AGR (*Aglenchus agricola* Pp_2, *Paratylenchus bukowinensis* Pp_2, *Bitylenchus dubius* Pp_3, and *Helicotylenchus dihystera* Pp_3 are all ectoparasites) suggests their close relationship with cultured crops.

Relationships of Soil type with soil properties and Nematode and microbial communities: Soil type was also an important factor affecting the nematode and microbial communities and soil properties. Soil properties were best in CM, with a neutral pH and the highest C and N contents and C/N ratio, followed by CS and SS. C and N contents were twice as high in CM than SS, in agreement with the general soil classification (Renčo et al., 2020).

Soil type was more important than ecosystem type for both the nematode and microbial communities. For example, nematode abundance, number of nematode species, and microbial biomass or diversity positively correlated in the CM soil type in two out of three ecosystems studied. Significant effects of soil type on the composition of nematode communities have been documented in forests, grasslands and in agricultural land. The populations of bacterivores (mainly *A. nanus*, *Eucephalobus striatus*, and *C. propinquus*) and fungivores (*A. avenae* and *F. vulgaris*) and microbial biomass in our study were highest in CM with aerobic conditions, a neutral pH, and a high humus content beneficial to microbial activities and associated nematodes. In contrast, the abundances of bacterivores and fungivores were low in SS because of its oxygen deficiency and acidic conditions. Cephalobidae bacterivores (*Acrobeloides*, *Acrobeles* and *Cervidellus*) were more abundant in a light sandy Regosol with a high pH, but disagreed with some , who reported that fungivores were more abundant in an acidic Cambisol.

The differences in soil properties, nematode communities, and microbial biomasses amongst the soil and ecosystem types suggest an obvious impact of environmental variables on biotic and abiotic soil characteristics (Renčo et al., 2020). The differences were larger amongst the soil types than the ecosystems. CM had the best soil properties, with a neutral pH and the highest C and N contents and C/N ratio and thus the highest number of species and diversity of nematode communities, as well as the MI, ΣMI, PPI, SI, and TD nematode ecological indices, and microbial biomass, richness, and diversity. The majority of the abiotic and biotic characteristics varied the most between CM and SS. The abiotic and biotic soil properties and their interactions were best in FOR, where the number of species and diversity of nematode communities, as well as the MI, ΣMI, PPI, SI, and TD ecological indices, and microbial biomass, richness, and diversity were highest. SM, C, and N contents and the C/N ratio were also highest in FOR, but the pH was lowest. C and N contents were twice as high in FOR than AGR but were similar to those in MEA, suggesting that established forests and natural meadows represent relatively stable environments, providing suitable conditions for soil microbial and nematode communities. C/N ratios and biological diversity were lower in the cultivated soils than in the natural ecosystems soils, likely due to periodic perturbation. This resulted in a lower abundance and diversity of nematode communities and microbial diversity. FOR and AGR generally differed the most. The soil properties, nematode communities, and microbial biomass were more similar in FOR and MEA. A multivariate analysis indicated that the abundance of most of the nematode guilds, total nematode abundance, number of nematode species, and microbial characteristics tended to be higher in the environment with a higher pH, the N and C contents. Sampling dates had a minor or no effect on most of the parameters, except the SM content, abundance of c-p2 nematodes, microbial richness, and several of the nematode ecological indices.

Soil Nematode Communities: The proportions of the different feeding groups in the soil nematode community vary between systems and seasons, and they are influenced by a variety of factors, including crop and soil management practices and the presence and abundance of natural enemies (Carmen Ugarte and Ed Zaborski, 2009). Management practices like tillage, crop rotation, and the use of organic amendments influence the physical and biological characteristics of the soil that influence the abundance of nematodes. Fungal-feeding, predatory, and omnivorous nematodes are very sensitive to soil disturbances and agricultural systems with fewer physical and chemical disturbances, such as pastures, hay fields, and orchards, tend to support larger populations of these nematodes than more frequently disturbed systems like vegetable- and row-crop fields. On the other hand, tillage and incorporation of organic residues increase the proportion of some bacterial-feeding nematodes, often offsetting declines in the numbers of other feeding groups and increasing the total abundance of nematodes. The wide variety of natural enemies that feed on or infect nematodes-predatory nematodes, predatory microarthropods, and nematode-trapping fungi, for example-may have a considerable impact on nematodes in agricultural systems.

Agricultural management may increase the abundance of soil nematodes, primarily through the increase in the abundance of bacterial-feeding nematodes associated with tillage and the incorporation of organic residues (Neher, 1999). Soil conditions in agricultural production systems can be improved by enhancing nutrient availability and providing habitat for beneficial soil organisms. Maintenance of large populations of bacterial-feeding nematodes with practices that promote N mineralization throughout the growing season may enhance crop productivity, but a surplus of mineral N is not desirable from the environmental point of view because of an increased risk of nitrate leaching. In an ideal production system, N supply would be synchronized with plant demand. On the other hand, cultural practices like tillage or cultivation may reduce the complexity of the soil food web. Thus, a decrease in the frequency and intensity of tillage may promote the conservation of predatory nematodes and contribute to improved farming system performance.

Soil properties associated with Phytonematodes: Based on observed associations between soil physical properties and nematodes, it was hypothesized that soil pH would be useful for determining where nematodes are likely to occur (Simon et al., 2018). However, lacking in most of the aforementioned studies on associations among cropping practices, soil proprieties, and nematode population densities were formal quantitative estimates of the nature, form, and strength of the associations. Such estimates would be useful for developing models to identify fields in which certain types of nematodes are most likely to be present and at high population densities in order to advance on-farm nematode research. Soil texture has also been shown to affect the distribution and population density of PPNs. Higher population densities of *X. americanum* were found in silty clay-loam soils when compared with silt loam soils. *Pratylenchus crenatus* was reported to occur in soils ranging from loams and silt-loams to light sandy soils, whereas *Pratylenchus penetrans* was found more frequently on soils with high sand content. Other studies reported that *P. penetrans* was favored by light colored silt-loam, whereas the population densities of other *Pratylenchus* species were positively correlated with fine silt. High population densities of *Criconemella* species were found in sandy soils; low population densities of *Tylenchorhynchus*, *Helicotylenchus*, and *Xiphinema* species in loams and sandy loams; high population densities of *Helicotylenchus* spp. in clays and silty clay soils and high population densities of *Tylenchorhynchus* species in coarse silt or dark colored silty clay-loam.

The overall null hypothesis was that there were no, or at best very weak, functional relationships between nematode population densities and soil type and region, cropping history, and soil properties. It was hypothesized that sandy soils were likely to have nematode population densities similar to those in high silt or clay content soils, and for any given soil type (textural class), fields under continuous corn was likely to have similar population densities to fields under crop rotation away from corn. Logistic regression modeling approaches were used to formally test these hypotheses and to quantify associations among the aforementioned covariates and the population densities of different morphological groups of nematodes. Firstly, binary logistic regression models were fitted to the nematode incidence data to estimate the odds of each morphological group being present, as influenced by the covariates, then ordinal logistic regression models were fitted to the population density data to estimate the odds (and probabilities) of lance, lesion, spiral, and pin nematode populations exceeding critical damage thresholds.

Particle Size and Texture: A subsample of the soil from each field was used to determine particle size distribution and texture based on the Pipette Method. Individual samples were air-dried and soil aggregates were crushed and passed through a sieve of 2 mm mesh size. Particle size distribution was then analyzed (Simon et al., 2018). In brief, 10 grams of each sample were placed into a sedimentation bottle containing a dispersing solution (sodium hexametaphosphate and sodium carbonate) and deionized water. The bottles were sealed and placed onto a shaker overnight at 120 oscillations per minute, after which more deionized water was added. After the prescribed stirring and settling intervals, an aliquot of the soil remaining in suspension, i.e., clay, was placed into a preweighed crucible and dried in an oven at 105°C for 24 h. After cooling in a desiccator at room temperature, the weight of the crucible + clay was recorded and percent clay calculated. The sand fraction of each sample was estimated by passing the remaining mixture through a 53 μm mesh sieve, rinsing the sand trapped in the sieve into a beaker, and then determining its dry weight. After determining the clay and sand contents, percent silt was calculated as 100 – (% clay + % sand). Textural classes were assigned to soil samples according to the USDA classification system. Most of soils (260) were classified as silt-loam.

Soil pH and Electrical Conductivity: A 1:1 mixture of soil and deionized water was placed into a plastic cup, stirred for 10 s, and allowed to stand for 30 min. Prior to running the test on the samples, the pH meter was calibrated using pH 4 and pH 7 buffer solutions. The electrode was then placed directly into the soil suspension and allowed to stabilize before reading the pH. The electrical conductivity (EC) of each sample was determined in a similar manner with an EC meter. The meter was first calibrated against a standard solution of known EC (1.41 millisiemens, mS) before being used to measure the conductance of each sample in microsiemens per centimeter (μS/cm). During both the pH and EC tests, the electrodes were rinsed with deionized water between samples to minimize cross contamination.

Soil characteristics and Cropping practices associated with the presence of plant-parasitic nematodes: The observed presence of PPNs in the 425 fields varied among the morphological groups (Simon et al., 2018). Based on results from binary logistic regression analyses, the effects of cropping sequence, tillage, soil region, silt content, pH, and EC on the log-odds of nematodes being present varied among morphological groups. Cropping sequence had a significant association with the presence of spiral and pin nematodes ($P < 0.05$), a marginally significant association with the lance nematode ($P = 0.083$), but was not associated with the presence of the other seven morphological groups. Tillage was significantly associated with the presence of three of the 10 groups (lance, stunt, and stubby), whereas soil region had a significant association with six groups, namely lesion, dagger, stunt, pin, ring, and stubby-root ($P < 0.05$). The continuous covariates silt content, pH, and EC, also had statistically significant associations ($P < 0.06$) with at least two of the morphological groups. The presence of spiral, lance, and stunt nematodes was associated with three of the factors evaluated, whereas the presence of cyst, dagger, and ring nematodes was associated with one of the covariates evaluated. The presence of the other morphological groups was associated with two factors.

Associations between soil region and the presence of PPNs depended on the nematode. With the exception of the dagger and ring nematodes, the odds of PPNs being present in one or more of soil regions 1, 2, 3, 4, or 5 were greater than the odds of them being present in soil region 6 (the reference soil region). Lance, stunt, pin and stubby-root nematodes were more than twice as likely to be present in soil region 3 as in region 6. The odds of stunt or spiral nematodes being present in region 5 were more than 7 times the odds of them being present in region 6. Soil region 4 was 2.9 times more likely to have lance nematodes, 3.9 times more likely to have pin nematodes and 6.5 times more likely to have spiral nematodes than region 6. Contrastingly, dagger and ring nematodes were respectively 2 to 3 and 5 to 15 times more likely to be present in soil region 6 than in any of the other soil regions. Lance nematodes were also more (5.5 times) likely to be present in soil region 6 than in region 1 and stunt nematodes were more than twice as likely to be present in soil region 6 as in regions 2 or 4. There were significant positive relationships between log-odds and soil pH for spiral and cyst nematodes ($P < 0.05$) (Simon et al., 2018).

The effects of silt content and EC of the soil on the probability of nematode population densities reaching moderate-higher risk levels also varied among nematodes. Slopes for relationships between the slit content of the soil and log-odds were negative for both the lesion and lance nematodes, whereas the slope for the relationship between EC and log-odds was positive for the lesion nematode but negative for the pin nematode. For both the lesion and lance nematodes, the predicted probability of population densities being in the moderate-high risk category decreased as the silt content of the soil increased, at fixed levels of other covariates. The predicted probability of population densities of the lesion nematode being in the moderate-high risk category increased as the EC of the soil increased, but the probability of pin nematode population densities being in that same risk category decreased with increasing EC.

When the effect of soil texture on the population density of lesion, lance, spiral, and pin nematodes was more directly investigated through the use of silt content as a continuous covariate, silt content had a significant association with population densities of the lesion and lance nematodes, but not the pin or spiral nematodes (Simon et al., 2018). For the former two nematodes, there was a significant negative relationship between soil silt content and the probability of population density in the moderate-high risk category, meaning that as soil silt content increased, the odds of lesion and lance nematode populations reaching the proposed moderate-high risk levels decreased. High population densities of *Pratylenchus brachyurus* were reported in soil with relatively low silt content (6% silt), suggesting that the lesion nematode may be less abundant in soils with relatively high silt content. However, these results do not support what was discussed above regarding possible effects of soil region on population density of lesion, spiral, and pin nematodes. This apparent disagreement could be due to the fact that other soil-region-specific factors likely had greater effects on nematode population than silt content or those factors interacted with silt content to affect population density. These results showed that although the probability of the lesion and lance nematode populations reaching moderate-high risk levels decreased as silt content increased, the relationship between predicted probability and silt content (slopes and heights of the regression lines) varied with soil region, and was influenced by tillage for the former and soil pH for the latter nematode.

Soil region, cropping practices, and soil pH and EC may all be good indicators of the presence and population density of one or more important groups of phytonematodes.

Bibliography

Agrios, G. N., 2005. Plant Pathology. 5th Edition. Academic Press. https://doi.org/10.1016/C2009-0-02037-6

Amal Ahmed Mohammed Al-Ghamdi, 2021. Relationship between Nematodes and Some Soil Properties in the Rhizosphere of Banana Plants. *Intl. Lett. Natu. Sci.*, **82**:1-12

Amanda Bennett, 2023. The soil food web and nutrient cycling. Agriculture and Horticulture Development Board, Middlemarch Business Park, Siskin Parkway East Coventry, CV3 4PE

Arthur, E., Schjønning, P., Moldrup, P. and de Jonge, L. W., 2012-03-01. Soil resistance and resilience to mechanical stresses for three differently managed sandy loam soils. *Geoderma,* **173–174**: 50–60.

Bais, H.P., Weir, T.L., Perry, L.G., Gilroy, S. and Vivanco, J.M., 2006. The role of root exudates in rhizosphere interactions with plants and other organisms. *Ann. Rev. Pl. Biol.*, **57**:233–266

Baker, K. and Cook, R. J., 1974. Biological Control of Plant Pathogens. New York, NY: WH Freeman and Company, 433 pages.

Barrios, E., 2007. Soil biota, ecosystem services and land productivity. *Ecol. Econ.,* **64:** 269-285

Bonanomi, G., Alioto, D., Minutolo, M., Marra, R., Cesarano, G. and Vinale, F., 2020. Organic amendments modulate soil microbiota and reduce virus disease incidence in the TSWV-tomato pathosystem. *Pathogens,* **9:** 379

Brady Nyle, C. and Ray R. Weil, 2010. Elements of the Nature and Properties of Soils. Upper Saddle River, NJ: Pearson Prentice Hall, Print.

Brussaard, L., Peter C. de Ruiter and George G. Brown, 2007. Soil biodiversity for agricultural sustainability, *Agri. Ecosyst. & Environ.,* **121**: 233-244

Carmen Ugarte and Ed Zaborski, 2009. Soil Nematodes in Organic Farming Systems. *eOrganic article.* Oregon State University

Carrión, V.J., Perez-Jaramillo, J.E., Cordovez, V., Tracanna, V., de Hollander, M., Buck, R.D., Mendes, L.W., van Ijcken, W.F.J., Expósito, R.G., Elsayed, S.S., Mohanraju, P., Arifah,

A., van der Oost, J., Paulson, J.N., Mendes, R., van Wezel, G., Medema, M.H. and Raaijmakers, J.M., 2019. Pathogen-induced activation of disease-suppressive functions in the endophytic root microbiome. *Science,* **366**:606–612.

del Pilar Rodríguez Guzmán, M., 2022. Soil Biodiversity and Root Pathogens in Agroecosystems. *IntechOpen.* doi: 10.5772/intechopen.99317

Dixon, G. R., 2002. Interactions of Soil Nutrient Environment, Pathogenesis and Host Resistance. Proc. 6th Conf. EFPP 2002, Prague. *Plant Protect. Sci.,* **38** (Special Issue 1): 87–94

Dixon, G.R. and Tilston, E.L., 2010. Soil Microbiology and Sustainable Crop Production, 197 DOI 10.1007/978-90-481-9479-7_6, © Springer Science+Business Media B.V. 2010

Dobson, A., Lafferty, K. and Kuris, A., 2005. Parasites and Food Webs. *In: Ecological Networks: Linking Structure to Dynamics in Food Webs*. Pascual, M. and J. A. Dunne (Eds.).Oxford University Press. P 119- 135

Fierer, N., Nemergut, D., Knight, R. and Craine, J.M., 2010. Changes through time: integrating microorganisms into the study of succession. *Res. Microbiol.,* **161**: 635-642.

Fierer, N., Wood, S.A. and Bueno de Mesquita, C.P., 2021. How microbes can, and cannot, be used to assess soil health. *Soil Biol. Biochem.,* **153:** 108111.

Gade, R.B. and Hiware, C.J., 2017. Studies on the Soil Nematode Diversity in Relation to the Soil Parameters from Sugarcane Fields of Aurangabad District Maharashtra State, India. *Intl. IJRASET* **5**:1632-1640.

Garbeva, P., Postma, J., van Veenand, J. A. and van Elsas, J. D., 2006. Effect of above-ground plant species on soil microbial community structure and its impact on suppression of Rhizoctonia solani AG3. *Env. Microbiol.*, **8**: 233-246

Gebremikael, M., Steel, H. and Buchan, D., 2016. Nematodes enhance plant growth and nutrient uptake under C and N-rich conditions. *Sci. Rep.,* 6: 32862

Gilligan, C.A. and van den Bosch, F., 2008. Epidemiological models for invasion and persistence of pathogens. *Ann. Rev. Phytopathol.,* **46**:385–418

Golldack, J., Augustin, C., Lentzsch, P. & Werner, A. 2004. Pathozones of genetic subtypes of *Gaeumannomyces graminis* in cereals. *Soil Biology and Biochemistry*, **36**: 145–154.

Grace C, Lynch MB, Sheridan H, Lott S, Fritch R, Boland TM. 2019Grazing multispecies swards improves ewe and lamb performance. *Animal* **13**: 1721-1729

Gregory, P. J., Bengough, A. G., George, T. S. and P. D. Hallett. 2013. Rhizosphere Engineering by Plants: Quantifying Soil–Root Interactions. American Society of Agronomy. DOI: 10.2134/advagricsystmodel4

Gu G, et al., 2018 Agricultural practices influence *Salmonella* contamination and survival in pre-harvest tomato production. *Front. Microbiol.,* **9:** 2451.

Hartmann, M., 2023. Six, J. Soil structure and microbiome functions in agroecosystems. *Nat Rev Earth Environ* **4:** 4–18

Heather L. Reynolds, Alissa Packer, James D. Bever and Keith Clay, 2003.. Grassroots Ecology: Plant-Microbe-Soil Interactions as Drivers of Plant Community Structure and Dynamics. *Ecology,* **84:** 2281-2291

Hiltner, L., 1904. Uber neure Erfahrungen und problem auf dem gebeit der bodenbackteriologie und unter besonderer berucksichtigung der grundungung und brache. *Ark Deut Landwirsch Ges,* **98**:59–78

Huang, K., Zhang, J., Tang, G.M., Bao, D., Wang, T.G. and Kong, D.P.,2023. Impacts and mechanisms of biochar on soil microorganisms. *Plant Soil Environ.,* **69**: 45–54.

Huber, D.M., 1980. The role of mineral nutrition in defense. *In: Plant Disease – An Advanced Treatise*, Vol. V. How Plants Defend Themselves, Horsfall, J.G. and Cowling, E.B.(Eds). Acad. Press, London: 381–406.

Jacobsen, C.S. and Hjelmsø, M.H., 2014. Agricultural soils, pesticides and microbial diversity. *Curr. Opin. Biotechnol.,* **27:** 15-20

Janvier, C., Villeneuve, F., Alabouvette, C., Edel-Hermann, V., Mateille, T. and Steinberg, C., 2007. Soil health through soil disease suppression: which strategy from descriptors to indicators? *Soil Biol. Biochem.,* **39**, 1-23.

Jia Z, Ishihara R, Nakajima Y, Asakawa A, Kimura M 2007. Molecular characterization of T4-type bacteriophages in rice field. *Environ. Microbiol.*, **9:** 1091–1096.

Joanne B. Emerson, 2019. Soil Viruses: A New Hope. 2019. Vol.4, AID - 10.1128/msystems.00120-19. *mSystems,* **4**: 10.1128/msystems.00120-19

John Idowu, Nicole Pietrasiak and Mikaela Hoellrich, 2023. Soil Biological Processes. Guide A-153. College of Agricultural, Consumer and Environmental Sciences, New Mexico State University.

Kinkel, L.L., Bakker, M.G. and Schlatter, D.C., 2011. A coevolutionary framework for managing disease-suppressive soils. *Ann. Rev. Phytopathol.,* **49:** 47-67

Kurt E. Williamson, Jeffry J. Fuhrmann, K. Eric Wommack, Mark Radosevich, 2017. Viruses in Soil Ecosystems: An Unknown Quantity Within an Unexplored Territory. *Ann.Rev. Virol.,* **4:** 201-219

Lehmann, J., Bossio, D.A., Kögel-Knabner, I. and Rillig, M.C., 2020. The concept and future prospects of soil health. *Nat. Rev. Earth Environ.,* **1:** 544-553.

Leigh Archer, U.A., 2020. *Antibiotics in crop production.* https://edis.ifas.ufl.edu/hs1366 (accessed on 30 November 2020).

Liu, X., Liu, L., Gong, J., Zhang, L., Jiang, Q., Huang, K. and Ding, W., 2022. Soil conditions on bacterial wilt disease affect bacterial and fungal assemblage in the rhizosphere. *AMB Express*, **12**:110.

Ludwig, M., Wilmes, P. and Schrader, S., 2018. Measuring soil sustainability via soil resilience. *Sci. Total Environ.,* **626:** 1484-1493

Mahdi M. Al-Kaisi, Rattan Lal, Kenneth R. Olson and Birl Lowery, 2017. Chapter 1 - Fundamentals and Functions of Soil Environment. Mahdi M. Al-Kaisi, Birl Lowery (Eds.), P.1-23. Soil Health and Intensification of groecosytems, Academic Press

Makoto Kimura, Zhong-Jun JIA, Natsuko Nakayama and Susumu Asakawa, 2008. Ecology of viruses in soils: Past, present and future perspectives. *Soil Sci & Pl.Nutrn.*, **54**: 1-32

McDearis Regan, 2022. Native soil virus abundance, composition, and mobility under natural and managed agricultural conditions. PhD diss., University of Tennessee, https://trace.tennessee.edu/utk_graddiss/7133

Mendes, R., et al.. 2011 Deciphering the rhizosphere microbiome for disease-suppressive bacteria. *Science,* **332:** 1097-1100.

Mendes, R., Kruijt, M., de Bruijn, I., Dekkers, E., van der Voort, M., Scheider, J. H. M., Piceno, I. M., DeSantis, T. Z, Andersen, G. L., Bakker, P. A. H. and J. M. Raaijmakers. 2011. Deciphering the Rhizosphere Microbiome for Disease-Suppressive Bacteria. *Science,* **332**:1097-1100.

Michel, V. V. and Mew, T. W., 1998. Effect of a soil amendment on the survival of *Ralstonia solanacearum* in different soils. *Phytopathology,* **88:** 300-305

Mishra, B.B., 2015. Soil Classification: Issues and Opportunities for Indian System. The 26th Dr. S.P. Raychaudhuri Memorial Lecture, 19th October. 2015, NBSS & LUP, Nagpur. *J.Indian Soc. Soil Sci.,* **63** (Supplement): 41-52.

Moussart, A., Wicker, E., le Delliou, B., Abelard, J.M., Esnault, R., Lemarchand, E., Rouault, F., le Guennou, F., Pilet-Nayel, M.L., Baranger, A., Rouxel, F. and Tivoli, B., 2009. Spatial distribution of *Aphanomyces euteiches* inoculums in a naturally infested pea field. *Eur. J. Plant Pathol.,* **123**:153–158

Neher, D. A., 1999. Soil community composition and ecosystem processes: Comparing agricultural systems with natural ecosystems. *Agroforestry Syst.*,**45**: 159–185

Nehl David, Allen Stephen and Kochman Joseph, 2003. Fusarium wilt of cotton: a fatal fungal affliction? *Microbio. Australia,* **24:** 8.

Otten, W., Gilligan, C.A., Watts, C.W., Dexter, A.R. Hall, D., 1999. Continuity of air-filled pores and invasion thresholds for a soil-borne fungal pathogen, *Rhizoctonia Solani. Soil Biol.& Biochem.,* **31:** 1803–1810.

Otten, W., Hall, D., Harris, K., Ritz, K., Young, I.M. and Gilligan, C.A., 2001. Soil physics, fungal epidemiology and the spread of *Rhizoctonia solani. New Phytologist*, **151**: 459–468.

Otten, W., Harris, K., Young, I.M., Ritz, K. and Gilligan, C.A., 2004. Preferential spread of the pathogenic fungus *Rhizoctonia solani* through structured soil. *Soil Biol.& Biochem.,* **36**: 203–210.

Otten, W. and Gilligan, C. A., 2006. Soil structure and soil-borne diseases: using epidemiological concepts to scale from fungal spread to plant epidemics. *European J. Soil.Sci.,* **57**: 26-37

Penton, C. R, Gupta, V. V. S.R., Tiedje, J. M., Neate, S. M., Ophel-Keller, K., Gillings, M., Harvey, P., Phan, A. and D. K. Roget, 2014. Fungal Community Structure in Disease Suppressive Soils Assessed by 28S LSU Gene Sequencing. *PLoS ONE,* **9**(4): e93893. doi:10.1371/journal.pone.0093893

Qi Gaofu, Ma Gaoqiang, Chen Shu, Lin Changchun and Zhao Xiuyun, 2019. Microbial Network and Soil Properties Are Changed in Bacterial Wilt-Susceptible *Soil Applied & Environ. Microbiol.,* **85:** e00162-19

Renčo, M., Gömöryová, E. and Čerevková, A., 2020. The Effect of Soil Type and Ecosystems on the Soil Nematode and Microbial Communities. *Helminthologia,* **57**:129-144.

Robert McSorley, 2019. Soil-inhabiting nematodes. UF/IFAS, Univ. of Florida.

Roberts and Alison, G , 2014. Plant Viruses: Soil-borne. In: eLS. John Wiley & Sons, Ltd: Chichester. DOI: 10.1002/9780470015902.a0000761.pub3

Sachidanand, B., Mitra, N.G., Vinod Kumar Richa Roy and Mishra, B.B., 2019. Soil as a Huge Laboratory for Microorganisms. *Agri Res& Tech., Open Access J.,* **22:** 556205.

Saha, S., Loganathan, M., Rai, A.B., Singh, A. and Garg, R., 2016. Role of microbes in soil health improvement. *SATSA Ann. Tech. Issue*, **20**:1-2.

Samaddar, S., Karp, D.S., Schmidt, R., Devarajan, N., McGarvey, J.A., Pires, A.F.A. and Scow, K., 2021. Role of soil in the regulation of human and plant pathogens: soils' contributions to people. *Philos. Trans. R. Soc. Lond. B. Biol. Sci.*, **27**: 376 (1834):20200179.

Schlatter, D., Kinkel, L., Thomashow, L., Weller, D. and Paulitz, T., 2017. Disease suppressive soils: new insights from the soil microbiome. *Phytopathology,* **107:** 1284-1297.

Shade, A., Peter, H., Allison, S. D., Baho, D. L., Berga, M., Bürgmann, H., Huber, D. H. Langenheder, S., Lennon, J. T., Martiny, J. B. H., Matulich, K. L., Schmidt, T. M. and Handelsman. J., 2012. Fundamentals of microbial community. Resistance and Resilience. *Front. Microbiol.*, Article 417, 19 pages. doi: 10.3389/fmicb.2012.00417

Simon, A. C. M., Horacio D. Lopez-Nicora, Terry L. Niblack, Elizabeth A. Dayton, David Tomashefski, and Pierce A. Pau, 2018. Cropping Practices and Soil Properties Associated with Plant-Parasitic Nematodes in Corn Fields in Ohio. *Pl. Disease*, **102**: 2519-2530.

Sun, Y.N., He, Z.W., Cao, X.Y., Zhao, Q., Liu, F., Jia, Z.H., Song, S.S. and Zhang, L.P., 2021. Involvement of receptor-like protein kinase PBL28 in the regulation of root growth by N-decanoyl-homoserine lactone in Arabidopsis. *Plant Physiology J.,* **57**: 1271–1280.

Terrence Bell, Kristy Borrelli, Sarah Isbell, Suzanne Fleishman, Laura Kaminsky and Mara Cloutier, 2021. Understanding and Managing Soil Microbes. PennState Extension. The Pennsylvania State University, Univ.Park, PA 16802

Thakur Rajneesh, Verma Shalini, Gupta Shivani, Negi Gulshan and Bhardwaj Paresh, 2022. Role of Soil Health in Plant Disease Management: A Review. *Agri. Rev.,* **43**: 70-76..

Trivedi, P., Leach, J.E., Tringe, S.G., Sa, T. and Singh, B.K., 2020. Plant–microbiome interactions: from community assembly to plant health. *Nat. Rev. Microbiol.,* **18:** 607-621.

USDA, 2016. Soil Biology Primer Photo Gallery. Natural Resources Conservation Service - Soils. Soil and Water Conservation Society, U.S. Department of Agriculture. Retrieved 14 August 2016

van Bruggen, A.H.C. and Semenov, A.M., 2000. In search of biological indicators for soil health and disease suppression. *Appl. Soil Ecol.,* **15:** 13-24.

van Elsas, J.D., Garbeva, P. and Salles, J., 2002. Effects of agronomical measures on the microbial diversity of soils as related to the suppression of soil-borne plant pathogens. *Biodegradation,* **13**:29–40

Walker, P. J., Siddell, S. G., Lefkowitz, E. J., Mushegian, A. R., Adriaenssens, E. M., Dempsey, D. M., Dutilh, B. E., Harrach, B., Harrison, R. L., Hendrickson, R. C., Junglen, S., Knowles, N. J., Kropinski, A. M., Krupovic, M., Kuhn, J. H., Nibert, M., Orton, R. J., Rubino, L., Sabanadzovic, S.... Davison, A. J., 2020. Changes to virus taxonomy and the Statutes ratified by the International Committee on Taxonomy of Viruses. *Arch. Virol.,* **165:** 2737–2748.

Wang, H., Li, X., Li, X., Wang, J., Li, X., Guo, Q., Yu, Z., Yang, T. and Zhang, H., 2020. Long-term no-tillage and different residue amounts alter soil microbial community composition and increase the risk of maize root rot in northeast China. *Soil Tillage Res.,* **196:** 104452.

Wen, K., Ortmann, A.C. and Suttle, C.A., 2004. Accurate estimation of viral abundance by epifluorescence microscopy. *Appl. Environ. Microbiol.,* **70:**3862–67

Xiang, L., Liu, S.J., Ye, S.J., Yang, H.L., Song, B., Qin, F.Z., Shen, M.C., Tan, C., Zeng, G.M. and Tan, X.F., 2021. Potential hazards of biochar: the negative environmental impacts of biochar applications. *J. Hazardous Mat.,* **420:** 126611

Yang, J., Liu, P., Zhong, K. *et al., 2022.* Advances in understanding the soil-borne viruses of wheat: from the laboratory bench to strategies for disease control in the field. *Phytopathol. Res.,***4:** 27. https://doi.org/10.1186/s42483-022-00132-2

Yifan Cao, Linda S. Thomashow, Yu Luo, Hangwei Hu, Xuhui Deng, Hongjun Liu, Zongzhuan Shen, Rong Li and Qirong Shen, 2022. Resistance to bacterial wilt caused by *Ralstonia solanacearum* depends on the nutrient condition in soil and applied fertilizers: A meta-analysis. *Agri. Ecosyst. & Environ.,* **329:** 107874

Yilmaz, P., Kottmann, R., Field, D., Knight, R., Cole, J.R, et al., 2011. Minimum information about a marker gene sequence (MIMARKS) and minimum information about any (x) sequence (MIxS) specifications. *Nat. Biotechnol.,* **29**:415–420

Zhang, F., Liu, S., Zhang, T., Ye, Z., Han, X., Zhong, K., et, al., 2021. Construction and biological characterization of an infectious full-length cDNA clone of a Chinese isolate of wheat yellow mosaic virus. *Virology,* **556:**101–109.

2

Nature and Importance of Rhizosphere and Rhizoplane

2.1 Roots and the Rhizosphere

The rhizosphere, the region of soil surrounding plant roots in which soil and microbes closely interact–plays an important role in soil carbon cycling. The rhizosphere is only about 1-2% of Earth's soil volume but can store up to 30-40% of Earth's total soil organic matter. Plants secrete organic compounds such as amino acids or enzymes known as root exudates, which are key in recruiting and selecting relevant beneficial microbes to colonize in the rhizosphere (Julie Bobyock, 2023). Studying these microbial-plant interactions is critical to better understand the relationship between plant growth and the role soils play in keeping carbon from the atmosphere. It is the region that is a few distances (2-80 mm) extended from the root system. Rhizosphere zone is the region of intense microbial activity, and it is isolated from the bulk soil that often called as Edaphosphere or Non-rhizosphere.

Plant growth results from interaction of roots and shoots with the environment. The environment for roots is the soil or planting medium, and its importance is often not appreciated beyond the physical and nutritional aspects. Roots do provide structural support for the plant, and they absorb water and nutrients, but they also produce plant growth-regulating substances, such as cytokinins. The cortex of the root is largely starch storage tissue, and represents a transition zone from the soil to the stele. The cortex also provides a niche to be occupied by mycorrhizal fungi. It is not as generally appreciated, however, that roots also support the growth and functions of a complex of microorganisms that can have a profound effect on the growth and survival of plants. These microbes interact in the soil around the root that is influenced by root exudates; this zone is called the rhizosphere. The effects of rhizosphere microorganisms on plant growth or health often are ignored unless the microbes weaken or kill the plant through disease. There are, however, beneficial interactions that can occur and be enhanced if actively managed by the grower. Establishment and maintenance of rhizosphere microbial populations is a dynamic process affected by age and kind of plant, nature and treatment of soil, environmental factors, foliar applications of chemicals, and interactions of microorganisms. It is important to remember that selective or preferential development of rhizosphere populations occurs in the early stages of plant growth and root development when root exudation is maximal but, thereafter, is altered and maintained at some microbial equilibrium by both microbial and plant activity, with both being influenced by edaphic and other environmental factors.

Roots may favor the colonization of certain microbes in the rhizosphere. Among them are mycorrhizal fungi (except possibly in the case of soilless mixes), which form symbiotic associations with roots by colonizing cortical tissues. These fungi also produce a hyphal network extending into the soil. Mycorrhizae benefit the plant in several ways, but primarily by aiding in absorption of water and nutrients, and especially immobile nutrients such as P. As this association forms, chemical and physical changes occur in the rhizosphere due to altered host physiology as well as to the physical and chemical presence of mycorrhizal fungus hyphae in the rhizosphere and beyond (Linderman, 1986). These changes were considered to be profound and permanent enough that he coined the term 'mycorrhizosphere' to describe the microbial ambience around mycorrhizae. The influence of root exudates and microbial interactions on components of the soil microflora and fauna other than mycorrhizal fungi may also be profound, although possibly more transient, and those microbes could have major effects on the growth and development of plants. Rhizosphere microbes are exposed to a selective pressure, which causes some organisms to be favored and thus to develop higher populations than those that are not favored. Those that dominate are most likely to have the greatest impact on plant growth. The selection process draws from the available background of organisms that have survived previous selection pressures. They probably represent a mixture of beneficial, detrimental, and neutral organisms relative to plant growth. If the prior treatments had eliminated the pathogens, and the soil or medium had been amended with beneficial organisms, then the selection process could only draw from a mixture of beneficial or neutral organisms. This, then, is the concept of active management of rhizosphere microorganisms: eliminate detrimental organisms by soil treatment, then reconstitute the microflora by inoculation with selected beneficial organisms to fill the biological void created by soil treatment. It is important that the void be filled by inoculation before it is filled by nonselected contaminants, including real or potential pathogens.

Among the more important microbial interactions that we seek to favor by microbial management is the suppression of pathogenic organisms by introduced biocontrol agents. In most instances, the key to successful exploitation of introduced biocontrol agents is to establish them at the infection court before the pathogen is introduced. Frequently, the agent produces metabolic inhibitors that suppress or kill the pathogen, or it may actively parasitize the pathogen. In the case of root pathogens, the rhizosphere and/or rhizoplane (root surface and surface cells) must be occupied heavily by the antagonist in order to achieve protection from disease agents. This means that the soil medium and/or the plant propagule itself must be preinoculated with the antagonist in order for it to grow and flourish prior to pathogen introduction. In order for an antagonist or any other beneficial organism to best establish itself in a soil medium, the microbial competition must be reduced by some pretreatment.

The region where the soil and roots make contact is called 'rhizosphere'. The microbial population of rhizosphere is much higher than that of root free soil.

- Bacterial growth in rhizosphere is enhanced by nutritional substance like vitamins, amino-acids, etc. released from the plants.
- The roots of higher plants are surrounded by both living as well as non-living environment.
- The organic and inorganic substances of soil make up the non-living environment.
- The organism that are present in the soil constitute the living environment.
- Rhizosphere is the region of soil immediately surrounding the roots of plant.

Structure of the rhizosphere: Structure of rhizosphere consists of three zones based on their proximity to the root system (Fig.8).

1. **Inner zone**: It is also termed as 'Endorhizosphere'. It is very close to the root. It includes the portion of cortex and epidermis, in which the microorganisms may occupy between the vacant space known as Apoplastic space.
2. **Rhizoplane**: It is the root system itself.
3. **Outer zone**: It is also known as 'Exorhizoshere'. This layer is adjacent to the portion of the epidermis.

The inner rhizosphere has a larger microbial population than the outer rhizosphere as in this region the interaction between microorganism and roots are most pronounced.

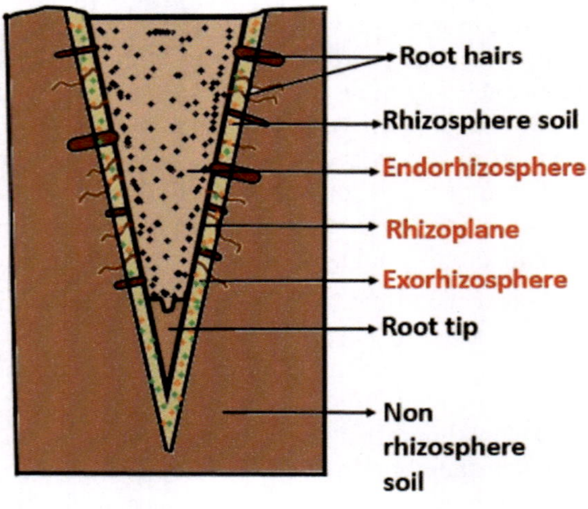

Fig. 8: Rhizosphere structure

Microbial activity in Rhizosphere Zone: The whole process inside the rhizosphere zone is explained below in the diagram.

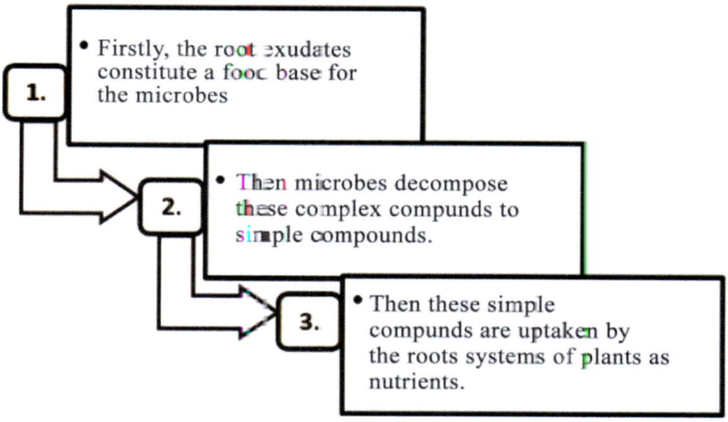

Rhizosphere Concept and Historical Background: The root system of higher plants is associated not only with soil environment composed of inorganic and organic matter, but also with a vast community of metabolically active microorganisms. As living plants create a unique habitat around the roots, the microbial population on and around the roots is considerably higher than that of root free soil environment and the differences may be both quantitative and qualitative.

Rhizosphere: It is the zone/region of soil immediately surrounding the plant roots together with root surfaces, or it is the region where soil and plant roots make contact, or it is the soil region subjected to influence of plant roots and characterized by increased microbial.

Rhizoplane: Root surface along with the closely adhering soil particles is termed as rhizoplane.

2.2 Historical Background: Term "Rhizosphere" was introduced for the first time by the German scientist Hiltner (1904) to denote that region of soil which is subjected to the influence of plant roots (Anonymous, 2018). The concept of "Rhizosphere Phenomenon" which shows the mutual interaction of roots and microorganisms was came into existence with the work of Starkey et al (1929), Clark (1939) and Rauath and Katznelson (1957).

- N. V. Krassinikov (1934) found that free living nitrogen-fixing bacteria, *Azotobacter* were unable to grow in the wheat rhizosphere.
- **Starkey (1938)** examined the rhizosphere region of some plant species and demonstrated the effect of root exudates on the predominance of bacterial population in particular and other soil microorganisms in general in the rhizosphere region. Thus, he put forth the concept of "Rhizosphere effect / phenomenon" for the first time.
- **F E Clark (1949)** introduced / coined the term "Rhizoplane" to denote the root surface together with the closely adhering soil particles.

- **R. I. Perotti (1925)** suggested the boundaries of the rhizosphere region and showed that it was bounded on one side by the general soil region (called as Edaphosphere) and on the other side by the root tissues (called Histosphere).
- **G. Graf and S. Poschenrieder (1930)** divided the rhizosphere region into two general areas i.e. outer rhizosphere and inner rhizosphere for the purpose of describing the same site of microbial action.
- **H. Katznelson (1946)** suggested the **R:S ratio** i.e. the ratio between the microbial population in the rhizosphere (R) and in the soil (S) to find out the degree or extent of plant roots effect on soil microorganisms. R: S ratio gives a good picture of the relative stimulation of the microorganisms in the rhizosphere of different plant species.
- **R: S ratio** is defined as the ratio of microbial population per unit weight of rhizosphere soil (R), to the microbial population per unit weight of the adjacent non-rhizosphere soil (S)
- **G. Lochhead and H. Katznelson (1940)** examined in detail the qualitative differences between the microflora of the rhizosphere and micro flora of the non-rhizosphere region and reported that gram-negative, rod shaped and non-spore forming bacteria are abundant in the rhizosphere than in the non-rhizosphere soil
- **Thom and H. Humfeld (1932)** found that corn roots in acidic soils yielded predominantly *Trichoderma* while roots from alkaline soils mainly contained *Penicillium.*
- **M J. Timonin (1940)** reported some differences in the fungal types and population in the rhizosphere of cereals and legumes. R: S ratio of fungal population was believed to be narrow in most of the plant species, usually not exceeding 10.
- **E. A. Peterson and others (1958)** reported that the plant age and soil type influence the nature of fungal flora in the rhizosphere, and the number of fungal population gradually increases with the age of plant.
- **M. Adati (1932)** studied many crops and found that though actinomycetes were relatively less stimulated than bacteria, but in some cases the R: S ratio of actinomycetes was as high as 62.
- **R. Venkatesan and G. Rangaswami (1965)** studied the rhizosphere effect in rice plant on bacteria, actinomycetes and fungi and reported that (i) for actinornycetes R: S was more (ranging from 0 to 25) depending on the age of plant roots and the dominant genera reported were Nocardia, (ii) R:S ratio reduced with the depth of soil.
- **E. A. Gonsalves and V. S. Yalavigi (1960)** reported the presence of greater number of algae in the rhizosphere
- **J. W. Rouatt et al** reported positive rhizosphere effect on protozoa, but a negative effect on algae in wheat plants.

2.3 Composition of the rhizosphere: However, the definition of the rhizosphere has been evolving as other physical, chemical and biological properties have been discovered. The rhizosphere is highly influenced by the roots of plants that promote intense biological and chemical activities. The organisms that coexist in the rhizosphere present a variety of interactions between them and also with plants. These interactions can affect the growth of a wide range of crops, which is why rhizospheres are very important as substitutes for chemical fertilizers and pesticides.

Characteristics of the rhizosphere: It is thin and subdivided into three basic zones. Structurally, the rhizosphere is approximately 1 mm wide and has no defined edges. Despite this, three basic zones in the rhizosphere have been described.

i. **The Endorizosphere:** It consists of root tissue and includes the endodermis and cortical layers.

ii. **The Rhizoplane:** It is the surface of the root, where the soil particles and microbes adhere. It is formed by the epidermis, the cortex and the layer of mucilaginous polysaccharides.

iii. **The Ectorizosphere:** It is the most external part; that is, the soil that is immediately adjacent to the root.

In the rhizosphere, different compounds are released. During the growth and development of a plant, a variety of organic compounds are produced and released by exudation, secretion and deposition. This causes the rhizosphere to be rich in nutrients, compared to the rest of the soil. Root exudates include amino acids, carbohydrates, sugars, vitamins, mucilages and proteins. The exudates act as messengers that stimulate the interactions between the roots and the organisms that inhabit the soil.

Change the pH of the soil around the roots: The environment of the rhizosphere generally has a lower pH, with less oxygen and higher concentrations of carbon dioxide. However, exudates can make the soil in the rhizosphere more acidic or alkaline, depending on the nutrients that the roots are taking from the soil. For example, when a plant absorbs nitrogen in ammonium molecules, it releases hydrogen ions that will make the rhizosphere more acidic. In contrast, when a plant absorbs nitrogen in nitrate molecules, it releases hydroxyl ions that make the rhizosphere more alkaline.

Microbiology: As mentioned above, the rhizosphere is an environment with a high density of microorganisms of several species. For better understanding, the microorganisms of the rhizosphere can be classified into three large groups, depending on the effect it causes on plants:

i. **Beneficial microbes:** This group includes the organisms that promote the growth of the plant directly - for example, by providing necessary nutrients to the plant - or indirectly, by inhibiting harmful microbes through various mechanisms of resistance. In the rhizosphere there is a constant competition for resources. Beneficial microbes limit the success of pathogens with several mechanisms: the production of biostatic compounds (which inhibit the growth or multiplication of microorganisms), the competition for micronutrients or stimulating the immune system of the plant.

ii. **Commensal microbes:** In this category are most of the microbes that do not harm or directly benefit the plant or the pathogen. However, it is likely that commensal microbes affect to some extent any other microorganism, through a complex network of interactions that would cause an indirect effect on the plant or the pathogen. Although there are specific microorganisms that are able to protect the plant (directly or indirectly) against pathogens, its effectiveness is largely influenced by the rest of the microbial community.Thus, commensal microorganisms can compete effectively with the other microorganisms exerting an indirect effect on the plant.

iii. **Pathogenic microbes:** A wide range of pathogens transmitted by the soil can affect the health of plants. Before infection, these harmful microbes compete with many other microbes in the rhizosphere for nutrients and space. Nematodes and fungi are the two main groups of plant pathogens transmitted by the soil. In temperate climates, pathogenic fungi and nematodes are agronomically more important than pathogenic bacteria, although some bacterial genera (*Pectobacterium, Ralstonia*) can cause substantial economic damage to some crops. Viruses can also infect plants through the roots, but require vectors such as nematodes or fungi to enter the root tissue.

Importance

i. **Attracts beneficial microorganisms:** The high levels of moisture and nutrients in the rhizosphere attract a much greater number of microorganisms than other parts of the soil. Some of the compounds secreted in the rhizosphere promote the establishment and proliferation of microbial populations, much higher compared to the rest of the soil. This phenomenon is known as the rhizosphere effect.

ii. **Offers protection against pathogenic microorganisms:** The cells of the roots are under continuous attack of microorganisms, which is why they have protection mechanisms that guarantee their survival. These mechanisms include the secretion of defense proteins and other antimicrobial chemicals. It has been determined that the exudates in the rhizosphere vary according to the stages of plant growth.

iii. **Protects the roots from drying:** Several studies suggest that the soil of the rhizosphere is significantly more humid than the rest of the soil, which helps protect the roots from drying out. The exudates released by the roots at night allow the expansion of the roots in the soil. When the perspiration is resumed with daylight, the exudates begin to dry and adhere to the soil particles in the rhizosphere. As the soil dries and its hydraulic potential decreases, the exudates lose water in the soil.

The rhizosphere is the narrow region of soil or substrate that is directly influenced by root secretions and associated soil microorganisms known as the root microbiome**.** Soil pores in the rhizosphere can contain many bacteria and other microorganisms that feed on sloughed-off plant cells, termed *rhizodeposition* and the proteins and sugars released by roots, termed root exudates**.** This symbiosis leads to more complex interactions, influencing plant growth and competition for resources (Walker, et al., 2003). Much of the nutrient cycling and disease suppression by antibiotics required by plants, occurs immediately adjacent to roots due to root exudates and metabolic products of symbiotic and pathogenic communities of microorganisms (Fig.9). The rhizosphere also provides space to produce allelochemicals to control neighbours and relatives. The *rhizoplane* refers to the root surface including its associated soil particles which closely interact with each other. The plant-soil feedback loop and other physical factors occurring at the plant-root soil interface are important selective pressures in communities and growth in the rhizosphere and rhizoplane. The term "rhizosphere" was used first in 1904 by the German plant physiologist Lorenz Hiltner to describe how plant roots interface with surrounding soil. The prefix *rhiza-* comes from the Greek, and means "root". Hiltner postulated the rhizosphere was a region surrounding the plant roots, and populated with microorganisms under some degree of control by chemicals released from the plant roots.

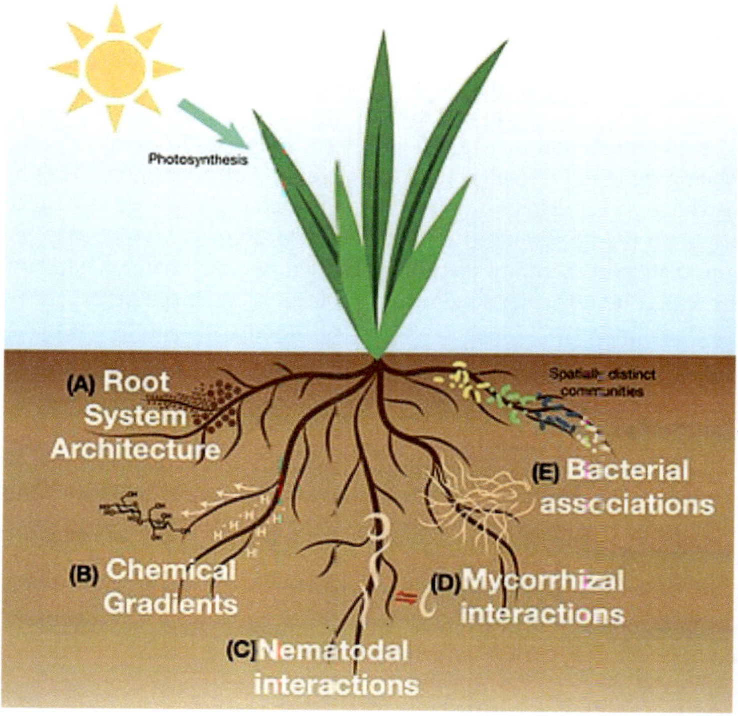

Fig. 9: Some rhizosphere processes in the soil- **(A)** Root system architecture is concerned with structural features of the root and responds to with environmental stimuli. **(B)** The rhizosphere produces photosynthetically fixed carbon that exudes into the soil and influences soil physicochemical gradients. **(C)** Free-living or parasitic nematodes interact with the rhizosphere via signaling interactions. **(D)** Mycorrhizal fungi create intimate relationships with the roots and engage in nutrient exchange. (E) Bacterial composition is distinct upon different parts, age, type of the roots.

The rhizosphere serves as the microbial seed bank where microorganisms transform organic and inorganic substances in the rhizosphere into accessible plant nutrients as plants harbor diverse microorganisms such as fungi, bacteria, nematodes, viruses, and protists among others (Dlamini et al., 2022). Although, the pathogenic microbes initiate diseases by infiltrating the protective microbial barrier and plants' natural defense systems in the rhizosphere. Whereas, the process is often circumvented by the beneficial microorganisms which antagonize the pathogens to instill disease resistance. The management of plant health through approaches focused on disease prevention is instrumental to attaining sustainable food security, and safety. Therefore, an in-depth understanding of the evolving and succession of root microbiomes in response to crop development as discussed in this review opens up new-fangled possibilities for reaping the profit of beneficial root–microbiomes' interactions toward attaining sustainable plant health.

Soil is one of the last great scientific frontiers and the rhizosphere is the most active portion of that frontier in which biogeochemical processes influence a host of landscape and global scale processes. A better understanding of these processes is critical for maintaining the health of the planet and feeding the organisms that live on it (Morrissey *et al*., 2004). There is a small but concerted effort under way to harness the root system of plants in an attempt to increase yield potentials of staple food crops in order to meet the projected doubling in global food demand in the next 50 years. These efforts are being done in the face of a changing global climate and increasing global population which will inevitably require more productively grown food, feed and fiber on less optimal (and often infertile) lands; a condition already encountered in many developing countries. Meeting the global challenges of climate change and population growth with a better understanding and control of rhizosphere processes will be one of the most important science frontiers of the next decade for which a diverse, interdisciplinary trained workforce will be required.

The rhizosphere is a dynamic region governed by complex interactions between plants and the organisms that are in close association with the root. The composition and pattern of root exudates affect microbial activity and population numbers, which in turn have an impact on the nematodes and microarthropods that share this environment. Beneficial or harmful relationships exist between rhizosphere organisms and plants, which ultimately affect root function and plant growth (Kennedy and de Luna, 2005). In addition, the rhizosphere may include organisms that do not directly benefit or harm plants but clearly influence plant growth and productivity. A better understanding of the soil–root and soil–seed interface is needed to manage microorganisms, increase plant growth, and reduce the impact of plant production and agriculture on the environment. The benefits of studying the rhizosphere include the use of plant growth-promoting organisms and the suppression of plant diseases and weeds using biocontrol agents. Rhizosphere organisms can also be used to enhance the formation of stable soil aggregates and as bioremediation agents of contaminated soils. With greater understanding of the ecology and biota in the rhizosphere, this zone of increased nutrients, biotic activity, and interactions can be manipulated to improve plant productivity and environmental quality.

The plant rhizospheric zone is a niche of microbial activity based on the resistance of the root system against phytopathogens. The region provides a variety of benefits to plant health both directly and indirectly (Dlamini et al., 2022). It is generally recognized that soil microbes serve as a primary nutrient reservoir for plants through their active participation in nutrient cycle processes such as nitrogen cycling, organic matter decomposition and mineral weathering, all of which influence plant growth. Thus, plant roots and microbe interactions are critical for plant participation in the ecosystem's functioning. Understanding ecosystem responses to the changing environments require knowledge of microbial diversity distribution and functions. More so, root exudates contain an extensive variety of chemicals, including amino acids, carbohydrates, siderophores, and enzymes, that can reinforce the activities of the microbial community and significantly change the structure of the rhizospheric microbial community, contingent on the plant species and genotype. Plant-associated microbial communities promote plant development and resilience to diverse abiotic and biotic stressors. Soil microbial species through their roles in carbon sequestration and biogeochemical cycling play important roles in the mitigation of climate change that presently constitutes a global concern. Improved ability to predict functionality from dynamic aberrations in root microbial composition is critical for building "designer" microbiome interventions and enhancing agricultural production and sustainability. This will necessitate a more comprehensive genomic and functional classification of plant-linked microbes, their genetic factor, and plant-microbe interactions for their exact useful roles.

Beneficial, commensal, or pathogenic relationships exist between specific microorganisms and their plant hosts. Plant microbiomes, in many instances, show a beneficial role at the community level, protecting plants from pathogens, while also boosting growth, health, and providing plants with an adaptive edge. Therefore, the interactions between plants and their microbiota are intricate and constantly changing. This interaction and interdependence on the abiotic and biotic components of the soil environment can be explained by the plant-soil feedback mechanism.

The definition of the Rhizosphere: In 1904 the German agronomist and plant physiologist Lorenz Hiltner first coined the term "rhizosphere" to describe the plant-root interface, a word originating in part from the Greek word "rhiza", meaning root (McNear Jr, 2013). Hiltner described the rhizosphere as the area around a plant root that is inhabited by a unique population of microorganisms influenced, he postulated, by the chemicals released from plant roots. In the years since, the rhizosphere definition has been refined to include three zones which are defined based on their relative proximity to, and thus influence from, the root (Fig.10). The endorhizosphere includes portions of the cortex and endodermis in which microbes and cations can occupy the «free space» between cells (apoplastic space). The rhizoplane is the medial zone directly adjacent to the root including the root epidermis and mucilage. The outermost zone is the ectorhizosphere which extends from the rhizoplane out into the bulk soil. As might be expected because of the inherent complexity and diversity of plant root systems, the rhizosphere is not a region of definable size or shape, but instead, consists of a gradient in chemical, biological and physical properties which change both radially and longitudinally along the root.

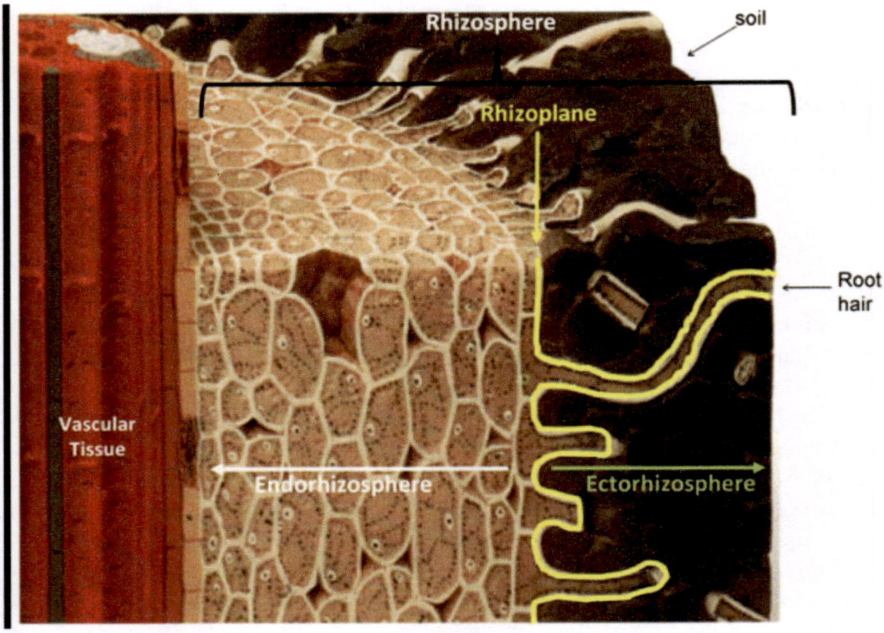

Fig. 10: Schematic of a root section showing the rhizosphere structure

2.4 Rhizosphere Soil Environment

Plant roots have strong effects on the physical environment in the rhizosphere. Shifts in the physical environment, which are influenced by root manipulation, have significant impacts on the availability and form of substrates used for microbial metabolism. In turn, this strongly affects the prevalence of different microbial functions, as well as their total biomass within the rhizosphere (*MicrobeWiki*). This interplay goes both ways, as manipulation by microbes also plays a strong role in the ways that plant roots interact with, benefit from, and influence the soil environment around them.

Water Relations: The availability of water is a major component in determining the diversity and prosperity of soil organisms. Water availability is quantified by the water potential that describes the transport and energy potential state of water in different forms and locations (Philippot et al.,2013). Plant transpiration, via leaf stomata, is a mechanism for connecting the soil-plant-atmosphere continuum that is driven by a vapor pressure gradient between two physical locations (I.E. the plant root and surrounding soil or the leaf interior and the atmosphere). As water transpires across the leaf surface, tension is created within the leaf that is ultimately transferred down to the root system via adhesion and cohesion of water through the vascular tissues of the plant. This can drastically alter the water potential of the rhizosphere, resulting in large fluctuations in the availability of water and solutes for microbes. This fluctuation is dramatically greater than that of the bulk soil, where changes in water potential are driven predominantly by evaporation that occur at much slower rates. This fluctuation in available water can put pressure on the diversity and total biomass of microbes in the rhizosphere, by selecting for those with better mechanisms for water uptake.

On average, 50% of soils consist as pore space, composed of air and water with differing concentrations of solutes (Scow, 2015). The constituents of the soil solution provide the substrates necessary for microbes to thrive, ultimately determining the types of biotic and abiotic interactions taking place. Within the rhizosphere, solute activity is significantly higher than that of the bulk soil. One reason is that plant roots secrete ions and organic compounds that interact with organisms and other molecules in the environment around them. Additionally, plant roots uptake solutes in the soil solution, causing changes in ion concentrations that drive the rate and direction of chemical processes within the rhizosphere. As a result, the reinforced microbial population carries out enzymatic processes, such as the transformation of carbon and nitrogen species, at higher rates than in the bulk soil.

Soil Physical Properties: The relative ratio of sand, silt, and clay has large impacts on the availability of water, air, and nutrients for microbial functions. The movement of roots through the soil profile occurs more rapidly in sandy soils than in clayey soils. This is due to sand having larger pore spaces between granules, allowing nutrient and water percolation to occur more rapidly. Increased porosity also allows for plant secreted compounds to extend further away from the root membrane. Therefore, the larger the granule size, the further the associated rhizosphere and microorganisms will extend into the surrounding soil (Kent and Triplett, 2002). This means that larger granules require microbes to travel further distances to attain nutrients or engage in interactions with plant roots. This is especially true in soils with very low water potential. Additionally, soil structure of the rhizosphere is also affected by the increased interactions of microorganisms. In addition to the binding effects of plant mucilages, other root carbon depositions also stimulate soil bacteria biofilm production. These biofilms are composed of extracellular carbon polymers that act as an aggregating mechanism for mineral particles and organic matter. The rate and composition of bacterial exopolymers is highly dependant on the composition of root-secreted compounds, although in general have been shown to be significantly higher in the rhizosphere than in the bulk soil.

Mycorrhizal fungi have also been shown to increase aggregation through the physical constriction of fungal hyphae, as well as through the secretion of their own exopolymer deposits. Further, the sorptive forces of plant roots and mycorrhizae provide an attractive force that promotes mineral alignment and provides a "backbone" for stabilizing aggregate surfaces. In addition, the decomposition of roots and hyphae also increases soil aggregation by the delivery of new organic material, as well as by producing larger pores and preferential flow pathways for air and water. The relationship of roots with associated mycorrhizae can alter both the decomposition rate as well as the final content of the decomposition products.

Soil Chemistry: The effects of pH extend to almost every characteristic of soil that are affected by multiple interactions within the rhizosphere. An example is respiration, a reaction that leads to carbon dioxide (and eventually to bicarbonate/carbonic acid) generation. In addition to the respiration of the roots themselves, the carbon-rich environment of the rhizosphere promotes high levels of respiration by other macro and microorganisms, to a far greater extent than what is occurring in the bulk soil. Increased microbial activity also increases microbial transformations of chemical species. These transformations result in increased acidity, such as nitric acid production during nitrification and sulphuric acid production during sulphur reduction. The influencing effects of pH in the rhizosphere are critical in supporting a biologically diverse microbial community (Scow, 2015). Manipulation of the pH by plant roots is constantly occurring and, while it is commonly believed that plants manipulate the soil pH in order to change the availability of nutrients, it also undoubtedly influences the abundance and diversity of associated microorganisms. Microbes are also essential in performing reduction-oxidation (redox) reactions in soil. These redox reactions provide energy for the microbes while also changing the valence of ions in solution. The changes in valence are important for plants, which require essential nutrients to be available, often in inorganic forms in order to be absorbed. Increases in microbial populations, associated with the rhizosphere, also result in increased redox reactions. Additionally, a study by Yang & Crowley (2001) showed that variations in different iron valence concentrations resulted in quantitative differences between rhizosphere microbial populations. While we have been aware of how the rhizosphere promotes increased microbe populations for some time now, we are only beginning to understand how the redox state of various ions is affecting these same populations.

Rhizodeposits and microbial exopolymer secretion also increase the amount of molecular chelation and complexation in the rhizosphere compared to the bulk soil. These complexes provide different functions, depending on the composition. Some of these functions include reducing or increasing solubility, altering the charge, and reducing toxicity of molecules. The changes in solubility are essential for allowing plant roots and mycorrhizae access to highly immobile nutrients, such as phosphate and iron.

2.5 Rhizoplane

The rhizoplane is the region where the root surface is in contact with soil and corresponds to the inner limit of the rhizosphere. At the rhizoplane level, plants exchange elements with the surrounding soil and the rhizoplane can therefore be considered as the region that drives nutrient movement and transformation in the rhizosphere. The rhizoplane differs in many respects from the bulk soil due to the far larger supply of substrates derived from the roots, with far greater microbial cell densities and reduced levels of diversity (Philippot et al., 2013). This is likely to result in completely different interaction profiles among microorganisms which may affect rhizosphere biogeochemistry. While the diversity of microorganisms associated with the rhizosphere and on the rhizoplane is getting increasing attention, knowledge on the spatial organisation of this diversity is still scarce. Rhizoplane simply refers to the root surface including the strongly adhering soil particles.

The rhizoplane, the innermost boundary of the rhizosphere, is the surface of the plant root including root hairs, and is a hot spot of plant–microbe activity (Knudsen and Dandurand, 2007). Bacteria and fungi on the rhizoplane are favorably positioned to intercept root exudates, and rhizoplane sites are points of root interaction with plant pathoens as well as beneficial microbes.

Rhizoplane Microbiome

The collective communities of plant-associated microorganisms are referred to as the plant microbiome or as the plants' other genome. In this context, plants can be viewed as superorganisms that rely in part on their microbiome for specific functions and traits. In return, plants deposit their photosynthetically fixed carbon into their direct surroundings, that is, spermosphere, phyllosphere, rhizosphere, and mycorrhizosphere, thereby feeding the microbial community and influencing their composition and activities (Rodrigo Mendes et al., 2013). To date, the interplay between plants and microorganisms has been studied in depth for various leaf pathogens, symbiotic rhizobia, and mycorrhizal fungi. However, for the vast majority of plant-associated microorganisms, there is limited knowledge of their impact on plant growth, health, and disease. Hence, deciphering the plant microbiome is critical to identify microorganisms that can be exploited for improving plant growth and health.

Plants are colonized by an astounding number of (micro)organisms that can reach cell densities much greater than the number of plant cells. Also, the number of microbial genes in the rhizosphere outnumbers by far the number of plant genes (Fig. 11). Microbial communities play a pivotal role in the functioning of plants by influencing their physiology and development. While many members of the rhizosphere microbiome are beneficial to plant growth, also plant pathogenic microorganisms colonize the rhizosphere striving to break through the protective microbial shield and to overcome the innate plant defense mechanisms in order to cause disease. A third group of microorganisms that can be found in the rhizosphere are the true and opportunistic human pathogenic bacteria, which can be carried on or in plant tissue and may cause disease when introduced into debilitated humans. Although the importance of the rhizosphere microbiome for plant growth has been widely recognized, for the vast majority of rhizosphere microorganisms no knowledge exists. To enhance plant growth and health, it is essential to know which microorganism is present in the rhizosphere microbiome and what they are doing. The main functions of rhizosphere microorganisms and how they impact on health and disease were reviewed and also discussed the mechanisms involved in the multitrophic interactions and chemical dialogues that occur in the rhizosphere. Several strategies to redirect or reshape the rhizosphere microbiome in favor of microorganisms that are beneficial to plant growth and health were highlighted by Rodrigo Mendes et al., (2013).

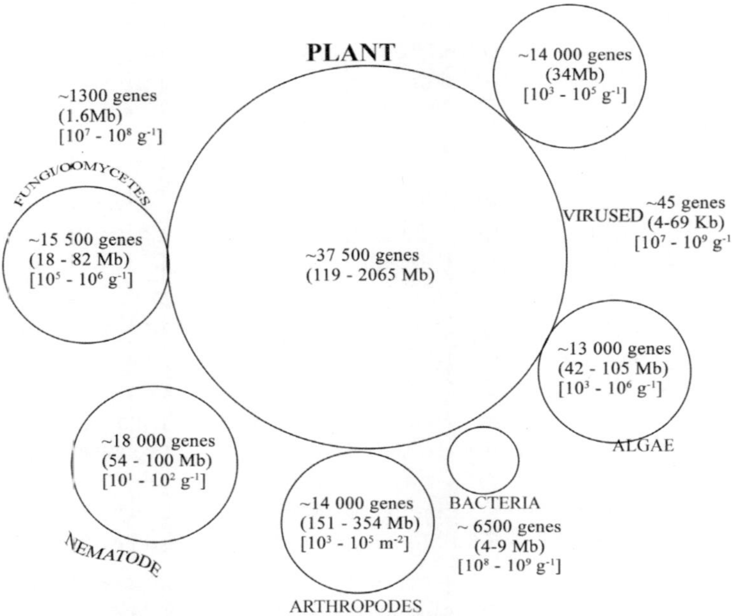

Fig. 11: Overview of (micro)organisms present in the rhizosphere zoo. The circle's size, except for VIRUSES, is a measure of the average number of genes in the genomes of representative species of each group of organisms; the size (or size range) of their respective genomes is indicated between parentheses. For each of these (micro)organisms, the approximate numbers for their abundance are indicated between square brackets. The endophytic microorganisms, including endosymbionts, are not included. The species selected to illustrate the composition of the rhizosphere microbiome and used to calculate the number of genes and genome sizes are: PLANT: *Glycine max, Populus trichocarpa, Zea mays, Oryza sativa, Arabidopsis thaliana,* and *Vitis vinifera;* PROTOZOA: *Dictyostelium discoideum;* VIRUSES: Pseudomonas phage 73, *Fusarium graminearum* dsRNA mycovirus-4, *Agrobacterium* phage 7-7-1, *Rhizoctonia solani* virus 717; ALGAE: *Chlorella variabilis* and *Chlamydomonas reinhardtii*; BACTERIA: *Pseudomonas fluorescens, Bradyrhizobium japonicum, Rhizobium leguminosarum, Bacillus cereus, Bacillus amyloliquefaciens, Burkholderia cenocepacia*, and *Streptomyces filamentosus*; ARTHROPODES: *Metaseiulus occidentalis, Acromyrmex echinatior,* and *Solenopsis invicta*; NEMATODES: *Caenorhabditis elegans* and *Meloidogyne hapla*; FUNGI/OOMYCETES: *Laccaria bicolor, Nectria haematococca, Piriformospora indica, Verticillium dahliae, Metarhizium anisopliae, Fusarium oxysporum, Sporisorium reilianum, Phytophthora sojae, Phytophthora parasitica, Aphanomyces euteiches, Phytophthora cinnamomi,* and *Pythium ultimum*; ARCHAEA: *Candidatus* Nitrosoarchaeum koreensis.

Huanglongbing impairs the rhizosphere-to-rhizoplane enrichment process: Among the three root-associated layers (i.e., rhizosphere, rhizoplane, and endorhiza), the rhizoplane is a key component serving a critical gating role that controls microbial entry into plant roots. The microbial communities colonizing the three layers are believed to be gradually enriched from the bulk soil inoculum. However, it is unknown how this enrichment process, particularly the rhizosphere to rhizoplane step, is affected by biotic stresses, such as disease. In a study, Zhang et al., (2017) addressed this question using the citrus root-associated microbiome as a model.

They identified the rhizosphere-to-rhizoplane-enriched taxonomic and functional properties of the citrus root-associated microbiome and determined how they were affected by Huanglongbing (HLB), a severe systemic disease caused by *Candidatus* Liberibacter asiaticus, using metagenomic and metatranscriptomic approaches. Multiple rhizoplane-enriched genera were identified, with *Bradyrhizobium* and *Burkholderia* being the most dominant. Plant-derived carbon sources are an important driving force for the enrichment process. The enrichment of functional attributes, such as motility, chemotaxis, secretion systems, and lipopolysaccharide (LPS) synthesis, demonstrated more active microbe-plant interactions on the rhizoplane than the rhizosphere. It was observed that HLB impaired the rhizosphere-to-rhizoplane enrichment process of the citrus root-associated microbiome in three ways, by decreasing the relative abundance of most rhizoplane-enriched genera; by reducing the relative abundance and/or expression activity of the functional attributes involved in microbe-plant interactions; and by recruiting more functional features involved in autotrophic life cycle adaptation, such as carbon fixation and nitrogen nitrification in the HLB rhizoplane microbiome. The data showed that inoculation of *Burkholderia* strains isolated from the healthy citrus root-associated microbiome could trigger the expression of genes involved in induced systemic resistance in inoculated plants.

HLB causes decreased relative abundance and/or expression activity of rhizoplane-enriched taxonomic and functional properties, collectively resulting in impaired plant host-microbiome interactions. Manipulation of the citrus root-associated microbiome, for instance, by inoculating citrus roots with beneficial *Burkholderia* strains, has potential to promote plant health. These results provided novel insights for understanding the contributions of the community enrichment process of the root-associated microbiome to the plant hosts.

Microbial populations and activities in the rhizoplane of rock-weathering desert plants: Dense layers of bacteria and fungi in the rhizoplane of three species of cactus (*Pachycereus pringlei, Stenocereus thurberi, Opuntia cholla*) and a wild fig tree (*Ficus palmeri*) growing in rocks devoid of soil were revealed by bright-field and fluorescence microscopy and field emission scanning electron microscopy (Puente et al., 2004). These desert plants are responsible for rock weathering in an ancient lava flow at La Purisima-San Isidro and in sedimentary rock in the Sierra de La Paz, both in Baja California Sur, Mexico. The dominant bacterial groups colonizing the rhizoplane were fluorescent pseudomonads and bacilli. Seven of these bacterial species were identified by the 16S rRNA molecular method. Unidentified fungal and actimomycete species were also present. Some of the root-colonizing microorganisms fixed in vitro N(2), produced volatile and non-volatile organic acids that subsequently reduced the pH of the rock medium in which the bacteria grew, and significantly dissolved insoluble phosphates, extrusive igneous rock, marble, and limestone. The bacteria were able to release significant amounts of useful minerals, such as P, K, Mg, Mn, Fe, Cu, and

Zn from the rocks and were thermo-tolerant, halo-tolerant, and drought-tolerant. The microbial community survived in the rhizoplane of cacti during the annual 10-month dry season. This study indicates that rhizoplane bacteria on cacti roots in rock may be involved in chemical weathering in hot, subtropical deserts.

Fungal assemblages in the rhizosphere and rhizoplane of grasses: Fungal communities associated with roots play an important role in nutrient cycling, supporting plant growth and the biocontrol of plant diseases. Experiments were conducted in 2004-2006 to isolate and characterize, based on their morphological features, rhizosphere and rhizoplane fungi from perennial grasses of the subfamily Panicoideae growing in forests of the Western Ghats in India (Vasanthakumari et al., 2011). Fungal species were isolated on potato dextrose, czapeck dox and water agar, in different locations and seasons. The results obtained on PDA were used for detailed analysis since most fungi occurred in high percentages. While certain grasses harbored diverse fungal species, others supported only a few species. Most fungi were isolated during winter followed by the rainy and summer seasons. The species richness, diversity and evenness of fungal assemblages in the rhizosphere and rhizoplane depended on the grass species and season. Ascomycetes were isolated in large numbers in most grass species. Species of *Aspergillus, Chaetomium, Penicillium* and *Trichoderma* occurred frequently. Certain others and non-sporulating fungi were grass species-specific. Most fungal species colonized the middle of the root more than the root tip or root base. Results suggested that perennial grasses harbor diverse fungal communities whose potential could be tapped for producing secondary metabolites and managing plant diseases.

Biocontrol microbes from the rhizoplane of *Decalepis arayalpathra*: Microbial communities present in the rhizoplane of plants play an essential role in both plant growth and physiological functioning. These microbes are attracted to the volatile compounds released by plants as a by product of their metabolism (Srivastava and Singh, 2020). Present investigation was designed to isolate and identify microbes from the rhizoplane regions of the tissue culture raised medicinal plant, *Decalepis arayapathra* KMA 05 clones. Two fungal species, i.e., *Cladosporium* sp. F42 (FJ755822) and *Epicoccum nigrum* (KR094452) were isolated from the rhizoplane. In antagonistic activity assay, *E. nigrum* was found to inhibit the growth of *Cladosporium* sp., thus displaying its biocontrol potential. The GC–MS analysis of fungal extracts showed bioactive molecules that can act as anti-microbial, anti-cancerous, and plant growth-promoting molecules. Moreover, exclusive and high concentration of PPDHMP from the *E. nigrum* supports for its involvement in *Cladosporium* growth inhibition. Thus the identified microbes can be exploited as biocontrol agents and plant growth-promoting microbes for *D. arayalpathra*.

Molecular mechanism of rhizoplane endobiotic bacteria in alleviating abiotic stress in plants: Bacteria present in rhizoplanes possess a significant proportion of genes that are engaged in Plant Growth Promoting (PGP) Traits under abiotic stress. Such traits are including ACC deaminase, IAA biosynthesis, phosphatase, siderophore synthesis, antioxidant enzymes, and exopolysaccharide production. Genes encoding

Na^+-transport and K^+/H^+ antiporter, have been reported in the rhizoplanes, meant to regulate the salt toxicity in plants and to resist hyperosmotic stress (Harleen Kaur et al., 2023). Various genes encoding glycine betaine transporter, betaine-aldehyde dehydrogenase, and betaine/carnitine/choline transporter were detected in the PGP bacteria genome, thereby helping the plants to survive under harsh conditions. The genes responsible for the synthesis and transportation of compatible solutes were predicted. The KLBMP 4941 genome also has a series of genes encoding cold shock and heat shock proteins, POX, SOD, Glutathione Peroxidases, and CAT, which could protect the cell from oxidative stress.

A transcriptome analysis of two endophytic bacteria (*A. endophyticus* SYSU 333,322 and *Nocardiopsis alba* SYSU 333,140) treated *Arabidopsis thaliana* plants was performed under salt stress. It was observed that these endophytic bacteria upregulated the expression of genes AT5G44610 responsible for shaping epidermal and cortical cells. Gene AT5G52300 was induced in response to water deprivation, gene AT1G04620 encoding for 7-hydroxymethyl chlorophyll *a* reductase was activated to reduce chlorosis in the plant, Gene AT3G46960 responsible for potassium ion uptake was found to be up-regulated. Scientific evidence was provided that the halotolerant *Dietzia natronolimnaea* STR1 strain defends wheat plants from salt stress by stimulating the expression of a group of genes accountable for salinity tolerance in plants. They noticed up-regulated expression of transcriptional factors (TaABARE, TaOPR1, and TaWRKY), ion transporters (TaNHX1, TaHAK, and TaHKT1) and SOS pathway-related genes (SOS1 and SOS4) indicated that halotolerant PGPR-mediated salinity tolerance is a complex phenomenon that involves modulation of ABA-signaling, SOS pathway, ion transporters and antioxidant machinery for protecting the plant from oxidative damage caused due to excessive salt concentrations. Similarly, many bacterial species possess mechanisms that make them resistant or tolerant to heavy metals. Analysis of the LK11 genome revealed the presence of a czc operon in the chromosome and plasmids. The czc operon was found comprised of three structural genes, czcA, czcB, and czcC, as well as two regulatory genes, czcD, and czcR. This operon was previously found to confer resistance to three heavy metals, namely cobalt, zinc, and cadmium.

Isolation and characterization of the Rhizoplane bacteria from a Legume: Two bacterial strains that amplified part of the nifH gene, RP1p and RP2p, belonging to the genus *Enterobacter* and *Serratia*, were isolated from the rhizoplane of *Lupinus albescens* (Adriana Giongo, et al., 2010). These bacteria are Gram-negative, rod-shaped, motile, facultative anaerobic, and fast-growing; the colonies reach diameters of 3–4 mm within 24 h of incubation at 28 °C. The bacteria were also able to grow at temperatures as high as 40 °C, in the presence of high (2–3 % w/v) NaCl concentrations and pH 4 -10. Strain RP1p was able to utilize 10 of 14 C sources, while RP2p utilized nine. The isolates produced siderophores and indolic compounds, but none of them was able to solubilize phosphate. Inoculation of *L. albescens* with RP1p and RP2p strains resulted in a significant increase in plant dry matter, indicating the plant-growth-promoting abilities of these bacteria.

Two bacteria with plant growth-promoting abilities and carrying the nifH gene were isolated from the rhizoplane of *L. albencens* where a significant increase in the dry matter production of the analyzed plants was observed in comparison with non-inoculated plants. Plant growth-promoting (PGP) activities have been reported for a series of bacterial species including *Pseudomonas, Azospirillum, Azotobacter, Klebsiella, Enterobacter, Alcaligens, Arthobacter, Burkholderia, Bacillus* and *Serratia.* Although the mechanisms by which PGPB promote plant growth are not fully understood, these bacteria can improve plant growth by N2 fixation, phytohormone and siderophore production, phosphorus solubilization, and disease control. Moreover, the root-surrounding environment, the rhizosphere, is a dynamic soil site of nutrient turnover. Bacteria around plant roots perform a wide range of metabolic activities and are able to make use of a wide range of low molecular mass organic compounds and of some more complex compounds as C and energy sources. These microorganisms play major roles in nutrient transformation and element cycling and influence the availability of these nutrients for plant uptake. Various soil microorganisms live in the plant rhizosphere and rhizoplane due to the rich nutrient availability. Although bacteria such as *Rhizobium, Azotobacter* and *Azospirillum* enhance plant growth as a result of their ability to fix-N2, bacteria with mechanisms such as phytohormone production in the rhizosphere and other PGP activities can contribute to improve the ability of the host plant to survive in extreme environments.

Rhizosphere-to-Rhizoplane enrichment of *Cellvibrio*: The rhizoplane poses different challenges for bacteria to colonize from the rhizosphere niche. On one hand, the plant immune response exerts stronger effects in the rhizoplane than in the rhizosphere; on the other hand, the richer nutrients in the rhizoplane attract microbes, leading to fierce competitive interactions among the microbes (Reinhold-Hurek et al., 2015). Common strategies used by microbes against other competitors include limiting resources such as iron and producing antimicrobial compounds. Interestingly, the functional category "Defense mechanisms" was overrepresented in the gene group that showed higher expression level in the rhizoplane, indicating that Bin79 encountered stronger selective forces in the rhizoplane. CirA, a critical receptor for the ferri-siderophore complex, exhibited higher expression level in the rhizoplane compared with the rhizosphere. The only bacterioferritin (Bfr)-encoding gene identified in the Bin79 genome, PROKKA_00691, was more strongly induced in the rhizoplane compared with the rhizosphere. Bfr is activated by iron and can store iron at concentrations much higher than the feasible solubility limits of iron (III) in bacterial cells. Sequestration of iron in the immediate vicinity represses the growth of competitive microbes, thus benefiting the iron-storing bacteria in the competitive microenvironment. Multiple genes involved in multidrug resistance and efflux pumps, such as *nepI*, *emrAB*, *acrAB*, and *tolC*, showed higher expression levels in the rhizoplane. The energy-dependent efflux pumps are extremely important for bacteria to survive in the presence of antibiotics and toxic compounds. AcrAB–TolC is the main multidrug efflux machinery in *Escherichia coli* to endow the bacteria with resistance to a broad spectrum of antibiotics.

Maintaining integrity of the plant cell walls is critical for plant health, however, a previous study showed that *Cellvibrio*, which is recognized by its robust ability to degrade plant cell walls, was enriched from the citrus rhizosphere to the rhizoplane (i.e., the root surface) (Zhang et al., 2020). The mechanisms underlying the rhizosphere-to-rhizoplane enrichment of *Cellvibrio* were investigated through genome-centric metagenomics and metatranscriptomics analyses. A near-complete metagenome-assembled genome representing a potentially novel species of *Cellvibrio* was recovered herein designated Bin79, with genome size of 5.71 Mb across 11 scaffolds. Differential gene expression analysis demonstrated that plant cell wall degradation genes were repressed, whereas genes encoding chitin-degrading enzymes were induced in the rhizoplane compared with the rhizosphere. Enhanced expression of multi-drug efflux genes and iron acquisition- and storage-associated genes in the rhizoplane indicated mechanisms by which Bin79 competes with other microbes. In addition, genes involved in repelling plant immune responses were significantly activated in the rhizoplane. Comparative genomics analyses with five related *Cellvibrio* strains showed the importance of gene gain events for the rhizoplane adaptation of Bin79. Overall, this study characterized a novel *Cellvibrio* strain and indicates the mechanisms involved in its adaptation to the rhizoplane from meta-omics data without cultivation.

The genus Cellvibrio comprises Gram-negative, aerobic, flagellated rod-shaped, and cellulolytic bacteria. Members of this genus have been identified from diverse environments, such as soil, plant root-associated habitats, spring water, and the gut of grass-feeding snails. Cellvibrio, showed significantly higher relative abundance in the rhizoplane microbiome, a niche characterized by more intimate host–microbe interactions, compared with the rhizosphere microbiome, and the mechanisms underlying their rhizosphere-to-rhizoplane enrichment were further revealed at the whole microbial community level. Recent developments on MAG (Metagenome-Assembled Genome) recovery and analysis methods enabled us to perform genome-centric analyses on the structure and function of the microbiome, which can provide more detailed and accurate information to reveal the mechanisms underlying niche adaptation of microbes compared with whole community-based analyses.

An organic hydroperoxides (OHPs) resistance enzyme-encoding gene *ohr*, and its regulator-encoding gene *ohrR*, exhibited higher expression levels in the rhizoplane. Production of OHPs is an important component of plant defenses and OHPs are highly toxic to bacteria. The expression of *ohr* is specifically induced by OHPs and the gene product is critical for bacterial survival in the presence of OHPs.

Spatial colonization of microbial cells on the rhizoplane: The rhizoplane is the region where the root surface is in contact with soil and corresponds to the inner limit of the rhizosphere. At the rhizoplane level, plants exchange elements with the surrounding soil and the rhizoplane can therefore be considered as the region that drives nutrient movement and transformation in the rhizosphere. The rhizoplane differs in many respects from the bulk soil due to the far larger supply of substrates derived from the roots, with far greater microbial cell densities and reduced levels of diversity (Raynaud, Xavier et al., 2017). This is likely to result in completely different interaction profiles among microorganisms which may affect rhizosphere biogeochemistry. While the diversity of microorganisms associated with the rhizosphere and on the rhizoplane is getting increasing attention, knowledge on the spatial organisation of this diversity is still scarce. It was therefore aimed at investigating the spatial arrangement of microbial rhizoplane colonization to increase our understanding of potential interaction dynamics within soil-microbe-plant interfaces. To study the spatial distribution of microbial cells on roots we cultivated rice plants in water-logged paddy soil. Root samples were taken three months after germination. After removing adhering rhizosphere soil the root samples were chemically fixed and prepared for CARD-FISH (Schmidt & Eickhorst, 2014). For hybridization, the oligonucleotide probes EUB I-III were applied to cover the majority of bacteria colonizing the rhizoplane. Root segments were then subjected to confocal laser scanning microscopy where triplicate image stacks of 10 μm thickness (0.5 μm layer distance) were acquired per region of interest (ROI). ROIs were defined as distances from the root tip (0, 5, 10, 15 mm) and corresponded to the root tip, elongation zone, and zone of maturation. Image stacks were processed using ImageJ software to extract microbial cells spatial coordinates, as well as other features of the root (e.g. root cell walls). For all the images analysed, we found that microbial cell distributions were not distributed randomly and strongly associated to root cell walls. The spatial organization of root cell walls could be used to simulate microbial cell distribution that have similar spatial properties compared to the microscopic data. Root cell walls thus appear as a strong determinant for microbial cell colonization of the rhizoplane.

Rhizoplane microbiota of superior wheat varieties: The rhizoplane portion of the root includes the root epidermis and the outer cortex. Soil particles and various microorganisms adhere to this area. Plant roots secrete vast amounts of organic carbon, which promotes the establishment and growth of bacteria. These bacteria respond to the plant, interact with each other, and affect their environment, including both chemical and physical properties of the soil in and around the rhizosphere, which ultimately affects plants. There are more bacteria present on the rhizoplane than on the rhizosphere (Ayesha Siddiqa et al., 2016). However, most previous studies have focused on the rhizosphere, rather than the rhizoplane. Bacteria present in the rhizoplane and rhizosphere can be beneficial to the plant (e.g., by pathogen suppression) as well as detrimental (e.g., by the depletion of nutrients). The microbial community that thrives in the rhizoplane/rhizosphere is dependent not only on the age and health of the plant, but also on its lineage. Moreover, the bacterial composition also differs between soil types and climates. Most importantly, microbial communities that flourish in the rhizosphere differ from those found on bulk soil. Hence, these communities exhibit what has been termed the “rhizosphere effect”. In studies on the rhizoplane of the plant *Oryza sativa,* neither fungi nor actinomycetes were observed on the root surface; rather, the entire surface of the basal region of the root was covered with bacteria. The bacteria found on the rhizoplane were largely distributed as aggregates. The most prevalent bacterial species were *Bacillus* and *Pseudomonas* (60-40%, respectively). Comparisons of microbial diversity among plant varieties with different yields may reveal hidden roles and subsequent effects on plant yield.

Bacterial load (CFU/mL) declined, and the succession of bacterial diversity occurred as the plants aged. Most auxin-producing bacteria and the highest concentrations of auxin (77 μg/mL) were observed during the second sampling point at the tillering stage (Ayesha Siddiqa et al., 2016). The Seher variety harbored the most auxin-producing as well as phosphate-solubilizing bacteria. Most of the bacteria belonged to *Bacillus* and *Pseudomonas*. *Planomicrobium, Serratia, Rhizobium, Brevundimonas, Stenotrophomonas,* and *Exiguobacterium* sp. were also found. These results suggested that the rhizoplane microbiota associated with higher-yield plant varieties have better plant growth-promoting abilities as compared to the microbiota associated with lower-yield plant varieties.

A substantial diversity of microbes was isolated from the rhizoplane of three different varieties of wheat (Seher, Faisalabad, and Lasani) at different stages of plant growth. Microbial diversity, microbial load, as well as plant growth-promoting abilities of microbiota (auxin production, phosphate solubilization, and HCN production) varied among the three wheat varieties, showing the importance and possible role of microbiota in the growth and fruit yield of the host varieties. These parameters also varied among the growth stages of plant varieties, suggesting the selection and succession of microbiota during plant growth. The 16S rRNA gene analysis revealed that most plant growth-promoting bacterial isolates in this study belonged to the genera *Bacillus (B. stratosphericus, B. cereus, B. subtilis, B. safensis,* and *B. marisflavi)* and *Pseudomonas (P. geniculata). Planomicrobium chinense, Serratia nematodiphila, Rhizobium pusense Brevundimonas diminuta, Stenotrophomonas rhizophila,* and *Exiguobacterium aurantiacum* were also found. In various studies, bacteria belonging to the genera *Pseudomonas, Bacillus,* and *Rhizobium* have been reported to have a powerful role in phosphate solubilization. The results of this study show that higher yielding plant varieties carry a comparatively larger number of plant growth-promoting bacteria with comparatively better plant growth-promoting abilities as compared to the microbiota of low-yielding plant varieties. However, other plant varieties should also be checked to confirm this hypothesis. Moreover, isolated bacteria need to be inoculated in low-yielding varieties in order to confirm their role in fruit yield.

Rhizosphere and Rhizoplane bacteria isolated from Subtropical region: Rhizosphere is a zone between the root surface and the soil adjacent to the roots. The bacteria inhabiting in the rhizosphere are called rhizobacteria. They live on the surface of roots in the soil and form a barrier to the root infecting parasites (bacteria, fungi, nematodes etc) (Abd El-Ghany et al., 2015). Rhizobacteria excrete antibiotic substances, thereby protecting the roots from plant parasites by the toxic effects. The boundary between rhizoplane and rhizosphere is very thin and therefore this habitat is largely considered as a continuum. On quantitative basis, the number of bacterial isolates was less in rhizoplane region than rhizospheric region (Subtropical region Jazan, Saudi Arabia). Bacterial isolates from soil of Sorghum bicolor field were *Microbacterium barkeri, Pseudomonas alcaligenes. Brevundimonas vesicularis, Pseudomonas syringae* pv *aceris* and *Pseudomonas syringae* pv *atrofaciens*. All bacterial isolates *M. barkeri* , *P. alcaligenes, B. vesicularis, P. syringae* pv *aceris* and *P.syringae* pv *atrofaciens* were present in the rhizosphere and not found on the root surfaces (rhizoplane) exept *P. alcaligenes* and *B. vesicularis*. *P. alcaligenes* can coexist as a rhizosphere and rhizoplane bacterium represented with 24.6 and 40.7% occurrence Frequency respectively while occurrence frequency of *M. barkeri* was

15.4 %. Future studies will determine the potential application of these isolates in plant growth promotion for enhancing plant productivity.

All bacterial isolates *M. barkeri* , *P. alcaligenes, B. vesicularis, P. syringae* pv *aceris* and *P.syringae* pv *atrofaciens* were present in the rhizosphere and not found on the root surfaces (rhizoplane) exept *P. alcaligenes, B. vesicularis*. Rhizosphere soil contained a great spectrum of bacteria whereas there were no bacteria isolated from the rhizoplane. The association between microorganisms and roots can be beneficial, neutral or harmful, but often the effects depend on the soil conditions. Microflora in rhizosphere soil is higher than the soil without rhizosphere indicates the influence of living roots in the soil. One such a physiological group, the ammonifying bacteria related to the capacity of mineralizing the nitrogenous materials by bringing about a rapid decomposition of organic nitrogen in the rhizosphere. Bacteria in the rhizosphere influence plant growth because they affect soil chemical properties and interact with plant roots, where the influence can be beneficial, neutral, or deleterious.

P. syringae was isolated as a Gram negative and rod shaped bacterium in rhizoplane, although many reports stated that *P. syringae* causing disease in most of the plant species. The present study demonstrated that *Pseudomonas alcaligenes* can coexist as a rhizosphere and rhizoplane bacterium represented with 24.6 and 40.7% occurrence Frequency respectively and can be used as a biofertilizer. Previeowisly studies demnostrated that inoculation of plants with *P. alcaligenes* significantly increased shoot dry mass, pod number, and content of chlorophyll, nitrogen, phosphorus and potassium in plants. Several studies found that *Pseudomonads* may improve plant growth by suppressing parasitic and nonparasitic root pathogens through the production of biologically active substances or the conversion of unavailable minerals and organic compounds into forms that are available to plants.

Rhizosphere, rhizoplane and phyllosphere bacteria and fungi as plant growth promoters in rice: The rice plant represents a habitat for diverse microorganisms, those that colonize the aerial parts (phyllosphere), the root surface (rhizoplane) as well as the zone around the root (rhizosphere). The phyllosphere comprises the aerial parts of plants and is dominated by the leaves. Most studies on the identity of organisms in the phyllosphere have focused on bacteria and, to a lesser extent, fungi (Mwashasha Rashid Mwajita1 et al., 2013). Potential beneficial interactions of phyllosphere bacteria with rice plants, such as plant growth promotion by bacterial nitrogen fixation or plant hormone production have been studied. Rhizoplane is the root surface zone where microorganisms attach themselves using surface structures such as flagella, fimbriae or cell surface polysaccharides. The boundary between rhizoplane and rhizosphere is very thin and therefore this habitat is largely considered as a continuum. The rhizosphere is a thin layer of soil immediately surrounding plant roots. This is an extremely important and active area for root activity and metabolism. A large number of microorganisms such as bacteria, fungi, protozoa and algae coexist in the rhizosphere.

Many soil micro-organisms are able to solubilize the unavailable phosphorus, increase uptake of nitrogen and also synthesize growth promoting hormones including auxin.

The aim of this study was to isolate and characterize phyllosphere, rhizoplane and rhizosphere micro-organisms from Kenyan rice with growth promoting habits (Mwashasha Rashid Mwajita1 et al., 2013). In this study whole plant rice samples were collected from different rice growing regions of Kenya. 76.2%, over 80% and 38.5% of the bacterial isolates were positive for phosphate solubilization, nitrogenase activity and IAA production whereas 17.5% and 5% of the fungal isolates were positive for phosphate solubilization and IAA production respectively. Hence these micro-organisms have potential for utilization as bio-fertilizers in rice production. Bacteria are the most abundant among them.

The bulk soil (rhizoplane) had the highest number of micro-organisms as opposed to the rhizosphere and phyllosphere. This is because the phyllosphere is a short-lived environment, as opposed to the rhizoplane which comprises the area in the soil around plant roots. The number of bacterial isolates from the rhizoplane and rhizoshere were more than that of the fungal isolates. There was a wide range of variation in nitrogenase activity among the 130 isolates tested in accordance with the regions and samples. Nitrogen fixing bacteria are present in the rhizosphere, rhizoplane as well as the phyllosphere. One of the important components of the nitrogen cycle in a range of ecosystems, is nitrogen fixation associated with roots of grasses. The rhizosphere is known to be a biologically active zone that contains soil-borne microbes where the biological, chemical and physical characteristics influence the roots. A large number of microorganisms such as bacteria and fungi coexist in the rhizosphere and bacteria are the most abundant among them.

Microbial colonization of cucumber rhizoplane: The ability of microorganisms belonging to the microbiological preparation "Embiko®" to colonize the rhizoplanes and rhizospheres of the Competitor and Phoenix Plus types of cucumber in vitro was investigated (Rzhevskaya, et al., 2013). The objects of study were the cultures of the lactic homofermentative streptobacteria *Lactobacillus plantarum* 20 and *L. casei* 6 and the homofermentative lactic streptococcus *Lactococcus lactis* 4/6, the yeast *Saccharomyces cerevisiae* 75 and the microbiological preparation "Embiko®" which includes the above-mentioned microorganisms. Germinated seeds were placed aseptically in biological test-tubes with starvation agar, where a suspension of the microbiological consortium of microorganisms had been added before. The ability of bacteria to colonize the root zone of plants was assessed visually by the intensity of formation of bacterial microcolonies on the surface of the roots of the seedlings and on the crushed micropreparations. The strain of *S. cerevisiae* colonized the entire volume of the agar along the entire length of the root, in the apical part of the root colonization was shown to be less active. With increasing duration of cultivation the intensity of colonization of the root zone by microorganisms was increased – colonies became larger. In various areas of the root the diameter of the cloud colonies *S. cerevisiae* was different in size. The strains of *L. casei* 6 and *L. plantarum* 20 colonized all the root zones, forming a cloud of small colonies around them. The strain of *L. lactis* 4/6 did not form colonies in the starvation agar and didn't colonize the root surface of the cucumber seedlings.

The microbiological preparation "Embiko®" colonized the root throughout its length, gradually narrowing in the apical zone. When inoculated with a pure culture of isolated strains the cloud was composed of monotypic colonies and looked homogeneous. When inoculated with the microbial consortium the cloud of the colonies looked heterogeneous, colonies of different sizes and colors were clearly visible. Under microscopy the preparations of the roots of the cucumber seedlings inoculated with the microbial preparation "Embiko®" yeast cells and cells of the lactic acid bacteria were found. This study of the ability of microorganisms from the preparation "Embiko®" to colonize the rhizoplanes and rhizospheres of roots of cultivated plants in vitro showed that the different strains of microorganisms form clouds of colonies around the roots which were distinct in turbidity and size: the strain of *L. plantarum* – almost transparent, and *S. cerevisiae* – very dense. The lowest growth rate of microorganisms was observed at the apex of the root, the highest – in the zone of root hair. Clearly, root exudates of plants are the main source of carbon and energy for the inoculated bacteria. The results indicated that the investigated microbial consortium has a promising potential to inoculate plants in order to stimulate their growth and development.

Bacterial community characterization of the rhizobiome of plants cultivated in desert soils: The microbial community associated with roots known as "rhizobiome" was proposed to be assembled in two steps, the rhizosphere is colonized by a subset of the bulk soil community and the rhizoplane and the endosphere are colonized by a subset of the rhizosphere community. All those plant-associated microorganisms together establish a core microbiome in the rhizobiome (Barraza et al., 2020). The core microbiome consisting of bacterial groups is common to the species. The rhizosphere (defined as microbes living in the zone around root is rich in nutrients due to the accumulation of plant exudates, and these contain amino acids and sugars that provide a nitrogen and carbon source required for colonizing bacteria. This induces the bacterial community to colonize the rhizoplane and rhizosphere. The rhizoplane microorganisms (defined as root surface adhered microbes can metabolize essential inorganic nutrients (P, S, Mg, Mn, and Cu) directly from minerals, fixing atmospheric nitrogen, and then promoting plant growth. Finally, some bacteria from the rhizoplane can colonize intercellular spaces in several internal plant tissues without causing any harm or disease, and some bacteria enhance plant growth and increase plant resistance to pathogens; those bacteria are collectively known as endophytic bacteria (defined as microbe living within plant tissue in the endosphere.

The plant *Solanaceae* family is one of the most important for global agriculture and nutrition. Within this plant family, two plant species stand out for their economic importance and for human consumption, which are tomato and chili pepper. Moreover, those plants support diverse and characteristic microbial communities that are uniquely suited to the plant habitat and intimately connected to plant health. The main objective of this work is the bacterial community characterization in the rhizobiome of tomato and chili pepper, cultivated in arid environments. Five crop fields located in the south of the peninsula of Baja California, Mexico, were sampled. Total DNA was extracted from rhizosphere, rhizoplane, and endophytic root compartment and sequenced by Illumina MiniSeq platform technology applied to 16S rRNA gene V3 region. They could obtain 1,195,426 total reads and 1,725,258 total reads for tomato and chili pepper samples, respectively. The analysis of the bacterial community structures confirmed that the two plant species showed differences in their microbial community structures. Nonetheless, the microbial community structures were directly and equally influenced by the crop field localization of each plant species. Interestingly, it was determined that in both plant species, the *Proteobacteria* was the main phylum. It was concluded that several bacterial families are part of the core rhizobiome (28 OTUs) for both tomato and chili pepper, but the most abundant were the *Pseudomonadaceae* family and the *Pseudomonas* genus, which most probably play a pivotal role in the microbial ecology to benefit both crop plants.

The bacterial community structures for all types of samples analyzed and for chili pepper were dominated by the *Proteobacteria* and *Firmicutes* phyla, and the main classes were *Alphaproteobacteria*, *Betaproteobacteria*, *Gammaproteobacteria*, and *Bacilli*. Interestingly, the family relative abundances showed specific differences for each type of sample: *Xanthomonadaceae*, *Bacillaceae*, and *Burkholderiaceae* were among the most abundant families for rhizoplane samples; *Aerococcaceae*, *Clostridiaceae*, and *Chromobacteriaceae* for rhizosphere samples; and *Thermoanaerobacteraceae* for root endophytic compartment samples. However, several abundant families were shared between the type of samples, such as *Enterobacteriaceae*, *Pseudomonadaceae*, and *Rhizobiaceae*. Other previous reports have found that the most abundant bacterial phyla for rhizoplane, rhizosphere, and root endophytic compartment samples were *Proteobacteria*, *Bacteroidetes*, *Actinobacteria*, and *Firmicutes*.

2.6 Host Exudates

Exudate is a term used to describe the fluid that is produced by plants as a result of photosynthesis. It can be found in the leaves, stems, flowers, fruits, and seeds of plants, as well as in soil and water. Exudates are considered complex mixtures of plant origin with solid or semi solid consistency such as resins, oleoresins, balsams and gums. The proportion of constituents may vary with climate, season and other factors. Plants produce a wide array of secretions both above and below ground. Known as mucilages or exudates, they are secreted by seeds, roots, leaves and stems and fulfil a variety of functions including adhesion, protection, nutrient acquisition and infection (Andrew F. Galloway et al., 2020). Plants exude a variety of substances, including sugars, proteins, carbohydrates, lipids, amino acids, vitamins, minerals, water-soluble compounds (such as vitamins A, C, D, E, K, Mg, P, Rb, Selenium, Thiamine, Riboflavin, Niacin and Vitamins B6, B12, Calcium, Iron, Magnesium, Phosphorus, Potassium, Sodium, Zinc, Copper, Manganese and Zirconium) and organic acids.

Exudation is also a by-product of the photosynthetic process, which is the process by which plants use sunlight to convert carbon dioxide (CO_2) into sugars and other organic compounds. The process is called "photosynthesis" because it involves the conversion of light energy into chemical energy. Photosynthesis is an energy-intensive process that takes place over a period of several days, but the amount of energy that plants produce during this process varies from plant to plant, depending on the type of plant and the conditions in which it is grown. For example, some plants are more efficient at converting sunlight into energy than others. In general, the more energy a plant uses to produce a particular chemical compound, such as sugar, protein, starch, fat, or oil, then the less energy it will use during the rest of its life cycle. This means that the plant will produce more of these compounds as it ages, resulting in a longer life span for that particular plant. However, this is not always the case. Some plants may be able to use up to twice as much energy during their life cycles as they do at the beginning of their lives, so they may live longer than they would have if they had used up all their energy in their first few years of life.

Structure and Constituents: Plant exudate structure and constituents can be quite varied and often plant-species-specific and include high-molecular-weight molecules (e.g. sugars, lipids, and proteins) and low-molecular-weight signalling molecules (e.g. organic and amino acids) (Marnie E. Rout, 2014). Sugars and amino acids comprise the majority of plant root exudates and they serve a variety of functions including as antimicrobials, allelopathic molecules, and pathogen/herbivore defences. Many of these exudates also serve as an energy source for the microbiome; prokaryotes can utilize many different sources of carbon, including many of the plant exudates. For example, in the invasive grass *Sorghum halepense*, the exudate sorgoleone is the dominant chemical exuded from root hairs and has well-known allelopathic properties. Further work using mineralization kinetics confirmed that rhizosphere microbes in invaded soils were capable of utilizing this allelopathic molecule as a carbon source. The multiple functions of this exudate reiterate the importance it serves as multiple currencies of the plant microbiome. Studies of the feedbacks between the influence of the microbiome and plant production of this and other allelochemical exudates are important, yet currently unknown, aspects of the functional contribution of the plant microbiome.

Roots function to uptake nutrients and chemical signalling molecules from the rhizosphere while simultaneously depositing nutrients and chemical signalling molecules into this same space. These exudates serve as a chemical currency to the plant, in turn providing a wide range of services to other components of the plant microbiome (e.g. as a nutrient substrate, a signalling molecule eliciting protective response, or other microbiome ecosystem service). The chemical components released by plants often are unique to a genus or family and can range in concentration and composition due to a wide variety of factors. Terpenoids, flavonoids, and isoflavonoids comprise many of the plant's antimicrobial defences. Isoprenoids are the most diverse group of plant metabolitic compounds both structurally and functionally. As primary metabolites, they regulate cellular processes. For example, the biological molecules that are the end result of isoprenoid metabolites are essential aspects of photosynthesis (as phytopigments) and seed emergence (as gibberellic and abscisic acids). As secondary metabolites, they provide a range of benefits to plants in the form of defences from pathogens and niche expansion through allelopathic molecules.

Environmental factors influencing root exudate composition and quantity can include elevated levels of CO_2, drought, and nutrient deprivation (particularly nitrogen and phosphorus). All of these factors can have profound impacts on the phytochemistry released as exudates. For example, increased carbon allocation to roots has been documented in CO_2-fertilization experiments, resulting in shifts in exudate composition and concentration that varied according to plant species. These species-specific impacts can result in increased yield, in no net productivity increase, or can be detrimental to plant growth and production. An example of this was shown in positive biomass responses in rye and clover to CO_2 fertilization, while maize showed no net biomass benefit. However, maize demonstrated increased exudation of several amino acids under CO_2 fertilization. These findings are not surprising considering that the C4 photosynthetic pathway facilitates growth under high levels of CO_2; however, the impacts of increased release of amino acids into the rhizosphere by the C4 grass (maize) might play a role in a multitude of feedbacks between other plants and microbes. The extent of this latter aspect is largely unknown and is a major reason why many scientists are studying the plant microbiome. Not all environmental stressors elicit the same responses from plant roots, and nutrient deprivation has been shown to have an inhibitory effect on root exudation. As an example, maize plants grown under either nitrogen (N) or phosphorus (P) limitations exuded decreased quantities of amino and organic acids from roots. An additional influence on exudate composition and quantity is exerted by the microbiome, whose members can (through their exudates) confer plant protection in addition to plant-growth promotion. We know less about the ways in which the microbiome utilizes plant exudates to up- or down regulate microbial gene expression, but research has shown clear roles of their involvement in mediating pathogen attack to plants.

2.6.1 Characteristics of plant exudates: Plant exudates have specific physical and biological characteristics depending on the species. The roles of plant exudates are diverse, and include defense against pests and diseases and allelopathic activity. Latex is a milky emulsion composed of small organic compounds suspended in a liquid dispersion medium. Not all latexes are whitish, there are some which are transparent or have a distinct color, such as yellow, orange, brown, or red (Abarca, Klinkhamer, & Choi, 2019). Most latexes are sticky and viscous after excretion, and even coagulate and form clots. However, latexes excreted from leaves of the mulberry tree (*Morus* spp.) and the oleander tree (*Nerium oleander* L.) are not viscous. The chemical properties of latex are related to their functions. Latexes act as the first line of defense against herbivores and pathogens, and it also plays a role in environmental stress conditions. The endophytes living inside the host plant are also believed to contribute via the latex to the production and biotransformation of diverse metabolites related to the plant defense against pests and diseases.

i. Resins: This term should not be mistaken with the synthetic resins result of the polymerization of long chains of low molecular weight organic compounds. Resins are exudates produced in specialized surface glands such as glandular hairs or internal ducts. They serve the protection against insects and pathogens (Henni Vanda et al., 2021). The characteristics of resins include stable, inert, amorphous, nonvolatile, insoluble in water, but soluble in organic solvents. Plant root exudates are produced as attractants and repellants in the rhizosphere. The compounds also may regulate the soil microbial community, control the herbivores, facilitate beneficial symbioses, change the chemical and physical properties of the soil, and inhibit the growth of competing plant species.

Exudates refer to those obtained from plants in their natural state or insects that parasitize plants. They are characterized by their insolubility in water, soluble in alcohol and ether, often non crystallizable, soften or melt at moderate temperature and burn at smoky flame. Usually, resins are terpenes oxidized from the volatile oils, result from their water insolubility. They taste bitter. Its chemical composition is not varied, some are acidic and alkali combinations that give a soapy consistency as in the case of rosin. A natural pure resin is distinguished to a prepared or derivative resin. The natural resin is obtained from plant exudate. A resin prepared can be made of an extraction by alcoholic percolation of a drug, in order to obtain its active resinous constituents, then treating it with excess water to precipitate the components which are washed and dried. For example the resin of Podophyllum peltatum L. A preparation of resin can also be obtained from a natural oleoresin by drying its volatile oil with warming. Ex: Rosin ((2*E*)-3-Phenylprop-2-en-1-yl β-D-glucopyranoside) (Fig. 12).

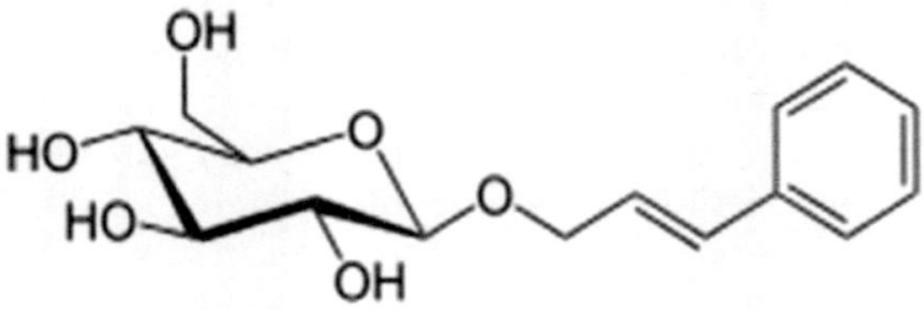

Fig. 12: Rosin

ii. **Oleoresins**: They are mixtures of resins with volatile oils, usually obtained by incision of the trunk of trees that contain it. The turpentine and the copaiba are natural oleoresins recognized by the USP-NF. The same way should be distinguished between oleoresins and prepared oleoresins, as the latter are concentrated of liquid preparations by percolation of the plant that naturally contains the resin and oil together. Suitable solvents are used (acetone, ether or alcohol) then the percolation is concentrated until remove the solvent. Examples are Aspidium, Capsicum and Ginger. Chemical structures of oleoresin ingredients capsaicin, capsanthin and capsorubin are provided in Fig.13

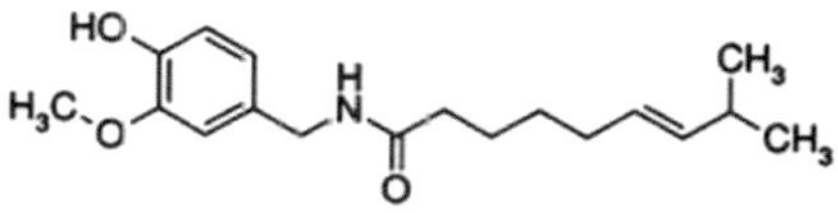

Capsaicin

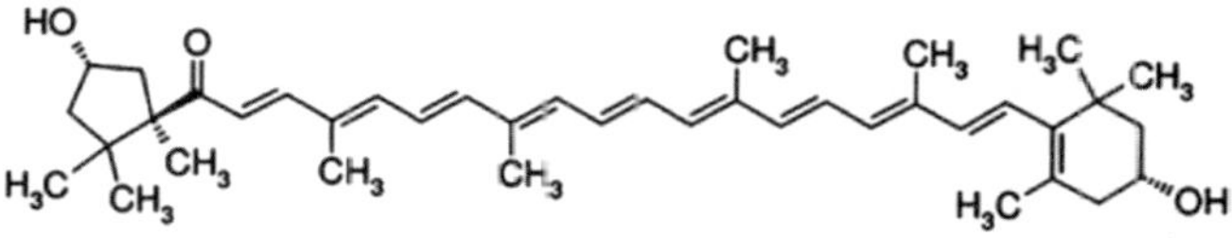

Capsanthin

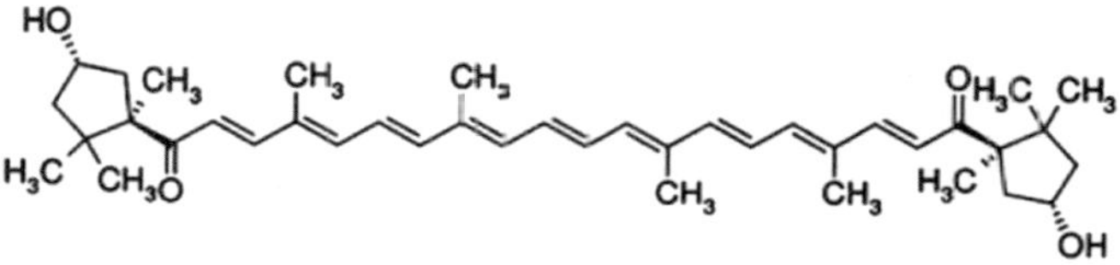

Capsorubin

Fig. 13: Capsaicin, Capsanthin and Capsorubin

iii. Gum Resins: Gum-resins are a mixture of both gums and resins and possess the properties of both the groups. They contain traces of essential oils. These are usually derived from the plant growing in dry and arid regions. Some of the commonly used gum-resins are asafoetida, myrrh, salai, guggul etc. The resins and gums are stored in the plant part called Old Xylem. Other than that, the by- products of metabolism like gums, latex, resins, and oils that are not useful to the plant are stored in plant parts like barks, stems, leaves, etc. Eventually, these parts are shed off, getting rid of the waste. Gum resin is typically collected by tapping into a tree. Guggul or Indian Myrrh is the yellowish gum-resin produced by the stem of the guggul tree (*Commiphora* spp.) Plant Sources include *Commiphora mukul* and *C.wightii.* Chemical structure of guggulsterone [4,17(20)-pregnadiene-3,16- (Fig.14).

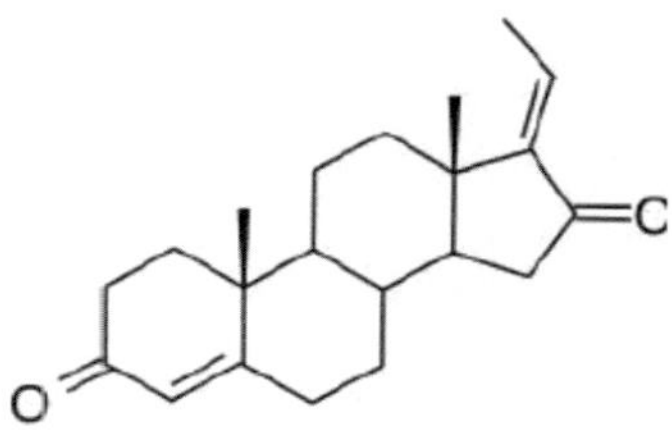

Fig. 14: Guggulsterone

iv. Balsams: Balsams have been used since ancient times, due to their therapeutic and healing properties; in the perfume industry, they are used as fixatives, and in the cosmetics industry and in cookery, they are used as preservatives and aromatizers. They are generally defined as vegetable material with highly aromatic properties that supposedly have the ability to heal diseases, not only of the body, but also of the soul. When

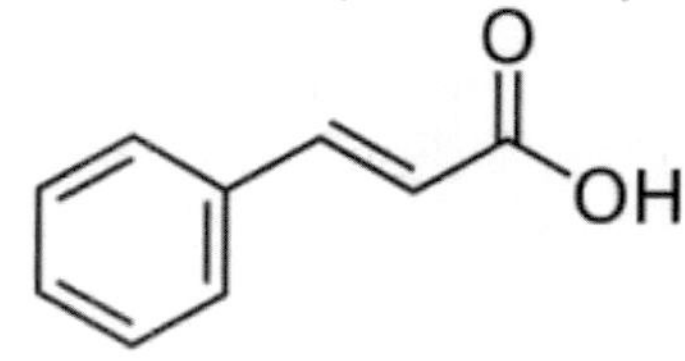

Fig. 15: Tolu Balsam

viewed according to this concept, many substances can be considered balsams. A more modern concept is based on its chemical composition and origin: a secretion or exudate of plants that contain cinnamic and benzoic acids, and their derivatives, in their composition. The most common naturally-occurring balsams (i.e. true balsams) are the Benzoins, Liquid Storaque and the Balsams of Tolu and Peru (Fig.15). Many other aromatic exudates, such as Copaiba Oil and Canada Balsam, are wrongly called balsam. These usually belong to other classes of natural products, such as essential oils, resins and oleoresins. Despite the understanding of some plants, many plants are still called balsams. This article presents a chemical and pharmacological review of the most common balsams.

v. Plant gums: They are present in large quantities in a variety of plants, with different structural and metabolic functions. Gums are considered to be pathological products as a result of plant injury or due to drought conditions. Some highly viscous gums, the mucilages, on the other hand, are water-insoluble, and produced as normal metabolites in epidermal cells of leaves, seed coats, and barks. Both gums and mucilages are translucent amorphous substances, consisting of polymers of a monosaccharide or mixed monosaccharides combined with uronic acids, they may also contain hydrophilic molecules, and when combined with water will form gels. The gums and mucilages also play a role in the response to the various forms of stress for the plant or plant cells. *Drosera* exudates fit best the class of gums. The exudate is transparent, just like drops of glue on the surface of tentacles, and act as trap to catch insects. The trapping mechanism involves two types of glands: stalked glands and sessile glands. Stalked glands specifically secrete a sweet mucilage to attract, capture, and digest the prey, while sessile glands particularly absorb the nutrients. There are different opinions about how the insects are attracted to the plants. Some researchers believe that it is due to the red color of the tentacles that contrast with the background, while other researchers believe that the insects are interested in nectar as a source of nutrients, which supports the idea that sugars play an important role in the trapping mechanism. Sugars found in the sticky fluid have a particular ratio that may form a NADES that act as a solvent of biomaterials.

Gums are hydrocolloids derived from plant exudates, seeds and seaweed extracts. These compounds are not digested in the upper intestinal tract and are resistant to the human digestive enzymes, being fermented in the large gut. This fermentation promotes the stimulation of the endogenous microbiota and the production of SCFA (María Ciudad-Mulero et al., 2019). Therefore, gums are used in food production as a source of DF with prebiotic effects and are also used for their functional properties such as, improve food texture, retard starch retro-gradation, improve moisture retention and enhance the overall quality of the products during storage (Qasem et al., 2017). Plant exudates are one of the main sources of gums, highlighting guar gum, gum arabic, gum tragacanth, karaya gum, etc.

Guar gum (Fig.16) is a galactomannan isolated from the seed of *Cyamopsis tetragonolobus* (guar). Due to its thickener properties, it is used as food additive. Guar gum has prebiotic properties and it can improve bowel transit. It also shows hypoglycemic and hypolipidemic effects (Tungland & Meyer, 2002). In this sense, according to European Commission (2012), Guar gum has approved the following health claim: "Guar gum contributes to the maintenance of normal blood cholesterol levels." This claim may be used only for food that provides a daily intake of 10 g of guar gum.

The exudate from the acacia tree (*Acacia* Mill.) is known as gum arabic. It is a mixture of a complex arabinogalactan polysaccharide with a glycoprotein. For its stabilizing and emulsifying properties, gum arabic is used by the food industry as additive. It has bifidogenic effect and hypolipidemic properties (Tungland & Meyer, 2002). Generally, the plants rich in gums are not used as food, but they are used as food additives. The most important gums in food belong to different genus of *Leguminosae* family.

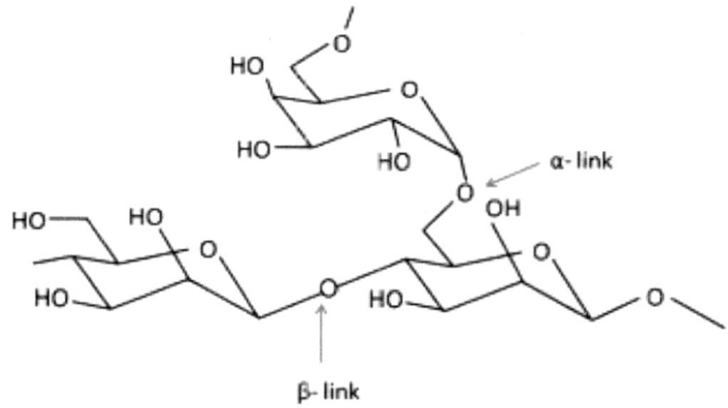

Fig. 16: Chemical structure of Guar gum

Haruna et al., (2017) discussed the natural plant exudates and mucilages, their sources, properties and uses and the potential applications. Research in natural polymeric materials has witness growing interest and attention. This is attributable to a number of factors which include their relative abundance, low cost, biodegradable nontoxic, and ecofriendly profile. They are polymers that are mostly plant in origin with a vast applications. Polysaccharides hydrocolloids including gum and mucilage are abundant in nature and commonly found in many higher plants. These polysaccharides constitute a structurally diverse class of biological macromolecules with a broad range of physicochemical properties which are widely used for various applications in pharmacy, medicine, food and other non- food applications. In recent years those polymers derived from plants have evoke tremendous interest because of their industrial applications as diluent binders, thickening agents, smoothening, emulsifiers, gelling agents and stabilizers. This increasing research in this group of these plant materials are clear indications of their increasing importance. The fact for increase in importance of natural plant based materials is that plant resources are renewable and if cultivated or harvested in a sustainable manner, they can provide a constant supply of raw materials. To that effect, more understanding of their nature, physicochemical properties, sources, uses be gained so that can be used for wider applications.

2.6.2 Plant exudates may stabilize or weaken soil: Soil physical conditions, particularly in the rhizosphere, are continually modified by the release of plant root exudates and microbial metabolites (Naveed et al., 2017). Plants potentially benefit from this modification of the rhizosphere because of improved physical conditions for root penetration, and nutrient and water uptake by the roots. However, the effect of biological exudates on soil physical properties might depend on their physicochemical characteristics. The quantity and physicochemical characteristics of root exudates are determined by the plant species, the age of an individual plant and external factors such as biotic and abiotic stresses (e.g. soil structure, presence of microorganisms and nutritional status). In general, root exudates are composed of an array of compounds such as carbohydrates, amino acids, organic acids, phenolic compounds, fatty acids, sterols, vitamins, enzymes and inorganic molecules. Among these, sugars (often polysaccharide derived), amino acids and organic acids are usually released in the largest quantities.

Plant exudates could either gel or disperse soil depending on their chemical characteristics. Barley (*Hordeum vulgare* L. cv. Optic) and maize (*Zea mays* L. cv. Freya) root exudates were collected using an aerated hydroponic method and compared with chia (*Salvia hispanica* L.) seed exudate, a commonly used root exudate analogue (Naveed et al., 2017). Sandy loam soil was passed through a 500-μm mesh and treated with each exudate at a concentration of 4.6 mg exudate g^{-1} dry soil. Two sets of soil samples were prepared. One set of treated soil samples was maintained at 4°C to suppress microbial processes. To characterize the effect of decomposition, the second set of samples was incubated at 16°C for 2 weeks at −30 kPa matric potential. Gas chromatography–mass spectrometry (GC–MS) analysis of the exudates showed that barley had the largest organic acid content and chia the largest content of sugars (polysaccharide-derived or free), and maize was in between barley and chia. Yield stress of amended soil samples was measured by an oscillatory strain sweep test with a cone plate rheometer. When microbial decomposition was suppressed at 4°C, yield stress increased 20-fold for chia seed exudate and twofold for maize root exudate compared with the control, whereas for barley root exudate decreased to half. The yield stress after 2 weeks of incubation compared with soil with suppressed microbial decomposition increased by 85% for barley root exudate, but for chia and maize it decreased by 87 and 54%, respectively. Barley root exudation might therefore disperse soil and this could facilitate nutrient release. The maize root and chia seed exudates gelled soil, which could create a more stable soil structure around roots or seeds.

There is growing evidence to suggest that certain compounds present in root exudates are involved in engineering the rhizosphere by dispersion and gelling of soil (Deng *et al.*, 2015), modulation of water and nutrient availabilities and attraction of rhizobacteria. The anions of organic acids in the rhizosphere may be adsorbed by soil mineral particles, thereby increasing the net negative charge of clays that would cause particles to disperse. Mucilages and other polysaccharides present in root exudates, which can function as stabilizing materials, might offset this effect. The pH of the exudates and thus rhizosphere will also affect soil structural stability if the change in pH results in dissolution or precipitation of stabilizing material such as Al or other polyvalent ions. Most research to investigate soil stabilization by biological exudates has measured aggregate stability. Although soil aggregate stability is a relevant test for studying the resistance of soil to erosion, it does not provide a quantitative measure of particle bonding that underpins the formation of soil structure. Fracture tests on dry soil discs or notched bars have quantified increased particle bond energy resulting from root exudate compounds, but such dry conditions are rarely met in reality. Formation of the soil structure occurs when soil is wet and particles are mobile; therefore, rheological measures of wet soil movement under stress are more relevant physically to understanding how mechanical stresses from root growth and exudation affect soil structural development. The mobility of soil when wet can be described with rheological measurements that provide controlled oscillatory shear stresses to quantify time-dependent flow under stress (i.e. viscosity and yield stress).

The physical characteristics of some of the exudates were in accord, like the chemical characteristics, with the limited body of research that provides comparable data (Naveed et al., 2017).. The viscosity of *Capsella* seed exudate measured agreed well with the zero-shear viscosity of chia seed exudate at similar concentrations in the present study. Zero-shear and infinite-shear viscosities for scleroglucan (a fungal exudate) were 10 times greater than for chia seed exudate at similar concentrations. Large standard errors observed for barley and maize root exudates were probably because of (i) a sharp decrease in viscosity with increasing shear stress near the inflection point and (ii) water-insoluble material present in root exudates. The source of this water-insoluble material could be the cells that sloughed off the root caps into the water during exudate collection. The variation in viscosities between different exudates could be attributed to polysaccharide sugars in the exudates (i.e. more polysaccharide sugars in the exudates resulted in greater viscosities. Greater zero-shear and infinite-shear viscosities for chia seed exudate are consistent with the presence of large amounts of sugars. The viscosity of the barley and maize root exudates represented relatively weak resistance against shear and would present only slight differences in capillary behaviour. Thus resistance to flow in soil pores because of increased viscosity of the soil solution might be minimal, although particles might gel to surrounding soil.

Exudates rich in organic acids, such as the barley root exudate in this study, have a net dispersing effect on soil in contact with roots. However, with microbial decomposition this dispersing effect decreased and the soil became more stable. The exudates rich in sugars, maize root and chia seed exudates in this study, have a net stabilizing effect on soil in contact with roots from the onset. There is considerable scope to extend this research to explore whether dispersion from particular root exudates might release physically protected carbon, causing a priming effect. Oxalic acid in root exudates liberates physically protected organic matter in soil, but they did not measure the physical mechanism. Moreover, root exudates from different genotypes of the same crop are known to cause large differences in microbial community structure and function of the rhizosphere (Pieterse *et al.*, 2016). Perhaps there is an opportunity to breed crops that could potentially manipulate the rhizosphere physically to improve resource availability and resist abiotic stresses.

2.7 Plant exudates and Root Penetration

Plant exudates greatly affect the physical behaviour of soil, but measurements of the impact of exudates on compression characteristics are missing. A number of studies have adopted an approach of upscaling rhizosphere conditions by mixing plant exudate compounds with soil to form repacked samples that are large enough for measurements (Oleghe et al., 2017). These have found a large impact of plant exudates on soil physical behaviour, which varies between plant species, seeds and roots. Exudates are often more viscous and have a lower surface tension than water. This will have a large impact on the capacity of plants to capture water from soils.

Oleghe et al., (2017) provided these data and explore how plant exudates may enhance the restructuring of compacted soils following cycles of wetting and drying. Two soils were amended with Chia (*Salvia hispanica)* seed exudate at 5 concentrations, compacted in cores to 200 kPa stress (equivalent to tractor stress), equilibrated to −50 kPa matric potential, and then compacted to 600 kPa (equivalent to axial root stress) followed by 3 cycles of wetting and drying and recompression to 600 kPa at −50 kPa matric potential. Penetration resistance (PR), compression index (C_C) and pore characteristics were measured at various steps. PR decreased and C_C increased with increasing exudate concentration. At 600 kPa compression, 1.85 mg exudate g^{-1} soil increased C_C from 0.37 to 0.43 for sandy loam soil and from 0.50 to 0.54 for clay loam soil. After 3 wetting-drying cycles the clay loam was more resillient than the sandy loam soil, with resilience increasing with greater exudate concentration. Root growth modelled on PR data suggested plant exudates significantly eased root elongation in soil. Plant exudates improve compression characteristics of soils, easing penetration and enhancing recovery of root induced soil compaction.

Exudate and soil properties: The chia seed exudate consisted of 40.7 g 100 g^{-1} carbon, 1.1 g 100 g^{-1} nitrogen and the carbon nitrogen ratio was 37. It had a pH-H_2O of 6.9 at 9.2 mg g^{-1} concentration.. The soil texture was sandy loam for the soil sampled from south Bullion and clay loam for the soil sampled from north Bullion. Total carbon content for the sandy loam soil was 2.25 g 100 g^{-1} and for clay loam soil was 2.95 g 100 g^{-1}. The soil pH_$Cacl_2$ at 1:5 soil to water was 5.48 for sandy loam soil and 5.15 for the clay loam soil.

2.7.1 Plant exudate impact on soil compression characteristics: The compression index (C_C) measures soil mechanical resistance to compression, with larger values indicating less resistance of soil to compression. C_C for 600 kPa compression increased by 17% for the sandy loam soil and 9% for the clay loam soil between 0 and 1.85 mg g^{-1} exudate amendment. Three cycles of wetting and drying, followed by recompression to 600 kPa had contrasting effects on C_C between soils. Both soils followed the same trend with increasing exudate concentration as observed for the soils before wetting and drying, but C_C had an overall drop of 5% for the sandy loam soil and increased by 7% for the clay loam soil ($P < 0.001$) (Oleghe et al., 2017).

Plant exudates obtained from *Salvia hispanica* seed coatings were found to greatly improve mechanical conditions for root growth, quantified from compression characteristics and penetration resistance. The decrease in penetration resistance of both sandy loam and clay loam soils with increasing exudate concentration demonstrates that exudates decrease soil resistance to local deformation. Similarly, an increase in compression index for both sandy loam and clay loam soils with increasing exudate concentration means the exudate used eased soil compression. A considerable decrease in shear strength and cone penetration resistance for loam and clay soils with increasing organic matter contents was reported. A decrease was found in cone penetration resistance with increasing soil organic matter and water content of a glacial till. In addition to easing mechanical conditions for root growth, exudates also enhanced the resilience of soil to a 600 kPa compression used to simulate axial root growth stresses. After 3 cycles of wetting and drying, increasing exudate concentration decreased the penetration resistance and increased the compression index. For a growing root that is transpiring water from soil, this improved resilience in the presence of exudates indicates potential structural re-arrangement of rhizosphere soil over time, creating better physical conditions for root elongation. Penetration resistance of a compacted soil decreases far more in the presence of roots than in fallow soil in a sandy loam soil. The capacity of plant roots to restructure compacted soils is well reported, driven by a combination of roots fracturing soil, enhancing cycles of wetting and drying, producing biopores and secreting exudates.

The possible mechanisms driving the changes in compression behaviour of soil as a result of exudation could be the amount of water retained by the exudates and hence effective stress, a lubricating effect of exudates that may decrease interparticle friction and the role of exudates in the dispersion, aggregation and hence pore structure development of soil. A more porous soil would be expected to be more compressible, but pore structure interactions with soil mechanical behaviour were not found for the sandy loam soil that we studied. After 600 kPa stress, none of void ratio, air ratio or water ratio for the sandy loam soil were correlated with penetration resistance or compression index. However, the clay loam soil after 600 kPa stress had a positive correlation between void ratio and compression index, and a negative correlation between void ratio penetration resistance (Oleghe et al., 2017). In this soil a more open pore structure therefore had the expected impact of decreased mechanical resistance. As pore structure did not influence the compression characteristics of the sandy loam soil, a lubricating effect of exudates was possibly the major driver.

Plant exudates eased soil compression and improved the mechanical resilience of compacted soils; the latter possibly having a large positive impact on rhizosphere physical conditions. The modelled increases in root elongation rate in soil, which was 40% faster in the sandy loam than the clay loam, are likely to impact on the capacity of roots to explore deep and disperse soil regions for resources. The physically quantified data generated from this study will be useful for models of how plant exudates may influence root growth and impact soil pore structure. Future research with model root exudates that vary in chemistry, real root exudates and plants with contrasting exudation properties could identify favourable exudate characteristics that improve the capacity of roots to grow in and restructure degraded soils. Such understanding would benefit practical applications of biological tillage by plants, selecting species in crop rotations to improve soil physical conditions and in crop breeding to improve the capacity of roots to grow through and restructure soils.

2.7.2 Biological properties and Pharmacological potential of Plant exudates: Traditional medicines derived from various plants species have been frequently used in daily health care system in developing countries since a long time (Licá et al., 2018). Knowledge obtained from ethnopharmacological studies have been employed to cure several diseases in traditional system of medicine and also known as the efficient means for drug discovery. The resurgence of public interest for plant based natural compounds due to their lesser side effects and better compatibility with the human body has resulted in finding out alternative ways to obtain bioactive compounds, which could be effective and safe as compared to synthetic drugs. The widespread use of synthetic drugs has resulted in development of resistance against the pathogens. It has become one of the world's most serious public health problems. It is difficult to eradicate some pathogenic bacteria and in several cases, use of more than one antibiotic is required, which may leads to side effects and other problems.

Natural products, mainly from plants represent potential alternative to treat various diseases caused by micro-organisms. The use of plants for therapeutic purposes is based on popular and scientific knowledge. These are also acts as suppliers of isolated active substances, such as total or purified extracts. Phytotherapeutic remedies are cost-effective with minimal toxicity and reduced health risks. These are also readily available in the market as compared to synthetic drugs. Among various plant- derived natural bioactive compounds, exudates employed routinely for primary health care in developing countries. These are the complex mixtures of large and small molecules, including carbohydrates, proteins, amino-acids, volatile compounds or inorganic ions released from healthy plants during the process of exudation. Exudates are known to possess several medicinal properties and have also been used as ingredients in medicines, food, perfumes and cosmetics. They may vary greatly among genera, between species or within a genus, and their function in plants is not fully understood.

Certain legumes exude specific flavonoids and isoflavonoids, which activate genes responsible for nodulation and promote chemotaxis. Other members of family Poaceae exude carbohydrates and amino acids that are energy sources and nutrients for microorganisms thus increasing the population of plant growth promoting bacteria in rhizosphere. Exudation of some organic acids, such as malic, pyroglutamic, succinic and fumaric acid and amino acids including; L-aspartic acid, L-glutamic acid, L-isoleucine, L-leucine and L-lysine by tomato plants influence flagellar motility of *Pseudomonas fluorescens*. The substances released during exudation may exert a chemo-attractant effect on microorganisms present in soil or even prevent the seed from being colonized by phyto-pathogens. Ethnopharmacological studies have indicated the use of plant exudates in traditional medicines but there are only a few scientific records or merits of these products.

2.7.3 Different types of exudates released from plants: Depending on the physical and chemical characteristics, exudates are generally classified as resins, sap, latex or gums (Licá et al., 2018). Additionally, there are other plant exudates such as seed exudates and root exudates. Latex, also known as milk sap is a cytoplasmic exudate of specialized cells, called laticiferals. The chemical composition of latex acts as a defense to fight against fungi and viruses through its constituents. More than 20,000 plant species from over 40 families exude latex, which accounts for about 8.9% of all angiosperm plants. The bestknown example is latex obtained from rubber, which has solids content over 50% of the weight of latex. Latex contains a variety of chemicals, such as terpenoids, alkaloids, rubber, cardenolides as well as various proteins and enzymes; such as proteases, chitinases, and glucosidases. Some chemical compounds in latex, such as morphine (an alkaloid from poppy latex) and cardenolides (from milkweed latex) show apparent toxicity against animals, including insects. Such toxic chemicals are suggested to have defense roles. Some of the species that produce latex include, *Asclepias syriaca* L. (milkweed), *Hevea brasiliensis* (Willd. ex A. Juss.) Müll. Arg. (rubber tree) and *Lactuca sativa* L.

The resins are solid or semisolid amorphous materials, generally comprising a complex blend of organic compounds called terpenes. They are insoluble in water but soluble in certain organic solvents. They can occur in almost any organ or tissue of the plant species. Important families that produce resins are Burseraceae, Dipterocarpaceae, Leguminosae (mainly, Caesalpinioideae), Styracaceae and two families of conifers, each with an important resin producing genus, viz. Araucariaceae (Agathis) and Pinaceae. Vegetable gums are solids consisting of mixtures of polysaccharides, which are water soluble or absorb water and swell to form a gel or gelatin, when placed in water. They are translucent and amorphous substances often produced by plants as a protection after an aggression. Many plants growing in semi-arid conditions produce gummy exudates in large quantities, when cortex is assaulted. This is to seal cutting and prevent dehydration. Gums can be obtained from the shells, or even from seeds, such as guar gum of *Cyamopsis tetragonoloba* (L) Taub and carob gum (or carob tree) of *Ceratonia siliqua* L., both obtained from the seed endosperm. Other natural gums are mucilages produced by algae (e.g. alginates, agar, carrageenan) and bacteria (dextran, xanthan). Sap is a fluid transported in xylem cells (vessel elements) or phloem sieve tube elements of the plant. These cells transport water and nutrients throughout the plant. Sap should not be confused with latex, resin or cell sap; it is a different substance, produced separately and possesses different components and functions. Xylem sap consists primarily of a watery solution of hormones, mineral elements and other nutrients. Transport of sap in xylem is characterized by its movement from roots towards leaves. Phloem sap consists primarily of sugars, hormones, and mineral elements dissolved in water.

Exudates released from plants, consist of complex mixtures of organic and inorganic molecules that have been used in traditional medicine from several years. They may vary among genera, species or within a genus and mainly include latex, sap, gums, resins, seed or root exudates (Licá et al., 2018). Plant exudates are known to possess several biological activities including, antimicrobial, anti-inflammatory, antioxidant, wound healing and anti-nociceptive. Exudates oozed out from plants have also been used as ingredients in medicines, food, perfumes and cosmetics. The present review provides brief overview about the exudates released from plants, their biological properties and beneficial effects for human beings. Due to the presence of various compounds, different methodologies and procedures have been employed for their collection and analyses. Literature studies suggest that plant exudates have extensive therapeutic potential for curing diseases with minimal toxic effects. This aspect could be taken into account in prospective studies regarding the search of new products derived from plant exudates with pharmaceutical value.

2.7.4 Exudates of plants and Sticky mucilages: Mucilages are generally polysaccharide-rich and often occur in the form of viscoelastic gels and in many cases have adhesive properties (Andrew F. Galloway et al., 2020). In some cases, progress is being made in understanding the structure–function relationships of mucilages such as for the secretions that allow growing ivy to attach to substrates and the biosynthesis and secretion of the mucilage compounds of the Arabidopsis seed coat. Work is just beginning towards understanding root mucilage and the proposed adhesive polymers involved in the formation of rhizosheaths at root surfaces and for the secretions involved in host plant infection by parasitic plants. In this article, knowledge on plant exudates and mucilages was summarised within the concept of their functions in microenvironmental design, focusing in particular on their bioadhesive functions and the molecules responsible for them. Attention was drawn to areas of future knowledge need, including the microstructure of mucilages and their compositional and regulatory dynamics.

Land plants have a great ability to adapt to a diverse range of environments. Being sessile and unable to evade unfavourable conditions, their survival depends critically on their ability to sense their environment and, if possible, create favourable local conditions. To meet these demands, plants have developed a range of strategies, including morphological adaptations, long-distance communication via volatiles or synthesis of protective metabolites. Plants have also evolved ways to bioengineer favourable microenvironments in their immediate surroundings by the secretion or exudation of mucilages from various surfaces. Mucilage, as a matrix of high-molecular-weight compounds, is secreted as a viscoelastic gel that is often polysaccharide-rich. The wider term exudates encompasses mucilages but can also include compounds of lower molecular weight and more soluble high-molecular-weight polysaccharide and protein components that may not contribute to gel-like structures. Depending on the plant species, exudates and mucilages can be secreted by almost any plant organ, and seemingly from all clades of the angiosperms and in a few cases multiple exudate types with different origins have been described for the same species or genus (e.g. seed coat, haustorial and stem exudates for parasitic dodder (*Cuscuta* spp.). Mucilages and other exudates provide an effective means to execute a variety of functions beyond the confines of their tissues and organ surfaces. Classic examples include the capacity of climbing plants to climb up vertical surfaces using sticky tendrils or roots. Other types of exudates are produced by leaves or by stems and have become known as extrafloral nectars. Below ground, the roots are equally adept in producing mucilages that have an impact on the properties of their immediate soil environment. Each type of secretion essentially consists of a unique blend of molecules that, depending on its location, serves a specific set of functions. These functions range from attracting beneficial microbiota and insects, modifying soils for enhanced nutrient and water uptake or surface anchorage for climbing to attachment to hosts before their infection.

Some molecules within mucilages have bioadhesive properties, making them sought-after materials for biotechnological, biomedical or agricultural applications (George & Suchitra, 2019). Important breakthroughs have been made in the last few years in investigations of the sticky adhesives used by English ivy and by sundew and a current focus on molecular genetic investigations of seed coats and the rhizosheath underpins the potential of this field. With this review we first highlight the wealth of the different sticky mucilage types that are produced by plants. The current state of the art regarding the analysis of the molecules was elaborated lending bioadhesive plant secretions their specific functions and point out where future efforts could be focused.

Exudates and mucilages enable plants to exert influences on their immediate surroundings, creating microenvironments that can be favourable for their growth. Some molecules present in these mucilages possess naturally adhesive properties that have potential biotechnological and biomedical value in the form of glues or wound covers – although the precise bioadhesive molecules present in the secretions have rarely been identified. These adhesives could also be used to prevent or even reverse the process of soil erosion, thus potentially increasing our ability to produce sufficient food for growing populations. Their application as blueprints for natural-based soil conditioners could boost the abundance of soil aggregates and water retention, thus preventing its degradation. However, to date, most of the insight into the molecular regulation, biochemical composition and structure–function relationships of mucilages has been based on a few key model species. In the case of the most prominent model, the thale cress *A. thaliana*, it does not have an ecological or economical value *per se* and can at best serve as a guide to look for similar genes, molecules and networks. This requires good genomic knowledge on other species with more prominent or potentially more valuable mucilages. A very promising approach that has yielded major breakthroughs in recent years is the use of techniques enabling visualisation of nanostructures (Huang *et al.*, 2016). Such techniques will potentially allow novel insights into generation of mucilages and exudates, particularly when combined with molecular approaches. Together, the benefits from translational approaches using plant mucilages and exudates as a basis for biotechnological applications give ample incentive to learn more about the common and the unique molecules that are part of the different plant mucilages and to understand their synthesis, properties and functions.

2.8 Plant Exudates and Mucilage Diversity

Exudates and mucilages can be categorised based on their location and on their potential to influence the microenvironment of the plant in various ways (Fig. 17). Some mucilages are highly conspicuous such as the droplets on the glandular hairs produced by carnivorous plants to capture insects and supplement their diet with nitrogen (Andrew F. Galloway et al., 2020). Others are less obvious or even hidden from sight as they are produced below ground. The following sections give a brief overview of their diversity.

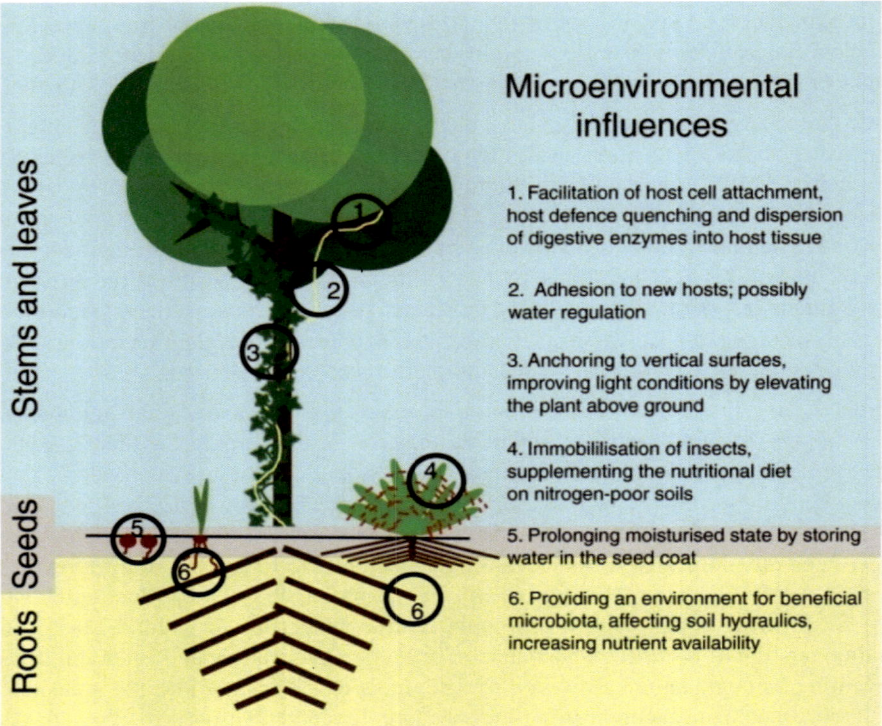

Fig. 17: Overview of plant mucilage types. Schematic representation of the location of plant mucilage production (indicated by circles) and the corresponding specific tasks of the sticky secretions. Above-ground organs are shown against a blue background, soil surface is shown on a grey background and the soil body with the root systems is shown in yellow. The ways in which they can shape the microenvironment of the plant are listed to the right.

Aerial Adhesive Mucilages facilitating climbing motion

Some climbing plants stick to surfaces of almost any nature with adhesive structures (Andrew F. Galloway et al., 2020). A study of the attachment organs of climbing fig (*Ficus pumila*) and English ivy (*Hedera helix*) identified clusters of adventitious roots as the organ from which an adhesive mucilage that robustly secures the plant to vertical surfaces is secreted. Virginia creeper (*Parhenocissus quinquefolia*), by contrast, produces adhesive discs at the end of short tendrils for climbing. A detailed compositional analysis has been so far only performed for *H. helix* and revealed that the major components of the mucilage secreted by English ivy are pectic rhamnogalacturonan-I (RG-I) polysaccharide domains that are held together by a nanospherical arabinogalactan protein (AGP) molecule. This multipolysaccharide architecture is supported by calcium-driven electrostatic interactions between the acidic residues of the AGP and RG-I domain. The AGP nanoparticles are highly uniform and evenly spread over the attachment area. The firm attachment is further supported by cellulose microfibril-dependent shape changes of adventitious root hairs at the attachment interface, accentuating the possibility that structural and chemical features have been fine-tuned together to form an attachment organ that meets the special needs of climbing plants. Whether the polysaccharide composition of the adhesive mucilages and their nanostructures are conserved amongst climbing plants or whether each climber has its own 'signature cement' is an intriguing question that should be addressed in the future.

Extrafloral nectars secreted by stems, leaves or aerial roots to provide protection and nutrition: Many land plant species secrete mucilage through their stems as extrafloral nectar (Andrew F. Galloway et al., 2020). Despite their name, extrafloral nectars are not involved in attracting pollinators but rather entice invertebrates such as ants or wasps to protect the plants from herbivorous predators or to supplement their diet with nitrogen by trapping insects or cultivating microbial symbionts. Sundew (genus *Drosera*) or butterworts (genus *Pinguicula*) possess trichomes with terminal glands, which produce the mucilage in small droplets on their upper leaf surface while pitcher plants secrete larger volumes of their mucilage in special leaves formed like a cone. So far, only the mucilage of *Drosera* has been subjected to an analysis of its polysaccharide composition and was shown to contain predominantly high-molecular-weight substances that resemble pectins. The architecture of the mucilage appears to have a higher order nanostructure that changes depending on the presence or absence of tensile forces. A hypothetical polygonal polysaccharide-based nanonetwork was proposed based on atomic force microscopy images. These nanostructures can be transformed into a parallelogram shape upon application of a tensile force, which would explain the high tensile strength reached by the sundew mucilage (Huang *et al.*, 2015). *Myo*-inositol (MI) was identified in the lower molecular weight fraction as the predominant nonpolysaccharide organic component. MI seems to be unique for sundew and could not be detected in other plant mucilages that served as control.

The parasitic giant dodder (*Cuscuta reflexa*), which infects shoots of other land plants, was also found to possess a stem exudate in the form of droplets whose sticky nature helps it to facilitate contact with host surfaces. It has been hypothesised that the dodder secretes excess sugars with it in order to secure the flow of less abundant organic or inorganic compounds and maintain an osmotic balance. The composition of this exudate has not yet been examined.

2.9 Haustorial Mucilage

In addition to the extrafloral 'nectar', *Cuscuta* also produces additional types of secretions that are crucial for succeeding with its infection (Andrew F. Galloway et al., 2020). Being shoot-parasitic, *Cuscuta* species wrap around the stems of other plants (their hosts) and withdraw their nutrients by forming physical and physiological nutrient bridges, called haustoria. To penetrate the plant cell wall barrier, the parasite must firmly attach to their hosts and push the haustorium into the host tissue using a combination of mechanical force and chemical tissue maceration. To provide a counter-force for the penetration process, the upper haustorium (sometimes also termed appressorium) exudes a viscous mucilage from club-shaped epidermal cells that form at the interface with the host (Vaughn, 2002). Immunological analyses revealed that this mucilage is mainly composed of pectins, extensins and AGPs. In addition, another mucilage was observed surrounding the developing endophytic part of the *C. reflexa* haustorium. This mucilage is predominantly visible during early infection stages and is difficult to analyse without interference from host activities, explaining why it has not yet been characterised. Besides *Cuscuta*, the root parasite *Orobanche crenata* was also found to produce a pectin-rich mucilage at the interface between itself and its hosts. *O. crenata* forms similar host attachments as *C. reflexa* but with the difference that it connects to the host root system below ground. Interestingly, the mucilage that the parasite secretes and that may help the parasite to infect susceptible hosts appeared to fill host xylem vessels in resistant hosts and ultimately led to the death of the parasite. Future investigations need to focus on comprehensive analyses of the haustorial mucilage composition in these two and other species, and also link possible differences in composition to differences in infection strategies, host specificity and host responses. In this context, and with the tensile strength of the infection organ in mind, a nanostructural analysis of the mucilages involved in parasitic plant infection would be very interesting.

i. Seed coat mucilage: The production of a coat of gelatinous material derived from cell wall polysaccharides is fairly widely found in seeds of land plants. A prominent example of seed coat mucilage is the viscin produced by hemiparasitic mistletoe species (Andrew F. Galloway et al., 2020). This mucilage protects the seeds during their passage through the guts of birds and ensures that the seed is firmly attached to the branches of its host trees when excreted by the birds. The viscin mucilage of *Phoradendron californicum* has been subject to extensive biochemical characterisation and was found to consist predominantly of highly branched xylans, arabinans, pectic rhamnogalacturonans and xyloglucans. A comparative study involving *P. californicum* and two other Viscaceae species indicated that compositional differences, mainly in the type of neutral sugars, exist between the species and that these may be specific enough to serve as taxonomic markers.

In the model land plant *Arabidopsis thaliana*, specialised mucilage secretory cells (MSCs) in the outer integument produce large amounts of mucilaginous hydrophilic polysaccharides in rings around a central columella that is compositionally similar to a secondary cell wall. The mucilaginous rings in each cell dehydrate during seed maturation but rapidly expand when in contact with water during imbibition of the mature seed, accompanied by rupturing and rapid degradation of the outermost cell walls. The released mucilage envelope can be divided into an inner and an outer layer and this organization provides coherence and stability. The outer layer is a poor adhesive due to its higher content of unbranched pectic RG-I that stretches out into the soil, whereas the inner layer has strong adhesive properties and contains pectins in a matrix of cellulose scaffolds that span this layer and connect it to the seed coat. Linkage analyses and the investigation of mutants impaired in several cell-wall-associated proteins pointed towards a significant contribution of proteins, including AGPs of type I, in the seed coats. The analysis of flax mucilage revealed that many of these components are located in distinct domains or layers (Miart *et al.*, 2019), suggesting a highly sophisticated and complex regulatory network in charge of seed coat production. High-throughput screens for phenotypic differences in the mucilaginous seed coats of *Arabidopsis* mutants have helped to identify specific genes that secure proper seed coat production. In particular, these studies have highlighted the contribution of transcription factors to MSC differentiation and have led to a proposed hierarchical network involving three distinct tiers or levels of regulation.

Seed coats are a potential source for carbon and therefore tentatively a source for nourishing the developing seed, although some studies suggest that this resource is not utilised. Rather, the seed coat mucilage seems to enhance water availability to the seed and make it less vulnerable to short-term dry spells. It provides a barrier to the environment and keeps the seed moist due to its water-holding capacity. It is further implicated in the mediation of germination under waterlogged conditions, in soil seed bank maintenance and in seed dispersal (Tsai *et al.*, 2017). The specific set of functions can vary between species. It is plausible that seed coat mucilage could also be involved in the recruitment of beneficial soil microbiota (as hypothesised for root bioadhesives), which is a crucial foothold for plant growth, although experimental evidence for this has yet to be generated.

ii. Root mucilage secretions: While disputed, most data agree that plants invest a considerable amount of their resources into releasing exudates from roots and estimates that between 10% and 40% of photosynthate ‘currency’ is spent on root exudates have been put forward (Andrew F. Galloway et al., 2020). Root exudates, including a great complexity of both low- and high-molecular-weight components, influence zones of soil at root surfaces known as rhizospheres. The release of polysaccharide-rich mucilage from root tips is well established and this may lubricate roots to ease penetration through deeper layers of soil and may also be involved in forming a protective barrier. It is proposed that bioadhesive mucilage components of exudates are important factors, along with root hairs, in the formation of cylinders of soil around roots known as rhizosheaths. Rhizosheaths could enable plants to sustain and increase nutrient and water uptake from the soil. Rhizosheath bioengineering by some grass species during periods of drought has been observed, where the grasses increased the thickness of their rhizosheaths. This increase in structural integrity was thought to be caused by a combination of mechanistic force exerted by root hairs as well as increased mucilage production. Xyloglucan, an otherwise abundant cell wall component, has been demonstrated to be released by plants and is a soil-binding factor that could enhance water infiltration and aeration.

Adhesiveness is likely to be only one functional aspect of root high-molecular-weight exudates in the bioengineering of soil. There will be highly complex interactions between the structurally diverse molecules released from plant roots and the soils and soil microbiomes, which themselves are also highly heterogenous. The capacity for putative bioengineering to alter the physical microenvironments will be considerable. It appears that this will involve differing impacts on soil aggregation through varied viscosities and surface properties, on drying/wetting cycles and the potential for hydrophobicity of mucilage after drying and binding properties in general, including the sequestering of heavy metals (Kroener *et al.*, 2018). Similarities between some of these aspects of root mucilage properties and impacts and those of other plant secretions will be a fruitful area for future research in conjunction with detailed structural characterisations.

Regarding the mechanisms involved in root mucilage secretion, the leading hypothesis is that mucilage could be secreted from continually lysing epidermal cells on the root body. However, other reports indicate that higher amounts of mucilage could be secreted than what would be predicted from lysing cells. Approaches using hydroponics, where roots are not subjected to penetrative forces, have also detected continual secretion. An important future goal will be to untangle the polysaccharide secretions from root tips and the possible secretion from root hairs that promote rhizosheath formation. Difficulties in collecting pure enough samples for immuno- and physicochemical analyses have been one limiting factor. The isolation of root-derived high-molecular-weight polymers from hydroponic media has provided a more reliable and scalable sampling method but uses an artificial environment removed from soil. Nevertheless, it is a valuable approach to understanding the formation of rhizosheaths that will hopefully in the future be supported by molecular genetic approaches.

iii. Molecules in the mucilage conferring bioadhesive properties: Due to the range of molecules present in mucilages and due to their varying accessibility, a variety of methods have been used to isolate them (Andrew F. Galloway et al., 2020). Despite the fact that the composition of each type of mucilage has yet to be fully deciphered, some mono- and polysaccharides as well as glycoproteins that are typically also key architectural components within the plant cell wall appear to be very common in the plant mucilages. AGPs and pectins (RG-I) are the key components in most mucilages, appearing in aerial, haustorial, rhizal and seed coat mucilages. The adhesive properties of both groups of molecules are well documented, particularly with their use as emulsifiers in industry. Xyloglucan along with β-1,3-glucans have recently been shown to have soil-binding properties and are target polymers for our understanding of rhizosheath formation. Nanostructure analysis of climbing plant and insectivourous plant mucilages has revealed highly ordered structures that contribute to extreme tensile strengths. An application of the techniques used to visualise these structures to other types of mucilages promises to be a revealing undertaking, and can uncover whether there are tentative common structures related to adhesion or other common functions, despite the differences in the mucilage compositions.

iv. Molecular mechanisms of the secretion of mucilage: The molecular mechanisms of mucilage secretion have been mainly investigated using the *Arabidopsis* seed coat as a model, where special secretory cells are used to produce and exude the mucilage components (Andrew F. Galloway et al., 2020). Polysaccharides are formed either through monosaccharide cytosolic synthesis or within the Golgi apparatus by glycosyl-transferase enzymes. The exception are glycoproteins, in which the protein domain is formed within the rough endoplasmic reticulum and is attached to the polysaccharide domain within the Golgi apparatus (Voiniciuc *et al.*, 2018). How these two domains localise together remains unknown. The key proteins involved in the production of cellulose, pectin and hemicellulose moieties of mucilage include Cellulose Synthase-like (CSL) A2, Galaturonosyl Transferase-like (GATL) 5, Galaturonosyl Transferase (GAUT) 11 and Rhamnose Biosynthesis (RHM) 2. The products of these enzymes are packaged into vesicles and are exported to the plasma membrane where they join the apoplast pathway for secretion. Following the secretion of mucilage, Pectin Methylesterase Inhibitor (PMEI) 6, beta-Xylosidase (BXL) 1 and beta-Galactosidase (BGAL) 6 play a role in assembling mucilage to stabilise its final form, although how they do this remains unclear. Before mucilage biosynthesis, the epidermal cell walls of the seed coat undergo major changes to accommodate such levels of secretion. For other mucilages, the principal pathways of production and shuttling to the apoplastic compartment are probably the same as for the seed coats although in some cases (e.g. in root mucilage) sloughing and lysing of cells contribute to mucilage production.

2.10 Plant Exudates and Soil

Soil physical conditions, particularly in the rhizosphere, are continually modified by the release of plant root exudates and microbial metabolites. Plants potentially benefit from this modification of the rhizosphere because of improved physical conditions for root penetration, and nutrient and water uptake by the roots (Naveed et al., 2017). However, the effect of biological exudates on soil physical properties might depend on their physicochemical characteristics. The quantity and physicochemical characteristics of root exudates are determined by the plant species, the age of an individual plant and external factors such as biotic and abiotic stresses (e.g. soil structure, presence of microorganisms and nutritional status). In general, root exudates are composed of an array of compounds such as carbohydrates, amino acids, organic acids, phenolic compounds, fatty acids, sterols, vitamins, enzymes and inorganic molecules. Among these, sugars (often polysaccharide derived), amino acids and organic acids are usually released in the largest quantities. There is growing evidence to suggest that certain compounds present in root exudates are involved in engineering the rhizosphere by dispersion and gelling of soil, modulation of water and nutrient availabilities and attraction of rhizobacterial. The anions of organic acids in the rhizosphere may be adsorbed by soil mineral particles, thereby increasing the net negative charge of clays that would cause particles to disperse. Mucilages and other polysaccharides present in root exudates, which can function as stabilizing materials, might offset this effect. The pH of the exudates and thus rhizosphere will also affect soil structural stability if the change in pH results in dissolution or precipitation of stabilizing material such as Al or other polyvalent ions

Root exudates and microbial metabolites can have a marked effect on soil stability and its resistance to disruption from both mechanical and hydraulic stresses. There is considerable evidence suggesting that root exudates improve soil stability. Improved aggregate stability was found for only certain biological exudates they studied, with large differences between soils containing swelling or non-swelling clay minerals. They observed improved tensile strength and aggregate stability for soil treated with xanthan, polygalacturonic acid and dextran, but not for lecithin. Compared with the number of studies that reported an increase in soil stability from treatment with plant-derived biological exudates, those that showed the dispersion of soil with exudates are rare. Most research to investigate soil stabilization by biological exudates has measured aggregate stability. Although soil aggregate stability is a relevant test for studying the resistance of soil to erosion, it does not provide a quantitative measure of particle bonding that underpins the formation of soil structure. Fracture tests on dry soil discs or notched bars have quantified increased particle bond energy resulting from root exudate compounds, but such dry conditions are rarely met in reality. Formation of the soil structure occurs when soil is wet and particles are mobile; therefore, rheological measures of wet soil movement under stress are more relevant physically to understanding how mechanical stresses from root growth and exudation affect soil structural development. The mobility of soil when wet can be described with rheological measurements that provide controlled oscillatory shear stresses to quantify time-dependent flow under stress (i.e. viscosity and yield stress) (Markgraf et al., 2012).

Plant exudates could either gel or disperse soil depending on their chemical characteristics. Barley (*Hordeum vulgare* L. cv. Optic) and maize (*Zea mays* L. cv. Freya) root exudates were collected using an aerated hydroponic method and compared with chia (*Salvia hispanica* L.) seed exudate, a commonly used root exudate analogue (Naveed et al., 2017). Sandy loam soil was passed through a 500-µm mesh and treated with each exudate at a concentration of 4.6 mg exudate g^{-1} dry soil. Two sets of soil samples were prepared. One set of treated soil samples was maintained at 4^0C to suppress microbial processes. To characterize the effect of decomposition, the second set of samples was incubated at 16^0C for 2 weeks at −30 kPa matric potential. Gas chromatography–mass spectrometry (GC–MS) analysis of the exudates showed that barley had the largest organic acid content and chia the largest content of sugars (polysaccharide-derived or free), and maize was in between barley and chia. Yield stress of amended soil samples was measured by an oscillatory strain sweep test with a cone plate rheometer. When microbial decomposition was suppressed at 4^0C, yield stress increased 20-fold for chia seed exudate and twofold for maize root exudate compared with the control, whereas for barley root exudate decreased to half. The yield stress after 2 weeks of incubation compared with soil with suppressed microbial decomposition increased by 85% for barley root exudate, but for chia and maize it decreased by 87 and 54%, respectively. Barley root exudation might therefore disperse soil and this could facilitate nutrient release. The maize root and chia seed exudates gelled soil, which could create a more stable soil structure around roots or seeds.

2.11 Complex Interactions in the Rhizobiome that Improve Plant Health

The rhizobiome harbors many complex interacting organisms that drive functional activities that have a positive impact on plant health. Although not considered because they are out of the scope of this review, plant–plant interactions, which involve allelochemicals, are one of the predominant interactions in the soil (Oluwaseyi Samuel Olanrewaju and Olubukola Oluranti Babalola, 2022). It is worthwhile mentioning that allelopathy originates in the rhizosphere through the roots of one plant impacting the roots of another plant. These interactions are exclusive between plants alone. We will consider other important interactions that are specific to the rhizobiome such as interactions between the various microbes that exist in the rhizosphere. Bacteria, fungi, viruses, and archaea, among others, are present in the rhizosphere. These organisms interact together and their interaction has an effect on the rhizobiome.

Quorum signaling/sensing (QS) involves the communication and coordination of microbe activities using molecules that act as signals. These molecules must be at a sufficient concentration to induce a response from the receiving microbe(s). Signaling can occur between pathogen-beneficial microbes, pathogen-pathogen microbes, and beneficial-beneficial microbes. QS is required for successful adaptation by both beneficial and non-beneficial microbes to the host environment. Among the key mediators of QS, especially in gram-negative bacteria, are *N*-acyl-l-homoserine lactones (AHLs). They are produced by bacteria and diffuse through the membranes into the rhizosphere, and plants have been reported to produce metabolites that mimic the actions of AHLs. Canavanine, a toxic arginine analogue to many bacteria except rhizobia, is present in root exudates of some legumes. It is used for selective purposes by legumes to control their rhizosphere bacterial populations. They use it to select for beneficial rhizobia and remove other bacteria genera from their rhizosphere. Another mimicking compound that is secreted by *Combretum albiflarum* is catechin, which is reported to interfere with the signaling of *Pseudomonas aeruginosa* PAO1. Catechin is a flavonoid used by the plant as the first line of defense against invading pathogens. Some PGPM use this mechanism of signal disruption in controlling pathogens. By disrupting the signaling between pathogens, they can disrupt the expression of virulence genes. In addition, rhizosphere QS can be disrupted by abiotic factors such as alkaline pH (Rana et al., 2020).

The secretion of AHLs by PGPM can also induce systemic resistance in the host plants. For example, *Serratia liquefaciens* MG1 and *Pseudomonas putida* IsoF secret AHLs that initiates induced systemic resistance (ISR) against *Alternaria alternate* in tomato plants using the SA and ethylene-dependent pathways. For PGPM to exert their effects on host plants, they must be able to successfully colonize their host plants and the activities of these PGPMs are influenced by the exudates from the roots which is termed the 'rhizosphere effect". Plants provide a carbon source which informs a chain of signals for the microbes and subsequently for microbe colonization.

Living organisms in the rhizosphere: The rhizosphere is a centre of intense biological activity due to the food supply provided by the root exudates. Bacteria, actinomycetes, fungi, protozoa, slime moulds, algae, nematodes, enchytraeid worms, earthworms, millipedes, centipedes, insects, mites, snails, small animals and soil viruses compete constantly for water, food and space (Rebecca Lines-Kelly, 2005). Soil chemistry and pH can influence the species mix and functions of microbes in the rhizosphere. Most soil microorganisms do not interact with plant roots, possibly due to the constant and diverse secretion of antimicrobial root exudates. However, there are some microorganisms that do interact with specific plants. These interactions can be pathogenic (invade and kill roots and plants), symbiotic (benefit plant growth), harmful (reduce plant growth), saprophytic (live on dead roots and plants) or neutral (no effect on plants). Interactions that are beneficial to agriculture include mycorrhizae, legume nodulation, and production of antimicrobial compounds that inhibit the growth of pathogens.

2.12 Plant-microbe Cross-talk and Rhizosphere-Effect

Rhizosphere, the interface between soil and plant roots, is a chemically complex environment which supports the development and growth of diverse microbial communities. The composition of the rhizosphere microbiome is dynamic and controlled by multiple biotic and abiotic factors that include environmental parameters, physiochemical properties of the soil, biological activities of the plants and chemical signals from the plants and bacteria which inhabit the soil adherent to root-system (Shyamalina Haldar and Sanghamitra Sengupta, 2015**).** Recent advancement in molecular and microbiological techniques has unravelled the interactions among rhizosphere residents at different levels. Insights into the underlying structural principles of indigenous microbial population and the key determinants governing rhizosphere ecology provided directions for developing techniques for profitable applicability of beneficial microorganisms in sustainable agriculture and nature restoration.

Rhizosphere-Effect: 'Rhizosphere effect', a term coined by Starkey (1938), is defined by collective processes occurring at the root-soil interface of a plant and includes root exudation, microbial activity, genetic exchange, nutrient transformation and gradient diffusion (Shyamalina Haldar and Sanghamitra Sengupta, 2015). In living plants, organic carbon released by plant roots is decomposed to CO_2 in a mechanism known as rhizosphere priming effect (RPE). Although the estimate of plant carbon economy is still controversial, approximately one-third to half of total assimilated carbon is allocated to below-ground, of which 15-25% is exuded from the roots into the soil to induce fast carbon-turnover in the rhizosphere. Due to intensive carbon uptake by the roots, other nutrients in the rhizosphere are strongly limited. In contrast, in a root-free soil, all the nutrients except carbon are unlimited. The surplus of easily available carbon and strong nutrient limitation, together, make the rhizosphere milieu strongly different from that of the root-free zones. The increased microbial number and activity in the rhizosphere compared to those in bulk soil are mainly due to the release of organic carbon by the plant roots. The substances released by plant roots are collectively termed as "rhizodeposits." These compounds include (i) water-soluble ions and low-molecular-mass compounds such as mono-saccharides, amino acids and organic acids which are lost passively along a concentration gradient, (ii) high-molecular-mass compounds such as carbohydrates, proteins and lipids which are actively transported along an electrochemical gradient, (iii) insoluble mucilage composed of polysaccharides and polygalacturonic acid, (iv) an array of secondary metabolites such as antimicrobial compounds, nematicides and flavonoids and (v) remnants of the dead and lysed root-cap and border cells. Of these, sugars and amino acids are thought to be released in the greatest quantities.

The rhizosphere effect is the influence of plant roots on the development of soil microorganisms as a result of the physical and chemical alteration of soil and the release of root secretions and exudates within the rhizosphere (Anupama Sapkota, 2022). The rhizospheric effect is observed on the basis of the microbial biomass of the rhizosphere when compared to the biomass of the bulk soil. The rhizosphere effect on soil microbial population can be measured by comparing the population density [colonies forming units (CFU)] between the rhizosphere soil (R) and the bulk soil (S), for which the "R/ S ratio" is employed. The rhizosphere effect is higher for bacteria > fungi > actinomycetes > protozoa. The microorganism diversity is higher near to the rhizoplane, which then decreases with an increase in distance from the rhizoplane. The interaction between plant nutrients in soil and plant exudates modifies the microclimate of the rhizosphere. The rhizosphere effect is a result of the interaction between the plant root and the microbial community of the region, where both factors influence each other. In the rhizosphere, microbial activity influences the plant root, and the plant root secretions influence the microbial biomass.

Measurement of rhizosphere effect: Influence of plant roots on the growth and development of microorganisms in the nearby soil is called 'rhizosphere effect'. It is a stimulus which can be measured by finding out the R:S ratio. Rhizosphere effect is normally measured by plating technique. It is always greater for bacteria than any other microorganism. The R:S ratio is defined as the ratio of microbial numbers/ unit weight rhizosphere soil to the population in a unit weight of the adjacent non-rhizosphere control soil.

$$\text{R:S ratio} = \frac{\text{No. of organisms/gm of soil of Rhizosphere}}{\text{No. of organisms/ gm of non-rhizosphere soil}}$$

As the distance from rhizosphere increases the R:S ratio decreases.

Plant-Microbe and Microbe-Microbe Interactions: The plant–microbe interaction in the rhizosphere may be categorized as associative, symbiotic, neutralistic, or parasitic. The positive interactions include symbiotic and associative interactions with beneficial microbes, such as endo- and ecto-mycorrhizal fungi, nitrogen-fixing bacteria and plant growth-promoting rhizobacteria (PGPR) whereas negative interactions include association with parasitic plants, pathogenic bacteria, fungi, oomycetes, nematodes and invertebrate herbivores (Shyamalina Haldar and Sanghamitra Sengupta, 2015). Symbiosis between legume plants and Rhizobia in the soil is of particular importance in agriculture and much research has been focused on characterization of the molecular mechanisms that establishes species-specific collaboration. Legume-Rhizobia interactions are mediated by host-specific flavonoids secreted in the root-exudates. These molecules belong to a diverse family of aromatic compounds derived from plant's secondary metabolism. Depending on the host and the bacterium, flavonoids activate a series of transcriptional events culminating in the production of main rhizobial nodulation signals called Nod factors (NF) or lipo-chito oligosaccharides (LCOs). Nod factors induce diverse developmental processes in the plant partner leading to morphological changes in legume root hairs, infection thread formation, nodule development and symbiotic nitrogen fixation. Rhizobium sp. is a vast group of soil-borne rhizobia with other plant growth promoting activities such as siderophore production and disease suppression.

Translating Knowledge into Technology: Studies in rhizosphere research have undoubtedly improved our ability to steer the knowledge into technological applications in agriculture, ecological engineering and nature restoration. The applicability of PGPR in agriculture has been steadily increased as they offer the opportunity to replace chemical fertilizers, pesticides and mineral nutrients (Shyamalina Haldar and Sanghamitra Sengupta, 2015). Cultivation of plant genotypes that can thrive on biotic and abiotic soil resources may also serve as an alternative in sustainable agriculture. This involves engineering plants with a higher level of resistance to pathogens via diverse mechanisms. For example, plants expressing bacterial acyl-homoserine lactonase and producing 5'-O-glucosyltransferase at an increased rate were shown to harbor enhanced resistance to plant pathogen, *Pectobacterium carotovorum*. Recent studies on the rhizosphere ecology of non-cultivated plant species have increased an awareness of reconstructing soil structure by ecological engineering of soil biota. Rhizoremediation can provide a cost effective and environmentally sustainable "Green Technology" for the breakdown of petroleum contaminants from the soil. Carboxylates comprise a significant component of the root exudate mixture and it is hypothesized that these compounds may enhance biodegradation of petroleum hydrocarbons by promoting microbial activity by providing energy source. Future research involving the use of advanced techniques such as SIP combined with powerful metagenomic approaches will greatly improve our ability to understand microbial–carboxylate dynamics and aid in the improvement of rhizoremediation as a viable remediation strategy.

2.13 Rhizosphere Microbiome

The microbial population in the rhizosphere consists of different groups of microorganisms like bacteria, fungi, parasites, nematodes, viruses and algae (Anupama Sapkota, 2022). The microbial population in the rhizosphere is known as the rhizosphere microbiome and the microbial population in such an area much higher than the bulk soil. In the rhizosphere, there is a microbial population distinct from the rest of the soil. Bacteria in the rhizosphere are larger and have higher proportions of Gram-negative and denitrifying bacteria than those in the bulk soil. Rhizosphere fungal populations, abundant in both pathogenic and mycorrhizal species, can be 10 to 20 times higher than those in the non-rhizosphere. Protozoa and other microfauna also thrive in the rhizosphere because that is where food is most plentiful. The type and population of microorganisms in the rhizosphere are highly influenced by the type of plant grown on the soil. Microbes in the bulk soil often experience long periods of nutrient deprivation; they have different survival strategies in dealing with starvation and stress. The rhizosphere bacterial community is recruited from the main reservoir of microorganisms present in the soil. Next to the recruitment of specific soil microbes into the rhizosphere microbiome, plant roots also influence specific functions of the microbiome.

Plant Growth Promoting Rhizobacteria (PGPR): Plant Growth Promoting Rhizobacteria (PGPR) are a group of bacteria that enhances plant growth via various plant growth-promoting substances as well as biofertilizers (Anupama Sapkota, 2022). PGPR act as biofertilizers are efficient soil microbes for sustainable agriculture and hold great promise in the improvement of agriculture yields. These bacteria enhance crop growth and can help in the sustainability of the safe environment and crop productivity. The rhizospheric soil contains diverse types of PGPR communities, which exhibit beneficial effects on crop productivity. Some common examples of PGPR genera exhibiting plant growth-promoting activity are *Pseudomonas*, *Azospirillum, Erwinia, Mycobacterium, Azotobacter, Bacillus, Burkholderia, Enterobacter, Rhizobium, Mesorhizobium, Flavobacterium*, etc. PGPR have different roles in promoting soil health as well as crop health. These bacteria function as biofertilizers, biocontrol agents, and biological fungicides. The function of PGPR as biofertilizers is by virtue of the production of growth-stimulating phytohormones such as indole-3-acetic acid (IAA), gibberellic acid (GA3), zeatin, ethylene, and abscisic acid. Besides, these bacteria also promote plant nutrition as bacteria like *Rhizobium* act as phosphate solubilizing bacteria increase the availability of accumulated phosphate, increase the efficiency of biological nitrogen fixation and render availability of iron and zinc through the production of plant growth-promoting substances. PGPR also protects plants against pathogens by direct antagonistic interactions between the biocontrol agent and the pathogen, as well as by induction of host resistance.

Factors influencing the growth and activities of microbiome: Microbial growth and activities in soil depend on resources (carbon and other mineral nutrients) available and the physiochemical conditions of their habitat (Anupama Sapkota, 2022).

i. **Nutrients:** The majorities of soil microorganisms are presumably chemoorganoheterotrophic and use organic compounds as carbon and energy sources. The humic fraction of soil organic matter provides the stable microbial nutrient base for microorganisms; however, humic substances have incredibly complex structures and are resistant to decomposition. In addition, humus is associated with mineral particles and forms organo-clay complexes. Therefore, humus only serves as a slow-release source of carbon and energy for oligotrophic microorganisms. In the rhizospheric area, other forms of nutrients are released by the plant in the form of root secretions and root exudates.

ii. **Physiochemical factors of soil**: The physical factors of soil like soil moisture, soil atmosphere, and soil temperature also affect the growth and activities of microorganisms in the soil and the rhizosphere region. Soil water affects not only the moisture available to organisms, but also the soil aeration status, soil temperature, and a variety of soil chemical reactions. Soil temperature affects the rates of physical, chemical, and biological processes in the soil. Soil pH influences microorganisms indirectly by changing the chemical forms, solubility, and availability of chemical compounds.

iii. **Interactions:** Different soil organisms frequently interact with each other; the interactions can be positive, negative, or neutral. Positive interactions (commensalism, synergism, and mutualism) enhance the abilities of populations to survive within a particular habitat; negative interactions (competition, amensalism, and predation/ parasitism) limit the population growth. Interactions like symbiosis and mycorrhiza favor the growth of organisms and their activities in the soil.

Positive and negative effects of rhizospheric microbes on plants: Rhizospheric microorganisms play an important role in the ecological fitness of the plant and the soil (Anupama Sapkota, 2022). Important microbial processes like plant protection, growth promotion, production of antibiotics, geochemical cycling, and plant colonization take place in the rhizosphere. Rhizospheric microorganisms increase the supply of mineral nutrients from the soil to the plant. Another group of microorganisms in the rhizosphere stimulate plant growth indirectly by preventing the growth or activity of plant pathogens. These microorganisms are responsible for direct growth promotion by the production of phytohormones. Plant growth-promoting rhizobacteria act as biofertilizers by enhancing phytochrome production, phosphate solubilization, and siderophore production.The capacity of rhizospheric organisms to synthesize anti-fungal metabolites such as antibiotics, fungal cell wall-lysing enzymes, or hydrogen cyanide suppresses the growth of fungal pathogens.

Negative effects: One of the most important negative effects of rhizosphere microorganisms is a competition where the microorganisms compete with the plants for water, nutrient, and space. Some of the members of the rhizosphere microbiome might act as plant pathogens, resulting in different forms of plant diseases. Competition between the microorganisms in the microbial community in the rhizosphere can even result in the loss of beneficial microorganisms.

2.13.1 The Effect of the Rhizosphere on Bio-geochemical Cycling: Plant roots modulate biogeochemical cycles both indirectly and directly. Analysis of the stability of organic carbon revealed that mineralization of ancient soil organic matter is stimulated by root-derived carbon. Conversely, it was recently demonstrated that 50–70% of the humus build-up in boreal forest ecosystems is derived from roots and root-associated microorganisms, highlighting the role of mycorrhizal fungi in sequestering carbon in the soil (Laurent Philippotet al., 2013). Nitrogen cycling in the rhizosphere has also received considerable attention, as nitrogen is the nutrient most often limiting to plant growth. It is argued that roots affect several abiotic and biotic factors, such as oxygen pressure and carbon and nitrogen availability, which in turn influence nitrogen transformations by soil microorganisms. For example, plant roots were reported to increase denitrification rates in the rhizosphere up to 22-fold. It is generally accepted that roots promote denitrification because they consume oxygen, thereby increasing the anaerobic volume of the soil. The higher carbon availability in the rhizosphere is another important factor stimulating denitrification and emissions of the greenhouse gas N_2O. Unravelling the plant traits that affect microbial guilds involved in nitrogen-cycling emissions holds great potential for breeding new cultivars that could decrease nitrogen losses through leaching and greenhouse gas emissions.

Nitrification can be either inhibited or stimulated in the rhizosphere, which can lead to shifts in the relative amount of ammonium and nitrate available for nitrogen nutrition in the plant. As plant species-specific nitrogen preferences might influence species coexistence and dominance, the ability to control nitrification can be an important trait for invasive plants. This was shown by the increased nitrification rates in soil invaded by exotic grasses, with doubled rates in Californian soil in part owing to increased abundance and changes in the composition of ammonia-oxidizing bacteria. By contrast, the Australian savannah-invasive species Andropogon gayanus, which prefers ammonium over nitrate as a nitrogen source, inhibits nitrification but stimulates ammonification. Modelling the preferences of plants for ammonium versus nitrate indicated that the ability to control nitrification modifies the patterns of coexistence between plants. Some studies reported that invasive plants could also modify denitrification and nitrogen fixation. Together, these studies show that the plants can both cause and benefit from altered microbial nitrogen transformation, which is of importance for ecosystem functioning and plant community structure.

Rhizosphere Ecological Engineering and Nature Restoration: Recent studies on the rhizosphere ecology of non-cultivated plant species has increased awareness that the building of soil structure by soil biota (ecological engineering) and the restoration of former natural ecosystems to conserve biodiversity depend on interactions between plants and soil biota (Laurent Philippotet al., 2013). Traditionally, plant–soil relationships were considered mainly from the perspective of resource utilization and competition. However, it is becoming more evident that pathogenic and mutualistic–symbiotic organisms, and decomposers in the soil biota, influence plant community diversity and succession. Further progress is to be expected if we can understand how rhizosphere interactions influence the legacy effects of soil biotic conditions and how these legacy effects influence priority effects in plant community composition. Thus, current drivers of plant community composition often seem to be reflections of rhizosphere interactions that took place in the previous plant generation or in even earlier generations. A key problem is how these historical events or legacy effects can be managed during nature restoration or ecological engineering. For example, the colonization and proliferation of soil-borne fungi might be a factor in the transition of bacterium-dominated to more fungus-dominated soil food webs, which influences ecosystem responses to drought events. Also, the removal of introduced exotic plant species might not be sufficient to restore the original plant communities of invaded sites, as the exotics might have changed the community composition of soil biota and other soil characteristics. Therefore, enhancing our knowledge of the rhizosphere interactions of successive plant species and of the ecological consequences during vegetation development is pivotal to making ecological engineering and nature restoration more effective.

Although considerable progress has been made in our understanding of the microbial ecology of the rhizosphere, rhizosphere ecologists face several major challenges. These include the development of new crops and cropping systems to produce sufficient biomass for food, feed, fibre and bioenergy at low environmental costs.

Production methods need to focus on efficient recycling of nutrients and effective control of pests and pathogens. The acknowledgement that above-ground pest control is influenced by below-ground interactions in the rhizosphere has opened up new avenues for integrated crop management. The restoration and conservation of above-ground biodiversity, as well as the prevention of exotic plant invasions, require insight into the microorganisms and fundamental processes in the rhizosphere. Studies using crop-based systems are now increasingly being complemented by studies on non-cultivated plant species, which will provide insight into the rhizosphere ecology of the co-evolved communities in natural ecosystems. In these systems, associated microbial communities play a major part in plant adaptation to biotic and abiotic stresses. This might account for the conservation of rhizodeposit release in all plant species on all continents, when these photosynthates could have been used instead for plant flowering and seed dissemination (Lambers, et al., 2009). Thus, the promotion of plant growth and health by the rhizosphere microbiota balances the cost represented by rhizodeposits.

Better knowledge of how plants differ in their patterns of rhizodeposition, both in quality and quantity, according to genotypes, traits, environmental parameters and cropping systems is a challenging field of research to promote beneficial interactions in the rhizosphere (Laurent Philippotet al., 2013). In this context, progress in understanding signalling not only between plants and microorganisms but also among the plants themselves, through microorganisms transporting signalling compounds to neighbouring plants, is another important challenge. Indeed, it was recently demonstrated that the mycorrhizal mycelial network can function as a messaging system, allowing uninfested plants to invoke herbivore defences before attack by aphids. There is a need for integrative studies in soil microbiology, and a holistic consideration of the various mechanisms at play in the rhizosphere would undoubtedly improve prediction and management of the rhizosphere microbiota. Despite these challenges, it is clear that 'going back to the roots' of natural plant communities holds great promise to further improve the sustainability of crop production for food, feed, fibre and fuel.

2.13.2 Rhizomicrobiome - Significance of plant beneficial and plant pathogenic microbes: Microbial communities play a pivotal role in the functioning of plants by influencing their physiology and development. While many members of the rhizosphere microbiome are beneficial to plant growth, also plant pathogenic microorganisms colonize the rhizosphere striving to break through the protective microbial shield and to overcome the innate plant defense mechanisms in order to cause disease (Rodrigo Mendes et al., 2013). A third group of microorganisms that can be found in the rhizosphere are the true and opportunistic human pathogenic bacteria, which can be carried on or in plant tissue and may cause disease when introduced into debilitated humans. Although the importance of the rhizosphere microbiome for plant growth has been widely recognized, for the vast majority of rhizosphere microorganisms no knowledge exists. To enhance plant growth and health, it is essential to know which microorganism is present in the rhizosphere microbiome and what they are doing. This review covers the main functions of rhizosphere microorganisms and how they impact on health and disease; discussed the mechanisms involved in the multitrophic interactions and chemical dialogues that occur in the rhizosphere and highlighted several strategies to redirect or reshape the rhizosphere microbiome in favor of microorganisms that are beneficial to plant growth and health.

Rhizosphere organisms that have been well studied for their beneficial effects on plant growth and health are the nitrogen-fixing bacteria, mycorrhizal fungi, plant growth promoting rhizobacteria (PGPR), biocontrol microorganisms, mycoparasitic fungi, and protozoa. Rhizosphere organisms that are deleterious to plant growth and health include the pathogenic fungi, oomycetes, bacteria, and nematodes. A third group of microorganisms that can be found in the rhizosphere are the human pathogens. Over the past decade, there is an increasing number of reports describing the proliferation of human pathogenic bacteria in and on plant tissues. Understanding the processes that shape and drive the composition and dynamics of the rhizosphere microbiome is therefore an essential step not only to safeguard plant productivity but also to safeguard human health. This review has focussed on the frequency, diversity, and activities of plant beneficial ('the good'), plant pathogenic ('the bad'), and human pathogenic ('the ugly') microorganisms in the rhizosphere and how they impact on health and disease. It should be emphasized that the use of the terms 'the good', 'the bad', and 'the ugly' is arbitrary as a specific microbial species may be beneficial or deleterious depending on its abundance. Hence, this anthropogenic terminology is merely used to facilitate the discussion of the complex rhizosphere microbiome environment. Specific attention is given to mechanisms involved in multitrophic interactions and chemical dialogues that occur in the rhizosphere.

In addition to a comprehensive phylogenetic analysis of the rhizosphere microbiome, there is a strong need to go beyond cataloguing microbial communities ('collecting stamps') and to determine which microorganisms are active during the different developmental stages of plant' root growth and which functions and pathways are displayed in time and space (Rodrigo Mendes et al., 2013).A variety of molecular approaches to study gene expression in the rhizosphere was discussed. It is apparent that most of our current knowledge of genes and functions expressed in the rhizosphere is still based on studies with reporter genes. Despite their limitations, reporter genes do enable the evaluation of how specific members of the rhizosphere microbiome perceive their habitat in terms of chemical, physical, and biological stimuli. An elegant promotertrapping strategy, referred to as in vivo expression technology (IVET), was adopted to identify *Pseudomonas fluorescens* genes with elevated levels of expression in the rhizosphere. Genes induced in *P. fluorescens* during rhizosphere colonization were genes involved in nutrient acquisition, stress response, and secretion. To go beyond the 'one-gene-at-a-time' approach whole genome transcriptome profiling was used to evaluate the effects of root exudates from two sugar beet cultivars on gene expression in *Pseudomonas aeruginosa*. In addition to genes previously identified in plant–microbe interactions (i.e. metabolism, chemotaxis, type III secretion, it was showed that 104 genes were significantly altered in response to both root exudates and that the majority of these genes were regulated in response to only one of the two exudates. Recently, a whole genome microarray was also used to study endophytic colonization of rice by *Azoarcus* sp. BH72.

'Omics' approaches that enable the identification of gene transcripts, proteins, or metabolites have been developed to provide a more detailed insight into the genes and functions expressed in the plant microbiome. A metaproteogenomic approach was first reported for bacterial communities in the phyllosphere of Arabidopsis, soybean, and clover plants. For the rhizosphere, a recent metaproteomics study revealed complex interactions between plants and rhizosphere microorganisms in different cropping systems. The development of stable isotope probing (SIP) to track plant-derived carbon into microbial nucleic acids has provided exciting new insights into the metabolically active rhizobacterial populations.

Plant beneficial microbes are found within the rhizosphere, roots, and root surfaces. Many of these rhizosphere microbes can communicate with plants, and some form mutually beneficial relationships with plants. These rhizosphere microbes can improve plant health and soil nutrient cycling by regulating root exudation, alleviating plant stresses, and influencing other soil biota. The abundance and type of rhizosphere microbes can be influenced by any of these diverse soil conditions: temperature, moisture, and other organisms (e.g., plant species and other biota) as well as the presence of root and soil chemicals, atmospheric gases, crop species, and applied management practices. Learning how the rhizosphere can improve crop production and regulate soil quality, as well as the interactions between management practices and environmental variables, will help identify the next generation of management systems for sustainable agriculture.

2.13.4 Rhizosphere plant pathogens: Soil borne plant pathogens cause major yield reductions in the production of food, feed, fiber, and fuel crops. Two main groups of soil borne plant pathogens are the nematodes and the fungi, including the true fungi and the fungal-like oomycetes. In temperate climates, plant pathogenic fungi, oomycetes, and nematodes are agronomically more important than plant pathogenic bacteria, although some bacterial genera (i.e. Pectobacterium, Ralstonia) can cause substantial economic damage in some crops (Rodrigo Mendes et al., 2013). For bacterial pathogens that infect plants via roots, *Agrobacterium tumefaciens, Ralstonia solanacearum*, *Dickeya dadanthi* and *D. solani*, and *Pectobacterium carotovorum* and *P. atrosepticum* are among the top 10 most notorious. Also, viruses can infect plants via the roots but require vectors such as nematodes or zoosporic fungi to enter the root tissue. Compared with our understanding of the role of rhizodeposits in the communication between symbionts and plants, information on root exudates that activate and attract soil borne plant pathogens is more scarce and fragmented. The limited knowledge of the communication between plants and root pathogens is largely due to a poor understanding of the complex physical–chemical conditions in soil and rhizosphere environments. Hence, expanding the analytical skills to elucidate the chemistry of rhizodeposits and their spatiotemporal production and distribution patterns, collectively termed 'ecometabolomics' will be important to resolve the dialogues between pathogens and plant roots.

The processes and chemical cues that are important for soilborne pathogens to colonize the rhizosphere and to infect the root tissue are briefed.

i. Fungi and Oomycetes: For germination, growth and establishment in the rhizosphere, fungal, and oomycete pathogens depend on several different cues from the host plant. Dormancy of fungal spores can be triggered by a range of factors, including changes in abiotic conditions (i.e. pH) and root exudates. Phenolic compounds like p-hydroxybenzoic, gallic, coumaric, cinnamic, ferulic, salicylic, and sinamic acids in root exudates stimulated, at low concentrations, conidial germination of pathogenic fungi; when concentrations increased, an inhibitory effect was observed. Similar effects of four phenolic acids from cotton root exudates were found on germination of *Verticillium dahliae* spores (Rodrigo Mendes et al., 2013). Also, alkaloids from roots of *Veratrum taliense* (Liliaceae) were shown to inhibit growth of *Phytophthora capsici* and *Rhizoctonia cerealis*. Soil type and soil microorganisms greatly affected the composition of alkaloids in roots and shoots of *Jacobaea vulgaris*, in particular retrorsine and retrorsine N-oxide. Both alkaloids inhibit mycelium growth of several plant-associated fungi, including *F. oxysporum, Fusarium sambucinum*, and *Trichoderma* sp.. Based on these results it was postulated that the effect of microorganisms on the alkaloid composition of plants could have other ecological consequences as these changes may attract specialist herbivores aboveground while deterring generalists. The fungal root pathogen *Gaeumannomyces graminis* var *avenae* (Gga) can detoxify avenacin A-1 and infect oat roots, whereas Gga-mutants that lack the detoxifying hydrolase, designated avenacinase, were more sensitive to avenacin A-1 and were no longer able to infect. On the plant side, avenacin-deficient mutants showed compromised resistance to several pathogens. Recent studies further suggested that avenacin or avenacin intermediates may also elicit other processes in the plant such as callose deposition.

In contrast to root-infecting fungi, oomycete pathogens produce motile zoospores that swim toward the plant root to initiate infection. Next to the effects of specific compounds in seed and root exudates on zoospore behavior, the work pointed to electrotaxis as a key root-targeting mechanism for zoospores. Plant roots generate external electrical currents due to the flow of protons and other ions into and out of growing and wounded regions. In a series of experiments, the profile of endogenous electrical fields generated by plant roots coincided with the sites where electrotactic species of zoospores accumulated. Induced or imposed electrical fields were capable of overriding local chemical cues in the rhizosphere that either mediate attraction or repulsion. Electrotaxis is an important cue for zoosporic pathogens to selectively colonize living rather than dead roots, thereby maximizing their survival rate. Whether electrotaxis also plays a role in the chemotactic responses of soil borne pathogenic fungi and bacteria is, to our knowledge, not known yet.

ii. Nematodes: Most nematodes in soil are free living, but some feed on the root exterior (migratory ectoparasitic), some penetrate and move in the root interior (migratory endoparasitic), while others develop a feeding site in the root where they reproduce (sedentary endoparasites). For plant parasitic nematodes other than cyst or polyphagous root knot nematodes, it is critical to exploit chemical gradients to find their host plant (Rodrigo Mendes et al., 2013). Their sensory apparatus enables them to orientate, move, and locate nutrient sources. In the physico-chemically complex soil matrix, volatile as well as water-soluble compounds are important cues for nematode foraging. Volatile compounds have been suggested to play a major role in long-range chemotaxis, whereas water-soluble compounds were proposed to be more suitable for short-range chemotaxis. Most studies to date have reported on plant-derived compounds that attract nematodes, but also nematode repellent compounds have been identified like a-terthienyl, inositol, and cucurbitacin A. Among the volatiles emitted by plant roots, CO_2 is the main so-called long-distance kairomone for root location by plant parasitic nematodes, with a theoretical action-radius of up to 1 m for a single root and more than 2 m for a complete root system.

CO_2 most likely serves as a 'response activator' that alerts the entomopathogenic nematodes to the general presence of other organisms and may enhance their responsiveness to more specific cues. Besides CO_2, many other compounds from different chemical classes induce chemotaxis in nematodes such as 2,4-dihydroxy-7-methoxy-1,4-benzoxazin-3-one (DIMBOA), glutamic and ascorbic acid (Rasmann et al., 2012). Knowledge of the chemical cues that attracts nematodes and other pathogens can and has been exploited to attract these pathogens to nonhost crops. Classic example of these trap crops is *Asparagus officinalis* that attract a wide range of nematodes which, after being lured in, are killed by defensive compounds such as glycosides.

Future Prospects: Although the importance of the rhizosphere microbiome in the functioning of plant ecosystems has been widely recognized, traditional approaches to unravel these functions are limited in their capacity and for the vast majority of rhizosphere organisms, no knowledge exists. Coupling traditional approaches with advanced nextgeneration sequencing techniques to assess organismal or community ecology and physiology will bring new insights to understand microbial life in the rhizosphere(Rodrigo Mendes et al., 2013). Identification of the exudates, signals, and key players in the rhizosphere microbiome will provide chemical and microbial markers to elucidate whether and how plants recruit and stimulate beneficial (micro) organisms. Unravelling the rhizosphere microbiome also holds potential to improve crop protection and to uncover numerous yet unknown soil microorganisms, functions, and genes for diverse applications.

Given that several antagonistic traits are shared by different rhizosphere microorganisms, there may be a certain level of functional redundancy among members of the core microbiome. In this respect, one could also design a 'minimal microbiome'. Analogous to the concept of the minimal genome, the minimal microbiome would comprise the minimal set of microbial traits needed to fulfil a specific ecosystem service, in this case protection of plants against soil borne diseases. One may argue that control of different pathogens on different crops requires a different subset of antagonistic microorganisms. This is probably true for the different groups of soil borne plant pathogens, that is, bacteria, fungi, oomycetes, and nematodes. However, designing a core microbiome for each of these pathogen groups separately may be feasible. This assumption is based on several observations: studies on natural disease-suppressive soils have pointed to common players and identical mechanisms and genes in the suppressiveness of soils to different fungal pathogens; also, the onset of natural disease suppressiveness of soils follows a similar pattern for various fungal pathogens, suggesting that similar processes, mechanisms, and microorganisms may be required for the transition of a soil from a conducive to a suppressive state. Knowledge of the shifts in microbial community composition and activities during this transition phase will provide the basis to select potential candidate members of the core microbiome and to set their initial densities to jump start disease suppressiveness.

2.14 Rhizomicrobiome Signalling

2.14.1 Rhizosphere: A pool of Plant–Microbe Signaling: The rhizosphere serves as a hotspot for diverse microbial activity. It is an intricate ecosystem comprising nutrient-rich soil that surrounds the plant roots, which provides a pool for plant–microbe communication. The term "Rhizomicrobiome" is defined as a microbial community that is present in the rhizosphere (Jamil et al., 2022). A variety of microorganisms reside within the rhizosphere, including bacteria, fungi, nematodes, protists, and invertebrates. Most of the microbiome studies within the context of rhizosphere signaling have been focused on the bacteria, fungi and nematodes that make up the major portion of the rhizosphere microbiome. However, viruses that infect bacteria, which are known as phages, have an influence on the dynamics of the rhizosphere microbiome. The most abundant known phages in the soil virome include *Microviridae, Siphoviridae*, and *Podoviridae*. The diversity and amount of identified viruses in the rhizosphere is much less, i.e., 2700 less than bacterial abundance, which amounts to around 10^{10} species per g of soil. The complete biological entity comprising a plant and its associated microbial community, i.e., the symbionts (facultative and obligate) and microbes that have parasitic relationships with the host plant, is referred to as a "holobiont". The cascade of intricate chemical signals develops a communication within the rhizosphere that facilitates various mechanisms in a holobiont, i.e., root–root interactions, biofilm formation, nutrient acquisition, microbiota development, and resistance against pathogens. Different types of interactions between microbes and plants are depicted in Figure 18.

As per the "rhizosphere effect", this region is associated with a lesser microbial diversity but a higher bacterial abundance compared to bulk soil. The recruitment of microbes within the rhizosphere directly depends on the soil properties, soil type, and plant metabolites. Different plant genotypes and physicochemical soil properties develop specific environments for the selection of a promising microbiome (Sasse et al., 2018). Plant-derived metabolites or root exudates play a significant role in the shaping of the rhizomicrobiome by modifying the soil chemistry around the plant roots and serving as substrates for the growth of specific microbiota. The root exudates differ both in quality and quantity depending on the type, nutritional status, and growth stage of the plant. These root exudates, including both primary and secondary metabolites, contain 10–16% of the total plant nitrogen and 11% of the photosynthetically fixed carbon that actively tunes the microbiota in the microbial reservoir that is within the vicinity of the plant roots. Another factor that affects the rhizomicrobiome is the cellular response that is shown either by the microbes or the plants, which results in transformation, catabolism, and resistance to the chemical that is being sensed. The release of root exudates into the rhizosphere by plants selects the desired microbial community by attracting it while deterring harmful communities, which in turn allows plants to be adaptable. This selection process for microbiota is known as “niche colonization”.

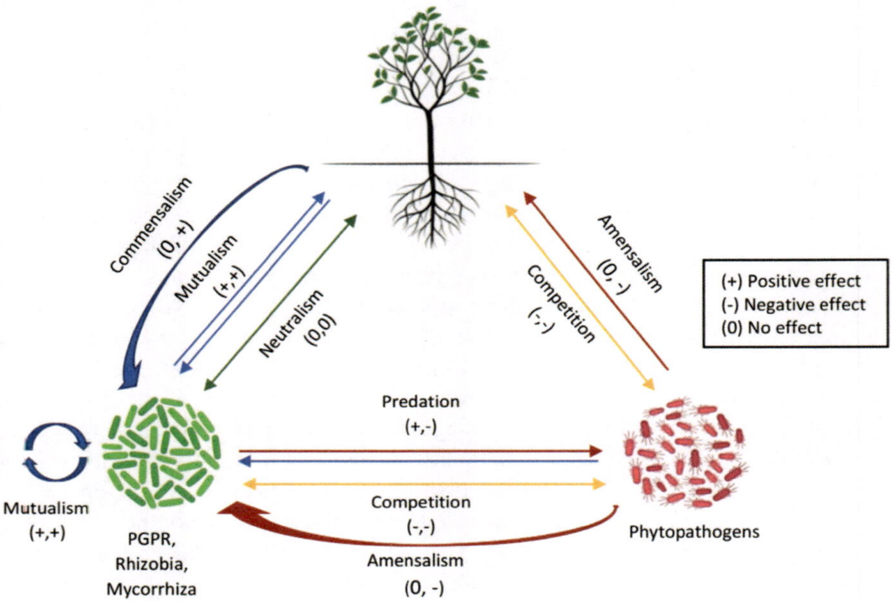

Fig. 18: Schematic representation of interactions between rhizo-microorganisms and plants

Signaling in the rhizosphere can be categorized into two major types based on the direction of the communication, i.e., the inter-and intraspecies microbial signaling and the interkingdom signaling between microbes and plants.

i. Inter- or Intraspecies Signaling among Microorganisms: Microbes in the rhizomicrobiome interact with each other by producing signaling molecules that adjust their gene expression. Inter- or intraspecies communication among microbes occurs via the quorum sensing mechanism, which involves cell density-dependent coordination (Jamil et al., 2022). Quorum sensing (QS) is the cell-to-cell signaling process that takes place through the synthesis, release, and detection of chemical signals. These chemical signals are also known as autoinducers (AIs) and they regulate the gene expression of some bacterial functions upon the recognition of the signal by the recipient, i.e., biofilm formation, adhesion, motility, propagation, virulence, metabolism and symbiotic association.

Quorum sensing signals in bacteria are categorized into two groups: acyl-homoserine lactone (AHLs or AI-1), which is found in Gram-negative bacteria, and autoinducer peptides (AIPs), which are found in Gram-positive bacteria; and autoinducer type 2 (AI-2), which has the traits of both AHLs and AIPs and is found in both Gram-negative and Gram-positive bacteria. N-acyl homoserine lactone (N-AHL) has been majorly reported in various Gram-negative bacteria, including *Burkholderia* sp., *Pseudomonas syringae*, *Pseudomonas putida*, *Pseudomonas chlororaphis*, *Erwinia* sp., and *Serratia* sp. As well as AHLs, a diverse range of signals have been reported in Gram-negative bacteria, i.e., fatty acid methyl esters, 2-alkyl-quinolones, furanone, and γ-butyrolactones. Another type of QS bacterial signals is antibiotics, which have been reported for intra- or interspecies communication. A recent study revealed some antibiotic signals in a bacterial strain that resided in the tobacco rhizosphere, *Lysobacter capsica*. The strain possessed biocontrol potential due to its production of antifungal and antibiotic signals, i.e., cyclic lipodepsipeptides and cyclic and polycyclic tetramate macrolactams. Fungi also produce quorum sensing signals that communicate with bacterial species, e.g., γ-heptalactone, tyrosol, γ-butyrolactone, dodecanol, and farnesol (Clinton and Rumbaugh, 2016). Fungal and bacterial interaction through QS signals creates a competition for existence among the associated microbes in the rhizosphere in terms of infecting the host.

Another important class of signaling molecules is volatile organic compounds (VOCs). Microbial VOCs are synthesized and released for long-distance communication among a microbial community and also in microbe–plant interaction (Bitas et al., 2013). VOCs are low molecular weight lipophilic compounds (100–500 Da) that are produced by various bacterial and fungal species through distinct metabolic pathways that are specific to the species genotype. The VOCs of bacterial origin include alkanes, ketones, alkene, terpenoids, and sulfurs[l]. VOCs play an important role in microbe–microbe communication by serving as antimicrobial QS signaling molecules and influencing microbial activity, i.e., virulence, stress resistance, and biofilm formation. These attributes are exemplified as a low concentration of nitric oxide (NO) influences the biofilm formation in bacterial species such as *P. aeruginosa*, *B. licheniformis*, and *S. marcescens*. Apart from this, VOC signals have been reported to regulate plant growth (root architecture and hormonal signaling) and plant immunity against biotic and abiotic stresses[l]. Additional studies will further unravel other signaling mechanisms that are associated with microbial VOCs. Hence, these intricate signaling mechanisms among rhizospheric microorganisms have a key role in the shaping of the rhizomicrobiome by recruiting specific microbes through inter- or intraspecies communication.

ii. **Interkingdom Signaling:** Microbes and plants interact with each other via interkingdom signaling, which can influence plant growth by either inducing or suppressing gene expression. Interkingdom signaling can be subdivided into two categories based on the stimulus direction: microbe–plant signaling and plant–microbe signaling.

2.14.2 Microbe–Plant Signaling: In microbe–plant signaling, microbes produce and emit signals that induce symbiotic interactions with the plant. Microbial signals that are of rhizosphere origin can trigger definite changes in the plant transcriptome. As in plants, phytohormones are also produced by PGPR (Jamil et al., 2022). These plant growth-stimulating signals can regulate the developmental processes of plants and can also provide plants with resistance to abiotic and biotic stresses. Many PGPRs, including *Azospirillum* spp., *Bacillus amyloliquefaciens*, *B. muralis*, *B. thuringiensis*, *Rhizobium* spp., and *Pseudomonas* spp., have been reported to produce a diverse group of phytohormones, e.g., auxin, abscisic acid, salicylic acid, cytokinin, gibberellin, and strigolactones. A recent study in Nature reported three newly isolated strains of *Phoma* spp. in the rhizosphere of *Pinus tabulaeformis*. These strains secreted stress the resistance substance abscisic acid (ABA) when under drought **stress,** which triggered drought resistance mechanisms in the pine tree and also stimulated its antioxidant activities.

Most of the literature on microbe–plant communication has been focused on beneficial interactions, such as the induction of plant growth and plant defenses against biotic and abiotic stresses. Rhizospheric microbes that have plant-friendly associations include mycorrhiza, rhizobia, plant growth-promoting bacteria, and fungi (PGPR or PGPF) (Venturi and Keel, 2016). Microbial signals are recognized by plants as microbe-associated molecular patterns (MAMPs), i.e., flagellin, chitin, and lipopolysaccharides, via pattern recognition receptors (PRRs) that trigger a local defense through a hormonal signaling network, which in turn produces immune responses.

Quorum sensing signals have also been reported in interkingdom communication. Bacterial QS signals that are perceived by plants elicit various plant responses, e.g., AHLs (N-butanoyl homoserine lactone and N-hexanoyl homoserine lactone), which aid in the development of symbiotic association with plants and enhance root growth by modifying the hormonal levels in the plant. Another bacterial QS molecule, DSF, stimulates innate immunity (the detection of harmful microbes via PRRs) in different plants (Kakkar et al., 2015). As well as causing physiological changes in plants, bacterial QS molecules also provoke the plants into secreting molecules that mimic the QS molecules of pathogenic microbes. For example, the p-coumaric acid that is produced by garlic, the isothiocyanate sulforaphane that is produced by broccoli, the curcumin that is produced by turmeric , and the patulin that is produced by fruits (apple, banana, pear, grape, etc.). Studies on AHL mimicry have shown that the mimicry molecule of bacterial AHL (AHL analog) specifically interferes with QS regulatory pathways and can either stimulate or inhibit the gene expression of the original QS receptor or host, as illustrated in figure 19**.**

This type of interference in microbial QS signaling for attenuating pathogenicity is known as quorum quenching (QQ). A recent study identified a quorum quenching mechanism in a novel bacterial strain, Acinetobacter sp. strain XN-10, which degraded the QS molecules of the AHL family that was acting against the QS-mediated pathogenicity. The strain suppressed the pathogenicity of *Pectobacterium carotovorum* subsp. *carotovorum* (Pcc), thereby protecting tissue maceration in potatoes, cabbage, and carrots. Another study reported a quorum quenching defense mechanism against *Pseudomonas aeruginosa* infection in plants. Upon the interaction of the plant with *P. aeruginosa*, the plant-secreted QQ molecule (rosmarinic acid) mimicked and competed with the QS pathogenic signal (C4-HSL) and stimulated QS-mediated responses that provide disease protection to the plant, i.e., virulence factors and biofilm formation.

Rhizospheric bacteria have been reported to produce antimicrobial signaling compounds. These compounds stimulate systemic resistance in plants against phytopathogens by altering hormonal pathways, e.g., the di-acetyl phloroglucinol (DAPG) that is produced by *Pseudomonas* sp. It is common that microbially produced VOCs can induce plant growth and resistance to biotic stress. Plants perceive and respond to various VOC signals of PGPR or PGPF origins, such as undecanone and heptanol. For example, 2-heptanol and 2-undecanone, which are produced by *B. subtilis* and *B. amyloliquefaciens*, promote the growth of *Arabidopsis thaliana* when cultivated with these PGPR strains.

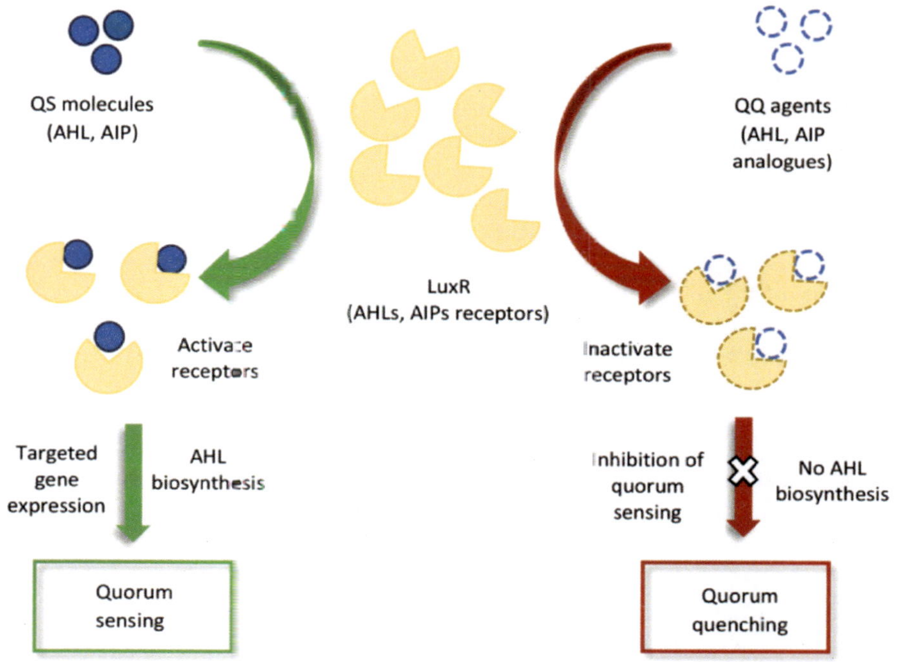

Fig. 19: Schematic representation of quorum sensing (QS) and quorum quenching (QQ) inhibition of signal perception pathways. The binding of QQ agents to the LuxR receptors either inactivates quorum sensing receptors or reduces the quantity of receptors in the QS molecules for targeted gene expression

2.14.3 Plant–Microbe Signaling: Plants serve as residences for microbial communities, which include mutualists, commensals, and parasitic microorganisms. Plants respond to the microbial signals by secreting chemical compounds, which are known as root exudates. These include low molecular weight compounds (organic acids, sugar, aliphatic acids, fatty acids, amino acids, flavonoids, and secondary metabolites) and high molecular weight compounds (proteins and mucilage) (Jamil et al., 2022). The plant–microbe interaction through the secretion of phytochemicals affects the biology of the rhizosphere. The secreted chemicals attract rhizospheric microbes toward the plant roots and develop either pathogenic or symbiotic interactions with them. The most studied plant–microbe interaction is the symbiotic relationship of legumes with nitrogen-fixing bacteria. The cascade of signals that is produced upon plant–microbe interaction forms root nodules. Rhizospheric microbes feed on these nodules and in turn provide the plant with an available form of nitrogen. Plant signals for nodule formation have been extensively studied over the last decade.

A recent work demonstrated pathogenic plant–microbe interaction through a broccoli plant induced defense system, which released root exudates upon interaction with fungal species. The root exudates (isothiocyanates (ITC) and glucosinolates (GLS)) showed antifungal potential against the rhizospheric microbes *Pseudomonas syringae, Sphingomonas suberifaciens*, and *Fusarium oxysporum*. The chemical complexity and specificity of the root exudates depend on the genotype of the plant. The chemical signals of root exudates attract specific microorganisms for interaction. For example, cucumber secretes citric acid, which attracts *Bacillus amyloliquefaciens*, whereas bananas secrete fumaric acid, which attracts *B. subtilis* N11, but both interactions result in biofilm formation. Therefore, specific root exudates recruit specific microbial communities within the rhizosphere. Phytochemicals are also capable of stimulating the QS signaling mechanism in microbes, e.g., the flavonoids that are produced by legumes upregulate the gene expression of AHL genes in rhizobia (Pérez-Montaño et al., 2014).

Plants also release volatile organic compounds (VOCs) from different parts: leaves, flowers, fruits, and roots. These include terpenoids, fatty acids, phenylpropanoids, and amino acids that constitute approx. 1% of the total plant secondary metabolites. VOCs can easily cross plant membranes and be released into the atmosphere or soil. In soil, they attract root colonizing pathogens and inhibit their growth (Rasmann, et al., 2005). Plant secretions, including root exudates, volatiles, and strigolactones, are widespread signaling compounds that bind with bacterial receptor proteins and elicit a response in microbes that regulates gene expression. The signaling molecules that have been identified so far are known to influence the developmental and defense processes of plants. These phytochemical compounds are responsible for shaping or adapting the rhizomicrobiome due to their specificity and complexity. However, many of these chemical compounds and their target functions are yet to be explored. The rhizosphere signaling mechanisms that are involved in the interactions between microorganisms and between the microorganisms and plants are represented in Figure 20.

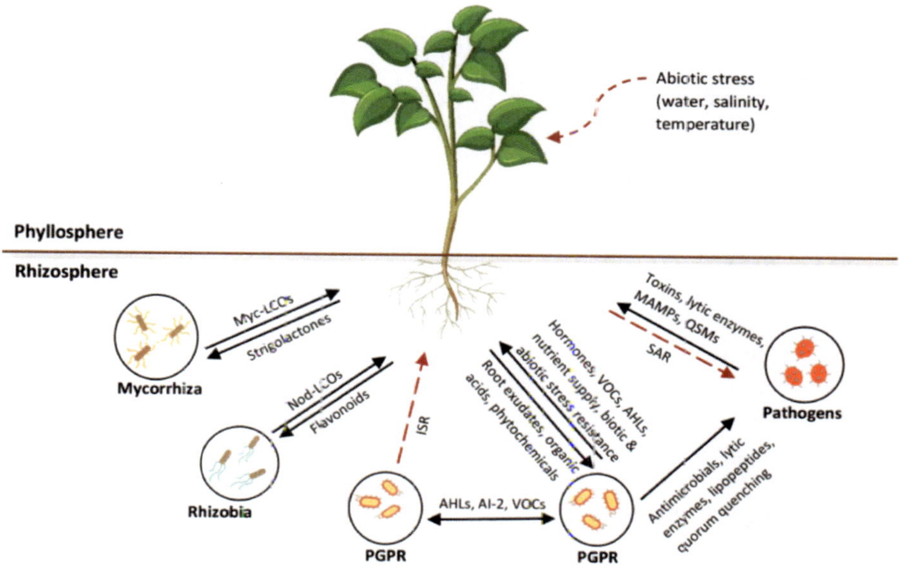

Fig. 20: Overview of rhizosphere communication in intra- or interspecies signaling among microorganisms and interkingdom signaling between microbes and plants. Myc, mycorrhizal; LCOs, lipo-chitooligosaccharides; Nod, nodulation; VOCs, volatile organic compounds; Ais, autoinducers; AHLs, N-acyl homoserine lactone; QSMs, quorum sensing molecules.

2.15 The Rhizosphere Microbial Complex in Plant Health

The rhizobiome has been proven to be important to crop production, and the impact of the rhizobiome on plant health cannot be overemphasized. Being rich in diverse complex microbial interactions, the rhizosphere has become a major force in recent plant growth promotion studies. The upsurge in next-generation sequencing applications with the various "omics" technologies is helping to unearth information relating to rhizosphere impact on plant growth. Explaining the complex interactions between and across microbial species present in the rhizosphere is important to further enhance our understanding of their mechanistic and mutualistic functions. Knowledge from this can be used in rhizosphere biome engineering for improved plant growth and yield in the face of the various biotic and abiotic challenges.

The rhizosphere is a chemically complex environment that harbors a strikingly diverse microbial community. The past few decades have seen a rapid growth in the body of literature on plant–microbe–microbe interactions and plant health (Jane Chepsergon and Lucy N Moleleki, 2023). Thus, the aim of this paper is to review current knowledge on plant–microbe–microbe (specifically bacteria) interactions in the rhizosphere and how these influence rhizosphere microbiomes and impact plant health. This article discusses how the plant recruits beneficial rhizosphere bacteria and how competition between rhizosphere bacteria and mechanisms/weapons employed in bacteria–bacteria competition shapes rhizosphere microbiome and in turn affects plant heath. The discussion mainly focuses on interference competition, characterized by production of specialized metabolites (antibacterial compounds) and exploitative competition where a bacterial strain restricts the competitor's access to nutrients such as through secretion of siderophores that could allude to cooperation. Understanding mechanisms employed in bacteria–bacteria and plant–bacteria interactions could provide insights into how to manipulate microbiomes for improved agricultural outcomes.

Being rich in diverse complex microbial interactions, the rhizosphere has become a major force in recent plant growth promotion studies. The upsurge in next-generation sequencing applications with the various "omics" technologies is helping to unearth information relating to rhizosphere impact on plant growth (Oluwaseyi Samuel Olanrewaju and Olubukola Oluranti Babalola, 2022). Explaining the complex interactions between and across microbial species present in the rhizosphere is important to further enhance our understanding of their mechanistic and mutualistic functions. Knowledge from this can be used in rhizosphere biome engineering for improved plant growth and yield in the face of the various biotic and abiotic challenges. Many microbes establish beneficial interactions with plants including endophytes.

2.15.1 Rhizosphere protists as key determinants of plant health: Plant health is intimately influenced by the rhizosphere microbiome, a complex assembly of organisms that changes markedly across plant growth. However, most rhizosphere microbiome research has focused on fractions of this microbiome, particularly bacteria and fungi. It remains unknown how other microbial components, especially key microbiome predators-protists-are linked to plant health. Wu Xiong et al., (2020) investigated the holistic rhizosphere microbiome including bacteria, microbial eukaryotes (fungi and protists), as well as functional microbial metabolism genes. These communities and functional genes were investigated throughout the growth of tomato plants that either developed disease symptoms or remained healthy under field conditions. It was found that pathogen dynamics across plant growth is best predicted by protists. More specifically, communities of microbial-feeding phagotrophic protists differed between later healthy and diseased plants at plant establishment. The relative abundance of these phagotrophs negatively correlated with pathogen abundance across plant growth, suggesting that predator-prey interactions influence pathogen performance. Furthermore, phagotrophic protists likely shifted bacterial functioning by enhancing pathogen-suppressing secondary metabolite genes involved in mitigating pathogen success.

Despite the fact that the microbiome consists of diverse groups, most research aiming to understand the role of the microbiome in plant health or disease suppression has focused on bacteria and fungi. A whole-microbiome view to decipher the main microbial determinants and their potential interactions that determine plant performance is currently missing. As such, a more complete microbiome analysis is needed to identify the microbial groups and potential interactions that help predicting plant health. In particular, protists that steer the taxonomic and functional composition of the rhizosphere microbiome through trophic predator-prey interactions have so far rarely been included in microbiome analyses linked to plant performance.

Protists, especially microbial-feeding phagotrophs, have various functions within the rhizosphere. For instance, some of these phagotrophs can directly prey on plant pathogens. Studies using model protists have shown that protists control microbiome diversity and structure leading to plant growth promotion. These changes are at least partly explained by the fact that protists feed selectively on microbial prey taxa, which differs between protistan species. Through this selective predation, protists can, for instance, increase those bacteria that produce pathogen-suppressive secondary metabolites (Mazzola et al., 2009). Yet, all these studies investigating potential links of protists with plant performance were carried out under artificial laboratory or greenhouse conditions focusing on one or few protistan species. As such, we have yet to identify the links between a complex diversity of protists, the microbiome and plant performance, especially in agricultural systems under field conditions (Wu Xiong et al., 2020).

Protists and their interactions with other microorganisms are also subject to change throughout plant growth (Hassani et al., 2018). Yet, the composition of the microbiome is often investigated only once during plant growth, usually at the time of plant maturity or after disease has already developed. Such approaches make it difficult to disentangle causality between plant health and potentially underlying characteristics in microbial communities, especially for diseased plants that host high amounts of pathogens. Recently, it was shown that bacterial communities at plant establishment can predict plant health at maturity. The findings bridge evidence from laboratory or greenhouse studies focusing on single protist model species to the community level in agricultural systems under field conditions, showing that protists affect bacterial communities and their functioning through predation, leading to changes in plant performance. Compared with diseased plants, healthy plants hosted higher relative abundances of phagotrophic protists, potentially plantbeneficial bacteria, and secondary metabolite genes likely implicated in pathogen suppression 5 weeks after plant establishment—the time point when pathogen symptoms first developed in diseased plants. Moreover, phagotrophic protists negatively correlated with bacteria that positively linked with a pathogen-suppressing gene coding for non-ribosomal peptides across plant growth. This finding might also contribute to pathogen suppression. However, the interaction between plants and the rhizosphere microbiome is a complex and dynamic process. Future experiments are needed to further examine how plants affect bacterial, fungal, and protistan communities and their interactions and how those changes in the soil microbiome in turn affect plant performance. Together, we propose that predation-induced shifts in microbiome composition and functioning are likely involved in controlling pathogen development and therefore plant health.

2.15.2 Rhizosphere Bacteriophages in plant health: The composition and diversity of viruses associated with the soil rhizosphere are largely unexplored. Bacteriophages play an important role in the ecology and evolution of the host-associated microbiome and are directly linked to host fitness (Pratama et al., 2020). The rhizosphere virome might alleviate plant responses to abiotic stress, thereby influencing plant resistance and resilience to climate change scenarios.

Phages play a key role in nutrient turnover and lysis of bacteria in terrestrial ecosystems. While phages are abundant in soils, their effects on plant pathogens and rhizosphere bacterial communities are poorly understood. Pratama et al., (2020) used metagenomics and direct experiments to causally test if differences in rhizosphere phage communities could explain variation in soil suppressiveness and bacterial wilt plant disease outcomes by plant-pathogenic *Ralstonia solanacearum* bacterium. Specifically, two hypotheses were tested viz., that healthy plants are associated with stronger top-down pathogen control by *R. solanacearum*-specific phages (i.e. 'primary phages') and that 'secondary phages' that target pathogen-inhibiting bacteria play a stronger role in diseased plant rhizosphere microbiomes by indirectly 'helping' the pathogen.

Using a repeated sampling of tomato rhizosphere soil in the field, it was shown that healthy plants are associated with distinct phage communities that contain relatively higher abundances of *R. solanacearum*-specific phages that exert strong top-down pathogen density control. Moreover, 'secondary phages' that targeted pathogen-inhibiting bacteria were more abundant in the diseased plant microbiomes (Yang et al., 2023). The roles of *R. solanacearum*-specific and 'secondary phages' were directly validated in separate greenhouse experiments where we causally show that phages can reduce soil suppressiveness, both directly and indirectly, via top-down control of pathogen densities and by alleviating interference competition between pathogen-inhibiting bacteria and the pathogen. Phage communities had an important indirect role in bacterial wilt disease progression by affecting the strength of competitive interactions between pathogen-inhibiting bacteria and *R. solanacearum*. Several inhibitory bacteria have been linked to a reduction in bacterial wilt disease incidence and it was previously shown that *Bacillus* and *Pseudomonas* bacteria isolated from the same field showed a higher *R. solanacearum* inhibition in vitro when they originated from healthy versus diseased plant rhizosphere microbiomes (Wei et al.,2019). In line with this, also the isolated *B. tequilensis* bacterium showed a clear *R. solanacearum* growth inhibition both in vitro and in planta, in addition to two other inhibitory bacteria, *S. maltophilia* and *E. ludwigii.* While the mechanism of inhibition remains unclear, we found that healthy plant microbiomes contained a relatively higher abundance of bacterial genes associated with secondary metabolism and antibiosis. In combination with inhibition assay data, pathogen growth suppression via secretion of antimicrobial compounds is hence the most plausible explanation for the results, even though the potential importance of other factors, such as iron-scavenging siderophores or resource competition cannot be excluded.

Findings demonstrated that soil suppressiveness, which is most often attributed to bacteria, could be driven by rhizosphere phage communities that regulate *R. solanacearum* densities and strength of interference competition with pathogen-suppressing bacteria. Rhizosphere phage communities are hence likely to be important in determining bacterial wilt disease outcomes and soil suppressiveness in agricultural fields.

2.15.3 Root-Microbe Communication: Root-microbe communication is another important process that characterizes the underground zone. Some compounds identified in root exudates that have been shown to play an important role in rootmicrobe interactions include flavonoids present in the root exudates of legumes that activate *Rhizobium meliloti* genes responsible for the nodulation process (Travis S. Walker et. al, 2003). Although the studies are not yet conclusive, these compounds may also be responsible for vesicular-arbuscular mycorrhiza colonization. In contrast, survival of the delicate and physically unprotected root cells under continual attack by pathogenic microorganisms depends on a continuous "underground chemical warfare" mediated by secretion of phytoalexins, defense proteins, and other as yet unknown chemicals. The unexplored chemodiversity of root exudates is an obvious place to search for novel biologically active compounds, including antimicrobials. For instance, rosmarinic acid (RA) was identified in the root exudates of hairy root cultures of sweet basil (*Ocimum basilicum*) elicited by fungal cell wall extracts from *Phytophthora cinnamoni*. Basil roots were also induced to exude RA by fungal in situ challenge with *Pythium ultimum*, and RA demonstrated potent antimicrobial activity against an array of soil-borne microorganisms including *Pseudomonas aeruginosa*. Similar studies with *Lithospermum erythrorhizon* hairy roots reported cell-specific production of pigmented naphthoquinones upon elicitation, and other biological activity against soil-borne bacteria and fungi. Given the observed antimicrobial activity of RA and naphthoquinones, these findings strongly suggest the importance of root exudates in defending the rhizosphere against pathogenic microorganisms. Moreover, the aforementioned studies complement earlier research that mainly focused on the regulation and production of these compounds by providing valuable insights into the biological importance of RA and shikonin.

Both Gram-negative and -positive bacteria, including important plant pathogenic bacteria such as *Erwinia* spp., *Pseudomonas* spp., and *Agrobacterium* spp., possess quorum-sensing systems that control the expression of several genes required for pathogenicity. Quorum sensing is a form of cell-cell communication between bacteria mediated by small diffusible signaling molecules (autoinducers); these are generally acylated homo-Ser lactones (AHLs) for Gram-negative bacteria and peptide-signaling molecules for Gram-positive bacteria. Upon reaching a threshold concentration at high population densities, an auto-inducer then activates transcriptional activator proteins that induce specific genes. Thus, intercellular signals enable a bacterial population to control the expression of genes in response to cell density AHL-producing transgenic tobacco plants restored pathogenicity to an avirulent AHL deficient *Erwinia carotovora* mutant. Root exudates from pea seedlings were found to contain several bioactive components that mimicked AHL signals in well-characterized bacterial reporter strains, stimulating AHL-regulated behaviors in some strains while inhibiting such behaviors in others.

2.16 Root Exudates

Root exudates are in the form of carbohydrates, including sugars, and proteins. Amazingly, their presence wakes up, attracts, and grows specific beneficial bacteria and fungi living in the soil that subsist on these exudates and the cellular material sloughed off as the plant's root tips grow. All this secretion of exudates and sloughing-off of cells takes place in the rhizosphere, a zone immediately around the roots, extending out about a tenth of an inch or a couple of millimeters (1 millimeter = 1/25 inch). The rhizosphere, which can look like a jelly or jam under the electron microscope, contains a constantly changing mix of soil organisms, including bacteria, fungi, nematodes, protozoa, and even larger organisms. All this "life" competes for the exudates in the rhizosphere, or its water or mineral content. At the bottom of the soil food web are bacteria and fungi, which are attracted to and consume plant root exudates. In turn, they attract and are eaten by bigger microbes, specifically nematodes and protozoa (amoebae, paramecia, flagellates, and ciliates). The eat bacteria and fungi primarily for carbon to fuel their metabolic functions. Anything they don't need is excreted as wastes, which plant roots are readily able to absorb as nutrients. The reason for intensive microbial activity is due to the phenomena called 'Root exudation'. During unfavourable conditions, the organic and inorganic compounds diffuse out of the root, in the form of root exudates. Therefore, root exudates form a network between the plants and microorganisms. Root exudates basically include the root secretions and root diffusates.

2.16.1 Classification and role of Root exudates: Based on chemical nature, root exudates are of three types:

i. **Organic compounds**: These are the compounds that consist of carbon molecule as a backbone, which can be covalently linked to the other molecules like hydrogen, nitrogen etc.

   Ex: Proteins, organic acids, vitamins, amino acids, sugars, flavonoids, nucleotides, enzymes etc.

ii. **Inorganic compounds**: These are the compounds, which consist of carbon or hydrogen, but not together. Ex: Water, anions and gases like carbon, oxygen, nitrogen etc.

iii. **Miscellaneous compounds**: Other than organic and inorganic compounds, some other substances are also released by the plant, which can impose a negative effect and called as miscellaneous compounds.

   Ex: Auxins, glycosides, saponins, hydrocyanic acids etc.

Role of Root Exudates

- It provides nutrient supply to the microbes.
- Root exudates protect the plant from desiccation.
- It provides moisture to the soil, which favours the growth of microorganisms.
- By releasing some defence proteins and antimicrobial chemicals, it also protects the root against biotic stress, i.e. pathogens.
- It induces systemic resistance against abiotic stress like temperature, salt concentration, pH.
- By mucilage secretion, it helps in soil aggregation.
- It can change the chemical and physical properties of the soil through the factors like Osmotic pressure, Ionic balance and Redox potential.

Factors affecting Root Exudation

- High temperature and light
- Wilting of plant releases a high amount of amino acids
- A release of secondary metabolites by certain microorganisms
- Rhizospheric microflora that can affect the permeability and metabolism of root

Rhizosphere Microorganisms: Rhizosphere shows huge microbial diversity, which includes (Bacteria, fungi, actinomycetes, algae, protozoa and nematodes are common, in addition to the viruses, archaea, arthropods etc.,). Microorganisms can be classified into three types (beneficial, deleterious and neutral) based on their effect.

i. **Beneficial microorganisms**: These do not harm the root system and includes mycorrhizal fungi, protozoa, N_2 fixing bacteria etc.

   Role: Fixes nitrogen, Helps in solubilizing inorganic phosphate, iron etc., Helps in root nodulation, Colonize the root system, Promotes plant growth.

ii. **Deleterious microorganisms**: These harm the root system of soil and includes the pathogenic fungi, bacteria, nematodes etc.

   Role: Cause diseases in plant and invades nutrient.

iii. **Neutral microorganisms**: These show a neutral effect and includes actinomycetes, algae etc.

   Role: Helps in soil aggregation, Shows antagonistic property against other organisms.

2.16.2 Factors affecting Rhizosphere Microorganisms

- **Type of soil**: A microbial population is very high in sandy soil and least in humus soil.
- **The moisture content of soil**: Low moisture favours the growth of organisms.
- **Soil amendments**: It has no quantitative and qualitative effect on microbial growth and activity.
- **The pH of soil**: It is inversely proportional to the microbial growth, i.e. a decrease in pH results in an increase of the microbial population.
- **Plant species**: Microorganisms vary with different plant species respective to their rooting habits, composition etc. For instance, leguminous plants show more rhizospheric effect.
- **The proximity of root with hair**: Rhizosphere effect is more when there is a less distance between the soil and root. As the R:S effect increases, the number of organisms will also increase.
- **Age of plant**: It is an essential factor, which decides the degree and magnitude of rhizosphere effect. R:S effect is directly proportional to the age of the plant, i.e. it increases with the age of the plant.

- **Root excretion**: This process strongly influences the microbial activity by providing a sufficient amount of organic and inorganic matter to the rhizospheric microorganisms.

Root exudates include both secretions (including mucilage) which are actively released from the root and diffusates which are passively released due to osmotic differences between soil solution and the cell, or lysates from autolysis of epidermal and cortical cells. The organic compounds released through these processes can be further divided into high and low molecular weight (HMW and LMW, respectively). By weight, the HMW compounds which are those complex molecules that are not easily used by microorganisms (e.g. mucilage, cellulose) make up the majority of C released from the root; however, the LMW compounds are more diverse and thus have a wider array of known or potential functions. The list of specific LMW compounds released from roots is very long, but can generally be categorized into organic acids, amino acids, proteins, sugar, phenolics and other secondary metabolites which are generally more easily used by microorganisms. The cocktail of chemicals released is influenced by plant species, edaphic and climactic conditions which together shape and are shaped by the microbial community within the rhizosphere. There is still very little known about the role that a majority of the LMW compounds play in influencing rhizosphere processes.

Root exudate is one of the ways for plant communication to the neighboring plant and adjoining of microorganisms present in the rhizosphere of the root. The chemicals ingredients of the root exudates are specific to a particular plant species and also depend on the nearby biotic and abiotic environment (Keshav Prasad Shukla et al., 2011). The chemical ingredient exuded by plant roots include amino acids, sugars, organic acids, vitamins, nucleotides, various other secondary metabolites and many other high molecular weight substances as primarily mucilage and some unidentified substances. Through the exudation of a wide variety of compounds, roots may regulate the soil microbial community in their immediate vicinity, cope with herbivores, encourage beneficial symbioses, change the chemical and physical properties of the soil and inhibit the growth of competing plant species. Root exudates mediate various positive and negative interactions like plant-plant and plant-microbe interactions. The present review has been undertaken to examine the possible role of root exudates on the removal of the polluted matter and nourishing the neighboring microorganisms present in the rhizosphere of the root.

A survey of the literature exposes an extensive range of compounds exuding from intact and healthy roots; these include sugars, amino acids, peptides, enzymes, vitamins, organic acids, nucleotides, fungal stimulators, inhibitors and attractants, eelworm hatching and attracting factors and many miscellaneous compounds. Organic acids, sugars, amino acids, lipids, coumarins, flavonoids, proteins, enzymes, aliphatics and aromatics are examples of the primary substances found within rhizosphere of the root. Among them, the organic acids have received considerable attention due to their role in providing substrates for microbial metabolism and for serving as intermediates for biogeochemical reactions in soil. The field of rhizosphere biology has find out the relative importance of root exudates in mediating interactions with neighboring plants and microbes. Root exudation is an element of the rhizodeposition process, which is a major source of soil organic carbon released by plant roots. Upon assembling a challenge, roots typically respond by secreting certain small molecules and proteins.

Root secretions may play both positive and negative communication in the rhizosphere. The positive communication includes symbiotic associations with beneficial microbes, such as mycorrhizae, rhizobia and plant growth promoting rhizobacteria (PGPR). Negative interactions include association with parasitic plants, pathogenic microbes and invertebrate herbivores. The rhizospheric bacteria are responsible for the elimination of the contaminants while the roots are responsible for providing nutrients (root exudates) used by the microorganisms to proliferate. Selected microbes can degrade most environmental pollutants. The process stops when the microbe is starved of food. In order to determine that such microbes can have access to the best food source available in soil, workers have described an enrichment method for the isolation of microbes, which combine the properties of degradation of a selected pollutant and excellent root colonization. The process 'rhizoremediation' was illustrated to emphasize the roles of the root exudates and the rhizosphere competent microbes.

2.16.3 Chemical Nature and Composition of Root Exudates: Root exudates refer to a suite of substances in the rhizosphere that are secreted by the roots of living plants and microbially modified products of these substances. They consist of low-molecular-weight organic compounds that are freely and passively released root-cell material and mucilage associated with roots (Koo et al., 2005). Root exudates are a complex mixture of soluble organic substances, which may contain sugars, amino acids, organic acids, enzymes, and other substances (Table 5). The rhizosphere also contains cell lysates, which are released when root cortex cells are ruptured by mechanical damage, microbial grazing, fungal infection, or the emergence of lateral roots that break through the root epidermis. In addition to these substances, rhizodeposition includes the secretion of enzymes, such as acid phosphatase secreted in response to P deficiency.

Although the identified components differ in molecular weight, amount released, and biochemical role, the dominant exudate components have consistently been shown to be organic acids, sugars, and amino acids. Typically, the total organic acid concentration in roots is approximately 10–20 mmol l^{-1} (i.e., 1–4% of total dry weight) for corn (*Zea mays* L.), with the other main organic solutes, namely amino acids (10–20 mmol l^{-1}) and sugars (90 mmol l^{-1}), present in root cells. Higher concentrations are generally found in rhizosphere soil than in bulk soil. The carboxylic acids that play a significant role in complexing metal ions in soils are not utilized as rapidly as carbohydrates by the soil microorganisms. It appears that microorganisms preferentially utilize the carbohydrates in root exudates. Consequently, organic and amino acids may be the prevalent chemical substances in root exudates.

Table 5: Composition of root exudates

Compositions	**Substances identified**
Organic acid	Acetic, butyric, citric, glutaric, lactic, maleic, malic, malonic, oxalic, propionic, pyruvic, succinic, tartaric, valeric
Amino acid and amide	α-Alanine, β-alanine, arginine, asparagine, aspartic acid, cystine / cysteine, glutamine, glycine, histidine, lysine, methionine, phenylalanine, proline, serine /homoserine
Enzyme	Amylase, invertase, phosphatase, protease, polygalacturonase
Growth factor	*p*-Amino benzoic acid, auxins, biotin, choline, inositol, *n*-methyl nicotinic acid, niacin, pantothenate, pyridoxine, thiamine
Phenolic acid and coumarin	Caffeic acid, cinnamic acid, coumarin, ferulic acid, salycilic acid, syringic acid, vanillic acid
Sugar	Arabinose, fructose, fucose, galactose, glucose, maltose, oligosaccharide, raffinose, rhamnose, ribose, sucrose, xylose
Others	Nucleotide, flavonone, fatty acids, proteins, sterols, lipids, aliphatics, aromatics, carbohydrates

Typical organic acids found in root exudates are malic, malonic, acetic, citric, fumaric, succinic, lactic, tartaric, oxalic, and others. Organic acids of root exudates seem to have a major role in solubilization of mineral nutrients. Malic and citric acids are major root exudate components for some plant species under aseptic conditions. For example, in corn grown under sterile conditions, exudation of malic and citric acids increase significantly under nutrient deficiency. Compared with the organic acids, amino acids appear to play a less significant role in complexing or mobilizing plant nutrients. The proteinaceous amino acids are not readily detected in root exudate solutions. The nonproteinaceous amino acids (phytosiderophores) in the rhizosphere have been shown to enhance the mobility of plant micronutrients in soils.

In general, the component of carbohydrates compounds extracted by the root is arabinose, glucose, galactose, fructose, sucrose, pentose, rhamnose, raffinose, ribose, xylose and mannitol. Current evidence suggests that sugars such as glucose dominate root exudates (Keshav Prasad Shukla et al., 2011). All 20 proteinogenic amino acids, l-hydroxyproline, homoserine, mugineic acid, and aminobutyric acid are included in amino acids exudates. Acetic acid, succinic acid, laspartic acid, malic acid, l-glutamic acid, salicylic acid, shikimic acid, isocitric acid, chorismic acid, sinapic acid, caffeic acid, p-hydroxybenzoic acid, gallic acid, tartaric acid, ferulic acid, protocatacheuic acid, p-coumaric acid, mugineic acid, oxalic acid, citric acid, and piscidic acid are categorized in organic acids and phenolic compounds. Some of these compounds, especially the phenolics, influence the growth and development of surrounding plants and soil microorganisms. Naringenin, kaempferol, quercitin, myricetin, naringin, rutin, genistein, strigolactone and their substitutes with sugars are the flavonols.

Catechol, benzoic acid, nicotinic acid, phloroglucinol, cinnamic acid, gallic acid, ferulic acid, syringic acid are derivatives of lignin (Dayakar et al., 2009), sinapoyl aldehyde, chlorogenic acid, coumaric acid, vanillin, sinapyl alcohol, quinic acid, pyroglutamic acid. Coumarins (Umbelliferone), aurones (benzyl aurones synapates, sinapoyl choline), glucosinolates (cyclobrassinone, desuphoguconapin, desulphoprogoitrin, desulphonapoleiferin, desulphoglucoalyssin), anthocyanins (cyanidin, delphinidin, pelargonidin and their substitutes with sugar molecules), indole compounds (indole-3-acetic acid, brassitin, sinalexin, brassilexin, methyl indole carboxylate, camalexin glucoside), fatty acids (linoleic acid, oleic acid, palmitic acid, stearic acid), sterols (campestrol, sitosterol, stigmasterol), allomones (jugulone, sorgoleone, 5,7,4′-trihydroxy-3′, 5′-dimethoxyflavone, DIMBOA, DIBOA) and proteins and enzymes (proteins, lectins, proteases, acid phosphatases, peroxidases, hydrolases, lipase) are the compounds found on root exudates, which is isolated from various plant roots. The main carbohydrates secreted from rice identified were glucose, arabinose, mannose, galactose, and glucuronic acid and rice exudates may induce a higher chemotactic response for endophytic bacteria (*Corynebacterium flavescens* and *Bacillus pumilus*) than for other bacterial strains (plant growth promoting bacteria, *Azospirillum brasilense* and *Bacillus* sp.) present in the rice rhizosphere. Endophytic bacteria appear to be ever-present in most plant species, inducing beneficial effects, as incidental from their ability to promote plant growth to confer resistance against plant pathogens.

Factors affecting Exudation: The exudation of organic compounds by roots are influenced by either biotic (for example, soil microbial uptake) or abiotic processes. In some instances, our knowledge is sufficient to explain why exudation is affected by the root environment, but often our ignorance of the physiological processes involved in exudation precludes a correct explanation (Keshav Prasad Shukla et al., 2011). Some of the factors influencing exudation include plant species, root age, microbes, temperature, light etc.

The roots exude water and compounds broadly known as exudates. Root exudates include amino acids, organic acids, carbohydrates, sugars, vitamins, mucilage and proteins. The exudates act as messengers that stimulate biological and physical interactions between roots and soil organisms (Rebecca Lines-Kelly, 2005). They modify the biochemical and physical properties of the rhizosphere and contribute to root growth and plant survival. However, the fate of the exudates in the rhizosphere and the nature of their reactions in the soil remain poorly understood. The exudates have several functions, viz., defend the rhizosphere and root against pathogenic microorganisms ; attract and repel particular microbe species and populations ; keep the soil around the roots moist ; obtain nutrients ; change the chemical properties of the soil around the roots ; stabilise soil aggregates around the roots ; inhibit the growth of competing plant species.

2.16.4 Root exudates and Mineral nutrition: Plants respond to nutrient deficiency by altering root morphology, recruiting the help of microorganisms and changing the chemical environment of the rhizosphere. Components in root exudates assist plants in accessing nutrients by acidifying or changing the redox conditions within the rhizosphere or directly chelating with the nutrient. Exudates can liberate nutrients via dissolution of insoluble mineral phases or desorption from clay minerals or organic matter where they are released into soil solution and can then be taken up by the plant (David H Mcnear, 2013). The nutrients most limiting to plant growth are nitrogen and phosphorus. Even thought 78% of the Earth's atmosphere is composed of nitrogen (N_2 gas), it is in a form that is only utilizable by nitrogen-fixing organisms. As such, inorganic forms of N (NO_3- ,NH4 +) that can be used by plants are added to soils. The availability of nitrogen in most soils is low because of the leaching losses of soluble nitrate (NO_3-) with infiltrating rainwater, fixation of ammonium (NH_4 +) in clays and soil organic matter and bacterial denitrification. Plants respond differently depending on the form of nitrogen in the soil. Ammonium has a positive charge, and thus the plant expels one proton (H +) for every NH_4 + taken up resulting in a reduction in rhizosphere pH. When supplied with NO_3- , the opposite can occur where the plant releases bicarbonate (HCO_3-) which increases rhizosphere pH. These changes in pH can influence the availability of other plant essential micronutrients (e.g., Zn, Ca, Mg). Phosphate (PO_4 3-), the form of P used by plants, is highly insoluble in soils, binding strongly to Ca, Al and Fe oxide, and soil organic matter rendering much of the P unavailable to plants. Under P deficiency, plants have evolved special mechanisms to obtain PO_4 3- which depend on plant type (dicot vs. monocot), species and genotype. Plant roots can exude organic acids such as malic and citric acids into the rhizosphere which effectively reduced rhizosphere pH and solubilize P bound in soil minerals. Pigeon pea uses another mechanism by releasing piscidic acid in response to P deficiency which chelates Fe in $FePO_4$, releasing the PO_4 3-. Plants also liberate PO_4 3- from organic sources by releasing enzymes such as acid phosphatase.

Iron deficiency elicits a response from plants which generally differs depending on whether the plant is a dicot or monocot. Dicots respond to Fe deficiency by releasing protons into the soil environment and increasing the reducing capacity of the rhizodermal cells. In monocots, Fe deficiency triggers the release of phytosiderophores such as mugienic acid which is a non-proteinogenic amino acid with extremely high affinity for Fe. The phytosiderophore chelates strongly with Fe and is then brought back to the root via diffusion where plasma membrane transporters specific to the chelated Fe shuttle it into the cells. Plants using the former method of Fe acquisition are called Strategy I and those using the latter Strategy II.

2.16.5 Root exudation and Plant nutrition

i. Root exudation of primary metabolites is driven by facilitated diffusion but plants and microbes can control the root exudation process by modifying concentration gradients depending on their nutritional status: Root exudation occurs mainly at the root meristematic apex, where fluxes of metabolites into the soil are the highest, due to the lack of root endodermis differentiation. Concentration gradients are fundamental in driving root exudation as most of the primary metabolites efflux from root cells through membrane transporters that facilitate diffusion (Canarini et al., 2019). Soil microbes actively consume root exudates and keep concentrations of the respective substance low in the soil environment and at the same time can stimulate specific root exudation patterns. Plants, on the other side, can control concentration gradients by different mechanisms: they can adjust the supply rate through phloem loading and transport (source activity) or modify phloem unloading and sink activity, including the inhibition of root meristematic activity or up/downregulation of efflux transporters. The functioning of efflux carriers can also be controlled by rapid post-translational mechanisms (e.g., phosphorylation), which suggest that a fast control of root exudation is possible. The exudation flux ultimately will reflect the nutritional status and strategy of the plant or of the microorganisms (soil environment). While plant source-sink dynamics are increasingly recognized as being fundamental to understanding C allocation and nutrient uptake, idea that plant-associated micro-organisms affect plant carbon allocation by affecting concentration gradients that drive root exudation was supported. Hence, a novel conceptual framework was introduced for the control of root exudation that considers source-sink dynamics of plants and micro-organisms together as an integrated system.

ii. Changes in the concentration of primary metabolites at the root tip are sensed by the plant and signals are translated to modify root system architecture: Multiple examples of how important root metabolite concentrations can be increased or decreased inside the root tip were presented. Specifically, regulation of efflux transporter expression in roots, root exudate re-uptake, phloem loading, and modulation of meristematic activity are major mechanisms that integrate the root exudation process with the soil environment surrounding the root tips. Modulating this flux can change the temporal concentrations of solutes inside the root tip, which elicits responses in root system architecture to modulate root growth to environmental cues. Indeed, the three chemical groups mentioned above (sugars, amino acids and organic acids), comprising a vast proportion of the root exudates, can all be sensed in different ways and induce a range of responses in the root system architecture from lateral root formation to root meristem exhaustion, depending on the plant and soil nutritional status. Soil microbes can create specific patterns of metabolite concentrations around the root tip by preferentially consuming some metabolites and by stimulating plant exudation. Therefore, soil microbes can exert a strong control on plant nutrient sensing.

Figure 21 illustrates a summary of the concepts proposed for the role of root exudates in plant-microbe interactions and consequences for ecosystems (Canarini et al., 2019). Plant nutrient strategy determines different exudation patterns, while the soil physical and nutrient composition determines the microbial community and its nutrient demand. Root exudation is a flux with a bi-directional influence between plant and soil microbes. Indeed, while plants can regulate root exudation, soil micro-organism can enhance exudation rates and release molecules that affect root exudation patterns. Ultimately, root exudation has a significant effect on SOM decomposition and greenhouse gases (GHGs) release from soil. Also, plant-soil feedbacks can significantly affect plant performance and plant community composition.

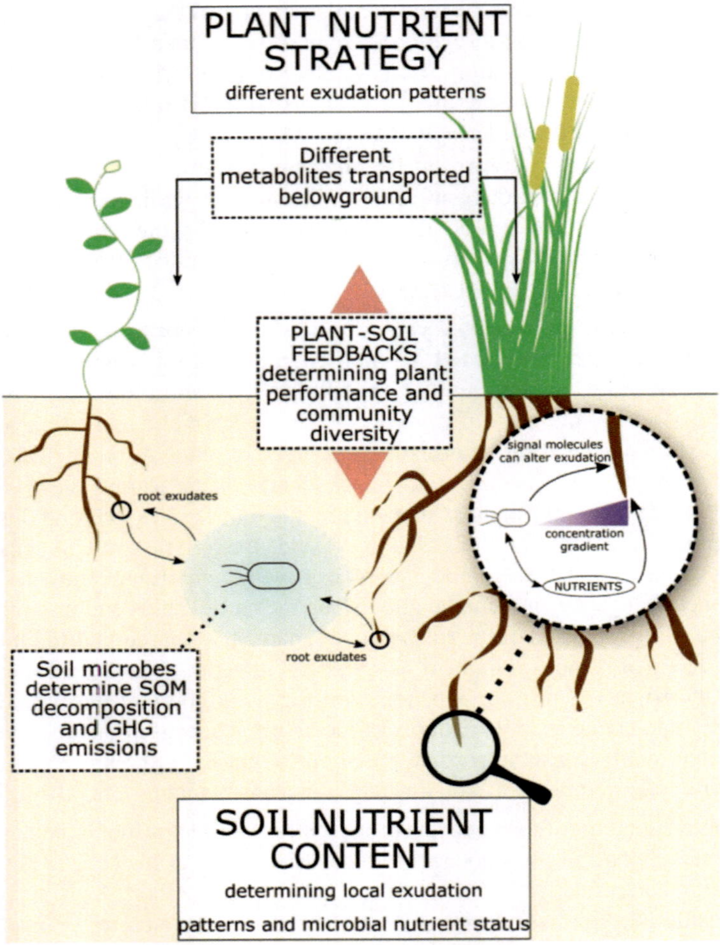

Fig. 21: Root exudates link soil nutrients to plant nutrient strategy

2.17 Dynamic Root Exudate Chemistry and Microbial substrate

The rhizosphere represents a critical hotspot for biogeochemical transformation that underlies the process of soil formation, carbon cycling and the ultimate productivity of Earth's terrestrial ecosystems. Within the rhizosphere, complex and dynamic interactions between plants and networks of organisms, particularly microorganisms, have been shaped by over 450million years of co-evolution (Kateryna Zhalnina et al.,2018). With such evolutionary optimization, it is not surprising that somewhat consistent patterns have emerged. The 'rhizosphere effect'1 describes the enrichment of microbial cells and activity near growing roots and has been shown to involve the 'selection' of phylogenetically related microorganisms, with plants of different species, geographical locations, climates and land management showing distinct rhizosphere microbiomes. Across systems, some consistent trends have also been observed, for example, an increase in the number of bacteria of the Alphaproteobacteria subphylum was reported in rhizosphere soil of various plants. Conversely, the abundance of Actinobacteria has been shown to decrease during late developmental stages. Several specific traits have been identified and associated with rhizosphere enrichment, such as the presence of secretory systems, adhesion, phage defence, iron mobilization and sugar transporty, the rhizosphere microbial community structure is the result of a complex series of interactions and feedbacks between plant roots, microorganisms and the physical and chemical environment of the soil. However, despite the growing number of studies demonstrating that plant development influences the composition of the soil microbiome and its functional capacity, relatively few studies have sought to understand the molecular and chemical basis of the role of dynamic plant exudation in the establishment of rhizosphere microbiota.

Plants exude a variable but substantial amount (11–40%) of photosynthesis-derived carbon, creating a diverse chemical milieu. Exuded compounds include sugars, amino acids, organic acids, fatty acids and secondary metabolites. The composition of root exudates is not a uniform or static property and varies depending on plant species, developmental stage, root traits, environmental conditions, nutrition and soil type, among other factors. Released compounds have been shown to attract beneficial microorganisms and influence the assembly of rhizosphere microbiomes that enhance the capacity of plants to adapt to their environment (Bulgarelli et al., 2013). It is not clear how the interaction between root exudate chemistry and microbial substrate preferences combines to influence rhizosphere community assembly and succession. Are there relationships between the dynamics of exudate composition and the growth of specific soil microorganisms? If so, can those relationships be predicted and generalized?

Interactions between roots, microorganisms and the soil matrix represent a suite of extremely complex processes. Acknowledging this complexity, the work presented here represents a defined suite of experiments designed to evaluate the potential for metabolic plant–microbial linkages in the rhizosphere of an annual grass in the absence of soil matrix effects. Kateryna Zhalnina et al., (2018) showed that programmed developmental processes in plants result in dynamic patterns of the chemical composition of root exudates. This chemical succession in the rhizosphere interacts with microbial metabolite substrate preferences that can be predicted from genome sequences. It was proposed that the combination of these plant exudation traits and microbial substrate uptake traits contributes to a metabolic synchronization that underlies microbial community assembly patterns observed in the rhizosphere.

Interactions between plant roots and soil microorganisms are critical for plant fitness in natural environments. Given this co-evolution and the pivotal importance of plant–microbial interactions, it has been hypothesized, and a growing body of literature suggests, that plants may regulate the composition of their rhizosphere to promote the growth of microorganisms that improve plant fitness in a given ecosystem. Using a combination of comparative genomics and exometabolomics, Kateryna Zhalnina et al., (2018) showed that pre-programmed developmental processes in plants (Avena barbata) result in consistent patterns in the chemical composition of root exudates. This chemical succession in the rhizosphere interacts with microbial metabolite substrate preferences that are predictable from genome sequences. Specifically, they observed a preference by rhizosphere bacteria for consumption of aromatic organic acids exuded by plants (nicotinic, shikimic, salicylic, cinnamic and indole-3-acetic). The combination of these plant exudation traits and microbial substrate uptake traits interact to yield the patterns of microbial community assembly observed in the rhizosphere of an annual grass. This discovery provides a mechanistic underpinning for the process of rhizosphere microbial community assembly and provides an attractive direction for the manipulation of the rhizosphere microbiome for beneficial outcomes.

Some functions that root exuded compounds conduct in the soil

i. **Sugars:** Sugars serve as some of the most important carbon sources available for soil microorganisms. Many forms of sugars provide accessible carbon to soil borne microbes. Sugars are also highly abundant in the environment, as they are the building blocks of all polysaccharides like cellulose and hemi-cellulose - which both give rise and structure to plant cell walls; starches - which function in the plant as food storage tissues; as well as several other organic polymers.

ii. **Amino Acids:** As root exudates, these compounds can promote microbial growth and activity, soil fertility, and can bind with micronutrients to make them more available to plants (Adler and Key, 2022).

iii. **Organic Acids:** A broad class of acids, and similar to AAs above, organic acids play a role in the physiochemical process of mineralization/solubilization of poorly available minerals. Organic acids also contribute to the carbon cycle and can aid plants in detoxification of metals from soils (Dakora et al., 2002). Specifically, some organic acids (nicotinic, shikimic, salicylic, cinnamic, and indole-3-acetic) can serve as the preferred carbon source for certain beneficial rhizosphere bacteria (Pantigoso et al., 2021).

iv. **Vitamins:** Vitamins like Biotin (Vitamin B7) can help with plants under high pH environments (e.g. carbonate stress) and wild mustard plants were seen to exude this vitamin in high amounts when under carbonate stress (Wang et al., 2020).

v. **Tannins (and other HMW compounds):** Tannins, one example of a root exuded high molecular weight compound (HMW), can help plants with nitrogen (N) use efficiency as well as the retention of N in agroecosystem soils.

Why do plants make root exudates?: Root exudation is a primary form of communication of plants with their rhizosphere microbiome, facilitating several responses such as nutrient absorption and competition, plant to plant signaling and the recruitment of beneficial microorganisms (Canarini et al., 2019).

Plant nutrition: Phenolic acids from root exudates, as one example, are commonly exuded compounds from N-fixing leguminous plants. These phenolics are essential signals to nodule-forming Rhizobiaceae bacteria, and it is in these nodules where atmospheric N is converted to ammonia, a plant-available form of N. Organic acids from root exudates can also convert unavailable forms of Ca-, Fe-, and Al-phosphates into plant usable forms of these nutrients, which is particularly important in field soil systems where large proportions of applied phosphates can be "locked up" by the presence of these calcium, iron, and aluminum ions and unavailable to the plant (Meng et al., 2021). Additionally, plants growing in low-nutrient environments can use root exudates to signal soil microbes to aid in nutrient procurement.

Microbial recruitment: Plants change their root exudate patterns to recruit microbes that help accommodate their developmental needs, and these needs change as the plant ages. For example, arabidopsis plants secrete mostly sugars during early growth but as the plant matures, exudation compounds switch from mostly sugars to mostly amino acids and phenolics (Chaparro et al., 2013). It was also demonstrated that applying root exudates from arabidopsis to soils in the absence of plants resulted in presence of highly similar microbiota as observed when arabidopsis plants were growing in those soils (Broeckling et al., 2008). Interestingly, anywhere from 5% to 21% of all photosynthetically fixed carbon is transferred to the rhizosphere in the form of root exudates (Pantigoso et al., 2021).

Exudate re-uptake: Plants can also re-uptake their exudates to prevent wasting of energy. This occurrence is best studied by using radio-labeled isotopic carbon 13 (13C), as this unique form of carbon can be tracked through the plant body and soils in controlled systems. As an example, tomato plants were seen to re-uptake their root exudates when growing in P deficient growing systems (Tiziana et al., 2021).

Beneficial plant-plant signaling: Some plant root exudates can induce defense responses in neighboring plants when the exuding plant is under attack. This can signal to neighboring plants that can respond to the signals to start producing defense compounds to lower their susceptibility to the attacking pathogen. These signals can sometimes tell neighboring plants to initiate production of volatile compounds that attract root shielding microorganisms or insect predators of plant enemies.

Detrimental plant-plant signaling: Examples of negative plant to plant communication by root exudates could be resource competition, chemical interference, or parasitism. Allelopathy is one example of chemical interference, and is characterized as when plants directly exude phytotoxins in the soil to limit growth of a close growing, but unrelated plant species. Also, parasitic plants (plants that can't photosynthesize or produce chlorophyll) such as the ghost plant (*Monotropa uniflora*) use exudates from their hosts to associate onto their roots (Bias et al., 2006).

2.17.1 Evolutionary Divergences in root exudate composition: Given the potential for root exudation to improve nutrient acquisition by plants, numerous studies have investigated variation among species for root exudate composition (Bowshe et al.,

2016). Many species respond to nutrient deficiency with increased exudation of organic acids and sugars, although this response varies both qualitatively and quantitatively among species. Increased exudation of sugars and organic acids under nutrient deficiency is often expected to be adaptive, given the influence of these metabolites on soil nutrient availability and microbial community structure. Therefore, infertile soils may generally favor species with either inherently high exudation of organic acids and sugars, or a high capacity to increase exudation of these metabolites in response to nutrient deficiency. However, it has also been suggested that exudation of organic compounds in general should be inherently higher in fast-growing species characteristic of fertile soils, due to their high nutrient demand and capacity to sustain rapid metabolic activity via high rates of nutrient assimilation. For example, higher exudation in species native to fertile soils may contribute to the faster rates of nutrient cycling observed in the rhizosphere of such species in common garden studies. Studies directly comparing species native to low nutrient soils and those native to high nutrient soils are therefore needed to clarify how root exudate composition (i.e. metabolite identity and relative quantity) has differentiated across environments, and to shed light on the role of root exudation in adaptation to soil fertility.

Among the studies that have examined variation among species for root exudate composition in relation to native soil characteristics, few have done so in a phylogenetically-informed framework. Controlling for species' phylogenetic relatedness is an important consideration in comparative studies in order to remove the potential confounding influence of species' shared evolutionary histories on inferences of adaptive differentiation. Similar trait shifts in response to local environmental conditions in multiple phylogenetically-distinct lineages indicates convergent evolution, and provides evidence for the adaptive significance of that trait (Pineda-García et al., 2011). Thus, while comparative analyses of root exudate composition across species native to contrasting environments are useful for generating testable hypotheses, studies which incorporate species pairs from multiple lineages are particularly informative for making adaptive inferences.

Plant roots exude numerous metabolites into the soil that influence nutrient availability. Although root exudate composition is hypothesized to be under selection in low fertility soils, few studies have tested this hypothesis in a phylogenetic framework. Bowshe et al., (2016) examined root exudates of three pairs of *Helianthus* species chosen as phylogenetically-independent contrasts with respect to native soil nutrient availability. Under controlled environmental conditions, seedlings were grown to the three-leaf-pair stage, then transferred to either high or low nutrient treatments. After five days of nutrient treatments, gas chromatography-mass spectrometry was used for analysis of root exudates, and detected 37 metabolites across species. When compared in the high nutrient treatment, species native to low nutrient soils exhibited overall higher exudation than their sister species native to high nutrient soils in all three species pairs, providing support for repeated evolutionary shifts in response to native soil fertility. Species native to low nutrient soils and those native to high nutrient soils responded similarly to low nutrient treatments with increased exudation of organic acids (fumaric, citric, malic acids) and glucose, potentially as a mechanism to enhance nutrition acquisition. However, species native to low nutrient soils also responded to low nutrient treatments with a larger decrease in exudation of amino acids than species native to high nutrient soils in all three species pairs. This indicates that species native to low nutrient soils have evolved a unique sensitivity to changes in nutrient availability for some, but not all, root exudates. Overall, these repeated evolutionary divergences between species native to low nutrient soils and those native to high nutrient soils provide evidence for the adaptive value of root exudation, and its plasticity, in contrasting soil environments.

2.17.2 Unifying diversities in the rhizosphere- Role of root exudates: Although rhizosphere microorganisms differ in traits and phylogeny, they co-exist and operate together despite their divergencies. The complexity of the soil and its richness in nutrients makes it a favorable environment for various forms of life to exist (Oluwaseyi Samuel Olanrewaju and Olubukola Oluranti Babalola,, 2022). Microorganism interactions, co-existence, and "divergence-overlook" in the rhizosphere are controlled by signals released from plants through the roots into the rhizosphere. These signaling compounds are called exudates. The root systems of plants acquire water and nutrients from the soil while also releasing low and high-weight molecular compounds.

Root exudates which include secondary metabolites, volatile organic compounds, sugars, hormones, and primary metabolites among others, act as catalysts or activators of interactions in the rhizosphere thereby shaping the rhizosphere microbial communities. These exudates are not always passively released and release activators such as pathogen attack, and nutrient and water deficiency may stimulate the release of exudates. Although research into the molecular mechanism of root exudate release is on-going, it is believed that these exudates are crucial to the recruitment of the rhizobiome. The means by which exudates diffuse through the soil and target required microbes remains unresolved. Further, it is not yet established how exudates signal the appropriate microbes to meet the plant's needs. For example, there are microbes that are good biocontrol agents, but inefficient growth promoters and *vice versa*. In a situation where exudates are released because of pathogen attack, the question of how the signals attract appropriate biocontrol agents to help the plant in suppressing the pathogens arises. Such questions require answers before a full understanding of plant–microbe interactions is achieved.

Root exudates majorly serve as signals to recruit microbes to the rhizosphere irrespective of microbe species. The only requirement of the exudates is that the plant's needs are met. Different conditions lead to the release of different exudates from plant roots. The microbes are unified in their response to the exudates to help the plant. The plant in turn provides nutrients for the microbes for their survival. Signal molecules that plants use in seeking help from the rhizosphere microbes have been reported in some studies. Studies on this aspect of plant growth promotion by microorganisms are still in its early stage. The signals are in the form of flavonoids, jasmonic acids, strigolactones and hormones such as auxins and gibberellic acids.

Factors affecting Root-Exudation: The quantity and composition of rhizodeposits are influenced by a variety of biotic and abiotic factors and vary in time and space with respect to the position on the root (Shyamalina Haldar and Sanghamitra Sengupta, 2015). The host-dependent factors include genotype and age of the plant while the abiotic components include different environmental factors, grazing activities and anthropometric habits. The wide quantitative and qualitative variation in root exudation of different plant species suggests that it is determined by plant genetic makeup. In a study with eight accessions of *Arabidopsis thaliana*, a unique combination of exudates was found to be secreted from the plants of different cultivars grown simultaneously under uniform conditions. This gave rise to genotype-specific rhizobacterial communities at six weeks post-germination. The cultivar-specific variation in root exudation is mainly regulated by the root system architecture (RSA). The zone immediately behind the root tip is considered to be the major site of exudation although different sites have been recorded to be active in different plant species. Furthermore, different parts of the root system exude different compounds. For example, asparagine and threonine are exuded from meristem and root elongation zone; glutamic acid, valine, leucine and phenylalanine are secreted from root hair zone; while aspartic acid is exuded from the whole root. Use of ^{14}C to identify the sites of exudation from wheat roots has shown that non-diffusible materials are released from both primary and lateral root tips and diffusible materials are released from the whole length of roots.

Plant age vis-a-vis developmental stages plays a pivotal role in characterizing root-exudation pattern by its impact on RSA. Exudation is positively correlated with root growth. Seedlings produce the lowest amounts of root exudates which gradually increases until flowering and decreases again at maturity. Plant developmental windows seem to influence the chemistry of root-exudates as well. Secretion of phenolic compounds and defense-related proteins by Arabidopsis roots at flowering stage has been correlated with the age-related resistance of the plant. Accumulation of pathogenesis-related1 (PR1) in the apoplast of tobacco plant during its transition to fowering stage in response to infection by *Phytopthora parasitica* and interaction of Xa-21 resistance gene product with the *X. oryzae* Ax21 effector molecule during the vegetative growth of rice in response to infection by *Xanthomonas oryzae* pv. *oryzae* are other well studied examples of developmentally regulated plant defense mechanisms against fungal pathogens.

The environmental factors affecting root exudation include soil characteristics such as soil texture, pH, and moisture content in addition to climate, temperature, light and humidity. High soil moisture leads to hypoxia which may cause an increased anaerobic respiration rate, accumulation of ethanol, lactic acid and alanine in the rhizosphere. The presence or absence of particular minerals or toxic metals in the soil affects the composition of root-exudation. For example, citric, malic and oxalic acids are secreted to detoxify aluminum; secretion of phenolic compounds is increased in phosphorus deficient soils. Secretion of signaling molecules like flavanones and flavones is enhanced in nitrogen limiting conditions (Abdel-Lateif et al., 2012). Mineral-deficiency enhances the production of elicitors that influence root-exudation. Potassium deficiency increases jasmonic acid mediated defense responses. Lower temperatures and light reduce the secretion of root exudates. For example, the exudation of tannins and phenolic compounds in Vicia faba was greatly reduced at 4 °C. The root exudation process follows diurnal rhythms. Flavonoid content in the root exudates from *Alnus glutinosa* (L.) is elevated under light conditions. The presence of neighboring plant species and the rhizosphere microflora also affect the root exudation. Outcome of glucosinolates activation in Arabidopsis was regulated by the presence of neighboring plants. Production of glucosinolates is increased in a high-density environment having large number of Arabidopsis plants grown together compared with a low-density range.

2.17.3 Influence of plants on rhizosphere microbes: The rhizosphere represents tremendous complex biological system owing to the following chemical activities of plants. Plant roots release a wide variety of materials to their surrounding soil like alcohols, ethylene, sugars, amino-acids, organic acids, vitamins, nucleotides, polysaccharides, enzymes, etc. These materials favour the growth of microorganisms. Death of plant roots produce a rich source of carbohydrates and their derivatives, which support the growth of many useful bacteria around the plant roots, e.g., nitrogen fixing bacteria, sulfur oxidising bacteria, etc.

Influence of microorganisms: Microbial population in the rhizosphere which produces beneficial effects on plant growth are described as Plant Growth Promoting Rhizobia (PGPR). They are a wide range of root colonizing bacteria which promote plant growth. Microbial population in the rhizosphere benefit the plants in the following ways:

- Some bacteria and fungi make N_2 available to plants as nitrates or inorganic form.
- Sulfur oxidising bacteria make sulfur available as sulfates.
- Heterotrophic metabolism makes carbon available as CO_2, for photosynthesis.
- High concentration of CO_2 in the rhizosphere also increases solubility and availability of calcium and thereby increase calcium uptake by plant roots.
- Some microorganisms (phosphate solubilisers) also produce acids which convert insoluble rock phosphate to the soluble form and make it available as nutrient.
- Some bacteria synthesize phytohormones like auxins, gibberellins, etc. which stimulates the plant growth.
- Some microorganisms can remove hydrogen sulfide (H_2S) which is toxic to the plant roots.
- Rhizosphere fungi and bacteria produce siderophores, low molecular weight organic molecules that are able to complex with ferric iron and supply it to the plant cells.
- Similarly, organic chelating agents produced by microorganisms make manganese compounds available to the plants.
- *Azotobacter* and other bacterial spp. produce antifungal substances which protect the plant against microbial invasion.

- *Agarobacter, Rhizobium, Corynebacterium* etc. produce cytochyanins which maintains plant growth.
- Thus, the microflora of rhizosphere is very beneficial to plants.
- The rhizosphere is the most active factory for transformation of element into living constituents and returning essential elements to soil upon death and decomposition of microorganisms.
- The hydrolytic forms of microorganisms like cellulose digester transform the plant material into humus, glucose and other valuable food material which can be used both by plant as well as microorganisms.
- A very heavy growth of microorganisms absorb, nitrogen, sulfur, phosphorus, potassium and other elements in soluble form which might otherwise be removed from the soil by rain and drainage.
- Thus, higher plants act as a food manufacturer and storage house for microorganisms of soil rhizosphere and microorganisms act as collectors, processors and treasurers of food for higher plants.

Root Exudates and Microbial Colonization: The structural and functional diversities of rhizosphere microbial communities of a wide variety of plants including Arabidopsis, Medicago, maize, pea, wheat and sugar beet are shaped by plant developmental stages. For example, rhizosphere microbial communities in early development of Arabidopsis were different from that observed in the bulk soil and this difference decreased with plant age (Shyamalina Haldar and Sanghamitra Sengupta, 2015). A strong correlation has been documented between compounds released from the roots at different stages of plant development and the expression of microbial genes involved in metabolism of specific compounds. Metatranscriptomic analysis of the rhizosphere microbiome of *Arabidopsis thaliana* (genotype Pna-10) revealed that eighty-one unique transcripts were significantly expressed at different stages of plant development. It was surmised from these studies that the blend of compounds and phytochemicals in the root-exudates were differentially produced at distinct stages of plant development and orchestrated the rhizobiota.

Components of root exudates are diverse and the researchers have only begun to dissect the role on individual compound. Carbohydrates, organic acids and amino acids stimulate positive chemotactic responses in bacteria. The capacities of different microbial species to compete for these substrates and utilize vary. In the rhizosphere of maize, exudation of the benzoxazinone DIMBOA (2,4-dihydroxy-7-methoxy-1,4-benzoxazin-3-one) resulted in an increase of population density of Pseudomonas putida strain with plant beneficial characteristics. In A. thaliana, active exudation of phytochemicals mediated by ABC (ATP-binding cassette) transporters was demonstrated. Phenolic compounds in the root exudates were suggested to send specific signals for soil bacteria. Besides chemo-attraction, root exudates regulate the adherence, colonization and formation of biofilm in the rhizosphere among the bacterial isolates. Many soil-borne organisms follow the concentration gradients of sugars and amino acids towards the root. This leads to differential proliferation of microorganisms along different compartments of rhizosphere. Since quantity and composition of root exudates respond rapidly to subtle environmental changes, the selective pressure exerted on microbial communities varies spatiotemporally. Rhizosphere pH also shapes microbial ecology in the root environment. Root-induced pH changes have a dramatic impact on key biogeochemical processes such as (i) dissolution/precipitation of soil minerals and (ii) adsorption/desorption of ions, and hence bioavailability of either anions (e.g. phosphate) or cations (e.g. copper). Further understanding of rhizosphere biogeochemistry is required to quantify the driving role of higher plants in ecosystem services such as soil formation and nutrient cycling.

2.17.4 Root exudates v/s disease suppressiveness of rhizosphere microbiomes: Terrestrial plants can affect the growth and health of adjacent plants via interspecific interaction. Zhou et al., (2023) studied the mechanism by which plant root exudates affect the recruitment of the rhizosphere microbiome in adjacent plants-with implications for plant protection-using a tomato (*Solanum lycopersicum*)–potato-onion (*Allium cepa* var. *agrogatum*) intercropping system. It was showed that the intercropping system results in a disease-suppressive rhizosphere microbiome that protects tomato plants against Verticillium wilt disease caused by the soilborne pathogen *Verticillium dahliae*. Second, 16S rRNA gene sequencing revealed that intercropping with potatoonion altered the composition of the tomato rhizosphere microbiome by promoting the colonization of specific *Bacillus* sp. This taxon was isolated and shown to inhibit *V. dahliae* growth and induce systemic resistance in tomato plants. Third, a belowground segregation experiment found that root exudates mediated the interspecific interaction between potatoonion and tomato. Moreover, experiments using split-root tomato plants found that root exudates from potatoonion, especially taxifolin-a flavonoid compound-stimulate tomato plants to recruit plant-beneficial bacteria, such as *Bacillus* sp. Lastly, ultra-high-pressure liquid chromatography–mass spectrometry analysis found that taxifolin alters tomato root exudate chemistry; thus, this compound acts indirectly in modulating root colonization by *Bacillus* sp. The results revealed that this intercropping system can improve tomato plant fitness by changing rhizosphere microbiome recruitment via the use of signaling chemicals released by root exudates of potatoonion. This study revealed a novel mechanism by which interspecific plant interaction modulates the establishment of a disease-suppressive microbiome, thus opening up new avenues of research for precision plant microbiome manipulations.

The results supported the hypothesis that tomato plants can perceive chemical signals released by root exudates of potatoonion to structure a disease-suppressive rhizosphere microbiome. This occurs through induced changes in the tomato root exudates upon interspecific interaction. These results indicate that-to some extent-the ability of plants to actively recruit root-beneficial bacteria can be activated by the presence of specific neighboring plants. Previous studies have shown that plant diversity tends to correlate with increased resistance against soilborne pathogens through the enhancement of plant-beneficial bacteria and revealed a mechanism was by which this phenomenon operates. The study provided new avenues for research by illustrating that harnessing the plant microbiome for beneficial outcomes can benefit from exploiting plant interspecific interactions in diversified agricultural systems.

Plants can actively recruit plant-beneficial bacteria by releasing specific molecules via root exudates. This is even more pronounced when plants are exposed to biotic or abiotic stresses (e.g., nutrient deficiency, pathogen infection) (Liu et al., 2021). Tomato– potatoonion intercropping resulted in the modification of tomato root exudates (i.e., tomato plants exposed to potatoonion root exudates) and enhanced the ability of tomato plants to attract B56 and structure a disease-suppressive microbiome. Therefore, the ability to detect cues from the neighboring plants and chemically respond to them (i.e., morphological and biochemical plasticity) can confer benefits associated with stress responses in plants.

It was found that root exudates play an important role in the interaction between tomato and potatoonion plants. Plant–plant communication can also be mediated by several other mechanisms (e.g., volatile organic compounds) (Zhou et al., 2023). For example, the plant-beneficial bacterium *Bacillus amyloliquefaciens* was shown to trigger tomato plants to release specific volatile organic compounds, which act as signals of intraspecific plant–plant communication resulting in altered rhizosphere microbiomes of neighboring tomato plants. Therefore, testing the role of other mechanisms may improve our understanding of the relationship between plant–plant interaction and rhizosphere microbiome assembly. It should also be noted that certain plant species may possess a diverse array of bioactive secondary metabolites that can also be involved in belowground plant–plant interactions, all of which can be explored for their potential roles in mediating plant–biotic interactions.

Bibliography

Abd El-Ghany, T.M., Masrahi ,Y.S., Alawlaqi, M.M. and Mohamed A. Al Abboud, 2015. Rhizosphere and Rhizoplane Bacteria Isolated from Subtropical Region of Jazan in Saudi Arabia. *J. Biol. Chem. Res.*, **32:** 934-944.

Abdel-Lateif, K., Bogusz, D. and Hocher, V., 2012. The role of flavonoids in the establishment of plant roots endosymbioses with arbuscular mycorrhiza fungi, Rhizobia and Frankia bacteria. Plant *Signal Behav.,* **7:** 636-41.

Adler, L. and Key, M., 2022. Amino Acids in Agriculture. https://impellobio.com/blogs/inoculants/aminoacids

Adriana Giongo, Anelise Beneduzi, Adriana Ambrosini, Luciano Kayser Vargas, Marcos Roberto Stroschein, Flávio Luiz Eltz, Maria Helena Bodanese-Zanettini and Luciane Maria Pereira Passaglia, 2010. Isolation and Characterization of two plant growth-promoting bacteria from the rhizoplane of a legume (Lupinus albescens) in sandy soil. *R. Bras. Ci. Solo,* **34:**361-369

Andrew F. Galloway, Paul Knox and Kirsten Krause, 2020. Sticky mucilages and exudates of plants: putative microenvironmental design elements with biotechnological value. *New Phytologist,* **225** : 1461-146

Anonymous, 2018. Rhizosphere Concept and its Historical Background. *Soil Microbiol.*, AGRIINFO.IN. MAY 20, 2018

Anupama Sapkota, 2022. Rhizospheric microorganisms and effects, PGPR and Mycorrhiza. *Microbe News,* Jan 14, 2022

Ayesha Siddiqa, Yasir Rehman and Shahida Hasnain, 2016. Rhizoplane microbiota of superior wheat varieties possess enhanced plant growth-promoting abilities. *Front. Biol.*, **11:** 481–487

Bais, H,P,, Weir, T.L., Perry, L.G., Gilroy, S. and Vivanco, J.M.,2008. The role of root exudates in rhizosphere interactions with plants and other diversity. *Appl. Environ. Microbiol.*, **74**:738–44

Barraza, A., Caamal-Chan, M.G., Castellanos, T. *et al*, 2020. Bacterial community characterization of the rhizobiome of plants belonging to *Solanaceae* family cultivated in desert soils. *Ann. Microbiol.*, **70:** 34.

Bitas, V., Kim, H.S., Bennett, J.W. and Kang, S., 2013. Sniffing on Microbes: Diverse Roles of Microbial Volatile Organic Compounds in Plant Health. *Mol. Plant-Microbe Interact.*, **26:** 835–843.

Bowsher, A.W., Ali, R., Harding, S.A., Tsai, C.J. and Donovan, L.A., 2016. Evolutionary Divergences in Root Exudate Composition among Ecologically-Contrasting *Helianthus* Species. *PLoS ONE*, **11**(1): e0148280

Bulgarelli, D., Schlaeppi, K., Spaepen, S., van Temaat, E. V. L. and SchulzeLefert, P., 2013. Structure and functions of the bacterial microbiota of plants. *Ann. Rev. Plant Biol.*, **64:** 807–838

Canarini, A., Kaiser, C., Merchant, A., Richter, A. and Wanek, W., 2019. Root Exudation of Primary Metabolites: Mechanisms and Their Roles in Plant Responses to Environmental Stimuli *Front. Plant Sci.*, 21 https://doi.org/10.3389/fpls.2019.00157

Clinton, A. and Rumbaugh K.P., 2016. Interspecies and Interkingdom Signaling via Quorum Signals. *Isr. J. Chem.*, **56**: 265–272.

David H Mcnear, 2013. The rhizosphere-roots, soil and everything in between. The rhizosphere-roots, soil and everything in between. Nature Education Knowledge. 4.

Deng, W., Hallett, P.D., Jeng, D.-S., Squire, G.R., Toorop, P.E. and Iannetta, P.P.M., 2015. The effect of natural seed coatings of *Capsella bursa-pastoris* L. Medik. (shepherd's purse) on soil-water retention, stability and hydraulic conductivity. *Plant and Soil*, **387:** 167–176.

Dlamini, S.P., Akanmu, A.O. and Babalola, O.O., 2022. Rhizospheric microorganisms: The gateway to a sustainable plant health. *Front. Sustain. Food Syst.*, **6:**925802

George, B. And Suchitra, T.V., 2019. Plant-derived bioadhesives for wound dressing and drug delivery system. *Fitoterapia* **137**: 104214.

Harleen Kaur and Ashish Sharma, 2023. Role of endophytic bacteria in regulating phsiological and molecular aspects of plants under abiotic stress. *In: The Role of Growth Regulators and Phytohormones in Overcoming Environmental Stress*

Haruna, S., Sidi-Aliyu, Bala Baba and Abdulhamid, 2017. Plant gum exudates (Karau) and mucilages, their biological sources, properties, uses and potential applications: A review. *Bayero J. Pure & Appl. Sci.*, **9**: 159.

Hassani, M.A., Durán, P. and Hacquard, S., 2018. Microbial interactions within the plant holobiont. *Microbiome*, 6:58

Henni Vanda and Young Hae Choi, 2021. Eutectic Solvents and Stress in Plants. *In: Advances in Botanical Research*. 2021

Huang, Y., Wang, Y., Tan, L., Sun, L., Petrosino, J., Cui, M.Z., Hao, F. and Zhang, M., 2016. Nanospherical arabinogalactan proteins are a key component of the high-strength adhesive secreted by English ivy. *Proc. Nat. Acad. Sci. USA*, **113**: 3193–23102.

Jamil, F., Mukhtar, H., Fouillaud, M. and Dufossé, L., 2022. *Rhizosphere.* In:*Encyclopedia.* https://encyclopedia.pub/entry/22952

Jane Chepsergon and Lucy N Moleleki, 2023. Rhizosphere bacterial interactions and impact on plant health. *Curr. Opin. Microbiol.*, **73**: 102297

Julie Bobyock, 2023. EESA Scientists Close Rhizosphere Microbiome Knowledge Gaps. *Earth & Environ.Sci.,* July 20, 2023

Kakkar, A., Nizampatnam, N.R., Kondreddy, A., Pradhan, B.B. and Chatterjee, S., 2015. *Xanthomonas campestris* cell–cell signalling molecule DSF (diffusible signal factor) elicits innate immunity in plants and is suppressed by the exopolysaccharide xanthan. *J. Exp. Bot.*, **66:** 6697–6714.

Kateryna Zhalnina, Katherine B. Louie , Zhao Hao, Nasim Mansoori, Ulisses Nunes da Rocha, Shengjing Shi, Heejung Cho, Ulas Karaoz, Dominique Loqué, Benjamin P. Bowen, Mary K. Firestone, Trent R. Northen and Eoin L. Brodie, 2018. Dynamic root exudate chemistry and microbial substrate preferences drive patterns in rhizosphere microbial community assembly. *Nature Microbiol.,* www.nature.com/naturemicrobiology

Kennedy, A.C. and de Luna, L.Z., 2005. Rhizosphere. *In: Encyclopedia of Soils in the Environment,* 2005

Kent, A.D. and Triplett, E.W., 2002. Microbial communities and their interactions in soil and rhizosphere ecosystems. *Ann. Rev. Microbiol.,* **56:** 211-236

Keshav Prasad Shukla, Shivesh Sharma, Nand Kumar Singh, Vasudha Singh, Kirti Tiwari and Sphoorti Singh, 2011. Nature and role of root exudates: Efficacy in bioremediation. *African J. Biotechnol.,* **10:** 9717-9724

Knudsen, G.R. and Dandurand, L.M., 2007. Analysis of Spatial Patterns of Rhizoplane Colonization. *In: The Spatial Distribution of Microbes in the Environment.* Franklin, R.B., Mills, A.L. (Eds.) Springer, Dordrecht.

Koo, B..J.,Adriano, D.C., Bolan, N.S. and Barton, C.D., 2005. Root Exudates and Microorganisms. Encyclopedia of Soils in the Environment. Daniel Hillel, Editor (s): Elsevier,2005, 421-428.

Kroener, E., Holz, M., Zarebanadkouki, M., Ahmed, M. and Carminati, A., 2018. Effects of mucilage on rhizosphere hydraulic functions depend on soil particle size. *Vadose Zone J.,* **17**: 170056.

Lambers, H., Mougel, C., Jaillard, B. and Hinsinger, P., 2009. Plant–microbe–soil interactions in the rhizosphere: an evolutionary perspective. *Plant Soil,* **321:** 83–115

Laurent Philippot, Jos M. Raaijmakers, Philippe Lemanceau and Wim H. van der Putten, 2013. Going back to the roots: the microbial ecology of the rhizosphere. *Microbiol.*, **11;**789-799

Licá, I.C.L., Soares, A.M.D.S., de Mesquita, L.S.S. and Malik, S., 2018. Biological properties and pharmacological potential of plant exudates. *Food Res. Int.,* **105:**1039-1053.

Linderman R.G.,1986. Managing Rhizosphere Microorganisms in the Production of Horticultural Crops. *Hort.Sci.,* 21(6): December 1986

Liu, H., Li, J., Carvalhais, L.C., et al., 2021. Evidence for the plant recruitment of beneficial microbes to suppress soil-borne pathogens. *New Phytol.,* **229**:2873–2885

María Ciudad-Mulero and Patricia Morales, 2019. Functional Food Ingredients from Plants. *In: Advances in Food and Nutrition Research,* 2019

Markgraf, W., Watts, C.W., Whalley, W.R., Hrkac, T. and Horn, R., 2012. Influence of organic matter on rheological properties of soil. *Appl. Clay Sci.,* **64:** 25–33

Marnie E. Rout, 2014. Genomes of Herbaceous Land Plants. In: *Advances in Botanical Research,* 2014

Mazzola, M., de Bruijn, I., Cohen, M.F. and Raaijmakers, J.M., 2009. Protozoan-induced regulation of cyclic lipopeptide biosynthesis is an effective predation defense mechanism for Pseudomonas fluorescens. *Appl. Environ. Microbiol.*, **75**:6804–6811

McNear, Jr., D. H.,2013. The Rhizosphere - Roots, Soil and Everything In Between. *Nature Edu. Knowledge* **, 4:1**

Meng, X., Chen, W., Wang, Y., Huang, Z., Ye, X., Chen, L. and Yang, L., 2021. Effects of phosphorus deficiency on the absorption of mineral nutrients, photosynthetic system performance and antioxidant metabolism in Citrus grandis. *PLoS One,* https://doi.org/10.1371/journal.pone.0246944.

Mwashasha Rashid Mwajita, Hunja Murage, Akio Tani and Esther M Kahangi, 2013. Evaluation of rhizosphere, rhizoplane and phyllosphere bacteria and fungi isolated from rice in Kenya for plant growth promoters. *Springer Plus,* **2:**606

Naveed, M., Brown, L., Raffan, A., George, T., Bengough, A., Roose, T., Sinclair, I., Koebernick, N., Cooper, L. and Hallet, P. 2017. Plant exudates may stabilize or weaken soil depending on species, origin and time. *European J.Soil Sci.,* **68:** 806-816

Oleghe, E., Naveed, M., Baggs, E.M. *et al.,* 2017. Plant exudates improve the mechanical conditions for root penetration through compacted soils. *Plant Soil,* **421:** 19–30

Oluwaseyi Samuel Olanrewaju and Olubukola Oluranti Babalola, 2022. The rhizosphere microbial complex in plant health: A review of interaction dynamics. *J.Integr. Agri.,* **21:** 2168-2182

Pantigoso, H.A., He, Y., DiLegge, M.J. and Vivanco, J.M., 2021. Methods for Root Exudate Collection and Analysis. *Methods Mol. Biol.,* **2232**:291-303

Pérez-Montaño, F., Alias-Villegas, C., Bellogín, R.A., del Cerro, P., Espuny, M.R., Jiménez-Guerrero, I., López-Baena, F.J., Ollero, F. and Cubo, T., 2014. Plant growth promotion in cereal and leguminous agricultural important plants: From microorganism capacities to crop production. *Microbiol. Res.,* **169:** 325–336.

Philippot, L. et al., 2013. Going back to the roots: the microbial ecology of the rhizosphere. www.nature.com. *Nat. Rev.Microbiol.*, 23 Sept. 2013.

Pieterse, C.M.J., de Jonge, R. and Berendsen, R.L., 2016. The soil-borne supremacy. *Trends in Pl. Sci.,* **21:** 171–173

Pineda-García, F., Paz, H. and Tinoco-Ojanguren, C., 2011. Morphological and physiological differentiation of seedlings between dry and wet habitats in a tropical dry forest. *Plant Cell and Environ.,* **34:** 1536–1547.

Pratama, A. A., Terpstra, J., de Oliveria, A. L. M., and Salles, J. F., 2020. The role of rhizosphere bacteriophages in plant health. *Tre.in Microbiol.,* **28:** 709-718.

Puente, M.E., Bashan, Y., Li, C.Y. and Lebsky, V.K., 2004. Microbial populations and activities in the rhizoplane of rock-weathering desert plants. I. Root colonization and weathering of igneous rocks. *Plant Biol., (Stuttg),* **6**:629-642

Rana, S., Bhawal, S., Kumari, A., Kapila, S. and Kapila, R.,2020. pH-dependent inhibition of ahl-mediated quorum sensing by cell-free supernatant of lactic acid bacteria in *Pseudomonas aeruginosa* PAO1. *Microbial Pathogenesis*, **142**, Article 104105

Rasmann, S., Ali, J.G., Helder, J. and van der Putten, W.H., 2012. Ecology and evolution of soil nematode chemotaxis. *J. Chem. Ecol.*, **38:** 615–628

Rasmann, S., Köllner, T.G., Degenhardt, J., Hiltpold, I., Toepfer, S., Kuhlmann, U., Gershenzon, J. and Turlings, T.C.J., 2005. Recruitment of entomopathogenic nematodes by insect-damaged maize roots. *Nature,* **434**: 732–737

Raynaud, Xavier, Eickhorst Thilo, Nunan, Naoise, Kaiser, Christina, Woebken, Dagmar, Schmidt, Hannes, 2017. Spatial colonization of microbial cells on the rhizoplane. 19th EGU General Assembly, EGU2017. Proc.from the Conference held 23-28 April, 2017 in Vienna, Austria., p.18837

Rebecca Lines-Kelly , 2005. The rhizosphere. Soil Biology Basics. State of New South Wales

Reinhold-Hurek, B., Bunger, W., Burbano, C.S., Sabale, M. and Hurek, T., 2015. Roots shaping their microbiome: Global hotspots for microbial activity. *Ann. Rev. Phytopathol.,* **53:** 403–424.

Rodrigo Mendes, Paolina Garbeva and Jos M. Raaijmakers, 2013. The rhizosphere microbiome: significance of plant beneficial, plant pathogenic, and human pathogenic microorganisms. FEMS *Microbiol. Rev.,* **37:** 634–663

Rzhevskaya, V., Teplitskaya, L. and Oturina, I., 2013. Colonization of rhizoplane of cucumber roots by microorganisms which are components of the microbial preparation "Embiko®". *Regulatory Mechanisms in Biosyst.,* **4:** 63-70.

Sasse, J., Martinoia, E. and Northen, T., 2018. Feed Your Friends: Do plant exudates shape the root microbiome? *Trends Plant Sci.,* **23:** 25–41.

Scow, K., 2015. Soil Microbiology class notes. Winter 2015, University of California, Davis.

Shyamalina Haldar and Sanghamitra Sengupta, 2015. Plant-microbe Cross-talk in the Rhizosphere: Insight and Biotechnological Potential. *The Open Microbiol.J.,* 9:1-7

Srivastava, S. and Singh, V.P., 2020. Characterization of biocontrol microorganisms from the rhizoplane of *Decalepis arayalpathra* and screening of secondary metabolites. *Vegetos,* **33:** 352–359

Tiziana, R., Fabio, T., Toury, P., Silva, C., Stefano, C. and Tanja, M., 2021. Tomato plants reuptake root exudates and alter carbon isotope fractionation under phosphorus deficiency. vEGU21, the 23rd EGU General Assembly, held online 19-30 April, 2021,id. EGU21-9459

Travis S. Walker, Harsh Pal Bais, Erich Grotewold and Jorge M. Vivanco, 2003. Root Exudation and Rhizosphere Biology. *Plant Physiology,* **132**: 44–51

Tsai, A.Y.L., Kunieda, T., Rogalski, J., Foster, L.J., Ellis, B.E. and Haughn, G.W., 2017. Identification and characterisation of *Arabidopsis* seed coat mucilage proteins. *Plant Physiology,* **173**: 1059–1074.

Vasanthakumari, M.M. and Shivanna, M.B., 2011. Fungal assemblages in the rhizosphere and rhizoplane of grasses of the subfamily Panicoideae in the Lakkavalli region of Karnataka, India. *Microbes Environ.*, **26**:228-36.

Vaughn, K.C., 2002. Attachment of the parasitic weed dodder to the host. *Protoplasma,* **219**: 227–237 Miart, F., Fournet, F., Dubrulle, N., Petit, E., Demailly, H., Dupont, L., Zabijak, L., Marcelo, P. and Boudaoud A, Pineau C *et al*. 2019. Cytological approaches combined with chemical analysis reveals the layered nature of flax mucilage. *Front. Plant Sci.,* **10**: Article 684.

Venturi, V. and Keel, C., 2016. Signaling in the rhizosphere. *Tre.Plant Sci.,* **21:** 187–198

Voiniciuc, C., Pauly, M. and Usadel, B., 2018. Monitoring polysaccharide dynamics in the plant cell wall. *Plant Physiology,* **174**: 2590–2600.

Walker, T. S., Bais, H. P., Grotewold, E. and Vivanco, J. M., 2003. "Root exudation and rhizosphere biology". *Plant Physiology,* **132:** 44–51

Wang, Y., Wang, M., Ye, X., Liu, H., Takano, T., Tsugama, D., Liu, S. and Bu, Y., 2014. Biotin plays an important role in Arabidopsis thaliana seedlings under carbonate stress. *Plant Science,* 300, https://doi.org/10.1016/j.plantsci.2020.110639

Wei, Z, et al., 2019. Initial soil microbiome composition and functioning predetermine future plant health. *Sci. Adv.,* **5:** https://doi.org/10.1126/sciadv.aaw0759.

Wu Xiong, Yuqi Song, Keming Yang, Yian Gu, Zhong Wei, George A. Kowalchuk, Yangchun Xu, Alexandre Jousset, Qirong Shen and Stefan Geisen, 2020. Rhizosphere protists are key determinants of plant health. *Microbiome,* **8**:27

Xie, J., Song, L., Li, X., Wu, Q., Wang, C., Zu, H., Cao, Y. and Qiao, D., 2013. Isolation and identification of oleaginous endophytic fungi. *African J. Microbiol. Res.*, **7:** 2014-2019

Yang, K., Wang, X., Hou R. *et al.*, 2023. Rhizosphere phage communities drive soil suppressiveness to bacterial wilt disease. *Microbiome*, **11:** 16 https://doi.org/10.1186/s40168-023-01463-8

Zhang, Y., Xu, J. and Wang, E., 2020. Wang N. Mechanisms Underlying the Rhizosphere-To-Rhizoplane Enrichment of *Cellvibrio* Unveiled by Genome-Centric Metagenomics and Metatranscriptomics. *Microorganisms,* **8**:583.

Zhang, Y., Xu, J., Riera, N. *et al.*, 2017. Huanglongbing impairs the rhizosphere-to-rhizoplane enrichment process of the citrus root-associated microbiome. *Microbiome,* **5:** 97

Zhou, X., Zhang, J., Khashi, U., Rahman, M., Gao, D., Wei, Z., Wu, F. and Dini-Andreote, F., 2023. Interspecific plant interaction via root exudates structures the disease suppressiveness of rhizosphere microbiomes. *Mol. Plant.*, **16:** 849–864.

3

Soil and Root Inhabiting Fungi

3.1 Soil Inhabiting Fungi

There's a wild and wonderful world that remains hidden for most of us, at least most of the time. It's an amazing ecosystem filled with fascinating creatures interacting with one another to create an intricate, dynamic web of life. It's right under our feet: the soil ecosystem! Many organisms make up this ecosystem, and some of the most important ones are the fungi. Healthy soil is alive and teeming with an array of fungus species, each playing a vital role in its environment. Healthy soils are alive with an especially diverse array of fungal species, and many are essential to the health of the plants growing in that soil. In fact, a soil devoid of fungi is a dead soil, and is unlikely to support thriving plants. Some soil fungi are single-celled and visible only through a microscope. Others form large, underground networks of filaments that can cover an area the size of a football field or larger. Most of us are familiar with the above-ground parts of some soil fungi, the mushroom.

Fungi are one of the key microbial groups in terrestrial ecosystems that enabled colonization of land by plants and facilitated development of soil that supports most of the biota on Earth. The kingdom Fungi is one of the most diverse groups of life with an estimated 1.5–6 million species that represent heterotrophic mutualists, pathogens, and saprotrophs (Tedersoo et al., 2017). The 70,000–100,000 currently recognized species are distributed among 156 orders, 46 classes, and 12 phyla. Fungi have traditionally been identified and classified based on morphological characters of fruiting bodies and living cultures. Similar to bacteria and archaea, merely <1% of fungal species have been cultivated with established protocols, which renders large taxonomic groups undescribed and virtually unknown to science. Roughly 80% of all soil-inhabiting fungal taxa cannot be identified at the species level, and 20% cannot be reliably assigned to known orders.

Fun facts about Fungi: Genetically speaking, fungi are more closely related to animals than to plants.In our scientific classification system, they belong to their own kingdom: Kingdom Fungi. Humans have just scratched the surface when it comes to identifying and understanding this vast kingdom. It's estimated that there are 2 to 3 million species, yet fewer than 150,000, less than 1%, have been identified and described. Fungi are found worldwide, including in some of the most extreme environments, such as in deserts and in areas with high salt concentrations. Some species have been exposed to and survived intense radiation such as that encountered during space travel. According to the Guinness Book of World Records, the largest living organism on the planet is a single specimen of honey mushroom (*Armillaria ostoyae*), which covers 2,385 acres (an area equal to 1,350 soccer fields) in the Malheur National Forest in Oregon. Most of its tissue is made up of underground mycelium, a network of filament-like hyphae.

Soil fungi are microscopic plant-like cells that grow in long threadlike structures or hyphae that make a mass called mycelium. The mycelium absorbs nutrients from the roots it has colonised, surface organic matter or the soil. It produces special hyphae that create the reproductive spores. Some fungi are single celled (Ex: yeast) (Abigail Jenkins, 2005). Fungi have many different structures but they can act in similar ways and thus are not as plant specific in their needs as some soil bacteria such as Rhizobia.

Distribution of Soil fungi: Fungi are found wherever there is hard, carbon-rich woody organic matter. This could be dead rotting trees in a forest, leaf litter on the surface of orchard soils, or plant roots. Mycorrhizal fungi are found naturally in all soils. Techniques to determine their presence usually focus on indirect methods or look at the colonization of plant roots and are therefore not that reliable (Nilshi, 2021). It is difficult to get mycorrhizal fungi to grow outside their natural state, but staining techniques and microscopy have been useful in identifying mycorrhiza from soil and plant samples. Fungi tend to dominate over bacteria and actinomycetes in acid soils as they can tolerate a wide pH range. Fungi can survive in the soil for long periods even through periods of water deficit by living in dead plant roots and/or as spores or fragments of hyphae. A gram of garden soil can contain around one million fungi, such as yeasts and moulds. Fungi have no chlorophyll, and are not able to photosynthesis; besides, they can't use atmospheric carbon dioxide as a source of carbon, therefore they are chemo-heterotrophic, meaning that, like animals, they require a chemical source of energy rather than being able to use light as an energy source, as well as organic substrates to get carbon for growth and development.

Many fungi are parasitic often causing disease to their living host plant, although some have beneficial relationships with living plants. In terms of soil and humus creation, the most important fungi tend to be saprotrophic, that is, they live on dead or decaying organic matter, thus breaking it down and converting it to forms which are available to the higher plants. A succession of fungi species will colonize the dead matter, beginning with those that use sugars and starches, which are succeeded by those that are able to break down cellulose and lignins. Fungi spread underground by sending long thin threads known as mycelium throughout the soil; these threads can be observed throughout many soils and compost heaps. From the mycelia the fungi is able to throw up its fruiting bodies, the visible part above the soil (e.g., mushrooms, toadstools and puffballs), which may contain millions of spores. When the fruiting body bursts, these spores are dispersed through the air to settle in fresh environments, and are able to lie dormant for up to years until the right conditions for their activation arise or the right food is made available. The range of fungi known to occur in soil is very wide, from chytrids to agarics, from saprophytes to root parasites, from parasites of amoebae to parasites of man. Interest in fungi occurring in soil has been great and considers that soil has probably been studied more extensively than any other natural habitat of fungi.

Lifestyles of Soil fungi: The lifestyles adopted by fungi in soil fall into two types, viz., Ruderals and Mycorrhizal fungi (Nilshi, 2021). Ruderals take advantage of the flushes of nutrients usually associated with rainfall. The water moving through soil carries with it dissolved organic molecules flushed from plant surfaces, surface litter and dead microbes. The fungi respond immediately, and grow actively while soil remains moist. Members of the order Mucorales are widely considered to be ruderals. The genera *Mucor* and *Rhizopus* are common in soil and air, and Pilobolus and Pilaira on dung. Their sporangiospores germinate rapidly in response to available energy. The mycelium develops rapidly when nutrients are available. Sporangiospores or other reproductive propagules are formed very soon after initiation of the mycelium. At the first sign of depletion of nutrients, or competition, the mycelium starts to collapse. Each individual is "alive" for a short time. Ruderals capture the readily available resources quickly. However, they appear unable to utilize complex carbohydrates because they lack the appropriate array or activity of enzymes. Energy is rapidly transformed into reproductive units, not hyphae. Sporangiospores appear to have a relatively short life, thus the fungi do not survive for long or at high densities in stable, low nutrient, highly competitive habitats. These fungi usually sporulate rapidly, and exist through dry periods as asexual spores. Common genera include *Absidia, Aspergillus, Chaetomium, Fusarium, Mortierella, Mucor*, and *Penicillium.* These are the fungi that are commonly isolated from soil using soil dilution techniques.

Mycorrhizal fungi subsist almost entirely by tapping into the roots of plants for their organic energy. The fungi are extremely common in soil, though quantification of the fungi is extremely difficult. Up to 5 m of living hyphae of arbuscular mycorrhizal fungi can be extracted from 1 g soil. The same quantity of soil may reveal 1,000 colonizing propagules of the same fungi, and maybe the same number of spores of ruderals. As ruderals are defined as organisms that form huge numbers of survival units, one might argue that mycorrhizal fungi in these soils should be classified as ruderals. The lifestyle of mycorrhizal fungi is however, much more stable. While ever the host is alive, the fungi have access to organic carbon, and their proliferation is comparatively measured. Indeed, they probably exude simple organic molecules that are a suitable source of carbon for some other microbes. AM fungi are ubiquitous yet lack clear methods of widespread dispersal found in ruderals. The life style is very different.

Finally, a group of fungi exist as hyphae in soil. Spores of these fungi are extremely difficult to find in soil, and some- may never sporulate. While some of the fungi have the potential to cause disease, it is likely that for much of their lifecycle, these organisms metabolize and grow slowly. They are usually associated with organic fragments, which they slowly degrade. Their isolation from soil requires special techniques. However, once in culture the fungi can grow rapidly utilizing a wide range of sources of complex carbon. These fungi exhibit some characteristics of combative and stress tolerant fungi. Fungi do not appear to readily degrade all forms of complex organic carbon. Humus is the name given to the complex of organic materials that may be resident in soil for decades to centuries. Humus consists of a suite of mixed polymers of aromatic and aliphatic compounds, largely derived from lignin (plant) and melanin (all organisms including fungi). Some 60% of organic carbon in soil may be humus.

Humus is resistant to degradation because it covalently bonds the reaction sites to metals and clay minerals making the sites unavailable to enzymic attachment. Humus also bonds to various organic pollutants such as polyaromatic hydrocarbons and alanines (and shares a number of similarities) making both unavailable for degradation. Further, humus is often trapped within soil aggregates and therefore has an exposed diameter that effectively prevents bacteria from accessing the complex. Humus breaks down more readily when the soil is exposed to oxygen such as by cultivation, when soil is acidic and when the moisture level is around 20%. The fungi play an important role in soil, degrading complex sources of organic carbon, some of which may be organic pollutants. While degradation is part of the normal cycle of carbon and other elements, loss of carbon from soil, especially agricultural soil, is responsible for as much as 20% of the increased CO_2 in air, contributing to global heating (Nilshi, 2021). Indeed, a reduction of fungal degradation of organic carbon leading to an increase of humus in soil may be one mechanism whereby both improved soil quality and carbon sequestration may be achieved.

Common genera of Soil fungi: Both the generic composition and the size of the flora vary with the type of soil and with its physical and chemical characteristics. Most species are mesophilic in their temperature relationships; however, a few thermophilic strains can be demonstrated in normal soil, during heating of rotting composts. Among the terrestrial thermophiles are species of *Aspergillus, Humicola, Mucor, Chaetomium* etc (Nilshi, 2021). The dominance of one or the other group is frequently related to factors like season, vegetative cover, type of soil, application of inorganic fertilizers etc. Some of the common fungi found in soil include *Amanita, Tricholoma, Torrendia, Descomyces, Thelephora, Verticillium, Phytophthora, Rhizoctonia,* and *Pythium.* Most isolates are placed in one of the two classes, Hyphomycetes or Zygomycetes. The list of classes is derived from the scheme of classification proposed by Answorth.

i. **Hyphomycetes:** Lack of sexual stages. Form a mycelium that has spores on special branches (the sporophores) or that bears no spores. Examples: *Aspergillus, Botryotrichum, Botrytis, Cladosporium, Curvularia, Cylindrocarpon, Epicoccum, Fusarium, Fusidium, Geotrichum, Gliocladium, Myrothecium, Paecilomyces, Penicillium, Rhizoctonia, Scopulariopsis, Stachybotrys, Stemphylium, Trichoderma, Trichothecium, Verticillium.*

ii. **Coelomycetes:** Lack of sexual stages. Spores in pycnidia. Examples: *Coniothyrium, Phoma.*

iii. **Zygomycetes:** The sexual resting spores are zygospores. Examples: *Absidia, Cunninghamella, Mortierella, Mucor, Rhizopus, Zygorhynchus.*

iv. **Pyrenomycetes:** Spores in the sexual stage are ascospores. Examples: *Chaetomium, Thielavia.*

v. **Oomycetes:** Possess biflagilate motile cells known as zoospores. Spores in the sexual state are oospores. Example: *Pythium.*

vi. **Chtridiomycetes:**Uniflagilate spores and produce oospores. Commonly called chytrids.

vii. **Hymenomycetes:**Sexual spores are basidiospores. Commonly called basidiomycetes.

viii. **Acrasiomycetes:**Have a free-living amebal phase, the abebae combining to dorm a pseudoplasmodium.

Ix. **Hymenomycetes:** Sexual spores are basidiospores. Commonly called basidiomycetes.

x. **Acrasiomycetes:** Have a free-living amebal phase, the abebae combining to dorm a pseudoplasmodium. The term yeast has no taxonomic validity, but the group includes fungi that exist primarily as unicellular organisms that reproduce by budding or fission. The genera of soil yeasts most frequently isolated are *Candida, Cryptococcus, Debaryemyces, Hansenula, Lipomyces, Pichia, Pullularia, Rhodotorula, Saccharomyces, Schizoblastosporium, Sporobolomyces, Torula, Torulaspora, Torulopsis, Tricjosporan* and *Zygosacchatomyces.*

Fungi groups: There are three functional groups of fungi.

i. **Decomposers**: Decomposers or saprophytic fungi convert dead organic matter into fungal biomass (i.e., their own bodies), carbon dioxide and organic acids. They are essential for the decomposition of hard woody organic matter. By consuming the nutrients in the organic matter they play an important role in immobilising and retaining nutrients in the soil. The organic acids they produce as by products help create organic matter that is resistant to degradation. Fungi are capable of degrading cellulose, proteins and lignin, some of which are highly resistant to breakdown.

ii. **Mutualists**: These fungi develop mutually beneficial relationships with plants. They colonise plant roots where they help the plant to obtain nutrients such as phosphorus from the soil. Their mass hides roots from pests and pathogens, and provides a greater root area through which the plant can obtain nutrients. For instance, mycorrhizal fungi are the best known of the mutualists. Mycorrhiza means fungus root, and mycorrhizal fungi grow inside plant roots. Up to 5m of living hyphae of mycorrhizal fungi can be extracted from 1g of soil. The four groups of mycorrhizal fungi are arbuscular, ectomycorrhizal, ericoid and orchid. Arbuscular mycorrhiza (VAM) are the most common form of mycorrhiza, especially in agricultural plant associations. This fungi has arbuscles which are growths formed inside the plant root that have many small projections going into the cells. About 150 arbuscular mycorrhiza species are known. Most plants (90%) have some sort of association with these fungi except for groups such as the Cruciferae family (Ex: mustard, canola, broccoli), Chenopodiaceae (eg spinach, beets, saltbush) and Proteaceae (banksia, macadamia).

iii. **Pathogens:** This group includes the well known fungi such as *Verticillium, Phytophthora, Rhizoctonia* and *Pythium.* These organisms penetrate the plant and decompose the living tissue, creating a weakened, nutrient deficient plant, or death. The pathogenic fungi are usually the dominant organism in the soil. Soils with high biodiversity have been shown to suppress soil-borne fungal diseases. Suppression mechanisms include the suite of native organisms out-competing the pathogenic organisms, physically protecting roots and providing better nutrition to the plant.

Roles of fungi: Fungi perform important functions within the soil in relation to nutrient cycling, disease suppression and water dynamics, all of which help plants become healthier and more vigorous.

i. **Decompose woody organic matter**: Along with bacteria, fungi are important decomposers of hard to digest organic matter. They use nitrogen in the soil to decompose woody carbon rich residues low in nitrogen and convert the nutrients in the residues to forms that are more accessible for other organisms.

ii. **Increase nutrient uptake:** Mycorrhizal fungi are well known for their role in assisting plants in the uptake of phosphorus. Ectomycorrhizal fungi can benefit plants by promoting root branching and increasing nitrogen, phosphorus and water uptake due to their large surface area and internal cellular mechanisms.

iii. **Improve plant resilience:** The sheer size and mass of fungal hyphae help decrease plant susceptibility to pests, diseases and drought.

iv. **Improve soil structure**: Fungal hyphae bind the soil particles together to create water-stable aggregates which in turn create the pore spaces in the soil that enhance water retention and drainage.

Beneficial Fungi: Here are 5 things beneficial fungi in soil.

1. **They Increase Nutrient Uptake:** The formation of mycorrhizae increases the root's surface area up to a thousandfold. The hyphae also help to increase the soil volume, as well as the nutrients, for easier access by plants. More importantly, they provide the plant with phosphorus in a form that is more usable. Mycorrhizal fungi can break down any element that is tightly bound to the soil, returning the minerals to the plants in a usable form.

2. **Provides Help to Nitrogen-Fixing Bacteria:** Rhizobia, a nitrogen-fixing bacteria, requires a substantial amount of phosphorus to fix nitrogen. However, the bacteria live and work on the coarse roots of legumes that may not be able to extract nutrients like phosphorus. Making sure that you have mycorrhizal fungi will increase the availability of fixed nitrogen and phosphorus in the soil.

3. **Increases the Availability of Water and Boosts Drought Tolerance:** Fungal hyphae have a microscopic diameter that enables them to harvest water from very tiny soil pores not accessible by the roots of plants. If you add more nutrients to the soil, the plants' ability to survive periods of drought will increase. Mycorrhizal fungi are, therefore, a helpful solution to water restrictions.

4. **Glomalin - Carbon Sink and Super Glue:** Glomalin, a super sticky substance, comprises 30 to 40 % carbon and is only produced by arbuscular mycorrhizal fungi (AMF). It accounts for 27 % of the carbon stores in the soil.

Role in nutrient cycling: Soil fungi play a crucial role in nutrient cycling in terrestrial systems, due to the primary role they play as decomposers, mediated by a large repertoire of catabolic activities. They are particularly involved in cycling C, N, and P, but have roles in most of the other soil elemental cycles. Fungi are primarily involved in the decomposition of plant material, the lignocellulosic components of which are relatively recalcitrant to bacteria (Ritz, 2005). There is typically a succession of fungal (and other microbial) species on new substrate when it enters soil, based upon the initial decomposition of simpler compounds and a subsequent degradation of more complex polymers. Cellulases are produced by a wide variety of fungi across the taxonomic classes; these include the brown rots, so-called because their activity results in a darkening of wood as the cellulose is decomposed and lignin (which they are not able to decompose) remains. Within fungi, polyphenol oxidases (mainly laccases), involved in the decomposition of lignin, are largely confined to Basidiomycetes and Ascomycetes. These include the white rots, species which normally also have strong cellulase activity and render wood a light color during decay.

As well as invoking the enzymatically mediated degradation of polymers, many fungi also produce a variety of compounds such as organic acids or siderophores, which solubilize or immobilize essential or toxic metals, with consequences for both their own growth and the growth of proximal organisms. Fungi also act to immobilize nutrients in soil through sequestration in mycelia, which are further released following death and lysis or attack by pathogens or grazers. Eucarpic fungi play a significant role in physically transporting nutrients through the soil fabric via translocation within mycelia, thus short-circuiting other diffusion and mass-flow transport mechanisms, both in saprotrophic and mutualistic (mycorrhizal) contexts.

Role of Soil fungi: Fungi are an important part of the microbial ecology. The majority of fungi decompose the lignin and the hard-to-digest soil organic matter, but some fungi consume simple sugars. Fungi dominate in low pH or slightly acidic soils where soils tend to be undisturbed (James J. Hoorman, 2016). Fungi break down the organic residues so that many different types of microbes can start to decompose and process the residues into usable products. Approximately 80 to 90 percent of all plants form symbiotic mycorrhizae fungi relationships by forming hyphae networks. The hyphae are about 1/60 the diameter of most plant root hairs and assist the plant in acquiring nitrogen, phosphorus, micronutrients and water in exchange for sugar produced by the plant. This mutually beneficial relationship is called a mycorrhizae network.

Hyphae interact with soil particles, roots, and rocks forming a filamentous body that promotes foraging for soil nutrients. These networks release enzymes into the soil and break down complex molecules that the filaments then reabsorb. Fungi act like natural recycling bins, reabsorbing and redistributing soil nutrients back to plant roots. Most hyphae are either pure white or yellow and are often misidentified as plant hair roots. The rhizosphere is an area next to the root dominated by soil microbes where many chemical and biochemical process occur. Soil fungi make up 10 to 30 percent of the soil rhizosphere. Generally there are fewer individual fungi than bacteria but fungi dominate the total biomass due to their larger size in a healthy soil. Fungi biomass in the soil ranges from the equivalent of two to six cows in a healthy soil or 1,100 to 11,000 pounds of biomass. Fungi prefer slightly acidic conditions, low disturbance soils, perennial plants, internal nutrient sources directly from the plant, and highly stable forms of organic residues with high carbon to nitrogen (C:N) values and slower recycling time. Bacteria dominate in highly disturbed ecosystems with fast nutrient recycling, low C:N values, prefer annual plants, and external nutrient additions outside the plant. Bacteria are single-celled organisms and need a film of water to survive, while fungi are multi-celled organisms that grow rapidly and in great lengths in the soil (feet or meters). This allows fungi to bridge gaps in the soil so as to transport nutrients relatively far distances back to the plants (Lowenfels & Lewis, 2006).

Taxonomy and Functional Groups of Fungi: There are at least 70,000 different species of fungi identified but it is estimated that there may be 1.5 million species worldwide. Genetically, fungi evolved a billion years ago and are closely related to plants and animals (James J. Hoorman, 2016). Membrane bound organelles present in each cell are similar to those found in insects, plants and animals. Fungi have 80 percent or more of the same genes as humans (Dick, 2009). There are four major groups of soil fungus: *Zygomycota, Ascomycota, Basidiomycota* and *Deuteromycota. Zygomycota* are less than 1,000 species and are mostly common bread molds. *Ascomycetes* have about 30,000 species and are mostly yeasts used in baking. *Basidiomycetes* include most mushrooms, toadstools and puffballs, while *Deuteromycota* include the lichens and the mycorrhizal fungus. Fungi are classified as heterotrophs so the carbon source originates from the decomposition of organic compounds or residues. Decomposers are also called saprophytic fungi which decompose cellulose and lignin in the soil. Sugar fungi called Zygomycetes decompose the simple sugars but most fungi decompose the more recalcitrant or hard-to-decompose organic residues high in cellulose, hemicellulose, lignin or cell walls. Some of the byproducts of this decomposition may turn to humus and remain in the soil for thousands of years (Lowenfels & Lewis, 2006).

Pathogenic fungi cause many agricultural root diseases including *Phytophthora, Rhizoctonia, Phythium* and *Verticillium*, and downy mildew Ascomycetes fungi are microscopic in size and dominate in agricultural soils and grassland while the Basidiomycetes have large fruiting bodies or mushrooms that dominate in high residue and forested soil (Dick, 2009). Some fungi help to control diseases and predators including a nematode trapping fungi that feeds on insects and can be used as biological controls. Mutualistic mycorrhizal fungi form a beneficial relationship with plants. "Mycorrhizae grow within the root cells and are commonly associated with grasses, row crops, vegetables, and shrubs. Some plant species like the *Cruciferae* family (e.g., cabbage, broccoli, mustard and canola) and the *Chenopodiaceae* family (e.g., lambsquarters, spinach, beets, and oilseed radish) do not form mycorrhizae associations."

3.2 Ecological Plant-Microbe Interactions

The microbes and plants together regulate many soil processes including the carbon cycle and nutrient recycling. The microbial species diversity and the total microbial population determine the ability of plants to obtain soil nutrients like nitrogen, phosphorus and micronutrients (James J. Hoorman, 2016). Plant diversity and abundance may change the entire soil ecosystem through the release of root exudates that attract or inhibit the growth of specific organisms. These carbon-rich substances can range from less than 10 percent to as much as 20 percent of a plant's total carbon production (Sylvia et al., 2005). Plants secrete large amounts of specific carbon compounds into the soil to be used by the microbes as a food and energy source and to enhance and improve soil structure. By encouraging certain microbial species to grow, the microbes in return supply nutrients to the plant roots and encourage the microbes to protect the plant from pathogenic microbes. Plants feed, raise and encourage certain microbes just like farmers raise and feed plants and livestock for food and fiber.

Many plants have a preferred fungus to bacteria ratio (F:B ratio). Most vegetable crops prefer more bacteria with a F:B ratio of 0.3 to 0.8 F:B which is optimal for carrots, lettuce, broccoli and cole crops. Tomatoes, corn and wheat like an F:B ratio of 0.8 to 1:1. Lawns prefer a F:B ratio of 0.5 to 1:1 and on forested soils, trees grow better with a F:B ratio of 10:1 where the soil is more acidic. Many conventionally tilled agricultural soils have a F:B ratio of 0.1 to 0.3 with soils that are high in nitrogen, low carbon, neutral pH, and with disturbed soil conditions which promotes weed production. Annual crops prefer lower F:B ratios and perennials prefer a higher F:B ratio. Many plants cultivate certain species of both bacteria and fungus to increase nutrient extraction from the soil.

Fungi benefit most plants by suppressing plant root diseases and fungi promote healthier plants by attacking plant pathogens with fungal enzymes. Fungi also use antagonism to reduce competition by producing antibodies, which suppress other microorganisms from growing. They produce many vitamins which promote plant growth. By competing with other fungus for nutrients, beneficial fungi prevent pathogenic and disease-causing organisms from getting established. Beneficial fungi along with some bacteria may also form protective webs and nets around roots and even leaves to protect the host plant. The fungus *Trichoderma* protects plant roots from attack by harmful microorganisms. Fungi also protect plants by supplying a protective sheath to supply both water and phosphorus to the plant roots during droughts (James J. Hoorman, 2016).

Hyphae must be in close contact with living or dead organic soil residues to absorb nutrients, so they usually grow in association with other soil microorganisms. They compete aggressively for scarce nutrients, and competition usually results in a succession or change in microbial populations as nutrients are absorbed or depleted. Initial colonizers absorb simple sugars, amino acids, and vitamins from plant parts such as fruits, seeds and vegetables, and are classified as "sugar fungus." The dominance of these fungi is short-lived because waste products accumulate quickly. Cellulose degraders appear next and they are the most diverse fungi and competitive. For example, degradation of straw with high C:N ratio (80:1) requires that fungus parasitize or decompose other fungi to obtain nitrogen for growth and enzyme production. Degradation of lignin is stimulated by low nitrogen. Lignin makes up 60 percent of the total mass of humus, but the low number of fungal species that can degrade lignin reduces competition.

Fungi generally reproduce asexually by spores (microscopic parts similar to plant seeds). Spore dispersal occurs in a variety of ways including triggers, jet propulsion, springs, and scents depending on the environment (Lowenfels & Lewis, 2006). Fungi spores may live 50 years in an inactive vegetative state as a spore. This allows the fungi to survive and remain viable by staying inactive until environmental conditions improve. Microbes spend most of their time hungry and looking for nutrient resources just to survive but fungus may remain viable for years. Most fungi survive by continual growth and translocation of nutrients in wide networks as they search for food or they survive through the production of resistant spores during times of stress. Natural soils tend to suppress germinating fungal spores, especially when nutrients are limiting. The longevity of fungi has not been measured in many species but their open-ended growth suggests that they have a longevity measured in millions of years, because they are basically the same organism. For example, fairy fungal rings grow in ever widening circles, much like rings on a tree, and are measured in decades and centuries instead of days and weeks for most microbes. One fungus in Michigan's Upper Peninsula covered 20 hectares or almost 42 acres, connected itself to most of the trees and weighed 10,000 kilograms or 11 ton and was estimated to be over 1,500 years old (James J. Hoorman, 2016).

To conclude, most soil fungi decompose recalcitrant organic residues high in cellulose and lignin. Fungi carbon use efficiency is about 40–55 percent so they store and recycle more C (10:1 C:N ratio) and less N (10 percent) in their cells than bacteria. While fungi are smaller in number, they equal or exceed the biomass of bacteria due to their greater size. Fungi are more specialized but need a constant food source and grow better under undisturbed soil conditions. Fungi and bacteria recycle soil nutrients and generally have a symbiotic relationship with most plants.

A survey by Gams (2007) was concerned only with filamentous fungi living in the soil layer. The observed diversity of soil fungi largely depended on the method used and the numbers of isolates obtained. Particle-plating usually yielded higher numbers of taxa than dilution plating. The Centraalbureau voor Schimmelcultures (CBS) preserves a great diversity of soil fungi. The CBS database contained 2,210 species of soil fungi in 2001, an estimated 70% of the known species available in culture. Thus, the current estimate for described culturable soil fungi was approximately 3,150 species, many of which have a cosmopolitan distribution. Adding the ca 150 spp. of nonculturable Glomerales result in 3,300 species of currently known soil fungi. Molecular studies in such groups as *Fusarium*, *Trichoderma*, *Penicillium* and *Aspergillus* were finding a number of more narrowly distributed cryptic species. Thus the number of species of soil fungi was expected to be considerably higher than the 3300 species currently known.

3.3 Global Diversity and Geography of Soil Fungi

There are many reports on biodiversity and influence of different species of fungi on various soil locations were available at global level which indicates the soil fungi are ubiquitous and inhabit different types of soils. From various reports, the fungi commonly present in different type of soils found as *Aspergillus niger, A. flavus, A. clavatus, A. terreus, Curvularia lunata, Pencillium janthinellum, P. citrinum, Rhizopus nigricans, Trichoderma viride, T. harzianum, Fusarium. oxysporium, F. moniliforme, Mucor, Cladosporium, Verticillium, Acremonium, Chaetomium globosum, Corynespora cassicola, Eurotium chevelari, Botrycis cineea, Helminthosporium* sp. etc. Among the species of fungi reported in soil by various authors, the *Aspergillus* is the dominant genus followed by *Penicillium* species, several species of hyphomycetes and Zygomycetes are frequently occurring in various soil samples (Jayaraman et al., 2018).

Fungi play major roles in ecosystem processes, but the determinants of fungal diversity and biogeographic patterns remain poorly understood. Using DNA metabarcoding data from hundreds of globally distributed soil samples, Leho Tedersoo et al., (2014) demonstrated that fungal richness is decoupled from plant diversity.The plant-to-fungus richness ratio declines exponentially toward the poles. Climatic factors, followed by edaphic and spatial variables, constitute the best predictors of fungal richness and community composition at the global scale. Fungi show similar latitudinal diversity gradients to other organisms, with several notable exceptions.These findings advance our understanding of global fungal diversity patterns and permit integration of fungi into a general macroecological framework.

Fungi comprise some 100,000 described species (accounting for synonyms), but the actual extent of global fungal diversity is estimated at 0.8 million to 5.1 million species. Globally, the biomass and relative proportions of microbial groups, including fungi, co-vary with the concentration of growth-limiting nutrients in soils and plant tissues. Such patterns suggest that the distribution of microbes reflects latitudinal variation in ecosystem nutrient dynamics. Richness of nearly all terrestrial and marine macroorganisms is negatively related to increasing latitude,a pattern attributed to the combined effects of climate, niche conservatism, and rates of evolutionary radiation and extinction. Although morphological species of unicellular microbes are usually cosmopolitan, there is growing evidence that the distribution of microorganisms is shaped by macroecological and community assembly processes. Only a few of these biogeographic processes have been demonstrated for fungi at the local scale. Despite their enormous diversity and importance in ecosystem function, little is known about general patterns of fungal diversity or functional roles over large geographic scales.

Taxonomic and functional diversity: Pyrosequencing analysis of global soil samples revealed 1,019,514 quality-filtered sequences that were separated into 94,255 species-level OTUs (supplementary materials). Altogether, 963,458 (94.5%) sequences and 80,486 (85.4%) OTUs were classified as Fungi. Most other taxa belonged to animals (Metazoa, 3.3%), plants (Viridiplantae, 3.1%), alveolates (Alveolata, 2.8%), and amoebae (mostly Rhizaria, 1.3%). Kingdom-level assignment of 3.8% OTUs remained elusive. The fungal subset included 35,923 (44.6%) OTUs that were represented by a single sequence; these were removed from further analyses in order to avoid overestimating richness based on these potentially erroneous sequences (Dickie, 2010). The remaining 44,563 nonsingleton fungal OTUs in our data set numerically correspond to approximately half of the described fungal species on Earth. For comparison, there are currently 52,481 OTUs based on 98.0% similarity clustering of all fungal ITS sequences in publicly available databases. Global soil sampling revealed representatives of all major phyla and classes of Fungi. Of fungal taxa, Basidiomycota (55.7%), Ascomycota (31.3%), Mortierellomycotina (6.3%), and Mucoromycotina (4.4%) encompassed the largest proportion of sequences, whereas the most OTU-rich phyla were the Ascomycota (48.7%), Basidiomycota (41.8%), Chytridiomycota (2.3%), and Cryptomycota (syn. Rozellida; 2.1%). Except for the recently described phylum Cryptomycota, the relative proportions of major phyla correspond to the proportional distribution of taxa described and sequenced to date (www.indexfungorum. org). Below the phylum level, ~6% of all fungal OTUs could not be assigned to any known class of fungi. Further clustering of unidentified fungal sequences at 70% sequence similarity revealed 14 distinct taxonomic groups comprising >7 OTUs, suggesting that there are several deeply divergent class-level fungal lineages that have not yet been described or previously sequenced.

The classification revealed that 10,801 (24.2%) fungal OTUs exhibited >98% sequence similarity, and 33.8% exhibited >97% similarity, to preexisting ITS sequences in public databases. This is consistent with, reporting 48% of OTUs amplified from Alaskan soils with >97% similarity to any database sequences. In our study, only 4353 fungal OTUs (9.8%) were matched to sequences from herbarium specimens or fully described culture collections at >98.0% sequence similarity. Although many type collections are yet to be sequenced, the paucity of matches to database entries indicates that a majority of soil-inhabiting fungal taxa remain undescribed. These results highlight the current lack of data from understudied tropical and subtropical ecosystems. The phenomenon of high cryptic diversity and low success in naming OTUs at the genus or species level have been found in other groups of soil microbes and invertebrates, emphasizing our poor overall knowledge of global soil biodiversity. The main fungal phylogenetic and functional groups were present in all ecosystems, but their relative proportions varied severalfold across biomes. The ratio of Ascomycota to Basidiomycota OTUs was highest in grasslands and shrublands (1.86) and tropical dry forests (1.64) but lowest in the temperate deciduous forests (0.88). Chytridiomycota, Cryptomycota, and Glomeromycota were relatively more diverse in the grasslands and shrublands, accounting for 4.6, 3.6, and 1.4% of OTU richness, respectively. The relative OTU richness of Mortierellomycotina and Mucoromycotina (including most fast-growing molds but also some plant symbionts) peaked in the tundra biome (4.8 and 2.7%, respectively), but their abundance was lowest in tropical dry forests (1.0 and 0.6%, respectively). Archaeorhizomycetes, a recently described class of Ascomycetes from a boreal forest, was most diverse in tropical moist and montane forests, particularly in northern South America and New Guinea.

Among all fungal taxa, OTUs assigned to saprotrophs, EcM mutualists, and plant pathogens comprised 19,540 (43.8%), 10,334 (23.2%), and 1770 (4.0%), respectively. Other trophic categories contained in the northern temperate deciduous forests but accounted for a relatively low proportion (11.9%) in grasslands and shrublands, reflecting the paucity of host plants in these ecosystems. Similarly, the proportion of EcM fungal taxa was lowest in northern South America (8.0%), where AM trees often dominate. Plant pathogens were relatively more abundant and diverse in lowland tropical moist (6.2%) and dry (6.3%) forests.

A compelling study regarding patterns of biodiversity of fungi, carried out at a scale unprecedented to date for fungal biogeographical studies was presented. The study demonstrates strong global biogeographic patterns in richness and community composition of soil fungi. What concerns us with the study is what we don't see (Christopher W. Schadt and Anna Rosling, 2014). Unfortunately this study underestimates fungal diversity of one key group of soil fungi due to reliance on a single primer with known flaws.

The Internal Transcribed Spacer (ITS) region of the ribosomal RNA operon has been a de facto barcode for fungal taxonomy and ecology for 25 years. While these primers have been valuable for describing fungal diversity both taxonomically and ecologically, unfortunately like almost any single primer, they cannot capture all diversity in the fungal kingdom in an unbiased fashion. The authors correctly acknowledge in the main text that the Tulasnellaceae and Microsporidia will not be amplified with their chosen primer set. The authors have also taken considerable care to include variants of the forward ITS3 primer to increase the representativeness of amplicon pools and describe these variants in their supplemental methods. However, it was previously shown describing this new class of fungi and cited by the authors, that the ITS4 reverse primer has at least two mismatches to all known Archaeorhizomycetes. These mismatches have been shown to result in at least a ten-fold underrepresentation of Archeaorhizomycetes in artificial communities using known DNA template amounts from isolates and as a consequence other taxa in the artificial community also appeared to be more abundant than they were.

The Archaeorhizomycetes are a widespread, highly diverse, and ancient class of the Ascomycota documented in over 100 studies of soil fungal communities. These fungi were first identified from 28S rRNA gene sequences from tundra soils in 2003 (6) and subsequently shown to be globally distributed in soil rRNA gene studies. Using perfectly matched LSU primers Archaeorhizomycetes are shown to be dominant in many soils, typically comprising >10% of rRNA gene sequences recovered and in one case >90%. Extensive studies of soil fungal communities were performed in Alaska using primers unbiased against amplification of Archaeorhizomycetes. In Alaska at least 53 putative species level OTUs of Archaeorhizomycetes could be identified, with diversity ranging from 2 – 25 OTUs per sample. Several of these putative species-level groups were among the most abundant OTUs across Alaskaand comprised at least 5% of the species richness of Alaskan soil fungal communities. In their current study, Tedersoo and colleagues identify only 2 OTUs belonging to the Archaeorhizomycetes in one out of two Alaskan sites studied, and thus Archaeorhizomycetes are reported in their paper to represent only approximately 0.3% of the fungal diversity an Arctic Tundra samples. In contrast previous reports of alpine and arctic tundra samples revealed >10% overall relative abundance of the Archaeorhizomycetes class when suitable primers were used (Rosling et al., 2013). It should also be noted that as all studies of rRNA gene diversity are inherently semi-quantitative (based on relative abundances), a likely ten-fold underestimation of diversity and abundance of the class Archaeorhizomycetes would also effect the ratios of other functional and phylogenetic classes. The authors specifically discuss the patterns of Archaeorhizomycete diversity and abundance at the global scale in several instances within the paper, but do not acknowledge the likely bias created by mismatched PCR primer. Thus, Archaeorhizomycetes are not low diversity, low abundance soil fungi as described in Tedersoo et al., but in fact quite the opposite.

While it is apparent that the authors used a single ITS4 primer variant to facilitate inclusion of the barcode tag sequences for multiplexing samples, this no longer necessary as methods now exist for adding barcode tags post amplification. With the rapid advancement in parallelization and throughput of samples enabled by current sequencing technologies, the use of multiple primer sets or primer variants is achievable and advisable. The biases of PCR primers for studies of "unknown" organisms in bacterial and fungal ecology have been widely reported and the continued reliance on single primer sets and variants for such studies compromises our understanding of the diversity and biogeography of microorganisms. It is also incumbent upon microbial ecologists to interpret patterns conservatively, especially given known biases.

Community ecology: Variation partitioning analysis revealed that climatic, edaphic, and floristic variables (and their shared effects) are the strongest predictors for community composition of all fungi and most of their functional groups (Leho Tedersoo et al., 2014). However, the saprotroph community composition was most strongly explained by purely spatial variables. More specifically, PET and soil pH explained 2.4 and 1.5%, respectively, of the variation in total fungal community composition. PET contributed 3.8, 2.5, and 11.7% to community structure of saprotrophs, plant pathogens, and yeasts, respectively. Distance from the equator (1.3%) and soil pH (0.7%) were the strongest predictors of EcM fungal community composition, whereas mean annual temperature (4.0%) was the strongest predictor for animal parasites, and distance from the equator (3.5%) was the best predictor for mycoparasites. These results indicate that both environmental and spatial predictors generally have a minor influence on species-level composition of fungi at the global scale. Nonetheless, the significant globalscale pH effect in several groups of fungi is consistent with the substantial influence of pH on the phylogenetic structure of soil fungal and bacterial communities in both local and continental scales. The relatively stronger climatic and edaphic drivers of richness at the class and phylum level suggest that phylogenetic niche conservatism in fungal lineages is similar to cross-biome distribution patterns in vascular plants and protists.

Climatic variables explained the greatest proportion of richness and community composition in fungal groups by exhibiting both direct and indirect effects through altered soil and floristic variables. The strong driving climatic forces identified here open up concerns regarding the impact of climate change on the spread of disease and the functional consequences of altered soil microorganism communities (Van der Putten, et al., 2010). The observed abrupt functional differences between fungal communities in forested and treeless ecosystems, despite spatial juxtaposition, suggests that plant life form and mycorrhizal associations determine soil biochemical processes more than plant species per se. Loss of tree cover and shrub encroachment resulting from drying and warming may thus have a marked impact on ecosystem functioning both aboveand belowground. In addition to natural mechanisms, such as long-distance dispersal and migration over past land bridges, global trade has enhanced the spread of some non-native soil organisms into other ecosystems, where they sometimes become hazardous to native biota, economy, and human health. The results highlighted how little insight we still have into natural microbial distribution patterns, and this undermines our ability to appraise the actual role of humans in shaping these biogeographic processes. Even larger-scale sampling campaigns are needed to provide data for establishing natural distributions and building species distribution models, which will enable us to predict the spread and habitat suitability of non-native microorganisms.

3.4 Global Patterns in Endemicity and Vulnerability of Soil Fungi

Fungi are highly diverse organisms, which provide multiple ecosystem services. However, compared with charismatic animals and plants, the distribution patterns and conservation needs of fungi have been little explored. Leho Tedersoo et al., (2022) examined endemicity patterns, global change vulnerability and conservation priority areas for functional groups of soil fungi based on six global surveys using a high-resolution, long-read metabarcoding approach. It was found that the endemicity of all fungi and most functional groups peaks in tropical habitats, including Amazonia, Yucatan, West-Central Africa, Sri Lanka, and New Caledonia, with a negligible island effect compared with plants and animals. It was found that fungi are predominantly vulnerable to drought, heat and land-cover change, particularly in dry tropical regions with high human population density. Fungal conservation areas of highest priority include herbaceous wetlands, tropical forests, and woodlands. More attention should be focused on the conservation of fungi, especially root symbiotic arbuscular mycorrhizal and ectomycorrhizal fungi in tropical regions as well as unicellular early-diverging groups and macrofungi in general. Given the low overlap between the endemicity of fungi and macroorganisms, but high conservation needs in both groups, detailed analyses on distribution and conservation requirements are warranted for other microorganisms and soil organisms.

Comparative studies indicate that aboveground and belowground biodiversity is driven by different environmental predictors at local and global scales. This suggests differential responses of macro- and microorganisms to land use and climate change. Despite the vast global climatic gradients, soil pH appears to be the main driver of microbial diversity. For example, fungal richness peaks at weakly acidic soils, whereas bacteria and protists are highest in neutral soils. Another study that did not measure pH found that the fungal diversity peaked at lower temperatures, but dominant fungal species were most-influenced by mean annual temperature. As for plants and animals, soil fungal communities are likely vulnerable to global change drivers. For instance, high temperature stress (Masiak et al., 2021) and prolonged drought can alter fungal growth, functionality and community composition. Likewise, changes in land use that result in habitat fragmentation may lead to shifts in the prevalence of pathogenic, mutualistic and free-living fungal groups.

Compared with macroorganisms, information about the conservation needs of fungi and other microbes is very limited. While thousands of plant and animal species are listed as threatened on the International Union for Conservation of Nature (IUCN) global Red List, only 262 out of an estimated 2.2–3.8 million fungal species have been listed as such. The majority of these are from high-income countries in temperate regions and are from fungal groups that form conspicuous macroscopic fruiting bodies (Cao et al., 2021). However, the vast majority of fungi produce no, microscopic or inconspicuous fruiting bodies and are therefore difficult to survey using traditional morphology- and culturing-based identification methods, which have hampered assessments of their distribution. Consequently, we have no information for which parts of the world fungal conservation needs are the highest.

General findings: The recently generated Global Soil Mycobiome consortium data set along with data from five other global soil surveys and international nucleotide sequence databases were used to determine the diversity and endemicity of fungal functional groups, viz. AM fungi, ectomycorrhizal (EcM) fungi, non-EcM Agaricomycetes (mostly saprotrophic macrofungi), molds, pathogens, OHPs (mostly thermophilic saprotrophs), early-diverging unicellular lineages and yeasts (Leho Tedersoo et al., 2022). Compared with previous meta-analytical approaches, the cumulative data comprise the largest available globally standardized database based on directly comparable soil sampling and long-read molecular analysis protocols. Collectively, all data sets yielded 20,182,427 fungal reads composed of 905,841 “species”-operational taxonomic units (OTUs), each defined at <98% sequence similarity of the rRNA ITS barcode from all other OTUs. *Tomentella* (Basidiomycota), *Penicillium* (Ascomycota), and *Mortierella* (Mortierellomycota) were the most species-rich genera recorded. While endemicity patterns of non-EcM Agaricomycetes and AM fungi were similar to those shown for all fungi, different patterns were found for other functional groups (Fig. 22).

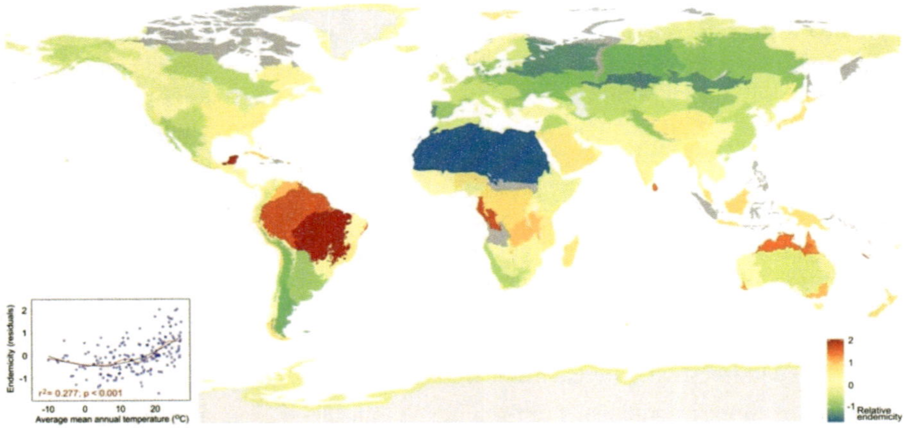

Fig. 22: Average endemicity of soil fungi across ecoregions. The inset graph indicates that endemicity has a positive and U-shaped relationship with mean annual air temperature, the best predictor of endemicity. Grey ecoregions were excluded because of insufficient data.

Endemicity of EcM fungi was related to high mean annual precipitation (R^2_{adj} =.147), with peaks in moist and dry tropical forests and temperate rainforests of Patagonia and New Zealand. Molds had relatively high endemicity in Amazonia, whereas pathogens and yeasts showed multiple endemicity hotspots. Molds (R^2_{adj} =.199) and pathogens (R^2_{adj} =.105) had relatively greater endemicity in strongly acidic or alkaline soils, indicating that extreme soil conditions may support unique soil biota, with limited effective dispersal across edaphically extreme habitats. Human footprint had a weak negative effect on endemicity of all fungi (R^2_{adj} =.018), pathogens (R^2_{adj} =.015) and OHPs (R^2_{adj} =.056), suggesting that anthropogenic habitat loss or homogenization may affect endemic species. European ecoregions had the lowest endemicity for all fungi (R^2_{adj} =.065), pathogens (R^2_{adj} =.086) and unicellular fungi (R^2_{adj} =.035) compared with those of other subcontinents. Averaged current aerial bioclimatic variables better explained endemicity compared with the ranges of those variables or bioclimatic variables of soil and LGM. Climate change since the LGM had a weak positive effect on endemicity of molds (mean diurnal range and overall climate change: R^2_{adj} =.073) and OHPs (bioclimatic variables climate isothermality and mean diurnal range: R^2_{adj} =.056).

The patterns in fungal endemicity were relatively consistent among the five individual endemicity indices and that, in agreement with our first hypothesis, they resembled endemicity patterns of vascular plants and animals, which exhibit major hotspots in wet tropical habitats. However, in a striking contrast to plants and animals, fungal endemicity showed no detectable relationship with insularity. This lack of insularity effect may reflect the greater long-distance dispersal capacity of fungal spores relative to plant propagules and animals. The negligible effects of soil nutrients on fungal diversity and endemicity (this study) suggest that nutrient availability may not limit fungal growth or promote speciation.

The literature abounds with hypotheses, including narrower niche breadth, more asymmetric biotic interactions (i.e., greater specialization to symbiotic partners), long-term climatic stability (LGM and earlier epochs), and more rapid evolution due to environmental energy (metabolic hypothesis), to explain the greater macroorganism richness and endemicity found in the tropics than elsewhere. The observed negligible effects of the LGM suggest that climatic stability is not an important driver of fungal endemicity, a pattern which contrasts with that of plants and animals. The greater phylogenetic diversity and mean neighbour taxonomic distance of fungi noted for the tropics reflect tropical origins for many lineages, as well as extensive radiation and rapid speciation of certain genera in higher latitude areas. On a global scale, plant diversity does not appear to be causally related to fungal diversity, but there is some evidence for stronger mutualistic plant–fungal interactions related to high rainfall. Pathogenic interactions warrant further research in this respect, given their major importance as regulators of plant diversity (Chen et al., 2019). Tropical soil fungi have relatively greater dispersal limitations and narrower distribution ranges (this study), suggesting that high local diversity may contribute to greater regional-scale endemicity.

3.4.1 Vulnerability to Global change drivers: Leho Tedersoo et al., (2022) evaluated the relative vulnerability of soil fungal functional groups by estimating the percentage of species occurring at their upper niche limits to three major global change drivers,land use (land cover change), heat (maximum monthly temperature), and drought (lowest quarterly precipitation). For all fungi taken together, predicted vulnerability to heat (best predictor: maximum monthly temperature; R^2_{adj} =.583) and drought (precipitation seasonality; R^2_{adj} =.456) were greatest in the drylands of tropical and subtropical latitudes. Vulnerability to land use change (best predictor: climate isothermality; R^2_{adj} = 145) peaked in the tropics. The overall additive global change vulnerability was, thus, the highest in densely populated and drier tropical and subtropical regions, especially in India and the sub-Saharan Sudanian savanna and was positively associated with temperature-related variables (cumulative R^2_{adj} =.382). Fungal functional groups had similar vulnerability patterns, which were mostly related to temperature. Among fungal groups, average vulnerability scores were highest for AM and EcM symbionts and unicellular fungi, but these scores differed only slightly across the global change drivers. The actual vulnerability was probably underestimated for biotrophic pathogens and EcM fungi because these groups associate with a limited number of plant species and are sometimes host-specific. Therefore, the loss of one of only a few key symbiotic partners may greatly reduce the biotic niches of specialist fungi.

Patterns of vulnerability in fungi are somewhat similar to those of terrestrial plants and animals, and in agreement with our second hypothesis, vulnerability peaks in drylands prone to desertification (Warren et al., 2013), Arctic/alpine areas (cold-adapted species) and regions with dense human populations. The relatively low vulnerability to heat in tundra-inhabiting fungi can be explained by their relatively high-temperature optima acclimation and poleward migration potentials, despite relatively greater predicted warming in Arctic ecosystems. Above certain tolerance thresholds, soil organisms may be physiologically constrained by increasing soil temperature and evaporation, lower soil water potentials and loss of oxygen due to greater respiration and faster decomposition, which result in hampered soil functioning and ecosystem multifunctionality. Open areas are predicted to increase due to climate change and human activities. This will further expose soil to solar radiation and result in erosion and the loss of plant hosts for fungi. Losses of soil carbon and nutrients, but also anthropogenic nitrogen deposition, may exacerbate fungal vulnerability. These drivers are expected to be more influential locally (Correia et al., 2021) and are partly accounted for by land cover change in this study. Furthermore, while here we calculated average vulnerabilities by summing the effects of individual drivers, global change impacts tend to be synergistic, so actual vulnerabilities may be much higher.

Soil fungi show strong endemicity patterns, which differ between functional groups and are driven by both climatic and edaphic factors. Fungal groups also differ strongly in their relative vulnerability scores to global change, which peak in heavily populated dry land areas. Unfortunately, these are the very areas most prone to further land degradation and desertification, with the potential loss of many species. Fungal endemicity and vulnerability patterns only partly mirror those of vascular plants and animals, with virtually no insularity effects detected among fungi-a pattern that may be ascribed to their more efficient dispersal mechanisms. Global conservation efforts should include fungal biodiversity surveys alongside assessments of soil health, below- and aboveground feedbacks and areas of highest conservation priority, to secure the protection of keystone host species and natural habitats. Furthermore, they should include the monitoring of regional fungal communities over time, to detect relevant changes and provide early warning signals of impending change. The analysis for soil fungi presented here strongly suggests that microorganisms, and soil organisms in general, also deserve detailed assessments of their geographic distributions and conservation needs.

3.4.2 Connecting the Multiple Dimensions of Global Soil fungal Diversity: As the largest contributor to global biomass after plants and bacteria, soil-inhabiting fungi play crucial roles in maintaining the health, productivity, and nutrient cycling of terrestrial ecosystems (Vladimir Mikryukov et al., 2023). Yet, our understanding of fungal diversity distribution across the globe is incomplete, hampering analysis and prediction of diversity-functioning relationships in changing environments. The relatively recent emergence of high-throughput molecular identification offers the tools for large-scale surveys of diverse fungal communities. Pioneering global studies, incorporating hundreds of samples and focusing on dominant phylotypes, have consistently reported a major role of precipitation in soil fungal alpha diversity (i.e., plot-scale diversity) and the existence of a latitudinal diversity gradient (the pattern of poleward declines of biodiversity intrinsic to many large groups of organisms). A recent metastudy involving heterogeneous data from a large number of soil samples (>3000) inferred higher fungal diversity in hyper-arid and Arctic habitats compared to much of the tropics, challenging the common understanding about the global distribution of fungal diversity and of biodiversity in general.

While the biogeographic patterns seen in plants and animals are explained by competing hypotheses involving interactions between various factors of diversity, the joint effect of key drivers—water and thermal energy supply-on the global distribution of fungal alpha diversity has not been assessed. There is also a notable lack of global data providing joint analysis of other diversity aspects, including beta diversity (spatial variability of community composition), gamma diversity (regional species richness), and phylogenetic diversity (incorporating the evolutionary component of community diversity). Considering these diversity aspects together is pivotal for comprehending the spatial organization of biodiversity and becomes an imperative in development of sustainable ecosystem management strategies and effective biodiversity conservation schemes (Socolar et al., 2016).

Vladimir Mikryukov et al., (2023) analyzed global patterns of fungal taxonomic and phylogenetic alpha, beta, and gamma diversities based on long-read DNA sequencing of ca. 4000 topsoil samples, collected during six global surveys following comparable sampling and molecular analysis protocols: the Global Soil Mycobiome consortium (GSMc), BIODESERT, MUSGONET and CLIMIFUN, GlobalAM and GlobalWetlands. The samples encompass a wide variety of land cover types, including different types of woody and herbaceous plant communities, deserts, and agricultural and urban environments. To increase the discriminatory power of the analysis compared to previous works, we use the full-length internal transcribed spacer (ITS) region of the ribosomal RNA (rRNA) gene as a taxonomic marker (Tedersoo et al., 2018). The resulting >14 million high-quality sequences are clustered into 800,000 operational taxonomic units (OTUs) using a 98% similarity threshold, a practical compromise for grouping sequences roughly corresponding to fungal species. Although the resulting OTUs may not always precisely represent distinct species, they serve as taxonomic proxies whose distribution patterns are likely similar to corresponding species.

How the multiple facets of soil fungal diversity vary worldwide remains virtually unknown, hindering the management of this essential species-rich group. By sequencing high-resolution DNA markers in over 4000 topsoil samples from natural and human-altered ecosystems across all continents, Vladimir Mikryukov et al., (2023) illustrated the distributions and drivers of different levels of taxonomic and phylogenetic diversity of fungi and their ecological groups. It was showed the impact of precipitation and temperature interactions on local fungal species richness (alpha diversity) across different climates. The findings revealed how temperature drives fungal compositional turnover (beta diversity) and phylogenetic diversity, linking them with regional species richness (gamma diversity). Fungi were integrated into the principles of global biodiversity distribution and present detailed maps for biodiversity conservation and modeling of global ecological processes.

While fungi exhibit immensely high functional diversity, they are, much like other organisms, commonly categorized by their most prominent roles in ecosystem functioning into pathogens, symbionts, and decomposers of dead organic matter. Leveraging the expert-curated FungalTraits database, species were categorized into functionally distinct, minimally overlapping groups: arbuscular mycorrhizal (AM) and ectomycorrhizal (EcM) fungi, molds, nonmycorrhizal Agaricomycetes (NMA;

mostly represented by saprotrophic macrofungi), and a group of species that are nonsymbiotically biotrophic on a wide variety of organisms (hereafter "pathogens"). We consider yeasts and nonyeast unicellular fungi that exhibit markedly different lifestyles from the predominant multicellular mycelial forms separately (Boekhout et al., 2022). Given their importance from a public health standpoint, opportunistic human pathogens (OHPs) were also distinguished.

The results of this multidimensional analyses indicated that, alongside other kingdoms, soil fungi exhibit peaks in alpha, beta, and gamma diversities in tropical (mostly mountainous) regions while decreasing toward the poles and abruptly dropping in hyper-arid regions(Vladimir Mikryukov et al., 2023). While water deficit in drylands is an evident limiting factor for alpha diversity, the models incorporating water-energy balance also explain the declines in local diversity in tropical hyper-humid soils-one of several unresolved issues in fungal biogeography. In addition, phylogenetic and species turnover are linked to gamma diversity by spatial variability in thermal energy supply. By integrating the distribution patterns of taxonomic, phylogenetic, and ecological group diversities, the mechanisms were outlined shaping soil fungal communities across environmental gradients and space and highlight the pivotal roles of EcM and AM fungi, molds, and nonyeast unicellular fungi in the mechanisms. One of the critical findings is that the predicted functional uniqueness of soil mycobiota across the majority of terrestrial ecoregions implies their role as functional units with different roles in global nutrient cycles. The potential cross-regional impacts of the loss of such functional units emphasize the imperative of collaborative conservation efforts that aim to preserve all types of biota to safeguard biosphere functional integrity. Considering the heightened vulnerability of endemic species to environmental changes, it was proposed, assessing the conservation value of local mycobiota based not only on its diversity but also on its compositional uniqueness. The models and high-resolution maps provided by our work represent the basis for conservation planning (of fungi or groups dependent on them) and for considering soil fungi in explanatory and predictive modeling of macroecological, biogeochemical, and epidemiological processes.

3.4.3 Meta-analysis of Global Fungal Biogeography: Considering the magnitude of ongoing global change, it is of high importance to determine how climate and other environmental factors affect the diversity and distribution of fungal communities (Větrovský et al., 2019). Fungal symbionts can benefit plants by ameliorating abiotic stressors associated with climate change, such as heat and drought, but the distribution of these symbionts may be driven by mechanisms other than climate. However, because the geographic distribution and environmental preferences of nearly all fungi remain unknown, their current status and the future threats to their existence are difficult to assess. Due to the importance of plant–fungal interactions, the ability to predict shifts in fungal distributions could help to understand or predict ecosystem-level changes.

The evolutionary and environmental factors that shape fungal biogeography are incompletely understood. Větrovský et al., (2019) assembled a large dataset consisting of previously generated mycobiome data linked to specific geographical locations across the world. This dataset was used to describe the distribution of fungal taxa and to look for correlations with different environmental factors such as climate, soil and vegetation variables. This meta-study identified climate as an important driver of different aspects of fungal biogeography, including the global distribution of common fungi as well as the composition and diversity of fungal communities. Fungal diversity was concentrated at high latitudes, in contrast with the opposite pattern previously shown for plants and other organisms. Mycorrhizal fungi appear to have narrower climatic tolerances than pathogenic fungi. It was speculated that climate change could affect ecosystem functioning because of the narrow climatic tolerances of key fungal taxa.

Climatic and edaphic variables as well as vegetation features were recently reported to influence the composition of fungal communities. Moreover, fungal diversity has been reported to respond to climate factors in drylands. It is less clear whether the global fungal biodiversity patterns correspond with higher diversities at low latitudes that were previously demonstrated for terrestrial macroorganisms and bacteria. The distribution of plant and animal species also demonstrates strong division into biogeographical regions. Although some biogeographic patterns were also observed for bacteria, other data show that dominant microbes are widespread and may thus exist in all areas of suitable environments. In addition to the climate, soil characteristics or vegetation, fungal diversity and community composition might also be shaped by other factors, such as dispersal limitations of certain taxa, although transcontinental dispersal of fungi has also been documented. Similarly, fungal community surveys have shown a mixed degree of regional endemism and identified groups with rather cosmopolitan distributions (Davison et al., 2015).

The current paradigm suggests that regional fungal assemblages are characterised by frequent endemic species occurrences and the distribution of these endemics is constrained by dispersal limitations. The dataset of this study does not allow us to address the question of endemism directly, however, most of the top SH were cosmopolitan species, with 29% found on all six continents, 20% found on five continents and 14% found on four continents (Větrovský et al., 2019). Only 8.5% of top taxa were constrained to a single continent. Of the 3943 frequent SHs that occurred in 10 or more samples, the share of taxa constrained to one continent was also low (23%), while 48% of SHs occurred on three or more continents and 8% occurred on all continents. In a global survey of arbuscular mycorrhizal fungi, endemism was also rare and limited to 7% of taxa found on only one continent while 34% of taxa occurred on all sampled continents and contrasted with the high frequency of endemism displayed by plants. Although mostly cosmopolitan, top taxa were frequently associated with samples from sites with similar climates.

The overall compositions of the fungal communities showed strong association with environmental variables, while separate clustering by continents was less clear. The limited share of potentially endemic species among top taxa may indicate that for the most common fungal species, environmental variables represent more important drivers of community composition than local speciation and dispersal barriers. This is also well documented by high similarity of fungal communities from similar environments, but different continents (e.g., tropical rain forests in South America,

Africa and Southeast Asia). In contrast to macroorganisms such as mammals and plants, the distribution of dominant fungi seems to be less constrained by geographic barriers, probably due to better dispersal abilities. Given that there are also a number of cases where endemism and dispersal limitations have been documented for populations and higher taxonomic groups and considering that the share of endemic species is likely higher for less common fungal taxa, future work should attempt to reconcile these observations and our findings with respect to frequently observed taxa.

3.5 Soil Fungal Diversity and Assembly along a Xeric Stress Gradient

Fungi are ubiquitously present in the environment and can tolerate extremely diverse environmental conditions, including the desiccation and temperature extremes of deserts. Arid soils are primarily dominated by the fungal phyla Ascomycota, Basidiomycota and Chytridiomycota (Vikram et al., 2023). Soil fungal communities generally show higher resistance to environmental disturbances than prokaryotic communities. Nevertheless, a number of climatic and edaphic variables have been reported to affect edaphic fungal diversity. Increasing aridity has particularly shown to decrease soil fungal abundance and diversity. In the Namib Desert, edaphic fungal communities are dominated by Ascomycota independently of the soil studied; i.e., dune vs gravel plain or rhizosphere/rhizosheath vs open soils. A recent diel shotgun meta-transcriptomics analysis of low-rainfall Namib Desert gravel plain soils also indicated that Ascomycota were the most active fungal taxa, particularly during the cooler night hours. Furthermore, edaphic fungal communities were found to assemble primarily through deterministic processes and particularly environmental filtering and to be strongly influenced by soil composition (particularly clay and sand content) and by soil NH_4^+ and total organic carbon contents. However, these studies compared very different Namib Desert edaphic systems (i.e., dune vs gravel plain vs salt pans vs riverbeds and/or open soils vs plant-associated rhizosphere/rhizosheaths) which can bias the results of such analyses.

The Namib Desert of south-western Africa is one of the oldest deserts in the world and possesses unique geographical, biological and climatic features. While research through the last decade has generated a comprehensive survey of the prokaryotic communities in Namib Desert soils, little is yet known about the diversity and function of edaphic fungal communities, and even less of their responses to aridity (Vikram et al., 2023). Soil fungal community diversity was characterized across the longitudinal xeric gradient across the Namib desert (for convenience, divided into the western fog zone, the central low-rainfall zone and the eastern high-rainfall zone), using internal transcribed sequence (ITS) metabarcoding. Ascomycota, Basidiomycota and Chytridiomycota consistently dominated the Namib Desert edaphic fungal communities and a core mycobiome composed of only 15 taxa, dominated by members of the class Dothideomycetes (Ascomycota), was identified. However, fungal community structures were significantly different in the fog, low-rainfall and high-rainfall zones. Furthermore, Namib Desert gravel plain fungal community assembly was driven by both deterministic and stochastic processes; the latter dominating in the all three xeric zones. Data was presented that suggest that the inland limit of fog penetration represents an ecological barrier to fungal dispersal across the Namib Desert.

Three phyla, namely, Ascomycota (41%), Basidiomycota (18%) and Chytridiomycota (9%), clearly dominated the Namib Desert edaphic mycobiome. The overall dominance of members of the phylum Ascomycota and the low relative abundance of Chytridiomycota in arid environment mycobiomes has also been reported (Vikram et al., 2023). Chytridiomycota represented 14% and 10% of the low-rainfall and high-rainfall soil sample communities, respectively, but were not detected in fog-zone communities. Similarly, Ascomycota were significantly more abundant in the high-rainfall zone than in fog zone communities (Tuckey's HSD, q = 0.18. In contrast, Basidiomycota abundances did not differ significantly across the three zones (ANOVA p > 0.05). Other fungal phyla (Rozellomycota, Mucoromycota, Mortierellomycota, and Calcarisporiellomycota) were identified at less than 2% relative abundance and showed no significant differences between the xeric zones. At the genus level, fog zone soils were significantly enriched in *Cladosporium* (Ascomycota; fog-low rainfall, $q = 0.005$; fog-high rainfall, $q = 0.005$), *Malassezia* (Basidiomycota; fog-low rainfall, $q = 0.003$; fog-high rainfall, $q = 0.004$) and *Alternaria* (Ascomycota; fog-low rainfall, $q = 0.033$; fog-high rainfall, $q = 0.038$) and significantly depleted in members of the *Monosporascus* (Ascomycota; fog-low rainfall, $q = 0.036$) and *Rhizophlyctis* (Chytridiomycota; fog-low rainfall, $q = 0.0004$; fog-high rainfall, $q = 0.035$) genera. The higher relative abundance of the genera *Cladosporium*, *Malassezia* and *Alternaria* in the fog zone parallels observations from the Atacama and Namib Desert, suggesting that the resident members of these fungal genera may be specifically adapted to low water-activity and higher soil salinity habitats. *Westerdykella* were significantly more abundant in high-rainfall zone soils (Ascomycota, q < 0.05).

Of the 2508 fungal ASVs identified, 73.2% (i.e., 1835/2508) were zone-specific; i.e., exclusively present in either the fog (156/2508; 6.2%), low-rainfall (557/2508; 22.2%) or high-rainfall (1122/2508; 44.7%) zone soils ; 600 fungal ASVs were shared between the low-rainfall and high-rainfall zones, while very few were shared between the other two zone pairs (fog/low-rainfall and fog/high-rainfall. The core mycobiome of the Namib Desert gravel plains; i.e., the fungal taxa common to all three zones, was only composed of fifteen ASVs (0.6% of the total ASVs but representing 6.7% of total rarefied sequences). This further confirms our contention that intense selective pressure occurs in the different xeric zone soils; i.e., deterministic niche-partitioning. All members of the core mycobiome were assigned to the Ascomycota Dothideomycetes (orders Pleosporales, Capnodiales, Dothideales and Hypocreales) and Basidiomycota Tremellomycetes (order Filobasidiales) classes. Dothideomycetes are known to be able to tolerate multiple environmental extremes - including those typical of hot deserts (i.e., high temperatures, high solar radiation fluxes and extreme dehydration), and Filobasidiales are dominant fungal taxa in Atacama Desert soils.

3.6 Global Diversity and Biogeography of phytopathogenic fungi in a changing world

Phytopathogenic fungi pose a major threat to global food security, ecosystem service delivery and human livelihoods. Plant diseases have a range of direct, quantifiable economic consequences for crop and forest management across a range of terrestrial environments (Li et al., 2023). For example, the persistent presence of these diseases would only leave enough food to feed ~1/3 of the world's population, and a handful of plant diseases in forests could reduce global CO_2 absorption annually by 230–580 megatons. Attempts to mitigate the effects of such diseases caused by phytopathogenic fungi, via chemical fungicides and biocontrol agents, are often locally ineffective and globally inadequate due to the complexity and diversity of phytopathogenic fungal communities and an incomplete understanding of the factors regulating them. Therefore, ascertaining the distribution and the environmental attributes that structure phytopathogenic fungal communities across the globe was recently considered to be a priority research direction.

Despite their ecological and agricultural importance, very few studies have focused on phytopathogenic fungal distributions at the global scale, but with limited taxonomic and environmental coverage, restricting our understanding of the role of climate in shaping these communities into the future. Given the tightly coupled relationships between climate and pathogen development, climate change has increased the prevalence and severity of some human, animal and plant diseases. While predicting the consequences of climate change remains challenging, models have already predicted that crop yields, carbon sequestration and pollination rates are expected to decrease, whilst evapotranspiration and tree mortality are expected to increase under future climate scenarios. In addition, as the geographic ranges of plant species shift in response to climate change and alien species are introduced, the potential of emerging novel plant pests and pathogens increases leading to increased incidence and severity of the diseases they cause (Wang et al., 2022). Considering the magnitude of global climate change, it is imperative to determine how a changing climate affects the distribution of phytopathogenic fungi, and potentially to use this new knowledge to inform policies to control the emergence of future plant diseases and maintain ecosystem functions and services.

Phytopathogenic fungi threaten global food security but the ecological drivers of their global diversity and biogeography remain unknown. Li et al., (2023) constructed and analysed a global atlas of potential phytopathogenic fungi from 20,312 samples across all continents and major oceanic island regions, eleven land cover types, and twelve habitat types. They showed a peak in the diversity of phytopathogenic fungi in mid-latitude regions, in contrast to the latitudinal diversity gradients observed in aboveground organisms. This study identified climate as an important driver of the global distribution of phytopathogenic fungi, and the models suggested that their diversity and invasion potential will increase globally by 2100. Importantly, phytopathogen diversity will increase largely in forest (37.27-79.12%) and cropland (34.93-82.51%) ecosystems, and this becomes more pronounced under fossil-fuelled industry dependent future scenarios. It was recommended improved biomonitoring in forests and croplands, and optimised sustainable development approaches to reduce potential threats from phytopathogenic fungi.

For the top ten most relatively abundant genera, seven (*Fusarium, Alternaria, Acrodontium, Trichoderma, Epicoccum, Erysiphe* and *Pseudocercospora*) were expected to increase in relative abundance, while the other three genera (*Fusicladium, Neoerysiphe* and *Plectosphaerella*) were predicted to show opposite trends. It is important to note that these predictions are based on observational data alone, with no mechanistic inference to drive these trends. Moreover, our projections are founded on a permanent climate-diversity/relative abundance relationship, and the projections may need to be amended if climate-diversity/relative abundance relationships change under future climate scenarios. For example, if in the future, extremely high temperatures experienced in some regions exceed the thermal limits for growth of many phytopathogenic fungi, the prediction will fail for those regions. The predictions may also be biased if major land use or vegetation type changes occur. Nevertheless, these predictions are consistent with previous research, which predicted an increased proportion of soil-borne pathogens (~0.8% to 2.3%) under warming conditions in global soils, and thus lends additional inference to support this global trend.

Limitations, conclusions and future perspectives: This study provided an omnidirectional understanding on the global diversity and biogeography of phytopathogenic fungi. It is important to acknowledge that using closed-reference data, and lacking a reference database fully resolved to species level, may obscure some of the overall diversity and abundance of phytopathogenic fungi (Li et al., 2023). However, the distributions and biogeographic patterns of phytopathogenic fungi presented here are still valuable, and provide a robust first-order approximation of their underlying ecology and susceptibility to climate change. In addition, CO_2 fertilisation effects were not considered in this study's predictions of future changes in phytopathogenic fungi, as the direct relationship between CO_2 concentration and phytopathogens remains ill-defined. Furthermore, without global data on the range distribution of plants, directly linking fungal pathogens with actual plant host distributions remains challenging. The observational data can only provide correlative insights into the distributions of plant pathogens, but a mechanistic understanding of current and future trajectories of pathogen biogeography will require detailed information about host plant ranges. Further research into the distribution of plant species and their niche ranges will be fundamental to facilitating the joint species distribution modelling that is needed to assess these trends in the future.

These findings highlight the importance of climate factors in determining the diversity and composition of phytopathogenic fungal communities across the globe. More importantly, the proposed model projections indicate a potential increase in phytopathogenic fungal diversity, and an increased global invasion risk especially under unsustainable development scenarios. Such increased fungal pathogen risks are likely to have direct consequences for the productivity and sustainability of managed and natural ecosystems, with direct implications for food production and carbon sequestration. The research suggests that low-emission-dependent climate scenario can mitigate the invasion risks caused by climate changes, shedding light to optimised sustainable development solutions that protects agriculture and forestry from the harmful impacts of phytopathogens.

3.7 Fungal Communities in soils under Global change

Plant pathogenic fungi, on the other hand, may opportunistically attack plant hosts that are under stress due to the rapid change in their environment (Desaint et al. 2021). Therefore, soil fungi, particularly plant associated guilds, mediate the effects of global change on natural vegetation and agricultural crops in multiple ways. In addition to direct effects, climate change can indirectly affect soil fungi through shifts in soil chemistry and vegetation structure (Baldrian et al. 2022). It is thus important to understand how global change affects soil fungi. Even though this question has been repeatedly addressed in many contexts and settings in the past, it is still difficult to give a general answer. Soil is the habitat with the highest fungal diversity and generalisations based on the observed response of individual species are difficult. This high diversity is associated with high levels of functional redundancy in the communities of saprotrophic as well as symbiotic fungi. Consequently, loss of some species may in theory be replaced by other taxa. However, the critical level of species loss with consequences for ecosystem processes remains largely unknown. Additionally, the diversity, and dependence on plant hosts, of fungal lifestyles (i.e., free-living saprotrophs, mutualistic symbionts and plant pathogens) affect fungal species responses to climate change.

Soil fungi play indispensable roles in all ecosystems including the recycling of organic matter and interactions with plants, both as symbionts and pathogens. Past observations and experimental manipulations indicate that projected global change effects, including the increase of CO_2 concentration, temperature, change of precipitation and nitrogen (N) deposition, affect fungal species and communities in soils. Although the observed effects depend on the size and duration of change and reflect local conditions, increased N deposition seems to have the most profound effect on fungal communities (Baldrian et al., 2022). The plant-mutualistic fungal guilds, ectomycorrhizal fungi and arbuscular mycorrhizal fungi, appear to be especially responsive to global change factors with N deposition and warming seemingly having the strongest adverse effects. While global change effects on fungal biodiversity seem to be limited, multiple studies demonstrate increases in abundance and dispersal of plant pathogenic fungi. Additionally, ecosystems weakened by global change-induced phenomena, such as drought, are more vulnerable to pathogen outbreaks. The shift from mutualistic fungi to plant pathogens is likely the largest potential threat for the future functioning of natural and managed ecosystems. However, our ability to predict global change effects on fungi is still insufficient and requires further experimental work and long-term observations.

Ecological guilds of fungi and global change: Global surveys of soil fungal occurrences in the GlobalFungi database show that members of various fungal guilds differ in the size of their climatic niche. Moreover, the level of dependence on vegetation varies from obligate biotrophs to free-living fungi (Lekberg et al., 2021). Due to this, global changes are expected to affect various ecological guilds of soil fungi (ECM fungi, AM fungi, ERM fungi, plant pathogens and saprotrophs) differently, affecting their relative share or community composition (Baldrian et al. 2022). These shifts may subsequently result in changes in various ecosystem processes such as decomposition rate or plant performance. Importantly, climate change-driven shifts in plant communities may lead to shifts in the host availability affecting those fungi that have a narrow host range. With increasing warming, some alpine communities have seen the replacement of forbs with deep rooted grasses and increasing nitrogen deposition can lead to reduced species richness though this effect depends on ecosystem characteristics, such as mean annual precipitation. Altered environmental conditions promote not only natural range shifts of plants species, but also enable naturalisation of alien plant species outside their native distribution range. Such events can affect local ecosystems and their fungal components in several ways: by competition for resources, by the introduction of novel fungal species (such as mycorrhizal symbionts or pathogens), or by selective recruitment of root-associating fungal species already present in the local pool by the alien plants. Because of all these factors, changes in local fungal communities are expected as has been already observed for plant introductions.

Due to the complex effects of N on soil chemistry and vegetation, and the fact that mutualistic mycorrhizal fungi mediate its transfer to plants, change in atmospheric deposition is perhaps the factor with greatest importance for guild composition of soil fungi. Indeed, nitrogen addition to 25 grasslands distributed across four continents led to the increase of fungal pathogens, although it did not significantly affect AM fungi and saprotrophs. These guild level responses were primarily mediated through nutrient-induced shifts in plant communities (Lekberg et al., 2021). On the other hand, no consistent shifts in guild composition were observed across N-supplemented forests in the USA. Among the various aspects of global change, changes in climate lead to severe ecosystem alterations. Forests are already facing increasing lengths of heat waves with unprecedented increases of temperature in high latitudes combined with long drought periods. This high level of climate stress likely increases the vulnerability of forests to disturbances including tree dieback and forest fires. These severe forest disturbances were shown to result in a shift of fungal communities from those dominated by ectomycorrhizal fungi in undisturbed forests to those dominated by saprotrophs in disturbed forests as a response to changes in primary productivity.

Plant pathogens: Plant pathogens may better cope with climate change than other fungal guilds (Chaloner et al., 2020). Conditions that affect pathogen overwintering and dispersal are of essential importance due to pathogen lifestyles, survival in soils, and outbreaks triggered by climatic and plant host signals. Global warming in areas with seasonal temperature variation has increased pathogen survival during winters and increased the length of vegetation seasons leading to faster pathogen spread or stronger outbreaks (Baldrian et al. 2022). Warming appears to be the most important driver of plant pathogen abundance. Climatic factors, especially the MAT and precipitation seasonality were the most important predictors of the relative abundance of plant pathogens across 235 global sites. Under future climate change and land-use scenarios, relative abundance of plant pathogens is predicted to increase. A nine-year warming experiment in a dryland on the Iberian peninsula showed higher relative share of pathogens, higher relative abundance of Alternaria and higher absolute abundance of Alternaria in warmed plots. While the increase in relative abundance, or sporulation, of plant pathogens may increase the risk of a disease outbreak, direct causal links may be difficult to find. It is possible that negative responses of mycorrhizal fungi and neutral or positive responses of pathogens to climate change can subsequently manifest in negative responses of vegetation. More importantly, climatic events seem to be predictive factors of fungal disease outbreaks with high humidity and high temperature being the most common factors (Romero et al., 2022). Pathogens may also use the opportunity to attack weakened host communities such as forest ecosystems after dieback caused by drought or heat stress.

In natural systems, pathogens appear to be more abundant in resource-rich environments and nutrient addition (e.g. fertilisation) has been linked to increased disease incidence in plants which may increase the risk of pathogen spread or outbreaks at elevated atmospheric N deposition (Baldrian et al. 2022). The effect of CO_2 increase on pathogens is less clear, however, concentrations of spores of several pathogens were increased by elevated atmospheric CO_2 (eCO_2) in a Populus tremuloides plantation in air and litter. Although the responses of fungi were not uniform, significant increases were found in the potential pathogenic genera *Alternaria, Cladosporium* and *Fusarium.* Plant pathogen community composition may not intrinsically affect ecosystems because it is often individual taxa that cause disease outbreaks. The effects of global change on individual plant pathogen taxa may thus be more important than the guild-level effects. Based on historical observations of higher *Alternaria* spp. spore concentrations at warm temperatures, spore concentrations are predicted to increase with warming in the United Kingdom and future climate models suggest increased prevalence of *Alternaria* brassicae in North Germany. In several instances, eCO_2 increased spore production by *Alternaria* spp. several-fold. Considering disease severity, both warming and eCO_2 has been shown to increase *Alternaria* leaf spot severity on rocket, cauliflower and cabbage.

Global change effects on ecosystems are highly context dependent and there are undoubtedly ecosystems where changes will be more pronounced. Where global change relieves existing limitations, such as the coldest or N-limited areas, novel limitations will arise, such as increased desertification or induced P-limitation, respectively. Unfortunately, these systems are rarely the subject of research. Experimental manipulations in underexplored systems are thus most welcome. Although the experiments combining multiple factors are relatively frequent (Yang et al., 2021), they are in most cases applying unrealistic treatment intensities and so far too rare to allow generalisations. Since global change factors act in combination and their effects are not simply additive, it would be more than welcome to see results of long-term manipulations based on complex predictions of multiple global change factors for given localities. Since it will never be possible to perform manipulations everywhere, long term collection of observational data is needed that would help to describe trends in the soil mycobiome. Global and regional initiatives intending to capture all available types of fungal community data, combined with paired environmental metadata, across time have the potential to scale our understanding of global change effects on soil fungi to a global level (Baldrian et al. 2022).

3.8 Impact of Abiotic Conditions on Fungal Diversity

The Soil fertility play a prominent role in biological sustenance in turn fungi are significant participants in bridging biotic and abiotic interactions between the components in the environment, and they involve in breakdown of organic/inorganic materials during bioleaching/bioconversion leading to bioavailability and thus recycling of nutrients. Local environmental factors (abiotic), such as the chemical and physical properties of the soil, greatly influence the composition of existing fungal communities and determine the diversity of soil fungi (Sindhushri Chauhan et al., 2023). A correlation study of the micro fungi found in ecologically two different regions (Karnataka state, India) of Uttara Kannada and Chitradurga districts was conducted. One is being part of India's Western Ghats, a biologically fertile ecosystem and the other being a dry, arid region of Deccan plateau with scanty rain fall. Soil samples were collected from various geographical locations in order to study and evaluate the role that certain soil physicochemical features play in fungi diversity. According to our findings, soil attributes (fertility) are related to the composition, richness, and diversity of fungi in soils and thus soil's physico-chemical properties assessed were linked to fungal diversity measurements. This led us to indicate the role of abiotic factors affecting the fungal diversity in soil and their prominent role in soil fertility.

The soil sampling has provided inclusive information about the biological species diversity particularly of the fungal isolates playing effective role in soil fertility. The first indications are with the pH of the soil that consistently was around 5.0 (mean average ± 5.607) which is indicative of efficient fungal colonization as it's the nominal growth pH of predominantly fungal species than the bacterial growth and activity. There was a clear difference in the % Organic Carbon (OC) content and Nitrogen (N), Phosphorus (P) and Potassium (K) by more than 40% in total giving an additional indication of the function of fungal diversity effectively playing the role in overall enhancement of the soil fertility index. These variables could include the purposeful movement of microorganisms independently and associated species, through the soil profile by water or bioturbation causing change in geochemistry, organic composition of the soil.

The concentration of hydrogen ions (H+) in the soil solution a factor attributed to the microbial activity of biodegradation and mineral leaching, determines the pH of the soil, there by indicates how acidic or alkaline the soil. Lower pH values are associated with higher H+ ion concentrations, and vice versa. The importance of pH in preserving soil fertility cannot be overstated. Electrical Conductivity/Conductance (EC) is a positive indicator for crops since it aids in nutrient absorption. Low EC values are shown to be suitable for plant growth, indicating better fertility (Jain et al., 2015). One of the important chemical indicators of soil quality is the amount of soil organic matter, often known as soil organic carbon. It influences soil porosity and fosters water and gas exchange processes. The findings demonstrated that the levels of OC in the soil samples from Chitradurga and Uttara Kannada differed significantly and were much greater when compared. Soil OM has a significant function to play in preserving the physical, chemical, and biological qualities of the soil as well as the crop productivity and output. The status of the available organic carbon may be connected to the variation in N content. Low organic matter content, little rainfall, and little vegetation in these soils all contribute to faster decomposition and removal of organic matter, which results in low nitrogen status.

When comparing soil samples from forests and barren areas, it was discovered that the amount of potassium in the former was substantially higher. There was an adequate supply of potassium in both areas (Sindhushri Chauhan et al., 2023). An increase in K might be the result of soil saturation, which widened clay minerals and released previously fixed K, as well as huge fertilizer storage, which caused these to dissolve in floodwater. A higher osmotic pressure in the plant is caused by more potassium in the soil, which increases its ability to absorb water. Phosphorus levels in soil samples from forests in both zones were noticeably higher than those from arid areas. Greater organic matter concentration may be the cause of higher phosphorus levels. Compared to soils with low organic content, soils rich in organic matter provide plants with organic phosphates. In contrast to soils with maximal leaching, soils with less leaching impact have higher levels of phosphorus, thus, giving an understanding of the climatic and ecological conditions playing an important scientific role in modulating the microbial (fungal) diversity and also maintaining the bioavailability and recycling of nutrients. The detailed characterization of the fungal species diversity, their biochemical properties and physiological role in colonizing the soil environments are in due course and would be the extended outcomes of this presently reported work. This would yield clear, detailed insights into the prominence of wildly adopted fungal species and their diversity in regulating bioconversion, bioavailability and recycling of nutrients affecting soil fertility.

An estimate of the lower limit of global fungal diversity: It has been suggested that the species diversity of fungi can be estimated using the species diversity of vascular plants in the same region (John Paul Schmit and Gregory M. Mueller, 2007). In small nature preserves with ongoing fungal surveys, there is a high ratio of fungi to plants, and the overlap between similar, extremely well studied preserves is low. Schmit et al., (2005) documented that tree species diversity was a statistically significant predictor of macrofungal species diversity through a meta-analysis of plot-based diversity studies in which plant and macrofungal diversity data were provided. Their findings support the use of plants as a surrogate taxon for at least some groups of fungi.

John Paul Schmit and Gregory M. Mueller (2007) conservatively estimated that there is a minimum of 712,000 extant fungal species worldwide, but it was recognized that the actual species richness is likely much higher. This estimate was calculated from the ratio of fungal species to plant species for various ecologically defined groups of fungi in well-studied regions, along with data on each groups' level of endemism. These calculations were based on information presented in the detailed treatments of the various fungal groups published in this special issue. The intention was to establish a lower boundary for the number of fungal species worldwide that can be revised upward as more information becomes available. Establishing a lower boundary for fungal diversity is important as current estimates vary widely, hindering the ability to include fungi in discussions of ecology, biodiversity and conservation. Problems inherent in making these estimates, and the impact that additional data on fungal and plant species diversity will have on these estimates were discussed.

There is considerable variation in species diversity and overlap factors among these groups (Macrofungi, Lichens, Lichenicolous fungi, Acuatic fungi, Soil-inhabiting fungi, Terrestrial plant-associated fungi, Arthropod-associated fungi, Microsporidians). Approximately half of the recorded fungal names are of lichens and macrofungi with the other half corresponding to microfungi. In total, it was estimated that there is a minimum of 712,285 fungal species worldwide, over 82% of which are microfungi associated with terrestrial plants. Approximately 13% of these species have already been described, but only 7% of the plant-associated microfungi have been described.

Some groups of fungi, such as endolithic fungi and most that are pathogens of organisms other than plants or arthropods, are not included in this estimate. While these groups of fungi are not currently known to be speciose, their exclusion makes our estimate even more conservative. Arthropod- and terrestrial plant-associated microfungi are clearly the two groups whose species diversity is hardest to assess at this time. For example, a single recent study added over 20% to the number or known yeast species by surveying yeasts found in beetles (Boekhout, 2005). Arthropod- and terrestrial plantassociated fungi are also the two groups that could most influence fungal diversity estimates. Both groups are clearly very diverse, but comprehensive data on their geographic distributions and host/substrata specificity have not yet been compiled. Distribution data, used in conjunction with estimates based on ratios of fungal species to host species, for these fungi are necessary to more accurately estimate the true diversity of fungi.

Research needs: Additional fungal surveys, especially in undersampled tropical regions, and rigorous monographic treatments to resolve species circumscriptions and distributions are crucial to improve our knowledge of species diversity (John Paul Schmit and Gregory M. Mueller, 2007). However, other research needs also exist. While nearly 80,000 fungal species have been described, information about them, their geographic distribution, and plant or insect hosts is widely scattered. Compiling the already available information into publicly available databases would enable researchers to make significant progress in understanding fungal diversity and biogeography. While ratio and other types of estimates may be useful, they are not based on well-proven ecological principles. Rather they are based on the reasonable, but as yet untested, notion that there is some causal link between the species diversity of various groups at various geographic and taxonomic scales. Studies at small scales are needed to elucidate if this is so, and to determine if there are consistent ratios between plants, arthropods, and fungi over various geographic scales.

3.9 Potential Links Between Wood-Inhabiting and Soil Fungal Communities

Advances in metabarcoding approaches using high-throughput sequencing (HTS) platforms enable detailed analysis of community composition and have been harnessed to reveal a more complete picture of fungal diversity in a wide range of habitats, including soil and deadwood (Witoon Purahong et al., 2019). Recent studies have demonstrated that diverse taxonomic and ecological functional groups of WIF colonize deadwood. The functional groups identified include saprotrophs, plant pathogens, endophytes, animal endosymbionts, mycoparasites, mycorrhizae, and lichenized fungi. Although several studies have investigated factors related to WIF community assembly, connections between the WIF community and soil fungi, and dispersal mechanisms responsible for their colonization between these compartments, are still unclear. In forest ecosystems, fungi (in spore or mycelium form) can be transported or dispersed from one place to another by for example (a) wind, (b) splash dispersal, (c) water dispersal, (d) animal dispersal, (e) seed-borne fungi (i.e. endophytes), or the (f) fungal mycelium network.

Wood-inhabiting fungi (WIF) are pivotal to wood decomposition, which in turn strongly influences nutrient dynamics in forest soils. However, their dispersal mechanisms remain unclear. Witoon Purahong et al., (2019) hypothesized that the majority of WIF are soil-borne. For this reason, the presented research aimed to quantify the contribution of soil as a source and medium for the dispersal of WIF to deadwood using high-throughput sequencing. Effects of tree species (specifically *Schima superba* and *Pinus massoniana*) on the percentage of WIF shared between soil and deadwood were tested in a Chinese subtropical forest ecosystem. The taxonomic and ecological functional group affiliations of the fungal community shared between soil and deadwood was also assessed. The results indicated that soil is a major route for WIF colonization as 12%–15% (depending on the tree species) of soil fungi were simultaneously detected in deadwood. We also demonstrate that tree species ($p < 0.01$) significantly shapes the composition of the shared soil and deadwood fungal community. The pH of decomposing wood was shown to significantly correspond ($p < 0.01$) with the shared community of wood-inhabiting (of both studied tree species) and soil fungi. The data suggested that a wide range of fungal taxonomic (Rozellida, Zygomycota, Ascomycota, and Basidiomycota) and ecological functional groups (saprotrophs, ectomycorrhizal, mycoparasites, and plant pathogens) may use soil as a source and medium for transport to deadwood in subtropical forest ecosystem. While 12%–62% of saprotrophic, ectomycorrhizal, and mycoparasitic WIF may utilize soil to colonize deadwood, only 5% of the detected plant pathogens were detected in both soil and deadwood, implying that these fungi use other dispersal routes. Animal endosymbionts and lichenized WIF were not detected in the soil samples. Future studies should consider assessing the relative contributions of other possible dispersal mechanisms (e.g. wind, water splash, water dispersal, animal dispersal, and mycelial network) in the colonization of deadwood by soil fungi.

There are four hypotheses for a shared occurrence of fungal OTUs between the deadwood and the soil, via., a) the fungal OTUs migrated from the soil to the wood, (b) the fungal OTUs migrated from the wood to the soil, (c) the fungal OTU migrated from somewhere else to both the wood and the soil, and (d) the fungal OTUs are ubiquitous, generalist organisms that were present in both the soil and the wood. Witoon Purahong et al., (2019) were not able to quantify the relative important of each hypothesis but we can assess the overall contribution of soil as source and medium for transport of wood-inhabiting fungi to deadwood. It was expected that the number of fungal OTUs migrating from the initial deadwood to the soil to be low, as we did not detect any of the fungal endophytes in the fungal community shared between soil and deadwood. Fungal OTUs that use soil as source and/or medium to colonize deadwood could follow any of the four migratory patterns, because they must be transported to, survive in, and/or colonize the soil. As the bark was removed from deadwood samples before homogenization and DNA extraction, fungi attached to the bark that had not penetrated the inner tissue of the deadwood (i.e. soil- or wind-dispersed fungal spores that had attached to the bark) should have been excluded from our analysis. Deadwood-inhabiting fungal communities in boreal forests interact along the decay gradient of Norway spruce logs, and a relatively high proportion of the total fungal community is present in both soil and deadwood. WIF locally influence the soil-inhabiting fungal communities at all decay stages because certain WIF only occur in the soil under specific decaying logs, but it is impossible to determine how many soil fungal OTUs colonized the deadwood due to the study's experimental set-up. To enable this, a long-term, time-series analysis (from initial decay to late decay stages) of the fungal OTUs present in both soil and deadwood would be required.

Soil-inhabiting fungi can integrate soil physical indicators: The role of soil fungi in soil physical properties and in soil water retention is rarely considered when searching for new agro-environmental indicators, even though i) fungi represent the majority of soil microbial biomass in volume, ii) they are the major player in the C cycle by amply exceeding bacteria in soil respiration and iii) they contribute to soil structures and moisture retention through a series of biochemical and biophysical mechanisms (Manici et al., 2019). Indeed, most fungi can absorb water from atmospheres at a lower relative humidity than those occurring at the permanent wilting point, this suggests a more important role than that hypothesized to date, especially in arid soils. Most of the evidence on positive correlation between soil fungi, soil water retention capacity and soil hydraulic properties so far p rovided concerns mycorrhizal fungi which play a role in shaping hydraulic properties of soils through multiple direct and indirect mechanisms. Indeed, these root symbionts of vascular plants can interconnect soil and host plants through hyphae by providing a series of benefits such as uptake of mineral nutrients and water; furthermore they improve soil aggregation and porosity through the enhancement of soil organic matter accumulation. To the contrary, little has been investigated on the effect of free-living filamentous fungi on soil water and hydraulic properties. Free-living saprophytic soil fungi, belonging to the Ascomycota and Basidiomycota divisions, play a dominant role from decomposition of plant residues to cycling processes, thus representing the largest part of soil microbial biomass. Saprophytic fungi mediate the formation of soil structure by the binding effect of the hyphal structure, a plastic structure having a growth habit based on mycelium extension and branching therefore absorbing nutrients, growing through air and bridging soil pores.

A study on the relationships between fungi and physical indicators of the soil applying a multivariate approach was performed in an olive growing area affected by a recent outbreak in the Apulia region, caused by the bacterium *Xylella fastidiosa* (Manici et al., 2019). The study was made with a view to improving the decisional tools for selecting disease-decimated groves to be replaced with new olive trees. Relative quantities of rDNA-PCR amplicons targeting soil fungi, ascomycetes and basidiomycetes, and a series of soil physical indicators, which accounted for a proportion of water and air in the soil including total organic carbon or structure stability index, were assessed in twelve olive orchards at six sites across the *Xylella* infected area in the Salento region. Multiple correlation and canonical correspondence analysis led to identification of a series of soil physical and fungal indicators which were linearly correlated. Further discriminant analysis showed the selected set of variables was able to effectively discriminate sites and classes of plant available water capacity, whilst it discriminated poorly for soil texture. Even though soil fungi showed a relatively low weight compared to soil physical indicators based on the classification function from discriminant analysis, they showed a discriminant ability close to that of soil physical indicators. Therefore, as far as agro-environments on the verge of desertification are concerned, such as those in this study, soil fungi can be effectively associated to soil physical indicators for multivariate analysis of physical soil properties.

Amplicon quantity from ITS4, targeting total soil fungi, was the only fungal indicator able to discriminate both sites and sampling locations within groves. This suggests that total fungi performed well in estimating not only the difference between agro-environments but also differences in soil tillage. In fact, the lower under-canopy values may be due to additional tillage and other practices, which are commonly applied to clean soil before harvesting. Furthermore, total fungi were also linearly correlated with SPIs, thus showing the best performance as soil fungal indicator. As far as ascomycetes and basidiomycetes are concerned, the former were also found to be linearly correlated with SPIs, while the latter greatly reduced linear correlation between the sets of SPIs and the fungal variables. The better performance of ascomycetes or microfungi as indicators, may be due to their better adaptation to the environment under study; in fact, unlike the basidiomycetes, they are able to quickly colonize soil with spores and smaller mycelial structures; moreover they decompose holocellulose, which characterize herbaceous materials, in preference to lignin which instead is the main basidiomycetes substrate (Manici et al., 2019). The latter observation suggests that Ascomycetes composition may also be related to organic matter quality of soils in different sites, which reflects different composition and management of vegetal groundcover of olive groves. In conclusion, the high distance between the composition of both total fungi and ascomycetes, overall reflected geographic distance between the experimental sites, which further suggested their best performance as indicators in this agro-environment.

The practical values of this research include i) the identification of total fungal abundance and Ascomycota abundance as potential indicators of soil structural/porosity qualities; ii) the demonstration of a way for selecting soil management regimens and orchard sites for re-planting, in the context of olive cultivation. This study suggests that multivariate approach to generate ecological indicators by combining physical-chemical to microbial indicators may be effective for setting new land management tools. In addition, this approach appears particularly useful when selecting sites or land management in degraded Mediterranean areas, where soil management or microclimate can have induced differences in soil quality.

3.10 Biodiversity of Soil-inhabiting Fungi

Fungi are very successful inhabitants of soil, due to their high plasticity and their capacity to adopt various forms in response to adverse or unfavorable conditions (Frac et al., 2018). Due to their ability to produce a wide variety of extracellular enzymes, they are able to break down all kinds of organic matter, decomposing soil components and thereby regulating the balance of carbon and nutrients. Fungi convert dead organic matter into biomass, carbon dioxide, and organic acids (Fig. 23). Many species of fungi possess the ability to act as an effective biosorbent of toxic metals such as cadmium, copper, mercury, lead, and zinc, by accumulating them in their fruiting bodies. Though these elements may inhibit their growth and affect their reproduction. The diversity and activity of fungi is regulated by various biotic (plants and other organisms) and abiotic (soil pH, moisture, salinity, structure, and temperature) factors. Fungi can be found in almost every environment and can live in wide range of pH and temperature.

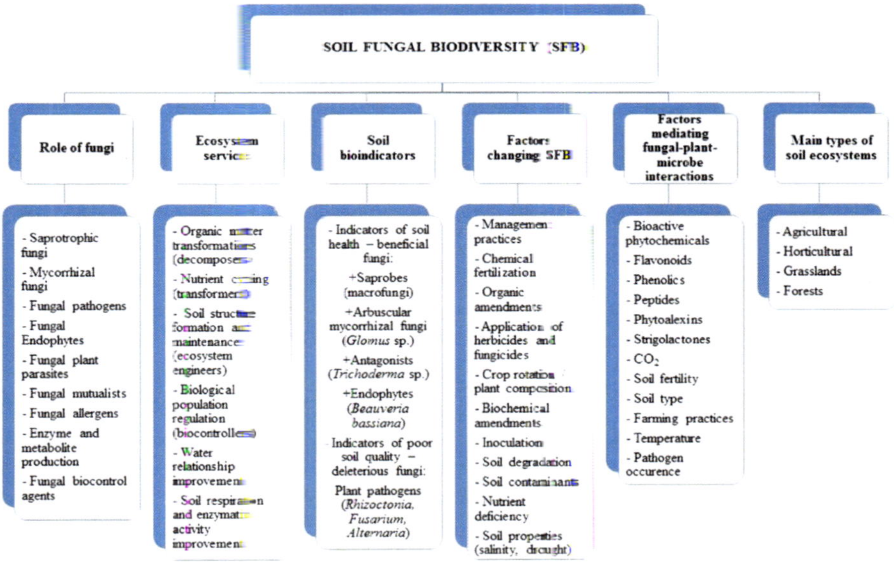

Fig. 23: Aspects of soil fungal biodiversity

Soil fungi categories: Soil fungi can be classified into three functional groups, viz., biological controllers, ecosystem regulators, and species participating in organic matter decomposition and compound transformations. Besides beneficial fungi, agri- and horticultural ecosystems contain also plant pathogens. The major groups of soil-borne root pathogenic fungi and oomycetes constitute of genera Fusarium, Verticillium, Rhizoctonia, Pythium, Phytophthora and many others, of global and local importance (Frac et al., 2018). The soil fungal diversity and methods of increasing it, particularly the populations of beneficial fungi within ecosystems should be used in practice for more sustainable plant production, decrease of chemical applications and protection of the soil environment. Ecosystem regulators are responsible for soil structure formation and modification of habitats for other organisms by regulating the dynamics of physiological processes in the soil environment. Biological controllers can regulate diseases, pests, and the growth of other organisms (Bagyaraj and Ashwin, 2017). For example, the mycorrhizal fungi improve plant growth by increasing the uptake of nutrients and protect them against pathogens. Fungal populations are strongly influenced by the diversity and composition of the plant community and in return affect plant growth through mutualism, pathogenicity and their effect on nutrient availability and cycling. Moreover, fungi participate in nitrogen fixation, hormone production, biological control against root pathogens and protection against drought. They also play an important role in stabilization of soil organic matter and decomposition of residues.

3.11 Methods and Recent Achievements in studies of Soil-Borne Fungi

The advent of next generation sequencing (NGS) has facilitated a sea-change in the analysis of soil and plant-associated fungal communities. Standardized pipelines for preparing rhizosphere soil samples for Illumina sequencing are widely available and in a relatively short time following sampling, files with millions of sequences can be generated. Important points to consider when preparing samples for NGS are: sufficient biological replication, sufficient sequencing depth, adequate coverage of target organisms (i.e., primer selection and DNA extraction), and the avoidance of contamination and bias (Frac et al., 2018). There are multiple pipelines for the analysis of fungal NGS data available, so it is not data-analysis that is problematic, but the interpretation of results. Alpha-diversity, representing either the number of species or diversity indices that account for evenness, was proposed as an indicator for robust, healthy soil. However, it is open to question whether absolute diversity or functional diversity should be emphasized. Usually the second step in analysis is to look at beta-diversity to see the effects of treatment/manipulation on the fungal community. Recently, it has been shown that fungal biodiversity in soils is strongly affected by plant community, soil moisture and the intensity of agricultural practices (Thomson et al., 2015). Studies identified key fungal species affected by soil treatments, but it is unknown if results obtained from studies conducted on one particular soil and ecosystem can be used to infer trends and identify key fungal groups at a global or continental scale.

Unlike the analysis of bacteria and archaea, where 16S rRNA is used as a barcode, fungi are usually identified based on the sequence of the Internal Transcribed Spacer (ITS) region allowing identification up to species level. Although in some cases, such as for soil-dwelling Fusaria, sequencing of additional genes, such as β-tubulin gene (β-Tub), and aminoadipate reductase gene (LYS2) was proposed to obtain the correct taxonomic identification at the species level, whereas others have suggested the use of

Translation Elongation Factor. Many researchers have in-house databases for functional classification of fungi but recently online resources for functional annotation of fungi have been made publicly available (Nguyen et al., 2016). Species level identification is , however only the first step from 'What is there?' toward the question 'What role does it play?' (Fig.24). Identification neither implies the microorganisms are alive and active nor does it describe their function. To unravel the function of the community, either (shotgun) metagenomics, metatranscriptomics or time-intensive culture based methods combined with functional tests must be used.

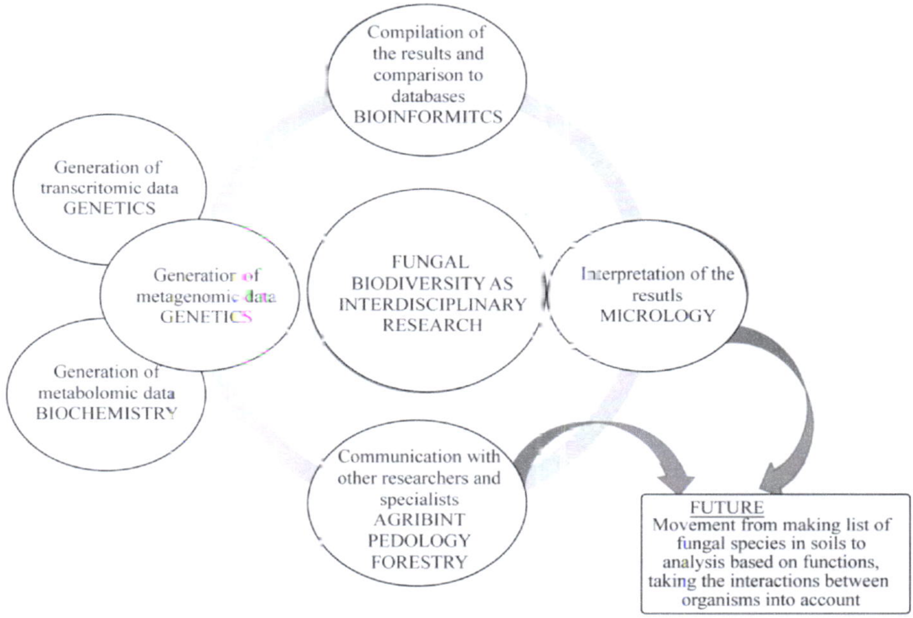

Fig. 24: Soil fungal biodiversity as an interdisciplinary research

A fast, but coarse alternative to molecular methods is MicroResp or community level physiological profiles (CLPP) approach, which gives an indication on substrate use of the total microbial community. This method, however, does not identify the species responsible for the process. Increasingly, especially in studies where plant community is included alongside NGS approaches, microorganisms are isolated from the soils and plant roots for further functional testing. To construct a more complete picture of a soil fungi community their interactions with other organisms must be taken into consideration. Strong linkage was proved between functional soil biodiversity and the function of the soil ecosystem (Morrien et al., 2017). Fungi interact with other soil organisms and thus changes in the fungal community have the potential to affect the function of the whole soil ecosystem. Analysis of the interactions in the soil can be achieved through indirect estimation of species interactions using co-occurrence networks or directly by using isotope tracers and/or gut content analysis.

3.11.1 Fungal Biodiversity and their functions in soil health of Agricultural, Horticultural, Grassland and Forest Ecosystems: In recent years the potential application of cultivating soil fungal biodiversity to improve soil quality and increase productivity of agricultural ecosystems has been highlighted as a new and very promising development in plant productivity, which may come be called 'the 2nd Green Revolution.' The implementation of such solutions may offer an alternative to the current overuse of fertilizers toward more sophisticated manipulations of plant productivity (Frac et al., 2018). Fungi participate in decomposition of organic matter and deliver nutrients for plant growth. Their role is very important in plant protection against pathogenic microorganisms as biological agents, which influences soil health. The assessment of fungal biodiversity as quality indicators cannot be limited only to the determination of biodiversity indexes, but also should include a structure analysis of fungal population in order to determine the functions they play in affecting soil quality and plant health. The use of different kinds of organic manure has a strong influence on soil health, through indirect effects (i.e., via changes in physicochemical characteristics) and a direct effect on soil fungal communities. Soil management is fundamental to all agricultural systems, and the reduction of soil degradation is a priority to sustain future production. This effect can be only achieved by taking soil fungal biodiversity into account. All cultural practices, such as the use of cover and rotational crops, composts and tillage systems, besides their known effects on soil-borne pathogens, are likely to affect also the other groups of soil fungi, especially beneficial fungal populations. It has long been known that the suppressiveness of soils can be enhanced by adding biopolymers such as chitin and its derivatives (i.e., chitosan). This suppressiveness is related to a change in the activity and structure of soil microorganisms.

Different tillage treatments can also impact soil fungi by soil disturbances that affect the functioning of fungal communities. Reduced tillage decreases the breakdown of hyphae causing fungal populations to remain more stable, retaining more nutrients and providing more suppressive effects against pathogenic microorganisms. Understanding and selecting the appropriate cultural practices, increasing fungal biodiversity, can prevent or decrease damage of root diseases and play a crucial role in the maintenance of soil quality and health. It should be taken into account that fungal diversity determines plant biodiversity, ecosystem variability, and productivity. Arbuscular mycorrhizal fungi (AMF) are the most important class of beneficial microorganisms in agri- and horticultural soils (Table 5). Significant increases in the yield of crop plants following inoculation with AMF have been observed in numerous experiments. The diseases of crop plants can be controlled by some antagonistic fungi such as *Glomus* sp. or *Trichoderma* sp. suppressing fungalpathogens. Species of *Trichoderma* (*T. asperellum, T. atroviride, T. harzianum, T. virens*, and *T. viride*) are frequently used in biocontrol and are known as biostimulants for horticultural crops. Other positive effects of fungi on soil quality and plant health include inoculation by microbial consortia of AMF together with plant growth promoting rhizobacteria (PGPR) and others such as N-fixing and P-solubilizing microorganisms. A synergistic, favorable impact of AMFs and PGPRs on horticultural plant growth and soil microbial diversity and activity has been reported. Plant pathogens under the major groups of soil-borne root pathogenic fungi, oomycetes and many others are of global and local importance. The soil fungal diversity and methods of increasing it, particularly the populations of beneficial fungi within ecosystems should be used in practice for more sustainable plant production, decrease of chemical applications and protection of the soil environment.

Table 5: Fungal community composition in different soil ecosystems and their function

Ecosystem	Fungal composition	Fungal function/reaction
Agricultural	*Agaricales* *Hypocreales*	Increase in drought-affected soils
	Sordariales *Capnodiales* *Eurotiales*	Reduction in soils affected by drought
	Mortierella *Fusarium* *Gibberella*	The dominant fungi in NPK treated soils
	Mortierella *Fusarium* *Schizothecium*	The dominant fungi in manure treated soils
	Ascomycota *Sordariomycetes* *Eurotiomycetes* *Dothideomycetes*	Key decomposers in agricultural soils. Increase after nitrogen fertilization
	Leotiomycetes *Helotiales*	Decline in N and P treated soils
	Cyphellophora *Penicillium* *Chloridium* *Trichoderma Acremonium*	Increase in fertilized soils
	Exophiala *Clonostachys* *Sarocladium* *Schizothecium* *Magnaporthe* *Phaeosphaeriopsis*	Decrease in N and P fertilized soils
Horticultural	*Glomus* *Gigaspora* *Scutellospora* *Acaulospora* *Entrophosporc*	Arbuscular mycorrhizal fungi (AMF) which improve plant growth by increasing phosphorus and the uptake of other nutrients
	Trichoderma such as:	Biostimulants and biocontrol agents, suppressing fungal pathogens like
	T. asperellum *T. atroviride* *T. harzianum* *T. virens* *T. viride*	Fusarium sp., Colletotrichum sp., or Phizoctonia sp.

Ecosystem	Fungal composition	Fungal function/reaction
Grasslands	*Hygrocybe* *Camarophyllopsis* *Dermoloma* *Entoloma* *Clavariaceae* *Geoglossaceae*	Saprotrophs, decomposers
	Glomus	Play a dominant role under different conditions of grasslands as AMF
	Agaricomycotina *Saccharomyceta* *Paraglomerales* *Chytridiales* *Archaeosporales* *Lobulomycetales* *Rhizophydiales* *Glomerales*	AMF, pathogens or decomposers
	Acaulospora *Glomus* *Scutellospora*	AMF important for soil and plant health, the species richness decrease in severely degraded grasslands
	Ascomycota *Glomeromycota* *Hypocreales* *Pezizales* *Dothideales (Aureobasidium)* Pleosporales such as: *Atternaria* *Cochliobolus/Bipolaris* *Phaeosphaeria* *Leptosphaeria* *Phoma*	Increase in Ascomycota and decrease in Glomeromycota after N and P addition into the soil. Glomeromycota phylum is composed almost entirely of AMF Most of them is known as plant pathogens
	Saprotrophic fungi	Decomposers which convert organic matter and produce enzymes
Forests	*Armillaria* *Phellinus* *Cronartium* *Arceuthobium*	Pathogen of trees

Ecosystem	Fungal composition	Fungal function/reaction
	Clonostachys candelabrum *Geomyces pannorum* *Penicillium adametzii* *P. commune* *P. daleae,* *P. janczewskii* *Trichoderma*	Antagonistic microbes suppressing soil-borne plant pathogens
	Basidiomycota	Ectomycorrhizal mutualists which protect plant families such as Pinaceae,
	Ascomycota	Fabaceae, Betulaceae, and Fagaceae
	Macrofungi	Biosorbents of toxic metals and compounds
	Tremellomycetes	The dominant fungal class in forest soil
	Dothideomycetes	

Fungal Biodiversity and their Functions in Soil Health of Grassland Ecosystems: Soil microorganisms, including fungi are an important component of grassland ecosystems due to their biochemical activity and engagement in nutrient cycling (Frac et al., 2018). Grasslands provide many forms of ecosystem services including: supporting, provisioning, regulatory, and cultural services. Importantly, the role of biodiversity has been established as fundamental in ensuring the performance of ecosystem functioning. Grazing activities influence soil fungal community structure by changing edaphic conditions and the vegetation biodiversity in plant communities. It has been proven that moderate grazing sustains plants diversity while heavy grazing results in species loss. Furthermore, plant-fungal interactions can inhibit biodiversity in grasslands due to the production of different root exudates such as enzymes, organic compounds, and polysaccharides. Plant pathogenic fungi also have a large impact on plant diversity in grasslands by limiting the abundance of their hosts, affecting biomass production. Fungal pathogens could affect nutrient cycling in grasslands reducing the abundance of dominant grasses and enhancing the growth of legumes. Soil fungal communities in grasslands can also be influenced by human activities and the components of long-term fertilization and other treatments. Unlike in agricultural soils, where ascomycetes dominate, in grasslands, basidiomycetes are major decomposers of dead organic matter.

Fungal Biodiversity and their functions in soil health of Forest Ecosystems: Knowledge of the soil chemical and physical properties has always been of interest to foresters to evaluate the capacity of sites and to increase forest productivity (Frac et al., 2018). Forest soils (including humus, litter, and coarse woody debris) are an important reservoir of microorganisms and soil biota that in turn influence carbon storage, soil structure, fertility, productivity, and plant/tree growth. Ectomycorrhizal associations are created by a specific group of plant families that includes the Pinaceae, Fabaceae, Betulaceae, and Fagaceae. A significant contribution by ectomycorrhizal mycelium to forest soil microbial biomass and by ectomycorrhizal roots to the production of extractable dissolved organic carbon, which is a carbon source for other microbes. During the processes of thinning, the transfer of nutrients from aboveground biomass to forest soil takes place. A higher concentration of nutrients comes from the green litter of thinned trees than litter returned to the forest floor after senescence or from the woody residue left on the ground after harvesting. Consequently, the quality and quantity of organic substrates presented to the soil fungal community by thinned and non-thinned forests may vary to a great extent. The community of soil microorganisms depends highly on organic matter as it provides a suitable environment and energy sources for them that are critical to maintain the nutritional quality and water-retaining capacity of forest soils (Jiménez-Morillo et al., 2016). Soil organic matter is of key relevance in maintaining soil resistance and stability, although it is uncertain how deterioration of soil properties and changes in fungal communities affect the functional stability of soils. Degradation of soil properties followed by deforestation may lead to decreases in soil fungal diversity and functional stability.

3.12 Fungal Community Composition, Diversity, Soil depth and Landscape Position

Fungal communities are highly diverse assemblages with an estimated 1.5 million species worldwide and hundreds of species within a single gram of soil (Daniel C. Schlatter et al., 2018). At the global scale, climatic conditions, soil characteristics and spatial patterns are believed to be the primary drivers of soil fungal communities. The immense diversity of fungal communities across the landscape, however, is due in part to their extensive variability at small spatial scales, where soil edaphic parameters (eg. water, temperature, C, N), plant species identity, and plant diversity drive variation in fungal community composition.Yet, with some exceptions the vast majority of work examining spatial variation in soil fungal communities has focused only on communities found within the top 10–20 cm of soil. Thus, we have only scratched the surface of our understanding of how fungal communities vary within the soil profile.

Soil edaphic characteristics are major drivers of fungal communities, but there is a lack of information on how communities vary with soil depth and landscape position in no-till cropping systems. Eastern Washington is dominated by dryland wheat grown on a highly variable landscape with steep, rolling hills. High-throughput sequencing of fungal ITS1 amplicons was used to characterize fungal communities across soil depth profiles (0 to 100 cm from the soil surface) among distinct landscape positions and aspects across a no-till wheat field. Fungal communities were highly stratified with soil depth, where deeper depths harbored distinct fungal taxa and more variable, less diverse fungal communities. Fungal communities from deep soils harbored a greater portion of taxa inferred to have pathotrophic or symbiotrophic in addition to saprotrophic lifestyles. Co-occurrence networks of fungal taxa became smaller and denser as soil depth increased. In contrast, differences between fungal communities from north-facing and south-facing slopes were relatively minor, suggesting that plant host, tillage, and fertilizer may be stronger drivers of fungal communities than landscape position (Daniel C. Schlatter et al., 2018).

Fungi residing in deeper layers of the soil profile contribute to key processes, such as soil formation, carbon and nutrient cycling, and xenobiotic degradation, and form intimate associations with plant roots where they can enhance water and nutrient uptake, provide protection from pathogens, or be pathogens themselves. Soil profiles are characterized by vertical gradients in carbon, nitrogen, pH and oxygen that frequently correspond with declines in microbial biomass and diversity. Further, deeper soil layers experience less extreme variation in moisture and temperature than surface soils. Although it is recognized that fungal communities are likely to vary throughout the soil profile, we have a poor understanding of the composition of deep-soil communities in agricultural systems and their diversity across the landscape.

Fungal community composition varies with soil depth and landscape slope/aspect Fungal communities varied primarily with soil depth (Adonis r2 = 0.25, P = 0.001), though members of the phylum Ascomycota were dominant among all depths (Daniel C. Schlatter et al., 2018). In general, the prevalence of fungal phyla was similar across the depth profile, though there was a distinct increase in the relative abundance of Basidiomycota at 10 cm below the soil surface (where the surface is 0 cm depth) versus at other strata (ANOVA F = 8.44, P<0.0001) that corresponded with a decrease in the prevalence of Ascomycota (ANOVA F = 3.25, P = 0.01). At the family-level, 22 groups varied significantly with soil depth (FDR adjusted p-values<0.05) where members of some families tended to be more abundant at the soil surface (0 cm; eg. Glomerellaceae, Teratosphaeriaceae, Mycosphaerellaceae, Pleosporaceae, Pezizaceae, Lasiosphaeriaceae), some in shallow soil layers (10 and 25 cm; eg. Cephalothecaceae, Clavicipitaceae, Nectriaceae, Trichocomaceae, Chaetomiaceae, Hypicreaceae and Mycotrichaceae), and others in the deepest strata (50, 75 and 100 cm; eg. Bionectriaceae). Among OTUs with relative abundances that differed significantly with soil depth, the soil surface was dominated by OTUs identified as Sordariales, *Fusarium, Mycosphaerella, Glarea* and *Humicola* species, upper soil layers (10 and 25 cm) were dominated by Ascomycota, Nectriaceae, Hypocreales, *Oidiodendron, Trichoderma* and *Penicillium* species and the deepest soils were dominated by OTUs related to *Bionectria, Ijuhya,* Hypocreales, *Mortierella* and *Ascomycota*. Thus, fungal community composition was highly stratified across soil depths, where some clades tended to dominate in shallow, intermediate or deep soil layers. Intriguingly, the variability in fungal community composition increased with depth (ANOVA F = 4.34, P = 0.001); communities deeper in the soil were more different from one another than those nearer to the soil surface.

It has been long understood that there is a succession of fungi on decomposing residue, since plant residue serves as a key resource base for fungi, stratification of fungal communities is likely to reflect different successional stages of residue decomposition. For example, the soil surface may have carbon derived from the standing crop residue and litter layer, and would have the greatest amounts of labile organic material. In parallel with studies of succession during decomposition, surface soil was dominated by many cellulolytic Ascomycetous fungi including those belonging to Sordariales, *Coniochaeta (Lecythophora), Fusarium, Mycosphaerella, Humicola* and Pleosporales, which are rapidly able to colonize wheat straw and are often found to be most abundant during early stages of decomposition. The initial cellulolytic colonists of wheat residue, however, are estimated to only utilize a small portion of the available carbon. In later stages of decomposition of polymers (eg. cellulose), other taxa, such as *Trichocladium* (Chaetomiaceae), which tended to predominate at 10 cm depth, are expected to be more competitive (Poll et al. 2010). Although some residue may be incorporated by higher disturbance hoe-type no-till seeders, the roots of wheat and other rotation crops may also be significant sources of carbon at this depth. As soil depth increases and root density decreases, those taxa that dominated deeper soil layers (50 and 100 cm), related to Hypocreales, *Mortierella, Ijuhya, Bionectria* and unidentified Ascomycota species, are likely to be those able to degrade older, recalcitrant carbon sources (eg. lignin). Moreover, anecic earthworms (Ex: *Lumbricus terrestris*), which are commonly present in no-till systems and can burrow to 3 m depth, may drag fresh surface residue into deep burrows, providing a significant source of C for microbial communities. Alternatively, fungi in deeper soil layers may act as pathogens or symbiotrophs in addition to saprotrophs as a means of obtaining nutrients, reflecting a shift in food web structures in deeper soil layers (Lamit et al. 2017). This is supported by the increase in the relative abundance of taxa assigned to trophic modes with pathogenic or symbiotrophic stages at deeper soil depths.

3.13 Phylotype Diversity Within Soil Fungal Functional Groups

Soil fungi are fundamental to plant productivity, yet their influence on the temporal stability of global terrestrial ecosystems, and their capacity to buffer plant productivity against extreme drought events, remain uncertain. Liu, et al., (2022) combined three independent global field surveys of soil fungi with a satellite-derived temporal assessment of plant productivity, and report that phylotype richness within particular fungal functional groups drives the stability of terrestrial ecosystems. The richness of fungal decomposers was consistently and positively associated with ecosystem stability worldwide, while the opposite pattern was found for the richness of fungal plant pathogens, particularly in grasslands. It was demonstrated that the richness of soil decomposers was consistently positively linked with higher resistance of plant productivity in response to extreme drought events, while that of fungal plant pathogens showed a general negative relationship with plant productivity resilience/ resistance patterns. This work provided evidence supporting the critical role of soil fungal diversity to secure stable plant production over time in global ecosystems, and to buffer against extreme climate events.

A survey by Gams, (2007) with filamentous fungi living in the soil layer revealed that the observed diversity of soil fungi largely depends on the method used and the numbers of isolates obtained. Particle-plating usually yields higher numbers of taxa than dilution plating. The Centraalbureau voor Schimmelcultures (CBS) preserves a great diversity of soil fungi. The CBS database contained 2,210 species of soil fungi in 2001, an estimated 70% of the known species available in culture. Thus, the current estimate for described culturable soil fungi is approximately 3,150 species, many of which have a cosmopolitan distribution. Adding the ca. 150 spp. of nonculturable *Glomerales* results in 3,300 species of currently known soil fungi. Molecular studies in such groups as *Fusarium*, *Trichoderma*, *Penicillium*, and *Aspergillus* are finding a number of more narrowly distributed cryptic species. Thus the number of species of soil fungi is expected to be considerably higher than the 3300 species currently known.

Soil fungi spread easily and most of them are regarded as having a cosmopolitan distribution. Even in tropical forests many taxa are similar to taxa found in temperate latitudes, and the numbers of species for a particular tropical soil are normally the same or even lower than those observed for soils in temperate regions. Needless to say that the species diversity in the litter layer is considerably higher and many novelties have resulted, e.g., from voluminous studies carried out in Ivory Coast. Thermophilic and heat-resistant species are largely cosmopolitan. The numbers of known species have roughly doubled within the last 25 years. In desert soils, these fungi can reach high proportions among the isolates. Among the 246 new fungal taxa described in the last 60 years from the Middle East, mainly desert and salt marsh soils, 53 species are soil fungi, many of which are adapted to high temperatures or high salt concentrations. Alkaliphilic and alkalitolerant fungi form another ecological group that is adapted to unusual substrata. In highly alkaline soils a range of fungi that grow well on agar media with pH 9.8, particularly of acremonium-like and Fusarium species, has been found in Indonesia. Unusual habitats also include deposits of burying rodents which have yielded several new *Penicillium* species.

3.14 Novel Clades of Fungi

For the last two decades, molecular discovery and characterization of fungi have rapidly outpaced traditional morphological description. Public sequence databases have accumulated internal transcribed spacer (ITS) barcodesrepresenting hundreds of groups of closely related fungal species with no taxonomic identity due to the paucity of relevant reference sequences and lack of phylogenetically informative ribosomal RNA (rRNA) genes (Tedersoo et al., 2017). Studies using a single molecular marker have shed light on several divergent but undescribed lineages of marine and terrestrial organisms among bacteria, protists and fungi. Analysis of multiple genetic markers obtained from vegetative tissues, single-cell genomics, or whole metagenome assays of the environment has improved the phylogenetic placement and classification for many of these previously unknown organisms, but many more remain overlooked. Because many of these lineages are not known from voucher material, the inability to name organisms only on the basis of sequence data hinders higher-level classification of fungi and other taxa.

Tedersoo et al., (2017) aimed to determine the phylogenetic placement of previously unclassified soil fungi by developing 452 taxon-specific primers targeting nuclear 18S (nSSU) and 28S (nLSU) rRNA genes in 263 ITS-based operational taxonomic units (OTUs) from global soil samples analyzed. Since the long 18S-ITS-28S rRNA gene sequences were generated by combining several amplicons from Sanger sequencing and 454 pyrosequencing, a multi-step quality control was performed to exclude any potentially artefactual entities. For the recovered novel soil fungal lineages, our purpose was to establish broad ecological niches for climatic and edaphic parameters and to determine geographic distribution together with endemicity patterns. It was hypothesized that tropical soils harbor relatively more enigmatic fungal lineages, because tropical habitats exhibit greater speciation but lower extinction rates, tropical forests harbor greater fungal richness and lower latitudes are relatively poorly covered by biodiversity and taxonomic research. Phylogenetic analyses revealed 37 major clades and seven single branches (singleton lineages) of previously unrecognized or unclassified fungi with distinct phylogenetic position that warrant at least order-level classification (Tedersoo et al., 2017). In the 18S rRNA gene and concatenated gene analyses, the clade GS01 was placed in a sister position to all remaining fungi, although the statistical support for this and most other early branching configurations remained poor.

Altogether, 11 clades (GS2–GS12) and three distinct branches (32%) of previously unclassified soil fungi were placed within Rozellomycota (Cryptomycota). These findings highlight that the remarkable phylogenetic diversity of Rozellomycota from aquatic ecosystems is also observed in terrestrial habitats. Unlike in recent analyses, Rozellomycota was separated from the phylum Aphelidea that accommodates the clade GS16, a large and well-supported group with no taxonomically characterized representatives. Other zoosporic phyla accommodated fewer undescribed fungal clades. Chytridiomycota harbored two distinct environmental groups, the clade GS13 with an unsettled position, and the clade GS14 in a sister position to Spizellomycetales. The clade GS15 formed a long branch within the Blastocladiomycota, albeit with low support (BS <70). Two clades of closely related soil fungi clustered with the enigmatic "chytrid" genus *Olpidium* that warrants a (sub)phylum of its own (Sekimoto et al., 2011). Taxonomically uncharacterized novel lineages of Chytridiomycota are particularly common in freshwater and marine environments.

Among the former zygomycetes, the clade GS19 formed a deep lineage at the base of Kickxellomycotina and Zoopagomycotina. Clades GS20, GS21, and GS22 were loosely associated with Endogonales (Mucoromycotina), whereas a single group (clade GS23) formed a monophyletic branch with Umbelopsidaceae (Mucoromycotina). Three class-level clades were related to the subphylum Pucciniomycotina of the Basidiomycota. Clades GS25 and GS26 represented successive sister groups to the remaining Pucciniomycotina, whereas the clade GS27 formed a sister group to Agaricostilbomycetes. The latter clade includes an 18S rRNA gene (Sanger) sequence from the voucher specimen RB1040 named as *Platygloea* sp. that appears distantly related to other Platygloeales and other Pucciniomycotina. Three novel clades (GS28–GS30) and branches were identified within the early-diverging Agaricomycetes, but their sister groups remained poorly resolved (BS <70). Multiple divergent sequences were also recovered in the orders Sebacinales, Trechisporales, Agaricales, Thelephorales, Hymenochaetales, and Atheliales. Within Ascomycota, the Taphrinomycotina subphylum included a well-supported sister group (clade GS31) to the Archaeorhizomycetes, a recently described class that is largely composed of environmental sequences (Rosling et al., 2011).

This study extended and illustrated previous findings that the soil habitat harbors thousands of undescribed fungal taxa, which was placed to >30 previously unrecognized well-supported fungal lineages. More importantly, these order- and class-level groups are distributed throughout the fungal tree of life and exhibit specific ecological preferences and/or biogeographic distribution patterns. To enable cross-communication of these major phylogenetic clades among research groups, a provisional naming system was proposed until their valid taxonomic description or matching with hitherto unsequenced species. These clade names were linked to fungal ITS and rRNA gene sequences in the UNITE database. Combining fluorescent probing and single-cell sequencing to cover nearly full-length rRNA genes will certainly improve our understanding about the ecophysiology and evolution of these enigmatic fungal clades.

Soil-inhabiting fungi can integrate soil physical indicators: The role of soil fungi in soil physical properties and in soil water retention is rarely considered when searching for new agro-environmental indicators, even though i) fungi represent the majority of soil microbial biomass in volume, ii) they are the major player in the C cycle by amply exceeding bacteria in soil respiration and iii) they contribute to soil structures and moisture retention through a series of biochemical and biophysical mechanisms (Manici et al., 2019). Indeed, most fungi can absorb water from atmospheres at a lower relative humidity than those occurring at the permanent wilting point, this suggests a more important role than that hypothesized to date, especially in arid soils. Most of the evidence on positive correlation between soil fungi, soil water retention capacity and soil hydraulic properties so far provided concerns mycorrhizal fungi which play a role in shaping hydraulic properties of soils through multiple direct and indirect mechanisms.

A study on the relationships between fungi and physical indicators of the soil applying a multivariate approach was performed by Manici et al., (2019) in an olive growing area affected by a recent outbreak in the Apulia region, caused by the bacterium *Xylella fastidiosa*. The study was made with a view to improving the decisional tools for selecting disease-decimated groves to be replaced with new olive trees. Relative quantities of rDNA-PCR amplicons targeting soil fungi, ascomycetes and basidiomycetes, and a series of soil physical indicators, which accounted for a proportion of water and air in the soil including total organic carbon or structure stability index, were assessed in twelve olive orchards at six sites across the *Xylella* infected area in the Salento region. Multiple correlation and canonical correspondence analysis led to identification of a series of soil physical and fungal indicators which were linearly correlated. Further discriminant analysis showed the selected set of variables was able to effectively discriminate sites and classes of plant available water capacity, whilst it discriminated poorly for soil texture. Even though soil fungi showed a relatively low weight compared to soil physical indicators based on the classification function from discriminant analysis, they showed a discriminant ability close to that of soil physical indicators. Therefore, as far as agro-environments on the verge of desertification are concerned, such as those in this study, soil fungi can be effectively associated to soil physical indicators for multivariate analysis of physical soil properties.

Free-living saprophytic soil fungi, belonging to the *Ascomycota* and *Basidiomycota* divisions, play a dominant role from decomposition of plant residues to cycling processes, thus representing the largest part of soil microbial biomass. Saprophytic fungi mediate the formation of soil structure by the binding effect of the hyphal structure, a plastic structure having a growth habit based on mycelium extension and branching therefore absorbing nutrients, growing through air and bridging soil pores. Furthermore, thanks to their fast and low nutrient-demanding hyphal growth, several ascomycetes have been found to explore soil highly efficiently thus mediating water translocation to new spaces.

Fungal plant pathogens and soil biodiversity: Soil-borne fungal plant pathogens have often proven difficult to manage with conventional methods of disease control. Nonetheless, researchers have characterized several naturally occurring "disease-suppressive" soils where crop loss from disease is less than would otherwise be expected. Suppressive soils can also result from the incorporation of various amendments into soil (Reeleder, 2003). In most cases, disease control in such soils has been shown to be biological in nature; that is, soil organisms appear to directly or indirectly inhibit the development of disease. Increased knowledge of the identity and functioning of these organisms may support the development of techniques that can be used to develop suppressiveness in soils that are otherwise disease-conducive. Populations of pathogens themselves have been shown to exhibit considerable genetic diversity; the ability of populations to respond to disease control measures should be considered when developing a management strategy. New molecular techniques can be exploited to better characterize soil communities, including the pathogens themselves, as well as community responses to various disease control options.

Historically, soil-borne fungal plant pathogens have proven difficult to manage; in many cases pesticides have not been effective nor has plant resistance been available. Disease management in these cases has relied, for the most part, upon crop rotation or cultivation practices that reduce pathogen populations or pathogen activity in soil. These tactics have been only partly successful. Broad-spectrum soil fumigants have been used effectively for disease management in certain high-value crops, however, the future availability of these products is uncertain. In this review, two aspects of biodiversity were discussed, the use and potential of soil biodiversity to manage these pathogens and the extent and impacts of diversity within populations of soil borne fungal pathogens.

3.15 Diversity of Soil Borne Plant Pathogens

Fungal plant pathogens are a subset of the total soil fungal population. Fungi are a complex group of eukaryotic organisms with polyphyletic origins. Their unique pattern of mycelial growth plus the utilization of parasexual modes of reproduction have together provided these microorganisms with tools that enable them to respond quickly to selection pressures (Reeleder, 2003). The development of agriculture in relatively recent times has likely altered selection pressures in cultivated soils sufficiently to select for more aggressive pathogens and, perhaps, new pathogens. Agricultural practices have resulted in ecosystems that are characterized by regular but massive disturbances. Changes in cropping sequences, tillage practices, the deployment of disease-resistant cultivars, the introduction of new crops, and pesticide use patterns all provide intense but often short-lived selection pressures that result in changes in soil fungal populations. Apart from the consequences for disease control, these processes promote variability within pathogen populations. Thus, as soil residents, the pathogens themselves contribute to soil biodiversity. Diversity within pathogen populations is important not only in terms of plant protection but also with respect to other roles that these fungi may have. In some cases, for example, pathogens and other soil fungi have industrial value as producers of useful secondary metabolites. The soil is likely a reservoir of a large number of heretofore unknown fungal-produced valuable bioproducts; this alone justifies more attention to these organisms. As colonizers of soil organic matter, many pathogens contribute to degradation processes in soil and recycling of nutrients and interact with other soil organisms. Some soil fungi that are regarded as weak plant pathogens or saprophytes in soil may also be implicated in human and animal diseases. Others can be employed as biocontrol agents for the management of weeds or plant diseases.

Measuring Diversity: Objective measures of diversity are required in order to both determine relationships between soil biodiversity and disease suppression and to compare effects of different crop cultural practices on disease management (Reeleder, 2003). The traditional approach in measuring diversity is taxon-based, i.e., species or other taxa in a given sample are enumerated. Recovery of representative taxa from soil, however, is problematic. Canadian researchers have documented fungal diversity and community assessments in both cultivated and non-cultivated soils, but it is recognized that most techniques recover only a small portion of the taxa present. Fatty acid analysis is one of several methods designed to overcome this difficulty. Molecular technologies (e.g., extraction of nucleic acids from soil, followed by their characterization and quantification), however, are becoming the tools of choice and should provide much more accurate measures of taxonomic diversity. It remains to be determined whether or not such measures will correlate in any way with disease-suppressive soil environments, or whether specific microbes or suites of microbes will be consistently and strongly linked to disease suppression. Further, there are technical and taxonomic issues that need to be addressed and standard extraction protocols that need to be developed. Without proper evaluation, procedures may be used that preferentially extract DNA from certain taxonomic groups, leading to inaccurate characterization of populations. Certain components in extraction solutions may also reduce total DNA yield. Such inefficiencies during extraction might effectively eliminate DNA of some organisms from the sample.

An alternative approach to taxon-based measures is to determine the functional diversity of a soil. In this case, diversity is expressed in an ecological sense rather than from a taxonomic perspective. Functional diversity assessment can be achieved, for example, by determining carbon substrate utilization (CSU) patterns in soil, by examining variation in enzyme production, or assessing secondary metabolite production in soil. Commercial kits are available for CSU assays and have been modified for studying fungi. Data from the assay of choice are then used to calculate diversity indices based on functional groups rather than taxa distribution. Functional diversity estimates may be of more use than taxon-based measures in assessing the potential disease-suppressiveness of a soil, but this has yet to be clearly demonstrated, in part perhaps because of technical limitations. CSU data, for example, may be quite dependent on incubation time and temperature, as well as dilution level. Because of the physiological plasticity of fungi and other microbes (Andrews 1992), assays may produce indices that are transient and not reflective of real constitutive differences between experimental treatments. In addition, functional diversity estimates need to encompass all soil organisms, including soil arthropods (Doles et al. 2001). CSU and other functional diversity assays are indirect measures of gene expression but this can also be determined directly. An extraction of RNA (as opposed to DNA) from soils, followed by the quantification of the mRNA corresponding to the gene in question, would provide such measures. However, practical use of such an approach awaits both the maturation of microarray and related technologies and, more importantly, the identification of genes critical to the functioning of pathways correlated with disease suppression.

3.16 Models for Studies sf Soil Biodiversity and its Impact on Soil Borne Fungal Pathogens

Because of their demonstrable effects on the security of food supply, soil-borne plant pathogens provide useful models for evaluation of the impact of soil biodiversity on agroecosystems (Reeleder, 2003). Developing a model system with measurable real-world outcomes would provide strong evidence of the benefits of soil biodiversity and may provide information that has more perceived value than "stand-alone" measures of biodiversity. Manipulation and monitoring of diversity parameters (pathogen biotypes, assemblages of microbes/ mycoparasites, mycophagic arthropods, soil texture, organic matter, nutrient/pesticide loads) may result in new methods to suppress pathogens. Use of molecular tools such as PCR and microarrays to quantify microbes and monitor gene expression will provide more information of greater detail than has previously been available. A number of well-known soil-borne plant pathogens of relevance to Canadian agriculture could provide suitable subjects for such models. Interactions between the pathogen *Sclerotinia sclerotiorum* (white mold of bean) and the soil environment would provide one example. Pathogens in the Oomycete genus *Pythium* (causal agents of root rot diseases in many crops) could provide useful targets in a second model system. Use of pathogens such as these in conjunction with naturally or amended soils known to be suppressive to them would provide an excellent starting point.

3.16.1 Fungal Biodiversity of agricultural field soil from Tamil Nadu, India: Fungi are an important component of the soil micro- biota typically of soil biomass than bacteria depending on soil depth and nutrient condition such as micro and macro nutrients. The saprophytic fungi represent the largest proportion of fungal species in soil and they perform a crucial role in the decomposition of plant structural polymers, such as cellulose, hemicelluloses and lignin thus contributing carbon and nitrogen source to increase the fertility of soil (Jayaraman et al., 2018). Mostly the distribution of fungi in agricultural soils is prominent and less is reported about the occurrence of fungi under natural soil condition. Soil microorganisms including fungi can have both positive and negative effects on plant growth and facilitate nutrient absorption by plants, promote plant growth or stimulate seedling development by producing hormone-like substances. Therefore, the soil is the most suitable environment and serves as a good culture medium for the growth of fungi. Generally, soils contain 105 to 106 cfu/g and 100 to 1500 g biomass per meter square. Some of the fungi have the ability to make rapid growth on organic substrates in dry, acid, coarse textured soil and thus to reach nutrients available to unicellular microorganisms by diffusion.

Soil microorganisms such as bacteria and fungi play an important role in soil fertility and promoting the plant growth. Sixty soil samples are collected of different agricultural crop fields such as groundnut field, sugarcane field, Plantain field, turmeric field, tapioca field and wasteland with Parthenium field in and around Thiruvannamalai, Tamil Nadu were investigated soil moisture content by oven drying method and diversity of fungi by using serial dilution plating method with standard Potato Dextrose Agar (PDA) for qualitative and quantitative pattern (Jayaraman et al., 2018). The moisture content of different soil samples collected from the above fields were found from 11.0% to 26.3%. Among the soil samples, the groundnut and turmeric field show higher moisture as 26.0% and 26.3% and the lowest level was observed in sugarcane and tapioca soil with 11.0% and 17.0% respectively. For analysis of fungi, 22 species were enumerated from the above soil samples in which *Aspergillus niger, A. flavus, A. terreus, A. fumigatus, Trichoderma* spp., *Cladosporium cladosporioides, Curvularia lunata, Gliocladium* sp., *Fusarium* sp. and white non sporulating fungus were remarkably present. Regarding the distribution of occurrence, Plantain filed show highest as 28% followed by groundnut field with 23%, Sugarcane field 17%, Tapioca and waste land soil each with 14% and the turmeric field soil has the lowest occurrence as 4% of fungal population. It is concluded that from the present study as the turmeric field soil has an antimicrobial effect which shows very low population of fungi in their soil environment when compared with other soil samples.

The fungi occurring in different type of soil samples like groundnut soil, sugar cane soil, plantain soil, Tapioca soil, Turmeric soil and Wasteland with Parthenium soil found with *A. niger* var. *japonicas, A. flavus, A.flavus* var *oryzae*,, *A. glaucs* sp.1, *A. glaucus* sp.2, *A.nidulans, A. terreus, A. candidus, A. fumigates, Penicillium citrinum, P. chrysogenum, Penicillium* sp., *Mucor* sp., *Rhizophus* sp., *Trichoderma virede, T. harzianum, Cladosporium cladosporioides, Curvularia lunata, Gliocladium* sp., *Fusarium* sp. and White non sporulating fungus. These results indicated that that the agricultural soil may contain various nutrients which are supporting for the growth and survival of wide variety of fungal species in the soil for prolonged period. The above results were in accordance with various reports, the fungi commonly present in diffferent type of soils found as *Aspergillus niger, A. flavus, A. clavatus, A. terreus, Curvularia lunata, Pencillium janthinellum, P. citrinum, Rhizopus nigricans, Trichoderma viride, T. harzianum, Fusarium. oxysporium, F. moniliforme, Mucor, Cladosporium, Verticillium, Acremonium, Chaetomium globosum, Corynespora cassicola, Eurotium chevelari, Botrycis cineea, Helminthosporium* sp. etc. On analysis of individual soil samples, in the Groundnut field shows *A.niger, A.japanicus, A.flavus, A.flavua* var *oryzae, A.glacus, Cladosporium* sp., *Cuvularia lunata*, and *Trichoderma sp.*. are the fungi commonly present in the order of dominance. At the same time the soil samples collected from the sugarcane field shows *A.niger, A.flavus, A.glacus amster, A.nidulans, Penicillium* sp. were present in significant level. But the mycoflora of the soil samples collected from the Turmeric field shows a unexpected results as very low percentage of fungal colonies with one each of Mucor and other nonsporulating white colony with poor growth.

The present study showed the versatile nature of fungal biodiversity from different agricultural soil samples with different population density which may be due to there between the soil types, crop cultivation, soil nutrients, soil pH, soil moisture and soil mycoflora. Soil environment of different areas of agricultural field with different crop cultivation was studied to know the relationship of fungal diversity and rhizosphere environment. The interaction between the presence of crop and soil fertility due to manure and pesticide addition and other chemicals present also were interpreted. There is no correlation was found with fungal population and the moisture content of soil samples from different environment.

3.16.2 Biodiversity and Ecology of soil fungi in a Temperate Coastal Dune System: Soil fungal communities were studied in an actively developing coastal dune system at Goeree Island, the Netherlands. A shore to inland sampling transect was laid out, extending from coastal brackish marshes to recently formed foredunes to older dune pastures to adjacent woodlands (Francesc X. Prenafeta Boldú et al., 2014). Soil samples from these biotopes were thoroughly characterized by analyzing physicochemical and microbial characteristics. Soil fungal community structure and composition were analysed by a combination of different phenotypic and genotypic methodologies (isolation of microfungi via a specialized soil washing technique and in situ observation of macrofungi, versus DGGE profiling and sequencing of multi-locus rDNA clone libraries). The results showed that fungal biomass tended to increase land-inwards along the gradient of maturity. The community structure was significantly correlated with progressive soil acidification land-inwards and with the exposure to brackish water in the coastal sites. Comparison between isolation and molecular datasets revealed that both methods were biased towards specific functional or phylogenetic groups. Most of the isolated fungi were common soil saprotrophic ascomycetes, while specialized fungi (biotrophic plant symbionts and pathogens, primary decomposers of recalcitrant organic matter, etc.) were only detected by molecular means. Phylogenetic specificity of PCR-based DNA profiling, on the other hand, strongly depended on primer selection. In spite of the relatively low number of common species that were identified among the isolated cultures and by clone library sequencing, as well as the potential biases of each characterization method, multivariate analysis on both isolation and molecular datasets yielded similar correlation patterns with the environment.

3.16.3 Diversity of soil fungi in different Land use types: The soils collected from five land use types located in the Tha Kum-Huai Raeng Forest Reserve, Trat province were studied with two objectives, to assess and compare the composition, abundance and diversity of soil fungi; and to relate the fungal diversity with leaf litter dry weight and soil physical and chemical properties (Pongsatorn Puangsombat et al., 2010). The five land use types were: secondary dry evergreen forest (DEF), Phayoong (*Dalbergia cochinchinensis*) plantation (PP), grassland (GL), rubber (*Hevea brasiliensis*) plantation (RP) and a pineapple field (PF). The soil collection was carried out in the rainy season by establishing a sample plot of 100×100 m in each land use type and nine soil samples were taken from 0- 10 cm soil depth. Before a soil sample was taken, the leaf litter covering the soil in a plot 1×1m was scraped and put into a plastic bag. The litter samples were dried and measured for dry weight. Each soil sample was analyzed for soil physical and chemical properties and isolated for soil fungi using the soil dilution plate method. The emerging fungal colonies were counted for numbers and identified to the species level based on morphological characteristics.

A total of 28 genera and 71 species of soil fungi were identified from the DEF, PP, GL, RP and PF soils. They belonged to the Class Zygomycetes (4 genera and 4 species), the Class Ascomycetes (4 genera and 5 species) and the Class Deuteromycetes (20 genera and 62 species) and most of them were organic matter decomposers. Some soil fungi appeared in all land use types, but some appeared only in a few land use types or in as little as one. The land use type which had the highest Shannon-Wiener's diversity index of soil fungi was PF, followed by DEF, PP, GL and RP, respectively. The highest similarity of fungal community composition between two land use types calculated by Sorensen's index of similarity occurred between DEF and PP, followed by PP and GL, and then DEF and GL, while the lowest similarity index value was between PF and RP. The analysis of variance of mean values of soil environmental factors, including leaf litter dry weight and soil chemical and physical properties, among the five land use types revealed that the soils were significantly different from each other. The relationships of soil fungi with all the mentioned soil environmental factors analyzed by the ordination method with canonical correspondence analysis (CCA) showed that some soil fungi related positively with the P, Ca and Mg contents in soils. It was recommended that the five land use types should be properly managed, so that there are no abrupt changes in their soil environmental conditions, in order to keep the existing diversity of soil fungi.

A total of 3,910 colonies of soil fungi were isolated from the five land use types. They were identified to 28 genera and 71 species, which belonged to the Class Zygomycetes (4 genera and 4 species), the Class Ascomycetes (4 genera and 5 species) and the Class Deuteromycetes (20 genera and 62 species). The numbers of fungal species identified from the soils in DEF, PP, GL, RP and PF were 33, 29, 38, 27 and 36, respectively. Table 1 lists the fungal classes, species and abundance levels of soil fungi (CFU/g wet weight soil) identified from each land use type. There was not only variation in the total number of soil fungi, but also in the variation of soil fungal species and abundance from one land use type to another. Some soil fungi appeared in all land use types (*Cuninghamella elegans, Gongronella butleri, Penicillium* sp.3, *Penicillium* sp.6, *Trichoderma hamatum, T. harzianum, T. oblongisporum* and unidentified sp. 1), but some appeared only in a few land use types (for example, *Mucor* sp., *Eupenicillium* sp., *Talaromyces* sp., *Aspergillus candidus, A. cervinus, A. japonicas, A. niveus, Cephalosporium* sp., *Chaetomella* sp.) or in as little as one (for example, *Mortiella pavisporum, Nectria* sp., *Aspergillus niger,* the *Aspergillus wentii* group, *Geotrichum* sp., *Gliocladium roseum, Heterocephalum* sp., *Leptographium* sp., *Penicillium oxalicum, P. striatrisporum* and *Penicillium* sp.16). There was a group of soil fungi which clearly had no relationship with the P, Ca and Mg contents in soils. They were: *Cunninghamella elegans* (s1), *Aspergillus niger* (s13), *Geotrichum* sp. (s23), *Leptographium* sp. (s27), *Penicillium adametzi* (s30) and an unidentified sp.2 (s70). These fungi had high abundance values and were only found in soil from RP.

3.16.4 DNA Metabarcoding in the Biodiversity of soil fungal communities in Agroecosystems: As pathogens, decomposers and plant mutualists soil fungi can affect plant and soil health in agro-ecosystems significantly. The impact of crop management practices on soil fungal communities is diverse and still poorly studied. The lack of knowledge is mainly related to their hidden life mode, high phenotypic diversity, the great heterogeneity of soil microhabitats, difficulty for culturing and species identification (Lazarova et al., 2019). Collecting data with high accuracy to detect effects relevant to ecosystem management is an ongoing challenge for soil ecological studies and biomonitoring. However, the combination of DNA-based identification methods and Next Generation Sequencing technology was recognized as a powerful tool to evaluate biodiversity in environmental samples, especially soil biodiversity. In the frame of the National Research Program "Healthy Foods for a Strong Bio-Economy and Quality of Life", soil microbiome diversity (fungi, prokaryotes and archaea) will be used as indicator for assessing soil and plant health, and ecosystem services in several agricultural ecosystems. Lazarova et al., (2019) presented some preliminary results on soil fungal communities evaluated by using amplicon DNA sequencing of the internal transcribed spacer 2 (ITS2) rDNA region. Multiple core soil samples were collected from 18 sampling plots in June 2019. Fungal diversity and community structure were evaluated at different taxonomic levels. The most common and abundant taxa at all sites were Fusarium Link 1809, Solicoccozyma X.Z.Liu, F.Y.Bai, M.Groenew. & Boekhout 2015, Cladosporium Link 1816, Alternaria Nees ex Wallroth 1816, and Mortierella E.Coemans 1863. The multivariate statistics (PCA) comparing the overall microbial composition revealed loose clusters linked to crops and localities. The study provides a new comprehensive overview of soil fungal communities (composition and diversity) from Bulgarian agro-ecosystems using high-throughput DNA sequencing.

3.16.5 Biodiversity of soil fungi on Integrated Pest Management Farming System: The greatest problem of modern agricultural practices is the use of chemical pesticides and fertilizers. It is noted that about 40% of the world's land surface is used for agriculture. The way this agricultural land is managed has a great influence on the global environment. Integrated Pest Management (IPM) farming system produces fewer negative externally affects, can effectively restore ecosystems and deliver well ecosystem services (Muhibuddin et al., 2011). Depending on methods applied and degree of production, agricultural practices impact biodiversity in the ecosystem and it also influences conservation practices within the ecosystem. The result showed that diversity of soil fungi on IPM was higher than diversity on conventional one. Trichoderma sp. and Acremonium sp. which are known as antagonist fungi were found on IPM but not on conventional one. Domination Index on IPM was lower than conventional field. Bacterial leaf blight (*Xanthomonas oryzae*) disease intensity showed no difference on both systems. The yield on IPM was higher (6.34 ton/ha) than conventional field (5.56 ton/ha). It has been found that rich biodiversity in agricultural environment improves productivity of agricultural systems. The IPM system gives more regard to the soil fungi biodiversity and considers it as an integrated system upon which the success of the agricultural production depends.

Fungal population known was higher on IPM than that on conventional field. On IPM was found 45 species (the total number was 14700 colonies per g soil) and on conventional was found 31 species (total number 8100 colonies per g soil). The fungi were classified based on their role: pathogen, decomposer, and unknown role. On IPM field was found 2 genera of pathogen i.e.*Cladosporium* sp. (1 species) and *Monilia* sp. (1 species). It was found 5 genusdecomposer fungi, i.e. *Trichoderma* sp. (1 species), *Acremonium* sp. (1 species), *Chaetomium* sp. (1 species), *Aspergillus* spp. (10 species), and *Penicillium* spp. (10 species). Twenty species of unknown fungi were also found. On conventional field was found 2 genus pathogens such as *Curvularia* sp. (1 species) and *Monilia* sp. (1 species). On this field also found 3 genus of decomposer fungi:*Chaetomium* sp (1 species), *Aspergillus* spp. (9 species), and *Penicillium* spp. (5 speces). There are 14 unknown spesies on this field. Some species fungi not found on conventional field (only found on IPM field) were *Trichoderma* sp., *Acremonium* sp., *Cladosporium* sp., *Aspergillus* sp. 1, *Aspergillus* sp. 6, *Aspergillus* sp. 7, *Penicillium* sp. 1, *Penicillium* sp. 2, *Penicillium* sp. 4, *Penicillium* sp. 5, *Penicillium* sp. 6, *Penicillium* sp. 7, *Penicillium* sp. 9, and unidentified fungus number 1, 4, 5, 6, 7, 10, 11, 12, and 14, while some species found on conventional, but not found on IPM, were *Curvularia* sp., *Aspergillus* sp. 8, *Aspergillus* sp. 10, *Penicillium* sp. 10, *Penicillium* sp. 11, and unknown fungi number 18, 19, and 21.

Diversity of the fungi on soil will make soil ecosystem more stable because there is a complementary cyclic between each other in that it can create a stable condition without domination by particular microorganism. Diversity of soil fungi on field IPM (index value = 3.50437118) was higher than the diversity of soil fungi in the area of conventional (index value field = 2.77). On field, 45 species of fungi IPM were obtained with the total of all the colonies as much as 14700 colony per g soil, whereas in conventional field, 31 kinds of mushrooms were obtained with a total of 8100 colonies colonies per gram soil.

Soil Fungal Biodiversity for Plant and Soil Health: Soil mycobiome (fungal microbiome) is essential, but still neglected, component of soil microbiome. Soil fungi are very important for agricultural, horticultural and forest ecosystems supporting functioning and environmental services to plant health, soil quality, fertility, and ecological stability. Soil functionality can be improved by managing the soil mycobiome (Frąc et al., 2023). However, fungi can also cause plant diseases, decrease yield, and affect human health. Therefore, there is a need to study the possibility of different mycobiome-based solutions and tools for sustainable and regenerative agriculture, including prediction of plant disease occurrence and control of mycobiome shift in soils. Metagenomic, metatranscriptomic, and metaproteomic methods combined with bioinformatic analyses may help to protect crops in modern agriculture. Moreover, there is a necessity to monitor environmental effects of newly developed microbial-based solutions. Although the knowledge on soil fungal community and mycobiomes is increasing, there is still a big black box concerning the future aspects of its influence on the environment.

Soil fungi play important ecological roles influencing plant health as symbionts, pathogens and decomposers. Soil fungal biodiversity is increasingly recognized as providing benefits to soil health as they participate in numerous ecosystem processes. Continued research on the identity, abundance and distribution of soil fungi, their various roles in soil microbiome community are thus fundamental to better understand all dimensions of fungal biodiversity, their impact on plant health as well as the prevention of diseases (Frąc et al., 2023). Monoculture farming is a fascinating testing ground for studies of soil mycobiome. The use of various cultivars, application of agronomic methods, pesticides, fertilizers and biological compounds at different stages of plant development causes changes in the structure and dynamics of soil microbiome. They can be rapid or subtle, depending on the applied factor. In this volume of Frontiers in Microbiology we have concentrated on mycobiome, which is much less popular in soil microbiome studies than bacteria. Mycobiome changes depend on the seasons and agricultural practices. Even short breaks in the monoculture, caused by intercropping, greatly increased soil mycobiome biodiversity. Similarly, the soil mycobiome dramatically changed after the use of cover crops, in the soil of continuous cucumber cultivation under plastic shed systems, mainly due to the increase of beneficial fungi. The study indicated positive sides of cover crops in creating plant-protective mycobiome in the soil.

3.17 Root Inhabiting Fungi

General features: In an ecological classification of the soil fungus flora, a number of groups can be distinguished, such as obligate saprophytes, root infecting fungi and fungal parasites of animals. Article by Garrett (1950) has been concerned with the ecological relationships between root infecting fungi and obligate saprophytes. The specialized plant parasites, together with mycorrhizal fungi (symbionts), have been grouped together as *root inhabiting fungi.* The remainder of the root infecting fungi, together with the obligate saprophytes, have been designated as *soil inhabiting fungi.*

The root inhabiting fungi are characterized by an expanding parasitic phase on the living host plant, and by a declining saprophytic phase in its absence. This restriction of the saprophytic phase of such root inhabitants is enforced by the competition of other soil micro-organisms, and they can therefore be considered as *ecologically obligate parasites*, inasmuch as their continued survival is dependent upon that of their host plants. With few exceptions, such fungi can readily be grown as saprophytes in pure culture and they are not, therefore, obligate parasites in the strict nutritional sense. The soil inhabiting fungi are characterized by ability to survive indefinitely as soil saprophytes. Some characteristics suggested as contributing to *competitive saprophytic ability* are a high growth rate, excretion and tolerance of antibiotic toxins, and a capacity for rapid and complete decomposition of dead plant tissues. Some distinguishing characteristics for the separation of these two groups of organisms have been listed. Root diseases caused by soil inhabiting fungi are most important in seedlings, and in older plants in which normal host resistance has been impaired by an adverse environment. Diseases of annual crops caused by such fungi are less amenable to control by crop rotation than are those caused by root inhabiting fungi.

Evolution of the host-parasite relationship is suggested as the determinant in the segregation of root inhabiting fungi. Competitive saprophytic ability is incompatible with such evolution, and is gradually lost, but the sheltered host-borne phase of the root inhabitant is prolonged. Evolution appears to have progressed furthest in the mycorrhizal fungi, as evidenced by their loss of competitive saprophytic ability, which has culminated in the nutritionally obligate parasitism of the vesicular-arbuscular endophytes (*Rhizophagus* spp.). Interference by other soil micro-organisms with the parasitic activity of a root inhabiting fungus is greatest at the root surface, where the soil micro-flora is greatly intensified, and is also changed in composition, by root excretion and by other activities of the living root. The rhizosphere and root surface microfloras thus appear to constitute the root's outermost barrier against invasion by pathogenic fungi.

In its invasion of the host root, a root inhabiting fungus is likely to be accompanied or followed by weak secondary parasites and obligate saprophytes. With progressive improvement in the relationship between host and primary parasite, gross disorganization of host root physiology and defence mechanisms is likely to diminish, and so the infected root may still be able to repel secondary invaders. Thus the ectotrophic mycorrhizal symbiosis is typically a closed association, and functions as a symbiosis for defence as well as for nutrition. The saprophytic survival of a root inhabiting fungus in dead infected host tissues is curtailed by the competition and antagonism of associated saprophytes, and is likely to be shortest under soil conditions optimum for microbiological activity.

3.17.1 Bioprospecting of Rhizosphere-Resident Fungi: Rhizosphere-resident fungi that are helpful to plants are generally termed as 'plant growth promoting fungi' (PGPF). These fungi are one of the chief sources of the biotic inducers known to give their host plants numerous advantages, and they play a vital role in sustainable agriculture. Today's biggest challenge is to satisfy the rising demand for crop protection and crop yield without harming the natural ecosystem (Mahadevamurthy Murali et. al, 2021). Nowadays, PGPF has become an eco-friendly way to improve crop yield by enhancing seed germination, shoot and root growth, chlorophyll production, and fruit yield, etc., either directly or indirectly. The mode of action of these PGPF includes the solubilization and mineralization of the essential micro- and macronutrients needed by plants to regulate the balance for various plant processes. PGPF produce defense-related enzymes, defensive/volatile compounds, and phytohormones that control pathogenic microbes' growth, thereby assisting the plants in facing various biotic and abiotic stresses. Therefore, this review presents a holistic view of PGPF as efficient natural biofertilizers to improve crop plants' growth and resistance.

Rhizosphere-resident plant growth promoting fungi (PGPF) are one of the effective eco-friendly management strategies of plant diseases, and they may also serve as an alternative tactic to boost the growth and defense mechanism in plants. The application of PGPF will reduce the use of chemical control to a minimum, and it protects plants against various biotic and abiotic stresses. Many PGPF species *Trichoderma, Talaromyces, Fusarium, Phytophthora, Penicillium, Rhizoctonia, Gliocladium, Phoma,* etc.) have been reported to date to boost plants' growth, and to enhance their innate immunity and other important secondary metabolites in plants (Mahadevamurthy Murali et. al, 2021). The above beneficial aspects rendered by PGPF are due to root colonization, the production of growth hormones, the solubilization of essential minerals, antagonistic properties, mycoparasitic and saprophytic resistance, and competition for space and nutrients, in addition to the induction of systemic resistance (ISR) in plants. Apart from these, PGPF prevents pathogen infection in plants, improves soil nutrient availability, and reduces ethylene in the native host via the production of 1-aminocyclopropane-1-carboxylate (ACC) deaminase and phytohormones. The overview of the beneficial importance of PGPF in the promotion of plant growth and resistance is represented in Figs.25 and 26.

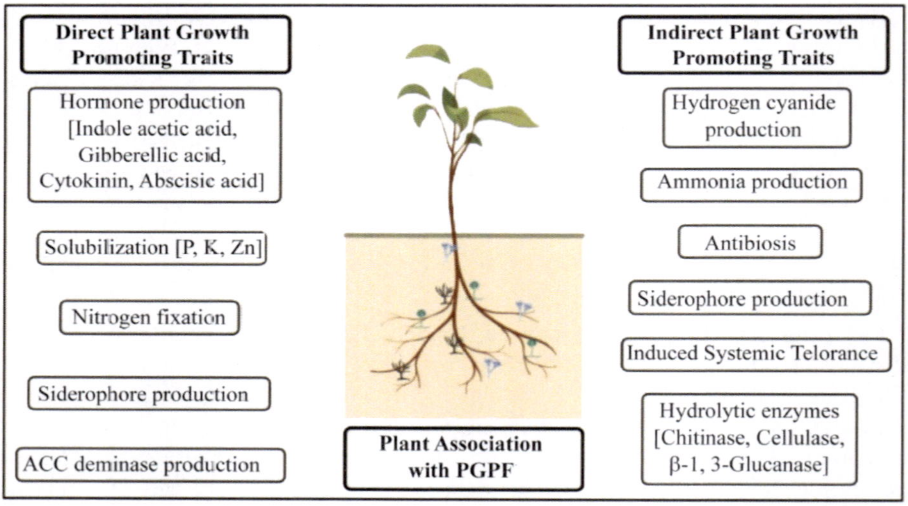

Fig. 25: Outlook on the beneficial characteristics of rhizosphere-resident plant growth promoting fungi (PGPF) for plant growth

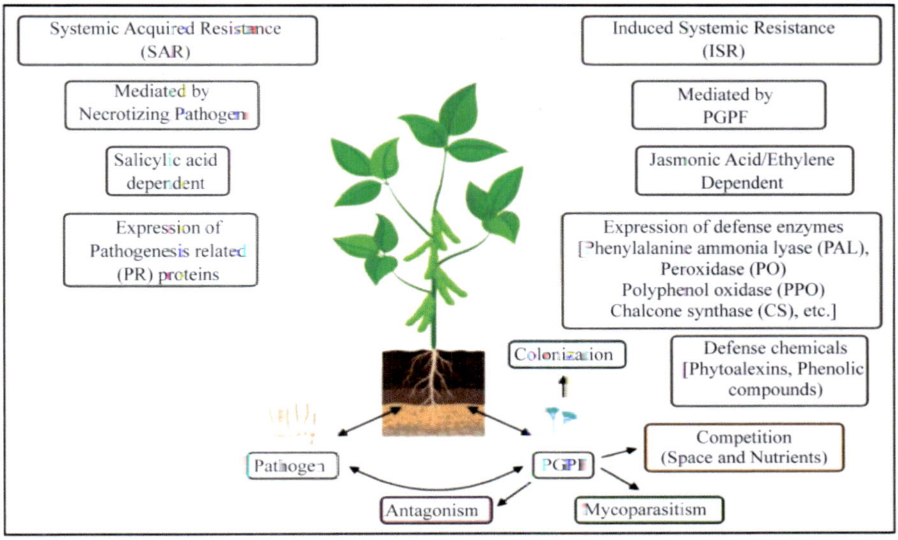

Fig. 26: Mechanism of resistance offered by PGPF against the invading pathogens

During ISR activation, PGPF treatment regulates the levels of defense signaling hormones such as jasmonic acid (JA) and/or ethylene (ET), along with the accumulation of certain proteins (thaumatin), and stimulates the production of defense-linked enzymes, chiefly GLU (glucanases) and CHI (chitinases), that prevent the growth and proliferation of pathogens either directly or indirectly. Before regulating defense signaling pathways, PGPF triggers the increased synthesis of secondary compounds (lignin, callose), along with the production of antioxidant enzymes that protect plants from various biotic and abiotic stresses (Aamir et al., 2019). Apart from the above-mentioned factors, the need for superior PGPF to reduce toxic chemical fertilizers has been highlighted as one of the key moves towards organic farming practices. Therefore, the review sheds light on the recent findings related to the application of rhizosphere-resident PGPF in improving plant growth, the elicitation of resistance, co-operation with the host, root colonization, and commercial formulation of byproducts (Fig. 27). The information produced from this review could be of great benefit to those looking for an organic method of farming primarily to achieve sustainability of crop production (Mahadevamurthy Murali et. al, 2021).

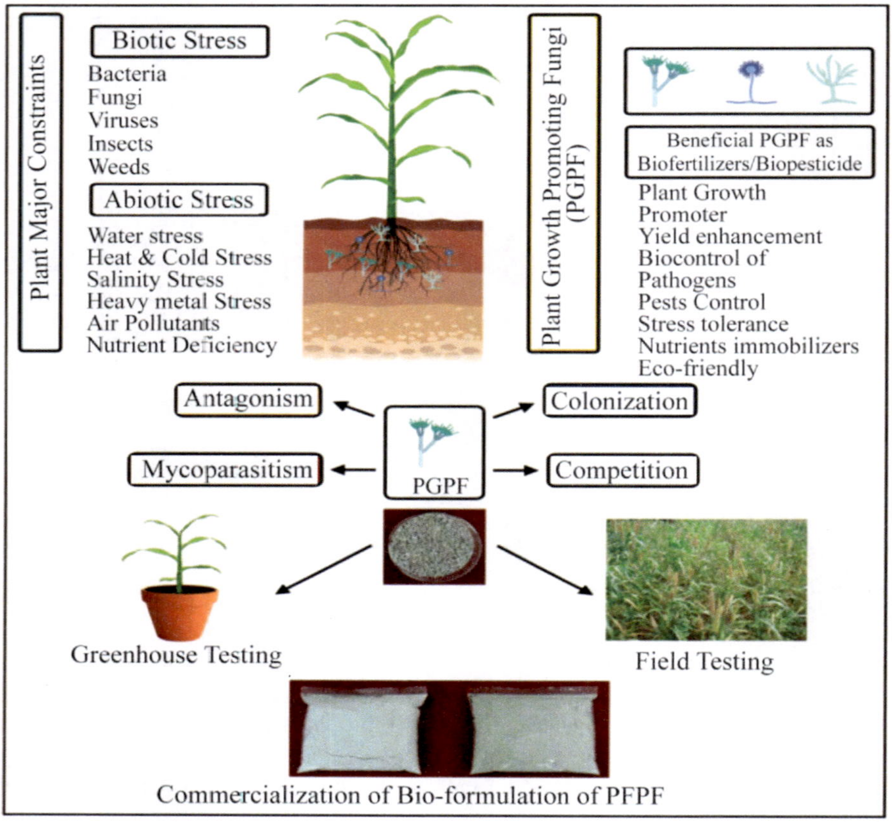

Fig. 27: Application of PGPF and its effect on the induced resistance in plants

Different groups of rhizosphere-resident fungi (*Trichoderma*, *Talaromyces*, *Rhizoctonia*, *Aspergillus*, *Penicillium*, *Fusarium*, etc.) tend to have a positive effect on augmenting the growth of various crop plants. In contradiction to this, *Penicillium* spp. (*P. janthinellim*, *P. citreonigrum* and *P. citrinum*) residing in the native rhizosphere hampered the growth of native zinnia and tomato plants. PGPF, such as sterile fungi, *Trichoderma* sp. and *Penicillium* sp., isolated from zoysiagrass specifically enhanced the growth of wheat plants of two different varieties. Thus, the reports suggested that rhizosphere-resident fungi exhibit unique specificity between host plants related to plant growth enhancement. Furthermore, non-sporulating fungi (SBF (sterile black fungus), SRF (sterile red fungus), and SDF (sterile dark fungus)) residing in the native rhizosphere soil of plants improved the growth and yield of corn, wheat, barley, chickpea, bromegrass, lupine, pea, clover, and ryegrass. It has been noted that the time taken by PGPF is one of the crucial characteristics in improving plant growth and yield, and this statement was validated, wherein PGPF (sterile fungi, *Trichoderma* sp. and *Penicillium* sp.) was reported to augment the growth of wheat plants in less than four weeks. However, in some plants, it has been stated that PGPF can take an even greater time to elicit plant growth.

The elicitors extracted from the PGPF have been employed to elicit plants' growth and resistance upon pathogen attack (Hammad et al., 2019). The elicitors include crude oligosaccharides, proteins, and sphingolipids, etc., which positively affect plant growth parameters. It has been noted that crude proteins extracted from PGPF improved the seed quality parameters predominantly in pearl millet and muskmelon plants compared to the control. Antagonism is one mechanism wherein rhizospheric fungi tend to antagonize pathogens' growth and development in plants. The antagonistic property of PGPF is suggested to be due to microbial predation, competition for nutrients, and antibiotic production, etc. The antagonism is suggested to be due to the production of lytic enzymes such as protease, chitinase and β- 1,3 glucanase. Various PGPF namely *T. harzianum*, *Phoma* sp., *F. equiseti* and *P. simplicissimum* have been reported to be antagonistic against *R. solani*, *P. irregulare*, *S. rolfsii*. *F. oxysporum*, *P. syringae* and *C. orbiculare*. Microbial predation mediated by PGPF involves the activation of the chitinase gene, which inhibits the growth of pathogens in an indirect manner. PGPF such as *Gliocladium virens* produce an antibiotic called Gliovirin, which has restricted the growth of *Pythium ultimum*. Induced resistance (IR) is generally demarcated as a phenomenon or a method of improving the plant's inherent immune system. It is elicited upon treatment with an inducer or elicitor which is active against biotic stresses imposed by bacteria, fungi, viruses, parasites, and nematodes. Plant treatment with *Phoma* spp., *Penicillium* spp., *F. equisti*, *Trichoderma* spp. and other non-sporulating fungi induced systemic resistance by suppressing the growth of some soil-borne and airborne pathogens in different crops.

The metabolites extracted from various fungi cause a great impression of imparting resistance in plants. The expression of elicitor-inducible PR proteins has been well correlated in imparting disease resistance (Gomes et al., 2017). Numerous glycoprotein elicitors have been extracted from *P. cinnamomic, Cladosporium fulvum*, *C. lindemuthianum* and *R. solani*, which have been shown to induce resistance in plants. PGPF applications are highly efficient in enhancing crop yield, crop quality, and the sustainable management of plant diseases. The treatment with PGPF will reduce the cost and pollution effect caused by chemical fertilizers. PGPF can be used as a substitute to reduce the use of agrochemicals in plant disease management. The use of PGPF improves plant performance and crop yield in agricultural fields. Information about PGPF and host plant interactions with the pathogen infection helps us understand the signaling mechanism involved in plant growth and resistance. The induction of resistance mediated by PGPF is linked with the production of phytohormones, defense-related enzymes, antibiotics and other signaling hormones, which play a crucial role in offering long-time protection and resistance during biotic and abiotic pressures (Mahadevamurthy Murali et. al, 2021). Thus, the treatment of PGPF increased crop yield production and reduced the disease-causing effect of pathogens in agricultural crop fields.

3.17.2 Root-inhabiting fungi in Alien plant species: The first step towards recognizing interactions between plants of alien origin and root-inhabiting fungi and thus better understand their invasion mechanisms is determining their presence in these plant species (Majewska et al., 2015). The importance of developing exhaustive checklists of the mycorrhizal status of non-native plants from various habitat types in different biogeographical regions has been highlighted. The presence of AMF in the roots of alien plants was examined in Asia, while similar research was carried out in Europe. In the latter research, 44 species of alien origin were studied, but the authors did not investigate the arbuscular mycorrhiza (AM) colonization rates and morphology, presence of fungal root endophytes and AMF species related to these plants. However, it has been shown that AM morphological type, the intensity of AMF colonization, AMF species identity and DSE may influence plant performance and, as a consequence, alien plant invasions. There was also a lack of data on the relationship between soil chemical properties and the presence and abundance of root-inhabiting fungi in alien plants in the sites they inhabit.

The objectives of this study were thus to determine the mycorrhizal status, AMF colonization rate, AM morphology and the presence of fungal root endophytes in 37 species of alien origin in Central Europe. Species from 32 genera and 17 families were chosen that were of different life forms and invasive status. chemical properties of soils and the AMF diversity in the soils collected from under those species were also studied. The hypothesis was that the mycorrhizal status of the investigated plants would depend on species identity. However, in view of the importance of mycorrhizae in plant nutrition, it was expected that the intensity of mycorrhizal colonization would relate to soil chemical properties being highest at sites featuring low nutrients. Furthermore, as for our study the alien plants were usually collected from semi-natural and anthropogenic habitats characterized by soil disturbance, we thus also predicted that AMF species found would be common and widely distributed both in Poland and around the world.

The first PCA axis (80.7 % of the total variance) distinguished between plant species with high values of mycorrhizal frequency, relative mycorrhizal root length and frequency of vesicles (left part of the diagram) and the species characterized by low mycorrhizal intensity or non-mycorrhizal (right part). The second axis (11.0 % of the total variance) separated species with high frequency of dark septate endophytes (upper part of axis 2), such as *Asclepias syriaca*, *Fraxinus pennsylvanica*, *Juglans regia* and *Parthenocissus inserta*. The plants grouped in lower left part of the diagram were associated with highest number of AMF species. This group comprises species belonging mainly to Asteraceae, such as *Aster lanceolatus*, *Aster novi-belgii*, *Erigeron annuus*, *Solidago canadensis* and *Xanthium albinum*. At the within-species and general (among-species) levels, no statistically significant correlations were found between the root colonization by AMF, DSE and *Olpidium* spp. and the soil chemical properties.

The study by Majewska et al., (2015) provided new records of the mycorrhizal status of nine species. The presence of AM in 26 plant species and the absence of mycorrhizal colonization in *Reynoutria japonica* were consistent with earlier observations. *Lupinus polyphyllus*, which was previously reported to be non-mycorrhizal, was found to form AM. The mycorrhizal status of plants under study seems to be determined by plant species identity. The species colonized by AMF are from genera and families that are considered mycorrhizal. The only two non-mycorrhizal plants, *R. japonica* and *Typha laxmannii*, belong to families which representatives are usually non-mycorrhizal or harbor weak and facultative AMF colonization. The most common fungal species found in our research, namely *Claroideoglomus claroideum*, *Funneliformis mosseae* and *Septoglomus constrictum*, are widely distributed both in Poland and around the world. Dark septate endophytes (DSE) were commonly found in the roots of the alien plants examined in our research with 32 of the 37 taxa being colonized.

3.17.3 Associations of Root-inhabiting fungi with Herbaceous plant species of temperate forests: The overwhelming majority of research on fungal interactions with plants in the forest ecosystems of the temperate climate zone focuses on ectomycorrhizal associations and no studies so far have compared the occurrence of root-inhabiting fungi in herbaceous plant species. Katarzyna Rożek et al., (2019) thus studied arbuscular mycorrhizal fungi (AMF) and fungal root endophyte colonization rates as well as AMF species richness and composition under 19 herbaceous plant species in temperate forests (southeast Poland) in relation to soil chemical properties. Seventeen species formed arbuscular mycorrhiza (AM), while 2 were non-mycorrhizal. The intensity of AMF colonization varied between species. Relative mycorrhizal root length (M_{AMF}%) ranged from 0% to 100%. AMF spore abundance ranged from 0 to 11.4 in 1 g of soils. Sixteen AMF species were recorded, both widespread (e.g. *Funneliformis constrictum*, *Claroideoglomus claroideum*) and rare (*Acaulospora cavernata*, *Entrophospora infrequens*).

The composition of AMF species related to the plants differed. Fungal root endophytes were recorded only in some plants; dark septate endophytes (DSE) in 13 species, while *Olpidium* spp. in 6 species. Moreover, DSE mycelia and Olpidium spp. sporangia were observed with low abundance, and their occurrence differed between particular plant species. Among soil chemical properties, only the concentration of available phosphorus was significantly negatively correlated with the M_{AMF}% parameter. In conclusion, several groups of root-inhabiting fungi were related to herbaceous plants; however, they occurred with varied frequency. AMF spore abundance and species richness differed as well; however, they persisted at a low level compared to other ecosystems. Nonetheless, significant negative correlation was detected between available P contents in soils and the intensity of mycorrhizal colonization, which suggests the importance of AM for the plants in sites with low P concentration.

In addition to AMF, there are other root-inhabiting fungal groups which affect plant fitness. Sometimes AMF are accompanied in roots by dark septate endophytes (DSE) and species of the genus *Olpidium* which also can play a significant role in ecosystem functioning. Detailed research on the relationship between the rate of root colonization by AMF, DSE, *Olpidium* spp. and soil chemical properties in the forest ecosystems of the temperate climate zone has, as yet, not been conducted. Therefore, the aim of this study was to analyze the intensity of fungal root colonization in 19 herbaceous plant species in relation to soil chemical properties as well as AMF species richness and composition related to these plants. It was hypothesized that due to the domination of ectomycorrhizal tree species in the studied forests, the availability of AMF propagules would be low, resulting in the low mycorrhizal frequency and colonization degree of herbaceous species. However, since mycorrhizal colonization of plants is dependent on soil fertility, it was expected that the extent of colonization would be higher in less fertile soils.

3.18 Root-Inhabiting and Rhizosphere Mycobiomes

The root mycobiome plays a direct role in plant productivity, and the study of its community composition allows for identification of organisms that influence plant health. To better understand the role of fungal community composition in crop productivity, the rootassociated mycobiomes of historically high and low yield sites of corn and wheat planted in rotation were characterized and analyzed along with soil physicochemical variables and crop yield (Wallace Marianna, 2022). In each field studied, root and rhizosphere mycobiomes reflected significant differences in fungal species composition. Several soil variables were found to be predictors of differences in composition of sample types including soil texture and pH. The genera *Fusicolla, Epicoccum,* and *Tetracladium* were positively correlated with yield, and *Neonectria, Pythium, Corynespora*, and *Mrakia* were negatively correlated. Identifying differences in the root-associated mycobiome of crops and changes in the soil environment could aid in the development of community management tools that maximize crop productivity.

ITS2, LSUA and LSUBG primers were used to amplify fungal DNA from corn rhizosphere soil, corn root, wheat rhizosphere soil and wheat root samples from historically and low yield sites of three fields that have a history of cornsoybean-wheat crop rotation. Significant differences in fungal species composition were found between each of the fields, requiring each field to be analyzed separately. Root samples exhibited lower alpha diversity compared to rhizosphere samples but there were no discernible trends between the alpha diversity of corn and wheat rhizosphere samples. The overall fungal community composition of crops (wheat or corn), types (rhizosphere soil or root) and sample types (high/low yield corn rhizosphere soil, high/low yield corn root, high/low wheat rhizosphere and high and low wheat root) were significantly different from one another in the same field. Species that differed in abundance between fields and sample types within each field that may be driving these differences include *Tausonia pullulans, Pythium biforme*, and *Linnemannia gamsii.* Fungal community composition in corn and wheat rhizosphere soils was found to be significantly correlated with yield measured at harvest. The indicator species analysis of rhizosphere soils in each field revealed some key species that may be driving the differences in community composition of high and low yield sites, including *Pseudeurotium bakeri, Cryptococcus watticus, Solicoccozyma terrricola*, and *Candida sake*, which were significantly associated with high yield sites, and *Leptodontidium camptobactrum* and *Soloacrosporiella*, which were significantly associated with low yield sites. Differences in fungal community composition of samples were correlated with soil physicochemical variables, including K ppm, Mn ppm, S ppm, Fe ppm, pH, and Mg ppm. Several soil physicochemical variables were significantly associated with yield of wheat and corn, with some soil characteristics differing between the two crops. Soil texture was an important parameter, along with % H, Fe ppm, soil moisture, pH, Ca ppm, CEC (cation exchange capacity), NO_3-N, Mn ppm, Mg ppm, and Al ppm. Genera were identified as being significantly correlated with yield measured at harvest. Notably, *Fusicolla, Epicoccum*, and *Tetracladium* were positively correlated with yield, and *Plectosphaerella, Neonectria, Dactylonectria, Aphanomyces, Pythium, Corynespora* and *Mrakia* were negatively correlated with yield. While there were some limitations to the study design, the results of this study emphasize the importance of comprehensive integrative analysis of soil fungi, soil characteristics, and yield.

Future studies should aim to sample multiple high and low yielding locations within a field to achieve a more comprehensive look at community composition and its relation to crop productivity, while reducing the impact of spatial dynamics. Identifying specific compositional shifts and differences in the root-associated mycobiome of crops, as well as the corresponding changes in the abiotic soil environment, could aid in the development of soil inocula or other community management tools that maximize crop productivity. These strategies can be integrated into current soil health management systems to increase agricultural yields and productivity in a low-input and sustainable manner.

Root-associated fungal communities are influenced more by soils than by root traits: Forest fungal communities are shaped by the interactions between host tree root systems and the associated soil conditions. How the soil environment, root morphological traits, and root chemistry influence root-inhabiting fungal communities in three tropical forest sites of varying successional status in Xishuangbanna, China were investigated (Aaron Hogan et al., 2023). For 150 trees of 66 species, we measured root morphology and tissue chemistry. Tree species identity was confirmed by sequencing rbcL, and root-associated fungal (RAF) communities were determined using high-throughput ITS2 sequencing. Using distance-based redundancy analysis and hierarchical variation partitioning, we quantified the relative importance of two soil variables (site average total phosphorus and available phosphorus), four root traits (dry matter content, tissue density, specific tip abundance, and forks), and three root tissue elemental concentrations (nitrogen, calcium, and manganese) on RAF community dissimilarity. The root and soil environment collectively explained 23% of RAF compositional variation. Soil phosphorus explained 76% of that variation. Twenty fungal taxa differentiated RAF communities among the three sites. Soil phosphorus most strongly affects RAF assemblages in this tropical forest. Variation in root calcium and manganese concentrations and root morphology among tree hosts, principally an architectural trade-off between dense, highly branched vs less-dense, herringbonetype root systems, are important secondary determinants.

The degree to which root functional (i.e. morphological and chemical) traits influence RAF communities in natural tropical forests is not well documented. Most of the studies that evaluate how root morphology and root-associated fungi (RAF) communities covary with variation in soil habitat often do so superficially. For instance, most studies either classify plant-host taxa generally as ectomycorrhizal (ECM) vs arbuscular mycorrhizal or use indirect measures of root functional traits from trait databases rather than via direct measurements. Thus, there is a need to directly measure and characterize root functional traits and root-inhabiting fungal communities across edaphic gradients to elucidate the role root functional strategies play in determining assemblages of fungal symbionts.

Multivariate community analyses applied to RAF communities of 150 host trees sampled across contrasting tropical forest habitats showed that the soil environment strongly shapes tree RAF communities in this tropical forest, overwhelming functional differences among root systems. This result aligns with previous studies in tropical forests, which also found the soil environment to be the main determinant of mycorrhizal communities, especially for ECM taxa. At least for tree roots in Asian Fagaceae–Lauraceae seasonal tropical forests, a secondary more stochastic environmental filter at the root system level exists (Yan et al., 2022), which selects fungal assemblages according to root morphological and tissue chemistry differences. the role of soils were first discussed in shaping plant RAF assemblages, and then the effects of root morphological traits and tissue chemistry on RAF communities were considered (Aaron Hogan et al., 2023).

In this sturdy, 5771 fungal were detected OTUs from 102 families on the roots of 150 trees from the three sites. The five most speciose fungal families were Glomeraceae (230 OTUs), Herpotrichiellaceae (104 OTUs), Mortierellaceae (89 OTUs), Clavariaceae (52 OTUs), and Agaricaceae (44 OTUs). One-third (34) of the families included in RAF communities contained five or more OTUs. Shannon's diversity of RAF was different among sites ($F(2,147) = 3.089$, $P < 0.05$) and was greater in RF than SE, with SR being statistically indistinguishable from both RF and SE. Fungal species richness was greater in the two secondary forests than in the primary RF site ($F(2,147) = 6.028$, $P < 0.01$), but species evenness showed the opposite trend ($F(2,147) = 13.440$, $P < 0.05$). FUNGuild was able to classify 57% of OTUs to a fungal guild, revealing differences in the relative abundance of RAF types among sites. The relative abundance of fungal guilds at SR resembled RF more closely than SE. For example, SE had a much higher relative abundance of ECM fungi and a lower relative abundance of saprophytic fungi than RF or SR. This pattern was also evident for ECM host trees among sites. Twenty fungal OTUs contributed ≥ 70% to differences RAF assemblages among sites. SIMPER values ranged from 0.0103 to 0.0439, which represent the relative contribution of each species to overall assemblage dissimilarity. These 20 fungal taxa were ordinated on the outskirts of the species space in the db-RDA, illustrating their influence on assemblage dissimilarity among the three sites.

Quantification of Root Fungi Quantification using Signature Fatty Acids: Both deleterious (pathogenic) and beneficial (mycorrhizal) fungi inhabit plant roots with strong impact on plant growth and health. Various methods have been used to quantify these fungi, such as disease index, staining techniques, serological/genetic/biochemical markers and indirect measurements of plant parameters (Sabine Ravnskov et al., 2019).

The objective of this work was to evaluate the possibility of using signature fatty acids to quantify root-inhabiting fungi in planta. Different fatty acid-based methods can be used to quantify fungi. Membrane bound phospholipid fatty acids (PLFA) can be used for biomass estimation and neutral lipid fatty acids (NLFA) for estimation of fungal energy reserves. The NLFA/PLFA ratio provides information on the physiological status of the fungus. The PLFA/NLFA method is, however, quite laborious, and if it is of minor importance to distinquish between PLFA and NLFA, the whole cell fatty acid (WCFA) analysis, which is much faster, can be used as an alternative to give information of root infection intensity.

Signature fatty acids have been used to quantify arbuscular mycorrhizal fungi (16:1ω5) and the pea root pathogen Aphanomyces euteiches (14:1ω9) in pot- and field experiments. In pot experiments, arachadonic acid (20:4) was further used to estimate root infection intensity of the plasmodiophorids Plasmodiophora brassica, causing club root in cabbage and analyses of the fatty acid profile of resting spores of the virus vector *Spongospora subterranea*, revealed high amounts of arachadonic acid. Signature fatty acids offer the possibility of measuring biomass, energy reserves and root infection intensity of various root-inhabiting microbes in planta. 14:1w9 and 16:1w5 seems to be highly specific to *Aphanomyces* and *Glomus*, respectively. It can be used in both field and greenhouse studies. 20:4 and 20:5 can be used to quantify root-inhabiting organisms among *Plasmodiophora*, *Spongospora*, *Aphanomyces, Pythium, Phytophthora* and *Glomus* under controlled conditions. No obvious fatty acid signatures for *Fusarium* and *Rhizoctonia* (Eumycota) in roots.

3.18.1 Spectra of root-associated fungi in submerged aquatic plants: Plant roots are inhabited by a diverse spectrum of rootassociated fungi (RAF). Besides mycorrhizal fungi, the RAF includes dark septate endophytes (DSE), aquatic hyphomycetes, hyaline fungal endophytes and other saprotrophic or pathogenic fungi. These fungi mostly belong to the phylum Ascomycota and Basidiomycota; however, also Zygomycota and Chytridiomycota are present (Petr Kohout et al.,2012). In contrast to the vast literature on mycorrhizal fungi, particularly the most widespread arbuscular mycorrhizal fungi (AMF) from the phylum Glomeromycota, our knowledge of the role of other RAF groups in plant ecophysiology and their importance in both terrestrial and aquatic ecosystem functioning is still very limited. In aquatic environments, the first described RAF was the AMF.

Similarly to plants from terrestrial ecosystems, aquatic species harbour wide spectra of root-associated fungi (RAF). However, comparably less is known about fungal diversity in submerged roots. Petr Kohout et al., (2012) assessed the incidence and diversity of RAF in submerged aquatic plants using microscopy, culture-dependent and culture-independent techniques. RAF of five submerged isoetid species collected in four oligotrophic freshwater lakes in Norway was studied. Levels of dark septate endophytes (DSE) colonization differed among the lakes and were positively related to the organic matter content and negatively related to pH. In total, 41 fungal OTUs were identified using culture-dependent and culture-independent techniques, belonging to Mucoromycotina, Chytridiomycota, Glomeromycota, Ascomycota as well as

Basidiomycota. Sequences corresponding to aquatic hyphomycetes (Ex:*Nectria lugdunensis, Tetracladium furcatum* and *Varicosporium elodeae*) were obtained. Eight arbuscular mycorrhizal taxa belonging to the orders Archaeosporales, Diversisporales and Glomerales were also detected. However, the vast majority of the fungal species detected (Ex: *Ceratobasidium* sp., *Cryptosporiopsis rhizophila, Leptodontidium orchidicola*, and *Tuber* sp.) have previously been known only from roots of terrestrial plants. The abundance and phylogenetic distribution of mycorrhizal as well as nonmycorrhizal fungi in the roots of submerged plants have reshaped our views on the fungal diversity in aquatic environment.

Besides the mycorrhizal fungi and the DSE, there are other groups of fungi capable of colonizing plant roots. One such group is aquatic hyphomycetes that are broadly defined as anamorphic fungi typically with relatively large branched (often stauroform) and scolecoform conidia occurring mainly in lotic waters. Their common substrate is submerged decaying tree leaves and woody debris, but they also enter healthy tissues as endophytes and act even as necrotrophic plant pathogens. The same fungi were also found in permanently or intermittently humid layer of forest plant litter or as endophytes of roots in nonaquatic habitats.

The most commonly isolated fungal taxa were: (i) an undetermined member of the phylum Ascomycota, tentatively named Ascomycota sp. 2 (from Lit. uniflora), (ii) isolates closely related to *Cryptosporiopsis rhizophila* (from Lit. uniflora and Lob. dortmanna) and (iii) isolates highly similar to *Leptodontidium orchidicola* (from Lit. uniflora). Additionally, fungi closely related to known species of aquatic hyphomycetes were isolated, including *Nectria lugdunensis, Tetracladium* sp. 1 (both from Lit. uniflora), *Tetracladium furcatum* and *Tricladium* sp. 1 (both from *S. aquatica*) (Petr Kohout et al.,2012). Only two fungal taxa were isolated from the roots of Isoe¨tes spp.: *Leohumicola minima* (from *I. echinospora*) and Ascomycota sp. 1 (from both Isoe¨tes species). The most diverse community of uncultured RAF was found in the roots of Lob. Dortmanna. Similarly to the results obtained from isolations, Helotiales constituted an abundant and diverse group of nonmycorrhizal RAF in culture-independent screening. However, there was a relatively low overlap in species assemblages detected using both techniques. The dominant sequences with affinities to the Helotiales matched well with known aquatic hyphomycetes. In addition to Ascomycota, several basidiomycetous fungi were also recorded, including the two OTUs (nos. 20 and 21) from both Isoe¨tes species with high similarities to Cantharellales. In phylogenetic analysis, these sequences clustered together with *Ceratobasidium* spp. from orchid or ericaceous roots. Additional components of the community of uncultured RAF were Chytridiales (OTUs 34, 35, 36), Endogonales (OTU 28), Rhizophydiales (OTU 29) and Monoblepharidales (OTU 30). Sequences of 5.8S rRNA genes of those taxa were phylogenetically analysed with similar sequences from the INSD.

In general, the diversity of nonmycorrhizal RAF is quite high, even within a single plant species. Because of their worldwide distribution and very high abundance, it has been predicted that endophytic fungi comprise a substantial component of global fungal diversity. There is a general consensus among researchers that a large proportion of fungal species still awaits description. Investigations into new habitats or host species will likely result in a discovery of new species and/or lineages of RAF.

Root-inhabiting fungi and Medicinal plants

Medicinal plants are an important part of the health system in many countries around the world. However, to date, <1% of global herbal species have been studied in terms of their interactions with soil-borne fungi; this figure includes arbuscular mycorrhizal fungi (AMF), which have been found to enhance the performance of medicinal plants. The aims of the present investigation were to examine overall mycorrhizal status, morphology of arbuscular mycorrhiza (AM), degree of root colonisation by AMF, and richness and composition of AMF species in relation to 40 plant species collected from the Garden of Medicinal Plants affiliated with the Institute of Pharmacology of the Polish Academy of Sciences in Kraków (Piotr Piszczek et al., 2019). In addition, the presence of root endophytes that frequently co-occur with AMF, such as dark septate endophytes (DSEs) and *Olpidium* spp., was examined. Correlations between fungal parameters and soil chemical properties were also tested. The survey showed the occurrence of AMF in 35 plant species. In 32 of these species, AMF were of the Arum morphotype. In the case of two plant species, the *Paris* type was found, while one plant represented the intermediate morphotype. DSEs were recorded within the roots of 11 species, *Olpidium* spp. in 10 species. Twenty-two common species of AMF were identified from spores isolated from soils from beneath the plant species. The most frequently occurring species were *Funneliformis constrictum*, *F. mosseae* and *Claroideoglomus claroideum*. No significant correlation was found between fungal parameters and soil chemical properties. The results of this study expanded our knowledge about the ecology of the plants under study. They may be also useful in further studies on the use of AMF inoculation to enhance production of secondary metabolites in these plants.

Root-inhabiting microbes identification by plant-derived carbon incorporation: To better understand the level of intimacy of root-inhabiting microbes such as arbuscularmycorrhizal fungi and bacteria, Philippe Vandenkoornhuyse et al., (2007) provided $^{13}CO_2$ to plants atatmospheric concentration during a 5-h pulse. They expected mi-crobes dependent on a carbon flux from their host plant to becomerapidly labeled. It was showed that a wide variety of microbesoccurred in roots, mostly previously unknown. Strikingly, the greatest part of this unsuspected diversity corresponded to activeprimary consumers. They found 17 bacterial phylotypes co-occurringwithin roots of a single plant, including five potentially newphylotypes. Fourteen phylotypes were heavily labeled with the ^{13}C. Eight were phylogenetically close to Burkholderiales, which encompass known symbionts; the others were potentially newbacterial root symbionts. By analyzing unlabeled and ^{13}C-enriched RNAs, differential activity in C consumption among these root-inhabiting microbes was demonstrated. Arbuscular mycorrhizal fungal RNAs were heavily labeled, confirming the high carbon fluxfrom the plant to the fungal compartment, but some of the fungipresent appeared to be much more active than others. The results presented here reveal the possibility of uncharacterized root symbioses.

The overall isotopic signature demonstrated that theroots of all three plant species were enriched in13C immediatelyafter the end of the pulse labeling. Carbon translocation was especially rapid to the roots of *T. repens*, consistent with thehigh AM mycelial carbon respiration that was recently demon-strated in this species. It is likely that demand by rhizobiain root nodules also contributed to the early accumulation of ^{13}C in clover roots. By contrast, ^{13}C built up in Agrostis roots overseveral days, although the results demonstrated that microbes still exploited a ^{13}C-rich pool of recent photosynthates. Cloning and sequencing of the various PCR products identi-fied 17 phylotypes (phylogenetically related sequences) of AMfungi colonizing the roots. By computing rare faction curves using a random resampling procedure and bootstrap from the 25 to 30 sequenced clones from each clone library, it was demonstrated that we have not underestimated the diversity ofAM fungal phylotypes in any of the RNA samples. The metagenomic approach using stable isotopes has allowed us to describe the community diversity of AM fungi and bacteria and to relate this diversity to a functional trait, the consumption of newly formed photosynthates. An extension of the approach would probably reveal the presence of additional micro-organisms, including archaea and a wider range of fungi. Because highly labeled RNAs were focused, it was concluded that the microorganisms involved in the process are closely linked witht heir host plant and are likely to be plant-dependent. However, results bring more questions than answers. There is much to learn about the ecology of microorganisms in roots, especially of thepotentially new bacterial groups described herein.

3.18.2 Root-inhabiting fungi in Alien plant species: The first step towards recognizing interactions between plants of alien origin and root-inhabiting fungi and thus better understand their invasion mechanisms is determining their presence in these plant species (Marta L. Majewska et al., 2015). The importance of developing exhaustive checklists of the mycorrhizal status of non-native plants from various habitat types in different biogeographical regions has been highlighted before.

In order to recognize interactions between alien vascular plants and soil microorganisms and thus better understand the mechanisms of plant invasions, Marta L. Majewska et al., (2015) examined the mycorrhizal status, arbuscular mycorrhizal fungi (AMF) colonization rate, arbuscular mycorrhiza (AM) morphology and presence of fungal root endophytes in 37 non-native species in Central Europe. The AMF diversity and chemical properties of soils from under these species was also studied. The plant and soil materials were collected in southern Poland. It was found that 35 of the species formed AM and their mycorrhizal status depended on species identity. Thirty-three taxa had AM of Arum-type alone. *Lycopersicon esculentum* showed intermediate AM morphology and Eragrostis albensis developed both Arum and Paris. The mycelia of dark septate endophytes (DSE) were observed in 32 of the species, while sporangia of *Olpidium* spp. were found in the roots of 10. Thirteen common and worldwide occurring AMF species as well as three unidentified spore morphotypes were isolated from trap cultures established with the soils from under the plant species. *Claroideoglomus claroideum*, *Funneliformis mosseae* and *Septoglomus constrictum* were found the most frequently. The presence of root-inhabiting fungi and the intensity of their colonization were not correlated with soil chemical properties, plant invasion status, their local abundance and habitat type. No relationships were also found between the presence of AMF, DSE and *Olpidium* spp. These suggested that other edaphic conditions, plant and fungal species identity or the abundance of these fungi in soils might have an impact on the occurrence and intensity of fungal root colonization in the plants under study.

This study provided new records of the mycorrhizal status of nine species. The presence of AM in 26 plant species and the absence of mycorrhizal colonization in *Reynoutria japonica* were consistent with earlier observations. *Lupinus polyphyllus*, which was previously reported to be non-mycorrhizal, was found to form AM. The mycorrhizal status of plants under study seems to be determined by plant species identity. The species colonized by AMF are from genera and families that are considered mycorrhizal. The only two non-mycorrhizal plants, *R. japonica* and *Typha laxmannii*, belong to families which representatives are usually non-mycorrhizal or harbor weak and facultative AMF colonization. The abundance of AMF in root systems cannot be simply taken as an indicator of their effects on plants. However, when the extent of root length colonized by AMF increases, plant growth and phosphorus content often increase. The majority of plant species in this study displayed a high mycorrhizal colonization rate. The invasion status was not directly related to the intensity of AMF colonization and among plants from the transformer category both highly mycorrhizal and non-mycorrhizal species were found. Because the group of plants in question was very heterogeneous, it is thus unlikely that, in general, mycorrhizal association could explain the success of their invasion. However, experiments are needed to determine the impact of AMF on plant performance and thus potential role of symbiotic fungi in the invasion of some species.

The most common fungal species found were *Claroideoglomus claroideum*, *Funneliformis mosseae* and *Septoglomus constrictum*. Dark septate endophytes (DSE) were commonly found in the roots of the alien plants examined in our research with 32 of the 37 taxa being colonized. However, their hyphae and sclerotia occurred in low abundance. These structures cannot be regarded as specialized interfaces for the transfer of nutrients between plant and fungus and it is thus probable that DSE do not influence plant performance through direct contact with roots (Newsham, 2011). This was the first report on the presence of *Olpidium* spp. in all the alien plant species under study with the exception of *Helianthus tuberosus*. *Olpidium sporangia* was observed in the roots of ten species, including the most common invasive plants, namely *Bidens frondosa*, *Echinocystis lobata*, *Helianthus tuberosus*, *Heracleum sosnowskyi* and *Impatiens parviflora*. As some of the *Olpidium* species may transmit viruses which can cause serious plant diseases, there is a possibility that the fungi spread with hosts during their invasions and may be a threat to native flora. However, this supposition requires confirmation. The vast majority of the examined alien plant species were AMF hosts. The monocultures formed by these species probably do not have such detrimental effects on AMF abundance in soils as is usually observed for non-mycorrhizal invasive species. However, the significant alterations in AMF species diversity, community structure and functional dynamics caused by mycorrhizal alien plants may occur.

The presence of root-inhabiting fungi and the intensity of their colonization were not correlated with soil chemical properties, plant invasion status, life form, their local abundance and habitat type. No relationship was also found between the occurrences of all fungal groups (Marta L. Majewska et al., 2015). These suggest that other environmental factors, plant and fungal species identities as well as the abundance of these fungi in soils, might have an impact on their occurrence and intensity of root colonization in the plants under study.

3.18.3 Root fungi in Non-orchidaceous Vascular Lithophytes: Based on their existence and dependence on rock, lithophytes are classified as epilithic or epipetric and endolithic forms. Epilithic forms grow superfi cially on rock surfaces (orchids, ferns, algae, bryophytes), whereas endolithic forms grow embedded in rocks (algae, fungi). Further, plants growing exclusively on rocks are known as obligate lithophytes and those that growing on rocky as well as on other substrates are known as the facultative lithophytes (Marimuthu Chinnathambi and Perumalsamy Priyadharsini, 2016).

Plant roots in natural ecosystems are colonized by a diverse group of fungi among which the most common and widespread are arbuscular mycorrhizal (AM) and dark septate endophyte (DSE) fungi. Though AM and DSE fungal associations are well reported for terricolous plant species, they are rather poorly known for lithophytic plant species. Marimuthu Chinnathambi and Perumalsamy Priyadharsini (2016) examined AM and DSE fungal association in 72 non-orchidaceous vascular plant species growing as lithophytes in Siruvani Hills, Western Ghats of Tamilnadu, India. Sixty-nine plant species had AM and 58 species had DSE fungal associations. AM fungal association in 42 and DSE fungal association in 53 plant species was reported for the first time. There were significant differences in total root length colonization and root length colonized by different AM and DSE fungal structures among plant species. In contrast, the differences in AM and DSE fungal colonization among plants in various life-forms and lifecycles were not signifi cant. AM morphology reported for the first time in 56 plant species was dominated by intermediate type AM morphology. A significant negative relationship existed between total root length colonized by AM and DSE fungi. These results clearly suggested that AM and DSE fungal associations are widespread in lithophytes.

A great diversity exists in inter- and intracellular distribution of AM fungal structures within plant roots. Based on the distribution of the fungal structures within plant roots, the AM colonization patterns have been termed as Arum-, Paris- and intermediate types. A group of predominantly ascomycetous fungi belonging to the order Helotiales also colonizes plant roots either individually or along with AM fungi. These fungi termed as dark septate endophyte (DSE) fungi form dematiaceous, regularly septate hyphae with microsclerotia and/or moniliform cells in plant roots. Although these fungi have been recorded in roots of more than 1000 plant species growing in a range of habitats, their occurrences in many plant taxa and habitats are yet to be examined. As lithophytic habitats are characterized by extreme environmental conditions and plants in stressed habitats depend more on association with endophytic microorganisms for their growth and survival, it was hypothesized that the incidence of endophytic root fungal associations in lithophytes are high.

The majority of the plant species investigated (81%) had DSE fungal colonization characterized by regularly septate hyaline/melanized hyphae bearing moniliform cells or microsclerotia (Marimuthu Chinnathambi and Perumalsamy Priyadharsini, 2016). Of these, co-occurrence of DSE association along with AM was evident in 97% of these plant species. DSE fungal structures were absent in roots of *Alstonia venenata* (Apocynaceae), *Argyreia populifolia, Merremia umbellata* (Convolvulaceae), *Costus speciosus* (Costaceae), *Agrostistachys borneensis, Phyllanthus urinaria* (Euphorbiaceae), *Smithia bigemina* (Fabaceae), *Parahemionitis cordata* (Hemionietidaceae), *Lygodium flexuosum* (Lygodiaceae), *Oplismenus burmanni, Setaria verticillata* (Poaceae), *Drynaria quercifolia, Pleopeltis macrocarpa* (Polypodiaceae), *Cheilanthes swartzii* (Pteridaceae). Roots of two of the three nonmycorrhizal plant species (*Murdannia nudiflora, Eriocaulon robustum*) were colonized by DSE fungi.

Occurrence of DSE fungal association has been reported in a wide range of terricolous plant species belonging to diverse plant families (Mandyam & Jumpponen 2015). Nevertheless, reports on the occurrence of DSE fungal association in non-terricolous plant species are limited. DSE fungal association were found in 82% of the plant species examined, which included *Eriocaulon robustum* and *M. nudiflora* the two of the three nonmycorrhizal species. DSE fungi could enable plants to survive extreme arid conditions prevailing in nonterricolous habitats by enhancing the drought tolerance mechanisms and nutrient capture. The results indicated that AM and DSE fungal associations are widespread among plants growing on rocks like those of saxicolous plant species. This is the first report of AM and DSE fungal associations for several plant species. It was speculated that the dual association of AM and DSE fungi in lithophytes could improve plant adaptability and survival under rapidly changing conditions of rock surfaces as shown for some harsh habitats. Despite their common occurrence, the roles of these fungal associations on plant adaptations to grow and survive on rock surfaces are not known unlike the role of bacteria. Experimental studies are, therefore, necessary to ascertain the role of AM and DSE fungal associations on growth and nutrition of plants growing on rock surfaces.

3.18.4 Influence of Root inhabiting fungal endophytes: Ecological consequences of climate warming-induced range-expanding plant species have received relatively little attention, despite the increasing incidents of intra-continental migration of native plant species to higher altitude and latitude due to climate change (Stefan Geisen et al., 2021). Similar to invaders from inter-continental origin, plants that expand their range within a continent may be less exposed to negative feedback from their soil microbiome, and host lower numbers of enemies in the new range. While experiments to decipher interactions between range-shifting plant species and soil microbiomes have focused on overall plant–soil feedback patterns, the roles of specific pathogens and mutualists have received less attention. In general, the majority of the plant-associated microbiome components, such as diverse root-inhabiting fungi and oomycetes in the following termed endophytes, remains functionally unknown. Under some conditions endophytes can shift, from plantneutral to mutualistic or pathogenic. Translating this ability to the symbiosis with invasive plants, endophytes may promote the spread of invasive plant species.

The mycobiome (fungal microbiome) influences plants from seed germination to full maturation. While many studies on fungal-plant interaction studies have focused on known mutualistic and pathogenic fungi, the functional role of ubiquitous endophytic fungi remains little explored. Stefan Geisen et al., (2021) examined how root-inhabiting fungi (endophytes) influence range-expanding plant species. endophytes were isolated from three European intra-continental range-expanders and three congenerics that are native both in the range expander's original (southern Europe) and new (northern Europe) range. To standardize the collection, endophytes were obtained from all six plant species growing under controlled conditions in northern (new range of the range expander) and southern (native range of the range expander) soils. The effects of all isolates were cultivated, molecularly identified and tested on seed germination, and growth of seedlings and older plants. Most of the 34 isolates could not be functionally characterized based on their taxonomic identity and literature information on functions. Endophytes affected plant growth in a plant species–endophyte-specific manner, but overall differed between range-expanders and natives. While endophytes reduced germination and growth of range-expanders compared to natives, they reduced seedling growth of natives more than of range-expanders. It was concluded that endophytic fungi have a direct effect on plant growth in a plant growth stage-dependent manner. While these effects differed between range expanders and natives, the effect strength and significance varied among the plant genera included in the present study. Nevertheless, endophytes likely influence the establishment of newly arriving plants and influence vegetation dynamics.

A limitation of the present study is the relatively low number of pairs tested. In order to determine the generality of these results, further studies are needed with additional plant species and from a variety of ecosystems. Furthermore, despite the notion that endophytes might hardly be affected by variation among plant genotypes, future studies should include plant seeds originating from a variation of populations rather than from one as was done in the present study. Another important challenge will also be to perform studies under semi-natural and natural conditions, including the stress conditions that may occur in the field. Studies that determine plant growth after a single inoculation or single sampling point might uncover only part of the ecologically relevant interactions between plants and (soil) biota.

3.18.5 Influence of Soil on Root inhabiting fungal endophyte communities: Symbiotic fungal endophytes are known to enhance plant fitness by conferring abiotic and biotic stress tolerance, and increasing biomass. Most symbiotic root-associated fungal endophytes (RAFE) enhance absorption of water and minerals and in return get direct access to host carbohydrates (Deepak Bokati et al., 2016). RAFE are less understood, especially in cereal crops. Studies examining the soil's effect on RAFE community of cereal grains such as wheat and maize have been rare; most of the published research investigated the identity and abundance of fungal colonists in (or on) the shoots of these crops. Therefore, additional information about how soil influences the structure and composition of the RAFE community within wheat and maize would provide the scientific community with much needed information about how RAFE communities are structured and how these economically important and cosmopolitan grasses respond to changes in their soil environment.

Root-associated fungal endophytes are vital component of root microbiome as some mitigate their host's abiotic and biotic stress. Deepak Bokati et al., (2016) characterized root-associated fungal endophytes in cereal grains and their progenitors grown on two different soil-types. It was aimed at determining how clay and desert soil affects the colonization of root fungal community. Both culture-dependent and culture-independent methods were employed to identify endophytes that successfully colonized greenhouse-grown host plants. The Internal Transcriber Spacer region of fungal ribosomal DNA was utilized for identification purposes. This study revealed soil as a prominent factor influencing the composition of microfungal communities inhabiting the roots of maize and its conspecific progenitor, teosinte (*Zea mays* subsp. *parviglumis*). Similar results were found in wheat and its progenitor (*Triticum monococcum* subsp. *monococcum*). The multidimensional comparisons of Morisita-Horn similarity values of fungal colonists of various host plant taxa indicated that soil plays a primary role in shaping the root fungal community; a secondary effect was plant host identity, even when the plant host is a conspecific. Future studies focused on characterizing root endophytes in other cereal grains, and studying the effect of edaphic factors on fungal colonization, can ultimately contribute to crop productivity.

A total of 270 and 210 root segments were assessed from maize and wheat plants, respectively (Tables 6 & 7). In both grasses, *Fusarium* spp. and *Alternaria* spp. were most prevalent. Grasses grown in clay soil had larger Shannon diversity values than those grown in desert soil (6 & 7) (Deepak Bokati et al., 2016). Grasses grown in the same soil-type harbored similar fungal taxa.

Table 6: Culture-dependent data of maize (*Zea mays* subsp. *mays*) and teosinte (*Zea mays* **subsp.** ***parviglumis***)

Plants	# of plants	Total morphotypes	Mean (SD) number of morphotypes/plants	# of species	Shannon diversity indices	Morphotypes represented by the most numerous sequences
Dmaize	7	8	1 (2)	4	0.90	Fusarium spp. (62%) Paraconiothyrium spp. (25%) Alternaria spp. (13%)
Cmaize	7	38	5 (2)	5	0.96	Fusarium spp. (54%) Alternaria spp. (37%) Exserohilum spp. (6%)
Dteosinte	7	5	1 (1)	3	0.50	Fusarium spp. (83%) Darksidea spp. (17%) N/A
Cteosinte	7	25	5 (3)	4	1.11	Alternaria spp. (56%) Fusarium spp. (12%) Chaetomium spp. (12%)
Total	28	44	19 (15)	9	0.87	Fusarium spp. (44%) Alternaria spp. (39%) Chaetomium spp. (4%)

Table 7: Culture-dependent data of *Triticum aestivum* subsp. *aestivum* and *Triticum monococcum* **subsp.** *monococcum*

Plants	**# of plants**	**Total morphotypes**	**Mean (SD) number of morphotypes/plants**	**# of species**	**Shannon diversity indices**	**Morphotypes represented by the most numerous sequences**
Daestivum	9	48	5 (4)	8	1.71	*Fusarium* spp. (*42%*) *Darksidea* spp. (*19%*). *Cochliobolus* spp. (*14%*)
Caestivum	9	24	3 (3)	7	2.04	*Alternaria* spp. (*32%*) *Fusarium* spp. (*23%*) *Periconia* spp. (*13%*)
Dmonococcum	9	11	1 (1)	4	1.55	*Darksidea* spp. (*50%*) *Embellisia* spp. (*30%*) *Cochliobolus* spp. (*20%*)
Cmonococcum	9	15	2 (2)	9	2.02	*Fusarium* spp. (*27%*) *Alternaria* spp. (*20%*) *Periconia* spp. (*30%*)
Total	36	97	28 (17)	18	1.83	*Fusarium* spp. (*28%*) *Alternaria* spp. (*13%*) *Periconia* spp. (*12%*)

The study documented soil as the prominent factor determining the composition and structure of microfungal communities inhabiting the roots of all grasses; that is, plants grown in the same soil-type harbored similar microfungal communities. Fungal colonization of host plants at the early developmental stage may provide a crucial competitive advantage to these plants, potentially improving their ability to cope with stress and/or utilize limited resources. The effectiveness of microfungal communities to influence host plant fitness might not depend solely on the level of colonization or fungal diversity, but also on the timing of colonization. Early colonization of symbiotic endophytes might play a vital role in enhancing absorption of nutrients and water and mitigating desiccation tolerance, thus conferring accelerated development. Even though plants grown in the same soil harbored similar fungal taxa, they were typically in different concentrations. For example, based on culture-independent data, a greater number of *Fusarium* spp. were observed in *Dmaize* (78%) than in *Dteosinte* (53%); this difference might have been the result of the host's role in microfungal colonization, as found in other studies (Bais et al., 2006). Since this was a greenhouse experiment, plants were provided with optimal growth conditions. However in farms or natural ecosystems those plants might not have such moderate growing conditions, a situation that could alter fungal colonization and fungal community development. It would be especially interesting to investigate root fungal community in arid lands because many studies have found that some symbiotic fungal endophytes tend to selectively colonize in stressful environmental conditions.

Bibliography

Aamir, M., Kashyap ,S.P., Singh, V.K., Dubey, M.K., Ansari, W.A., Upadhyay, R.S., Singh, S., 2019. *Trichoderma erinaceum* Bio-priming Modulates the WRKYs Defense Programming in tomato against the *Fusarium oxysporum* f. sp. *lycopersici* (Fol) Challenged Condition. *Front. Plant Sci.*,**10**:911.

Aaron Hogan, J., Michelle A. Jusino, Matthew E. Smith, Adriana Corrales, Xiaoyang Song, Yue-hua Hu, Jie Yang , Min Cao, Oscar J. Valverde-Barrantes and Christopher Baraloto, 2023. Root-associated fungal communities are influenced more by soils than by plant-host root traits in a Chinese tropical forest. *New Phytologist,* **238:** 1849–1864

Abigail Jenkins, 2005. State of New South Wales Department of Primary Industries. Soil Biology Unit, US Department of Agriculture

Bagyaraj, D. J. and Ashwin, R., 2017. Soil biodiversity: role in sustainable horticulture. *Biodivers. Hortic. Crops,* **5**: 1–18.

Bais, H. P., Weir, T. L., Perry, L. G., Gilroy, S. and Vivanco, J. M., 2006. The role of root exudates in rhizosphere interactions with plants and other organisms. *Ann.l Rev. Pl. Biol.,* **57:** 233–266

Baldrian, P., Bell-Dereske, L., Lepinay, C., Větrovský, T. and Kohout, P., 2022. Fungal communities in soils under global change. *Studies in Mycol.*,**103**: 1–24

Boekhout, T., Amend, A. S., El Baidouri, F., Gabaldón, T., Geml, J., Mittelbach, M., Robert, V., Tan, C. S., Turchetti, B., Vu, D., Wang, Q.M. and Yurkov, A., 2022. Trends in yeast diversity discovery. *Fungal Divers.,* **114:** 491–537

Boekhout,T., 2005. Gut feeling for yeasts. *Nature*, **434**:449–451

Cao, Y., Wu, G. and Yu, D., 2021. Include macrofungi in biodiversity targets. *Science*, **372:** 1160.

Chaloner, T.M., Gurr, S.J. and Bebber, D.P., 2020. Geometry and evolution of the ecological niche in plant-associated microbes. *Nature Commun.,* **11**: 2955.

Chen, L., Swenson, N. G., Ji N., Mi, X., Ren, H., Guo, L. and Ma, K., 2019. Differential soil fungus accumulation and density dependence of trees in a subtropical forest. *Science*, **366:** 124–128.

Christopher W. Schadt and Anna Rosling, 2014. Global diversity and geography of soil fungi: Minus one widespread group. Biosciences Division, Oak Ridge National Laboratory, Oak Ridge TN, USA.

Correia, M., Espelta, J. M., Morillo, J. A., Pino, J. and Rodríguez-Echeverría, S., 2021. Land-use history alters the diversity, community composition and interaction networks of ectomycorrhizal fungi in beech forests. *Journal of Ecology*, **109:** 2856–2870.

Daniel C. Schlatter, Kendall Kahl, Bryan Carlson, David R. Huggins and Timothy Paulitz, 2018. Fungal community composition and diversity vary with soil depth and landscape position in a no-till wheat-based cropping system. *FEMS Microbiol. Ecol.,* **94:** fiy098

Davison, J. et al., 2015. Global assessment of arbuscular mycorrhizal fungus diversity reveals very low endemism. *Science*, **349:** 970–973

Deepak Bokati, José Herrera, Ravin Poudel,et al., 2016. Soil Influences Colonization of Root-Associated Fungal Endophyte Communities of Maize, Wheat, and Their Progenitors. *J. Mycol.,* Article ID 8062073: 9

Dick, R., 2009. Lecture on Soil Fungus in Soil Microbiology, Personal Collection of R. Dick, The Ohio State University School of Environment and Natural Resources, Columbus, OH.

Dickie,. I. A., 2010. Insidious effects of sequencing errors on perceived diversity in molecular surveys. *New Phytol* **188**: 916–918

Doles, J. L., Zimmerman. R. J. and Moore, J. C., 2001. Soil microarthropod community structure and dynamics in organic and conventionally managed apple orchards in Western Colorado, USA. *Appl. Soil Ecol*., **18:** 83–96

Frac, M., Hannula, S.E., Belka, M. and Jedryczka, M., 2018. Fungal Biodiversity and Their Role in Soil Health. *Front. Microbiol*., **9**:707

Frąc, M., Jędryczka, M. and Hannula, E.S., 2023. Editorial: Soil fungal biodiversity for plant and soil health, volume II. *Front. Microbiol*.,**14**: 1170312.

Francesc X. Prenafeta Boldu, Richard C. Summerbell, Wietse de Boer , Henricus T.S. Boschker and Walter Gams, 2014. Biodiversity and ecology of soil fungi in a primary succession of a temperate coastal dune system. *Nova Hedwigia,* **99:** 347–372.

Gams,W., 2007. Biodiversity of soil-inhabiting fungi. *Biodiv.& Consn*.,**16.** 69-72.

Garrett, S. D., 1950. Ecology of the Root Inhabiting Fungi. *Biol. Rev*.,**25**: 220-254

Gomes, E.V., Ulhoam C.J., Cardozam R.E., Silvam R.N. and Gutierrez, S., 2017. Involvement of *Trichoderma harzianum* Epl-1 protein in the regulation of *Botrytis* virulence-and tomato defense-related genes. *Front. Plant Sci.,***8**: 880

Hammad, R. and Elbagory, M.,2019. Using plant growth-promoting fungi (PGPF), as a biofertilizer and biocontrol agents against *Tetranychuscucurbitacearum* on Nubian watermelon (*Citrullus lanatus* L.). *J. Adv. Microbiol.,***16:**1–15.

Jain, S., Middha, R., Juneja S.K, et al., 2015. Comparative physicochemical analysis of restored and unrestored soils of village Thooni Ram Laxmanpura of Chaksu block, Jaipur, Rajasthan. *Int. J. Curr. Res.,* **7:**15706-15710.

James J. Hoorman, 2016. Role of Soil Fungus. Ohioline, Ohio State University Extension

Jayaraman, P., Shalini,S., Saraswathi, K., Vadamalai, K. and Logambal, R., 2018. Comparative Studies on Fungal Biodiversity of agricultural field soil from Thiruvannamalai District, Tamil Nadu, India. *Int.J.Curr.Microbiol.App.Sci.,* **7:** 4259-4273

Jiménez-Morillo, N. T., González-Pérez, J. A., Jordán, A., Zavala, L. M., de la Rosa, J. M., Jiménez-González, M. A., et al., 2016. Organic matter fractions controlling soil water repellency in sandy soils from the Doñana National Park (Southwestern Spain). *Land Degrad. Dev.,* **27:** 1413–1423.

John Paul Schmit and Gregory M. Mueller, 2007. An estimate of the lower limit of global fungal diversity. *Biodivers. Conserv.*,**16**:99–111

Katarzyna Rożek, Kaja Rola, Janusz Błaszkowski and Szymon Zubek, 2019. Associations of root-inhabiting fungi with herbaceous plant species of temperate forests in relation to soil chemical properties. *Sci. Total Environ.*, **649**:1573-1579

Lamit, L.J., Romanowicz, K.J., Potvin, L.R, et al., 2017. Patterns and drivers of fungal community depth stratification in Sphagnum peat. *FEMS Microbiol. Ecol.*,:93 doi: 10.1093/femsec/fix082. PMID: 28854677.

Lazarova, S., Peneva, V., Mladenov, A., Lozanova, L., Bozhanova, V., Georgiev, S., Uzundzalieva, K., Bileva, T., Todorova, M. and Grozeva, N., 2019. Biodiversity of soil fungal communities in agro-ecosystems using DNA metabarcoding. *ARPHA Conf. Abstr*., **2:** e46553.

Leho Tedersoo, Mohammad Bahram, Sergei Põlme, Urmas Kõljalg, Nourou, S., Yorou, Ravi Wijesundera, et al., 2014. Global diversity and geography of soil fungi. *Science*, **346:** 12566

Leho Tedersoo, Vladimir Mikryukov, Alexander Zizka, Mohammad Bahram, Niloufar Hagh-Doust and Sten Anslan, et al., 2022. Global patterns in endemicity and vulnerability of soil fungi. *Global Change Bio.,* **28:** 6696-671

Lekberg, Y., Arnillas, C.A., Borer, E.T, et al., 2021. Nitrogen and phosphorus fertilization consistently favor pathogenic over mutualistic fungi in grassland soils. *Nat. Commun.,* **12:** 3484.

Li, P., Tedersoo, L., Crowther, T.W. *et al.,* 2023. Global diversity and biogeography of potential phytopathogenic fungi in a changing world. *Nat Commun.,* **14:** 6482

Liu, S., García-Palacios, P., Tedersoo, L. *et al.,* 2022. Phylotype diversity within soil fungal functional groups drives ecosystem stability. *Nat. Ecol. Evol.,* **6:** 900–909

Lowenfels, J. and Lewis, W., 2006. Teaming with Microbes: A Gardener's Guide to the Soil Food Web, Chapter 3: Bacteria, Timber Press: Portland, Oregon

Mahadevamurthy Murali, Banu Naziya, Mohammad Azam Ansari, Mohammad N. Alomary, Sami AlYahya, Ahmad Almatroudi, M. C. Thriveni, Hittanahallikoppal Gajendramurthy Gowtham, Sudarshana Brijesh Singh, Mohammed Aiyaz, Nataraj Kalegowda, Nanjaiah Lakshmidevi and Kestur Nagaraj Amruthesh, 2021. Bioprospecting of Rhizosphere-Resident Fungi: Their Role and Importance in Sustainable Agriculture. *J. Fungi (Basel),* **7**: 314.

Majewska, M.L., Błaszkowski, J., Nobis, M., Rola, K., Nobis, A., Łakomiec, D., Czachura, P. and Zubek, S., 2015. Root-inhabiting fungi in alien plant species in relation to invasion status and soil chemical properties. *Symbiosis*, **65:**101-115.

Manici, L.M., Castellini, M. and Caputo, F., 2019. Soil-inhabiting fungi can integrate soil physical indicators in multivariate analysis of Mediterranean agroecosystem dominated by old olive groves. *Ecol. Indicators*, **106**:105490

Marta L. Majewska, Janusz Błaszkowski, Marcin Nobis, Kaja Rola, Agnieszka Nobis, Daria Łakomiec, Paweł Czachura and Szymon Zubek, 2015. Root-inhabiting fungi in alien plant species in relation to invasion status and soil chemical properties. *Symbiosis*, **65**:101–115

Misiak, M., Goodall-Copestake, W. P., Sparks, T. H., Worland, M. R., Boddy, L., Magan, N., and Newsham, K. K., 2021. Inhibitory effects of climate change on the growth and extracellular enzyme activities of a widespread Antarctic soil fungus. *Global Change Biol.,* **27:** 1111–1125.

Morrien, E., Hannula, S. E., Snoek, L. B., Helmsing, N. R., Zweers, H., de Hollander, M., et al., 2017. Soil networks become more connected and take up more carbon as nature restoration progresses. *Nat. Commun.*, **8**:14349

Muhibuddin, A., Addina, L., Abaci, A. and Ahmad, A., 2011. Biodiversity of Soil Fungi on Integrated Pest Management Farming System. *AGRIVITA J. Agri. Sci.,* **33:** 111-118.

Newsham, K.K., 2011. A meta-analysis of plant responses to dark septate root endophytes. *New Phytol.*, **190**:783–793

Nguyen, N. H., Song, Z., Bates, S. T., Branco, S., Tedersoo, L., Menke, J., et al., 2016. FUNGuild: An open annotation tool for parsing fungal community datasets by ecological guild. *Fungal Ecol.,* **20:** 241–248.

Nilshi, M., 2021. Soil Fungi: Distribution, Life Style and Benefits. Soil Science, Soil management.

Petr Kohout, Zuzana Sy´korova´, Martina C˘tvrtlı´kova´, Jana Rydlova, Jan Suda, Martin Vohnı´k and Radka Sudova´, 2012. Surprising spectra of root-associated fungi in submerged aquatic plants. *FEMS Microbiol. Ecol.*, **80:** 216–235

Philippe Vandenkoornhuyse, Stéphane Mahé, Philip Ineson, +5, and J. Peter W. Young, 2007. Active root-inhabiting microbes identified by rapid incorporation of plant-derived carbon into RNA.*PNAS,* **104:** 16970-16975

Piotr Piszczek, Karolina Kuszewska, Janusz Błaszkowski Anna Sochacka-Obruśnik, Anna Stojakowska and, Szymon Zubek, 2019. Associations between root-inhabiting fungi and 40 species of medicinal plants with potential applications in the pharmaceutical and biotechnological industries. *Appl.Soil Ecol.,***137***:* 59-77

Poll, C., Brune, T., Begerow, D., et, al.,2019. Small-scale diversity and succession of fungi in the detritusphere of rye residues. *Microb. Ecol.*,**59**: 130–40.

Pongsatorn Puangsombat, Uthaiwan Sangwanit and Dokrak Marod, 2010. Diversity of Soil Fungi in Different Land Use Types in Tha Kum-Huai Raeng Forest Reserve. *Trat Province. Kasetsart J.,* (*Nat. Sci.,*) **44**: 1162 - 1175

Reeleder, R. D., 2003. Fungal plant pathogens and soil biodiversity. *Can. J. Soil Sci.,* **83:** 331–336.

Ritz, K., 2005. Fungi. *In: Encyclopedia of Soils in the Environment*, Nutrient Cycling.

Romero, F., Cazzato, S., Walder, F., et al., 2022. Humidity and high temperature are important for predicting fungal disease outbreaks worldwide. *New Phytologist.*, **234**: 1553–1556.

Rosling, A., Cox, F., Cruz-Martinez, K., Ihrmark, K., Grelet, G.A., Lindahl, B.D., Menkis, A. and James, T.Y., 2011. Archaeorhizomycetes: unearthing an ancient class of ubiquitous soil fungi. *Sci.*, **333:** 876–879.

Rosling, A., Timling, I. and Taylor, D. L., 2013. Distribution and abundance of Archaeorhizomycetes. P 333–349. *In: Genomics of soil- and plant-associated fungi.* B. A. Horwitz, P. K. Mukherjee, M. Mukherjee and C. P. Kubicek (Eds.,), Springer, Heidelberg, Germany

Sabine Ravnskov, Birgit Jensen, Sabine Ravnskov, Birgit Jensen, Steen Lykke Nielsen and John Lar Steen Lykke Nielsen and John Larsen, 2019. Quantification of Root Fungi Quantification of Root Fungi Using Signature Fatty Acids Using Signature Fatty Acids. Department of Crop Protection, Research Centre Department of Crop Protection, Research Centre Flakkebjerg Flakkebjerg Danish Institutes of Agricultural Sciences, DK-4200, Slagelse Slagelse, Denmark.

Schmit, J.P., Mueller, G.M., Leacock, P.R., Mata, J.L., Wu, Q.X. and Huang, Y.Q., 2005. Assessment of tree species richness as a surrogate for macrofungal species richness. *Biol. Conserv.*, **121**:99–110

Sekimoto, S., Rochon, D., Long, J.E., Dee, J.M. and Berbee, M.L., 2011 A multigene phylogeny of Olpidium and its implications for early fungal evolution. *BMC Evol. Biol.,* **11:**331.

Sindhushri Chauhan, Parashuram Ganure and Chethan J Dandin, 2023. Impact of abiotic conditions on fungal diversity and comparative analysis of soil quality of two distinct locations. *AGBIR*,.39: No.1

Socolar, J. B., Gilroy, J. J., Kunin, W. E.and Edwards, D. P., 2016. How should β-diversity inform biodiversity conservation? *Trends Ecol. Evol.,* **31**: 67–80

Stefan Geisen, Freddy C. ten Hooven, Olga Kostenko, L. Basten Snoek and Wim H. van der Putten, 2021. Fungal root endophytes influence plants in a species-specific manner that depends on plant's growth stage. *J.Ecol.*,**109:**1618–1632

Tedersoo, L., Bahram, M., Puusepp, R. *et al.,* 2017. Novel soil-inhabiting clades fill gaps in the fungal tree of life. *Microbiome,* **5:** 42.

Tedersoo, L., Tooming-Klunderud, A. and Anslan, S., 2018. PacBio metabarcoding of fungi and other eukaryotes: Errors, biases and perspectives. *New Phytol.,* **217:** 1370–1385

Thangavelu Muthukumar, Marimuthu Chinnathambi and Perumalsamy Priyadharsini, 2016. Root fungal associations in some non-orchidaceous vascular lithophytes 2016. *Acta Botanica Brasilica*, **30**: 407-421.

Thomson, B. C., Tisserant, E., Plassart, P., Uroz, S., Griffiths, R. I., Hannula, E. S., et al., 2015. Soil conditions and land use intensification effects on soil microbial communities across a range of European field sites. *Soil Biol. Biochem.,* **88**: 403–413.

Van der Putten, W. H., Macel, M. and Visser, M. E., 2010. Predicting species distribution and abundance responses to climate change: Why it is essential to include biotic interactions across trophic levels. *Philos. Trans. R. Soc.,* **B 365**: 2025–2034

Větrovský, T., Kohout, P., Kopecký, M. *et al.,* 2019. A meta-analysis of global fungal distribution reveals climate-driven patterns. *Nat. Commun.,* **10:** 5142

Vikram, S.,. Ramond, J.B., Ortiz, M., Maggs-Kölling, G., Pelser, K. and Cowan, D.A., 2023. Soil fungal diversity and assembly along a xeric stress gradient in the central Namib Desert. *Fungal Biol.*,**127**: 997-1003

Vladimir Mikryukov *et al.,* 2023. Connecting the multiple dimensions of global soil fungal diversity. *Sci. Adv.,* **9:** eadj8016 D

Wallace Marianna E., 2022. Root-Inhabiting and Rhizosphere Mycobiomes and Crop Yield of Corn and Wheat. Electronic Thesis and Dissertation Repository, 9028. The University of Western Ontario

Wang, C. Z. Et al.,2022. Occurrence of crop pests and diseases has largely increased in China since 1970. *Nat. Food,* **3:** 57–65

Warren, R., VanDerWal, J., Price, J., Welbergen, J. A., Atkinson, I., Ramirez-Villegas, J. And Lowe, J., 2013. Quantifying the benefit of early climate change mitigation in avoiding biodiversity loss. *Nature Climate Change*, **3:** 678–682.

Witoon Purahong, Katherina A. Pietsch, Helge Bruelheide, Christian Wirth, François Buscot, and Tesfaye Wubet, 2019. Potential links between wood-inhabiting and soil fungal communities: Evidence from high-throughput sequencing. *Microbiologyopen,* **8**(9): e00856.

Yang, Y., Li, T., Wang, Y., et al., 2021. Negative effects of multiple global change factors on soil microbial diversity. *Soil Biol. & Biochem.*,**156:** 108229

4

Types of Biocontrol Agents

Biological control is defined broadly as the "use of natural or modified organisms, genes, or gene products" to reduce the effects of pests and diseases. The many approaches to biological control can be categorized conventionally into regulation of the pest population (the classical approach); exclusionary systems of protection (a living barrier of microorganisms on the plant or animal that deters infection or pest attack) and systems of self-defense (resistance and immunization) (Anchal Sharma et al., 2013). The agents of biological control include the pest- or disease-agent itself (sterile males or a virulent strain of pathogens), antagonists or natural enemies, or the plant or animal managed or manipulated (immunized) to defend itself. Principles of plant health care are offered, know the production limits of the agro-ecosystem, rotate the crops, maintain soil organic master, use clean planting material, plant well-adapted, pest resistant cultivars, minimize environmental and nutritional stresses, maximize the effects of beneficial organisms and protect with pesticides as necessary.

Biological control is the control of one organism by another. This control may be expressed as either a longer population of the pest or as a restriction or prevention of the severity or incidence of pest damage without regard to the pest population (Cook and Baker, 1983). Biological control depends on knowledge of biological interactions at the ecosystem, organism, cellular, and molecular levers and often is more complicated to manage compared with physical and chemical methods. Biological control is also likely to be less spectacular than most physical or chemical controls but is usually also more stable and longer lasting. In spite of biological controls having been used in agriculture for centuries, as an industry biological control is still in its infancy. Biological control is now being considered for an increasing number of crops and managed ecosystems as the primary method of pest control. One reason for its growing popularity is its record of safety during the past 100 years considered as the era of modem biological control (Anchal Sharma et al., 2013). No microorganism or beneficial insect deliberately introduced or manipulated for biological control purposes has, itself, become a pest so far as can be determined, and there is no evidence so far of measurable or even negligible negative effects of biocontrol agents on the environment. The new tools of recombinant DNA technology, mathematical modeling, and computer technology combined with a continuation of the more classical approaches such as importation and release of natural enemies and improved germplasm, breeding, and field testing should quickly move bio-control research and technology into a new era.

Biological control was discovered by trial and error and then practiced in agriculture long before the term itself came into use (Baker and Cook, 1974). One example is the ancient practice of not growing the same crop species in the same field more frequently than every second or third year or even longer. Such crop rotation allows time for the pest or pathogen population in soil to decrease below some economic threshold because of the predatory, competitive, and other antagonistic effects imposed by the associated microflora and fauna. In other words, crop rotation allows time for the natural soil microbiota to sanitize the soil, especially with regard to the more specialized plant parasites and insect pests that are highly dependent on their host crop to maintain their populations. The era of modern biological control, involving the deliberate transfer and introduction of natural enemies of insect pests, was launched 100 years ego with the highly successful introduction of the vadalia beetle from Australia to California in 1888 to control the cottony cushion scale of citrus. In 1914, the German plant pathologist C. F. von Tubuef wrote a somewhat speculative article entitled "Biologische Bekampfung von Pilzkrankheiten der Pflanzen." This is apparently the first reference in the scientific literature to the term "Biologische Bekampfung" or "biological control" (Baker, 1987).

DeBach (1964) defined biological control as "the action of parasites, predators, or pathogens in maintaining another organism's population density at a longer average than would occur in their absence." This definition covers some highly successful biological controls of insect pests with natural enemies, but it does not accommodate some other highly successful controls accepted in other disciplines as examples of biological control. For example, citrus tristeza virus is controlled in Brazil by inoculating the citrus trees with a mild virus, which then protects the trees against the more severe strains. "Cross protection" was first shown by H. H. McKinney in 1929 to have potential for biological control of plant viruses. Plant pathologists refer to cross protection for control of plant viruses as biological control.

4.1 Need for Biological Control, Merits and Demerits

Beyond good agronomic and horticultural practices, growers often rely heavily on chemical fertilizers and pesticides. However, the environmental pollution caused by excessive use and misuse of agrochemicals, as well as fear mongering by some opponents of pesticides, has led to considerable changes in people's attitudes towards the use of pesticides in agriculture (Anchal Sharma et al., 2013). A concomitant increase in the proportion of pests and diseases resulted in the increased use of toxic chemical for their management. The number of species resistant to pesticides and fungicides is increased. In recent years after signing of the general agreement of trade and tariff of world trade organization more emphasis is given to the use of ecofriendly pesticide for crop production in view of their least toxic nature, low levels of disease resistance and low residue problems. However, Biological controls should be integrated with other control measures because different methods are effective at different times and locations under varying conditions.

Biocontrol agents are the organisms or microorganisms that control the plant pathogen. They play a key role in resisting pest population and prevent the crop from various diseases like leaf wilt, curling of disease, root rot disease, crown gall disease, etc. They kill the pest population before they spread the disease (Bhupendra Koul et al., 2022). Biocontrol agents are used at large scales as they do not cause any harm to the main crop and kill the pest population naturally. Many bacteria (*Bacillus* sp., *Pseudomonas* sp., etc.) and fungi (*Trichoderma* sp., *Candida* sp. etc.) are used as biocontrol agents. The major benefit of these biocontrol agents is that the hosts do not consume them, and thus host organism remains unaffected. The microbes used in both (entomology and plant pathology) to kill or control the pest population are named 'biocontrol agents' or 'biological control agents'. Moreover, the extracts or fermented products are taken from various natural sources can also act as biocontrol agents. These products or extracts can be used in any form (pure or mixture), having various effects on target microbes or pathogens. According to USRC (US Research Council), biocontrol agents are natural or modified organisms, genes or a gene product used to suppress the growth of the unwanted organism (pest/pathogen) and to enhance the growth and yield of the wanted organism (crops, microorganism, or insects).

Biological control agents reduce the activity of pathogens to damage the target organism, hence called 'natural enemies'. The utilization of fungi and bacteria is done for its various profit-oriented products to kill the pests, weeds, and to control diseases. It is found that their association enhances the productivity of the plant, and contributes to animal and human health (Savita and Anuradha, 2019). Thus, bacteria and fungi are extensively used as biocontrol agents to prevent the crop from diseases and enhance both the quality and quantity of food.

Merits of Biocontrol Agents

- Biological control is less costly and cheaper than any other methods.
- Biocontrol agents give protection to the crop throughout the crop period.
- They do not cause toxicity to the plants.
- Application of biocontrol agents is safer to the environment and to the person who applies them.
- They multiply easily in the soil and leave no residual problem.
- Biocontrol agents not only control the disease but also enhance the root and plant growth by way of encouraging the beneficial soil micro flora. It increases the crop yield also.
- Biocontrol agents are very easy to handle and apply to the target.
- Biocontrol agent can be combined with bio-fertilizers.
- They are easy to manufacture.
- It is harmless to human beings and animals (Environmentally safe.)

Demerits

- Pathogens will grow unmanageable in the end.
- There is always a potential that the predator may change to a new prey since it is only intended to handle one issue.
- This is a highly time-consuming operation, while other treatments, such as the application of pesticides, give benefits and that also immediately.
- The effects of biological agents may be seen in the long run. It requires a lot of patience. A successful system of biological control takes a lot of time and money to design.
- Predators can't survive without food, therefore eliminating all of their food sources isn't the solution because the population of harmful pests will only be reduced leading predators' reduction.
- The planning and development of an effective biological control system requires a significant investment of both time and resources.
- Can sometimes fail in its specificity

4.2 Bacterial and Fungal Biocontrol Agents Against Plant Diseases

A wide range of biological control agents (BCAs) are available for use, but further research is needed to identify potential microbes and their natural products with a broad-spectrum antagonistic activity to control crop diseases. Ayaz et al., (2023) reviewed to highlight the importance of biocontrol strategies for managing crop diseases. Furthermore, the role of beneficial microbes in controlling plant diseases and the current status of their biocontrol mechanisms was summarized. Various bacterial and fungal biocontrol agents against plant pathogens are listed in table 8. The review also covered the challenges and the need for the future development of biocontrol methods to ensure efficient crop disease management for sustainable agriculture. A diverse range of bacterial genera have demonstrated great potential as biocontrol agents for various plant diseases. Fungi have also been shown to play a significant role in preventing important diseases in major crops. Research on fungal strains as biocontrol agents for plant disease control has also received considerable attention. These fungi-based biocontrol agents exhibit a significant antagonistic activity against a variety of soil and airborne plant pathogens, making them potential biopesticides for field or greenhouse studies (Zubair, et al., 2021).

Table 8: Various bacterial and fungal biocontrol agents against plant pathogens

Plant Species	Biocontrol Agents	Pathogens	Mode of Action
Citrus fruit	*Bacillus megaterium*	Blue mold	In vitro antagonistic activity against post-harvest disease
Wheat	*Bacillus subtilis* 26DCryChS	*Stagonospora nodorum* Berk	Antimicrobial metabolites (surfactants showed antifungal activity against *S. nodorum* disease)
Brassica campestris L	*Bacillus thuringiensis*	*Sclerotinia sclerotiorum*	Suppressing *S. sclerotiorum* growth by inducing systemic resistance
Cotton/black root rot	*Paenibacillus alvei* K-165	*Thielaviopsis basicola*	K-165 inhibited *T. basicola* growth *invitro* through antibiosis and significantly reduced root discoloration and hypocotyl lesions on cotton seedlings
Tomato and Soybean	*Bacillus velezensis* DMW1	*Phytophthora sojae* and *Ralstonia solanacearum*	Antimicrobial metabolites (fengycin, iturin, and bacillomycin) demonstrated antagonistic activity in vitro and in pot experiments
Rice	*Bacillus atrophaeus* FA12 and *B. cabrialesii* FA26	*Xanthomonas oryzae* pv. *oryzae* (Xoo)	In vitro, antagonistic activity against various fungal pathogens significantly reduced Xoo lesions in greenhouse conditions
Rice	*Bacillus thuringiensis* GBAC46	*Aphelenchoides besseyi*	In vitro antagonistic activity through various proteins (Cry31Aa, Cry73Aa, and Cry40ORF) and in greenhouse conditions
Maize	*Pseudomonas protegens* Pf-5	*Pantoea ananatis* DZ-12	Antimicrobial pyoluteorin showed strong antagonistic activity against *P. ananatis* in vitro and in vivo
Wheat and Maize	*Bacillus Subtilis* ATCC6633	*Fusarium graminearum* and *Fusarium verticillioides*	Antimicrobial mycosubtilin showed a strong antagonistic activity against *F. graminearum* and *F. verticillioides* in vitro and in vivo
Tomato	*Bacillus atrophaeus* GBSC56	*Meloidogyne incognita*	Antimicrobial VOCs showed nematicidal activity and also produced ROS in nematodes
Rice	*Bacillus* spp. GBSC56, SYST2, and FZB42	*Aphelenchoides besseyi*	Antimicrobial VOCs of *Bacillus* spp. showed the strongest nematicidal activity and accumulated ROS as well as promoted rice growth

Plant Species	**Biocontrol Agents**	**Pathogens**	**Mode of Action**
Soybean and Rice	*Pseudomonas parafulva* JBCS1880	*Xanthomonas axonopodis* pv. *glycines*, and *Burkholderia glumae*	Strong antagonism and antibacterial activity against *Xanthomonas axonopodis* pv. *glycines* and *Burkholderia glumae*
Rice	*Pseudomonas putida* BP25	*Magnaporthe oryzae*	BP25 showed strong biocontrol activity against blasts caused by *M. oryzae*
Pepper	*Bacillus licheniformis* BL06	*Phytophthora capsici*	BL06 effectively reduced pepper *Phytophthora* blight severity in vitro and pot experiments
Wheat	*Bacillus atrophaeus strain* TS1	*Fusarium graminearum*	TS1 was found as a potential biocontrol agent to inhibit *F. graminearum* under low temperatures
Tomato	*Bacillus amyloliquefaciens* FZB42	*Sclerotinia sclerotiorum*	Antimicrobial potential (fengycin-induced systemic resistance in tomatoes against *S. sclerotiorum*)
Rape Seed and Tabaco	*Bacillus amyloliquefaciens* EZ1509	*Sclerotinia sclerotiorum*	*Bacillus* strain EZ1509 showed a strong antifungal activity against *S. sclerotiorum* and also led to the development of new biopesticides
Tomato	*Streptomyces* sp. AN090126	*Ralstonia solanacearum* and *Xanthomonas euvesicatoria*	*Streptomyces* sp. AN090126 can combine with antibiotics effectively control different bacterial plant diseases
Tomato	*Paecilomyces lilacinus*	*Meloidogyne javanica*	*P. lilacinum* is used as a biocontrol agent to control *M. incognita* and as a better alternative against chemical nematicides
Pineapple	*Purpureocillium lilacinum*	*Meloidogyne javanica*	The application of *P. lilacinum* significantly reduced nematode egg and egg mass production, reducing root galling damage in pineapple
Onion	*Trichoderma asperellum*	*Sclerotium cepivorum*	*T. asperellum* BCC1 exerts efficient biocontrol against *S. cepivorum* and activates onion systemic defenses against *S. cepivorum* under greenhouse conditions
Okra	*Trichoderma virens*	*Meloidogyne incognita*	*T. virens* observed a reduction in second-stage juveniles' hatching periods tested in vitro

Plant Species	Biocontrol Agents	Pathogens	Mode of Action
Carrot	*Pochonia chlamydosporia*	*Meloidogyne incognita*	*P. chlamydosporia* reduced nematode galls and also decreased juvenile 2 nematodes in vitro and pot experiment methods
Mango	*Trichoderma asperellum T8a*	*Colletrotrichum gloeosporiodes*	*T. asperellum* T8a plays a role in biological control against *C. gloeosporioides* and controlling anthracnose disease in mangoes
Beans	*Trichoderma asperellum*	*Sclerotinia sclerotiorum*	*T. asperellum* the reduced disease severity index and antagonistic activity against *S. sclerotiorum* in field trials of beans
Cabbage	*Trichoderma hamatum*	*Sclerotinia sclerotiorum*	*T. hamatum* LU593 reduced apothecial production, decreased disease severity index, and could potentially control *S. sclerotiorum* disease in cabbage

Biocontrol agents employ diverse mechanisms to protect plants against pathogenic invasion. Using one or a combination of processes, they may interact with the pathogen directly or indirectly to reduce plant disease (Ayaz et al., 2021). BCA in the rhizosphere region competes for space and resources, and interferes with pathogens' pathogenicity through different substances such as lipopeptides, biosurfactants, bacteriocins, volatiles, and enzymes that have antimicrobial effects by slowing the development or metabolic activity of pathogens. BCA could disrupt pathogens' quorum sensing (QS) by inhibiting the production of signal molecules that launch infections. One example is producing QS inhibitors that can break down QS signal molecules, such as chitinases, pectinases, and lactonases. These inhibitors prevent pathogen invasion and lead to reduced plant disease symptoms. In addition to direct interactions, BCAs may protect plants indirectly by triggering a defense response or promoting plant growth. As a result, the host undergoes a wide range of biochemical and molecular defensive processes that serve as a defense mechanism against various pathogens. Furthermore, beneficial microbes may promote plant growth by enhancing nutrients and water uptake or by producing chemicals such as hormones for maintaining plant fitness. Many processes take part in the complex interactions between pathogens, beneficial microbes, and plants. Therefore, identifying the mechanisms responsible for biocontrol is a great challenge. Hence, understanding the mechanism of action behind a BCA's protective effect will make it easier to optimize biological control; establish ideal conditions for the interaction between the BCA, the pathogen, and the host; and develop suitable formulations and application techniques to improve plant health and sustainable agriculture.

4.2.1 Bacteria for disease management: Many studies have investigated the role of beneficial bacteria in promoting plant growth and disease resistance in crops. At present, many bacteria from various genera, including *Bacillus, Paenibacillus, Agrobacterium, Bradyrhizobium, Acinetobacter, Azospirillum, Azotobacter, Pseudomonas, Rhizobium* and *Streptomyces,* have been documented as biocontrol agents to control various diseases in major crops (Ayaz et al., (2023). Phytopathogens, mainly bacteria, fungi, and nematodes, cause serious plant diseases, which present a significant barrier to sustainable crop health and yield. Crops are highly affected when infected by pathogenic bacteria such as *Pseudomonas savastanoi, Xanthomonas axonopodis*, and *Ralstonia solanacearum*, although inoculation with *Bacillus* spp. suppresses pathogen growth and reduces disease severity in many infected crops. The use of beneficial bacteria can improve plant growth and control various bacterial, fungal, and nematode diseases without harming the environment, as listed in Table 8.

Beneficial bacteria create biofilms and secondary metabolites, such as surfactin, iturin, bacillomycin, and fengycin, which reduce the plant pathogen population by establishing plant–microbial interactions in the rhizosphere region. *Bacillus* spp. attach to mycelial cell walls and deform hyphae through the production of extracellular enzymes such as chitosanase, protease, glucanase, and cellulase. Lipopeptides, including fengycin, iturin, pumilacidin, mixirin, and surfactin, are antifungal peptides that function against pathogenic fungi in rhizospheres. Many bacteria, particularly *Bacillus*, *Pseudomonas,* and *Burkholderia* spp. are known to suppress nematodes in various plants by affecting nematode behavior such as feeding and reproduction. Previous studies have shown that biological treatment with *Bacillus* isolates was effective in controlling a root–knot nematode infestation. They have also been reported to reduce nematode populations in infested roots and soil. *Bacillus* spp. have also been reported to stimulate induced systemic resistance in plants against various pathogens through increased defense-related enzyme activity, such as polyphenol oxidase, peroxidase and phenylalanine ammonia-lyase (PAL), as well as root exudates and modifications such as amino acids and polysaccharides (Jacob et al., 2020).

Bacillus and *Pseudomonas* spp. have been reported to be effective against various phytopathogens in major crops. The presence of several biosynthetic gene clusters responsible for the synthesis of secondary metabolites was found in *Bacillus velezensis* isolates. These chemically varied bioactive metabolites may be used as a resource for the development of new drugs. The recent studies on bioactive substance synthesis, chemical compositions, interesting bioactive gene clusters, and biological applications of *B. velezensis* and related *Bacillus* species will be effective at preventing plant diseases to ensure sustainable agriculture. There are numerous uses for antimicrobial compounds made by *B. velezensis* in the management of plant pathogens. *B. velezensis* uses the lipopeptides surfactin, iturin, and fengycin to produce antagonistic effects against *R. solanacearum* and *Fusarium oxysporum*. *B. velezensis* QST713 is used industrially as a biocontrol agent to protect *Agaricus bisporus*, an edible mushroom, in a compost micromodel and inhibit *Trichoderma aggressivum* f. *europium*, which causes green mold disease. To carry out their actions, they seem to principally rely on antibiosis and the development of systemic resistance in a variety of plant species. Additionally, *B. subtilis* has been identified as a key player in the control of different crop diseases. *B. subtilis* IBFCBF-4 is a potential biocontrol agent for watermelon *Fusarium* wilt. An important example of an exogenous endophyte for preventing banana *Fusarium* wilt in China is the *B. subtilis* R31 strain isolated from *Dendrobium orchid* leaves (Li et al., 2021). Thus, extensive research should be conducted to develop efficient formulation technology and to investigate various natural products with the determination of their antimicrobial activity in vitro as well as in the field to manage different crop diseases efficiently.

4.2.2 Fungi for diseases management: Beneficial fungi produce large quantities of bioactive compounds that can be used as agrochemicals for crop protection. The development of fungal strains as BCAs for plant diseases has drawn a lot of attention because it has been found that many beneficial fungi inhibit the growth of various plant pathogens (Ayaz et al., 2023). *Trichoderma*, *Aspergillus*, and *Penicillium* are among the most popular fungal genera used as BCAs against both bacterial and fungal plant diseases. Fungi, such as *Gliocladium* and *Saccharomyces,* have also been reported to possess an antagonistic activity against a variety of pathogens. Most fungal endophytes, which are diverse microorganisms, live asymptomatically in the internal plant tissues. Endophytic fungi interact closely and intricately with their hosts through mutualism and in rare cases, parasitism. They are crucial for protecting their hosts from pathogenic bacteria and pests because of their capacity to produce a large variety of structurally varied and physiologically active secondary metabolites. Endophytic fungi are a rich mine for discovering novel secondary metabolites with a wide range of potential agricultural applications. Numerous metabolites with various chemical structures, including terpenoids, alkaloids, steroids, peptides, isocoumarins, benzopyranones, and quinones, have been identified in endophytic fungi. The identification of these metabolites provided a solid chemical basis for the development of agrochemicals that may have antibacterial, antifungal, herbicidal, nematicidal, insecticidal, and other agricultural uses (Risoli, et al., 2022). Hence, there is a large variety of fungi that interact with plants in rhizospheric and endophytic associations. Different antimicrobial products produced by biological control agents, i.e., bacteria and fungi, and their impact on plant pathogens along with plant growth parameters are briefly summarized in figure 28.

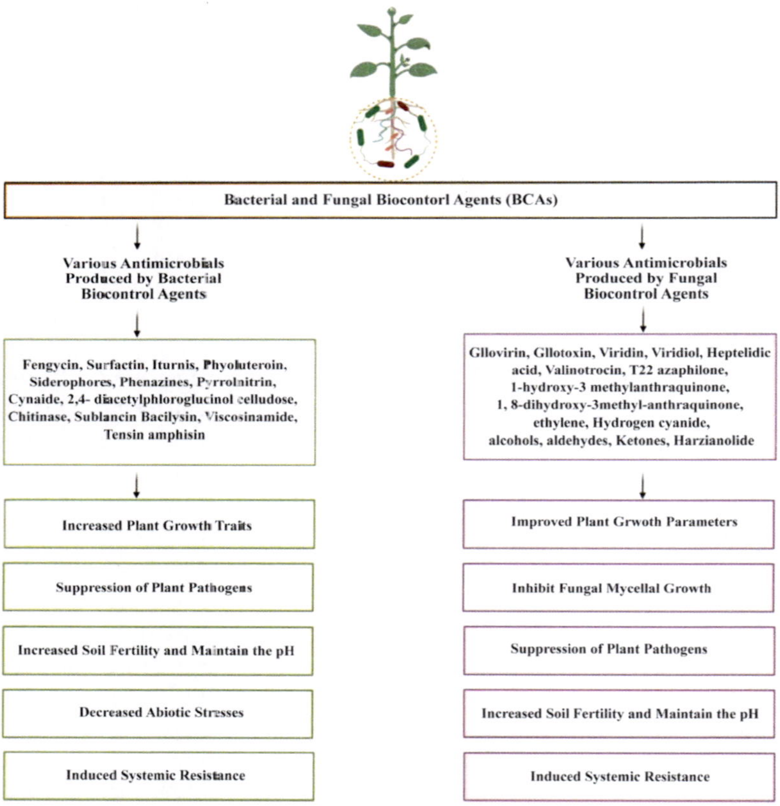

Fig. 28: Different antimicrobial products produced by biological control agents, i.e., bacteria and fungi, and their impact on plant pathogens and plant growth parameters. The antimicrobial products produced by bacteria and fungi, such as the secondary metabolites fengycin, surfactin, and bacillomycin D, induce systemic resistance and also have a broad-spectrum antagonistic activity against different diseases. The volatile organic compounds (VOCs) produced by microbes improved plant growth traits and reduced disease indexes, both directly and indirectly. The arrows show the different antimicrobial metabolites produced by bacteria and fungi along with their benefits. The figure was created with BioRender software (https://www.biorender.com/, accessed on 13 September 2023).

Several fungi, especially *Trichoderma*, are famous for their broad-spectrum antagonistic activities against various phytopathogens. *Trichoderma* is a widespread fungal genus found in soil as residents, saprotrophs, plant symbionts, and mycoparasites. This genus contains filamentous fungi that have been extensively investigated and employed as biocontrol agents against plant pathogens in agriculture (Degani and Dor, 2021). The direct and indirect control potential of BCAs against phytopathogens has been highly studied in the last decade to ensure efficient plant disease management. It has been demonstrated that *Trichoderma* controls insect pests directly through parasitism and the synthesis of insecticidal compounds, repelling metabolites and antifeedant chemicals. The indirect approaches include attracting natural enemies, stimulating systemic defense mechanisms of plant, and parasitizing insect−symbiotic microbes. *Trichoderma* use in agriculture is thus effective at fighting insect pests and plant pathogens, presenting it as a promising future option for the advancement of sustainable agriculture. This fungus has shown a strong antagonistic potential against more than 80% of plant pathogens. *T. aggressivum* f. *europaeum* TAET1 showed a strong antagonistic activity against *Botrytis cinerea*, *Sclerotinia sclerotiorum*, and *Mycosphaerella melonis* by affecting mycelial growth. Additionally, it completely stopped *S. sclerotiorum* sclerotia germination. *Trichoderma asperellum* TaspHu1, isolated from *Juglans mandshurica*. Rhizosphere soils in China has been shown to enhance resistance in tomato seedlings against leaf spot disease caused by *Alternaria alternata*. It has been clearly shown that *Trichoderma* has a significant role in inhibiting plant pathogens indirectly in the rhizospheric region through enhanced plant immunity. Thus, *Trichoderma* spp. can be employed in combination with several other alternative pest management techniques for sustainable agriculture.

4.3 Factors Affecting Biocontrol of Plant Diseases and Selection of Potential BCAs:

Despite decades of research, biological control remains relatively insignificant in managing plant health compared with chemical control. It has already been documented that biocontrol of plant pathogens under laboratory conditions (in vitro or in planta) is more effective than in open-field trials (Ayaz et al., 2023). BCAs face greater challenges transferring from the controlled environment of a laboratory experiment to the harsh conditions encountered in the field. BCA field efficacy may be equal to or better than that of synthetic pesticides, however, this could change over time and across regions. In other words, a BCA that inhibits or controls disease in the lab is unlikely to be efficient in open-field trials. This is a result of all of the complex interactions that take place between the host, pathogen, antagonist, and environment. For instance, the host undergoes a sequence of changes or mutations that alter its physical and chemical features. The efficiency of the antagonist is also affected by pathogenic behaviors. The antagonistic potential of BCA may change as a result of changes in the population and environment or from the existence of microbial colonizers in the natural environment. *Trichoderma* is the most extensively reported fungal BCA, but its broader application has been hampered by its unpredictable nature in the field. To create and put into practice *Trichoderma*-based agricultural production and preservation methods, it is essential to fully understand how it interacts with plants, other microbes, and the environment. The factors limiting the efficacy of BCA are given below.

i. **Plant species influence the biocontrol activity of BCAs:** Different plant species and genotypes have a significant impact on the effectiveness of the biocontrol potential of microbes. Several studies have been conducted that show the effect of plants on BCAs' biocontrol activities. The degree of rhizosphere colonization, the production of antibiotics by BCA, and the stimulation of induced systematic resistance vary among different plant species (Umer et al., 2021). In the rhizosphere region, the interaction between BCAs and pathogens is mostly influenced by root exudate excretion, water and mineral absorption, and other surrounding microbes. The expression of plant genes is reported to be involved in the rhizosphere enrichment process that provides fitness to the plants by recruiting microbes from the surroundings. The complexity behind the specificity of BCAs for specific hosts must be uncovered with in-depth research to control different plant diseases in a well-organized manner.

ii. **Pathogen influence the biocontrol activity of BCAs:** The behavior of pathogens toward BCAs is one of the most crucial factors in plant disease control. Each pathogen interacts with the host differently due to genetic variety and ecological fitness diversity. It is crucial to emphasize that the pathogen behaves differently from the BCAs because it is both harmful and vulnerable to antagonist activity. Depending on the complexity of their mode of action, phytopathogens may display a wide range of sensitivities to BCAs (including extremely low sensitivity). In a few generations, some pathogens can adjust to the selection pressure imposed by BCAs. The idea that biological control lasts longer than chemical control is a widespread misconception. According to research on pest management in fields, this assumption may not always be true. Numerous pests have developed tolerance to one or more Bt toxins, while the codling moth Cydia pomonella has established resistance to th***e*** *C. pomonella granulovirus (*Lahlali, et al., 2022).

   The durability of biological control of plant diseases, in contrast with pest management, has received little attention. Similar to single-mode chemical fungicides, populations of pathogens could change into forms that are resistant to BCAs. This would make it less likely that BCAs will be able to reduce plant diseases for a long time. Therefore, a weak spot in the pathogen life cycle must be found as an opportunity window for successful biocontrol. BCAs should enter the window of opportunity and stop the progression of pathogens. For example, the window of opportunity for some unspecialized necrotrophic pathogens is to obstruct the nutrient uptake required for growth. If BCAs are abundantly present near the pathogen spores, the depletion of local nutrients from the disease spore could hinder or stop germination.

iii. **Biocontrol agents and their specific nature***:* The ability of BCAs to adapt to local biotic and abiotic environmental conditions is the primary cause of their reduced field efficiency. Therefore, to completely understand this phenomenon, it is important to study the distribution pattern of BCAs in the rhizosphere (Ayaz et al., 2023). To obtain pertinent biocontrol results, additional suitable native BCA strains should be collected and examined. The mode of action of

BCAs, their selectivity against plant pathogens, and their resistance to harsh climatic conditions play a significant role in their success. The efficacy of the biocontrol agent improves when an optimal association exists. Therefore, the application of BCAs at the right time is crucial for effective biocontrol. If the antagonist is applied before the infection becomes established, biocontrol will be effective. Furthermore, it is crucial to understand how BCAs function to attain the best disease control effect. Numerous characteristics of fluorescent *Pseudomonas* make them excellent BCAs. These features comprise (i) the ability to efficiently colonize roots, hypocotyls, tubers, and other parts of the plant; (ii) the convenience of cultivation in the lab; (iii) the synthesis of various secondary metabolites; (iv) being able to utilize various kinds of organic compounds, which are usually found in exudates from roots and seeds; and (v) compatibility with common pesticides and biological agents. Hence, future studies should focus on identifying novel strains with potent biocontrol characteristics by investigating unexplored microbial diversity. This could involve extensive sampling and screening of different habitats in order to identify microorganisms with a broad-spectrum antagonistic activity against various phytopathogens.

iv. **Environmental stresses impact on BCA activity**: The microbial makeup of the soil and its environment has a significant impact on the efficacy of biological control measures. Soil biology research should be able to figure out the unique qualities of different species, especially those that live in the rhizosphere of plants, by figuring out what each possible microbe brings to the biocontrol process. Similarly, the ecological study should investigate all biotic and abiotic factors that have a significant impact on BCAs as crucial components of plant health. Many BCAs are extremely susceptible to changes in the biotic and abiotic environment. It is not always possible to transfer their efficiency from the laboratory to the field (Barratt et al., 2018). For instance, although many *Pseudomonas* and *Bacillus* BCAs function well in trials, they cannot effectively handle disease in a variety of field conditions. Therefore, it is important to select a BCA that has stable efficacy under various environmental conditions, such as soil texture, moisture, temperature extremes, or competition. Furthermore, knowledge about the ecological and biological components of the soil can improve the efficacy and success of BCAs in the control of plant diseases. Finally, investigating how biocontrol strategies can contribute to enhancing crop resilience in the face of climate-change-induced stressors could be an important avenue. Understanding the interactions between biocontrol agents, pathogens, and changing environmental conditions could guide future research.

4.4 Challenges in Establishing Beneficial Microbes as BCAs

The biological control strategy was initially employed at the end of the 19th century. Displaying an array of advantages, the literature has also mentioned many limitations of this biocontrol strategy for controlling plant diseases (Stenberg et al., 2021). As the method offers an effective alternative to chemical pesticides for controlling pests and plant diseases, there are still various challenges to overcome. These challenges are briefly discussed below (Ayaz et al., 2023).

i. The journey of biocontrol agents from Lab to Field: Researchers are screening potential microbes as part of their work to create biological pesticide remedies that are very effective for controlling plant diseases in agriculture. These strains are usually chosen based on factors such as disease resistance, host range, availability, formulation, mass production, and farmer practice. PGPR performance can be assessed by considering geographical locations, host crop species, soil types, and environmental factors. The growth of BCAs is typically easier to observe in controlled environments such as greenhouses. The preference of most researchers at this stage might be associated with the reliability of the environment. The performance of BCAs in greenhouse trials can provide important theoretical and practical support for field utilization. As a result, the feasibility and efficacy of PGPR for commercial horticulture production, disease management, and field climate change conditions may be ensured (Ali et al., 2022).

The stability of BCAs is affected by the method of formulation, shipment, and storage environment. To achieve high levels of BCA strategy success, it is important to improve the formulation technology, extend the storage duration of the BCA product, optimize the production of selected microbial strains, and achieve large-scale application through low-cost manufacturing companies. Many scientists are looking for a solution to increase the shelf-life of PGPR by lowering the storage temperatures and/or changing the combination of additive mixtures. During the formulation and field efficacy trials, private sector partners and licensed laboratories conduct the essential environmental and human health assessments, as well as quality assurance for the ultimate commercialization of biopesticides. Before utilizing biopesticides on a large scale in the field, government authorities must grant permission. Typically, the biopesticide registration and regulation portfolio includes modified versions of synthetic pesticides as well as risk evaluation. This includes ecotoxicological and toxicological tests, as well as studies into their mechanism of action and host spectrum. The majority of these requirements are difficult for authorities to achieve because producing effective biopesticides while maintaining acceptable safety and consistency standards for commercialization might be a difficult task.

ii. Limited number of registered biopesticides and lack of awareness: Despite their well-documented efficacy, BCAs now account for less than 5% of the crop protection sector's commercial value (He et al., 2021). The identification, characterization, and registration of potential microbes takes more time and requires academic–industry collaborations. Furthermore, using natural resources to manage diseases (i.e., BCAs) raises several ethical and legal issues that might affect the biodiversity of an area. In this situation, new BCA species and populations have been restricted from entering specified countries. However, commercial applications of PGPR in protected environments such as greenhouses are significantly easier due to the availability of more isolated and controlled environments and they have potentially less detrimental ecological impacts. Another challenge that has risen as a result of the extensive use of PGPR-based biocontrol is regulatory concern. At present, every country has its own regulatory system, which varies widely. For example, high development costs for new commercial BCAs have been recognized as an obstacle to the BCA industry's development in Australia. To facilitate the registration and commercialization of novel BCAs and their products. BCA registration requires extensive cooperation among governmental institutes, universities, and industry sectors. The number of programs that provide financial and ecological benefits plays a vital role in biopesticide registration. Hence, local usage and global marketing commercialization must fulfill international legislation. The International Organization for Biological Control (IOBC) was established to bring together scholars, scientists, and professionals from a wide range of industries and fields to identify barriers and offer recommendations to overcome them.

Farmers might experience little or no financial gain when compared with chemical insecticides, which are more predictable and reliable. Farmers are discouraged from using the biocontrol strategy if they lack of expertise and knowledge about successful technological advances. Positive initiatives, which could include community talks, training workshops, and free conferences, might increase awareness regarding the use of BCAs in certain farming areas (Kumar et al., 2018). PGPR-based biocontrol provides substantial advantages in terms of decreasing pesticide use in agriculture. Biocontrol management has a direct impact on farmers' expenses and income, but it also has an indirect impact on their financial benefits due to its effect on farmland biodiversity and environmental sustainability. However, farmers have not fully utilized commercial biopesticides due to a lack of information and awareness. Therefore, there is a need to strengthen the idea of biological management in the farming community to restore their trust in the use of biopesticides.

iii. Biopesticides Commercialization and Legislative Procedure: Despite their growing popularity in crop disease management, biocontrol agents currently represent only 1% of agricultural control techniques, whereas synthetic-based pesticides account for 15% in the agriculture sector. The commercialization of biopesticides is a multi-step procedure that faces numerous challenges. Before being licensed for commercialization, similar to synthetic pesticides, BCAs are subjected to risk evaluations. European Regulation (EC) No. 1107/2009 specifies standards based on risk assessment in plant protection product marketing. European Communities (EC) Regulation No. 540/2011 specifies microbes that have been certified to be used as biological controls in Europe. As a result, commercial authorizations for biopesticides (and synthetic pesticides) can be obtained after an extensive procedure. However, the current trend of reducing the use of synthetic pesticides and simplifying the regulatory process for low-risk products may allow BCAs to be commercialized worldwide.

4.5 Biotechnological and Omics Techniques for Biocontrol Strategy Improvement

A large amount of microbial genomic data are easily accessible and have opened a new chapter for the discovery of potential BCAs (Ayaz et al., 2023). Integrating advanced omics and metagenomic techniques could provide a deeper understanding of the interactions between beneficial microbes and pathogens. Studying the genomic, transcriptomic, and proteomic profiles of microbes, pathogens, and plants during biocontrol interactions could reveal key mechanisms and factors influencing disease suppression. It is expected that by using biotechnological approaches, researchers will be able to develop potential BCAs for efficient control of multiple crop diseases. Based on the information we have collected, we have done our best to summarize the recent advancements in biocontrol strategies in light of prominent emerging biotechnological and omics techniques.

i. Biotechnological techniques linked with Biocontrol Strategy: Using biotechnological techniques and synthetic biology provides an interesting direction for biocontrol strategy improvement. Through advanced techniques in microbial engineering, researchers could create microorganisms with strong phytopathogen suppression features, including increased synthesis of antimicrobial substances and efficient colonization of plant surfaces. There are many examples where BCAs are successfully modified for desired traits. Using a genetic engineering tool, introducing glucanase gene to *Trichoderma,* resulted in increased resistance to diseases such as *Pythium*, *Rhizoctonia*, and *Rhizopus*. Transferring a *Serratia* gene expressing the chitinase enzyme to the *Pseudomonas* endophyte also reduced *Rhizoctonia solani* disease in beans. The biosynthetic locus phlACBDE from strain CPF-10 was cloned into a mini-Tn5 transposon, and then the chromosome of *Pseudomonas fluorescens* P32 improved tomato tolerance to *R. solanacearum* bacterial wilt and wheat resistance to *Gaeumannomyces graminis* var. *tritici*. Several microorganisms associated with plants can produce plant hormones such as auxin, ethylene, and cytokinins, all of which play important roles in plant growth and disease resistance (Ayaz et al., 2021). The pathways responsible for these hormones are engineered to express them in other species. Heterologous expression of the IAA synthesis pathway in *Bacillus* spp. significantly increased IAA production, efficiently enhancing plant growth in harsh conditions. The banana endophytes *Kosakonia* sp. S1 and *Enterobacter* sp. E5 produced ACC deaminase, which enhanced banana plant development and improved tolerance to *Fusarium* wilt. Hence, researchers might be able to formulae a BCA that will efficiently mitigate the harsh conditions in the open field and maintain plant health. Using tools such as CRISPR/Cas, we can insert mutations into specific sections of the genome with high precision and efficiency. In addition, mutations can be made in multiple genes at the same time, enabling the determination of the role of various genes in biocontrol. Engineered BCA could help detect plant physiological variations caused by biotic stresses, besides delivering desirable characteristics.

ii. Biocontrol Strategy in the Era of Multi-Omics: An effective approach to better understanding the interactions of BCAs with the host plant, pathogen, and environment is now offered through an influx of new generations of molecular technologies (Larena et al., 2023). The name "omics" refers to these emerging technologies that involve next-generation sequencing (NGS) technology, proteomics, metabolomics, genomics (including its derivatives pangenomics and metagenomics), and transcriptomics. The use of diverse omics approaches could ultimately accelerate the development of BCAs against plant diseases. Omics also provide insight into the complex molecular mechanisms that support plant–pathogen interactions, aiding in the identification of functional elements in pathogens involved in disease.

Exploring endophytes' role in plant disease protection is an exciting area of omics research. Endophytes are microorganisms that reside within plant tissue without causing visible symptoms and are vital for plant health. At present, most studies have focused on the function of one or a few endophytes rather than complete communities. It has been reported that *Burkholderia phytofirmans* PsJN, through various extra-cytoplasmatic functional group elements (sigma factors and group IV), assist other bacteria in monitoring changes in their environment, such as extreme temperatures, and in altering their metabolism to survive in harsh conditions. This study used a dual omics technique, applying high-throughput DNA sequencing to identify *Burkholderia*, as well as other bacteria, and conducting critical metabolomics. Microorganisms that inhabit the soil are essential components of the ecosystem, playing crucial roles in plant growth, development, defense, and responses to stress (Crandall, et al., 2020). As a result, understanding the relationship between plants and the soil microbial community utilizing metagenomics will be useful in constructing agricultural systems for progressive sustainable agriculture production. The use of metagenomic techniques in the study of soils inoculated with organic manures would be beneficial for developing fertilization strategies to decrease the dependence on chemical fertilizers.

iii. Microbiome-based solution for plant disease management: Multi-omics and next-generation sequencing (NGS) reveal enormous microbial diversity across many plant species, as well as different phytopathogen antagonistic microbes. According to newly available data, a wide range of microorganisms effectively interact with plants in the rhizosphere region. The knowledge related to different factors that influence microbes in the rhizosphere region will be helpful for future efficient biocontrol strategies. The earliest plants grown on Earth, mosses, are notable for their extraordinary microbiological diversity and high level of antagonistic potential because of their ecology. In addition, medicinal plants have been determined to be a fascinating source of biodiversity. It has recently been discovered that archaea are a component of the plant microbiome, and their impacts on plants, as well as their biocontrol potential, remain unexplored. Hence, the use of different omics techniques will enable us to identify different untapped microbes from different regions that can be formulated further to improve the biocontrol strategy.

Furthermore, to maintain ecosystem diversity and health, integrated breeding and biocontrol measures are essential. These systematic approaches are essential to prevent further biodiversity loss and encourage long-term farming practices. Crop breeding has been highlighted as a major force in natural evolution. Currently, the breeding strategy is aimed at controlling diseases, increasing yield, and various growth features; however, in the past, only a few plant phenotypes were randomly selected, and changes in the plant microbiome were ignored (Marco et al., 2022). Although the use of PGP bacteria in the field appears to be a viable strategy for sustainable agriculture, efficacy has proven inconsistent, probably due to variations in environmental conditions, poor microbial colonization, and low long-term stability in the rhizosphere. These limitations could be minimized by genetic or genome engineering of active root colonizers or improved colonization with large phyto-microbiome subpopulations. Soon, consortiums of microbes and biocontrol products may be utilized to improve biodiversity associated with crops via microbiome engineering to accomplish specific microbiome outcomes (Ayaz et al., 2023).

iv. Microbiome Engineering: A Shining Approach in Biocontrol Strategy: It has already been documented that agriculture could benefit significantly from microbiome engineering. Therefore, a modified microbiome with essential features is required for crop disease management (Ayaz et al., 2023). Plant-associated beneficial microorganisms have enormous potential for commercial and sustainable agriculture. Plant microbiome engineering can be achieved in two ways: bottom-up methods involve isolating, altering, and reviving certain microbes, whereas top-down approaches involve synthetic ecology, which includes horizontal gene transfer to a variety of hosts in situ and then phenotyping the microbiome. The recent advancements in genome engineering tools, computational tools, genome-wide functional tools, and meta-omic tools can improve our ability to design microbes for biocontrol and biofertilization, as well as increase agricultural productivity and yield (Fig.29).

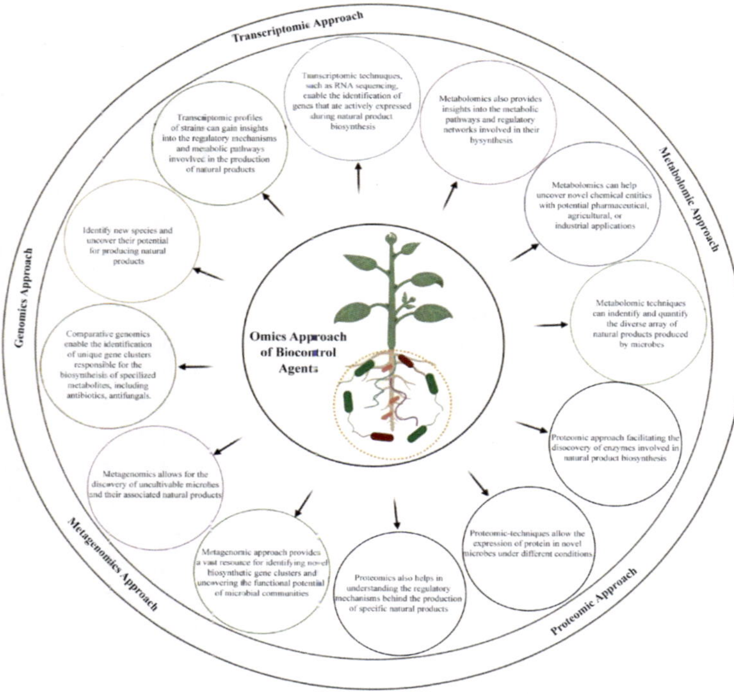

Fig. 29: Multi-omics approaches such as genomics, metagenomics, proteomics, transcriptomics, and metabolomics need to be used to identify potential microbial strains, a source of antimicrobial products. Omics also provides insight into the complex molecular mechanisms that support plant–pathogen interactions. The picture was made with the BioRender program (https://www.biorender.com/, accessed on 13 September 2023).

Current advances also report some devices that permit the evaluation of engineered microbiome function before field studies. For instance, lithographic 3D printing has made it possible to create small microscopic containers for arranging several bacterial species in virtually any 3D geometry. It will enable us to study their interactions at a microscopic level and compare them with those of native strains. The "tracking root interactions system" (TRIS), a microfluidic device developed to track root–bacteria interactions, is also beneficial for studying the engineered microbiome. From a practical point of view, it would be extremely helpful to develop microbiomes that are long-lasting, stress-resistant, and capable of increasing agricultural output. Finally, microbiome bioengineering is a fascinating choice for improving plant health. This approach, still in the early stages, will provide enormous benefits to biocontrol methods in the future.

4.6 Research Gaps and Future Direction

Biocontrol of plant diseases is an emerging strategy that still leaves many gaps for future research work. The journey of biocontrol agents from lab to field requires multiple efforts to reach a high peak. Further research could focus on discovering novel strains with potent biocontrol features by exploring untapped microbial diversity (Ayaz et al., 2023). This could involve extensive sampling and screening of various habitats to discover microbes with broad-spectrum antagonistic activity against different plant pathogens. By integrating modern omics and metagenomic approaches, researchers may be able to gain advanced knowledge of the relationships between beneficial bacteria and pathogens. The genomic, transcriptomic, and proteomic profiles of microorganisms and plants during biocontrol interactions may reveal critical mechanisms and factors affecting disease suppression. Using synthetic biology principles to design beneficial bacteria with better biocontrol capabilities could be a future avenue. Researchers could develop microorganisms with more disease-fighting properties, such as improved antimicrobial substance production or increased colonization of plant surfaces. In addition, nanotechnology-based biopesticides offer an environmentally responsible and efficient pest management. Biopesticides, in particular nano-biopesticides, have the potential to revolutionize global agriculture in terms of protecting crops for sustainable agriculture.

Modification of the plant microbiome to increase the growth and activity of beneficial microbes could be a promising approach. Upcoming research could look into fruitful ways to manipulate the composition and function of the plant microbiome to produce an environment suitable for biocontrol. The performance of BCAs in the greenhouse experiment can provide significant theoretical and practical support for field utilization. Future research could focus on large-scale trials to evaluate the practical effectiveness of biocontrol agents in different agroecosystems and under varying environmental conditions. Biocontrol could be incorporated into complete IPM programs that include a variety of techniques, such as cultural practices, resistant crop varieties, and biological control agents. Such integrated approaches when developed and optimized, could lead to more sustainable and reliable crop protection systems.

Future research should also look into the economic feasibility and societal consequences of utilizing biocontrol technologies on a broader scale. Analyzing the cost-effectiveness and potential benefits for farmers and the agriculture sector as a whole would be useful. It is critical for the widespread use of biocontrol strategies to spread knowledge about them among farmers, extension services, and policymakers. More efforts could concentrate on building educational programs and outreach campaigns to improve awareness and make biocontrol measures more practical. Considering biocontrol's potential to minimize dependency on chemical pesticides, further study could help build regulatory frameworks and policies that support the safe and effective use of biocontrol agents in agriculture. Finally, investigating how biocontrol methods can help enhance crop resilience in the face of climate-change-induced challenges could be a promising direction. Future studies could be motivated by knowledge of the relationship between biocontrol agents, diseases, and changing environmental conditions. Researchers can improve the subject of biocontrol to create more sustainable and effective crop disease management systems by focusing on the research gaps mentioned above.

4.7 Present Status, Historical Perspective and Future Prospects of Bio-Agents in Agriculture

Due to strong consumer demands to address food security and environmental safety has resulted in the lesser use of synthetic pesticides, the lowering of maximum residue limits and changes in the regulatory environment that favour more environmentally safe control options. Thus there is need to switch for more environmentally safe and ecologically sound pest control methods such as bioagents/ biopesticides (Nandani Shukla et al., 2019). Many microorganisms in biopesticides also deliver a number of benefits beside virulence to a target pathogen. For example, Trichoderma species are known to enhance the uptake of soil macro and micro nutrients by plants and specific isolates can deliver substantial plant growth benefits in the absence of a disease. There is recent evidence that endophytic microorganisms which provide protection against diseases, can also provide abiotic stress (drought and salt) tolerance attributes to certain crops. Benefits in addition to pest control can be an important selection criterion in screening programs to choose commercially attractive biocontrol isolates and open up a significant opportunity for marketing biopesticides as products with an added value. The interest in biopesticides is based on the advantages associated with such products which are inherently less harmful and less environmental load, ; designed to affect only one specific pest or, in some cases, a few target organisms, ; often effective in very small quantities and often decompose quickly, thereby resulting in lower exposures and largely avoiding the pollution problems ; and when used as a component of Integrated Pest Management (IPM) programs, biopesticides can contribute greatly.

Negative impact of chemicals has brought into focus the use of safer and effective alternative such as bioagents/biopesticides. Various microorganisms are currently being utilized as bioagents include, among others, *Trichoderma* spp., *Pseudomonas fluorescens, Bacillus* spp., *Ampelomyces quisqualis, Agrobacterium radiobacter,*

Pochonia chlamydosporia, *Purpureocillium lilacinum*, Mycorrhizae, nonpathogenic *Fusarium*, *Coniothyrium*, and atoxigenic *Aspergillus niger*. India has huge potential for the growth of industry as it losses crops of worth US$ 9259 million every year due to low consumption of pesticides in the country (https://www.mordorintelligence.com/).

Historical perspective: Biopesticide development for the control of plant diseases has undergone a similar transformation. During the early 20th century, studies of soil microbiology and ecology had led to the identification of many different microorganisms that act as antagonists or hyperparasites of pathogens and insect pests (Nandani Shukla et al., 2019). A number of these were shown to be useful in field-scale inoculations, but few were developed commercially because of the rapid adoption of chemical pesticides during that time period. Research into biological control of insect pests and pathogens slowed after the discovery of synthetic pesticide molecules during/after World War II. Renewed interest in biological pest management was stimulated following the demonstration by environmentalists and ecologists that broad and repeated application of these synthetic molecules could be ecologically harmful.

Commercial success stories from the 1980s and 1990s include products containing *Agrobacterium radiobacter* for the prevention of crown gall on woody crops and *Pseudomonas fluorescens* for the prevention of fire blight in orchards where the streptomycin had been overused and resistant pathogen populations were abundant. In the greenhouse and potting mix industry, products containing a variety of microbes that suppressed soilborne pathogens were introduced into the market. The genus *Trichoderma* was discovered by Persoon in 1794 but captured the attention only after Weindling and his associates showed that one species of the genus can kill other fungi and control plant diseases (Weindling 1932 and 1934). However, the first field success of biological control (target *Sclerotium rolfsii*) using *Trichoderma* was not until the 1970s. Applied and fundamental research on these fungi has continued since in biotechnology and agriculture.

Present scenario and Future projection: Presently, biopesticides cover only 2% of the plant protectants used globally; however its growth rate shows an increasing trend in past two decades. Agricultural biologicals have recorded double-digit sales growth and have accrued around US $2.3 billion in annual sales over the past few years (Nandani Shukla et al., 2019). Around two-thirds of US $2.3 billion is contributed by microbial formulations alone.

The global market of bioagents is expected to reach $4 billion by 2024 from $2 billion in 2016, growing at a CAGR of 8.8% from 2016 to 2024. Similarly, global investment in biopesticides was US$1.3 billion in 2011and is estimated to reach US $ 3.2 billion by 2017, with at 15.8 % compound annual growth rate from 2012 to 2017 (www.markets and markets.com) (Nandani Shukla et al., 2019). These formulations include live microbial cells and microbial active ingredients for seed treatment and foliar applications (www. naasindia.org). There are about 1400 biopesticides currently sold globally, and is estimated that the annual growth rate of the biopesticide sector is greater than that of synthetic pesticides (16% versus 3%) (www. naasindia.org). The

USA accounts for 40% of the global biopesticide use, followed by Europe (20%) and Oceania (20%) (www.naasindia.org). However, the usage of bioagents is only about 20% of that of synthetic fertilizers (www.fao. org).

In India, biopesticide industry is projected to grow at a CAGR of 20.2 % since 2010 -2020. Scope of Current market for pesticides was US$ 23.92 million in 2015 which represent only 4.2 % of the overall pesticide market. Currently, 34 microorganism have been included in the schedule of Gazette of India for registration as biopesticide with Central Insecticide Board, Faridabad, under section 9 (3B) and 9(3) of the insecticide act 1968 (Nandani Shukla et al., 2019). Over 150 Bio pesticide producing companies, 15 types of bio pesticides out of 227 pesticides are registered. Highest demand for bio pesticides was observed from West India - Maharashtra followed by South India. Microbial pesticides sale are dominated by *Trichoderma viride, Pseudomonas fluorescens* and *Bacillus thuringensis*. Central insecticide Board and registration Committee, Government of India have registered more than 970 Bio pesticide products. The biopesticides market has been segmented in different ways. On the basis of type, the biopesticides market is led by the bioinsecticides segment, followed by the bioherbicides, biofungicides, and bionematicides segments and others (sulfur, oil, insect repellent, moth control, and other biochemicals) respectively. The bioinsecticides segment is projected to be the fastest-growing type in the biopesticides market, due to the high crop loss by pests and diseases.

On the basis of crop type, classified as Grains & oilseeds, Fruits & vegetables, Others (turf, plantation, sugar crops, cotton, and ornamental crops). The market for fruit & vegetable crops accounted for the largest share and is also projected to be the fastest growing. This is mainly due to a high demand for fruits & vegetables by the growing population across the world (Nandani Shukla et al., 2019). Biopesticides market share from fruits & vegetables accounted for over 70% of the overall industry revenue (www. markets and markets.com). As many fruits & vegetables are eaten without proper processing, consumers demand for better crop safety processing. Pesticide residue is generally a concern among consumers in these crops than in row crops that are not consumed in raw form. This practice leads to a high increase in pressure on grocery stores and good marketers to offer pesticide-free fruits & vegetables. Grains & oil seeds will observe gains over 6% up to 2024 (www.naasindia.org). Oats, vegetables, grains, and oilseeds are majorly contributing crops. The products prevent the generation of pathogens in the yield and enhance crop productivity. Other segment includes pulses, turfs, forage, and greenhouse crops.

Varous modes of applications for biopesticides are foliar spray, soil treatment, seed treatment, and post-harvest. Foliar spray is highly used for applying biopesticides. To improve quality of crops and increase productivity of plants, foliar spray is effectively used to control pests on crops. The increases in acceptance of biological microbes and smart farming techniques have proven to be a major driving factor for the foliar application. These are marketed as dry and Liquid formulation (www. markets and markets.com). On the basis of Origin, classification include Microbial pesticides, Biochemical pesticides, Beneficial insects, Plant-incorporated protectants and lastly, by geography the biopesticides market is segmented on the basis of region into North America, Europe, Asia-Pacific, and the Rest of the World (RoW). The U.S, Mexico and Canada are covered under North America wherein Europe covers France, Germany, Italy, Spain and others. Asia pacific covers China, India, Japan and others and rest of the World (RoW) covers South America, Middle East and Africa. In 2015, North America accounted for the largest market share in the biopesticides market, followed by Europe and Asia-Pacific. U.S. and Canada constituted the largest country-level markets in the North American region in 2015. Increase in awareness about the benefits of biopesticides, among the cultivators and rise in population is leading to the growth of the market in this region. The biopesticides market in the Asia-Pacific region is projected to grow with investments from several multinational manufacturers. The market is projected to grow due to the increase in agricultural technologies and agricultural exports of key products. Moreover, extensive R&D initiatives have been undertaken for exploring the new varieties of biopesticides to be used on different pest management and soil fertility to increase the yield.

4.8 Research and Development in Biopesticide Based Pest Management

Out of all the biopesticides used today, microbial biopesticides constitute the largest group of broad-spectrum biopesticides, which are pest specific (i.e., do not target non-pest species and are environmentally benign) (Nandani Shukla et al., 2019). Over 200 microbial biopesticides are available in 30 countries affiliated to the Organization for Economic Co-operation and Development (OECD). There are 53 microbial biopesticides registered in the USA, 22 in Canada and 21 in the European Union (EU) although reports of the products registered for use in Asia are variable. Overall, microbial biopesticide registrations are increasing globally, the expansion of various technologies has increased the scope for more products and the change in the trend to develop microbial products is definitely on the rise.

i. **Fungal biopesticides**: Some of the most widely used species include *Trichoderma harzianum, Trichoderma viridae, Streptomyces griseoviridis, Pochonia chlamydosporia, Beauveria bassiana, Metarhizium anisopilae, Nomuraea rileyi, Ouroueocilium lilacinum, Paecilomyces farinosus, Verticillium lecanii, Glomus* spp., etc. Many of them have been commercialized globally (Nandani Shukla et al., 2019). *Trichoderma* is a fungicide effective against soil borne diseases such as root rot. It is particularly relevant for dryland crops such as groundnut, black gram, green gram and chickpea, which are susceptible to these diseases. Preparation of *Trichoderma* biopesticide is cheap and requires only basic knowledge of microbiology. This biofungicide is recommended as seed treatment, soil application, soil drenching, root dip technique etc for the control of seed and soil borne diseases. Many *Trichoderma* strains, mainly *T. harzianum, T. viride* and *T. virens*, have been identified as having potential applications in biological control and a partial list of genera of plant pathogenic fungi affected by *Trichoderma* includes: *Armillaria, Botrytis, Chondrostereum, Colletotrichum, Dematophora, Diaporthe, Endothia, Fulvia, Fusarium, Fusicladium, Helminthosporium, Macrophomina, Monilia, Nectria, Phoma,*

Phytophthora, Plasmopara, Pseudoperonospora, Pythium, Rhizoctonia, Rhizopus, Sclerotinia, Sclerotium, Venturia, Verticillium, and wood rot fungi. Recent studies also indicate potential use of *Trichoderma* strains in abiotic stress management i.e., drought and salt stress.

ii. **Bacterial biopesticides**:The bacteria that are used as biopesticides can be divided into four categories: crystalliferous spore formers (such as *Bacillus thuringiensis*); obligate pathogens (such as *Bacillus popilliae*); potential pathogens (such as *Serratia marcesens*); and facultative pathogens (such as *Pseudomonas aeruginosa*) (Nandani Shukla et al., 2019).Out of these, the spore formers have been most widely adopted for commercial use because of their safety and effectiveness. The most commonly used bacteria are *B. thuringiensis* and *B. sphaericus*. *B. thuringiensis* is a specific, safe and effective tool for insect control (Roy et al., 2007). It is primarily a pathogen of lepidopterous pests like American bollworm in cotton and stem borers in rice. When ingested by pest larvae, Bt releases toxins which damage the mid gut of the pest, eventually killing it. Main sources for the production of BT preparations are the strains of the subspecies kurstaki, galeriae and dendrolimus. Other species of bacteria have little impact on pest management though some commercial products based on *Agrobacterium radiobacter, B. popilliae, B. subtilis, Pseudomonas cepacia, P. chlororaphis, P.flourescens, P.solanacearum* and *P. syringae* are available.

iii. **Viral biopesticides**: Over 700 insect-infecting viruses have been isolated, mostly from Lepidoptera (560) followed by Hymenoptera (100), Coleoptera, Diptera and Orthoptera (40). About a dozen of these viruses have been commercialized for use as biopesticides (Nandani Shukla et al., 2019). The viruses used for insect control are the DNA-containing baculoviruses (BVs), Nucleopolyhedrosis viruses (NPVs), granuloviruses (GVs), acoviruses, iridoviruses, parvoviruses, polydnaviruses, and poxviruses and the RNA-containing reoviruses, cytoplasmic polyhedrosis viruses, nodaviruses, picrona-like viruses and tetraviruses. However, the main categories used in pest management have been NPVs and GVs. These viruses are widely used for control of vegetable and field crop pests globally, and are effective against plantchewing insects. Their use has had a substantial impact in forest habitats against gypsy moths, pine sawflies, Douglas fir tussock moths and pine caterpillars. Codling moth is controlled by *Cydia pomonella* GVs on fruit trees and potato tuberworm by *Phthorimaea operculella* GVs in stored tubers. Virus-based products are also available for cabbage moths, corn earworms, cotton leafworms and bollworms, beet armyworms, celery loopers and tobacco budworms. Baculoviruses are target specific viruses which can infect and destroy a number of important plant pests. They are particularly effective against the lepidopterous pests of cotton, rice and vegetables. Their large-scale production poses certain difficulties, so their use has been limited to small areas. They are not available commercially in India, but are being produced on a small scale by various IPM centres and state agricultural departments.

iv. **Nematode biopesticides**: Another group of microorganisms that can control pests is the entomopathogenic nematodes, which control weevils, gnats, white grubs and various species of the Sesiidae family (Nandani Shukla et al., 2019). These fascinating organisms suppress insects in cryptic habitats (such as soil-borne pests and stem borers). Commonly used nematodes in pest management belong to the genera *Steinernema* and *Heterorhabditis*, which attack the hosts as infective juveniles (IJs).

v. **Protozoan biopesticides**: Although they infect a wide range of pests naturally and induce chronic and debilitating effects that reduce the target pest populations, the use of protozoan pathogens as biopesticide agents has not been very successful. Protozoa are taxonomically subdivided into several phyla, some of which contain entomogenous species (Nandani Shukla et al., 2019). Microsporan protozoans have been investigated extensively as possible components of integrated pest management programmes.

Opportunities in India: Harmful impact of chemicals such as higher pesticides residues in food crops, specifically on grains and increasing pest resistance has brought into focus the use of safer and effective alternative such as bioagents/biopesticides (Nandani Shukla et al., 2019). Moreover, the area under organic crop cultivation is on the rise because of the growing demand of organic food, a result of increasing health consciousness among the people. This indicates that there is huge scope for growth of the biopesticide sector. Analysts believe that there would be a greater development in the biopesticides sector (Desai 1997). Due to its rich biodiversity India offers plenty of scope in terms of sources for natural biological control organisms as well as natural plant based pesticides. The richtraditional knowledge base available with the highly diverse indigenous communities in India may provide valuable clues for developing newer and effective biopesticide. The National Farmer Policy 2007 has strongly recommended the promotion of biopesticides for increasing agricultural production, sustaining the health of farmers and environment. It also includes the clause that biopesticides would be treated at par with chemical pesticides in terms of support and promotion. Increased adoption, further, depends on the following.

- Concrete evidences of efficacy of biopesticides in controlling crop damage and the resultant increase in crop yield
- Availability of high quality products at affordable prices
- Strengthening of supply chain management in order to increase the usage of biopesticides. In this regard, an efficient delivery system from the place of production (factory) to place of utilization (farm) of biopesticides is quite essential.

Future prospects: The biopesticide market will continue to grow in future due to increased pest resistance problem and high demand of safe and quality food products. However, there are many challenges that will need to be overcome (Nandani Shukla et al., 2019). Biopesticides clearly draw attention as safer alternative to manage pest and diseases while posing less risk to human being and the environment. In the US, biopesticides are monitored by Environmental Protection Agency which supports their registration, sale and distribution under the Federal Insecticide, Fungicide, and Rodenticide Act (FIFRA), as well as ensures a "reasonable certainty of no harm" under the Federal Food, Drug, and Cosmetic Act (FFDCA) to provide pesticide residue-free food and feed. Most of the times, it is the farmers who are affected by the problems of pesticide resistance and withdrawal of plant protection products, and yet they are policy takers rather than policy makers. Hence, a publicprivate sector approach to the development, manufacturing and sale of environment friendly alternatives to chemical pesticides for developing countries like India is the need of the day.

There is a need for well-defined selection criteria i.e. toxicity, production efficiency and safety of the product. Besides, most jurisdictions recognize that biopesticides are fundamentally different to synthetic pesticides. Concerted efforts to modify regulations will be crucial to boost development of new biopesticides. More research is required towards integrating biological agents into production system, improving capability of developing countries to manufacturing and commercialization of biopesticides. At the same time, it is also required to encourage public funded programmes, commercial investors and pesticide companies to take up biopesticideenterprises. Strict regulatory mechanisms to maintain the quality parameter and availability of the biopesticides at affordable cost in the developing countries must be followed. Thus, various aspects of biopesticides covering the current status, constraints, prospects and regulatory network towards their successful utilization for the benefit of people and environment need to be examined regularly. Progress has clearly been made, with many new biopesticides on the market and technological developments continuing to overcome impediments. However, in our opinion, they have not yet reached their potential, even though all predictions suggest biopesticides will outperform other pest control options in terms of market share increases in the near future. We see a bright future for biopesticides, if the research and industry groups can think bigger and act united, better communicate the positive messages about biopesticides, and demonstrate their ability to control pests effectively and economically.

4.9 Biocontrol Strategy- An Eco-smart Tool

Microbiologists, plant pathologists, and entomologists across the globe face a significant challenge as they work to find and develop environmentally friendly control agents against plant diseases & pests. Their goal is to reduce the widespread use of chemical pesticides, which would be an important step forward (Jaiswal et al., 2022). On the other hand, pesticides and biopesticides derived from beneficial microorganisms are among the most effective strategies for risk-free crop management during low to medium biotic stresses. Agricultural practices are being developed based on microbial bio-stimulants, which have fewer negative effects and are more efficient than synthetic agro-chemicals. In this context, one of the most important approaches in sustainable agriculture is the use of biocontrol microbes that can suppress phytopathogens. Simultaneously, it is critical to comprehend the role of these microbes in promoting growth and disease control, and their application as biofertilizers and biopesticides, the success of which in the field is currently inconsistent. Therefore, editorial is part of a special issue titled "Biocontrol Strategies: An Eco-smart Tool for Integrated Pest and Disease Management" which focuses on biocontrol approaches that can suppress the biotic stresses, alter plant defense mechanisms, and offer new eco-smart ways for controlling plant pathogens and insect pests under sustainable agriculture.

Challenges: Biological control will remain indispensable and play a significant role in modern agriculture. The growing concerns about the overuse of synthetic chemical pesticides and their residues, increased significance of insect pests and pathogens due to increased food demand, the withdrawal of several chemical pesticides, including soil fumigants, the appearance of new invasive species, and pesticide-resistant strains of pests, climate change, and specialized monoculture are all factors have contributed to the expansion of the biological control domain of plant protection under the sustainable agriculture goal (Jaiswal et al., 2022). However, bio-control agents (BCAs) have advantages over traditional crop protection (CPs) methods but are not yet ready to take their place. In many cases, the adaptability of BCA in a non-native environment is poor. Further, their efficacy against multiple pathogens/insect pests is also low. As a result, it hasn't been widely used (Shoham, 2020).

Another major challenge is the lack of adequate characterization of bio-agents coupled with the poor marketing strategy of bioagent-producing firms; for example, many PGPRs/biofertilizers projected as a biocontrol agent; low efficacy in non-native soils/environment work on adaptability, and its contains is lacking appropriate research on the efficacy, growth promoting activities; lacking response of bio-agents at the physiological and molecular level, and poor of characterization and product formulation of bio-agents (Jaiswal et al., 2022). The mild disease/pest suppression by these bio-agents may probably be due to their growth-promoting effect on plants. Therefore, this product confuses end users because they expect anti-pest activity from these products. As a result they are often ineffective, which helps diminish microbial biocontrol agents' reputation. Even though many strains blur the distinction between plant protection products and biopesticide/biofertilizers, strict regulation is required to ensure the efficacy of biopesticide microorganisms and prevent their misuse as plant protection products. Presently, only a few genera, species, and strains of BCAs (*Coniothyrium minitans*, *Gliocladium catenulatum*, *Pseudomonas chlororaphis* and spp., *Streptomyces griseoviridis* and *Streptomyces lydicus*, and *Trichoderma asperellum*, *T. atroviride*, and *T. harzianum*) are registered against some soil-borne pathogens. Similarly, *Bacillus firmus* and *Purpureocillium lilacinum* are BCAs approved for use against nematodes.

Experts have encountered great difficulties in developing various BCA products, in addition to the cost and scalability challenges of BCA. Many alternative solutions, such as those based on fermentation or pheromones, are prohibitively expensive to manufacture, providing customers with a little financial incentive to switch away from using known BCA-containing products. As a result, several companies are looking into novel ways to reduce production costs. Due to the low barriers to entry and high market attractiveness in this domain, hundreds of companies, ranging from major CP firms to many mid-tier firms, engage in the BCA and bio-stimulants industries. Many new businesses emerge due to the influx of venture capital, but they frequently lack the funds to register their company, develop their products, and enter the market (Shoham, 2020). To register a single strain for commercial usage, firms would have to conduct extensive, statistically significant efficacy trials for each crop/disease in each zone. This limitation has resulted in a dearth of products for the biological control of insect pests and diseases in Europe and Asia. As was previously stated, only a handful of these products have been approved for use by European growers.

Opportunities: Progressive farmers increasingly use the conservation and management of endangered species of biocontrol microorganisms, among other biologicals, to combat plant diseases (Jaiswal et al., 2022). The most successful approach to biological management for conservation objectives, is to concentrate on the most critical aspects of natural enemy ecology. There are two strategies to boost natural enemy effectiveness, viz., changing the habitat so that natural enemies benefit at the expense of pests or decreasing the detrimental effects of pesticides on natural enemies. Furthermore, the significance of biological control conservation in developing countries has been emphasized. Numerous microorganisms have been shown to be effective against soil-borne diseases and nematodes over the last 50 years. Among these are the active ingredients in at least one biopesticide that is already on the market. Even though several of these strains were developed a few years ago, none have achieved widespread commercial success due to competition from synthetic chemical fumigants, which are often more cost-effective, easier to apply, have a wider spectrum of activity, and are highly effective. Since the ban on methyl bromide and other chemicals, there has been a revived interest in microbial biocontrol agents against soil-borne diseases. These agents operate best in conjunction with other agronomic practices or resistant/tolerant plant varieties. The mechanism of action of microbial biocontrol agents against plant pathogens includes direct antibiosis, hyper-parasitism, resistance induction, and competition for space and nutrients.

In addition, researchers are investigating the role of non-pathogenic beneficial rhizobacteria in increasing plant resistance to pathogens, a process known as induced systemic resistance (ISR). Plant pathogen infection can result in systemic acquired resistance (SAR) (Aboulila et al., 2022). Some microorganisms (such as *Bacillus* spp.,*Pseudomonas* spp., *Acinetobacter calcoaceticus*; *Azotobacter* spp., *Azospirillum* spp., *Mesorhizobium*, *Bradyrhizobium*, *Burkholderia*) act as bio-stimulators by producing indole-acetic acid, nitrogen fixation, P-solubilizing, siderophore, HCN production, 1-aminocyclopropane-1-carboxylate (ACC) deaminase, degrading organic matter to improving the plant growths & yields, controlling disease & pest and maintaining soil health's. Furthermore, soil and plant microbiomes can act as inoculants, aid in nutrient absorption, biocontrol products, help protect plants from pests and diseases, or both. Some soil amendments may be required to ensure beneficial microbes' survival. Perhaps "probiotics" can be identified to maintain plant microbiomes healthy (Harutyunyan et al., 2022).

Research priorities: Basic biological research, particularly in taxonomy, ecology, and behavior, has tremendously aided procedures employed in the exploration, selection, and risk evaluation of biological control agents (Jaiswal et al., 2022). However, some questions remain unanswered in the field of biocontrol, such as the lacking efficacy in profiling plant-associated microbial bio-controlling agents, the lack of an overall understanding of a pathogen's biology, and the epidemiology of the resulting disease, which hinder the development of disease and pest management strategies. Therefore, scientists and researchers must keep the following goals in mind as they investigate emerging issues in the field in order to establish an eco-friendly bio-control strategy.

- Exploration of a new generation of biocontrol agents with higher efficacy, high productivity in fermenters, long shelf life and the ability to be stored at room temperature, and high compatibility with other control methods
- Standardizing of identification of BCA protocol against soil-borne disease & pest
- Population genetics research presents opportunities to better understand how the impact of biological control can be optimized.
- Improving microbiological control by integrating several strains of the same genus specie, or several genus-specie
- Increase our understanding of microbial biocontrol agents' potential against other soilborne pathogens beyond those listed on labels, as well as their potential use with carriers that can increase survival in soil, in order to demonstrate their environmental safety.
- Exploring cutting-edge genomic tools like CRISPR genome editing can reduce fewer desirable traits in biological control agents and insert new desirable characteristics such as insecticide resistance.
- Implementation of IPM strategies which include the use of microbial biocontrol agents with other management strategies
- Optimize and reduce the cost of production of BCA by improving the technologies of fermentation or use of low-cost carrier substrates for BCA
- To protect food and ornamental crops from pathogens, it is important to encourage and assist businesses in registering microbial products that meet the criteria of 'low-risk substances' either by expediting the registration of low-risk substances or by providing subsidies to farmers who choose low-risk substances).

How to improve the efficacy of microbial biocontrol agents?: Besides the specific indication given above for each group of biocontrol agents there are few ways to increase the efficacy of these tools. In the first instance wrong or bad uses/application of such products should be avoided (Ilaria Pertot et al., 2021). Most of the time, failure is due to the scarce knowledge on the way or conditions of application which results in a wrong use. Secondly we should distinguish failures caused by the application of low quality bio-inoculum. Often, inefficacy is due to the fact that fertilizers or compost containing biocontrol agents are used instead of registered plant protection products. This bad practice generated a lot of confusion around biocontrol agents.

The specific way to increase efficacy of these products are the following:

- Use in combination with agronomic practices in an IPM strategy: in this case the biocontrol agents are applied not just as substitution of a chemical soil fumigant but in combination of agronomic practices, for example after solarisation, anaerobic soil disinfestation or biofumigation;
- Verify the efficacy of the strain against the specific disease and the right conditions of application;
- Verify the compatibility between the strain and the crop or cultivar, in particular for rhizosphere colonizers;
- Give sufficient time to the biocontrol agent to reduce the inoculum of the disease;
- Apply under the right conditions;
- Calibrate the quantity per unit of soil to a level that can guarantee efficacy (often, the quantity per hectare is limited by the price of the product);
- Develop carriers or specific substrates to increase their efficacy;
- Improve production systems to reduce the cost of biocontrol agents;
- Develop multi-year strategies to control pathogens with biocontrol agents, especially for those supposed to reduce soil-borne inoculum of pathogens;
- Apply only when the inoculum is not too high;
- Identify more effective strains (most of the commercial strains have been isolated more than 20 years ago, most of the time without a targeted selection and wide comparison among strains).

Improve the cost/benefits: One of the major limiting factors in the wide use of microbial biocontrol agents is their cost compared to the chemical fumigants. The high cost is due to the cost of production (fermentation), the logistics (they comply should be kept at low temperatures and their shelf life is very short 0.5-2 years) and the cost of registration ((Ilaria Pertot et al., 2021).

Possible solutions can be found in

- Improving the production systems and the logistics
- Make them more attractive to companies for example by speeding-up the entrance to the market for those which fit with the 'low risk substance' by speeding up the registration process.
- Subsidies for growers to make them more familiar with these products

The effects on the whole soil-health: Commonly these biocontrol agents have no effect or only minor and transient effect on the soil microflora and microfauna, because many of them are highly specific against pathogens, their persistence is limited, they are biodegradable and do not bioaccumulate. Therefore if they are implemented in a good agronomic strategy they can be a sustainable tool to reduce the soil inoculum potential and improve soil and plant health.

4.10 Factors Affecting the Success of Bio-agents against Phytonematodes

The use of BCAs against plant parasitic nematodes (PPNs) is particularly timely given the negative impact of the use of chemical nematicides. However, BCAs are frequently slower acting, less effective, and more inconsistent than control normally achieved with chemicals. Therefore, a better grasping of the relevant edaphic and biological factors of BCAs is likely to give new directions in their use against PPNs (Abd-Elgawad and Askary, 2020). Such factors were reviewed in order to maximize their useful applications against PPNs. Continuous long-term research combining laboratory tests, followed by field experiments are still needed for more optimization of BCAs used in PPN. Such studies should exploit new technologies and resources that are becoming central to the development of sustainable systems for nematode management. For example, an integrated approach consisting of a lower dose of pesticide in conjugation with compatible bionematicide(s) coupled with appropriate cultural practice may be a good option needed to develop an economically viable and ecologically sustainable management of the nematodes. Improved and standardized sampling should be exercised. Genetic improvement of BCAs should address their efficacy and allocating markers of beneficial genes. Awareness raising for enhancing grower acceptance of biocontrol tactics should be better communicated.

Biological control agents (BCAs) are increasingly used against various plant-parasitic nematode (PPN) pests and offer a favorable alternative to hazardous chemical nematicides. Yet, their lack of efficacy, inconsistent field performance, and/or unfavorable economic factors have generally relegated them to a relatively small sector of pesticide market. Efficacy and biocontrol success can be boosted via holistic grasping of soil biological and ecological factors. Therefore, such factors were highlighted to give better directions for their use. Main points discussed currently are considered to affect the transmission success of these BCAs so that their use must be a way forward in crop protection/pest management. These included improved sampling, grasping BCAs interactions with soil biota and ecology, cost-effective use of BCAs, genetic manipulation for better PPN control, grower acceptance and awareness-raising of BCA techniques, and commercial application (Abd-Elgawad and Askary, 2020).

Commercialization of bionematicides against PPNs has experienced highs and lows. Examples of apparent commercial successes include the formulation of active ingredients such as *Purpureocillium lilacinum* products, BioAct WP, BioAct WG, MeloCon, and NemOut, against a wide range of PPNs on economically important crops; *Myrothecium verrucaria* products, DiTera DF, and DiTera WP, for use on a broad range of high-value fruit, vegetable, and ornamental crops with efficacy against a wide range of target PPN species; and *Bacillus firmus* I-1582 and Nortica 5WG, for control of PPNs on turf grasses. But for every success there have been numerous failures. In many cases, success was not achieved despite the nematode pests have shown promising susceptibility in laboratory or field plots, e.g., a bionematicide containing in vitro produced *Pasteuria* usage product, Econem, was an ineffective treatment for the management of *Belonolaimus longicaudatus* on golf course turf. The product only decreased *B. longicaudatus* populations on a single sampling date in one of eight field trials. Timper (2014) reported examples of how agricultural practices can enhance or decrease the biological control of PPNs and other soil borne pests. She stressed that the conservation biological control of PPNs to protect or enhance their suppression may not be effective in all field sites because they are linked to indigenous antagonists. The need and ability for scientists to publish negative data has been pointed out. Such data will enable the researchers to identify the deficiencies of the bionematicide, e.g., lack of efficacy, inconsistent field performance, and/or unfavorable economic factors. Thus, they can work on rectifying the shortcomings. This is especially timely with the current technological advances and major changes in the external environment, which have positively altered the outlook for biopesticides. Clearly, they are generally safe alternatives to hazardous chemicals. Nevertheless, the abovementioned deficiencies of the bionematicides have resulted in the failure of numerous biocontrol tactics against PPNs.

Fostering commercial demand of biocontrol agents: The remarkable operators for more commercial demands of a BCA are due to intrinsic, environmental, technological, societal, economic, and commercial items (Dutta et al. 2019). For instance, economic factors comprise the grower's sense to manage the PPN pests, the cost of biocontrol agent relative to other management options, the price of the commodity (e.g. per hectare), and the overall significance of the commodity in the agricultural market. Niche markets tend to be amenable to bionematicides use because both the crop value is high and the bionematicide product occupies a modest enough portion of the pesticide market for would-be rivals to shy away from registration costs and seek alternatives. Hence, low-value crops such as major row crops (e.g., maize, cotton, chickpea, and lentil) and some perennials (e.g., pigeon pea) are often unreachable for bionematicides marketing because the crop value is low and the market section is huge. These factors are interrelated, rendering more difficulty to solve the shortcomings. For instance, low BCA efficacy may result in a restricted product appeal, but effective cost application may adversely affect the efficiency of PPN control (Abd-Elgawad and Askary, 2020).

Successful biopesticides could be the compounds produced by the microbes, rather than a BCA itself. Improved packaging in certain delivery systems that target specific sites of pest occurrence can also enhance tolerance to environmental stresses, increase persistence, and offer cost-effective tactics (Abd-Elgawad 2019). Delivery requires proper mass production of the microbes and their bioactives, favorable bionematicide formulation, and application to the target. Because production and formulation of biopesticides are usually kept as trade secrets or patents, their further development is frequently hindered. Abd-Elgawad (2019) pointed out the need to use standardized procedures in developing such biopesticides in order to facilitate future reviews to be analytical, not just descriptive, and also offer building on them. Admittedly, BCAs occupy their own position as safe bionematicides, but for more pest control approaches, they are needed. Hence, the identification of research priorities for harnessing BCAs in sustainable agriculture in the context of grasping their relevant biology, ecology, interaction with other agricultural inputs, and mode of action is substantially required. The close examination of such targets to show the shortcomings and how to tackle or avoid them should widen the limited uptake of BCAs in pesticide markets. It can help in reducing BCA product costs and improving its efficacy and carryover effect which will stimulate their broader use in biocontrol. Therefore, the following section addresses the main topics currently considered to affect the transmission success of these BCAs to optimize their fitness, especially for lowvalue crops.

Environmental and biological factors: These factors are both overlapping and may be interacting with one another. The most important ones are soil texture, moisture, temperature extremes, predation or competition, and starvation (Abd-Elgawad and Askary, 2020). So, it is often difficult to control these factors in the field, but one solution resides in the addition of adjuvants to enhance their efficacy. P. lilacinum, with the adjuvant avermectin, is utilized effectively to control PPNs. Another technically advanced alternative is the use of BCA-filled capsules. Micro and nanomaterial-based pesticides are one of the formulations that are increasingly considered as a suitable alternative to traditional pesticides. Nanocapsules of lansiumamide B were used by the microemulsion polymerization method to improve the nematicidal efficacy of lansiumamide B against *Bursaphelenehus xylophilus* and J2 of *M. incognita*. Nanoformulated agricultural inputs were recently applied to control M. incognita on tomato, a tactic that should be further exploited.

Nevertheless, BCAs remain vulnerable to many intrinsic mortality factors in the plant rhizosphere that require further considerations. Such factors, singly or in combination, can result in an increased gradual decline in BCA numbers over the days following their application. Basically, BCAs rely mostly on their ability to proliferate or multiply in the rhizosphere intimately contacted with soil which should have suitable texture and structure for the added organism. So, soil texture and structure impact the growth and spread of microorganisms and their PPN hosts. The great bulk of soil (2500 t/ha) to plow depth (20 cm) forms difficulty for thorough BCAs incorporation and renders uneconomic broadcast treatments. The residual soil microflora competes with the introduced BCA for scarce energy sources and can significantly affect the performance of the BCA even when it is added to soil in a pre-colonized substrate (Abd-Elgawad and Askary, 2020). Soil moisture rarely limits the growth of most fungi but affects spore dispersal, especially of zoospores. Bacteria are more vulnerable to moisture levels than fungi but are unlikely to be damaged by levels that enable PPNs to survive. As for desiccation tolerance, BCAs have different thermal limits and optima depending on their species. Temperature commands the performance and persistence of biocontrol agents. P. lilacinus requires high soil temperatures which limits its application though it is easily produced in vitro, has rhizosphere competent, can attack the PPN eggs, and practically treat seed tubers for PPN control. Soil temperatures mostly range 10–15 °C in temperate regions for much of the growing season and 20–25 °C in warm climates at 10-cm depth. Such ranges have direct effects on the BCA growth and sporulation and on the PPN rate of development.

A strong positive correlation was found between biofumigation-induced *Globodera pallida* mortality and levels of microbial activity indicating that the soil microbes play a role in decreasing the nematode populations during the biofumigation. Such a biofumigation, due to the release of biocidal isothiocyanates (ITCs), indicates the suppressive actions of plants in family Brassicaceae on serious plant pathogens including PPNs. Conversely, due to the adverse effect of ITCs on a wide range of soil biota including beneficial flora and fauna, unregulated liberation of ITCs may result in disrupting soil food webs and exerting negative effect on nontarget BCAs' activities and survival (Dutta et al. 2019).

Considerations to develop cost-effective and efficient biocontrol programs

- Accurate evaluation of efficient biocontrol programs
- Synergistic combinations against plant-parasitic nematodes
- Developed techniques for BCAs against PPNs
- Registration of new BCAs against PPNs
- Genetic improvement for better and broader PPN control Molecular tools to utilize BCAs against PPNs
- Molecular tools for plant resistance
- Awareness-raising and grower acceptance of BCAs

4.11 Global Agriculture Biological Control Agents Market-Analysis and Insights

Global Agriculture Biological Control Agents Market, by Product Type (Weed Killers, Parasitoids, Predators, Bacteria, Fungi, Others), Application (Seed Treatment, Foliar Spray, Soil Treatment, Others), Crop Type (Grains, Fruits and Vegetables, Others), Country (U.S., Canada, Mexico, Germany, Sweden, Poland, Italy, U.K., France, Spain, Netherland, Belgium, Switzerland, Turkey, Russia, Rest of Europe, Japan, China, India, South Korea, New Zealand, Vietnam, Australia, Singapore, Malaysia, Thailand, Indonesia, Philippines, Rest of Asia-Pacific, Brazil, Argentina, Chile, Rest of South America, UAE, Saudi Arabia, Kuwait, South Africa, Rest of Middle East and Africa) Industry Trends and Forecast to 2028 (Anonymous, 2021) (Fig.30) (Source: Data Bridge Market Research Market Analysis Study 2021).

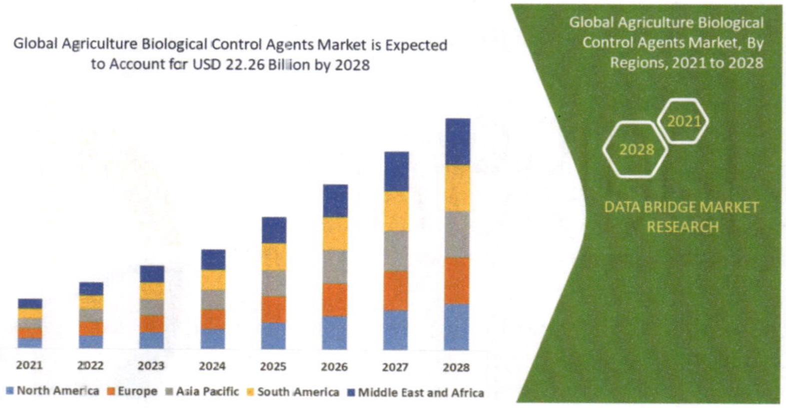

Fig. 30: Global Agriculture Biological Control Agents Market

Market Analysis and Insights

The agriculture biological control agents market is expected to witness market growth at a rate of 14.7% in the forecast period of 2021 to 2028 and is expected to reach USD 22.26 billion by 2028. Data Bridge Market Research report on agriculture biological control agents market provides analysis and insights regarding the various factors expected to be prevalent throughout the forecast period while providing their impacts on the market's growth (Anonymous, 2021). The rise in the demand for organically cultivated food products is escalating the growth of agriculture biological control agents market. Biocontrol agents can be referred to as the natural organisms such as predation, parasitism and other mechanisms for the controlling the plant pests. They play an essential role in controlling the plant pests like weeds, nematodes, mites. These biocontrol agents aid in the maintenance and balance the plant species such as insects, nematodes weeds and mites. Major factors that are expected to boost the growth of the agriculture biological control agents market in the forecast period are the rise in the utilization of biological products over the chemical based crop protection products.

Furthermore, the rise in the need for the sustainable agricultural practices is further anticipated to propel the growth of the agriculture biological control agents market. Moreover, the rise in the knowledge concerning the damaging effects of chemical pesticides is further cushion the growth of the agriculture biological control agents market. On the other hand, inadequate amount of shelf-life of biological control products is further projected to impede the growth of the agriculture biological control agents market in the timeline period. In addition, the research and developments in the field of biological products in the advancing countries for the market development will further provide potential opportunities for the growth of the agriculture biological control agents market in the coming years. However, the absence of knowledge amongst the growers might further challenge the growth of the agriculture biological control agents market in the near future.

This agriculture biological control agents market report provides details of new recent developments, trade regulations, import export analysis, production analysis, value chain optimization, market share, impact of domestic and localised market players, analyses opportunities in terms of emerging revenue pockets, changes in market regulations, strategic market growth analysis, market size, category market growths, application niches and dominance, product approvals, product launches, geographic expansions, technological innovations in the market. To gain more info on agriculture biological control agents market contact Data Bridge Market Research for an Analyst Brief, our team will help you take an informed market decision to achieve market growth.

Agriculture Biological Control Agents Market Scope and Market Size: The agriculture biological control agents market is segmented on the basis of product type, application, and crop type (Anonymous, 2021). The growth among segments helps you analyse niche pockets of growth and strategies to approach the market and determine your core application areas and the difference in your target markets.

- On the basis of product type, the agriculture biological control agents market is segmented into weed killers, parasitoids, predators, bacteria, fungi, and others.
- On the basis of application, the agriculture biological control agents market is segmented into seed treatment, foliar spray, soil treatment, and others.
- On the basis of crop type, the agriculture biological control agents market is segmented into grains, fruits and vegetables, and others.

Agriculture Biological Control Agents Market Country Level Analysis

The agriculture biological control agents market is analysed and market size, volume information is provided by country, product type, application, and crop type as referenced above (Anonymous, 2021). The countries covered in the agriculture biological control agents market report are the U.S., Canada, Mexico in North America, Germany, Sweden, Poland, Italy, U.K., France, Spain, Netherland, Belgium, Switzerland, Turkey, Russia, Rest of Europe in Europe, Japan, China, India, South Korea, New Zealand, Vietnam, Australia, Singapore, Malaysia, Thailand, Indonesia, Philippines, Rest of Asia-Pacific (APAC) in Asia-Pacific (APAC), Brazil, Argentina, Chile, Rest of South America as a part of South America, UAE, Saudi Arabia, Kuwait, South Africa, Rest of Middle East and Africa (MEA) as a part of Middle East and Africa (MEA).

Europe dominates the agriculture biological control agents market due to the increase in the initiatives taken by the government for the utilization of organic products. Furthermore, the overall agricultural land is dedicated to organic harvesting will further boost the growth of the agriculture biological control agents market in the region during the forecast period. Asia-Pacific projected to observe significant amount of the growth in the agriculture biological control agents market due to the sustainable form of agriculture practices. The country section of the agriculture biological control agents market report also provides individual market impacting factors and changes in regulation in the market domestically that impacts the current and future trends of the market. Data points such as consumption volumes, production sites and volumes, import export analysis, price trend analysis, cost of raw materials, down-stream and upstream value chain analysis are some of the major pointers used to forecast the market scenario for individual countries. Also, presence and availability of global brands and their challenges faced due to large or scarce competition from local and domestic brands, impact of domestic tariffs and trade routes are considered while providing forecast analysis of the country data.

Agriculture Biological Control Agents Market Share Analysis: The agriculture biological control agents market competitive landscape provides details by competitor (Anonymous, 2021). Details included are company overview, company financials, revenue generated, market potential, investment in research and development, new market initiatives, global presence, production sites and facilities, production capacities, company strengths and weaknesses, product launch, product width and breadth, application dominance. The above data points provided are only related to the companies' focus related to agriculture biological control agents market. The major players covered in the agriculture biological control agents market report are BASF SE, Valent BioSciences LLC, Certis USA L.L.C., Marrone Bio Innovations., BioBest Group NV, Syngenta AG, Corteva, Nufarm, United Phosphorus Ltd., Bayer AG, Novozymes, Koppert Biological Systems, Andermatt Biocontrol Suisse AG, Evogene, Lallemand, Inc., Chr Hansen Holding A/S, Precision Laboratories, LLC, Verdesian Life Sciences, Isagro SpA., and Symborg., among other domestic and global players. Market share data is available for global, North America, Europe, Asia-Pacific (APAC), Middle East and Africa (MEA) and South America separately. DBMR analysts understand competitive strengths and provide competitive analysis for each competitor separately.

4.12 Modes of Action of Biocontrol Agents

Microbial biological control agents (MBCAs) are applied to crops for biological control of plant pathogens where they act via a range of modes of action. Some MBCAs interact with plants by inducing resistance or priming plants without any direct interaction with the targeted pathogen. Other MBCAs act via nutrient competition or other mechanisms modulating the growth conditions for the pathogen (Köhl et al., 2019). Antagonists acting through hyperparasitism and antibiosis are directly interfering with the pathogen. Such interactions are highly regulated cascades of metabolic events, often combining different modes of action. Compounds involved such as signaling compounds, enzymes and other interfering metabolites are produced in situ at low concentrations during interaction. The potential of microorganisms to produce such a compound in vitro does not necessarily correlate with their in situ antagonism. Understanding the mode of action of MBCAs is essential to achieve optimum disease control. Also understanding the mode of action is important to be able to characterize possible risks for humans or the environment and risks for resistance development against the MBCA.

Preferences for certain modes of action for an envisaged application of a MBCA also have impact on the screening methods used to select new microbials. Screening of MBCAs in bioassays on plants or plant tissues has the advantage that MBCAs with multiple modes of action and their combinations potentially can be detected whereas simplified assays on nutrient media strongly bias the selection toward in vitro production of antimicrobial metabolites which may not be responsible for in situ antagonism. Risks assessments for MBCAs are relevant if they contain antimicrobial metabolites at effective concentration in the product. However, in most cases antimicrobial metabolites are produced by antagonists directly on the spot where the targeted organism is harmful. Such ubiquitous metabolites involved in natural, complex, highly regulated interactions between microbial cells and/or plants are not relevant for risk assessments. Currently, risks of microbial metabolites involved in antagonistic modes of action are often assessed similar to assessments of single molecule fungicides. The nature of the mode of action of antagonists requires a rethinking of data requirements for the registration of MBCAs.

4.12.1 Competition: *Microorganism competes* for space, minerals and organic nutrients to proliferate and survive in their natural habitats. This has been reported in both rhizosphere as well as phyllosphere. Competition has been suggested to a play a role in the biocontrol of species of *Fusarium* and *Pythium* by some strains of fluorescent pseudomonas. Competition for substrates is the most important factor for heterotrophic soil fungi (Anchal Sharma et al., 2013). Success in saprophytic ability (CSA) and inoculums potential of that species. Those fungi with highest number of propagules or the greatest mass of mycelia growth have the greatest competitive advantage. Competitive saprophytic ability is the summation of physiological characteristics that make for success in competitive colonization of dead organic substrates.

The majority of plant pathogens exploit nutrient sources in a much less specific way by degrading dead organic plant matter. Necrotrophic plant pathogenic bacteria, fungi and oomycetes kill and subsequently invade tissues of host plants and utilize the available nutrients as primary colonizers of these killed tissues (Köhl et al., 2019). Once necrosis has been induced by the pathogen, non-pathogenic microorganisms with saprophytic life style can potentially also colonize necrotic tissues so that a saprophytic competitive substrate colonization between different populations is common resulting in competition for nutrients and space. The principle competitive advantage of necrotrophic pathogens is that they are the first colonizers directly after they incited cell death. Non-pathogenic saprophytic endophytes being latently present in attacked host tissue may have a similar competitive advantage and may play an important role in competitive substrate colonization, e.g., of leaf lesions caused by necrotrophic pathogens.

The interaction with the host by killing and invading host tissue leading to damage in the infected crop is the most recognized stage of the life cycle of necrotrophic pathogens. However, most pathogen populations have another life style during significant parts of their life cycle when they live as saprophytes on necrotic plant tissues in soil, on crop residues, residues of non-hosts and on plant surfaces. During this stage pathogen populations survive, grow and spread independently of the host. The biology of such pathogens differs between species, some can develop completely independent from the host, other pathogen species will complete their life cycle only in the presence of the host during specific stages of their life cycle. Common for all necrotrophic pathogens during their saprophytic stage is that they depend on exogenous nutrients available in the environment, e.g., in colonized necrotic residues of host and non-host tissues. Successful host infection of most fungal necrotrophic pathogens also depends on exogenous nutrients during spore germination and formation of infection structures on host tissues. Also bacterial pathogens often depend on exogenous nutrients for multiplication to reach population levels sufficiently high to attack host tissues.

This dependency on exogenous nutrients during significant parts of their life cycle makes non-biotrophic pathogens vulnerable to nutrient competition. Consequently, highly competitive microorganisms are potential candidates for biological control using competition for nutrients and space as mode of action. To exploit this mode of action in disease control, detailed knowledge on the epidemiology is essential to identify stages where limitations of nutrients and space will affect pathogen development (Köhl et al., 2019). Typical situations are the presence of free nutrients in wounds of fruits which stimulate infection by various fruit pathogens, the presence of senescent floral tissues stimulating flower infection by *Botrytis cinerea* and the presence of dead host tissues on which the primary inoculum of pathogens is produced. An example of use of this efficient mode of action is wound protection of fruits from pathogen invasion by fast colonizing yeasts. Yeasts as single cell organisms are able to multiply rapidly under favorable conditions in wounds of fruits which are rich in nutrient supply. They can consume a broad range of carbohydrates such as disaccharides and monosaccharides but also various nitrogen sources. Spadaro and Droby (2016) reviewed competition processes between antagonistic *Pichia guilliermondii* and pathogenic *Penicillium digitatum, P. expansum, B. cinerea,* or *Colletotrichum* spp. in wounds of different fruits and *Aureobasidium pullulans* and *P. expansum* in apple wounds. Competition for carbohydrates in the carbohydrate rich wound environment in combination with competition for the limited amounts of nitrogen sources such as amino acids play the key roles in the antagonistic interactions.

Besides carbohydrates and nitrogen sources, restricted iron availability due to the low solubility of Fe^{3+} ions can be a limiting factor for microbial growth. Many microorganisms can produce a variety of low-molecular-weight siderophores with a high affinity for ferric iron (van Loon, 2000). Microbial strains with the ability to produce high amounts of siderophores with high affinity to iron play an important role in disease suppression and can be selected for biological control through competition for iron with pathogens that produce lesser amounts of siderophores with lower affinity for iron. This mechanism has been investigated in particular for isolates of Pseudomonas spp. and it has been demonstrated that siderophore mediated iron competition result in reduced pathogen populations in rhizospheres. Iron competition is also the mode of action of several fungal antagonists. For example, *Trichoderma asperellum* producing iron-binding siderophores controls *Fusarium* wilt. The yeast *Metschnikowia pulcherrima* transforms pulcherriminic acid and iron ions to the red pigment pulcherrimin. This process leads to iron depletion in media inhibiting development *of B. cinerea, A. alternata* and *P. expansum.*

Microbial biological control agents can also be targeted at the saprophytic stage of necrotrophic pathogens to outcompete the pathogen so that primary inoculum production on necrotic plant tissues is reduced or infections routes via senesced tissues to healthy tissues are blocked (Köhl et al., 2019). Examples are the application of *Microsphaeropsis ochracea* controlling *Venturia inaequalis* in apple, *Clonostachys rosea* controlling *B. cinerea* in roses and *Ulocladium atrum* controlling *Botrytis* spp. in various crops. Competitive antagonists may modulate growth conditions for the pathogen in the targeted niche not only through nutrient depletion but also by other mechanisms. Application of *Bacillus brevis* resulted in fast drying of leaf surfaces and reduced *B. cinerea* by 68% similar to the application of a standard fungicide in Chinese cabbage. Modulating leaf wetness periods by antagonists, e.g., via secretion of biosurfactants, may be a powerful mode of action for prevention of leaf diseases and diseases in stored products without any direct interaction between pathogen and antagonist. *A. pullulans* strains antagonistic to *Erwinia amylovora* causing fire blight in pome fruit strongly reduce growth of the bacterial pathogen through shifting the pH of the medium down to pH 4.0. Acidifying of the growth substrate may be an additional mode of action supporting antagonists during competition with bacterial pathogens.

4.12.2 Antibiosis: Antibiosis is defined as antagonism mediated by specific or non – specific metabolites of microbial origin, by lytic agents, enzymes, volatile compounds or other toxic substances. Antibiosis plays an important role in biological control. Antibiosis is a situation where the metabolites are secreted by underground parts of plants, soil microorganism, plant residues etc. It occurs when the pathogen is inhibited or killed by metabolic products of the antagonists. The products include the lyric agents, enzymes, volatile compounds and other toxic substances. Antimicrobial metabolites are secondary metabolites belonging to heterogeneous groups of organic, low-molecular weight compounds produced by microorganisms that are deleterious to the growth or metabolic activities of other microorganisms (Köhl et al., 2019). They are produced and released to the environment in small quantities by many microorganisms. Huge numbers of known antibiotics are produced by actinomycetes (8700 different antibiotics), bacteria (2900) and fungi (4900). Less than 1% of microscopically counted bacteria can be cultured on usual culture media. Approximately one-third of the bacterial divisions have no cultured representatives and are known only through rRNA sequences.

Microbial genome analysis revealed huge numbers of cryptic antibiotic gene clusters encoding still unknown antibiotics. Antimicrobial metabolites are often considered as the most potent mode of action of microorganisms against competitors allowing antibiotic producing microorganisms competitive advantages in resource-limited environments (Raaijmakers and Mazzola, 2012). Production of antimicrobial metabolites, mostly with broad-spectrum activity, has been reported for biocontrol bacteria belonging to *Agrobacterium, Bacillus, Pantoea, Pseudomonas, Serratia, Stenotrophomonas, Streptomyces*, and many other genera. In *Bacillus*, especially lipopeptides such as iturin, surfactin, and fengycin have been investigated, in Pseudomonas many antibiotic metabolites such as DAPG, pyrrolnitrin and phenazine have been studied. Many antibiotics are produced only when a microbial population reaches certain thresholds. This quorum-sensing phenomenon is well described for phenazine-producing Pseudomonas. Genomic information reveals that also these genera have the potential to produce many still unknown secondary metabolites with possible antimicrobial activity. Also fungal antagonists can produce antimicrobial compounds. For *Trichoderma* and closely related *Clonostachys* (former *Gliocladium*), 6-PAP, gliovirin, gliotoxin, viridin and many more compounds with antimicrobial activity have been investigated. Microorganisms producing antimicrobial metabolites with the potential to interfere with antibiotics in human and veterinary medicine must be excluded from use as MBCAs.

Several studies demonstrated that in vitro antagonism does not predict antagonism in complex assays including host plants which simulate the natural habitat situation under controlled or even in field situations. An example is the screening of *Trichoderma* isolates for their potential to control *R. solani.* Köhl (1989) tested 256 isolates belonging to *T. viride, T. hamatum, T. harzianum*, or *T. koningii* in dual cultures with *R. solani* and in pot experiments with lambs lettuce seeds planted in *R. solani* infested soil. Dual cultures on yeast dextrose agar revealed 192 antagonistic isolates. For these isolates, the average efficacy in reduction of damping off in the pot experiments was 61.2%. For the remaining 64 isolates, showing no in vitro antagonism, the average efficacy in pot experiments was similar with 59.7%. This example demonstrates that in vitro antagonism depends on the chosen conditions and by far does not explain the antagonistic potential of isolates. Also recent transcriptomic studies confirm that in vitro produced metabolites may not be expressed or play a minor role in situ.

Antibiosis observed on agar plates historically resulted in the development of pharmaceutical antibiotics. With similar expectations, results of agar plates often are translated to the control of plant pathogens in the field situation with antimicrobial metabolites seen as sole mode of action against competitors (Köhl et al., 2019). Production of 5 ng to 5 µg per gram of soil or plant tissue were reported depending on experimental conditions, strains used and type of produced antibiotic with exceptional higher values up to 180 µg per gram for a *Bacillus subtilis* isolate. Antibiotic concentration may be higher in certain microniches, but an important fraction of the antibiotics may be bound to the producing cells and may not diffuse in the habitat. Antibiotics are not stable in the soil environment. Half-life of antibiotics produced by biocontrol strains in soil was reported ranging between 0.25 and 5 days depending on biocontrol strain, antibiotic and experimental conditions. Such short life spans can be due to microbial decomposition but also to chemical and physical inactivation. Information on in situ concentration of antimicrobial metabolites produced by MBCAs against plant disease and their life span is hardly to be quantified and therefore often missing and not included in risk assessments on non-target effects. There is increasing evidence that antimicrobial metabolites have important functions for the producing microorganisms at subinhibitory concentrations. In other words: such compounds are characterized as being antibiotic because of their effect on microorganisms at high concentration under in vitro conditions although their function in the natural habitat is very different at the prevailing lower concentrations.

Lipopeptides of *Bacillus* and *Pseudomonas* are involved in the surface attachment of bacterial cells and biofilm formation by activating signaling cascades finally resulting in the formation of extracellular matrices which protect microorganisms from adverse environmental stresses. Some antibiotics, especially lipopeptides support the mobility of bacteria, most likely via changing the viscosity of the colonized surfaces (Köhl et al., 2019). Surface-active antibiotics allow bacteria to move to nutrient rich locations and also change the water dynamics on leaf surfaces which indirectly affects pathogen development. Other groups of antibiotics influence the nutritional status of plants. For example, DAPG-producing *Pseudomonas upregulates* the nitrogen fixation by plant growth-promoting *Azospirillum brasilense*, and redox-active antibiotics support mobilization of limiting nutrients such as manganese and iron. Antagonists with antimicrobial metabolites as mode of action can be selected using adequate bioassays if in situ production by living antagonists is envisaged or in vitro if the application of the formulated metabolites is envisaged. They generally have a broad host range and their activity depends on environmental conditions if their antagonistic activity depends on in situ production, thus on active growth. The risk for development of resistance against antimicrobial metabolites by pathogens can be considered as low in cases where metabolites are produced in situ. In cases where a single formulated microbial metabolite is applied on crops, the risk of development of resistance will be, depending on the genetics of the targeted pathogen and the stability of the metabolite in the environment, comparable to risk for chemical active substances.

4.12.3 Mycoparasitism / Hyperparasitism: Mycoparasitism or hyperparasitism occurs when the antagonist invades the pathogens by secreting enzymes such as chitinases, celluloses, glucanases and other lytic enzymes. Mycoparasitism is the phenomenon of one fungus being parasitic on another fungus. The parasiting fungus is called hyperparasite and the parasitized fungus as hypoparasite. In mycoparasitism, two mechanisms operate among involved species of fungi. This may be hyphal of interfungus interaction i.e., fungus-fungus interaction, several events take place which lead to predation viz., coiling, penetration, branching and sporulation, resting body production, barrier formation and lyses (Anchal Sharma et al., 2013).

This kind of interaction is often observed between fungi. For bacteria, hyperparasitism rarely has been reported. *Bdellovibrio bacteriovorus* is a predatory bacterium which has the unusual property to use cytoplasm of other Gram-negative bacteria as nutrients (Köhl et al., 2019). In initial research on biological control, *B. bacteriovorus* was tested in liquid cocultures with phytopathogenic bacteria belonging to the *Burkholderia cepacia* complex. Specific strains of *B. bacteriovorus* predated a broad host panel of the pathogen complex. Predation by *B. bacteriovorus* strains of other plant pathogenic bacteria such as *Agrobacterium tumefaciens, Xanthomonas vesicatoria, X. campestris pv. campestris, Erwinia carotovora* pv. *carotovora, Pseudomonas syringae* pv. *glycinea, P. syringae* pv. *tomato, P. marginalis,* and *Erwinia herbicola* was confirmed in similar tests. In biotrophic mycoparasitism, the hyperparasite depends on the living host fungus and gains nutrients from the host cells via haustoria without killing the host. Host and mycoparasitic fungus interact in a stable and balanced way. These often species–specific interactions may be important components in disease suppressiveness in ecosystems but are hardly to be exploited for commercial augmentative biocontrol because mass production of the hyperparasite depends on living host mycelium as substrate. Hyperparasites with a necrotrophic life style gain nutrients from dead host cells but also from other commonly available organic matter which allow mass production on artificial media making this group of hyperparasites much more favorable for commercial use as MBCA compared to biotrophic hyperparasites. Necrotrophic hyperparasites invade host spores or hyphal cells after killing such cells. Main mechanisms of parasitism is the excretion of CWDEs combined in some cases with excretion of secondary metabolites in close contact with the host cell leading to openings in the cell wall and subsequent disorganization of the cytoplasm. Cell wall degradation is typically caused by a range of chitinases, β-1,3-glucanases and proteases or, in case of hyperparasites of oomycota, cellulases. Such a necrotrophic hyperparasitism with invasion of killed host cells is frequently observed by microscopy and electron microscopy. The assumed nutrient transfer from the dead host cell to the invading fungus often has not been proven because it is technically difficult to investigate such processes, especially in the in situ situation in the field.

For some pathogen groups, researchers thoroughly investigated the phenomenon of hyperparasitism and found many antagonistic fungal species. For example, 30 hyperparasitic species against *Rhizoctonia solani* belonging to 16 genera have been reported (Köhl et al. 2019). Obligate biotrophic pathogens have been of particular interest for biocontrol using hyperparasites. There have been reports of eight hyperparasitic species of powdery mildews and 30 fungal species which show hyperparasitism against rust pathogens, including *Cladosporium uredinicola* against *Puccinia violae* and *Alternaria alternata* against *Puccinia striiformis* f. sp. *tritici*. In bioassays on rust-inoculated wheat seedlings, *A. alternata* germ tubes contacted with and penetrated into urediniospores of the pathogen at 24 hpi, and caused complete urediniospore collapse at 36–48 hpi. The most studied mycoparasites are belonging to the genera *Trichoderma* and *Clonostachys*. Antagonistic isolates of these genera vary in host range and individual strains mostly have a range of plant pathogenic hosts. They produce structures for attachment and infection, and kill their hosts by CWDEs, often in combination with antimicrobial secondary metabolites. These lytic enzymes are not constitutive but their production is triggered by complex signaling after recognition of the host. Surface compounds such as lectins from the host cell wall, surface properties and diffusible host-released secondary metabolites play important roles in the recognition and signaling pathways such as MAPK cascades, cAMP pathway and G-protein signaling (Karlsson et al., 2017).

Sclerotia of *Sclerotinia sclerotiorum*, microsclerotia of *R. solani,* individual urediniospores or pustules of rust pathogens and individual conidia or pustules of powdery mildew have been exposed to potential antagonist candidates and macroscopical and microscopical observations were made to find strains which invade the host structures, often accompanied with discoloration of these structures. Such studies generally are completed by assessments of the viability of invaded host structures, e.g., Zheng et al. (2017) confirmed that the viability of urediniospores from *A. alternata* treated pustules was only 25% whereas 80% of spores from untreated rust pustules were viable. Pathogens can develop resting structures such as endospores, chlamydospores, and melanised sclerotia with high resistance against hyperparasitism by naturally occurring antagonistic microorganisms. Pathogens can also repress synthesis of enzymes needed by the antagonist for hyperparasitic interactions. A considerable variation in susceptibility of *S. sclerotiorum* to the commercially applied hyperparasite *Coniothyrium minitans* has been observed in different regions in France.

Antagonists with hyperparasitism as mode of action can be selected using adequate bioassays. They generally have a narrow host range and their activity depends on environmental conditions because their antagonistic activity depends on active growth. The risk for development of resistance against hyperparasites by pathogens can be considered as low. Hyperparasitism acts through CWDEs which production is highly regulated by signaling from the potential host pathogen. Since these enzymes, ubiquitously produced in all ecosystems, are highly substrate specific and highly susceptible to rapid degradation, toxicological and eco-toxicological risks can be considered as very low and do not warrant a risk assessment.

4.12.3.1 Lytic enzymes: Lysis is the complete or partial destruction of a cell by enzymes. Lysis may be distinguished into two types, endolysis and exolysis. Endolysis (autolysis) is the breakdown of the cytoplasm of a cell by the cell's own enzymes following death, which may be caused by nutrient starvation or by antibiosis or other toxins. Endolysis does not usually involve the destruction of the cell wall. Exolysis (heterolysis) is the destruction of cell by the enzymes another organism. Typically exolysis is the destruction of the walls of an organism by chitinases, cellulases etc. and this frequently results in the death of the attacked cell.

An investigation by Tchameni et al., (2020) was aimed to evaluate the antimicrobial activities and involvement of extracellular lytic enzymes produced by four strains of *Trichoderma* in the inhibition of *Pythium myriotylum*. Antagonistic effects were tested by dual culture. Activities of lytic enzymes were evaluated from the filtrate of each strain after cultivation in selected media. Organic extracts were obtained from liquid media subsequent to the cultivation of *Trichoderma* in potato dextrose broth (PDB). Non-volatile organic compounds such as polyphenols and flavonoids were evaluated spectrophotometrically while volatile organic compounds (VOCs) were analyzed by gas chromatography coupled with mass spectrometry (GC-MS). The antimicrobial activity of the organic extracts was determined using the poisoning method. Results have shown that all the strains were antagonists against *P. myriotylum*. *T. erinaceum* (IT-58), *T. gamsii* (IT-62), *T. afroharzianum* (P8), and *T. harzianum* (P11) were found to produce cellulase, protease, and xylanase. Over 20 compounds were identified in each extract, including esters, lactones, and organic acids. The organic extracts also contained high amounts of polyphenolic compounds and flavonoids and significantly inhibited the mycelial growth of *P. myriotylum*. The minimal inhibition concentrations were 80 μg/μL, 40 μg/μL, 20 μg/μL, and 10 μg/μL, for extracts obtained from *T. erinaceum* (IT-58), *T. gamsii* (IT-62), *T. afroharzianum* (P8), and *T. harzianum* (P11), respectively. There was significant correlation between the production of total polyphenol and flavonoid content and the antagonistic effects of the tested strains.

Extracellular hydrolytic enzymes are a highly heterogenous group of enzymes, including lyases, esterases, glycosyl-hydrolases, and oxidoreductases. Microorganisms with degrading activities toward plant cell wall polysaccharides are the most source of hydrolytic enzymes for industrial applications. Although, large- scale industrial production of microbial enzymes as biopesticides in the management of phytopathogens has been considered by many researchers. Among rhizosphere microbes, various bacterial strains produce cell wall degrading enzymes, including chitinase, cellulase, proteases, lipases, and amylases, in response to phytopathogen attacks, which are fully discussed in the following sections (Table 9). These enzymes can either be applied indirectly by manipulating genes coding enzymes, purified enzyme proteins, or directly applying on phytopathogens.

Table 9: Extracellular lytic enzymes in the biocontrol of phytopathogens

Biocontrol bacteria	Hydrolytic enzyme	Target pathogen
Streptomyces isolate CT02	Chitinase	*Aspergillus niger, A. oryzae*
Streptomyces albus, S. flavogriseus, S. fumosus	Chitinase	*Fusarium. graminearum, Magnaporthe oryzae, Rhizoctonia solani, Botrytis cinerea, Puccinia* species
Streptomyces cellulosae	Chitinase	*Sclerotium rolfsii*
Pseudomonas putida, B. subtilis	Chitinase	*M. phaseolina*
Pseudomonas fluorescens	Chitinase	*Magnaporthe grisea*
Pseudomonas fluorescens, Enterobacter cloacae	Chitinase, cellulase, lipase, protease	*Phytophthora capsici*
Bacillus velezensis	Chitinase, protease, **β** Glucanase	*Verticillium daliae*
Bacillus velezensis NKG-2	Protease, Lipase	*Gaeumanomyces graminis* Var. *tritic., Gaeumannomyces graminis* var. *tritici*
Paenibacillus elgii HOA73	Chitinase	*B. cinerea*
B. subtills TD11	Chitinase, cellulase	*Colletotrichum, Aspergillus, Fusarium, Rhizoctonia.*
Bacillus velezensis NKG-2	Chitinase, **β**Glucanase, cellulase, amylase	*Alternaria alternata, Botrytis cirerea, Fusarium oxysporum, F. graminearum, Ustilaginoidea virens, Fulvia fulva*
Bacillus spp.	Chitinase, protease, glucanase	*Rhizoctonia solani, M. phaseolina*
Basillus spp.	Chitinase, **β**-1,3-glucanase, protease	*F. oxysporum* f. sp. *lycopersici*
B. cereus	Chitinase	*F. verticillioides*
Serratia plymuthica	Chitinase	*Pythium myriotylum*
B. aerius, Geobacillus, Thermodenitrificans	Chitinase	*P. capsici*
B. cereus, B. subtillis, Pantoea agglomerans	Chitinase	*Colletotrichum, Rhizoctonia, Aspergillus, Fusarium*
Bacillus simples, B. subtilis	Cellulase, chitinase, pectinase, xylanase	*Fusarium* spp.
B. velenzensis TSA32-1	Cellulase, protease	*F. fujikuroi, F. graminearum, Diaporthe actinidiae, A. alternata, Pythium ultimum*
B. subtilis	Cellulase, chitinase, glucanase	*Colletotrichum gloeosporioides* OGC1
Pseudomonas spp.	Cellulase, glucanase, xylanase	*Verticillium dahliae*

Biocontrol bacteria	Hydrolytic enzyme	Target pathogen
Bacillus pumilus	Protease, cellulase	*Fusicoccum aesculli, Phomopsis macrospora, Cytospora chrysosperma.*
B. cereus	Protease	*Bursaphelenchus xylophilus*
B. cereus BCM2	Protease, chitinase	*Meloidogyne incognita*
P. aeruginosa FG106	Protease, lipase	*Xanthomonas euvesicatoria* pv. *perforans, R. solani, P. infestans, A. alternata, B. cinerea, Clavibacter michiganensis* subsp. *Michiganensis, P. colocasiae*
P. putida ASU15	Lipase, protease, chitinase	*Uromyces appendiculatus*
B. cereus	α- amylase	*Rhizoctonia cerealis*
B. cereus SCB-1	α- amylase, cellulase, protease	*Fusarium, Alternaria, Curvularia, Neodeightonia, Saccharicola, Cochliobolus, Phomopsis*
Staphylococcus warneri, B. velezensis, B. megaterium, Caballeronia glebae, B. licheniformis	Amylase	*Ralstonia solanacearum*
One *Bacillus* and one *Pseudomonas isolate*	Amylase, protease	*R. solanacearum*
P. syringae, P. fluorescens, P. aeruginosa	α- amylase, protease	*R. solanacearum*

4.12.3.2 Hydrogen cyanide (HCN)**:** Many rhizobacteria produce hydrogen cyanide and this has been shown to play direct as well as indirect role in biological control of plant diseases and increasing the yields. The florescent pseudomonas themselves produce HCN and are able to suppress the pathogens. HCN is a broad-spectrum antimicrobial compound involved in biological control of root diseases by many plant-associated fluorescent pseudomonads. The HCN synthase is encoded by three biosynthetic genes (hcnA, hcnB, and hcnC), but little is known about the diversity of these genes in fluorescent *Pseudomonas* spp. and in other bacteria (Ramette et al., 2003). The partial hcnBC sequence was determined for a worldwide collection of biocontrol fluorescent *Pseudomonas* spp. Phylogenies based on hcnBC and deduced protein sequences revealed four main bacterial groups, but topological incongruences were found between hcnBC and rrs-based phylogenies, suggesting past lateral transfer of hcnBC among saprophytic root-colonizing pseudomonads. Three of the four groups included isolates from different countries and host plants. Yet, these groups corresponded to distinct, ecologically-adapted populations of HCN-producing biocontrol fluorescent pseudomonads, as indicated by high hcnBC distinctness ratio values and the differences in production levels of HCN in vitro found between groups. This is in accordance with previous results on catabolic properties and biocontrol abilities of these strains. HCN synthase gene diversity may thus reflect the adaptive radiation of HCN+ biocontrol fluorescent pseudomonads. Positive correlations were found between HCN production in vitro and plant protection in the cucumber/*Pythium ultimum* and tomato/*Fusarium oxysporum* f. sp. *radicis-lycopersici* pathosystems.

The high phenotypic plasticity developed by plants includes rapid responses and adaptations to aggressive or changing environments. To achieve this, they evolved extremely efficient mechanisms of signaling mediated by a wide range of molecules, including small signal molecules (Pablo Díaz-Rueda et al., 2023). Among them, hydrogen cyanide (HCN) has been largely ignored due to its toxic characteristics. However, not only is it present in living organisms, but it has been shown that it serves several functions in all kingdoms of life. Research using model plants has changed the traditional point of view, and it has been demonstrated that HCN plays a positive role in the plant response to pathogens independently of its toxicity. Indeed, HCN induces a response aimed at protecting the plant from pathogen attack, and the HCN is provided either exogenously (in vitro or by some cyanogenic bacteria species present in the rhizosphere) or endogenously (in reactions involving ethylene, camalexin, or other cyanide-containing compounds). The contribution of different mechanisms to HCN function, including a new post-translational modification of cysteines in proteins, namely S-cyanylation, was discussed.

Certain soil bacteria, algae, fungi, plants and insects possess the unique ability to produce HCN, which plays an important role in the biotic interactions of those organisms. In particular cyanogenic bacteria have been found to inhibit the growth of various pathogenic fungi, weeds, insects, termites and nematodes (Anju Sehrawat et al., 2022). Thus, the use of HCN-producing bacteria as biopesticides offers an ecofriendly approach for sustainable agriculture. The enzyme, HCN synthase, involved in the synthesis of HCN, is encoded by the *hcnABC* gene cluster. The biosynthetic regulation of HCN, antibiotics and fluorescent insecticidal toxins through the conserved global regulatory GacS/GacA system is elaborated in this review, including approaches that may optimize cyanogenesis for enhanced pest control. In addition, the effects of bacterially synthesized HCN on the production of indole acetic acid, antibiotics and fluorescent insecticidal toxins, 1-aminocyclopropane-1-carboxylate deaminase utilization and phosphate solubilization may result in the stimulation of plant growth. A more detailed understanding of HCN biosynthesis and regulation may help to elaborate the precise role of this compound in biotic interactions and sustainable agriculture.

Rijavec and Lapanje (2016) had re-addressed a new set of in vitro experiments, to test if HCN-producing rhizobacteria could inhibit the growth of phytopathogens. The level of HCN produced by the rhizobacteria in vitro does not correlate with the observed biocontrol effects, thus disproving the biocontrol hypothesis. A new concept was developed, in which HCN does not act as a biocontrol agent, but rather is involved in geochemical processes in the substrate (e.g., chelation of metals), indirectly increasing the availability of phosphate. Since this scenario can be important for the pioneer plants living in oligotrophic alpine environments, HCN producing bacteria was inoculated into sterile mineral sand together with germinating plants and showed that the growth of the pioneer plant French sorrel was increased on granite-based substrate. No such effect could be observed for maize, where plantlets depend on the nutrients stored in the endosperm. To support our concept, KCN and mineral sand were used and showed that mineral mobilization and phosphate release could be caused by cyanide in vitro. It was proposed that in oligotrophic alpine environments, and possibly elsewhere, the main contribution of HCN is in the sequestration of metals and the consequential indirect increase of nutrient availability, which is beneficial for the rhizobacteria and their plant hosts.

With in vitro experimental system using KCN, $FeCl_3$, and KH_2PO_4 it was demonstrated indirectly that cyanide interacts with iron, probably forming Fe-CN complexes. Unlike HCN, which is a weak acid and the product of the HCN-synthesis pathway, KCN is a salt, which upon dissociation increases the alkalinity of the solution and thus the solubility of metals (Rijavec and Lapanje, 2016). Judging by these results, the pH is not the predominant factor, as it is not consistently associated with KCN concentration and because cyanide interacts with iron at low and high pH. HCN interrupts the activity of metallo-enzymes and reacts with keto compounds and Schiff-base intermediates and as a result, it can inhibit the functioning of many enzymes or protein carriers and in specific cases probably can inhibits the growth of certain organisms.

4.12.3.3 Induced Systemic Resistance (ISR): ISR is the ability of an agent (a fungus, bacteria, virus, chemical etc.) to induce plant defense mechanisms that lead to systemic resistance to a number of pathogens. Inoculation of plants with weak pathogens or non-pathogens leads to induced systemic plant resistance against subsequent challenge by pathogens (Anchal Sharma et al., 2013). The mechanisms remain largely unknown but typically the induced resistance operates against a wide range of pathogens and can persist for 3-6 weeks. The biocontrol agents bring about induced systemic resistance (ISR) through fortifying the physical and mechanical strength of cell wall as well as changing physiological and biochemical reaction of host leading to the synthesis of defense chemicals against challenge inoculation of pathogens. Defense reaction occurs due to accumulation of PR proteins (chitinase, B-1, 3 glucanse), chalcone synthase, phenylalanine ammonia lyase, peroxidase, phenolics, callose, lignin and phytoalexins.

Induced plant defense mechanisms are triggered by stimuli recognized by specific recognition receptors. Typical recognized stimuli are PAMPs which induce defense pathways in the plant to increase host resistance against the recognized attacking pathogen. Resistance can be induced locally in the attacked tissue or spread via signaling through the plant or even to neighboring plants resulting in SAR. This type of induced resistance is a direct reaction to the stimulus of necrotizing pathogens (Köhl et al., 2019). Another type of induced resistance is ISR where the enhanced defensive capacity of the whole plant to multiple pathogens is induced by beneficial microbes. Both types of induced resistance decrease again in absence of the stimulus. Different to induction of resistance, priming of plants by stimuli leads to the sensitization for enhanced defense not only in the presence of the stimulus but also to a long lasting system of faster or stronger defense mechanisms in the future. Even transgenerational priming has been reported. Resistance inducing stimuli produced by microorganisms are called MAMPs. These specific molecular signatures are recognized by corresponding extracellular PRRs. After signal reception early responding molecular mechanisms can be measured in the plant cells within a few minutes. The typical pathways of plants induced by MAMPs resulting in ISR have been studied in detail.

The most studied MAMPs are the bacterial proteins flagellin and elongation factor EF-Tu. For both proteins the receptor systems are known and reactions are induced by subnanomolar concentrations (Köhl et al., 2019). Other examples for MAMPs with known receptors are glucan, chitin, xylan, e.g., produced by *Phytophthora megasperma* and *Trichoderma viride*. Many other MAMPs with still unknown receptors were listed such as proteinaceous MAMPs from bacteria, e.g., superoxide dismutase and 23-amino-acid peptide, or from oomycetes, e.g., pep13 trans-glutaminase, sterol-binding elicitins, and cellulose-binding lectins; lipophylic MAMPs, e.g., ergosterol and arachidonic acid; oligosaccharide MAMPs, e.g., N-glycosylated yeast peptides, cell wall components from glucan and chitin and peptidoglycans; and lipooligosaccharides produced by gram-negative bacteria. Common to all signaling compounds is that they induce reactions in the plant cells when present at millimolar to subnanomolar levels. Siderophores produced by iron-competing bacteria, antibiotics, such as DAPG and pyocyanin, biosurfactants and VOCs, such as 2R,3R-butanediol produced by B. subtilis GB03 (130) and a C13 volatile emitted by *Paenibacillus polymyxa* can act as elicitors inducing ISR. Mild viruses can also function as elicitors in plants. This principle is used for the biological control of Pepino mosaic virus in tomato which relies on mild variants of the Pepino mosaic virus. Products containing one or a combination of two mild virus isolates are commercially available.

It is expected that new approaches in "signalomics" based on metabolomics and metametabolomics will highlight the complex chemical communication within microbiomes and between microbiomes and plants including the contribution of signaling by released MBCAs to the continuous chemical cross-talk between organisms in the environment (Mhlongo et al., 2018). There is no selection pressure on the pathogen via resistance inducing MBCAs themselves or by released signaling compounds because they do not interact directly with the pathogen. The development of resistance against induction of resistance is thus not likely. However, the induced resistance itself causes selection pressure on the pathogen and pathogens may overcome induced resistance mechanisms. Screening for MBCAs with high potential in induction of resistance is complex because the mode of action depends on a sequence of events from establishment of the MBCA, release of signaling compounds, induction of a cascade of metabolic events to induce plant defense mechanisms and the response of the pathogen to this defense mechanism.

Using MBCAs for disease control through induction of resistance or priming relies on a complex sequence of events where the MBCA initially has to establish on the host, followed by the release of specific inducers which are recognized by specific receptors by the plant and there after triggering pathways within the host plant resulting in the onset of defense reactions or priming to make the plant ready for later challenges by pathogens. The first part of this sequence of events depends on the MBCA, the second part on the genetics and physiological status of the plant. The MBCA induces defense reactions of the plant through a broad variety of signaling compounds which are low molecular weight compounds, in some cases breakdown products of cell walls. They are commonly produced by microorganisms at very low millimolar to subnanomolar concentrations for chemical communication between microorganisms and with plants. These processes are ubiquitous in the environment and common wherever different microorganisms coexist with plants. It can thus be assumed that the production of signaling compounds by applied MBCAs will pose very low toxicological or eco-toxicological risks and generally do not warrant a risk assessment (Köhl et al., 2019).

4.12.3.4 Plant growth promotion: Biocontrol agents also produce growth hormones like, Auxins, Cytokinin, Gibberellins etc. These hormones suppress the deleterious pathogens and promote the growth of plants and simultaneously increase the yield. The studies on mechanism of growth promotion indicated that PGPR promotes plant growth directly by production of plant growth regulators or indirectly by stimulating nutrient uptake, by producing siderophores or antibiotics to protect plant from soil borne pathogens or deleterious rhizosphere organisms. *Pseudomonas* spp. may increase plant growth by producing gibberellins- like substances, mineralizing phosphates.

Plant growth promoting microorganisms (PGPMs) incorporated in a varied group of microbes which represent a wide range of genera. The PGPMs are portrayed by (i) root colonisation capacity, (ii) survivability and increase in microhabitats associated with the root surface (iii) competition with different indigenous microorganisms (iv) enlistment of resistance in host plants and (v) at last, plant growth development and advancement (Moh Tariq et al., 2020). The PGPMs incorporated in the plant-growth promoting rhizobacteria (PGPR) and the plant-growth promoting rhizofungi (PGPF) are frequently found to act as biocontrol agents (BCAs) and diminished plant pathogens. The utilisation of bacterial taxa in plant production has been evaluated previously for *Bacillus*, *Pseudomonas*, *Actinobacteria* and *Lactobacillus*. Furthermore, *Acetobacter, Azospirillum, Paenibacillus, Serratia, Burkholderia, Herbaspirillum* and *Rhodococcus* have also been shown to increase crop production. In this way, numerous BCAs are also involved in nutrient cycling and increment plant biomass (multiplication of shoots, roots and leaves). The rhizobacterial strains *Bacillus amyloliquefaciens, B. pumilis* and *Serratia marcescens* which demonstrated several *in vitro* PGPR attributes, like siderophore production, phosphate solubilisation, IAA production and antagonism towards fungal pathogens, could induce development in tea plants in terms of increment in the number of leaves, biomass and number of shoots in nursery just as field conditions. Soils with dynamic microbial ecologies and high organic matter commonly have lower fertiliser prerequisites than expectedly managed soils (Bender et al., 2016). Microorganisms that assist in plant nutrient procurement (biofertilisers) act through an assortment of mechanisms, including augmenting surface area accessed by plant roots, nitrogen fixation, P-solubilisation, siderophore production and HCN production.

Many types of PGPR produce cytokinins and gibberellins; however, the role of bacterially orchestrated hormones in plants and bacterial mechanisms of synthesis is not yet totally comprehended. A few strains of PGPR can advance generally a lot of gibberellins, prompting great increment in plant shoot growth (Jha and Saraf, 2015). Exudates are thought to comprise fundamentally sugars, amino acids and organic acids that are available at high concentrations in the cytoplasm of the plant, yet in addition incorporate smaller amounts of complex secondary metabolites such as flavonoids, terpenes and phenolic compounds that can attract explicit microorganism in the rhizosphere. PGPR colonisation brought about improved tissue health and physiology of plants, which produced more flowers and seeds.

Table 10: Mechanisms of bioagents against plant pathogens

Mode of action	Method for screening	Pathogen specificity	Risk of resistance1	Dependency on environmental conditions	Dependency on plant physiology	Use by distributors and end user
Induced resistance	Complex bioassay on plants	Specific to broad	Low	Low	High	Knowledge transfer needed
Competition	Simplified bioassays	Broad	Low	High	Low	Knowledge transfer needed
Hyperparasitism	Simplified bioassays	Pathogen specific interactions	Low	High	Low	Knowledge transfer needed
Antimicrobial metabolites produced in situ	Simplified bioassays	Specific to broad	Low	Moderate	Low	Knowledge transfer needed
Antimicrobial metabolites in product	In vitro assays	Broad	Moderate	Low	Low	Similar to use of fungicides
Helper strains2	Complex bioassays	Depends on MBCA	Low	Reduced	Reduced	Knowledge transfer needed
Assembled consortia combining different modes of action	In silico design followed by complex bioassays	Broad	Low	Low	Low	Knowledge transfer needed
Modulation of indigenous microbiota	Complex site-specific bioassays	Broad	Low	Medium	Low	Site-specific knowledge needed

’ Also depending on the specific evolutionary potential of targeted pathogen.2Applied in combination with MBCAs.

Pathogen populations thus can be limited by antagonistic microorganisms in very different ways. The nature of the mode(s) of action does not only determine how a pathogen population is affected by the antagonist. Also the characteristics of the MBCA depend on the exploited mode of action. Possible risks for humans or the environment, risks for resistance development against the biocontrol agent, its pathogen specificity and its dependency on environmental conditions and crop physiology may differ between different modes of action. Preferences for certain modes of action for an envisaged application of a biocontrol agent will also have impact on the screening methods used to select new antagonists (Köhl et al., 2011).

General modes of action of microbes has been provided in the table 10 (Köhl et al., 2019). Mechanisms of bioagents against plant pathogens is illustrated in fig.31 (Moh Tariq et al., 2020).

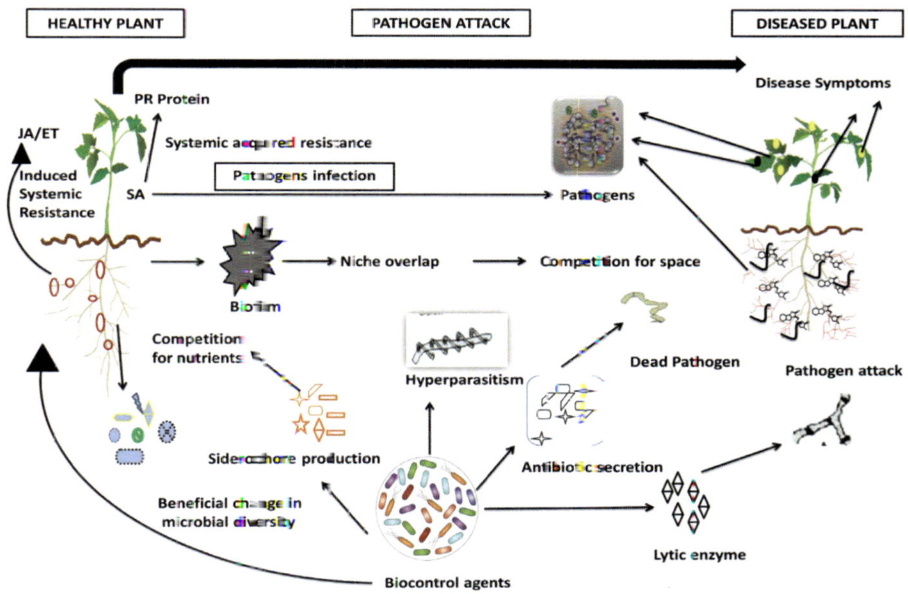

Fig. 31: Mechanism of biocontrol agent against plant pathogens

- The response of local and acquainted populations with various management practices
- The determinant factors of effective colonisation and articulation of biocontrol characteristics
- The components and dynamics of plant have host defence induction.

4.13 Identification and Development of New Microbial Strains

The microbial inoculants potentially replacing harmful pesticides are called biocontrol agents. Biological control, by definition, provides a non-chemical method for management of plant diseases by using other living entities, such as microorganisms

(Pirttilä et al., 2021). The biocontrol capacity of a microbe can result from production of antibiotic compounds, or enzymes capable of fungal cell wall lysis, depletion of iron from the rhizosphere, induced systemic resistance, and competition for niches with pathogens within the rhizosphere]. Production of one or more antibiotics is a mechanism most commonly associated with biocontrol ability. A number of biocontrol strains can also produce antifungal enzymes, for example chitinases, β1,3-glucanases, proteases, or lipases, with the capacity to lyse fungal cells. Synthesis of low-molecular mass siderophores that chelate iron in the soil near roots can inhibit the proliferation of fungal pathogens. For example, *Pseudomonas trivalis* strain BIHB 745 can produce siderophores and the siderophores pyochelin and pyoverdine have been identified in *P. fluorescens*. Many biocontrol strains can protect the host plant by out-competing phytopathogens for nutrients. They help the plant also by colonizing niches in the rhizosphere and preventing pathogens from infecting the plant.

Microbiological tools and biocontrol agents, which are bacteria and fungi capable of providing beneficial outcomes in crop plant growth and health, have been developed for several decades. Currently a selection of strains available as products for agriculture, predominantly based on plant-growth-promoting rhizobacteria (PGPR), soil, epiphytic, and mycorrhizal fungi, each having specific challenges in their production and use, with the main one being inconsistency of field performance (Pirttilä et al., 2021). With the growing global concern about pollution, greenhouse gas accumulation, and increased need for plant-based foods, the demand for biofertilizers and biocontrol agents is expected to grow. What are the prospects of finding solutions to the challenges on existing tools? The inconsistent field performance could be overcome by using combinations of several different types of microbial strains, consisting various members of the full plant microbiome. However, a thorough understanding of each microbiological tool, microbial communities, and their mechanisms of action must precede the product development. A brief overview of the available tools and consider various techniques and approaches that can produce information on new beneficial traits in biofertilizer and biocontrol strains was provided. Innovative ideas on how and where to identify efficient new members for the biofertilizer and biocontrol strain family was discussed.

4.14 Types of Microbes used as Biocontrol Agents

4.14.1 Plant-Growth-Promoting Rhizobacteria (PGPR): The most common biofertilizers and biocontrol agents currently in use belong to a group known as plant-growth-promoting rhizobacteria (PGPR) (Pirttilä et al., 2021). PGPR colonize the rhizosphere of many plant species, where they induce beneficial effects for the host, for example, increased plant growth and reduced susceptibility to diseases caused by plant pathogens, such as nematodes, fungi, bacteria, and viruses. The majority of the most-well known PGPR belong to the genera *Alcaligenes, Arthrobacter, Azospirillum, Azotobacter, Bacillus, Burkholderia. Enterobacter, Klebsiella, Pseudomonas, Rhizobium* and *Serratia.* Benefits of PGPR can include increased seed germination rate, root growth, yield, leaf area, chlorophyll content, nutrient uptake, protein content, hydraulic activity, tolerance to abiotic stress, shoot and root weights, and delayed senescence.

PGPR are, thus, often employed as biofertilizers. For example, *Azotobacter chroococcum* and *A. vinelandii* are used as nitrogen fertilizers worldwide. Many *Pseudomonas* species produce a broad spectrum of antibiotics, including pyoluteorin (PLT), pyrrolnitrin (PRN), 2,4-diacetylphloroglucinol, and hydrogen cyanide, and the strain *P. fluorescens* ZX produces and secrets lytic enzymes. Currently, there are several PGPR-derived products available for biocontrol based on, e.g., *Bacillus* strains, *Pseudomonas* strains, and *Streptomyces griseoviridis*. For example, *P. fluorescens* A506 controls fire blight on pome fruits by competing with the pathogen *Erwinia amylovora*. On the other hand, *P. chlororaphis* is effective on barley net blotch (*Drechslera teres*), barley leaf stripe (*Drechslera graminae*), and *Fusarium* pathogens. *B. subtilis* increases growth and salt tolerance in *Arabidopsis* and prevents the mummy berry fungus *Monilinia vaccinii-corymbosi* in blueberry, whereas *B. amyloliquefaciens* controls bottom rot on lettuce. *S. griseoviridis* K61 is effective against many fungal pathogens, such as *Fusarium oxysporum* f.sp. *lycopersici* and *Verticillium dahliae*, which are responsible of vascular wilt of tomato and *Verticillium* wilt in many crop plants.

PGPR as promising biocontrol agents: The rhizosphere is home to various microorganisms that provide steady PGPR supplies. The phytomicrobiome includes the bacterial population that colonize the rhizosphere, on the root surface, and between the root cortex cells (El-Saadony et al., 2022). Since plants can first colonize the terrestrial environments, PGPR have co-evolved with related plants; resulting in synergistic host plants' relationships (Cauca et al., 2018). The effects, methods, and possibility for successfully applying PGPR to agricultural plant production in controlled situations have been the subject of numerous studies. This is critical for developing more widely used methods of biological control that consider field settings.

Bacillus subtilis, Bacillus amyloliquefaciens and *Pseudomonas stutzeri* are among those shown to achieve success in root colonization as well as prevention of the pathogen *Phytophthora capsici* in cucumber. At the postharvest stage, *B. subtilis* can protect tomato fruits from infection by *Penicillium* sp. and *Rhizopus stolonifer*. Under greenhouse conditions, *B. amyloliquefaciens* isolates, diminish *Fusarium oxysporum* causing Fusarium wilt disease. In controlled situations, PGPR -as BCAsare effective, indicating their role in greenhouse production systems and their efficacy in commercial horticulture. It is not necessary to distinguish the indirect PGPR pathways for pathogen infection avoidance and plant growth promotion under abiotic stresses. In addition, PGPR with biocontrol activities that also enhance plant growth would be more effective in practice. Plant tolerance to abiotic conditions and resistance to phytopathogens causing plant diseases can be improved by PGPR. Some strains benefit plants' coping with stress and flourishing in abiotic environments. While most researchers have reported PGPR under these controlled conditions, few of them have investigated their effectiveness as BCAs, especially when combined with an abiotic stress. This is a critical factor in field biocontrol, and when climate change affects the ecosystem. The long synergism between PGPR and plant may deliver various benefits to the host plant (Fan et al., 2020).

Biocontrol mechanisms using PGPR: PGPR can enhance the availability of certain nutrients [phosphate solubilization and nitrogen (N2) fixation], or synthesize the phytohormones [indole-3 acetic acid (IAA), ethylene (ET), jasmonic acid (JA), gibberellic acid (GA), and cytokinins (CKs) (El-Saadony et al., 2022). PGPR colonizing a host plant can stimulate its growth through direct and indirect mechanisms. Direct mechanisms include the production of plant hormones, solubilization of phosphates, and increased uptake of iron. Indirect effects include antibiotics production, nutritional competition, parasitism, pathogen toxin inhibition, and induced resistance. The attitude of "PGPR" in creating phytohormones, molecules of signaling metabolites, and related substances describe how plants protect themselves from drought as an example of abiotic stress and salinity. PGPR may also alter the shape of the roots, resulting in increased root surface and improved root performance. In addition, PGPR can compete with other bacteria by colonizing rapidly and accumulating a greater supply of nutrients, preventing other organisms from growing (Abd El-Mageed et al., 2020). PGPR have different strategies to colonize, of which each is tied to a particular host. In general, pathogen infections can be suppressed by using antibiotics and antifungal metabolites; thus considered a well-known direct biological control strategy. Bacteriocins, antibacterial proteins, and enzymes are examples of antimicrobial peptides. Antibiotics are small antimicrobial molecules produced by PGPR that can inhibit the process of metabolic or growth activities of microbial pathogens. These antibiotics, which are mostly strain-specific, can target the ribosomal RNA (rRNA), alter the membrane structure, and damage the cell walls of bacterial pathogens.

Many bacteria produce bacteriocins, where some have a greater variety of inhibitory activities than others. Siderophores are specialized chelating agents of ferric iron that inhibit phytopathogens from gaining access to iron; thus, maintaining plant health particularly in iron-deficient environments. PGPR can manage various plant diseases by depriving pathogens of iron, thereby reducing disease development and generating extracellular siderophores. Bacteriocins, siderophores, and antibiotics have thus been identified as the three supreme operative approaches for potential biocontrol prior to in vivo applications. Several studies have investigated PGPR as a potential plant disease management tool to synthesize plant-beneficial metabolites such as siderophores. PGPR can indirectly increase crop stress tolerance. Signal chemicals, such as phytohormones and specialized signal molecules, enable plant-to-microbe and microbe-to-plant communication. The control and regulation of activities in the holobiont include the host plant and the "specific" phytomicrobiome (the plant-phytomicrobiome interaction). Two microbe-to-plant signals, lipochitooligosaccharides (LCOs) and thuricin17 (TH17), enhance stress tolerance in different plant species (Lyu et al., 2020). Resistance-inducing and antagonizing PGPR might be useful as new inoculants with combinations of different mechanisms of action, leading to a more efficient use for biocontrol strategies and plant growth promotion. PGPR can also produce volatile organic compounds (VOCs) that play a significant role in plant growth and induced systemic resistance (ISR) to pathogens. PGPR can trigger ISR as a strategy to improve disease resistance of plants. Roots colonization by arbuscular mycorrhizal fungi (AMF) and certain strains of non-pathogenic bacteria can improve plant resistance to biotic stresses.

ISR triggered by PGPF and plant growth-promoting fungi (PGPF) can be found in a wide range of plant taxa. Systemic acquired resistance (SAR); however, can be activated by a pathogen infection. Salicylic acid (SA) signaling is associated with the production of pathogenesis-related (PR) genes. Unlike SAR, ISR functions independently of SA, but requires responses to ET and JA. This can be achieved by the induction of defense-related gene expression; although is not always associated with induced PR proteins. PGPR are involved in diverse mechanisms to enhance plant growth and/or act as BCAs. Crop production promotion and disease management could be investigated together to ensure sustainability and cost-effectiveness of agricultural systems. Thus, effective PGPR strains can promote stress tolerance and nutrient absorption, plant development, and batting fungal/bacterial diseases. Thus, this appears to be a win-win situation to the PGPR strain and the host plant.

Challenges of employing PGPR as BCAs: PGPR-based biocontrol provides effective and long-lasting disease management. Europe and the United States are the most promising marketplaces for biocontrol products, followed by South America (Barratt et al., 2018). Although many PGPR have been tested in vitro and commercially proven as BCAs, new biocontrol products have been released from research activities carried out in the United States and Europe. In general, the market of BCAs and their products is growing; yet, it is not well-adopted compared with chemical pesticides as the most common crop management method. Before being publically accepted/registered as a commercial BCA, there are certain requirements/needs that have to be taken into consideration. As such, researchers should improve the efficacy of BCAs to manage certain disease(s). This can be achieved by having a BCA that has as many beneficial characteristics and mechanisms of action as possible. Such characteristics may include, but not limited to, the ability of the BCA to grow fast in vitro, produce a wide range of bioactive metabolites, possess high rhizosphere competence abilities, enhance plant growth performance, be environmentally safe, have the compatibility with other rhizobacteria/fungi, and be tolerant to abiotic stresses (Lyu et al., 2020). Successful colonization of root tissues and/or the rhizosphere is a critical component for any PGPR strain to be an effective BCA; thus, to perform well against plant pathogens. On the other hand, performance of the inoculated PGPR may vary, depending on the survival rate in the soil, crop compatibility, interaction with other local microbial species and the environmental factors. Survival and colonization are major components when identifying effective BCA isolates. In vitro antagonism experiments are often used to investigate the effect of bacterial isolates on certain diseases, prior to greenhouse and/or field trials (El-Saadony et al., 2022).

Performance of PGPR is generally assessed according to the geographical areas, soil types, host crop species, and under various environmental conditions. BCA growth is often easier to monitor under controlled conditions, i.e., greenhouses. The preference of this stage by most researchers could be attributed to the stability of environmental conditions. Greenhouse experiments evaluating the performance of BCAs under controlled conditions can provide strong theoretical and practical support for the application of PGPR in the field. Thus, this ensures the feasibility and efficacy of PGPR for commercial horticulture production, disease management and climate change conditions such as those found under field conditions. PGPR stability is also influenced by the method, formulations, transportation, and storage conditions. To achieve high levels of the BCA survival, one should improve the formulation technology, increase the shelflife of the BCA product, optimize the production of targeted microbial types and achieve low-cost production at large scales. Many scientists have attempted to extend the shelf-life of PGPR by decreasing the storage temperature and/or modifying the combinations of additives. Extensive research on the risks and benefits of BCAs is also required, because agricultural disease management approaches rely on this balance.

Due to the diverse modes of action, identification, characterization, the registration of promising PGPR strains take time and require academic-industry collaborations. Using natural sources (e.g., BCAs) to control pathogens also poses a set of legal and ethical issues that may threaten the local biodiversity (El-Saadony et al., 2022). In that regard, new species/populations of BCAs have been restricted from entering specific countries. For commercial uses, the application of PGPR in protected environments such as greenhouses is much easier, due to a more isolated and controlled environment delivery and potentially less negative ecological consequences. Another challenge that is linked to the widespread implementation of PGPR-based biocontrol is the regulatory problems. Currently, each country has its regulatory system that greatly vary among them. High development costs for new commercial BCAs, for example, have been identified as a barrier to the BCA industry's expansion in Australia. The high regulatory expenses of importing new BCAs into Australia is one of the most serious challenges. BCA registration requies tight coordination among governmental institutes, universities, and industrial sectors to facilitate the assessment and commercialization of new BCAs and their products. The shortage of programs for financial and ecological benefits can also be added as a challenging problem. For global marketing and local practical applications, commercialization should follow international legislation. The International Organization for Biological Control (IOBC) have gathered academicians, researchers and practitioners from different sectors/fields to identify the barriers and provide recommendations to overcome these limitation.

4.14.2 Rhizobacteria as BCAs: Rhizobacteria have gained attention when applied to grains, seeds, roots, and/or soils to help the plant grow and develop. Rhizobacteria are important for N2 fixation, promotion of plant growth, and biological control of plant pathogenic microorganisms. Recently, various microbial species are presently used in bacterization, containing *Azospirillum, Azotobacter, Bacillus, Rhizobium, Serratia, Stenotrophomonas, Streptomyces Acinetobacter, Agrobacterium, Alcaligenes, Arthrobacter, Bradyrhizobium, Frankia, Pantoea, Pseudomonas*, and *Thiobacillus* (El-Saadony et al., 2022). Many plant diseases associated with nematodial, bacterial and fungal infections have been reported to be managed by PGPR. The use of BCAs has been controversial in suppressing nematode populations because other soil microorganisms and the host plant can be adversely affected. To manage diseases associated with plant-parasitic nematodes a combination of biological management, nematicides, organic soil amendments, and crop rotation have been used. In vitro culture filtrates of a strain of *Pseudomonas* sp. can suppress juvenile mortality of *Meloidogyne javanica;* thus, considerably reduce root galls and nematode population, and enhance plant development and yield. Furthermore, inoculations with *Bacillus* spp. affect nematode behavior and feeding. *Pseudomonas striata*, *Pseudomonas fluorescens*, and *B. subtilis* strains also overturn the population of nematodes. Root-knot nematode (RKN) infestations have been successfully managed via biological management using *Bacillus* isolates. Few studies have reported the endophytic *P. fluorescens* and *Bacillus* spp. to promote systemic resistance in crops against nematodes owing to the increased activities of phenylalanine ammonia-lyase, polyphenol oxidase, and peroxidase, as defense-related enzymes for producing antagonistic chemicals and altering explicit root exudates such as amino acids and polysaccharides. *P. fluorescens* isolates increased defense enzymes in tomatoes resistant to RKN (Kavitha et al., 2013). In comparison with the control, the application of *P. fluorescens* and *Paecilomyces lilacinus* resulted in low nematode community in roots, tubers and soils.

Fungal species, such as *Alternaria, Aspergillus Penicillium*, and *Fusarium* producing mycotoxins, are harmful to green vegetables and cause post-harvest diseases. Mycotoxins, such as fumonisin, ochratoxins, aflatoxins and other toxins, are released in vegetables and fruits infected with the fungal pathogens, *Fusarium, Alternaria*, and *Aspergillus*. The use of biopriming and pelletizing techniques of *Serratia plymuthica* HRO-C48 alongside *Verticillium dahliae* in canola plants revealed a significant biocontrol; thus, providing evidence of the ability of BCAs to manage diseases comparable to chemical fungicides. *Bacillus* spp. produce a variety of compounds involved in the biocontrol of phytopathogens on various plants, including potato, rice, tomato, wheat, groundnut, brinjal, chickpea, and cucumber. *Bacillus* sp. BS061 isolate can mitigate the effect of *Botrytis cinerea* to reduce the occurrence of gray mold and powdery mildew diseases in strawberry and cucumber. *Pectobacterium carotovorum* SCC1 can manage soft rot disease in tobacco plants when conjugated with *B. subtilis* strain E4 and BTH fungicides. The root-knot and root rot pathogens are often suppressed when *Pseudomonas* spp. are used as BCAs. PGPR isolates had antifungal activities in mungbean plants against *Rhizoctonia solani*, *Macrophomina phaseolina, F. solani*, and *F. oxysporum*. These isolates can also be used to prevent fungal infections that cause root rot disease. The application of the bacterial BCA, *Pseudomonas aeruginosa*, can manage anthracnose in the chili pepper against the causal pathogen *Colletotrichum capsici*. *P. aeruginosa* can also induce systemic resistance of chili pepper to anthracnose (El-Saadony et al., 2022).

4.14.3 Fungi as BCAs Fungal BCAs are able to antagonize plant pathogens and protect their host plants. For example, several strains of *Trichoderma* have been developed as BCAs against the fungal pathogens *Penicillium, Fusarium, Aspergillus, Alternaria, Pythium, Rhizoctonia, Phytophthora, Pyricularia, Botrytis,* and *Gaeumannomyces* (El-Saadony et al., 2022). As a BCA, *Trichoderma* can suppress various air- and soil-borne plant pathogens; thus, can be conceivably used as biopesticides in greenhouse and/or field trials. Certain strains of nematophagous fungi can manage the populations of *Meloidogyne enterolobii* in an integrated pest management (IPM) approach. AMF could also protect crops against soil-borne pathogens, including RKN, albeit the unclear mechanisms of antagonism. The use of nematophagous and endoparasitic fungi has been deployed as antagonists to suppress RKN. The talc-based formulation of the fungal BCA, *Paecilomyces lilacinus*, was found to be more active in reducing the population of *Meloidogyne incognita* in soils cultivated with tomato plants. The efficiency of *P. lilacinus i*n controlling nematodes was observed in several horticultural crops, including tomato, okra, and capsicum. The most widely used BCA for plant-parasitic nematodes is the fungus *P. lilacinus*, which has shown an appropriate replacement to synthetic chemical control in pre- and post-planting applications. *P. lilacinus* infects eggs, juveniles and females of *M. javanica* by direct hyphal penetration. *P. lilacinus* can boost tomato yield while reducing the population of *M. incognita* in the soil and on the roots. RKN management can also be achieved by using *P. lilacinus* and *Bacillus firmus* either individually or in combination. However, the mixture of *P. lilacinus* and *B. firmus* applied in soils 2 weeks prior to tomato transplantation showed the best practice to control *Meloidogyne* spp.

Coating the seed with *Trichoderma viride* and *P. lilacinus* effectively reduced the nematode population in the soil. Species of *Aspergillus* and *Paecilomyces* were found to be antagonistic to M. incognita when compared to the single bio-agent treatment; thus, resulting in enhanced plant growth. A management technique was developed using the nematophagous fungus *Pochonia chlamydosporia* to manage RKN for the purpose of organic vegetable production. Okra seeds treated with *Trichoderma harzianum, T. viride, P. lilacinus, P. chlamydosporia*, and *P. fluorescens* at 20gkg−1 seed signifcantly reduced the nematode population in the soil and promoted plant growth development. *P. chlamydosporia* exhibited nematicidal effects against *M. incognita* on infected common bean under greenhouse condition. *Trichoderma* spp. synthesizing chitinases, lytic enzymes, proteases, and glucanases were found to manage vegetable crop diseases. *T. harzianum, T. viride*, and *T. hamatum* have nematocidal properties when they colonize the roots of host plant and enhance their growth performance. Because crop yield is mainly influenced by climatic conditions, agronomic factors, pests, and nutrient availability in the soil (Elrys et al., 2020), researchers must consider these factors in the selection of fungal BCAs.

Trichoderma spp. can prevent nematode penetration and development in plants through the regulation of metabolites. *M. incognita* and other RKN are more affected by the culture filtrate of *Trichoderma*. Species of *Trichoderma* can produce viridin, a nematicidal chemical. Gliotoxin and acetic acid, have also been reported as nematicidal substances in the culture filtrates of *T. virens* and *T. longibrachiatum,* respectively. In response to *M. incognita, T. polysporum* has the ability to synthesize cyclosporine, the peptide that has a nematicidal action. The efficacy of *P. lilacinus*, as bioagents or bioproducts in mixtures, significantly decreased the number of *M. incognita* on eggplant. It has been shown that different sources of N and carbon affect the growth and antagonism of *Trichothecium roseum* and *T. viride.* Although fructose and lysine were mostly effective against *T. viride*, rhamnose and glycine were more effective against *T. roseum*. There is an adverse effect of fungal culture filtrates on the egg hatching and juvenile mortality of RKN. Plants treated with the fungal BCA, *Lecanicillium muscarium,* decreased the number of galls in plants, eggs, juveniles (J2) and the reproduction factor (Rf) of *Meloidogyne hapla* compared to control plants (El-Saadony et al., 2022).

In addition, plant growth was greatly improved when treated with *L. muscarium. Trichoderma* and *P. lilacinum* isolates dramatically reduced nematode egg number and mass, minimized root gal injury, and improved plant root mass development when compared to control plants without the fungal BCAs. Overall, more than 30 genera and 80 species of fungi can parasite RKN. *Colletotrichum falcatum*, has antagonistic potential against the fungal pathogens *Penicillium citrinum, Botrytis cinerea*, and *Trichodermum glaucum*. Moreover, other fungi, such as *Ampelomyces speciosus* and *Acremonium alternatum*, have the ability to degrade the mycelia of fungal pathogens, indicating that not only rhizobacteria, but also fungi can serve as BCAs.

4.14.4 Mycorrhizal Fungi: Besides PGPR, mycorrhizal fungi are among the most used biofertilizers and act in biocontrol against root-damaging pathogens of the genera *Fusarium*, *Pythium*, and *Phytophthora*, as well as nematodes. Mycorrhizal fungi can improve the food-quality properties of crop plants by increasing antioxidant and vitamin contents of edible parts. Furthermore, they improve the physical properties of the soil by modifying the soil structure. For example, the hyphae can induce entanglement of soil particles with each other to create macroaggregates. The most common type used to improve agricultural crops is arbuscular mycorrhizal fungi (AMF), which penetrate the root cortical cells of a host plant and form highly branched structures, arbuscules. The glycoprotein of AMF, glomalin, is an important component of the mycorrhizal soil that mitigates formation of gas-water interfaces by reducing macroaggregate disruption during wetting and drying, and by preventing movement of water into soil pores.

The AMF are favored in improving agricultural crops, because they are naturally found in over 90% of plant species, and many agricultural practices, such as tillage and fertilization, can reduce abundance of AMF in the fields. As agricultural additives, they have been shown to increase grain yields depending on soil pH, plant species, and cultivar. For example, AMF strains can increase tomato yields by 26% and carrot yields by 300% by protecting host plants against nematodes. A meta-analysis suggests that C4 grasses, non-legume, and woody plants respond better to AMF than legumes and C3 grasses, especially when the soil community is complex and limited in phosphorus rather than nitrogen. For better predictivity of mycorrhizal inoculation, there is a MycoDB database consisting of meta-analysis data spanning > 10 years on plant productivity responses to mycorrhizal applications (Pirttilä et al., 2021).

Regardless of promising results with mycorrhizal inoculants, there are many technical challenges in their production. The up-scaling of mass production of mycorrhizal fungi is often hard, as some strains of the mutualistic symbionts are not cultivable in axenic conditions. With existing protocols, the mycorrhizal inocula may not be pure but contain fungal pathogens, therefore molecular tests are needed for analysis of the inocula composition. However, there are several biofertilizer and biocontrol products of mycorrhizal fungi available, based mainly on strains of *Glomus iranicum*, to improve crop absorption of water and nutrients, as well as tolerance towards nematodes. Other mycorrhizal inoculants contain spores of *Rhizofagus irregularis*, *Funneliformis mossae*, and *Claroideoglomus etunicatum* for improved fruit production (Aamir et al., 2020).

4.14.5 Endophytic Fungi: Regardless of numerous positive impacts on plants and potential in both biocontrol and biofertilization, endophytic fungi have been rarely developed as biotechnological tools in agriculture. The clavicipitaceous endophytes of grasses are an example of fungi that are used in biocontrol in agriculture. The tall fescue has been inoculated with *Epichloë coenophiala* for enhanced pest tolerance and ryegrass varieties inoculated with fungal endophytes are used in New Zealand and Australia with reduced pasture damage by insect herbivores. However, the use of clavicipitaceous endophytes is currently limited to pasture grasses only. Endophytic fungi internally colonize plant tissues and have the potential to act as control agents, such as biological agents or elicitors in the process of induced resistance and in attenuating abiotic stresses. In a review, Fontana et al., (2021) listed the mode of action of this group of microorganisms which can act in controlling plant diseases and describe several examples in which endophytes were able to reduce the damage caused by pathogens and adverse conditions. This is due to their arsenal of molecules generated during the interaction by which they form a kind of biological shield in the plant. Furthermore, considering that endophytic fungi can be an important tool in managing for biotic and abiotic stresses due to the large amount of biologically active substances produced, bioprospecting this class of microorganisms is tending to increase and generate valuable products for agriculture.

Endophytic fungi inhabit a similar ecological niche to that occupied by phytopathogens, thus being able to protect their environment and control them through competition, production of antagonistic substances, direct parasitism or even inducing resistance or tolerance. It is important to consider that some fungi which are endophytic for one plant species may be pathogenic for another species. In the same sense, the production of compounds such as antibiotics, for example, suggests that endophytes can control plant diseases. Both hosts and endophytes are benefited in the interaction among them. On the one hand, the microorganism benefits from protection, nutrition and shelter in the plant. On the other hand, endophytes also help their hosts by stimulating their growth, development, adaptation, and stress tolerance (Wang et al., 2015). Protection against diseases occurs by reducing the infection levels, as well as suppressing and reducing the growth of pathogens. In this sense, it is suggested that the presence of endophytes during the evolutionary process allowed the plants to grow better and be more resistant to insects, herbivorous animals and pathogenic organisms. The same can be inferred regarding adverse environmental conditions such as low humidity and/ or high temperatures.

The plants provide an environment in their interior for a high diversity of endophytic fungi. These microorganisms can colonize leaves, branches and roots, without causing damage to the hosts, systemically inhabiting the apoplast, vascular tissues and in some cases the cell interior. For example, in cacao grown in Bahia State, Brazil, it was observed that plants harbor endophytic fungi belonging to several groups, such as *Acremonium* spp., *Colletotrichum gloeosporioides*, *Fusarium*, *Gliocladium*, *Lasiodiplodia theobromae*, *Pestalotiopsis* spp., *Trichoderma* spp. and *Verticillium* spp.. *Fusarium* sp. and *Colletotrichum* sp. were also isolated as endophytes from cacao leaves in Panama. The most frequent endophytically isolated fungi include *Colletotrichum, Cladosporium, Fusarium* and *Xylaria* species. The diversity of plant endophytes from *Paeonia* spp. was recently analyzed and different genera of fungi were identified. The most abundant among them were *Fusarium, Phoma, Alternaria* and *Pestalotiopsis* (Yang et al., 2018). Other examples can also be found, such as *Coniothyrium* species isolated from the cortex of *Picea abies* branches, *Asteromella* fungus isolated from the inside of *Quercus emoryi* leaves, *Phoma* isolated from wheat leaves and *Aspergillus, Curvularia lunata, Fusarium, Penicillium* and *Trichoderma* isolated from sunflower seeds. In addition, there were a total of 60 isolates of endophytic fungi belonging to 16 different genera in the medicinal *Sceletium tortuosum* plant in South Africa, the most ubiquitous being the *Fusarium*, *Aspergillus*, *Penicillium* and *Phomopsis* genera. Yerba mate plants (*Ilex paraguariensis*) are also colonized by endophytic fungi, with the main ones being *Aspergillus*, *Penicilium*, *Acremonium*, *Fusarium* and *Colletotrichum.*

Most biological control endophyte (BCE) agents have frequently been found among different crops and are able to suppress important pathogens are listed below (Fontana et al., 2021).

Endophytic Fungi	**Plants**	**Fungi Targets**
Cladosporium tenuissimum	-	*Uromyces appendiculatus*
Trichoderma viride, T. harzianum, T. stromaticum, T. virens	-	*Rhizopus stolonifer*
Trichoderma viride	-	*Penicillium digitatum*
Trichoderma viride	-	*Phytophthora nicotianae*
Trichoderma viride	-	*Rhizoctonia solani*
Trichoderma viride, T. koningii	-	*Verticillium dahliae*
Fusarium oxysporum sensu lato	*Solanum lycopersicum*	*Phytophthora infestans and P. capsici*
Xylaria sp.	*Ginkgo biloba*	*Penicillium expansum and Aspergillus niger*
Heteroconium chaetospira	*Brassica oleracea*	*Verticillium dahliae*
Diaporthe helianthi	*Leuhea divaricata*	*Moniliophthora perniciosa*
Aspergillus, Penicillium and Trichoderma sp.	*Eucalyptus benthami*	*Botrytis cinerea*
Trichophyton sp., Chrysosporium sp., Candida pseudotropicalis, and Candida tropicalis	*Symphytum officinale*	*Sclerotinia sclerotiorum*
Colletotrichum gloeosporioides and Clonostachys rosea	*Theobroma cacao*	*Phythophthora sp. and Moniliophthora roreri*
Gliocladium catenulatum	*Theobroma cacao*	*Crinipellis perniciosa*
Diaporthe terebinthifolii	*Schinus terebinthifolius*	*Phyllosticta citricarpa*
Fusarium solani sensu lato	*Vitis labrusca*	*Botrytis sp.*
Aspergillus insulicola and A melleus	*Sesuvium portulacastrum*	*Pythium aphanidermatum*

Endophytic Fungi	Plants	Fungi Targets
Phyllosticta fallopiae	*Cornus officinalis*	*Alternaria alternata, A. arborescens, Botryosphaeria dothidea and Colletotrichum gloeosporioides*
Alternaria tenuissima	*C. officinalis*	*Alternaria alternata*
Alternaria alternata	*C. officinalis*	*Alternaria arborescens*
Botryosphaeria dothidea	*C. officinalis*	*Alternaria alternata, A. arborescens, Botryosphaeria dothidea and Colletotrichum gloeosporioides*
Colletotrichum gloeosporioides	*C. officinalis*	*Alternaria alternata*
Botryosphaeria berengeriana	*C. officinalis*	*Botryosphaeria dothidea*
Alternaria sp., Botryosphaeria ribis, Phoma medicaginis, Bionectria ochroleuca, Aureobasidium pullulans and Chaetomium spirochaete	*Vitis vinifera*	*Botrytis cinerea*
Ramularia pratensis, Phoma aliena and Fusarium acuminatum	*Vitis riparia*	*Botrytis cinerea*
		Bacteria targets
Xylariales sp.	*Distylium chinense*	*Clavibacter michiganensis, Xanthomonas citri pv. phaseoli var. fuscans and Pseudomonas syringae pv. lachrymans*
		Viruses targets
Paecilomyces variotii	*Nicotiana benthamiana and N. tabacum*	*Potato Virus X (PVX) and Tobacco mosaic virus (TMV)*
		Nematodes targets
Acremonium implicatum	*Solanum lycopersicum*	*Meloidogyne incognita*
Fusarium oxysporum sensu lato	*Musa spp.*	*Pratylenchus goodeyi*
Chaetomium globosum	-	*Meloidogyne incognita*
Daldinia cf. concentrica	*Olea europaea*	*Meloidogyne javanica*
Alternaria sp.	-	*Bursaphelenchus xylophilus*

*The possible mechanisms of action are commented on in the text. "-" means that the host plant was not identified in the cited reference. The Bold is applied to divide different kinds of plant pathogens.

Endophytic fungi can trigger innumerable mechanisms in the plant, providing protection against biotic and abiotic disorders. These fungi satisfactorily perform biological control against plant diseases with the potential to be used as a tool for bioprospecting new molecules and genetic modification of plants due to their potential for genetic modulation and interaction with the host. Tolerance to abiotic stresses can be obtained by an association of endophytes with the target cultures, presenting promising results and making it possible to grow plants in certain places where plants without association with the endophytic agent could have difficulties to developing. The secondary metabolites produced by endophytes exhibit important biological activity and can become valuable products. Thus, isolating and characterizing endophytic microorganisms from plants which have not yet been studied can enable discovering new species with the potential to produce substances of interest which can be used in the biological control of diseases, as elicitors in induced resistance and for inducing tolerance to abiotic stresses.

4.14.6 Rhizospheric Fungi: There are also several soil-associated fungi used in agriculture for increasing crop health and fitness. Members of rhizospheric and epiphytic *Trichoderma* spp. are used in many products due to their capacity to reduce abiotic and biotic stress on host plants, for example, by controlling many plant pathogens and nematodes (Vosátka et al., 2012). *Trichoderma viride* is antagonistic towards soil borne pathogens in genera *Fusarium*, *Sclerotium*, *Rhizoctonia*, and *Pythium*. *T. harzianum* can control the same fungal pathogens as *T. viride*, but also *Botrytis*, *Gaeumannomyces*, *Sclerotinia*, *Verticillium*, and wood-rot fungi. *T. harzianum* and *T. polysporum* are used together to increase the spectrum of activity against diseases caused by fungal pathogens and *Trichoderma* spp. can also be used in combination with mycorrhizal fungi.

The discovery of novel biocontrol agents requires the continuous scrutiny of native microorganisms to ensure that they will be useful on a regional scale. The goal of the present work was to discover novel antagonistic bacteria against *Fusarium oxysporum* f. spp. *lycopersici* race 3 (*Fol* R3) and *radicis-lycopersici* (*Forl*) causing *Fusarium* wilt disease and *Fusarium* crown and root rot of tomatoes, respectively (Md. Masudur Rahman Khalil et al., 2021). High-throughput liquid antagonism screening of 1,875 rhizospheric bacterial strains followed by dual confrontation assays in 96-well plates was used to select bacteria exhibiting > 50% fungal growth inhibition. In a second dual confrontation assay in 10-cm Petri dishes, bacteria showing > 20% *Fol* R3 or *Forl* growth inhibition were further screened using a blood hemolysis test. After discarding β-hemolytic bacteria, a seedling antagonistic assay was performed to select five potential antagonists. A phylogenetic analysis of 16S rRNA identified one strain as *Acinetobacter calcoaceticus* (*Ac*DB3) and four strains as members of the genus *Bacillus* (*B. amyloliquefaciens Ba*MA26, *Bacillus siamensis Bsi*DA2, *B. subtilis Bs*TA16 and *B. thuringiensis Bt*MB9). Greenhouse assays demonstrated that *Bs*TA16 and *Ac*DB3 were the most promising antagonists against *Fol* R3 and *Forl*, respectively. Pathogen biocontrol and growth promotion mechanisms used by these bacteria include the production of siderophores, biofilm, proteases, endoglucanases and indole acetic acid, and phosphate solubilization. These five bacteria exerted differential responses on pathogen control depending on the tomato hybrid, and on the growth stage of tomatoes. the use of an *Acinetobacter calcoaceticus* isolate (*Ac*DB3) to control *Forl* in tomato under greenhouse conditions was reported for the first time.

Rhizosphere-resident fungi that are helpful to plants are generally termed as 'plant growth promoting fungi' (PGPF). These fungi are one of the chief sources of the biotic inducers known to give their host plants numerous advantages, and they play a vital role in sustainable agriculture (Murali et al., 2021). Today's biggest challenge is to satisfy the rising demand for crop protection and crop yield without harming the natural ecosystem. Nowadays, PGPF has become an eco-friendly way to improve crop yield by enhancing seed germination, shoot and root growth, chlorophyll production, and fruit yield, etc., either directly or indirectly. The mode of action of these PGPF includes the solubilization and mineralization of the essential micro- and macronutrients needed by plants to regulate the balance for various plant processes. PGPF produce defense-related enzymes, defensive/volatile compounds, and phytohormones that control pathogenic microbes' growth, thereby assisting the plants in facing various biotic and abiotic stresses. Therefore, this review presents a holistic view of PGPF as efficient natural biofertilizers to improve crop plants' growth and resistance.

4.14.7 Mycoparasitic and Entomopathogenic Fungi: In addition to plant-associated fungi, there are several fungal products available based on mycoparasitism of plant pathogens. *Ampelomyces quisqalis* and *Fusarium proliferatum* are mycoparasites of fungi causing powdery and downy mildew on crop plants, respectively (Vosátka et al., 2012). *Beauveria bassiana* is an entomopathogenic fungus used to control pathogenic insects, such as white flies, thrips, mites, aphids, and their various developmental stages infesting numerous crop plants. Furthermore, *Chaetomium cupreum* protects plants from fungal diseases such as rust, early and late blight, leaf spot, and stem and tuber rot.

Current trends: A great majority of the microbiological tools currently available colonize the rhizosphere of a crop plant, consisting mainly of PGPR, epiphytic, mycorrhizal, and soil fungi (Fig.32). Consequently, their application often results in inconsistent field performances in agriculture.

These variations can be associated with poor competence of inoculants in the field conditions, plant genotype, interactions with the existing plant-host microbiome, and environmental conditions. The soil ecosystem is highly variable, with localized microenvironments that are affected by changing temperatures and humidity. Soil microbial communities can have a crucial role in survival of the incoming microbes. Depending on soil quality, environmental conditions, and the interactions between soil microbiome, the inoculant, and the host plant, the biofertilizer or biocontrol agent can have varying rates of success in increasing the crop yield. A constant search for new potent, stable strains and the effort for deeper understanding of mechanisms of action of biofertilizers and biocontrol agents will be the key for achieving dependable tools for improved and sustainable crop production.

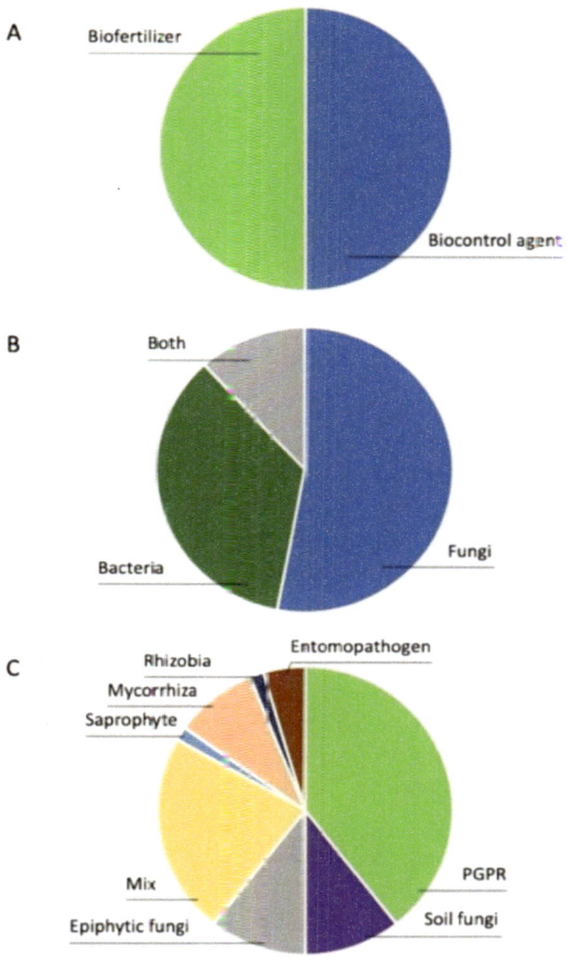

Fig. 32: Microbiological tools, available in 2021 for agriculture with published field test reports by (A) application, (B) organism, and (C) type

4.15 Identification of New Potent Microbial Traits

Agriculture, at present relies on a few microbial beneficial traits improving the growth of various plant crops. Biofertilizers typically help plant growth by fixing nitrogen, solubilizing phosphates, producing plant hormones, oxidizing sulfur, and decomposing organic compounds, whereas biocontrol agents induce systemic resistance, produce one or several antibiotic compounds or enzymes, deplete iron, and compete for niches with pathogens within the rhizosphere, all reviewed in detail several times in the past (Pirttilä et al., 2021). However, other yet-unexplored microbial traits beneficial for the host plant likely exist, typical for each strain. Furthermore, specific traits and concerted action of microbial communities can result in improved plant growth and stress tolerance, currently unstudied.

4.15.1 Genome Mining: Sequencing of genomes and analysis of gene functions could reveal new mechanisms on how microbes help plants thrive. The genome of the first commercially available biocontrol agent, *Bacillus amyloliquefaciens* GB03, was reported in 2014 and the genome of *B. velezensis* FZB42, which is considered a model for plant-growth-promoting and biocontrol rhizobacteria, was sequenced already in 2007. Since then, numerous bacterial strains found or suggested effective in biocontrol or biological fertilization of various crop plants have been sequenced. So far, genome mining of the strains has, however, focused on known traits, listed above. For example, the genome of *B. velezensis* FZB42 was reported to carry 13 gene clusters responsible for synthesis of predicted antimicrobial metabolites or volatile compounds. Sequencing of several other biocontrol strains of *B. velezensis*, WRN014, 9912D, M75, RC 218, and CC09 have revealed antimicrobial clusters in their genomes and biosynthesis genes of surfactin, bacillaene, fengycin, and bacillibactin were proposed responsible for the biocontrol activity of *B. atrophaeus* GQJK17. *B. methylotrophicus* B25, which enhances plant health, carries several genes responsible for secondary metabolism, such as nonribosomal peptide synthetases and polyketide synthases and *Paenibacillus polymyxa* strains SC2 and E681 possess genes required for biosynthesis of antibiotics such as polymyxin and fusaricidin. The genome of one of the best-known biocontrol strains, *P. fluorescens* Pf-5, was shown diverse with secondary metabolism genes and another *Pseudomonas* biocontrol strain, SH-C52, was reported equally versatile with biosynthesis of secondary metabolites, such as thanamycin, hydrogen cyanide, achromobactin- and ornicorrugatin-type siderophores, bacteriocin, arylpolyene, insect toxins Tcc2 and Tcc4, and three NRPS-based lipopeptides 2 (Van Der Voortet al., 2015).

Regardless of the high potential, analysis of genomes has rarely revealed new traits behind plant-growth promotion or protection against biotic or abiotic stresses. There are a couple of examples, however. In the biofertilizer strain *Trichoderma harzianum* t-22, genome mining revealed biosynthesis of tricholignan A, which is a natural product helping the plant in assimilating iron from the soil (Chen et al., 2019). In the plant-growth-promoting strain *Methylorubrum extorquens* DSM13060, which increases pine growth to the same extent as mycorrhizal fungi, genome analysis confirmed lack of plant hormone biosynthesis genes, but revealed an array of phospholipase A genes, potentially responsible for growth promotion. In *Micrococcus luteus* K39, phospholipase D, superoxide dismutase, and ferredoxin NADP reductase were suggested responsible for induced tolerance of oxidative stress, salinity, and drought in the host. A recent comparative analysis of an array of plant-associated bacterial genomes revealed 64 plant-resembling domains, including NLR class of intracellular innate immune receptors, which were hypothesized to interfere with plant immune functions. Such intracellular immune receptors could potentially be responsible of enhancing plant resistance by biocontrol strains as a new trait (Levy et al., 2018).

4.15.2 Molecular Plant–Microbe, Microbe–Microbe and Phage Interactions: A detailed analysis of interactions between the host plant and the microbe can reveal new beneficial traits in biofertilizers and biocontrol agents. For example, a study on gene expression of *B. velezensis* FZB42 suggests the involvement of regulatory small RNAs in the plant–microbe interaction (Fan et al., 2015). A detailed study on the beneficial compounds produced by the plant-growth-promoting strain *M. extorquens* DSM13060 revealed that bacterial metabolism of polyhydroxybutyrate plays an important role in stress tolerance of the bacterium and host. In the biocontrol strain *P. fluorescens* Pf0-1, the importance of histidine metabolism during the interaction with host plants has been discovered through gene-expression analyses. Furthermore, antimicrobial peptides are a group of less studied, but potentially important compounds associated with biocontrol, which have been discovered through studies on bioactive metabolites in both fungal and bacterial endophytes.

Microbe–Microbe Interactions: Another largely unknown field is the microbe–microbe interactions, and studies on such interactions would likely reveal completely new traits useful in biocontrol and biofertilization (Pirttilä et al., 2021). Once a microbe is inoculated into a plant, it affects the full holobiont, the host, and the associated microbiome. This is clearly seen with pathogens infecting host plants, as they create shifts in the structure of endophytic communities. Similar but less pronounced changes have been observed as the result of endophyte inoculation in potato. An example of microbe–microbe interactions as a mechanism in biocontrol is the endophytic community playing a role in control of citrus variegated chlorosis (CVC) by *Methylobacterium mesophilicum* in sweet orange (*Citrus sinensis* L.). The pathogen *Xylella fastidiosa* invades the xylem vessels of sweet orange, where the colonization is suppressed by resident endophytic community influenced by *M. mesophilicum* (Azevedo et al., 2016). This suggests that the performance of specific biocontrol agents can in reality be the consequence of a community response. Further information on microbe–microbe interactions is necessary for a full understanding of the action of biofertilizers and biocontrol agents. Therefore, analysis of the microbial community of the holobiont should accompany studies on beneficial effects of microorganisms in plants.

Phage Interactions: Bacteriophages or phages, are emerging as a new trait in biocontrol, replacing or accompanying bacterial biocontrol agents. Phage interactions are typical for a root microbiome, which suggests phages could be useful tools in biocontrol. For a successful tailored product, life cycle and host interactions of the pathogen need to be thoroughly understood (Holtappels et al., 2021). The antimicrobial mechanisms are based on lysis of bacterial cells caused by virulent phages or lytic enzymes. The phages could also be manipulated genetically to deliver specific lethal genes to the target organisms, or to sensitize them against traditional antimicrobials which, however, raises the problems of spreading antibiotic resistance and responsible usage of genetically modified organisms (GMOs) in the environment. Overall, phages may represent an "outside of the box" solution in biocontrol, and such unconventional thinking is necessary to create new innovations in agriculture.

4.15.4 How to identify New Potent Microbial strains?: A current and widely approved trend involves using multiple (2–5) microbiological tools in combination to achieve more effective and consistent results in agriculture. Plant-growth promotion can be enhanced by using several biofertilizer strains, both fungal and bacterial, together, complementing the effects of each other (Pirttilä et al., 2021). Similarly, microorganisms with biocontrol capacities can have a synergetic effect and stimulate the antagonism of other biocontrol agents against plant diseases (Sharma et al., 2017). However, when *B. velezensis* FZB42 was studied in the rhizosphere, the main mechanism of protecting the host plant was not the production of antibiotic compounds by single or several strains in combination, but the systemic resistance induced by the bacteria in the host plant. These results emphasize the importance of a healthy microbiome, which is behind the healthy phenotype in animals, as well as in plants. For example, a recent study showed that the composition of soil microbiome could predict the survival of a plant host against disease. Specifically, species belonging to genera *Pseudomonas* and *Bacillus* were important members of the protective microbiome (Wei et al., 2019). Microbiome engineering, i.e., rationally constructing microbiomes and inoculating them into crop plants could be a solution for sustainable agriculture of the future. Such attempts should not only focus on the root microbiome, but on the full plant microbiome including shoot endophytic communities.

i. Modification of Microbial Genomes: Genomic modification of microorganisms can provide more potent strains of biofertilizers or biocontrol tools (Pirttilä et al., 2021). Ideas and attempts to genetically enhance biocontrol agents have already been cultivated and tested. For example, *P. syringae* was modified early on to enhance frost tolerance of strawberry and potato by removing ice nucleation protein from the bacterial genome. *P. fluorescens* strains F113 and CHA0, as well as *P. putida* WCS358 have been genetically engineered for enhanced production of antibiotic compounds, 4-diacetylphloroglucinol (PHL) and pyoluteorin (PLT). Furthermore, a strain of *P. fluorescens* CHA0 has been modified to overproduce the phytohormone indole acetic acid (IAA). However, when tested in a microcosm, the *P. fluorescens* strain CHA0 overproducing PHL and PLT decreased the numbers of rhizobia, *Sinorhizobium meliloti*, and reduced nodulation in alphalpha. The IAA-overproducing strain of CHA0 increased root yield in natural soil, but reduced root growth in autoclaved soil. The development of genome editing using CRISPR/Cas system has revived the field of using genetically modified microorganisms to fight plant diseases. For example, development of non-pathogenic strains of fungal pathogens by CRISPR/Cas was recently proposed as an approach of biocontrol (Glandorf, 2019). However, fungal endophytes that are close, non-pathogenic relatives of plant pathogens are already typically an innate part of plant microbiomes, therefore, the approach is already naturally in use.

Overall, there are severe ethical aspects associated with GMOs as biofertilizers or biocontrol agents in agriculture. Some tests with genetically enhanced biocontrol agents have shown changes in the microbial community structure of the rhizosphere. More importantly, the GMOs would be released to the environment and, the transferred traits, such as biosynthesis genes of antimicrobial compounds, would with a very high likelihood spread across microbiomes, spreading antimicrobial resistance. Furthermore, restricting the access of the GMOs to the edible plant parts would need considerate testing. A deliberate release of genetically modified organisms into the environment is regulated by Council Directive 2001/18/EC, which repeals Council Directive 90/220/EEC in Europe. Nevertheless, genetic modification using CRISPR/Cas, for example, is an important tool in elucidating the mechanisms of plant–microbe interaction (Shelake et al., 2019).

ii. Bacterial Endophytes and Endosymbionts: Apart from the PGPR that can have an endophytic phase, or rhizobia of legumes, bacterial endophytes have rarely been the choice for biocontrol or biofertilizer products. However, the currently used microbiological tools that typically colonize plant rhizosphere, soil, root surface, or root apoplast, have demonstrated inconsistent field performances, especially when used as single inoculations. Systemic bacterial endophytes, which colonize all internal tissues within a plant, and endosymbionts, which colonize plant cells, would likely provide more persistent benefits to the host plant than rhizobacteria or apoplastic endophytes. They would be better protected from environmental conditions, such as abiotic stress, being well adapted to the targeted niche. Endophytic bacteria share many growth-promoting and biocontrol traits with PGPR, such as production of plant hormones and antibiotic compounds, induction of systemic resistance, or nitrogen fixation (Adeturji et al., 2019).

Potentially limiting the use of shoot endophytes in agriculture is their existence in the edible parts of the crop plant. However, this line of product development should be reconsidered (Pirttilä et al., 2021). The fear of introducing microorganisms into edible plant parts has no basis on the grounds of the fact that plant tissues are already naturally colonized by a variety of microorganisms, both completely harmless strains, as well as potential human pathogens. Selection of non-pathogenic microorganisms should pose no threat to human health, which can be ascertained through food security programs. Furthermore, several food products, such as fermented foods, are already manufactured with the help of an array of microorganisms, and they are considered healthy for the human gut. The shoot endophytic strains could be applied to host plants during micropropagation, as many crop plants are routinely propagated in vitro. For crops propagated through seeds, formulations similar to those prepared for rhizobacteria could provide a suitable application method. For example, the endosymbiont *M. extorquens* DSM13060 systematically colonizes the host plant after application of bacterial cells on germinating roots.

iii. Microorganisms of Plants Adapted to Extreme Growth Conditions: So far, the majority of biofertilizers and biocontrol agents have been identified from rhizosphere of crop plants. For example, among the PGPR, *Azotobacter chroococcum* has been isolated from wheat rhizosphere, *Azospirillum brasilense* strains from rice and wheat and *Pseudomonas fluorescens* Pf5 from rhizosphere of cotton seedlings (Pirttilä et al., 2021). Although some strains can be host-specific, many are generalists, and plants growing in extreme growth conditions are fruitful targets for finding new potent biofertilizers and biocontrol agents, because they often rely on their microbial partners in surviving in the challenging environment. For example, the growth-promoting endosymbiont *M. extorquens* DSM13060 has been isolated from Scots pine (*Pinus sylvestris* L.), which is a conifer species adapted to harsh growing conditions, poor levels of nutrients in the soil, drought, and low temperatures. The endophytic fungus *Curvularia protuberata* provides heat resistance in the tropical panic grass *Dichanthelium lanuginosum*, which can be transferred to crop species such as tomato, watermelon, and wheat. Similarly, *Piriformospora indica*, isolated from a desert plant, provides tolerance towards salt and biotic stress in several crop plants. New effective biocontrol agents are also likely to be discovered in resistant plants living in environments with a high pathogen reservoir.

iv. Microorganisms of Diverse Environments: When searching for new biofertilizers and biocontrol agents, a rational choice is to screen environments with high microbial diversity. With soils harboring the greatest reservoirs of microbes, land-use change is considered the most severe threat to biodiversity. A recent study showed that conversion of forest soils to arable land changes especially bacterial diversity, but also viral diversity, whereas Archaean diversity was less responsive. The bacterial diversity was reduced by 45–75%, depending on forest type, and correlated with loss of functional diversity in the soil (Muñoz-Arenas et al., 2020). There are similar reports on fungal endophytes. In a forest setting, the diversity of foliar fungal endophytes declines along a gradient of pine trees growing in old-growth forest, managed forest, and in nursery. Grapevine varieties grown on organically managed farms host richer fungal endophyte communities than the varieties grown under an integrated pest-management system. The bacterial endophytic communities respond in a similar manner to biodiversity of the environment. For example, using a green manure by *Erythrina poeppigiana,* O. F. Cook or *Inga* trees modify the endophytic microbiome of banana, compared to plants grown in monoculture. Therefore, environments rich with various plant species, such as forests, likely host a rich pool of microorganisms beneficial for plants. Discovery of new microbial strains that can be used as members of an engineered healthy plant microbiome, or as single inoculants in biocontrol and biofertilization, is more fruitful in forests than on agricultural land (Pirttilä et al., 2021).

Future Prospects: Further development and increased use of biofertilizers and biocontrol agents is necessary for sustainable food production in the future. Currently, we have ~50 functional strains that are applied in agriculture. The demand for both biofertilizers and biocontrol agents is expected to rise, which provides a good basis for further development (Pirttilä et al., 2021). The major problem regarding the current tools is the inconsistency of the products in field conditions. Using combinations of microbial strains that rely on different biocontrol or biofertilizer properties could alleviate the problem of inconsistency; this approach is already is in use to some extent. However, while the current tools are all based on the few well-known mechanisms of action, plant–microbe and microbe–microbe interactions are more multifaceted than that and need a thorough investigation.

Detailed understanding of each microbial strain, their mechanisms of action, and components of a healthy plant microbiome will likely provide a firm basis for developing reliable tools to enhance plant health and growth in varying field conditions. To achieve this, the following are needed.

i. A continuous search for new efficient microorganisms, enabling healthy plant cultivation in various field conditions, is pertinent. The potential new microbiological tools are likely to be found in plants adapted to extreme, in nutrient-poor growth conditions, in resistant plants under pathogen exposure, and in environments with high biodiversity, such as forests.

ii. Studies on microbial traits beneficial for plants should be extended beyond mechanisms already known. The research for new microbial traits should cover not only plant–microbe interactions, but also microbe–microbe interactions, including phage effects and transmission.

iii. The focus should be turned from the rhizosphere towards the plant holobiont, including strains occupying all plant organs, both bacteria and fungi. A high number of microbial inoculants should be combined in the biocontrol and biofertilizer products, with the aim of plant microbiome engineering.

iv. For enabling the production of such complex biological products, the bottleneck of formulation needs to be solved. Therefore, more efforts should be placed on the development of technologies for microbial culture and live preservation. With the recent enormous advances in molecular and microbiological techniques, a technological jump enabling microbial live preservation in various conditions should be more than feasible.

v. A thorough analysis of the capacity of the strains to enhance plant health and growth under several growth conditions should become a routine in biofertilizer and biocontrol agent discovery. There are frequent reports in the scientific literature on new potent strains identified, but usually they are only tested in the laboratory conditions for few of the well-known microbial traits, and in one or two plant-growth conditions (Nian et al., 2021). The effects of inoculants need to be tested under stressful conditions, along with an analysis of changes in the microbiome of the holobiont.

4.16 Potential Postharvest Biological Control Agents

Fresh fruits and vegetables are often washed and sanitized immediately after harvest and handled under low temperatures in controlled or modified atmosphere. This contributes to the low incidence of insect attacks (Ertwange, 2006). However, plant pathogens are favoured in most cases, leading to diseases. Biological control of pre- and postharvest diseases has been one of the most extensively studied alternatives and appears to be a viable technology. Research and development of biological control products for pre- and postharvest use has been on a fast track. Several commercial products are already available and others will be available in the near future. However, huge commercial successes in biological control are still rare. Currently, the main hurdle facing widespread use of postharvest biocontrol strategies is the decreasing efficacy and lack of consistency found when these methodologies are applied as standalone treatments under commercial conditions. Research into formulation technology might offer a new stimulus to help develop successful biological control agents from the vast number of strains isolated in the past decades. Combining biological, chemical, physiological, and other postharvest practices in an integrated control strategy can give excellent pest control during long term storage and transport of certain fruits. The opportunities of successful implementation of this technology is growing as information regarding various aspects related to formulation technology, application, microbial ecology, and genetics becomes available.

i. ***Bacillus* spp**: A natural epiphytic *Bacillus subtilis* isolated from avocados showed great promise as a biological control agent for pre- and postharvest pathogens (Irtwange, 2006). The organism is a natural inhabitant of the avocado leaf and fruit surfaces and dominates on these surfaces throughout the year. It has a natural ability to compete with other organisms by means of competition for space and nutrients. The organism is of particular importance due to its competitive ability against avocado fruit pathogens. The effective use of *Bacillus subtilis* and its advantages for man has been known earlier. Japan has been using *Bacillus subtilis* as an important ingredient in the fermentation of soybean, providing them a product sometimes referred to as natto, which is one of the most widely used food products in Japan. The organism has also been given GRAS (Generally Regarded As Safe) clearness in the USA and has been marketed as a natural biological control product in many countries. *Bacillus* spp. have been successfully tested in field trials, postharvest applications on the use of other *Bacillus* spp. in controlling avocado postharvest diseases has revealed a significant reduction of the severity of postharvest diseases such as anthracnose and fruit rot complex, and stem-end rot.

ii. ***Pseudomonas syringae***: *P. syringae* is a nutritionally versatile organism. There are 41 pathovars of *P. syringae* which are able to cause diseases on various plants. The antagonistic strain is their non-pathogenic counterpart and is antagonistic to pathogens on many plants. It can grow well on wounded plant tissue and can control a variety of diseases on different fruits, including pome fruits, banana, and citrus fruits as well as vegetables. The fruit's surface is a natural habitat for many yeasts which dominate fruit microflora, especially close to harvest (Irtwange, 2006). The yeasts grow rapidly on this substrate, which has a high concentration of readily available carbon sources, such as those contained in juices leaking from wounded fruits. *P. syringae* (strain L-59- 66 renamed as strain ESC- 11) can control blue mold caused by *P. expansum*, gray mold caused by *B. cinerea*, and Mucor rot caused by *Mucor* spp. on apple and pear. It can also control blue mold caused by *Penicillium italicum*, and green mold caused by *P. digitatum* on citrus fruit. Another strain (ESC-10) of this bacterium, superior in controlling decays of citrus fruit, has been isolated by EcoScience, Corp. Strain ESC-11 has been shown to reduce crown rot of banana, which is caused by a complex of fungi, including *Fusarium semitectum* and *F. moniliforme*, and it also reduced Fusarium dry rot on potato caused by *F. sambucinum*. This strain also prevented growth of the food borne pathogen, *Escherichia coli* O157:H7, in apple wounds. This pathogen can grow quickly on damaged apple tissue and consumption of unpasteurized apple cider contaminated with this bacterium has caused outbreaks of illnesses in recent years.

iii. **Yeast isolates:** Yeasts appear to be particularly promising biocontrol agents since production of antibiotics probably is not involved in their activity (Irtwange, 2006). Various observations suggest that competition for nutrients between yeasts and molds and parasitism are likely to be the main mechanisms of action. Many beneficial yeasts can effectively deplete the sugar occurring in fruit wounds (place of infection) and inhibit germination of mold propagules. Radioactively labeled sugar was taken-up more rapidly by beneficial yeast than by the mold propagules. It is postulated that the nutrient competition can play a significant role as a sole biocontrol mechanism, or it may weaken the mold propagules and predispose them to other mechanisms. Some examples of yeast isolates used as BCAs include *Cryptococcus infirmo-miniatus*, a biological control of naturally occurring yeast, to control decay of sweet cherry either alone or in combination with a fungicide and modified atmosphere packaging. Several postharvest diseases affect sweet cherry, including blue mold (*Penecillium expansum*) and brown rot (*Monilinia fructicola*).

C. infirmo-miniatus alone did not control brown rot in air-stored cherries, but it did control blue mold, probably because the longer germination time of blue mold conidia allowed the yeast time to grow and colonize wound sites to prevent infection. A single pre harvest application of the fungicide iprodione reduced brown rot, but better control was achieved when cherry fruit were also treated with a postharvest dip in the beneficial yeast suspension. The combination of preharvest fungicide and postharvest biocontrol reduced decay by 88-97%. Modified atmosphere packaging also reduced brown rot, with the greatest reduction when used in conjunction with treatment with the beneficial yeast. Combining all three treatments in an integrated control strategy reduced brown rot from 41% to less than 1%. This strategy gives excellent decay control during long-term storage and transport of sweet cherry fruit. Another yeast isolate from peach fruit is *Pichia membranefaciens*. It is suggested that *P. membranefaciens* at 5x108 cells per ml completely inhibited Rizopus rot in nectarine wounds artificially inoculated with 5x104 spores per ml at 25, 15, and 30o C storage conditions. Furthermore it is suggested that the use *P. membranefaciens* for biocontrol is compatible with several common postharvest practices including fungicide, calcium treatment, and cold storage. Picking wounds were inoculated with yeasts then inoculated with *Botrytis* spores.

Innovative biological control strategies should take into consideration the growing concern over contamination of the produce with human pathogens, as well as plant pathogens (Irtwange, 2006). The time is ripe to integrate biocontrol agents with one or more physical treatments such as heat treatments, controlled and modified atmospheres, natural biocides, and food-grade preservatives. Such an integrated approach will probably provide adequate control levels comparable to those achieved by chemical fungicides. To achieve this goal, the fruit and vegetable industry should adopt certain changes in packing and sorting lines required for successful implementation of an integrated control strategy. Molecular approaches may prove useful in developing biocontrol agents with enhanced biocontrol activity and set directions for full exploitation of the genetic potential of these antagonists. Yeast antagonists selected for their biocontrol activity appear to be very well adapted to growth and colonizing fruit surface and especially fruit wound sites. This feature suggests that these yeasts provide an excellent means of delivering bioactive compounds, such as fungal cell-wall-degrading enzymes, directly to areas where pathogen propagules are most likely to germinate and infect the tissue. Combining the ability of these yeasts to rapidly colonize wound sites with enhanced constitutive production of antifungal proteins may prove to be a useful strategy for improving biocontrol effectiveness.

4.17 Virus-based Bioagents and Phage Against Plant Diseases

Among BCAs, viruses, a major driver for controlling host populations and evolution, are somewhat underused, mostly because of regulatory hurdles that make the cost of registration of such host-specific BCAs not affordable in comparison with the limited potential market. Jeroen Wagemans et al., (2022) provided a comprehensive overview of the state of the art of virus-based BCAs against fungi, bacteria, viruses, and insects, with a specific focus on new approaches that rely on not only the direct biocidal virus component but also the complex ecological interactions between viruses and their hosts that do not necessarily result in direct damage to the host. Implementing viruses as BCAs for plant biotic stress is not new: The use of bacteriophages as BCAs in bacteria was first hypothesized a century ago. In fungi, the first example of a virus as a successful BCA was the use of the mycovirus *Cryphonectria hypovirus* 1 (CHV1), which was discovered to cause hypovirulence in the ascomycetous fungus. *C. parasitica* has since been deployed to interfere with chestnut blight disease in chestnut orchards and forests. In insects, the first examples of virus biocontrol were the uses of baculoviruses to control populations of lepidopteran insects. More surprisingly, viruses can also be used to control diseases caused by plant viruses: In this case, viral biocontrol relies on the use of attenuated strains of plant viruses to prime plants against an infection of more severe viral strains of the same virus species. This concept is known as cross-protection or preimmunization.

Phage biocontrol against plant-pathogenic bacteria: Among the low impact measures against phytopathogenic bacteria is the use of bacterial viruses or bacteriophages as biocontrol tools. Promising results and technical limitations have been reviewed elsewhere (Jeroen Wagemans et al., 2022). Even though there is clearly great potential for the implementation of phage biocontrol in crop production, there is currently only a limited number of commercially available phage cocktails on the crop protection market, especially in Europe, likely because, in the European Union, biologicals need to pass a regulatory process similar to that of synthetic pesticides.

Since 2005, the US Environmental Protection Agency (EPA) has authorized four different phage cocktails as active substances for bactericides, all from Omnilytics (part of Phagelux Agrihealth). The first cocktail, called AgriPhage, contains bacteriophages against *Xanthomonas campestris* pv. *vesicatoria* and *Pseudomonas syringae* pv. *tomato* and protects tomato and pepper plants from bacterial spot and speck in both the greenhouse and open field. Similarly, Omnilytics also developed cocktails against bacterial canker disease on tomatoes and citrus caused by *Clavibacter michiganensis* subsp. *michiganensis* and *Xanthomonas citri* subsp. *citri,* respectively, and against *Erwinia amylovora* fire blight on apple and pear trees. Unfortunately, no concrete details on the phage contents and expected symptom development reductions of the current cocktails are described in the literature, although early formulations of the *X. campestris* pv. *vesicatoria* phage cocktail reduced bacterial spot disease severity on tomato plants by an average of 17% (15, 39). A cocktail against *X. citri* subsp. *citri* containing at least some of the Omnilytics phages was able to reduce citrus canker disease severity by 59% on average in a set of five greenhouse experiments (14). In 2021, the EPA approved another phage product, XylPhi-PD, from the Otsuka Pharmaceutical subsidiary A&P Inphatec that protects grapevines in California from Pierce's disease, which is caused by *Xylella fastidiosa*. The company reported that consistent XylPhi-PD injection in the xylem reduced disease incidence by 57% in a four-site California field study.

Since 2018 the Hungarian government has authorized Enviroinvest to locally sell the phage cocktail Erwiphage Plus as a preventive bactericide against *E. amylovora*. This product is adapted yearly to reduce resistance development and can be applied only under strict conditions during the blossoming period, and this emergency authorization is approved yearly for only 120 days from mid-March to midJuly. The Erwiphage Plus cocktail contains two specific phage species, PhiEaH1 and PhiEaH2, which were isolated locally and successfully reduced the occurrence of fire blight cases in field experiments when no artificial infection was applied (https://www.erwiphage.com). Although not registered as a PPP, APS Biocontrol sells a phage product within Europe: Biolyse-PB reduces *Pectobacterium* soft rot of potato tubers packed for human consumption (Jeroen Wagemans et al., 2022).

4.17.1 Mycoviruses as putative biocontrol agents of plant-pathogenic fungi: Mycoviruses are obligate parasites prevalent in most economically important plant-pathogenic fungi. Mycovirus–fungus interactions are frequently associated with latent infections; however, mycovirus infection can also be beneficial or detrimental to the fungal host (Jeroen Wagemans et al., 2022). Mycoviruses can have many types of genomes: mainly positive- and negative-sense single-stranded RNA (ssRNA+/−) or double-stranded RNA (dsRNA) genomes and, much less frequently, single-stranded DNA (ssDNA) genomes, with the noticeable exclusion of dsDNA viruses. Their genomes are mono- or multi-segmented, naked, or encapsidated. The historical success of using mycovirus-infected hypovirulent *Cryphonectria parasitica* for the control of chestnut blight has stimulated additional studies on mycoviruses and their potential for biocontrol of plantpathogenic fungi in other pathosystems.

i. Hypovirulent Strains for the Biocontrol of Chestnut Blight: A successful case example of biocontrol of a fungal disease with a mycovirus is the use of *Cryphonectria* *hypovirus* 1 (CHV1) against its fungal host *Cryphonectria parasitica,* the causal agent of chestnut blight, a severe disease of chestnut with characteristic canker symptoms (necrosis of the bark and cambium) that can girdle the trunk and lead to death of the canopy. Comprehensive reviews have been written on the history of controlling this invasive species with hypoviruses and the molecular aspects of the hypovirus-fungus interaction. The practical aspects of this hypovirus-based biocontrol approach, the reasons behind some failed attempts at controlling this disease with mycoviruses in the United States, and some possible new approaches to overcome such failures were mainly focussed. Hypovirulence was first described by Antonio Biraghi in Italy in the early 1950s followed by approaches to develop the hypovirus-containing strain as BCA (Rigling and Prospero, 2017). The demonstration of its cytoplasmic localization, the presence of a dsRNA molecule, and, finally, its viral nature were the three discoveries that led to implementation of an active biocontrol strategy in both Europe and the United States. However, no mycovirus-based PPP is registered for commercial use. In Europe, hypovirulent isolates of *C. parasitica* have been used to treat chestnut blight disease.

ii. Mycoviruses as Biofungicides: The use of mycoviruses as BCAs has relied on the infected hypovirulent fungal strain to transmit the mycovirus to the virulent fungal strain by hyphal anastomosis. However, the formulation of mycoviruses in such ways that they could be used as external fungicides, similar to the commercial products based on other microorganisms such as phages, fungi, and bacteria, would be more promising for market success. In this way, field transmission barriers due to the fungal host vegetative incompatibility could be avoided. In this regard, there are a few examples of extracellular infections of fungi using mycoviral particles. In 2010, Yu and coworkers discovered the first mycovirus with an ssDNA circular genome associated with *Sclerotinia sclerotiorum* [*Sclerotinia sclerotiorum* hypovirulence associated DNA virus 1 (SsHADV-1)]. The treatment of Arabidopsis leaves and *Brassica napus* plants with SsHADV-1 virions as a virus-based fungicide was protective against *S. sclerotiorum* infection; the results were comparable with the application of carbendazim, a chemical fungicide frequently used for fungal control (Yu et al., 2013). Moreover, the isolation of ssDNA-infected strains from the treated leaves indicated that the inhibition of the infection was due to the conversion from virulent to hypovirulent variants by mycoviral infection. Furthermore, in rapeseed field trials, the application of SsHADV-1 particles also prevented secondary infections, reduced both the incidence and severity of Sclerotinia stem rot, and significantly improved seed yield.

The deployment of fungal isolates with virus-mediated hypovirulence had the advantage of limiting the environmental spread of the mycovirus, raising very little concern for off-target effects; nevertheless, such approaches that rely on the horizontal spread of the virus among pathogen populations with delayed efficacy are not suitable for annual crops but remain particularly useful for forest diseases in natural settings (Jeroen Wagemans et al., 2022). Recent new approaches seem to be better adapted to protect annual crops, using mycovirus-infected strains, but have the advantage of not relying on horizontal transfer of the virus for effective biocontrol.

A more recent mycovirus-based biocontrol approach relies on the discovery of the complex relationship of a mycovirus-infected hypovirulent strain with the plants that need protection. Three groundbreaking works have been published in the past year sharing the discovery that a mycovirus can switch the lifestyle of a fungal host from a pathogen to an endophyte and that the presence of the endophytic fungus in the plant hosts has positive effects on the growth of the plant and protection from the virus-free pathogen. Zhang and coauthors started from the observation that SsHADV-1 DNA could be found in rapeseed pods long after treatment with the hypovirulent strain, suggesting that mycovirus-infected S. sclerotiorum could colonize *B. napus*; they showed that mycovirus-infected S. sclerotiorum isolates formed compound appressoria and could grow on the surface and under the leaf epidermis without killing the cells (as normally done by virus-free isogenic isolates). Looking at mycovirus-caused differential expression patterns, the authors showed that the mycovirus suppressed the production of pathogenicity/virulence factors and provided the fungus with the ability to regulate plant defense and hormone function, stimulating plant growth and resistance to pathogen infection, including Botrytis cinerea (Zhang et al., 2020).

Virus-induced gene silencing and synthetic biology: Most mycoviruses that are found through metatranscriptomic data are not causing hypovirulence, because they coevolved with their host, which tamed their pathogenic potential. Nevertheless, there are biotechnological approaches that can make those viruses pathogenic to their fungal host, interfering with the fungal life cycle or specifically interfering with their virulence potential for causing disease on plants (Jeroen Wagemans et al., 2022). Virus-induced gene silencing (VIGS) is an approach in which a fragment of a target host gene is inserted into a viral genome (into a viral infectious clone). From CHV1 infectious clones, an expression vector was derived and the same site of insertion can be used to insert a fragment of a fungal host gene that would be downregulated during infection. Furthermore, the mycoviruses that can be used for biocontrol strategies can include those that result from the amplification and assembly of infectious clones from metasamples analyzed in silico, as recently demonstrated for an ssDNA virus amplified from a plant metagenome. Such infectious clones can be adapted to become virus-induced gene editing (VIGE) tools, increasing the potential for application (Oh et al., 2021).

4.17.2 Virus-based biocontrol tools for plant virus diseases: It is intriguing that plant virus diseases can be prevented by viruses themselves. The mechanism of virus-based biocontrol tools for virus diseases called cross-protection differs from what we have shown for the other organisms in which viruses are chosen because of their intrinsic parasitic properties and their ability to interfere with host homeostasis in an ecological sense (Jeroen Wagemans et al., 2022). To control virus-caused plant diseases with virus-based biocontrol tools, other paradigms better describe the rationale of the approach used, and the most fitting definition, sometimes used in the past, is that of preimmunization with a nonpathogenic (mild) isolate of the virus, as is the case of the live attenuated vaccine against several human viral diseases. In fact, virus-based biocontrol tools are mild strains of a virus that, when preinoculated to a host plant, can protect from subsequent infection of a second, severe isolate of the same virus. Cross-protection was first described more than a century ago with mild strains of tobacco mosaic virus protecting tobacco from strains causing a more severe yellow mosaic symptom. Since then, cross-protection has been used in open-field conditions to protect against important virus diseases caused by citrus tristeza virus (CTV) on citrus species, zucchini yellow mosaic virus on several cucurbit species, and papaya ringspot virus in papaya.

Some of the possible molecular mechanisms rely on reprogramming the adaptive immune response of the plant (antiviral silencing pathway), and therefore the analogy with the preimmunization is somewhat congruent with this approach. An appropriate framework for understanding cross-protection is that of the use of satellite viruses, satellite RNAs, and defective interfering RNAs as BCAs that offer some potential for tempering the helper-caused disease severity (Pyle et al., 2017). Nevertheless, the silencing pathway seems not to be involved in some specific cases of crossprotection because it also occurs in silencing deficient mutant plants. Cross-protection was also recently developed and employed on a wide scale in Europe for pepino mosaic virus using several mild strains) which were also granted distinct approvals from the European Food Safety Agency. A wealth of literature has dealt with the molecular mechanism of cross-protection. Most likely, different mechanisms are involved in different systems, of which some are protein mediated and others are RNA mediated. Different mechanisms of superinfection exclusion at the cellular level seem to be a common theme for the molecular mechanism of cross-protection. Regarding technological advances in developing mild strains, recent work relies on reverse genetics tools to produce mutants in vitro through site-directed mutagenesis without having to rely on natural mutants, a time-consuming and often limiting factor in cross-protection approaches.

Several examples show that viruses are viable as BCAs for limiting the impact phytopathogens have on crops. In any case, care should be taken to select viruses that have the expected properties in terms of applicability and safety (Jeroen Wagemans et al., 2022). As some examples show, off-target or unpredicted outcomes can occur. Targeted research is thus needed to fully understand the biology of the viruses that are applied. Moreover, an understanding of the effect of BCAs on the natural microbiome (soil and plant microbiome) is needed to further pave the regulatory pathway and ensure biosafety. Finally, social acceptance is an important aspect that should be investigated in more detail as a crucial component of getting the products to the agricultural community.

4.18 Novel Biocontrol Approaches and Mode of Action

Microbial biological control agents interact with the plant, the targeted pathogen and the resident microflora. Studies on the interactions with the resident microflora have been hampered in the past because of limitations of available methods. This changed drastically with arrival of Next Generation Sequencing (NGS) methods such as metagenomics and metatranscriptomics allowing to identify the composition and functions of the microbiome (Kohl et al., 2019).Further steps by adding information of metametabolomics and signalomics (Mhlongo et al., 2018) will complete the picture on interactions between introduced MBCAs and resident microbiota. As a result, a holistic in depth understanding on MBCA-microbiota interactions will support better timing, formulation and application of MBCAs and prevent failures. It is expected that three new developments will have significant impact on biological control of plant diseases. First, functional analysis will allow a "prebiotic approach". Application of specific compounds or complex substrates will modulate indigenous microbiota compositions with the aim to enhance microbial suppression of plant pathogens. Such a manipulation of resident microbiota toward disease suppression may be comparable to conservation biological control applied in insect pest control, e.g., via improving nutrient availability for beneficial insect populations by planting flower strips. Simple or complex substrates applied for such a prebiotic approach may not be considered as plant protection products. A second approach will be the selection and application of "helper" strains which have no biocontrol properties on their own but support MBCAs in establishment, survival and antagonistic activity in situ. A third expectation is that core microbiomes will be designed consisting of different strains of biological control species combining various modes of action. The transfer of tailor-made core-microbiomes will become a viable strategy for plant disease management.

The ecological considerations supporting the idea of assembled consortia are sound. However, practical considerations may hamper their introduction. Validation and optimization of in silico-designed consortia under ranges of relevant environmental conditions will be complex and will need substantial resources. In a commercial setting, development of mass production, down streaming and storage procedures separately for each individual consortium member will need substantially more resources and investments compared to production of single strain MBCAs. Registration of assembled consortia as plant protection products will add further difficulties. Regulations in the EU demand the risk assessment of each active ingredient before the product can be registered. In case of assembled consortia, costs will thus increase substantially. In this context, strategies to develop helper strains or to shape the indigenous microbiota may clearly have advantages above the use of assembled core consortia. On the other hand, an adapted legislation for novel disease control systems would benefit society as a whole as well as the environment.

MBCAs are functioning directly on the spot where the targeted organism is harmful, generally combining different modes of action to highly regulated cascades of events (Kohl et al., 2019). Current thinking on how to consider the mode of action during the risk assessment and registration procedure of MBCAs focuses on a single mode of action and potential risks of in vitro produced metabolites, very similar to the risk assessment of synthetic fungicides with a single compound as active ingredient. A rethinking is needed considering that the effectiveness of MBCAs in most cases is based on natural, complex, highly regulated interactions between microbial cells and plants on site but are not the results of a single action of a single metabolite. Toxicological and ecotoxicological risks of such complex processes of interaction can be considered as very low. Moreover, humans and other organisms have been and still are exposed to such processes in evolutionary terms and adverse effects are not known.

Since an antimicrobial action of a single metabolite is not relevant in many cases, the existing EU regulations may require such a rethinking in the registration of MBCAs as long as antimicrobial metabolites are not present in the formulated MBCA at relevant concentrations.

Conclusion and Perspectives: Biocontrol agents hold the potential to replace chemical pesticides. They can be employed as biopesticides, biofertilizers, and stimulators for plant development. The BCA employs various action mechanisms to control the pathogens of a plant which either involve a direct interaction of BCA with the pathogen, such as the production of hyperparasites, lytic enzymes or antibiotics, or the indirect interaction of BCA and pathogen, e.g., nutrient competition and spatial induction and plant resistance (Vanshree et al., 2022). As part of an integrated pest management (IPM) system, biocontrol agents can be created to reduce the environmental impact of chemical fungicides. Further investments are needed in biocontrol agents' research and development. Further research on dosage, formulation, impact on the action of BCA, and the effects of BCA on the native plant microflora of different environmental circumstances should be carried out. The mechanism of action of biocontrol agents is crucial to be fully understood in order to create more effective antagonists. The formulation processes must be improved so that formulations may be formulated with increased shelf-life. The encapsulation of microbial organisms is one of the potential techniques of the effective formulation. However, the most utilized encapsulation technique is still alginate-based beads. Therefore, efforts must be made to develop new techniques and enhance the present formulations and delivery methods. The interaction between the capsule matrix and physical–chemical soil or agroecosystem is also crucial to acquire understanding. Environmental factors play an important role to determine biocontrol agents' effectiveness. The environmental conditions under which the biocontrol agents function efficiently thus need to be studied. Factors leading to the rhizosphere colonization and the influence of BCA on the native land population must also be determined. The genes, gene products, and other signaling molecules that are responsible for the antagonistic activity of microorganisms are also crucial to determine. Effective techniques for mass manufacturing of BCA must also be developed and technological challenges linked to their mass production resolved and formulations improved so that gram-negative bacteria may be employed successfully.

- Microbial biological control agents use a great variety of mechanisms to protect plants from pathogens (Kohl et al., 2019).
- Important modes of action strengthen the resistance of the plant, e.g., induced resistance or priming, or modulate the local growth conditions for pathogen development, e.g., nutrient competition, but do not interfere directly with the pathogen.
- Hyperparasitism and secondary metabolites are directly affecting the targeted pathogen via highly regulated cascades of physiological events but not by a single constitutively produced metabolite.
- Secondary metabolites produced in vitro may have antimicrobial activity at high concentration but low amounts are produced in situ very locally during

interaction and metabolites have short life spans, often with functions such as signaling, very different from antibiosis.

- During the cascades of events a range of different compounds with different modes of action are used to outcompete the pathogen. Such events of signaling and interaction are common wherever microorganisms interact.
- The highly regulated in situ production of ubiquitous mechanisms commonly used in the microbial interplay makes the use of MBCAs a safe and sustainable technology.
- In situ produced compounds such as MAMPs, enzymes or secondary metabolites are not relevant for risk assessments so that detailed toxicology and ecotoxicological studies of these compounds are not relevant, and should not be required.
- The fear of antimicrobial metabolites produced by MBCAs after their release is not based on real risks but fed by the wrong perception on how biocontrol acts if studied under in vitro conditions.
- Better screening assays for finding the next generation of MBCAs are needed to measure the overall effect of the interplay of different modes of action.
- Multi-omics will help to further understand the complex events during microbial interactions in the environment.
- Current EU regulations on registration of MBCAs should allow a science-based differentiation between the majority of compounds involved in modes of action to be considered as safe and not relevant for detailed risk assessment and the limited number of cases relevant for risk assessments where secondary metabolites are present as active ingredients in MBCAs formulations at high concentrations.

Bibliography

Aamir, M., Rai, K.K. and Zehra, A., 2020. *Microbial Endophytes.* Elsevier; Amsterdam, The Netherlands: 2020. Microbial bioformulation-based plant biostimulants: A plausible approach toward next generation of sustainable agriculture; pp. 195–225

Abd El-Mageed, T. A., Rady, M. M., Taha, R. S., Abd El Azeam, S., Simpson, C. R. and Semida, W. M., 2020. Effects of integrated use of residual sulfur-enhanced biochar with effective microorganisms on soil properties, plant growth and shortterm productivity of *Capsicum annuum* under salt stress. *Sci. Hortic.*, **261:**108930

Abd-Elgawad and Askary, 2020. Factors affecting success of biological agents used in controlling the plant-parasitic nematodes. *Egyptian J.Biol. Pest Cont.*, **30**:17

Abd-Elgawad, M.M.M. 2019. Towards optimization of entomopathogenic nematodes for more service in the biological control of insect pests. *Egyptian J Biol. Pest Cont.*, **29**:77

Aboulila, A.A., 2022. Efficiency of plant growth regulators as inducers for improved systemic acquired resistance against stripe rust disease caused by *Puccinia striiformis* f. sp. *tritici* in wheat through up-regulation of PR-1 and PR-4 genes expression. *Physiol. Mol. Plant Pathol.*, **121**:10[illegible]382.

Adetunji, C.O., Kumar, D., Raina, M., Arogundade, O. and Sarin, N.B., 2019. Endophytic Microorganisms as Biological Control Agents for Plant Pathogens: A Panacea for Sustainable Agriculture. 10.1007/978-3-030-26657-8_1

Ali, Q., Ayaz, M., Mu, G., Hussain, A., Yuanyuan, Q., Yu, C., Xu, Y., Manghwar, H., Gu, Q. and Wu, H., 2022. Revealing Plant Growth-Promoting Mechanisms of Bacillus Strains in Elevating Rice Growth and Its Interaction with Salt Stress. *Front. Plant Sci*., **13**: 994902

Anchal Sharma, Diwevidi, V.D., Smita Singh, Khsitiz Kumar Pawar, Mahesh Jerman, Singh, L.B., Satish Singh and Deepak Srivastawa, 2013. Biological Control and its Important in Agriculture. *Intl. J. Biotechnol.& Bioengg. Res*., **4:** 175-180

Anju Sehrawat, Satyavir S. Sindhu and Bernard R. Glick,2022. Hydrogen cyanide production by soil bacteria: Biological control of pests and promotion of plant growth in sustainable agriculture. *Pedosphere*, **32**: 15-38

Anonymous, 2021. Market Analysis and Insights: Global Agriculture Biological Control Agents Market, Agriculture and Animal Feed, Upcoming Report, Jun 2021, Global, 350 Pages

Ayaz, M., Ali, Q., Farzand, A., Khan, A.R., Ling, H. and Gao, X., 2021. Nematicidal Volatiles from *Bacillus atrophaeus* GBSC56 Promote Growth and Stimulate Induced Systemic Resistance in Tomato against *Meloidogyne incognita*. *Int. J. Mol. Sci.,* **22***:* 5049

Ayaz, M., Li, C.H., Ali, Q., Zhao, W., Chi, Y.K., Shafiq, M., Ali, F., Yu, X.Y., Yu, Q., Zhao, J.T., 2023. Bacterial and Fungal Biocontrol Agents for Plant Disease Protection: Journey from Lab to Field, Current Status, Challenges, and Global Perspectives. *Molecules,* **28**:6735

Azevedo, J.L., Araújo, W.L. and Lacava, P.T., 2016. The diversity of citrus endophytic bacteria and their interactions with *Xylella fastidiosa* and host plants. *Genet. Mol. Biol.***,** **39**:476–491

Baker, K. F. and Cook, R. J., 1974. Biological Control of Plant Pathogens, W. H. Freeman and Co, San Francisco, California. 433 pp. (Book, reprinted in 1982, Amer. Phytopathol. Soc., St. Paul, Minnesota).

Baker, K. F., 1987. Evolving concepts of biological control of plant pathogens. *Ann. Rev. Phytopathol.,* **25**:67-85.

Barratt, B. I. P., Moran, V. C., Bigler, F. and van Lenteren, J. C., 2018. The status of biological control and recommendations for improving uptake for the future. *Biol. Control*, **63:** 155–167.

Bhupendra Koul, Manpriya Chopra and Supriya Lamba, 2022. Chapter 3 - Microorganisms as biocontrol agents for sustainable agriculture. *In: Relationship Between Microbes and the Environment for Sustainable Ecosystem Services*. Jastin Samuel, Ajay Kumar, Joginder Singh (Eds.), Volume 1, Elsevier, P 45-68,

Chen, M., Liu, Q., Gao, S., Young, A.E., Jacobsen, S.E. and Tang, Y., 2019. Genome mining and biosynthesis of a polyketide from a biofertilizer fungus that can facilitate reductive iron assimilation in plant. *Proc. Natl. Acad. Sci. USA.,* **116**: 5499–5504

Cook, R. J., and Baker, K. F.1983. The Nature and Practice of Biological Control of Plant Pathogens. Amer. Phytopathol. Soc., St. Paul, Minnesota. 539 pp

Crandall, S.G., Gold, K.M., Jiménez-Gasco, M.d.M., Filgueiras, C.C. and Willett, D.S., 2020. A Multi-Omics Approach to Solving Problems in Plant Disease Ecology. *PLoS ONE*, **15:** e0237975.

De Bach, P., ed. 1964. Biological Control of Insect Pests and Weeds. Reinhold, New York, New York. 844 pp.

Degani, O. and Dor, S., 2021. *Trichoderma* Biological Control to Protect Sensitive Maize Hybrids against Late Wilt Disease in the Field. *J. Fungi,* **7:**, 315

Dutta, T.K., Khan, M.R. and Phani, V., 2019. Plant-parasitic nematode management via biofumigation using brassica and non-brassica plants: current status and future prospects. *Curr. Pl. Biol*., **17**:17–32

Elrys, A. S., Abdo, A. I. E., Abdel-Hamed, E. M. and Desoky, E. S. M., 2020. Integrative application of licorice root extract or lipoic acid with fulvic acid improves wheat production and defenses under salt stress conditions. *Ecotoxicol. Environ. Saf.*, **190**:110144

El-Saadony, M.T., Saad, A.M., Soliman, S.M., Salem, H.M., Ahmed, A.I., Mahmood, M., El-Tahan, A.M., Ebrahim, A.A.M., Abd El-Mageed, T.A., Negm, S.H., Selim, S., Babalghith, A.O., Elrys, A.S., El-Tarabily, K.A. and AbuQamar, S.F., 2022. Plant growth promoting microorganisms as biocontrol agents of plant diseases: Mechanisms, challenges and future perspectives. *Front. Plant Sci.*, **13**:923880

Fan, B., Li, L., Chao, Y., Förstner, K., Jörg Vogel, J., Borriss, R. and Wu, X.Q., 2015. dRNA-seq reveals genomewide TSSs and noncoding RNAs of plant beneficial rhizobacterium *Bacillus amyloliquefaciens* FZB42. *PLoS ONE,* **10**:e0142002

Fan, D., Subramanian, S. and Smith, D. L., 2020. Plant endophytes promote growth and alleviate salt stress in *Arabidopsis thaliana. Sci. Rep.,* **10**:12740.

Fontana, D.C., de Paula, S., Torres, A.G., de Souza, V.H.M., Pascholati, S.F., Schmidt, D. and Dourado Neto, D., 2021. Endophytic Fungi: Biological Control and Induced Resistance to Phytopathogens and Abiotic Stresses. *Pathogens,* **10**:570

Glandorf, D.C.,2019. Re-evaluation of biosafety questions on genetically modified biocontrol bacteria. *Eur. J. Plant Pathol.,* **154**:43–51.

Gouda, S., Kerry, R. G., Das, G., Paramithiotis, S., Shin, H. S., and Patra, J. K., 2018. Revitalization of plant growth promoting rhizobacteria for sustainable development in agriculture. *Microbiol. Res.*, **206:** 131–140.

Harutyunyan, N., Kushugulova, A., Hovhannisyan, N. and Pepoyan, A., 2022. One health probiotics as biocontrol agents: one health tomato probiotics. *Plant,* **11**:1334.

He, D.C., He, M.H., Amalin, D.M., Liu, W., Alvindia, D.G. and Zhan, J., 2021. Biological Control of Plant Diseases: An Evolutionary and Eco-Economic Consideration. *Pathogens*, **10**: 1311

Holtappels, D., Fortuna, K., Lavigne, R. and Wagemans, J., 2021. The future of phage biocontrol in integrated plant protection for sustainable crop production. *Curr. Opin. Biotechnol.,* **68**: 60–71

Ilaria Pertot, Claude Alabouvette, Estefanía Hinarejos Esteve and Soraya Franca, 2021. Mini-paper – The use of microbial biocontrol agents against soil-borne diseases. Eip-Agri Focus Group Soil-Borne Diseases, *eip-Agri.,* European Commission

Irtwange. S. V., 2006. Application of Biological Control Agents in Pre- and Postharvest Operations". Agricultural Engineering International: *the CIGR E journal*. Invited Overview No. 3. Vol. VIII. February, 2006.

Jacob, J., Krishnan, G.V., Thankappan, D. and Amma, D.K.B.N.S., 2020. Endophytic Bacterial Strains Induced Systemic Resistance in Agriculturally Important Crop Plants. In *Microbial Endophytes*; Elsevier: Amsterdam, The Netherlands, pp. 75–105

Jaiswal, D.K., Gawande, S.J., Soumia, P.S. *et al.*, 2022. Biocontrol strategies: an eco-smart tool for integrated pest and diseases management. *BMC Microbiol,* **22:** 324

Jeroen Wagemans, Dominique Holtappels, Eeva Vainio, Mojgan Rabiey, Cristina Marzachì, Salvador Herrero, Mohammadhossein Ravanbakhsh, Christoph C. Tebbe, Mylène Ogliastro, María A. Ayllón and Massimo Turina, 2022. Going Viral: Virus-Based Biological Control Agents for Plant Protection. *Ann. Rev. Phytopathol.,* **60**: 21–42

Karlsson, M., Atanasova, L., Jensen, D. F. and Zeilinger, S., 2017. Necrotrophic mycoparasites and their genomes. Microbiol. *Spectrum*, **5**: FUNK-0016-2016.

Kavitha, P. G., Jonathan, E. L. and Sankari Meena, K., 2013. Induction of defence enzymes in tomato by plant growth promoting rhizobacteria, *Pseudomonas fluorescens* against root-knot nematode, *Meloidogyne incognita. Indian J. Nematol.*, **43**: 94–96

Köhl, J., Kolnaar, R. and Ravensberg, W.J., 2019. Mode of Action of Microbial Biological Control Agents Against Plant Diseases: Relevance Beyond Efficacy. *Front. Plant Sci.*, **10**:845.

Kumar, K.K., Sridhar, J., Murali-Baskaran, R.K., Senthil-Nathan, S., Kaushal, P., Dara, S.K. and Arthurs, S., 2018. Microbial Biopesticides for Insect Pest Management in India: Current Status and Future Prospects. J. Invertebr. *Pathol.*, **165**: 74–81

Lahlali, R., Ezrari, S., Radouane, N., Kenfaoui, J., Esmaeel, Q., El Hamss, H., Belabess, Z. And Barka, E.A., 2022. Biological Control of Plant Pathogens: A Global Perspective. *Microorganisms*, **10**: 596.

Larena, I., Espeso, E.A. and Veloso, J., 2023. Editorial: Impact of Novel Omic Technologies on Biological Control against Plant Pathogens. *Front. Microbiol.*, **14**: 1162422.

Levy, A., Gonzalez, I.S., Mittelviefhaus, M., Clingenpeel, S., Paredes, S.H., Miao, J., Wang, K., Devescovi, G., Stillman, K., Monteiro F., et al., 2018. Genomic features of bacterial adaptation to plants. *Nat. Genet.*, **50**:138–150.

Li, C., Cheng, P., Zheng, L., Li, Y., Chen, Y., Wen, S. and Yu, G., 2021. Comparative Genomics Analysis of Two Banana *Fusarium* Wilt Biocontrol Endophytes *Bacillus subtilis* R31 and TR21 Provides Insights into Their Differences on Phytobeneficial Trait. *Genomics,* **113:** 900–909.

Lyu, D., Backer, R., Subramanian, S. and Smith, D., 2020. Phytomicrobiome coordination signals hold potential for climate change-resilient agriculture. *Front. Plant Sci.*, **11**:634

Marco, S., Loredana, M., Riccardo, V., Raffaella, B., Walter, C. and Luca, N., 2022. Microbe-Assisted Crop Improvement: A Sustainable Weapon to Restore Holobiont Functionality and Resilience. *Hortic. Res.,* **9:** uhac160

Md. Masudur Rahman Khalil, Rosario Alicia Fierro-Coronado, Ofelda Peñuelas-Rubio, Alma Guadalupe Villa-Lerma, Rigoberto Plascencia-Jatomea, Rubén Félix-Gastélum and Ignacio Eduardo Maldonado-Mendoza, 2021. Rhizospheric bacteria as potential biocontrol agents against Fusarium wilt and crown and root rot diseases in tomato. *Saudi J. Biol. Sci.*, **28**: 7460-7471

Mhlongo, M. I., Piater, L. A., Madala, N. E., Labuschagne, N. and Dubery, I. A., 2018. The chemistry of plant–microbe interactions in the rhizosphere and the potential for metabolomics to reveal signaling related to defense priming and induced systemic resistance. *Front. Plant Sci.,* **9:**112

Moh Tariq, Amir Khan, Mohd Asif, Faryad Khan, Taruba Ansari, Mohammad Shariq and Mansoor A. Siddiqui, 2020. Biological control: A sustainable and practical approach for plant disease management. *Acta Agriculturae Scandinavica, Section B - Soil & Plant Sci.,* **70:** 507-524

Muñoz-Arenas, L.C., Fusaro, C., Hernández-Guzmán, M., Dendooven, L., Estrada-Torres, A. and Navarro-Noya, Y.E., 2020. Soil microbial diversity drops with land-use change in a high mountain temperate forest: A metagenomics survey. *Environ. Microbiol. Rep.,* **12**:185–194

Murali, M., Naziya, B., Ansari, M.A., Alomary, M.N., AlYahya, S., Almatroudi, A., Thriveni, M.C., Gowtham, H.G., Singh, S.B., Aiyaz, M., Kalegowda, N., Lakshmidevi, N. and Amruthesh, K.N., 2021. Bioprospecting of Rhizosphere-Resident Fungi: Their Role and Importance in Sustainable Agriculture. *J. Fungi (Basel),* **7:**314.

Nandani Shukla, Erayya Archana Negi Akansha Singh, B.C. Kabadwa, Roopali Sharma and Jatinder Kumar, 2019. Present Status and Future Prospects of Bio-Agents in Agriculture. *Int.J.Curr.Microbiol.App.Sci.*, **8**: 2138-2153

Nian, J., Yu, M., Bradley, C.A. and Zhao, Y., 2021. *Lysobacter enzymogenes* strain C3 suppresses mycelium growth and spore germination of eight soybean fungal and oomycete pathogens and decreases disease incidences. *Biol. Control.*, **152**: 104424.

Pablo Díaz-Rueda, Laura Morales de los Ríos, Luis C Romero and Irene García, 2023. Old poisons, new signaling molecules: the case of hydrogen cyanide. *J. Exp. Bot.*, **74:** 6040–6051

Pirttilä, A.M., Mohammad Parast Tabas, H., Baruah, N and, Koskimäki, J.J., 2021. Biofertilizers and Biocontrol Agents for Agriculture: How to Identify and Develop New Potent Microbial Strains and Traits. *Microorganisms,* **9**:817

Pyle, J.D. and Scholthof, K.B.G., 2017. Biology and pathogenesis of satellite viruses. *In: Viroids and Satellites.* A Hadidi, R Flores, J Randles, P Palukaitis (Eds.), pp. 627–36. Cambridge, MA: Acad. Press

Raaijmakers, J. M. and Mazzola, M., 2012. Diversity and natural functions of antibiotics produced by beneficial and plant pathogenic bacteria. *Ann. Rev. Phytopathol.*, **50:** 403–424

Ramette, A., Frapolli, M., Défago. G. and Moënne-Loccoz, Y., 2003. Phylogeny of HCN synthase-encoding hcnBC genes in biocontrol fluorescent pseudomonads and its relationship with host plant species and HCN synthesis ability. *Mol. Plant Microbe Interact.*, **16**:525-35

Rigling, D. And Prospero, S., 2017.*Cryphonectria parasitica*, the causal agent of chestnut blight: Invasion history, population biology and disease control. *Mol. Plant Pathol.*, **19**:7–20

Rijavec, T. and Lapanje, A., 2016. Hydrogen Cyanide in the Rhizosphere: Not Suppressing Plant Pathogens, but Rather Regulating Availability of Phosphate. *Front. Microbiol.,* **7**:1785.

Risoli, S., Cotrozzi, L., Sarrocco, S., Nuzzaci, M., Pellegrini, E. and Vitti, A., 2022. *Trichoderma*-Induced Resistance to *Botrytis cinerea* in *Solanum* Species: A Meta-Analysis. *Plants,* **11:** 180.

Sharma, P., Bora, L., Puzari, K., Baruah, A., Baruah, R., Talukdar, K., Kataky, L. and Phukan A., 2017. Review on bacterial blight of rice caused by *Xanthomonas oryzae* pv. *oryzae*: Different management approaches and role of *Pseudomonas fluorescens* as a potential biocontrol agent. *Int. J. Curr. Microibol. Appl. Sci.*, **6**:982–1005

Shelake, R.M., Pramanik, D. and Kim J.Y., 2019. Exploration of Plant-Microbe Interactions for Sustainable Agriculture in CRISPR Era. *Microorganisms,* **7:** 269

Shoham, J., 2020. The rise of biological products in the crop protection and plant nutrition markets. *Outlook Pest Manag.*, **31:**129–31.

Spadaro, D. and Droby, S., 2016. Development of biocontrol products for postharvest diseases of fruit: the importance of elucidating the mechanisms of action of yeast antagonists. *Trends Food Sci. Technol.*, **47:** 39–49.

Stenberg, J.A., Sundh, I., Becher, P.G., Björkman, C., Dubey, M., Egan, P.A., Friberg, H., Gil, J.F., Jensen, D.F., Jonsson, M., et al., 2021. When Is It Biological Control? A Framework of Definitions, Mechanisms, and Classifications. *J. Pest Sci.,* **94:** 665–676.

Tchameni, S.N., Cotârleț, M., Ghinea, I.O. *et al.,* 2020. Involvement of lytic enzymes and secondary metabolites produced by *Trichoderma* spp. in the biological control of *Pythium myriotylum*. *Int. Microbiol.,* **23:** 179–188

Timper, P.,2014. Conserving and enhancing biological control of nematodes. *J. Nematol.,* **46**: 75–89

Umer, M., Mubeen, M., Iftikhar, Y., Shad, M.A., Usman, H.M., Sohail, M.A., Atiq, M.N., Abbas, A. and Ateeq, M., 2021. Role of Rhizobacteria on Plants Growth and Biological Control of Plant Diseases: A Review. *Plant Prot.,* **5:** 59–73

Van Der Voort, M., Meijer, H., Schmidt, Y., Watrous, J., Dekkers, E., Mendes, R., Dorrestein, P.C., Gross, H. and Raaijmakers, J.M., 2015. Genome mining and metabolic profiling of the rhizosphere bacterium *Pseudomonas* sp. SH-C52 for antimicrobial compounds. *Front. Microbiol.,* **6**: 693

Vanshree, C.R., Muskan Singhal, Mansi Sexena, Mahipal Singh Sankhla, Kapil Parihar, Ekta B. Jadhav, Kumud Kant Awasthi and Chandra Shekhar Yadav. Chapter 11 - Microbes as biocontrol agent: From crop protection till food security. Jastin Samuel, Ajay Kumar, Joginder Singh (Eds.). Relationship Between Microbes and the Environment for Sustainable Ecosystem Services, Volume 1, Elsevier, P 215-237

Vosátka, M., Látr, A., Gianinazzi, S. and Albrechtová, J., 2012. Development of arbuscular mycorrhizal biotechnology and industry: Cur-rent achievements and bottlenecks. *Symbiosis,* **58**: 29–37.

Wang, X., Zhang, X., Liu, L., Xiang, M., Wang, W., Sun, X., Che, Y., Guo, L., Liu, G. and Guo, L., et al., 2015. Genomic and Transcriptomic Analysis of the Endophytic Fungus Pestalotiopsis Fici Reveals Its Lifestyle and High Potential for Synthesis of Natural Products. *BMC Genom.,* **16**:28.

Wei, Z., Gu, Y., Friman, V.P., Kowalchuk, G.A., Xu, Y., Shen, Q. and Jousset, A., 2019. Initial soil microbiome composition and functioning predetermine future plant health. *Sci. Adv.,* **5**:eaaw0759.

Yang, G., Li, P., Meng, L., Xv, K., Dong, F., Qiu, Y., He, L. and Lin, L., 2018. Diversity and Communities of Culturable Endophytic Fungi from Different Tree Peonies (Geoherbs and Non-Geoherbs), and Their Biosynthetic Potential Analysis. *Braz. J. Microbiol.,* **49**:47–58.

Yu, X., Li, B., Fu, Y., Xie, J., Cheng J, et al., 2013. Extracellular transmission of a DNA mycovirus and its use as a natural fungicide. *PNAS*, **110**:1452–57

Zhang, H., Xie, J., Fu, Y., Cheng, J., Qu, Z., et al., 2020. A 2-kb mycovirus converts a pathogenic fungus into a beneficial endophyte for brassica protection and yield enhancement. *Mol. Plant*, **13**:1420–33

Zheng, L., Zhao, J., Liang, X., Zhan, G., Jiang, S. and Kang, Z., 2017. Identification of a novel *Alternaria alternata* strain able to hyperparasitize *Puccinia striiformis* f. sp. *tritici*, the causal agent of wheat stripe rust. *Front. Microbiol.,* **8**:71

Zubair, M., Farzand, A., Mumtaz, F., Khan, A.R., Sheikh, T.M.M., Haider, M.S., Yu, C., Wang, Y., Ayaz, M., Gu, Q., et al., 2021. Novel Genetic Dysregulations and Oxidative Damage in *Fusarium graminearum* Induced by Plant Defense Eliciting Psychrophilic *Bacillus atrophaeus* Ts1. *Int. J. Mol. Sci.,* **22:** 12094.

5

Inoculum Potential and Density in Relation to Host and Soil Variables

5.1 Important Terms

i. Inoculum: The infective pathogen propagules coming in contact with the host constitutes the inoculum. The inoculum in the liquid state is a significant factor in anaerobic digestion in solid phase because it is very useful to convey the micro-organisms, the water of recirculation. Depending on the particular inoculant formulation, the inocula can be used for seed coating, for dipping seedlings, direct application to the furrow, or as foliar application. An appropriate inoculum consists of copious amounts of active microbes having the potential to convert the organic matter to biogas and the S/I ratio is essential for the rate of degradation of a recalcitrant substrate since it governs not only the biogas composition, digestion period and reactor. Spread of inoculum may occur by wind, water, infected plant material, insects, animals, birds, and people. Plant diseases may have more than one source of inoculum. The infection cycle of some economically important soil-borne plant pathogens involves a combination of primary infection from particulate inoculum residing in the soil and secondary infection as disease is spread from infected to susceptible hosts.

The infectivity of inoculum can be quantified by the pathozone profile which measures changes in the probability of infection when inoculum occurs at different distances from the host. For instance, the germinability of inoculum, the growth of the mycelial colony and the infectivity of mycelium at the surface of the host combine to dictate the shape of the pathozone profile for *Rhizoctonia solani* on radish. The ultimate shape of the pathozone depended on inoculum type and was particularly sensitive to changes in the density and distribution of the mycelium in the fungal colony. Mycelium from an infected radish plant grew much further and at a higher density than that from particulate inoculum (mycelial discs). This results in pathozone profiles that differed in shape. For particulate inoculum the profile rise and fall with distance whilst for an infected plant the decay is sigmoidal.

In fungi, the inoculum may include the spores, sclerotia (compact mass of mycelium) or fragments of the mycelium. In bacteria, phytoplasmas, protozoa, viruses and viroids, the inoculum is always their whole individuals. In nematodes, it may be adult nematodes, juveniles or eggs. In parasitic flowering plants, the inoculum may be plant fragments or seeds.

ii. Primary inoculum: The inoculum that survives dormant in the winter or summer and brings about original infections in the spring and autumn is called primary inoculum, and the infections it causes are called primary infections. A range of initial sources of inoculum is reputedly held responsible for primary and early outbreaks of potato late blight. Infected tubers (cull piles and volunteer potatoes), latently infected seed tubers and oospores are distinguished as primary sources of inoculum.

iii. Secondary inoculum: Primary inoculum starts the diseases. Secondary inoculum spreads the disease. Ex: Overwintering or oversummering pathogen or its spores that cause primary infection. In *Puccinia,* the urediospore is the primary inoculum.

The inoculum produced from primary infections is called secondary inoculum and it, in turn causes secondary infections. Generally, the amount of inoculum and prevalence of favourable environmental conditions determine the success of infection.

iv. Inoculation: It is the initial contact of the pathogen with the site of the plant where infection is possible. In simple terms, inoculation in microbiology is the process of introducing microbes into a culture media so that it reproduces there. Commonly, it is used in the introduction of vaccines, serum or any antigenic substance in the body so as to boost immunity against a particular disease. Inoculation is the study of introducing microorganisms into environments where they will grow and reproduce.

v. Inoculum Potential: It is the inoculum needed for successful infection. It is a function of inoculum density and their capacity. It is defined as the resultant of the action of environment, the vigour of pathogen to establish an infection, susceptibility of the host and amount of inoculum present or it is defined as the energy of growth of a parasite available for infection of a host at the surface of the host organ to be infected. It is the resultant of action of the environment, the vigour of the pathogen to establish an infection, the susceptibility of the host and amount of the inoculum present. The inoculum, which survives whether on perennial plants, plant debris or soil, or on the propagative planting material is carried to the host plants mostly by wind, water, insects or man. Only a tiny fraction of the potential inoculum produced actually lands on the susceptible host plants. The bulk of the produced inoculum lands on the things that cannot become infected. Some types of inoculum, in the soil, e.g., zoospores and nematodes may be attracted to the host plant by chemical substances like sugars and amino acids diffusing out of the plant roots. This process is known as chemotaxis. Vector transmitted pathogens are usually carried to their host plants with an extremely high efficiency. THE importance of inoculum potential of a pathogen in overcoming host resistance and causing infection is well recognized. This is particularly so with root-infecting pathogens which have to overcome not only host resistance but also competition from soil saprophytes. The importance to disease incidence of serious competition between the pathogen and saprophytes associated in general soil habitat or in the rhizosphere seems to be undetermined. A fungus like *Fusarium oxy-sporum* f. *vasinfectum* having a high degree of competitive saprophytic ability, can be expected to build up sufficient inoculum in soil by colonizing saprophytic substrates.

vi. Inoculum density: The density of inoculum refers to the number of cells or organisms that are introduced into a system for a specific purpose, such as cell expansion, phytoremediation, or disease resistance studies. It is a measure of the number of propagules of a pathogenic organism per unit area or volume. In the context of human-induced pluripotent stem cells (hiPSCs), the inoculum density significantly affects cell expansion, differentiation, and production of hiPSCs. In the study on the combined use of plants and bacteria for soil remediation, the inoculum density of an alkane-degrading bacterium influenced its colonization and hydrocarbon degradation activity. In the evaluation of soybean resistance to nematodes, increasing the initial density of *Heterodera glycines* inoculum resulted in an increase in nematode population. For clubroot in canola, the severity of disease reaction and gall formation varied depending on the inoculum density of the pathogen. In a study on Jurkat cell cultures, the growth rate was found to be dependent on the initial inoculum density. The inoculum density (ID) in a sample is estimated as number of microsclerotia per gram of dry soil, and it was recorded for each sample area studied.

Inoculum of a pathogen is its biomass available for infection. Inoculum density (ID), usually expressed per unit of volume or (dry) weight of soil, is the common measure to evaluate the effect of soil disinfestation. Interpretation of ID of a given pathogen depends on the sampling and detection procedures. For example, soil drying and sieving selects for relatively large surviving propagules, and ID only reflects the density of these surviving structures (Aad Termorshuizen and Jeger, 2014). Factors affecting the reliability of the prediction of future damage based on ID include: no more than one type of surviving propagule; absence of a saprophytic phase; strong sensitivity to fungistasis; high survival capacity; and availability of validated sampling and detection methods. The efficacy of soil disinfestation is often determined rapidly after finishing the disinfestation treatment. If that is the case, certain detection methods may lead to false positives. The perspective of a given soil disinfestation method not only depends on the relationship between ID and disease incidence/severity, but also on the initial ID prior to treatment. For a given soil disinfestation method there is, theoretically, a threshold initial ID, above which the method leads to a final ID that is higher than the damage threshold; if no other methods are available, a switch to resistant crops may be the only other control alternative. In soil disinfestation experiments, scientists can choose between a natural level of infestation or introduce inoculum. In the first case, spatial variation of inoculum has to be dealt with, while in the second case, care has to be taken that the introduced inoculum must have the same properties such as persistence and germinability as the inoculum occurring naturally.

5.2 Soil-borne Inoculum

Soil-borne diseases are huge problem for agriculture production. To the casual observer, the most immediate impression of a soil relates to its physical properties; colour, texture, horizons, compactness or strength, moisture status and perhaps temperature. More detailed inspection would reveal a diverse array of physical and chemical properties, many of which are inter-related. However, soil is more than just an inert medium utilised by plants as a stable base from which to obtain water and nutrients (Nehl David et al., 2003). Soil is teeming with microorganisms which interact with abiotic components, with plants and with each other. This ecosystem is primarily driven by carbon compounds derived directly, or indirectly, from plants. Depending on the environmental conditions, substrate availability and its competitive abilities, a soil-borne pathogen will be either dormant, parasitic or saprophytic. In each of these states a pathogen will encounter different obstacles to survival and dispersal.

A soil borne pathogen is a disease-causing agent which lives both in soil and in a plant host, and which will tend to infect undiseased plants which are grown in that soil. Soil is a reservoir for many plant pathogens and plants are under regular attack by these soil borne organisms. If inoculum levels are high enough and environmental conditions become favorable for infection, susceptible plants will develop disease. Soil borne pathogens are readily spread if infested soil or contaminated water moves into other fields or planting areas. Levels of soil borne pathogens, including bacteria, fungi, nematodes and some viruses can be reduced in the soil by appropriate treatments.

During the period between hosts, soil-borne pathogens are exposed to fluctuations in the soil environment. When the availability of host material results in continuous infection chains, then the required survival time is reduced. When the lack of substrate or host tissue leads to discontinuous infection chains then the probability of exposure to adverse biological and nonbiological conditions is increased. The greatest priority for a pathogen in dry soil, where most microbial activity ceases, is to survive desiccation and declining energy reserves due to cell maintenance and respiration. In contrast, a pathogen in a moist soil must contend with reduced oxygen availability, bacterial, fungal and viral hyperparasites, grazing by protozoans, nematodes or arthropods and volatile or soluble todc substances. Many plant pathogens of economic and ecological importance, including, fungi, bacteria, nematodes and viruses spend part, or all, of their life cycle in soil. Survival and dispersal of these pathogens requires different strategies to those of airborne pathogens because the soil environment presents physical, chemical and biological impediments. The availability of host tissue to pathogens depends upon its location in time and space.

Plant pathogens may also be spread from plant to plant via insect vectors or by soil animals such as nematodes. By far the most damage is caused by fungal pathogens which are responsible for a wilts, root rots, club rot and blight. The life cycle for many fungal plant pathogens can be split between above ground on the leaves and shoots (phyllosphere) and below ground in the soil. Many live as saphrophytes, not causing disease, for part of their life cycle whilst others are totally obligate pathogens. Fungal pathogens can be differentiated on the basis of their action on the host; they can either destroy the hosts' tissue (necrotrophs), such as damping-off caused by *Pythium*, live as parasites deriving some nutrient/energy from the host (e.g. downy mildew) for possibly long periods or, display properties of both (hemibiotrophs) such as apple scab (Ingram & Robertson, 1999). Toxins may be produced by the pathogen as well as chemicals which regulate the hosts' metabolism. Bacteria often take advantage of damage such as wounds or natural openings such as stomata to gain entry into the hosts' tissues. Some diseases cause swellings as a result of their infection to bring about distinct anatomical changes to the plant for example *Agrobacterium radiobacter* var. *tumefaciens*, a bacterial infection of sugar beet causes characteristic swellings to the root tissue brought on by the transfer of DNA from the invading bacterium to the host. For some plant diseases, soil can be considered a reservoir for the disease organism with the organisms residing in the soil until the right opportunity arises. However, depending on their life cycle and biology, the time that they can survive in the soil without a suitable host may be limited from a few days to many years.

Soil conditions play an important part in plant disease affecting the survival of the pathogen, movement through the soil to a potential host or interaction with an antagonist. The amount of water can affect some disease organism more than others. Above ground, very humid conditions tend to promote infection promote downy mildew. Below ground, high levels of water due to poor drainage promote anaerobic conditions which deplete the amount of oxygen available thereby reducing root health leading to greater susceptibility to infection and also to greater movement of pathogen.

5.3 The Biological Soil Environment and Inoculum Survival

The biological component of the soil includes an extensive flora and fauna ranging in size from bacteria to earthworms. It has been estimated that there are up to 10^9 micro-organisms representing up to 10^4 species in a gram of soil. The types of relationship that exist between the various components of this dynamic microflora vary from synergism and mutualism through no interaction to inhibition, antagonism, competition and parasitism. The biology and consequently Bulk soil differs greatly from that influenced by roots). Rhizosphere the chemistry of the soil, changes near roots (Nehl David et al. 2003). in the rhizosphere (zone of soil adjacent to and pH may differ from the surrounding soil by up to 2 pH units due to root exudation, which occurs by both passive leakage and active release. Root exudates support a dense and highly competitive community of micro-organisms, usually within 1 to 3 mm of the root, which is very different to that in non-rhizosphere soil. While root exudates play an important role in attracting root pathogens the microbial flora of the rhizosphere is potentially antagonistic to them. All root-infecting pathogens must pass through the rhizosphere, where microbial competition is intense. Although such competition is difficult to quantify, the carbon and nitrogen supplied by root exudates is important to successful infection of the host by many pathogens.

Soil micro-organisms increase their competitive saprophytic ability by producing substances which antagonise others or by sequestering substrates. Toxins include volatile and non-volatile by-products, lytic enzymes and synthesised antibiotics. Rhizosphere pseudomonads produce phenazine antibiotics and siderophores which increase their competitiveness and are involved in the suppression of take-all in wheat. Certain siderophores can also act as powerful antibiotics, thus increasing their function in competition. Most, if not all, soil-borne pathogens are subject to predation or parasitism by other organisms. Bacteria are infected by bacteriophages and consumed by protozoans and nematodes. Fungi are lysed by bacteria, grazed by nematodes and arthropods and colonised by hyperparasitic fungi. Nematodes are infected by bacteria, trapped and parasitised by fungi and consumed by predacious nematodes. Life in the soil is tough.

Survival of soil-borne inoculum: Survival effectively represents 'dispersal in time'. Inoculum of a pathogen must be able to survive the period between inoculum production on one host and either re-infection of the same host or infection of another susceptible host. This inter host or inter-crop period may vary from a few days to a few months or several years (Nehl David et al., 2003). The soil environment produces several obstacles to successful survival that must be overcome by soil-borne inoculum in the absence of susceptible hosts. The pathogen or its inoculum must be able to either avoid, compete with, or withstand antagonism or parasitism from other soil micro-organisms. The inoculum must also be able to tolerate extremes and fluctuations of temperature and moisture in the soil. The initial step in an infection chain or illness cycle is the means of survival. Primary infection and primary inoculum are terms used to describe the initial infection that results from the sources of pathogen survival (infected host as a reservoir of inoculum, saprophytic survival outside the host, dormant spores, and other structures in or on the host or outside the host) in the crop. After the disease has begun to affect the crop, the spores or other pathogen structures serve as sources of secondary inoculum and secondary infection, which spreads the disease across the field (Praveen Kumar Singh et al., 2023).

Sources of survival of pathogens

- The infected host serves as a source of inoculum or must survive in close proximity to living plants.
- Survival without the host as saprophytes.
- Surviving by using specialised resting structures either inside, outside, or on the host.
- Surviving in conjunction with nematodes, fungus, and insects

i. Infected host as reservoir of inoculum: The infected host is divided into the following categories: Seed: During the process of development and maturity in fruit or pod, seeds may become externally or internally infected by plant pathogens. The majority of seedborne diseases persist as long as the seed is still viable. For example: (i) The pathogen *Ustilago nuda tritici* infects the early seed of wheat by entering the stigma and style, where it then survives as mycelium. (ii) It has been demonstrated that *Pseudomonas syringae* can live for 20 years in dried tomato seed.

ii.**Collateral hosts:** Collateral hosts are those that are sensitive to the plant diseases of crop plants and offer suitable conditions for their growth and reproduction during the off-season. Alternate hosts are those that are wild hosts of the same families. During the noncropping season, weeds that persist and live allow the disease to continue to develop and multiply. In this way, the weed hosts facilitate the transition between two crop seasons. For instance, *Pyricularia grisea* (Teleomorph: *Magnaporthe grisea*), the fungus that causes rice blast disease, can persist during the off-season of the rice crop by infecting grass weeds like *Brachiaria mutica, Dinebra retroflexa, Leersia hexandra*, and *Panicum repens*. The conidia (inoculum) released by the weed host and spread by wind as soon as a new rice crop is planted infect the new rice crop (Praveen Kumar Singh et al., 2023).

iii. Alternative hosts or wild hosts from different families: These hosts play a less significant role than collateral hosts. However when a pathogen has a very broad host range (like *Sclerotium rolfsii*, *Rhizoctonia solani, Fusarium moniliforme*, etc.) and is adaptable to a wide range of environmental conditions, the disease's alternate hosts become a crucial part of its survival. For the heteroecious rust pathogens to complete their life cycle, these alternative hosts are crucial. For instance, the barberry bush *(Berberis vulgaris*), the alternate host of *Puccinia graminis tritici* (the pathogen that causes black or stem rust on wheat), coexists side by side with the cultivated host in temperate climates. This wild host, which belongs to a separate family, is crucial to the fungus' survival in such conditions.

iv. Self-sown crops: Early-sown, self-sown, and voluntary crops are all sources of various plant diseases. As an illustration, self-sown rice plants contain the pathogen (Rice tungro virus) and the vector (*Nephottetix virescens*).

Saprophytic survival outside the host: Several plant diseases may exist without growing vulnerable plants thanks to their capacity for saprophytic life. Typically, saprophytic life exists in or on the soil. There is a difference between soil invaders and soil dwellers. The former includes the fundamental fungal flora of the soil, whereas the latter are transient exotics. Fungi can exist without the cultivated host plant as saprophytes and fall into one of three categories for study.

- **Soil residents**: Organisms that live permanently in the soil as saprophytes in the absence of the host plant are referred to as soil residents. For instance, *Pythium, Rhizoctonia*, and *Sclerotium* species
- **Root occupants:** These highly specialised parasites live in soils alongside their hosts in close proximity. As long as the host tissue in which they are residing as parasites has not entirely disintegrated, the active saprophytic phase continues. Examples include cotton root rot, *Fusarium* species, and *Verticillium* (fungi that cause vascular wilt) (*Phymatotrichum omnivorum).*
- **Rhizosphere colonisers:** They are organisms that inhabit the dead substrates near the roots and remain there for an extended period of time. They are more resilient to soil antagonists. Example: *Cladosporium fulvum*, a tomato leaf mould.

v. Survival as dormant spores or specialized resting structures: Plant viruses do not have a resting phase and spread continuously along an infection chain. The phytopathogenic bacteria do not also create dormant spores or other similar structures. They continue to exist in either their active saprophytic stage on the remains of dead plants or their active parasitic stage in the current host (Praveen Kumar Singh et al., 2023). Many soil-borne plant pathogens produce resistant resting structures such as the large sclerotia formed by *Sclerotinia sclerotiorum* and *Sclerotium rolfsii,* microsclerotia formed by *V. dahliae*, chlamydospores produced by *Fusarium oxysporum* and *T. basicola*, oospores produced by *Phytophthora cinnamomi* and cysts produced by nematodes such as *Globodera* spp.and *Heterodera* spp (Nehl David et al., 2003). These resistant structures allow the pathogen to survive in the soil from season to season or sometimes over several seasons. If they are embedded in crop residues then the life of these resistant structures may be extended even further. While thickened walls serve as a physical barrier to desiccation and attack by antagonists, most resting structures also have chemical defences. The insoluble phenolic compounds known as melanins have a role in the defence systems of most organisms, including fungi. Melanins and their precursors can exhi.bit antibiotic activity but the major role for melanins is protection against the hydrolytic enzymes of antagonistic organisms. Hyaline spores or hyphae in soil are attacked and lysed quickly while melanised ones may survive for years. Consequently, most soil-borne fungi produce melanins. The rind of fungal sclerotia is pigmented and in most cases contains melanins. Plant pathogenic bacteria do not produce spores but many are tolerant of desiccation and survive for long periods under dry conditions. Survival of soil borne bacteria is enhanced by exudates (polysaccharides) which form a protective coating around micro-colonies.

vi. In conjunction with the following

Nematodes: They can exist as active parasites that feed on a living host, as well as dormant forms like eggs, cysts, and galls that grow inside the tissues of the host. These structures could be found in seed lots or in the soil.

Phanerogamic parasites: They use seeds to stay latent for a long time. For instance, *Orobanchae* seeds can last longer than 7 years in soil.

Fungi are the only organisms among plant pathogens that create spores, which are similar to nematode eggs, and other resting structures for their inactive survival. The following categories can be applied to these dormant structures of survival.

1. Soil borne fungi

a. Dormant spores, such as conidia (*Taphrina deformans* causes peach leaf curl), chlamydospores (*Fusarium* species causes wilt), oospores (Downy mildew fungi), and perithecia (*Venturia inaequalis* causes apple scab), etc.
b. Other dormant components, such as thickened hypha, sclerotia (Cottony Rot Fungus, *Sclerotinia sclerotiorum*), microsclerotia (*Verticillium*), rhizomorphs (*Armillaria mellea*), etc

The following elements affect pathogen survivability in soil.

Physical components (high temperature, irradiation, desiccation and anaerobiosis),

Chemical elements (antibiotics, compounds produced by competing bacteria that are toxic)

Biotic elements (parasitism, predation by microflora and microfauna)

2. Fungi from seeds

a. Externally seed-borne: Dormant spores on seed coat; examples include covered grains of barley, jowar grains, wheat bunts, etc.
b. Seed-borne mycelium that is dormant and found inside the embryo or beneath the seed coat. as in loose smut of wheat (*Ustilago nuda tritici*).
c. Moisture and temperature have an impact on the pathogen's ability to survive on or inside the seed.

vii. Structures of dormant fungi on active or dormant hosts - Examples: The fungal mycelium may be present in a dormant form in the affected twigs in the case of downy mildew of grapevine, powdery mildew of grapevine, apple, etc., or its oospores or perithecia may be entrenched in the tissues of the affected organs. In addition to using eggs, cysts, and larvae of plant parasitic nematodes as over-seasoning structures, parasitic phanerogams can also persist as seeds (Praveen Kumar Singh et al., 2023).

viii. Surviving in conjunction with nematodes, fungus, and insects: Several significant plant diseases could live and overwinter inside an insect's body. The corn wilt pathogen, *Xanthomonas stewartii,* is carried inside the body of the corn flea beetle, *Cheatocnema pulicaria*, which aids in overwintering. Nematodes or fungi found in the soil help plant viruses including wheat mosaic, tobacco necrosis, tobacco rattle, and tobacco ring spot viruses live in between crop seasons. The nematode *Xiphinema americanum* is connected to tobacco rings pot. The viruses are carried internally by the fungi *Polymyxa graminis* (Wheat soil borne mosaic and Barley yellow mosaic) and *Spongospora subterranea* (Potato mop top virus), which also spread the viruses through the dormant spore (Praveen Kumar Singh et al., 2023).

5.4 Impact of Inoculum Density of *Fusarium oxysporum* f. sp. *zingiberi* on Yield

Fusarium yellows, caused by the ascomycete fungus *Fusarium oxysporum* f. sp. *zingiberi* (*Foz*), is a devastating soil-borne disease of ginger. It has curtailed ginger production in Australia and around the world, leading to significant economic losses. An integrated approach is required to manage soil-borne diseases such as those caused by *Foz*. However, little is known about the influence of *Foz* inoculum on disease severity (Matthews et al., 2023). This study aimed to establish a minimum threshold level of spores per gram of soil required for plant infection and to develop and evaluate a pot inoculation method for screening large numbers of plants in a controlled environment. To achieve this, the dominant Australian ginger cultivar Canton was inoculated with 10^1, 10^3, 10^5, 10^6 and 10^7 microconidia g^{-1} soil. The inoculum density was positively associated with leaf and stem yellows, and rhizome discolouration, and negatively associated with root length and rhizome weight. The lowest threshold required for plant infection was 10^1 microconidia g^{-1} soil, which may provide an important basis for outbreaks of *Foz* in the field. This finding adds significantly to our knowledge of the impact of soil health on ginger production, thereby contributing to the integrated management of *Foz*. When used at a high dose, this method can facilitate reliable and accurate screening of *Foz*-susceptible ginger genotypes in a controlled environment.

Once the soil is infected, *Foz* is thought to survive as chlamydopores in the absence of a susceptible ginger host and hence can persist for years in the form of these resting spores in the soil (Soesanto et al., 2022). The inoculum concentrations in this study only refer to microconidia. It has not been established if macroconidia or chlamydospores would produce similar results. Chlamydospores, as long-term survival structures, may be the primary source of infection after an extended absence of ginger and it would be particularly beneficial to better understand the role of these other spores in infection cycles.

TEF-1α sequencing showed that UQ6848 (NCBI GenBank: OR425153), UQ6851 (OR425154), UQ6852 (OR425155) and UQ6853 (OR425156) shared 100% nucleotide identity with each other, as well as with three other known *Foz* isolates (*Foz* Goomboorian, *Foz* Eumundi, *Foz* BRIP44986). Their sequences differ from that of *Foz* BRIP39299 by three SNPs and a single base INDEL. The phylogenetic tree reconstructed using Bayesian inference produced a topology that separated the four species complexes of *Fusarium*. This phylogeny was presented, and its topology described in a previous study (Czislowski et al., 2021). Within the *Fusarium oxysporum* species complex (FOSC), all four isolates used in this study are clustered together in a subclade with known *Foz* isolates, including *Foz* Goomboorian, *Foz* Eumundi, *Foz* BRIP44986 and *Foz* BRIP39299. The sequences and the phylogenetic positions of these four isolates within FOSC confirmed their *Foz* identity (Matthews et al., 2023). Inoculum concentration had a discernible effect on the severity of Fusarium yellows in ginger cultivar Canton. *Foz*-inoculated plants could be easily differentiated from uninoculated controls by the development of yellowing symptoms typical of *Foz* infection. The degree of senescing leaves and tillers was positively associated with the inoculum dosage, with sporadic yellowing of stems and leaves observed at the low-to-medium inoculum concentration (10^1, 10^3, 10^5 microconidia g^{-1} soil) in comparison to the uninoculated controls which showed little-to-no yellows. *Foz* susceptibility was exacerbated at the high inoculum levels (10^6, 10^7 microconidia g^{-1} soil), where tiller abortions due to Fusarium yellows occurred at a relatively high rate and the whole root systems appeared discoloured

5.5 Influence of Soil-borne Inoculum of *Plasmodiophora brassicae* Disease Severity

Club root (*Plasmodiophora brassicae*) is the most serious soil-borne disease of brassica oilseed and vegetable crops world-wide, causing appreciable yield losses. Managing *P. brassicae* is one of the major challenges in rapeseed production and active measures must be taken, as the resting spores persist in the soil for up to 17 years, with an estimated half-life of 3.6 years (Wallenhammar et al., 2021). Spread of infected soil by machinery, wind and water has been demonstrated. The classic disease symptoms consist of an enlarged root that contains tens of millions of spores per gram of root. The resting spores are robust, with a well-developed mechanism involving various spore walls that provide protection against degradation by soil microorganisms. At high soil moisture and favorable temperature, the resting spores germinate, and zoospores are produced. These infect the root hairs and an intracellular primary phase takes place, followed by release of secondary zoospores. The second phase of infection takes place in rhizodermal cells of the root and the pathogen subsequently induces local hypertrophy caused by mis-regulated plant-derived hormonal pathways. The main part of the life cycle takes place well protected inside the root.

Use of resistant cultivars is considered the most effective tool in managing club root. Three club root-resistant commercial winter oilseed rape (OSR) cultivars and a susceptible 'Cultivar mix' were evaluated for disease severity index (DSI) and yield performance in field soils, selected for varying abundance of natural inoculum of *Plasmodiophora brassicae* (Wallenhammar et al., 2021). Seven field trials were carried out during 2017–2019 in winter OSR crops, and comparative bioassays were performed in a growth chamber. Substantial variation in clubroot infection between years was observed in the field trials. For Cultivar mix, a negative correlation ($y = -252.3\ln(x) + 58{,}897.6$) was found between inoculum density and seed yield in five trials, whereas no correlation was found for the resistant cultivars. In bioassays, Cultivar mix exhibited a significantly high correlation between DSI_b and number of gene copies g^{-1} soil ($R^2 = 0.72$). For resistant cvs., Mentor and Alister, correlation was $R^2 = 0.45$ and 0.58, respectively, indicating that resistance was under pressure. In field trials, DSI_f of the resistant cultivars was lower (<27). The recommendation is thus to use club root-resistant cultivars of OSR as part of Integrated Pest Management in situations where abundance of *P. brassicae* DNA exceeds 1300 gene copies g^{-1} soil.

There was also substantial variation in club root infection between years. There was a clear correlation between amount of soil inoculum (gene copies g^{-1} soil) as indicated by real-time qPCR and DSI_f for the susceptible Cultivar mix. DSI_f rapidly increased with increasing pathogen abundance above approximately 4000 gene copies g^{-1} soil (corresponding to 9000 spores g^{-1} soil). There was no such correlation for the resistant cultivars ($R^2 = 0.0192$), even though these cultivars were infected at a higher rate, particularly at inoculum densities >70,000 gene copies g^{-1} soil (corresponding to 160,000 spores g^{-1} soil). The DSI_f recorded was close to 30, indicating that cultivar resistance was under pressure and confirming our initial hypothesis. The club root-resistant cultivars showed partial, rather than complete, resistance, as found in earlier studies on summer oilseed turnip rape. The results also support recent findings in a greenhouse set-up that increasing inoculum load applied to the resistant cv. Mendel increases the infection rate (Zamani-Noor et al., 2021). Canadian studies on club root-resistant OSR cultivars have found increased inoculum loads (quantified by qPCR) after cultivation and a peak in *P. brassicae* DNA in the following year.

Disease development of club root is strongly influenced by environmental factors, with soil water content long being recognized as a major abiotic factor contributing to the development of clubroot. Disease symptoms can develop after 10–18 h in excessively wet soil. When soil moisture content exceeds 50% soil water-holding capacity, the disease develops very quickly. Disease assessments at the different field sites in this study clearly showed the influence of soil moisture on infection. Temperature is an abiotic factor of great importance for successful club root infection. Early studies found that the conducive temperature range for infection is 12–27 °C, with galls developing only above 9°C. More recent studies show that air temperature has a significant effect on club root incidence in infected soil, with 40-fold lower risk of club root infection at 12°C compared with 22°C and with higher club root infection at 25°C compared with 20°C, particularly at lower pH levels. Recent studies on the survival of *P. brassicae* spores in different temperature regimes show that spores buried in the soil at 4°C and 20°C show better survival than spores on the soil surface at −20°C and 30°C, also ultraviolet (UV) light has been shown to have a negative effect on spore viability (Zhar et al., 2021).

5.6 Rapid Quantification of soil borne pathogen communities

Rainfed experiments operated continuously for up to 84 years in semiarid eastern Oregon are among the oldest agronomic trials in North America. Disease incidence and severity had been quantified visually but quantification of inoculum density had not been attempted (Smiley et al., 2016). Natural inoculum of 17 fungal and nematode pathogens were quantified for each of 2 years on eight trials using DNA extracts from soil. Crop type, tillage, rotation, soil fertility, year, and their interactions had large effects on the pathogens. *Fusarium culmorum* and *Pratylenchus thornei* were more dominant than *F. pseudograminearum* and *P. neglectus* where spring crops were grown, and the opposite species dominances occurred where winter wheat was the only crop. *Bipolaris sorokiniana* and *Phoma pinodella* were restricted to the presence of spring cereals and pulse crops, respectively. *Helgardia* spp. occurred in winter wheat-fallow rotations but not in annual winter wheat. *Gaeumannomyces graminis* var. *tritici* was more prevalent in cultivated than noncultivated soils and the opposite generally occurred for *Rhizoctonia solani* AG-8. Densities of *Pythium* spp. clade F were high but were also influenced by treatments. Significant treatment effects and interactions were more prevalent in two long-standing (>50-year) annually cropped experiments (29%) than two long-standing 2-year wheat fallow rotations (14%). Associations among pathogens occurred mostly in an 84-year-old annual cereals experiment. This survey provided guidance for research on dynamics of root-infecting pathogens of rainfed field crops and identified two pathogens (*Drechslera tritici-repentis* and *P. pinodella*) not previously identified at the location.

Inoculum density of *F. culmorum* differed significantly for all possible interactions of main treatments as well as for the main treatment variables of crop and year pathogens. The density of inoculum for a broad group of pathogens in these stabilized ecosystems could not have been uniformly identified in any other way. Results of this survey provided guidance for subsequent research on pathogen dynamics and root disease complexes in semiarid temperate climates. A summary of pathogenic fungi and parasitic nematodes detected in this research was presented. With one exception, 2-year mean minimum and maximum inoculum densities for each pathogen averaged across replicates for individual treatments within each experiment. Data for experiment 7 is from a single year of comprehensive sampling during 2015. Inoculum densities for 6 of the 11 pathogens were significantly affected by one or more treatment effects or treatment interactions in at least five of the eight experiments. The pathogens that were frequently influenced by experimental variables included the fungi *F. culmorum, G. graminis* var. *tritici, Pythium* spp. clade F, and *R. solani* AG-8, and the nematodes *Pratylenchus neglectus* and *P. thornei.*

Pathogens for which significant treatment differences were detected included *G. graminis* var. *tritici, Phoma pinodella*, *Pythium* spp., and *Pratylenchus neglectus* (Smiley et al., 2016). The only significant interaction among treatments for the three fungal pathogens was the interaction of rotation phase–fertility treatment for *G. graminis* var. *tritici.* That interaction indicated that inoculum concentration did not differ greatly between the planted and fallow phases for the 0 N and pea vine treatments but did differ greatly between the two rotational phases for each of the other four fertility treatments. In each treatment, a much lower inoculum concentration occurred during the year following fallow compared with the year following wheat. Inoculum density of *G. graminis* var. *tritici* was significantly affected by main fertility treatments, with greater DNA densities in the 90 N+burn treatment than in the 0 N or 90 N treatments: 8.8, 0.4, and 0.2 pg/g of soil, respectively (HSD0.05 = 5.3). The density of *Pythium* spp. was also significantly affected by fertility treatment, with 38.9, 25.8, 15.9, 15.6, 12.9, and 12.5 pg/g in the manure, pea, 90 N, 45 N, 0 N, and 90 N+burn treatments, respectively (HSD0.05 = 17.3). The density of *Pythium* spp. was especially high in both rotational phases of the manure treatment.

The density of *Pratylenchus neglectus* was influenced by significant fertility treatment–rotation phase and fertility–year interactions. The fertility–phase interaction indicated nearly equal densities of *P. neglectus* in the pea treatment of each phase but higher densities following wheat than fallow for all other management treatments. Inoculum of *Phoma pinodella* was significantly influenced by the crop–tillage interaction and by each of the main treatment effects. Densities of both *Pratylenchus neglectus* and *P. thornei* were significantly affected by crop–tillage interactions, and the tillage–year interaction was also significant for *P. thornei*. Inoculum density of *F. culmorum* differed significantly for all possible interactions of main treatments as well as for the main treatment variables of crop and year.

5.7 Disease Potential of Soil- and Tuber Borne Inocula of *Colletotrichum coccodes* and Black Dot severity on Potato

Inoculum of *Colletotrichum coccodes*, the cause of potato black dot, is soil- or tuber borne. Understanding the disease potential of sources of inocula is crucial for developing disease management strategies and resistance screening techniques. Two hypotheses were tested in this study, viz., soil borne inoculum causes more disease than tuber borne inoculum and black dot severity is related to the concentration of soil borne inoculum (Nitzan et al., 2008). Trials were conducted in the greenhouse with standardized inoculum. Plants grown in infested soil had more sclerotia on roots than plants grown from infected tubers in three of four trials. In general, plants grown in infested soil produced fewer tubers and lower yields than the noninoculated plants. Plants grown from infected tubers produced similar numbers of tubers as the noninoculated plants in all trials, and had reduced yields in one of the four trials. Increasing concentrations of soil borne inoculum had a nonlinear association with disease development. Foliar symptoms, sclerotial density on roots, and sclerotial development on stems did not increase when soil inoculum exceeded 0.5 or1.7 g/ liters of soil. In this study, soil borne inoculum caused more disease than tuber borne inoculum and disease severity remained constant above a threshold of soil borne inoculum.

Interactions between fungal isolates and inoculum source were recorded in 2005 and 2006 for the length of the stem from the soil surface upward on which sclerotia were visible (sclerotia on aboveground stem). Isolate C55 produced significantly ($P < 0.05$) more sclerotia on aboveground stems when its inoculum originated from a combination of soil and seed tuber (soil+/seed+) or the soil (soil+/seed–) than from only the seed tuber (soil–/seed+) in both years. Isolate C59 produced more ($P < 0.05$) sclerotia on aboveground stems when its inoculum originated from only the soil (soil+/seed–) than from only the seed tuber (soil–/seed+) in both years. In addition, isolate C59 produced significantly ($P < 0.05$) more sclerotia on aboveground stems than isolate C56 when their inoculum originated from the soil both years. Data of sclerotial density on roots, the number of progeny tubers and the total weight of progeny tubers were combined for all three isolates of *C. coccodes* due to lack of interactions. Sclerotial density on roots was greater ($P < 0.05$) in the presence of soil borne inoculum (soil+/tuber+ or soil+/tuber–) than in its absence (soil–/tuber+) in three of four years.

The relationship between soil borne inoculum concentration and disease sever-ity was described by nonlinear regression more accurately than by linear regression (Nitzan et al., 2008). The nonlinear model that described the relationship with the smallest variation (mean square error of the residuals) and the best goodness of fit (R2) was the sigmoid model, indicating that soil borne inoculum concentration had a nonlinear, sigmoid relationship with the severity of disease. The disease progress curve increased as the concentration of soil borne inoculum increased from 0 to 2 g of rye grains per 1 liter of soil, after which the curve levelled out. The sclerotial density on the roots in-creased as the concentration of soil borne inoculum increased from 0 to 0.5 g/liter and then levelled out at inoculum concentrations ranging between 0.5 and 1.7 g/liter. The relationship of sclerotial density on the root to soil borne inoculum concentration was explained ($P < 0.05$) by the sigmoid model.

Soil borne and seed tuber inocula of *C. coccodes* differ. Soil borne inoculum consists mostly of sclerotia, which must germinate, produce appressoria, and penetrate host tissue for infection and colonization to occur. Tuber borne inoculum may be external, on the epidermis as sclerotia or internal, in the vascular bundles associated with the stolon end as hyphae. Internal seed-tuber inoculum possibly develops with the growing plant. Sclerotia on tubers germinate and produce appressoria infecting roots, stolons, and crowns through the external surface. The mode of infection of the later is similar to that of soil borne inoculum. Sclerotia on the surface of the tuber may germinate inward, into the tuber, to infect the plant. However, application of Prochloraz (mycelial growth inhibitor) reduced infection substantially, indicating that this mode of infection is not likely to occur. Soil-borne inoculum is potentially more aggressive than tuber borne, probably because it is often spread throughout the soil profile and infects the roots in multiple infection courts, resulting in more damage. Tuber-borne inoculum, on the other hand, probably acts as a single infection source.

5.8 Synergistic Effect of Organic And Inorganic Fertilization on the Soil Inoculum Density of the Soil Borne Fungal Pathogens

The increasing demand of food causes an excessive exploitation of agricultural lands, often inducing phenomena of soil sickness accompanied by the development of soil borne diseases. The use of residual biomasses together with inorganic fertilizers can be considered a good agricultural practice for controlling the inoculum density of soil borne phytopathogens since soil conditioners can release inorganic nitrogen, polyphenols and fatty acids that, especially in vitro, have demonstrated various degree of suppressiveness against such pathogens (Claudio Cocozza et al., 2021). Further, soil organic amendments can also modify the population of soil culturable bacteria and fungi that, in turn, can affect the soil borne diseases in several ways. With this study, the authors aim to evaluate the impact of the synergistic application of diferent biomasses and inorganic fertilizers on the soil inoculum density of *Verticillium dahliae* and *Phytophthora* spp. during two potato cycles under open-feld conditions. The biomasses used for the fertilization of the potato crop were olive pomace residues (OPR), composts from municipal solid wastes (CMW), spent mushroom compost (SMC), and livestock manure-based compost (BRX).

The inoculum density of *Verticillium dahliae* appeared inhibited by BRX due to its low C/N ratio that caused a quicker release of inorganic nitrogen with respect to the others soil conditioners. In contrast, OPR was conducive to the aforementioned soil borne pathogen since that biomass was characterized by a very high percentage of unsaturated fatty acids that, rather, stimulate the inoculum density of *V. dahliae*. Finally, polyphenols did not influence the same pathogen because they apparently turned into no toxic compounds very quickly. The inoculum density of *Phytophthora* spp. was reduced equally by all the biomasses used in combination with the inorganic fertilizers, regardless of their composition and quantity, mainly because of the development of general microbial suppression. Therefore, the chemical characteristics of the soil conditioners apparently did not affect the inoculum density of *Phytophthora* spp.

The higher nitrate content recorded at the end of each potato crop cycle is the results of the fertilization of potato and, since all plots received the same amount of inorganic fertilizers, the differences among treatments were possibly due to the quality of each biomass, rather than the quantity applied, that could have played a role in the mineralization and release of the mineral N. In particular, at the end of the experiment, soil nitrate content increased by about 152% after the amendment with BRX, while it increased only by 26% after the amendment with OPR. This result is apparently associated to the C/N ratio of biomasses, since it can influence their mineralization and the development of the beneficial microbial community. BRX and OPR showed C/N ratio of 12.3 and 92.4, respectively and since a positive correlation between N assimilation into microbial biomass and the C/N ratios of soil conditioners exists, the former soil conditioner induced the soil microbial community to immobilize less N and increased the rate of nitrification. Te faster mineralization of BRX, as indicated by the higher soil nitrate content at the end of each crop cycle and the negative correlation between *V. dahliae* ID and soil nitrate concentration ($r=-0.991$, $p=0.001$), determined a better control of this pathogen. Other parameters generally responsible to lower pathogens ID are the amount and composition of SCB and SCF, influenced by the nature of amendments (Claudio Cocozza et al., 2021).

The application of different biomasses with the same inorganic fertilization affected differently the ID of *V. dahliae* and *Phytophthora* spp. in open field. With respect to the former pathogen, the mineral N can be considered the most important parameter for its control, therefore one should select for that aim biomasses characterized by less recalcitrant organic matter and low C/N ratio, since their organic N can mineralize easier. Soil conditioners characterized by large quantity of unsaturated fatty acids and applied at high rate per hectare can be conducive to *V. dahliae* and can stimulate the population level of both culturable fungi and bacteria. Soils already infested by *V. dahliae* or cultivated with crops susceptible to that pathogen should not be amended with this type of biomasses. The polyphenols content of each soil conditioner apparently does not play an important role in controlling *V. dahliae* since they should be quickly transformed into no toxic compounds shortly after the amendment. *Phytophthora* spp. are always susceptible to the synergistic application of inorganic fertilizers and biomasses regardless of the composition and the quantity of the latter. The reasons could be ascribed to an increase of beneficial microorganisms able to control these pathogens, rather than chemical parameters; however, more studies in open fields are necessary to confirm these conclusions.

5.9 Influence of Inoculum Density on Disease Development

A better understanding of the influence of inoculum density on disease, as related to pathogen race, may lead to better prediction of disease development and to the development of more efficient disease management strategies. Artificial inoculation experiments were carried out at 25 ◦ C to determine the effects of inoculum density of *Fusarium oxysporum* f.sp. *ciceris* races 0 (Foc-0) and 5 (Foc-5) and susceptibility of chickpea cultivars P-2245 and PV-61 on development of *Fusarium* wilt. Foc-5 proved much more virulent than Foc-0. Increasing the inoculum density of *F. oxysporum* f.sp. *ciceris* caused an exponential reduction in disease incubation period and a monomolecular increase of disease incidence and the area under the disease intensity progress curve (Juan A. Navas-Cortes et al., 2000). The extent of these effects was highest in the most conducive 'P-2245'/Foc-5 combination and decreased in the less susceptible 'PV-61' and for the less virulent Foc-0, in that order. For 'P-2245'/Foc-5, the highest disease intensity was attained with 6 chlamydospores g−1 of soil, the lowest inoculum density in the study. One thousand chlamydospores g−1 of soil of the same race were needed to attain a comparable disease intensity in 'PV-61'. Twenty thousand chlamydospores g−1 of soil of Foc-0 were required for maximum disease intensity in 'P-2245'. The disease intensity curves were adequately described by the Gompertz model. Using this model, a response surface for disease intensity was developed, in which the model parameters are expressed as a function of both time from inoculation and inoculum density. This response surface confirmed that the final amount of disease intensity increases in a monomolecular relationship with increasing inoculum density and showed that the relative rate of disease progress increases exponentially with increasing inoculum density of the pathogen.

The role of initial inoculum density of soil borne fungi in disease development has been well documented in many pathosystems. Generally, for *Fusarium* wilt diseases, increasing the amount of initial inoculum enhances disease severity and reduces the time required for maximum disease development. Both the inoculum density and race of *F. oxysporum* f.sp. *ciceris* significantly influenced the quantitative development of

Fusarium wilt in the two chickpea cultivars. No disease developed in 'PV-61' grown in soil infested with Foc-0, in spite of the high inoculum densities used. For the remaining chickpea cultivar × pathogen race combinations, disease intensity (DII) for both the wilt and yellowing ('P-2245'/Foc-0) syndromes increased with the increase in inoculum density in soil. The incubation period (IP) of the disease decreased exponentially with the increase in inoculum density of both Foc-0 and Foc-5 in soil irrespective of the cultivar used, the shortest IP occurring for the most conducive interaction, i.e. cultivar P-2245 grown in soil infested with the wilt-inducing Foc-5. The IP in cultivar P-2245 inoculated with Foc-5 was always shorter than that in cultivar PV-61, and the difference between the IP in the two cultivars infected by Foc-5 decreased with the increase in inoculum density. Conversely, for disease reaction within a comparable range of inoculum density of either *F. oxysporum* f.sp. *ciceris* race, the highest IP always occurred in cultivar P-2245 grown in soil infested with the yellowing-inducing Foc-0 irrespective of inoculum density in soil.

In order to analyze *Fusarium* wilt development over time as influenced by inoculum density of *F. oxysporum* f.sp. *ciceris,* the nonlinear form of both the Richards and Gompertz models was fitted to the observed DII progress curves (Juan A. Navas-Cortes et al., 2000). The Richards model offered an improved fit to disease progression data compared to the Gompertz model, which lacked the shape parameter. However, for most of the DII progress curves studied, the estimated value of the shape parameter in Richards function was not significantly different from 1. A response surface for DII progression as a function of both time from inoculation and inoculum density was developed with a two-stage method. Increasing the inoculum density of *F. oxysporum* f.sp. *ciceris* influenced development of *Fusarium* wilt in chickpea through an exponential reduction of the time for disease onset (decreasing IP) and a monomolecular increase of the overall amount of disease (increasing disease incidence and SAUDPC). However, the extent of these effects was significantly influenced both by virulence of the pathogen race and susceptibility of the chickpea cultivar.

Results from the present study allowed the development of a response surface function for disease development over time as influenced by inoculum density. Such a surface is based on the Gompertz model, in which the model parameters are expressed as a function of the pathogen inoculum density. This response surface confirmed that in the chickpea / *F. oxysporum* f.sp. *ciceris* pathosystem the final amount of disease increases in a monomolecular relationship with increasing inoculum density, except in the most susceptible cultivar for which a low inoculum density of the most virulent race saturates the system. Similarly the response surface indicated an exponential increase in the relative rate of disease progress with increasing inoculum density of the pathogen, except for the least conducive 'P-2245'/Foc-0 combination. This relationship suggests a synergistic interaction within Foc-5 propagules at high inoculum densities for the rate of disease development over time, the effect of such an interaction being more pronounced in the most susceptible cultivar. As shown by results from this research, the benefits provided by such a cultural practice for *Fusarium* wilt management can be overridden if a cultivar too susceptible to the disease is used, or if soils where a highly virulent race of the pathogen prevails are present, or both.

5.10 A Conceptual Framework to Reduce Inoculum Potential and the Effect of Bulk Density

The inoculum potential of plant pathogens embodies their intrinsic characteristics to survive, reproduce, spread, and initiate infection of their hosts. Preservation of inoculum potential is critical to the continuity of the life cycle of plant pathogens, and thereby constitutes a lurking threat to crop production systems. Reducing inoculum potential is central to management of plant pathogens in order to minimize the negative impacts of these pathogens on crop health. An important tool used for reducing the inoculum potential of soil borne pathogens is soil disinfestation, which is achieved through a wide array of methods including fumigation, solarization, chemical treatment, and anaerobic soil disinfestations.

Soum Sanogo (2019) proposed the concept of "genetic soil disinfestation" as an additional innovative approach for managing soil borne pathogens. Although many components of "genetic soil disinfestation" include well known and familiar tools such as crop rotation, the concept of "genetic soil disinfestation" redefines cropping systems in a unified perspective with focus on using a genetic approach to optimize the attributes of hosts and nonhosts that significantly reduce the populations of soilborne plant pathogens and the efficiency of invasiness of these pathogens.

Genetic soil disinfestation, sensu stricto, is the use of host-specific resistance to reduce the inoculum potential in soil. The basic tenet of host resistance lies in the selective inclusion and exclusion of host-adapted pathogens from partaking in the infection of the host Soum Sanogo (2019). Utilization of host resistance may efficiently suppress the inoculum potential of selected pathogenic strains whereas other strains remain unaffected. The concept of genetic soil disinfestation redefines cropping systems with focus on using a genetic approach to optimize the attributes of hosts and nonhosts that significantly reduce the populations of soil borne plant pathogens and the efficiency of invasiness of these pathogens. Hosts and nonhosts with high capacity for genetic soil disinfestation should, among other things, resist invasion or colonization and saprophytic growth of pathogens. These attributes are quantifiable and can be used to carry out a benchmark evaluation of hosts and nonhosts for their efficiency in reducing the inoculum potential of soil borne pathogens. The implementation of the concept of genetic soil disinfestation does not necessitate any drastic changes in cropping systems. It will require a systematic planning of existing crop management schemes to include hosts and nonhosts based on their capacity in reducing inoculum potential. In order to assist agricultural practitioners in this endeavor, the research community must deploy effort in establishing parameters that are reliable and easy to measure for gauging the genetic capacity of candidate hosts and nonhosts for soil disinfestation. In all, the concept of genetic soil disinfestation provides a new unified framework for managing the inoculum potential of soil borne pathogens and opens a new direction of research to validate the concept and broadness of its applicability.

Effect of bulk density on the spatial organisation of *Rhizoctonia solani* in soil: A basic characteristic of eucarpic fungi is their mycelial growth form, which enables a highly effective spatial exploration of habitats. This is particularly important for saprotrophic and parasitic activity in soil (Kirsty Harris et al., , 2003). As soils are a complex three-dimensional physical framework, mycelia must extend through the heterogeneous networks of pores, to enable them to locate nutrients. Most studies with fungal growth in soil have been carried out with little attention to the soil structure within which the mycelia grow and microbial interactions occur. In this paper we report on growth patterns of *Rhizoctonia solani* introduced into a mineral soil.

The mycelial growth form of eucarpic fungi allows for a highly effective spatial exploration of the soil habitat. However, understanding mycelial spread through soil has been limited by difficulties of observation and quantification of fungi as they spread through this matrix. Kirsty Harris et al., (2003) reported on a study on the effects of soil structure by altering the soil bulk density, on the spatial exploration of soil by *R. solani* using a soil thin-sectioning technique. They quantified fungal densities in microscopic images (0.44 mm^2). At this scale, hyphae were either absent, or present as minor fragments, typically occupying less than 1% surface area of the thin section. From contiguous microscopic images, then produced large-scale (6.21 cm^2) spatial distribution maps of fungal hyphae. These maps were superimposed onto soil structural maps, which quantify the degree of porosity in each microscopic image. Alterations in soil structure by changing the bulk density are shown to affect the distribution of the fungus within the soil. The volume of soil explored by the fungus increased with increasing bulk density. This was associated with a shift from a few large pore spaces to more evenly distributed small-scale pores. Fungal hyphae were present in all porosity classes within each bulk density, including areas that contain less than 5% visible pore space. However, fungal hyphae were more often found in areas with a higher porosity, in particular at low soil bulk densities. The results show that soil structure is a major component in the spatial exploration of soil by fungi.

Mycelial exploration through soil is influenced by soil physical characteristics since hyphae must ramify through the complex heterogeneous network of pores. The effects of physical conditions on hyphal spread are difficult to ascertain because of the geometric complexity of the pore networks. Furthermore, soils are comprised of interdependent solid, liquid and gaseous phases, where altering one of the components could indirectly affect the other phases or associated processes. Many factors may modulate fungal growth in soils, such as nutrient availability, pH, aeration, and microbivory. Previous studies of the spread of fungi through soil have shown that soil physical conditions can also play a significant role in controlling the growth of mycelia. One of the principal limiting factors to fungal growth in soil is the availability of air-filled pore space. The connectivity and tortuosity of air-filled pore networks are dictated by the architecture of the soil and the interaction between the pore networks and water.

The growth of fungi through soil involves spreading through a heterogeneous three-dimensional framework of tortuous pores (Kirsty Harris et al., , 2003). Manipulation of the soil framework is one of the principal means by which microbial dynamics, in particular fungi, can be regulated at the small scale with effects that may be realised in the field. Regulation of microbial dynamics by this means occurs through the fragmentation in habitat space, water and substrate distribution and the spatial arrangement of pore pathways. Soil bulk density had a significant impact upon the spatial organisation of R. solani growing through the soil volume. The spatial distribution of hyphae was patchy at lower bulk densities and the volume of soil that contained hyphae increased with bulk density, between 1.2 and 1.4 Mg cm^3. Generalisation of these results must be done cautiously: the results are concerned with a central radial region of the fungal colony (20 mm) in which density of hyphae is expected to be high in non-sterile soil.

5.11 Role of Inoculum Density and Incubation Period on Saprophytic Behaviour of *Pythium ultimum, Sclerotium rolfsii* and *Rhizoctonia* spp.

The success of root infecting fungi depends on their potential to survive saprophytically in soil or on organic residues buried in soil, or on dead tissues colonized during the parasitic phase. During this period, these root fungi face fierce antagonism by other resident obligate saprophytes. The potential of a root parasite to survive saprophytically in soil in the absence of host depends on its intrinsic capacity known as competitive saprophytic colonization (Pandey et al., 1999). Knowledge regarding this aspect of the problem is important in understanding the ecology of root pathogens.

This article provided detailed information on saprophytic behaviour of these two pathogens as influenced by inoculum density and incubation period. The data presented indicate that in the control (soil only) baits were not colonised by *S. rolfsii*, but in case of *P. ultimum* a few pieces of both the substrates were colonized after 96 hours of burial of baits. This shows that *P. ultimum* is more prevalent than *S. rolfsii* in the agricultural lands of the IISR farm. Both *P. ultimum* and *S. rolfsii* colonised baits in the presence of resident saprophytic microflora, normally present in soil. Of the two baits used, sugar beet hypocotyles were better traps than the wheat straw. This is expected because sugar beet hypocotyle contains more nutrients than the wheat straw which contains mostly cellulose. The colonization of baits by the test fungi increased as the inoculum density increased.

The competitive saprophytic ability (CSA) of *Pythium ultimum* and *Sclerotium rolfsii* isolated from diseased sugar beet seedlings was studied. Each pathogen was multiplied on autoclaved sand - maize-meal in 500 ml conical flasks. The biological baits used were pieces of wheat straw and sugar beet hypocotyl. Wheat straw was cut into 1 cm long pieces and boiled in water for 10 minutes before use. The pieces of sugar beet hypocotyl were of the same size and used as such. *Pythium* species are not active competitors and their saprophytic activities are greatly restricted. Generally, these grow and proliferate only under circumstances where other organisms are either not present or have limited activity in soil environment. However, in the present study, *P. ultimum* was found to be a fast colonizer of biological baits; the colonization potential depended upon both the inoculum density and incubation period in soil. Even at 2% inoculum density 17.3% sugar beet baits were colonised in 72 hours.

The infection of a root or other subterranean organ of a susceptible host plant by a propagule of a soil-borne pathogen depends upon the distance of the propagule from the host. As distance increases, the probability of infection decreases, and beyond a certain distance the chance of infection is effectively zero. These relationships have been formalized in plant pathology into the terms pathozone and infection efficiency. The pathozone is defined as the volume of soil surrounding subterranean plant organs within which the centre of a propagule must occur if it is to have any chance of infecting the organ. Propagules outside this volume either do not respond to the presence of the host or, if they do, have insufficient reserves of nutrients to reach and infect the host. The pathozone is a more general term than rhizosphere or spermosphere and can be applied to any subterranean organ such as a root, seed, hypocotyl, epicotyl or mesocotyl. Exceptionally, it can also be applied to combinations of organs, when, for example, a seed and an epicotyl comprise an infection court.

Experiments involving direct placement of inoculum at various distances from an infection court were used to estimate the probability of infection and/or disease of seminal roots of wheat by *Gaeumannomyces graminis* var. *tritici* and of germinating seeds of radish by *Rhizoctonia solani* (Gilligan and Simons, 1987). Two concepts underlie the relationships under investigation. These are the pathozone, which is characterized by the furthest extent of inoculum from the host from which infection is recorded {z), and infection efficiency, which is characterized by two parameters alpha and beta. The asymptotic infection efficiency, beeta, is estimated by the proportion of propagules that cause infection when placed in contact with the host; alpha is a measure of the decline in infection efficiency of propagules of inoculum with increasing distance from the host. Three non-linear models are used to describe the profile of infection efficiency over distance. The sensitivity of the method of placement of inoculum in detecting treatment effects is demonstrated first by showing differences in response amongst two isolates of *G. graminis* and two times of observation. Maximal values of z for *G. graminis* ranged from 4 to 8 mm, depending upon the type of inoculum of *G. graminis* and the growth medium used. Values of z for *R. solani* and germinating radish seeds varied from 12 to 16 mm for propagules comprising fragments of chopped potato soil inoculum sieved to 1 to 2 mm. The amounts and nutritional basis of inoculum influence the probability of infection and the parameters alpha, beeta and z. Thus, for *G. graminis*, estimated values for alpha $^{-1}$, beeta and z were greater for inoculum comprising colonized millet grains than for colonized sand particles. When seedlings of radish were exposed to clumps of propagules of *R. solani*, the parameters beeta and z increased, while a decreased with increasing clumip density. Reduction in propagule size to 0.5 to 1.0 mm reduced the proportion of infected hosts. When clumps of seedlings of radish were challenged with clumps of inoculum of *R. solani*, the probability of infection of clumps increased with clump density, but there was no evidence for an effect on the probability of infection of individual seedlings within a clump. The relationship between the estimates of pathozone width and infection efficiency and models for infection of subterranean organs was discussed.

Role of inoculum density and incubation period on saprophytic behaviour of *Rhizoctonia solani* and *R. bataticola*: The success of root infecting fungi depends on their potential to survive saprophytically in soil or on organic residues buried in soil, or on dead tissues colonized during the parasitic phase (Pandey et al., 1999). During this period, these root fungi face fierce antagonism by other resident obligate saprophytes. The potential of a root parasite to survive saprophytically in soil in the absence of host depends on its intrinsic capacity known as competitive saprophytic colonization. Knowledge regarding this aspect of the problem is important in understanding the ecology of root pathogens. The competitive saprophytic ability (CSA) of two destructive root rot pathogens of sugar beet seedlings, viz., *Rhizoctonia solani* and *R. bataticola* was investigated using the Cambridge method by employing two types of baits (wheat straw and sugar beet hypocotyl pieces) in a sandy loam soil. The competitive saprophytic colonization (CSC) of these two baits by these two facultative parasites appreciably increased as the concentration of inoculum increased in the soil. The per cent colonization of baits was also influenced by the incubation period.

The regression analyses showed that the biological baits were not colonized by *R. solani* and *R. bataticola* in soil not inoculated with the test fungi. The colonization of the substrates by the test fungi increased directly with fungal density for both the test pathogens. The maximum colonization of the baits was found where fungal density was 100 per cent. At 2% fungal density, *R. bataticola* did not colonize pieces of wheat straw. However, the colonization started at higher levels of inoculum. The degree of colonization increased with increase in incubation period but in none of the cases one hundred per cent pieces were colonized by *R. bataticola*. The maximum percentage of pieces showing colonies of this fungus was 82.67 per cent. Like, *R. solani*, sugar beet hypocotyl baits were also suitable for *R. bataticola*. The degree of colonization of this bait was significantly higher than wheat straw. The maximum number of pieces was found to be colonized at 98-100 % inoculum level incubated for 144 h. Sugar beet hypocotyl can be a better biological bait than wheat straw since the per cent colonization of sugar beet hypocotyl was more than the wheat straw. The saprophytism of the fungus is highly influenced by the type of soil, organic matter content, quality and quantity of microbial population, kind of substrate and type of inoculum.

Sampling was carried out on soils around pistachio trees in various regions of Rafsanjan, Iran (Saberi-Riseh et al., 2004). Following isolation and identification of *Phytophthora* isolates, the predominant species was found to be *P. drechsleri* and used for further investigation. For studying saprophytic survival of the fungus, soils collected from different areas were combined and autoclaved. Sterile soil was divided into 10 parts and mixed with fungal inoculum at various concentrations of 0-9% (w/w) separately. Each soil part (100g) was placed in 15cm diameter plastic pot. Some soils in pots were supplemented with sterile wheat straw whereas others were mixed with pistachio leaves surface sterilized with 5% (v/v) sodium hypochlorite. After 3, 6 and 9 weeks of incubation, five leaves or straws samples were taken from each replicate and cultured on CMA-PARPH medium and the fungal colony formation was monitored. The experiment was performed using completely randomized design with factorial experiments including three factors (substrate type, inoculum density and time), 10 treatments (0-9 g inoculum levels) and nine replicates. The results showed that the type of substrate (wheat straw and pistachio leaf) was very important for the fungal saprophytic survival in that this was significantly greater for the pistachio leaves. Time was also considered another critical factor for the fungal survival. With passing incubation time, saprophytic survival of the fungus declined. Further, it was demonstrated that increasing inoculum density would result in longer survivability of *P. drechsleri* and maximum fungal survival on substrate was obtained when inoculum density was at 9% (w/w).

5.12 The Importance of Primary Inoculum to Crop Health, Food Security and Rice Sheath Blight Disease Development

In some epidemics that have devastating consequences, the primary inoculum plays an important role in both epidemic onset and intensification. This article documents the dynamics of such epidemics, and illustrates their importance using two examples: Huanglongbing of citrus and begomoviruses of tomato (Bergamin Filho et al., 2016). The latter disease is a major constraint to tomato production in Brazil, while the former has become a threat to global citrus production and farmers' livelihoods. In spite of their importance little is known of the characteristics of these diseases and their management. This is because classical botanical epidemiology considers two types of diseases: polycyclic diseases, where the inoculum that causes infections is produced during the epidemic in or on individual plants that had been previously infected in the course of that epidemic; or monocyclic diseases, where inoculum that causes infection is not produced in or on individual plants that had been infected in the course of the epidemic, but in the soil, on secondary hosts, or in infected crop plants of the same host in other fields. Diseases of the first type typically present a logistic disease progress curve and management is based on reducing the rate of infection, whereas diseases of the second type present a monomolecular disease progress curve and management is based on reducing the initial inoculum. This article deals with plant diseases that depart in their structure and behaviour from these two archetypes, because they borrow elements from both.

Bergamin Filho et al., (2016) addressed polycyclic diseases in which the primary inoculum has a continuous and dynamic role, and in which the secondary inoculum contributes to epidemic build-up, i.e., polycyclic diseases with continuous primary spread. This epidemiological structure generates less clear-cut disease progress curves, but usually follows a monomolecular dynamic. The focus on this type of disease is multifold because this more complex, combined, pattern is actually quite common, often leading to grave plant diseases epidemics, with impacts at the farm, community, and country scales ; and such epidemics are among the most difficult to manage. The analysis leads us to assess past errors and current courses of action. It allows us to recognize, in addition to the conventional tools for management with local effects, the critical importance of collective action. Collective management action – at the farm, community, or national scales – is congruent with the characteristics of many epidemics, because they also entail properties at successive and nested scales. The management of such epidemics needs to address both the primary and secondary inoculum. More importantly, these actions have to be performed in an area-wide, regional basis in order to be effective.

The two diseases, by contrast, highlight also differences which have consequences for the identification of management strategies. First, HLB pathogen and its vector have a comparatively narrower range of hosts than begomoviruses and their vectors. Second, perennial (citrus) and annual (tomato) crops, inherently confer strong differences from their lifespan: the development of resistant materials is obviously more difficult in the case of perennials, and management options associated with changes in host population density over time (e.g., consideration of a crop free period) and space (e.g., elimination of plants or plots) clearly correspond to contrasted situations in both cases. While the former feature (host range) raises difficulties in achieving efficient area wide management strategies for tomato virus diseases, the second feature corresponds to a more manoeuvrable landscape management for tomato.

The two diseases addressed in this article highlight the key commonalty between diseases exhibiting this type of epidemiological pattern: the importance of inoculum flow from outside the field or orchard to epidemic build-up calls for observations, measurement, understanding, and management that accounts for processes occurring outside the field under consideration. In other words, this corresponds geographically to linking field and landscape, and to linking farmer to community in terms of social and decision making processes. Such considerations have been outlined for livestock epidemics, where both the individual (individual farm) and the community (neighbouring farms) scales need to be considered to estimate optimum strategies to control epidemics. Due to both the key role of primary infection and the lack of effective measures to suppress it, HPR represents an important element for the management of polycyclic diseases with continuous primary spread. Availability of HPR and possibilities of rapid deployment are very variable according to the pathosystem considered. HPR prospects are good in the case of the tomato-begomoviruses pathosystem; but they are still quite remote in the case of citrus and HLB.

Impact of soil-borne inoculum on rice sheath blight disease development: The rice sheath blight (*Rhizoctonia solani* AG-1 IA) disease cycle usually starts from overwintered asexual primary inoculum, mainly floating sclerotia present in the soil, irrigation water or stubble or fungal mycelia. Fungal mycelia survive on rice debris, infected weeds and seeds (Maryam Khoshkdaman et al., 2020). During cold winter conditions, mycelia in plant debris lose viability after harvest. Seed borne inoculum is known to play a significant role in overwintering and long-distance dispersal of the sheath blight pathogen. Secondary infection (polycyclic phase) takes place in the upper part of the canopy by strands of mycelium that are produced on the surface of primary lesions and run on the surface of healthy leaves and sheaths to establish new lesions. The hypothesis is that the initial inoculum of the pathogen has great impact on disease development. However, determining the relationship between initial inoculum density and disease development for crop diseases which their initial inoculum is soil-borne and cannot be easily measured and counted like rice sheath blight, is difficult.

Primary inoculum density seems to be a major factor in sheath blight disease in rice outbreak. Maryam Khoshkdaman et al., (2020) determined the relationship between the primary inoculum density and type and the disease intensity, grain yield and yield loss. Field experiments were conducted in both years of 2017 and 2018 in Guilan province, Iran. Disease incidence and severity were significantly higher when the highest inoculum densities (mycelial and sclerotial) were tested. When sclerotia were applied as the primary inoculum, disease developed more quickly. Based on the results of the current study, in a temperate lowland rice system in Guilan province, sclerotia floating on the water surface after puddling can be the primary source of inoculum and play a major role in sheath blight epidemics whereas mycelia in plant debris probably lose their viability in winter. These results suggested that control of sheath blight disease in order to prevent sclerotia production and reduce the main disease inoculum can be a promising strategy for suppressing this disease in the rice fields of Guilan province.

Disease incidence and severity and grain yield loss were positively and linearly correlated with initial inoculum density of the pathogen. Rice grain yield was negatively and linearly correlated with the initial inoculum density. The sheath blight disease incidence showed more significant correlation with the sclerotial or mycelial inoculum density than the disease severity in both years. Yield loss also showed more significant correlation with the disease incidence than the disease severity in both years. Therefore based on the results of the regression analysis, sclerotial density is a suitable predictor of the disease incidence and the disease incidence is a suitable predictor of yield loss in the field conditions. Sheath blight disease in all sclerotial treatments developed much faster and higher than mycelial treatments. These results suggested that sclerotia floating on the water surface after puddling may be the main primary inoculum and play a more important role in sheath blight epidemics. Although the plots that were infected with *R. solani*– infested whole rice grain/ rice hull (as mycelial inoculum) showed sheath blight disease symptom, the disease was significantly lower compared to those plots infected with sclerotia. The findings suggested that in a temperate climate such as Guilan province, mycelia in plant debris lose viability due to the cold winter conditions prevailing after harvest.

Not only inoculum density, but also inoculum type (mycelia or sclrotia) were positively correlated with disease incidence, severity and yield loss in 2017 and 2018. Disease development was higher and faster when sclerotia were applied as primary inoculum. It could be concluded that in Guilan province with temperate climate, sclerotia floating on the water surface after puddling might be the main primary inoculum source and play a more important role in sheath blight epidemics. Mycelia in plant debris probably lose their viability due to the cold winter conditions and are less important in disease development.

5.13 Inoculum Levels on the Aggressiveness of *Fusarium oxysporum* f.sp. *ciceris*

An optimized inoculation method is required to screen the germplasm for disease resistance. Number of inoculation techniques has been reported for *F. oxysporum* f.sp. *ciceris* such as root inoculation by dipping in conidial suspension, seed infestation method and soil infestation method (Maitlo et al., 2016). Different methods of pathogen inoculation result in very different level of disease. Moreover, inoculums level is also proportional to the amount of disease produced.

To screen the plants against diseases suitable inoculation method together with appropriate inoculum density are essential. Measurements of propagules required for infection are also necessary for the development of potential disease risk analysis. To find out the best method for the pathogenicity test of chickpea wilt caused by *Fusarium oxysporum* f.sp. *ciceris* five inoculation methods viz., soil infestation, seed infestation, drenching, injection and spraying methods were evaluated by Maitlo, et al., (2016). The soil infestation, seed infestation followed by drenching method was highly effective in establishing the *F. oxysporum* f.sp. *ciceris* infection in chickpea plants. The plant inoculation with these methods produces higher plant mortality and higher infection as well as reduced plant growth. In order to determine the effective inoculum level of *F. oxysporum* f.sp. *ciceris* for the development of disease, chickpea plants were inoculated with 101 , 102 , 103 , 104 , 105 and106 spores g -1 of soil and found out that fourth inoculum level (104 spores g-1 soil) enough to produce the maximum disease determined by plant mortality, root infection and plant growth. The results indicated that up to certain level disease development (plant mortality, pathogen infection and plant growth) positively correlated with the increasing inoculum densities.

Inoculation method and inoculums level of *F. oxysporum* f.sp. *ciceris* influence the plant mortality and pathogen infection as well as plant growth of chickpea. Among five different inoculation methods for pathogenicity test, the soil infestation and seed infestation methods were highly effective in establishing the *F. oxysporum* f.sp. *ciceris* infection in chickpea plants. The disease was more pronounced in these methods of inoculation as compared to the drenching, spraying and injection methods. Either the plants inoculated through soil infestation, seed infestation or drenching showed significantly higher plant mortality and pathogen infection as well as reduced plant growth. The soil infestation method and drenching method may efficiently be used in greenhouses for inoculation with wilt pathogens, but have limitation for screening a large number of plants in the field soil. Technically sterilization of field is impossible otherwise there may present antagonistic organisms that may interfere with the normal pathogenicity procedure. Seed infestation methods produce equal plant mortality and infection as compared to soil infestation method, hence may be used for screening a large number of plants. Injection method was difficult due to the small diameter of stem and spraying method was not suitable for vascular wilt pathogen it rather suits for the pathogens causing foliar diseases. Direct introduction of the vascular wilt pathogen to the stem does not allow the activation of resistance mechanism present in the roots for soil borne pathogen getting roots as the main point of entry.

Different inoculum levels of *Fusarium oxysporum* f.sp. *ciceris* influenced the disease development. The results indicated that up to certain stage disease development (plant mortality, pathogen infection and plant growth) was positively correlated with the increasing inoculum levels. Significant correlation between number of lesion and shoot weight and between shoot weight and pods reported in kidney bean inoculated with *Rhizoctonia solani.* The impact of pathogen inoculation was more pronounced at initial levels, reached to peak at 4th or 5th inoculum level and then static. It means at the level of 10^4 or 10^5 spores g^{-1} of soil, the soil become fully saturated with the pathogen populations, so that any further increased in pathogen inoculum failed to cause more noticeable change in either disease development or on plant growth. Disease establishment and symptom expression were positively correlated with the pathogen inoculum, infection increase in response to increased inoculum density (Tooley et al., 2013). As the inoculum density increases, wilt disease symptoms appear earlier in susceptible cultivar, while at low level of inoculum densities, wilt disease of chickpea appears late and develop slowly. The effective inoculum levels to produce disease greatly varied with the pathogen race and variety under evaluation. At optimum temperature, disease development was greater at 250–1000 chlamydospores per g of soil as compared to the 25–100 chlamydospores per g of soil.

5.14 Inoculum dose–Disease Response Relationships

In areas with endemic Aphanomyces euteiches problems, the most recommended practice is disease avoidance based on determining inoculum potential of field soil indexing through greenhouse grow out tests in field soils (Chatterton et al., (2023). Inoculum potential is an index of potential disease activity of the soil dependent on pathogen infectivity and density and soil factors that can either inhibit or promote infection. Historically, inoculum potential was determined in a greenhouse bioassay by growing a susceptible pea cultivar in collected field soils under conditions that are conducive to disease development.

Chatterton et al., (2023) related oospore levels in autoclaved and non-autoclaved soils to ARR severity across soil types from the vast prairie landscape and to determine the relationship of measured DNA quantity of *Aphanomyces euteiches* using droplet digital PCR or quantitative PCR to the initial oospore inoculum dose in soils. These objectives support a future end goal of creating a rapid assessment method capable of categorizing root rot risk in field soil samples to aid producers with pulse crop field selection decisions. The ARR severity to oospore dose relationship was statistically significantly affected by the soil type and location from which soils were collected and did not show a linear relationship. For most soil types, ARR did not develop at oospore levels below 100/g soil, but severity rose above this level, confirming a threshold level of 100 oospores/g soil for disease development. For most soil types, ARR severity was significantly higher in non-autoclaved compared to autoclaved treatments, demonstrating the role that other pathogens play in increasing disease severity. There was a significant linear relationship between DNA concentrations measured in soil and oospore inoculum concentration, although the strength of the relationship was better for some soil types, and in some soil types, DNA measurement results underestimated the number of oospores. This research is important for developing a root rot risk assessment system for the Canadian prairies based on soil inoculum quantification, following field validation of soil quantification and relationship to root rot disease severity.

Although *A. euteiches* is the most destructive pathogen to pea roots, it often is detected as a complex with other soil borne pathogens. A number of *Fusarium* species were commonly isolated from pea roots in southern Alberta, Canada, and *F. avenaceum* and *F. solani* were the most aggressive among tested species (Safarieskandari et al., 2021).

Co-inoculation of *A. euteiches* with these two species, and the weakly aggressive *F. redolens*, resulted in statistically significantly higher disease severity ratings compared to single-species inoculation. Therefore, the synergistic interactions of *A. euteiches* with other soil pathogens may affect disease severity. In this context, the bioassay may be more predictive, in some cases, than DNA-based analyses of *A. euteiches* alone.

The finding that soil type influences oospore quantification has been described for related species, *Aphanomyces cochlioides* where oospore detection limits were higher in high clay soils and *Phytophthora medicaginis* where quantification was lower in sand than in soil (Bithell et al., 2021). In terms of the effect of autoclaving, dark brown silt loam, black silt loam, and dark brown clay loam soils had significantly different slopes between autoclaved and nonautoclaved treatments, even though oospores were added after autoclaving. Autoclaving can affect soil properties, causing, for example, a decrease in aggregation, a corresponding increase in the clay fraction, and more dissolved organic matter. The disease severity and oospore dose relationship may also have been affected by changes in soil properties due to autoclaving, but other soil sterilization procedures may also result in changes to soil properties and other sterilization equipment are not as readily available as an autoclave. It is possible that this differential effect was due to the different spiking events and random variation in oospore distribution when collecting 2 × 250 mg samples from each spiking event. Taken together, the results for the effects of soil texture and zone and autoclaving suggest that soil properties may affect the performance of both qPCR and ddPCR, although the exact nature requires further research. This is being tested with more replicates on a larger number of soil samples to determine if random variation in sampling is the biggest factor.

Inoculum density-Disease development relationship in *Verticillium* Wilt: *Verticillium dahliae* survives by means of microsclerotia in soil for more than 14 years and causes monocyclic diseases. Therefore, inoculum density of the pathogen in field soils at planting plays a critical role in the epidemiology of *Verticillium* wilt diseases. Fungicide treatments and soil fumigants cannot successfully control this disease, and artichoke cultivars with resistance to *V. dahliae* are not commercially available (Berbegal et al., 2007). The long survival of *V. dahliae* microsclerotia in soil and the broad host range of *V. dahliae* are limiting factors for the effectiveness of crop rotations as a control strategy of *Verticillium* wilt in many crops. Although long rotations with resistant or immune crops can reduce the amount of microsclerotia in soil, shortterm rotations do not result in significant reductions of viable microsclerotia. Consequently, quantitative information about the inoculum density in soil and its relationship with *Verticillium* wilt development would be useful for assessing the efficacy of control strategies. It also is essential for disease risk predictions based on the preplanting assessment of initial inoculum.

The relationship between inoculum density of *Verticillium dahliae* in soil and disease development was studied in 10 commercial artichoke fields. Inoculum density of *V. dahliae* varied between 2.2 and 34.2 microsclerotia (ms) g–1 of soil near planting. Artichoke plants were monitored for disease at the beginning and the end of each growing season. There was a significant correlation, which was best described by negative exponential models, between inoculum density and disease incidence, symptom severity, and recovery of the pathogen from the plants. Inoculum densities ranging from 5 to 9 ms g–1 of soil were associated with a mean percentage of infected plants of about 50%. Additionally, three fields were monitored in two consecutive growing seasons to evaluate the population dynamics of *V. dahliae* microsclerotia in soil and disease development. Numbers of microsclerotia per gram of soil decreased significantly by the end of the first growing season but slightly increased at the end of the second growing season (Berbegal et al., 2007). In these fields, symptom severity was greatest during the second growing season when high percentages of infected plants also were recorded.

The incidence of *Verticillium* wilts in herbaceous hosts is proportional to the pathogen inoculum density expressed as the number of viable microsclerotia per gram of soil; however, considerable variation occurs, with such a relationship depending upon crops and cultivars. Furthermore, the threshold inoculum density for *Verticillium* wilt in soil required to cause wilt seems to be crop dependent. Nevertheless, it often has been observed that similar amounts of inoculum did not always induce the same amount of infection. A lack of correlation between incidence of infection and soil inoculum density has been observed in cotton and other crops, such as potato.

This is the first study that characterizes the relationships between the ID of *V. dahliae* in soil and development of *Verticillium* wilt of artichoke. Information about these relationships is essential for disease risk assessment and development of disease management options. The perennial growth habit of artichoke also allowed the study of the dynamics of microsclerotia population in soil and progression of disease during two growing seasons (Berbegal et al., 2007) The relationship of *V. dahliae* microsclerotia density in soil and *Verticillium* wilts vary with crops. For example, in cotton, field inoculum densities of 19 ms g–1 can result in the infection of 50% of plants. In strawberry fields, only 0.3 ms g–1 can result in the infection of more than 5% of the plants. Low levels of *V. dahliae* ID in soil could become an important limiting factor for crop profitability. In artichoke fields at the Comunidad Valenciana Region, ID in soil ranging from 5 to 9 ms per g of soil were associated with a mean percentage of infected plants of about 50%.Therefore, for disease management, it is desirable to reduce ID in soils as much as possible. Information from this study could help to decide the nature of disease control strategies worth applying, as well as when and how long they should be used, based on the results of preplant analysis of soil samples for the quantification of microsclerotia.

Frequency of *V. dahliae* isolations from plants was higher in autumn than in spring in all fields of the study, suggesting a clear influence of seasons and, perhaps, temperature. This fact was observed earlier in artichoke and in other crops, such as olive, maple, and potato. Seasonality also may be influenced by anatomical and physiological changes in the plants. Several factors, such as microbial antagonism, could be involved in the quick loss in the number of viable microsclerotia from soil. Microsclerotia would germinate near the plants roots, leading to a decline in the population of viable propagules in soil until the incorporation of new inoculum from infected tissues at the end of the crop

5.15 Inoculum Density and Spatial Patterns of *Plasmodiophora brassicae* at the Field Level

Clubroot, caused by *Plasmodiophora brassicae*, is an important soilborne disease of the Brassicaceae. Knowledge of the spatial dynamics of *P. brassicae* at the field level and the influence of soil properties on pathogen spatial patterns can improve understanding of clubfoot epidemiology and management (Botero-Ramirez et al., 2021). To study the spatial patterns of *P. brassicae* inoculum density and their relationship to different soil properties, four club root-infested fields in central Alberta, Canada, were sampled in 2017 and 2019, and *P. brassicae* inoculum density, soil pH, and boron, calcium, and magnesium concentrations were quantified. Spatial autocorrelation of the inoculum density was estimated for each of the fields in both years with the Moran's *I* and semi-variograms. A Bayesian hierarchical spatial approach was used to model the relationship between *P. brassicae* inoculum density and the soil parameters. Patchiness of the pathogen was detected, with most patches located at the field edges and adjacent to the entrance. Infested patches grew in size from 2017 to 2019, with an average increase in diameter of 221.3 m and with this growth determined by the maximum inoculum density and active dispersal methods such as movement by machinery and wind. Soil pH, boron, calcium, and magnesium concentrations were not found to have an important effect on the inoculum density of *P. brassicae.*

As a soilborne pathogen, the physical and chemical conditions of the soil affect *P. brassicae* survival and infectivity, and therefore, club root development. Alkaline soils tend to be less favorable for clubroot and are associated with milder levels of the disease. This effect appears to result from reduced germination of the *P. brassicae* resting spores, declines in root hair infection, and inhibition of maturation of plasmodia, sporangia, and zoosporangia. As such, liming of the soil to increase alkalinity has often been suggested as a club root management strategy, particularly for vegetable Brassicas grown over smaller areas. Nonetheless, it has also been reported that acidic soils can negatively affect *P. brassicae* resting spore survival, likely to be due to stimulation of resting spore germination in absence of a host, possibly resulting in a more rapid decline in inoculum levels.

The impact of soil properties on *P. brassicae* and clubroot development is dependent on soil inoculum density, and consequently, under high inoculum concentrations, their effect is hidden. As a result, it has been proposed that the longevity and viability of *P. brassicae* inoculum is determined by soil type, pH, ion concentration and host susceptibility, which ultimately influence pathogen inoculum density at a particular site. Studies directly addressing the relationship between chemical soil properties and inoculum density of *P. brassicae* are, however, scarce. The development and implementation of improved practices for club root management require a better understanding of the epidemiology of this disease. From a practical perspective, spatial epidemiological studies can help to enhance knowledge of pathogen biology and ecology, which is essential for refined and more effective sampling methods and disease management programs. Soil borne diseases are characterized by aggregated spatial patterns, with dynamic patches that vary in size over time, and which may reflect local soil conditions that favor the disease or inhibit crop growth. Observations of in-field club root spatial patterns have indicated that disease incidence is higher at the field entrances or field margins, suggesting that disease spread occurs mainly through the movement of infested soil on farm machinery. However, while it is widely known that club root tends to have a patchy spatial pattern, detailed evaluations of within field spread over time or the effect of soil properties on inoculum density have not been conducted. The percentage of positive samples as well as the maximum inoculum density increased from 2017 to 2019, which at first glance suggested an expansion of the infested area in all fields from one year to the other. Maximum inoculum density variation was within the same order of magnitude in Fields 1 and 2, but increased by two orders of magnitude in Fields 3 and 4 between 2017 and 2019 (Table 11) (Botero-Ramirez et al., 2021).

Table 11: Minimum and maximum *Plasmodiophora brassicae* inoculum density quantified by qPCR in soil samples collected from four clubroot-infested fields located in Sturgeon and Westlock counties in central Alberta, Canada.

Field/Year	2017		2019	
	Minimum (Resting Spores/g Soil)	Maximum (Resting Spores/g Soil)	Minimum (Resting Spores/g Soil)	Maximum (Resting Spores/g Soil)
Field 1	na *	1.4×10^5	4.3×10^3	2.7×10^5
Field 2	1.1×10^4	1.7×10^7	6.7×10^3	9.9×10^6
Field 3	na	1.7×10^3	4.3×10^3	1.7×10^5
Field 4	5.4×10^3	1×10^5	5×10^3	3.2×10^7

*na, not applicable.

Fitted semi-variograms indicated spatial autocorrelation in all fields with a small nugget effect (error). Structural variance *(C/(C$_0$+C))* ranged from 77% to 100%, suggesting high spatial dependency of the inoculum density in all of the fields. In Fields 2 and 4, where semi-variograms for both years were fitted, the slope of the semi-variogram curve in the spatial range was higher in 2019 than in 2017, indicating an increase in the spatial autocorrelation. The patch diameter, measured by the spatial range, changed between years in all of the fields, with average patch growth of 221.3 m (Table 12).

Table 12: Semi-variogram parameters describing the spread of the log-transformed *Plasmodiophora brassicae* inoculum density in four clubroot-infested fields located in Sturgeon and Westlock counties in central Alberta, Canada.

Field	Year	Nugget (C_0)	Sill ($C+C_0$)	$C/(C+C_0)$	Range (m)
Field 1	2017 *	NA	NA	NA	40.0 †
	2019	0.01	0.20	0.96	289.2
Field 2	2017	0.12	0.72	0.83	346.1
	2019	0.00	4.00	1.00	634.9

Field	Year	Nugget (C_0)	Sill ($C+C_0$)	$C/(C+C_0)$	Range (m)
Field 3	2017 *	NA	NA	NA	40.0 †
	2019	0.00	0.10	1.00	77.7
Field 4	2017	0.00	0.912	1.00	113.6
	2019	0.58	2.49	0.77	422.9

*Identification of only one positive sample did not allow fitting of a semi-variogram, and hence parameters are not presented; NA, not applicable. † When only one positive sample was observed, half of the distance between sampling points was assumed as the range (40 m).

P. brassicae inoculum occurred mostly at the field margins, converging at the field entrance. Infested patches grew each year, with this growth determined by the maximum inoculum density and active dispersal methods. In the fields evaluated, wind and possible mechanical operations contributed to in-field dispersal of the pathogen. These results suggested that adoption of field sampling strategies based on the likely spatial patterns of the pathogen, with more intense sampling towards the field edges and scattered sampling at the center of the field, may be warranted. Nonetheless, further research is required to improve understanding of the underlying processes determining *P. brassicae* spatial patterns, limit further spread of the pathogen, and optimize in-field management practices.

5.16 Effects of Inoculum Density and Cultivar Susceptibility on *Rhizoctonia* Damping-Off and Crown & Root Rot

Rhizoctonia crown and root rot (RCRR) (*Rhizoctonia solani*) is influenced by soil temperature and moisture. A daily accumulation of 11 growing degree days (GDDs) was required for infection of 10-week-old sugar beet with increasing disease as temperature increased to an optimum of 22 GDDs per day (Jason R. Brantner and Ashok K. Chanda, 2021). At the optimum 22 GDDs per day, disease developed at soil moistures as low as 25% moisture holding capacity, with enhanced development as soil moisture levels increased. Although soil temperature and moisture are certainly important factors in disease onset, other factors including inoculum density and cultivar susceptibility likely play a role. With the soilborne pathogen *Verticillium dahliae* the higher the inoculum density, the earlier the disease onset in cauliflower. Similarly, for Fusarium wilt of chickpea, increasing inoculum density of *Fusarium oxysporum* f. sp. *ciceris* caused an exponential reduction in disease incubation period and an increase in disease incidence and area under the disease progress curve (Navas-Cortes et al., 2000). Most of the previous research with sugar beet and *R. solani* inoculum density has focused on documenting the effects of rotation crops on inoculum and subsequent sugar beet crop. The nonhost crop wheat reduced *Rhizoctonia* inoculum potential in the soil, whereas incorporation of host plant debris (sugar beet and maize) into soil increased *Rhizoctonia* inoculum potential and incidence of sugar beet rot.

Intercrops used for biocontrol of beet cyst nematode showed different levels of susceptibility to *R. solani* and affected disease on sugar beet in the subsequent year. With other crops, *R. solani* inoculum density has been shown to impact disease development. Higher inoculum densities of *R. solani* AG 8 and *R. oryzae* resulted in pre-emergence damping-off, stunting, and early crown root production in barley. Inoculum density of *R. solani* decreased emergence and increased root rot severity in both faba bean and canola. In greenhouse studies, increasing *R. solani* inoculum density resulted in increased damping-off and hypocotyl rot in dry beans at 14 days after planting (DAP) and decreased emergence and dry matter production and increased root rot severity in chickpea. In contrast to these studies, it was found that areas of diseased beet and high *R. solani* soil populations that developed in subsequent sugar beet crops did not coincide with previously selected disease areas, and monitoring of *R. solani* populations in the season prior to and during the early season of sugar beet did not provide a basis for forecasting disease in fields. Initiation of disease patches in these sugar beet fields was therefore governed by factors other than inoculum density. Several factors were involved in *R. solani* disease patch dynamics, including primary inoculum density as influenced by previous disease, saprophytic growth on host debris, sclerotial development, effects of soil micro flora, rate of growth for secondary infection, and plant density.

Sugar beet cultivar resistance has also been shown to reduce RCRR, increase sugar yield, and decrease inoculum in the soil compared with a susceptible cultivar. Disease severity in field trials was reduced by 39 to 46%, depending on the crop rotation, with a resistant cultivar compared with a susceptible cultivar. White sugar yield in the same trial was increased by 9 to 19% with the resistant cultivar compared with the susceptible cultivar. In field trials under natural infestations, it was found that disease severity of 6% for the susceptible cultivar compared with 1 and 1.5% for the intermediate and resistant cultivars. At these low disease levels, white sugar yield was still 8.5% higher for the susceptible and intermediate cultivars compared with the resistant cultivar. Under inoculated conditions, however, disease severity was more than twice as high and white sugar yield was 37% less for the susceptible cultivar than the intermediate and resistant cultivars. In *Cylindrocladium* black rot of peanut, onset of epidemics was delayed 1.5 to 2 weeks in a moderately resistant cultivar and further in highly resistant breeding lines compared with a susceptible cultivar. Cultivar resistance has been shown to increase as the age of plants increases in both field and growth chamber trials (Jason R. Brantner and Ashok K. Chanda, 2021).

Previous studies looking at inoculum density of *R. solani* in sugar beet have focused on measuring the effects of rotation crops on inoculum and subsequent disease. The importance of sugar beet cultivar resistance on disease level and sugar yield at harvest has also been studied. Unlike previous studies, this study used a range of inoculum densities and cultivar susceptibility levels to demonstrate the importance of *R. solani* inoculum density and sugar beet cultivar susceptibility on the temporal development of damping-off and early-season root rot, which lead to losses in plant stand, crown and root rot, and resulting sugar beet yield and sucrose quality under field conditions. Inoculum densities used in this trial were chosen to create mild, moderate, and severe disease pressure similar to that observed in commercial fields either in patches or across whole fields based on our experience infesting soils for field trials. Stand loss in sugar beet caused by *R. solani* AG 2-2 can occur throughout the growing season, but plants are more susceptible when young (Liu et al., 2019). Information on the temporal development of Rhizoctonia disease in field-grown sugar beet is critical for management decisions regarding the need for and timing of fungicide that can be applied in furrow and/or postemergence to protect against early-season stand loss and crown and root rot.

5.17 The effect of Aggregation of Pathogen and Biocontrol Microbe Inocula on Biocontrol Potential

One important ecological trait affecting microbial population dynamics is related to the spatial distributional properties of their propagules and their abilities of being spatially dispersed (Xu et al., 2020). The spatial distribution and dispersal of propagules have been shown to affect biocontrol efficacies for specific pathosystems. The initial distribution of introduced BCA propagules can largely influence their distance to pathogen propagules, affecting subsequent biocontrol outcomes. This is supported by a limited number of modelling studies, including a spatially explicit simulation study of *Ulocladium atrum* against *Botrytis cinerea,* spatial host heterogeneity in affecting biocontrol of fungal pathogens with mycoviruses and spatial distribution of mycoparasite propagules in soil on controlling soilborne pathogens. Furthermore, environmental heterogeneity, on both spatial and temporal scales, is the rule rather than an exception when foliar pathogens are concerned. Such spatio-temporal heterogeneity can affect pathogens and BCAs differentially, thus affecting biocontrol outcomes.

Xu et al.,(2020) focused on the effects of the spatial aggregation of initial pathogen and biocontrol propagules (inocula) under spatially homogeneous and heterogeneous conditions on biocontrol potential. The simulation showed that, as expected, increasing the biocontrol propagule density led to increased biocontrol potential. Under a given inoculum density, increasing spatial aggregation of BCAs is not only likely to reduce biocontrol potential but also to increase variabilities in biocontrol outcomes. A spatially random distribution of BCA propagules is most likely to result in best biocontrol outcomes. Spatial aggregation of BCAs is more important than spatial heterogeneity in influencing biocontrol potential. Thus, the present simulation study illustrates the importance of ensuring a close-to-homogeneous distribution of BCA propagules for maximising biocontrol potential. By the same reasoning, a faster host growth rate will reduce biocontrol potential if BCA cannot keep up with host growth in terms of coverage: increasing BCA aggregation essentially leads to decreased effective coverage.

A simple simulation approach was used to illustrate the importance of the spatial distribution properties of BCA propagules on potential biocontrol outcomes. The distance of each pathogen propagule to its nearest BCA propagule was used to represent biocontrol potential. As expected, increasing BCA abundance led to increased biocontrol potential. Under both homogeneous and heterogeneous conditions, spatial distributions of BCA propagules were the most important factor affecting biocontrol potential. The more spatially aggregated BCA propagules were the more variable and lower the biocontrol potential was. Unlike previous studies of biocontrol under spatially heterogeneous conditions (Kessel et al.,2005), the effects of initial aggregation of BCA and/or pathogen propagules on the biocontrol efficacy were not confounded with other dynamic aspects of BCA/pathogen development. Moreover, the present study investigated the joint effects of spatial heterogeneity on leaf surfaces and spatial distributional properties of pathogen/BCA propagules on biocontrol potential.

The distance between a pathogen and its nearest BCA propagule is likely to overestimate biocontrol potential, particularly under clustered situations. In the present study, many pathogen propagules could all share the same BCA propagule as their nearest BCA neighbour, especially when pathogen propagules were extremely aggregated. For instance, propagules within a sporulating lesion or bacterial biofilm are extremely aggregated. Bacterial biofilms can protect bacterial aggregates from adverse conditions and reduce their contact with potential antagonists (Nongkhlaw and Joshi, 2014). Therefore, under these circumstances, a single BCA propagule may not be able to suppress the highly aggregated pathogens, particularly if the biocontrol mechanism involved is competition or direct parasitism. It was suggested that reducing aggregation of BCA propagules whilst maintaining the overall population density is the most important first step to improve biocontrol potential in practice. One way to achieve this is to improve spray coverage. The coverage of wheat spikes by *B. amyloliquefaciens* post application was inadequate for protecting spikes from Fusarium head blight pathogens in the field. Commercial field spraying equipment for strawberry could result in not only large unprotected areas (30–70%) but also large variability in the coverage (Xu et al.,2020).

It was concluded that under a given inoculum density, increasing spatial aggregation of BCA propagules is not only likely to reduce biocontrol potential but also to increase variabilities in biocontrol outcomes. Thus, it is critically important to ensure a close-to-homogeneous distribution of BCA propagules in order to maximise biocontrol potential in commercial agriculture.

5.18 Inoculum Potential of *Sclerotinia sclerotiorum* Sclerotia

The soilborne fungus *Sclerotinia sclerotiorum* infects many important crop plants. Central to the success of this pathogen is the production of sclerotia, which enables survival in soil and constitutes the primary inoculum. This study aimed to determine how crop plant type and *S. sclerotiorum* isolate impact sclerotial production and germination and hence inoculum potential (Taylor et al., 2018). Three *S. sclerotiorum* isolates (L6, L17, L44) were used to inoculate plants of bean, carrot, lettuce, oilseed rape (OSR) and potato, and the number and weight of sclerotia per plant quantified. Carpogenic germination of sclerotia collected from different hosts was also assessed for L6. Production of sclerotia was dependent on both crop plant type and *S. sclerotiorum* isolate, with OSR and lettuce supporting the greatest number (42–122) and weight (1.6–3.0 g) of sclerotia per plant. The largest sclerotia were produced on OSR (33–66 mg). The three *S. sclerotiorum* isolates exhibited a consistent pattern of sclerotial production irrespective of crop type; L6 produced large numbers of small sclerotia while L44 produced smaller numbers of large sclerotia, with L17 intermediate between the two. Germination rate and percentage was greatest for larger sclerotia (4.0–6.7 mm) and also varied between host plants. Combining sclerotial production data and typical field crop densities suggested that infected carrot and OSR could produce the greatest number (3944 m^{-2}) and weight (73 g m^{-2}) of *S. sclerotiorum* sclerotia, respectively, suggesting these crops potentially contribute a greater increase in inoculum. This information, once further validated in field trials, could be used to inform future crop rotation decisions.

In addition to representing a huge reproductive potential and providing the primary inoculum for subsequent epidemics, these sclerotia enable the fungus to survive over winter (or for longer periods) in soil. *Sclerotinia sclerotiorum* sclerotia can then germinate carpogenically or myceliogenically, depending on environmental conditions, with the former resulting in the production of apothecia and airborne ascospores. This often represents the principal source of inoculum and hence allows infection of aboveground plant parts. Carpogenic germination of *S. sclerotiorum* occurs only in sclerotia that are brought close to the soil surface and, in temperate climates, a period of cold conditioning is also required to break dormancy. The number of sclerotia in the soil increases when successive susceptible crops become infected by *S. sclerotiorum*, leading to escalating disease problems in many cropping rotations. The situation is exacerbated by short rotations and pressure on land use, as *S. sclerotiorum* sclerotia are generally thought to survive for 4–5 years in soil under natural conditions, although this period can vary depending on factors such as depth, temperature and moisture. However, few researchers have attempted to investigate and quantify the effect of crop type and *S. sclerotiorum* isolate on the production of sclerotia and hence understand how these factors might influence subsequent inoculum pressure.

The data generated for the number of *S. sclerotiorum* sclerotia produced on infected plants in this study allowed estimation of the inoculum potential for different crops. Although infection of OSR and lettuce resulted in greater numbers of sclerotia per plant than the other crops, carrot was identified as potentially supporting the greatest returns of sclerotia to the soil when planting density in the field was taken into account, producing 2666–4794 sclerotia m^{-2} compared with 1026–2754 sclerotia m^{-2} in OSR. However, based on the weight of sclerotia produced per plant and crop density, both OSR and carrot constitute the greatest returns with 86 g and 83 g sclerotia m^{-2}, respectively. These are estimated figures and also assume that all plants are completely colonized by *S. sclerotiorum*, which is unlikely in a field situation. Nonetheless, these estimates appear to be consistent with some preliminary field data where sclerotia on the soil surface were counted in quadrats for crops with high levels of sclerotinia disease. Up to 3000 sclerotia m^{-2} were recorded for a carrot crop compared to up to 900 m^{-2} for OSR and 195 m^{-2} for potato (Young *et al.*, 2014). Data from a published field survey in the UK suggested that pea, potato and sunflower crops resulted in high returns of sclerotia to the soil of up to six sclerotia per kg soil, with 92 sclerotia m^{-2} being recorded on the soil surface of an infected pea crop. These crops were therefore considered a major threat to subsequent OSR plantings. However, subsequent tillage operations may redistribute sclerotia left on the soil surface throughout the soil profile and many will be killed by microorganisms or adverse environmental conditions. Only a small proportion of sclerotia that remain close to the soil surface can subsequently germinate and these are difficult to quantify. As stated in the introduction, just a few germinating *S. sclerotiorum* sclerotia are required to initiate disease due to the large numbers of ascospores produced by apothecia, and as few as two apothecia m^{-2} were shown to lead to 11% disease incidence in soybean.

Sclerotial production and germination by *S. sclerotiorum* is dependent on isolate, host plant and size of sclerotia. Both carrot and OSR potentially return more sclerotia by weight or number to the soil compared to the other plants tested here and hence potentially pose a bigger disease risk for subsequent susceptible crops (Taylor et al., 2018). However, further work is required to extensively quantify sclerotial production in the field in order to validate this. This information could then be used to develop optimum crop rotation strategies to reduce the impact of sclerotinia disease and maximize financial returns. This has been attempted for *S. sclerotiorum* using a modelling approach based on assumptions relating to potential yield loss for different combinations of susceptible and breaks crops over time. It was concluded that rotations selected to reduce build-up of sclerotia were financially justified.

5.19 Inoculum Potential of *Fusarium* species

i. *Fusarium verticillioides:* José da Cruz Machado et al., (2013) evaluated the effect of *Fusarium verticillioides* on the early stage development of maize seedling/plants, considering different inoculum potentials of this fungus in seeds, under controlled conditions. To obtain seeds with different inoculum potentials of that fungus, the osmo-priming technique as referred to in literature was used. By this technique, the non-germinated seeds of the same lot, exhibiting typical structures of *F. verticillioides*, were grouped into five categories according to a grading system that was established based on the visible area of the seeds covered by the fungal structures. Seeds of each category were sowed in soil substrate and plants were grown under controlled conditions, 25°C and 15 h light/day, for 25 days. All variables used, plant population (stands), speed of seedling emergence, and height/weight of emerged plants, were negatively affected by the increase of the inoculum intensity of the fungus initially present in the seeds. The regression equation showed high values of correlation coefficient (>0.85) for all variables, except for plant height which presented correlation coefficient of 0.77. The most severe effect of the fungus in this study occurred on the root system of the emerged plants as determined by the highest inoculum intensity of the pathogen on the planted seeds.

Fusarium verticillioides may be quite damaging to maize at the initial stage of the crop development depending on its inoculum potential present in the seeds. Plant population (stands) declined gradually in the proportion of the inoculum potential evaluated at 15 and 25 days after sowing a reduction of about 60% in such variable occurred when inoculum intensity of the fungus was higher than 75% according to the grading system used in this work. For both features initial and final stands, the correlation coefficients (R^2) were high, then explaining the exclusive effect of that pathogen on seed and seedling performance of maize under the circumstances of this experiment. Regarding the effects of that fungus on the root system, it was clear that the damages were also quite high with the increased inoculum potential in the seeds. The dry weight in this case was reduced to almost nil when seeds, used for planting, were covered by the fungal mycelium at proportions higher than 75%. It is also timely to register that, in any case, necrotic symptoms on the roots caused by *F. verticillioides* were seen (José da Cruz Machado et al., 2013). Except for the emergence speed index, the effects of *F. verticillioides* on the initial development of plants of maize for all variables considered were inversely proportional to the initial inoculum intensity (potential) in the seeds.

In general, a high proportion of seed lots in Brazil presents variable levels of such fungal species. Under normal weather conditions, crop damage caused by that organism is hardly observed, neither on seed germination nor on the developing crop in the field. However, under stress conditions, in which high and prolonged relative humidity occur in some seasons damages caused by that species can be expressive, as part of a complex of other organisms causing seed deterioration. Information on the single effects of *F. verticillioides* on seed performance and on the crop is hardly found in literature, except in cases mycotoxins are concerned. The presence of *F. verticillioides* in seed lots of maize may affect the performance of the emerged plants at variable intensity depending on the proportion of infected seeds with high inoculum potential. The highest inoculum potential in seeds, the most severe are the effects of the fungus on the development of the seeds and seedlings/plants. The osmo-priming technique used to screen seeds with different inoculum potentials of *F. verticillioides* proved to be useful and reliable as it has been demonstrated for other pathosystems. The inoculation of seeds by the soaking process, in which seeds are imbibed in conidial suspensions or by dry mixing seeds with inoculum, deals with the contamination process rather than infection. It has been also demonstrated that seeds of some species submitted to soaking treatment may be physically injured.

The pathogenic fungi may be associated with the seeds during development, harvest and storage, and it can be disseminated through them. This work evaluated the seeds performance of three hybrid corn cultivars, BM 840 PRO, BM 810 and BM 207, submitted to different inoculum potentials of *Fusarium verticillioides* (Tatiana Botelho Fantazzini et al., 2016). Seeds were uniformly distributed over the medium fungal colony developed on PDA (potato dextrose agar) containing mannitol in water potential -2.0 MPa. Seeds remained in this way for different periods of contamination (0, 24, 48, 72 and 96 hours). Before and after seven months of seed storage the evaluations were performed. Seeds treated with fungicide Derosal Plus® (carbendazim + thiram, 250 mL p.c. 100 kg-1 of seeds) were used as control. The increased of inoculum potential of *F. verticillioides* fungus on corn seeds caused reduction in physiological quality in the potentials 72 and 96 hours of exposure before and after storage. The highest *F.verticillioides* incidence is observed in non-stored seeds. *F.verticillioides* incidence on corn seeds accentuates the deterioration in storage.

F.verticillioides incidence in corn seed of the cultivar BM 840 PRO showed lower average 38% in inoculum potential of 24 hours after seed storage. But with increasing inoculum potential, these averages did not differ among studied periods, with values close to 100%. For the cultivar BM 810, it was found a significant difference among seeds non-stored with those stored in the potentials of 24 and 48 hours. There is a reduction of the physiological quality of the seeds inoculated with *F. verticillioides* in the range from 72 to 96 hours of exposure, both before and after storage. The highest *F. verticillioides* incidence is observed in the non-stored seeds. The *F. verticillioides* incidence on corn seeds accentuates the deterioration in storage.

ii. *Fusarium avenaceum* and *F. graminearum*: The increased occurrence of *Fusarium*-mycotoxins in Norwegian cereals over the last decade, is thought to be caused by increased inoculum resulting from more cereal residues at the soil surface as a result of reduced tillage practices. In addition, weather conditions have increasingly promoted inoculum development and infection by *Fusarium* species (Hofgaard et al., 2016). The objective of this work was to elucidate the influence of different tillage regimes (autumn plowing; autumn harrowing; spring plowing; spring harrowing) on the inoculum potential (IP) and dispersal of *Fusarium* spp. in spring oats. Oat residues from the previous year's crop were collected within a week after sowing for evaluation. IP was calculated as the percentage of residues infested with *Fusarium* spp. multiplied by the proportion of the soil surface covered with residues. *Fusarium avenaceum* and *F. graminearum* were the most common *Fusarium* species recovered from oat residues. The IP of *Fusarium* spp. was significantly lower in plowed plots compared to those that were harrowed. Plowing in either the autumn or spring resulted in a low IP.

Harrowing in autumn was more effective in reducing IP than the spring harrowing, and IP levels for the spring harrowed treatments were generally higher than all other tillage treatments examined. Surprisingly low levels of *F. langsethiae* were detected in the residues, although this species is a common pathogen of oat in Norway. The percentage of the residues infested with *F. avenaceum, F. graminearum, F. culmorum*, and *F. langsethiae* generally related to the quantity of DNA of the respective *Fusarium* species determined using quantitative PCR (qPCR). *Fusarium* dispersal, quantified by qPCR analysis of spore trap samples collected at and after heading, generally corresponded to the IP. *Fusarium* dispersal was also observed to increase after rainy periods. These findings are in line with the general understanding that plowing is a means to reduce the IP of *Fusarium* spp. in cereal fields. The main inoculum source for *F. langsethiae* remains unclear. These results will be useful in the development of forecasting tools to calculate the risk of *Fusarium* in cereals.

Tillage significantly influenced the IP of both *F. avenaceum* and *F. graminearum* at most locations, whereas no significant effects of straw removal were evident. However, significant interactions between straw removal and tillage regime were found within many fields. Therefore, a Tukey analysis was performed to identify significant differences in IP of *F. avenaceum* and *F. graminearum* between combined treatments (straw removal combined with tillage treatment). Generally, higher IPs were estimated for harrowed compared to plowed plots. Within individual locations, the highest IPs were found on spring-harrowed plots where the straw was retained after harvest. In fields where the IPs exceeded 10% in spring-harrowed plots, significant differences in IP were often recorded between harrowed and ploughed plots. Moreover, in some of these spring-harrowed plots, straw removal significantly reduced the IP of *F.*

avenaceum and *F. graminearum*. Hardly any significant effect of tillage on the IP was detected in fields where the *Fusarium* spp. were isolated from less than 25% of the residues sampled, as was the case for *F. graminearum* in Østfold and Solør in 2012 (Hofgaard et al., 2016). Similarly, hardly any significant effect of soil cultivation was detected when the different tillage treatments resulted in a maximum plant residue cover below 20%. The lowest IPs (<5%) were estimated for ploughed plots where the removal of straw residue or timing of tillage treatment had no apparent impact upon the *Fusarium* IP. No significant differences in IPs were found between spring and autumn ploughed plots.

Removal of straw residues generally reduced the IP in harrowed treatments, though the effect was not always significant. By contrast, the removal of straw residues did not significantly influence the IP on the ploughed treatments. The reason for the different effect of straw removal in harrowed vs. ploughed treatments is most probably the generally low amount of residues, and the correspondingly low IPs, in the plowed treatments. These results were in line with the general understanding of plowing as the best means of reducing the occurrence, and thereby the IP, of *Fusarium* spp. in cereal fields. The effect of tillage on FHB development may not always be evident, as weather factors play an important role in the dispersal and infection of *Fusarium* spp. (Prussin et al., 2014). According to this data, spring plowing appears to be the best option to both reduce the risk of soil erosion and at the same time minimize *Fusarium* diseases. Removal of cereal straw in autumn would be the best practice for reducing the IP of *Fusarium* spp. where harrowing is the preferred tillage system.

iii. *F. proliferatum*, *F. oxysporum* and *F. redolens*: Asparagus decline syndrome (ADS), one of the most important diseases affecting asparagus crops, causes important yield losses worldwide. *Fusarium proliferatum, F. oxysporum* and *F. redolens* are among the main species associated with ADS (de la Lastra et al., 2021). To explore their potential inoculum sources and the effectiveness of soil disinfestation practices for ADS management, molecular methods based on a quantitative real-time polymerase chain reaction (qPCR) were developed. qPCR-based molecular tools demonstrated advantages in the sensitive and specific detection and quantification of fungal pathogens in comparison with less-accurate and time-consuming traditional culture methods. *F. proliferatum, F. oxysporum* and *F. redolens* could be specifically detected and accurately quantified in asparagus plants, soil and irrigation water collected from asparagus fields with ADS symptoms by means of the designed TaqMan qPCR protocols. Furthermore, these molecular tools were successfully applied for evaluation of the efficacy of diverse soil disinfestation treatments. Chemical fumigation with dazomet and biosolarization with pellets of *Brassica carinata* contributed to a significant reduction in the inoculum densities of the three *Fusarium* species in treated soils, which was correlated with production increases.

The capability to accurately detect and quantify the main *Fusarium* species involved in ADS in plants, soil and water samples by means of qPCR will allow identification of high-risk fields that can be avoided or managed to reduce yield losses. Quantification of pathogen densities in the soil may also provide essential insights into the effectiveness of soil disinfestation methods for ADS management.

iv. *Fusarium acuminatum*, *F. avenaceum*: The objective of a study by Abigail J. Stack et al., (2022) was to assess the occurrence and pathogenicity of *Fusarium* species in an almond (*Prunus dulcis*) bare root tree propagation system with focus on asymptomatic host and other production materials to identify potential inoculum sources through genetic characterization of isolates. In addition to *F. acuminatum* and *F. avenaceum*, we included *F. brachygibbosum* and *F. californicum*, which were identified during the course of this study and shown to be capable of causing disease. Surveys of several impacted bareroot-propagated almond nurseries and pathogenicity tests of fungi isolated from symptomatic and asymptomatic trees identified *Fusarium acuminatum*, *F. avenaceum*, and to a lesser degree, *Ilyonectria robusta* and *Cylindrocarpon obtusiusculum*, as causal agents of the disease. It was also found in our pathogenicity assays that reduced bark moisture, imposed before inoculation (a stress that occurs during cold storage and transplanting), strongly increases size of necrotic lesions. Surveys to determine potential inoculum sources within the nursery production system were incomplete in the initial study.

Based on the MLST analysis (i.e., using *TEF1-a*, *ITS*, *TUB2*, *RPB1*, and *RPB2*), seven species of *Fusarium* were identified. Of the 77 isolates from Nursery A that were sequenced, 40 were *F. acuminatum* or *F. avenaceum* , five were *F. californicum*, and an additional 11 isolates were *F. brachygibbosum*. Of the 21 isolates from Nursery B that were sequenced, 10 were either *F. acuminatum* or *F. brachygibbosum*. Of the five isolates from nursery C, one was *F. brachygibbosum*. In addition to the isolates mentioned above, five isolates were *F. californicum* and two isolates were *F. solani*. Eighteen isolates of *F. equiseti* and one isolate of *F. proliferatum*, both nonpathogenic in almond, also were identified. The findings confirm asymptomatic infections of almond trees and their rootstocks and wheat residues by *F. acuminatum*. This study expands potential sources of *F. acuminatum* inoculum to include soil, metal and wood pallets, carts, and walls inside of cold-storage facilities. Recovery of *F. acuminatum* from air samples indicates that this pathogen can be dispersed aerially within a cold-storage unit. *F. avenaceum* was isolated from bareroot almond trees and a cart inside a cold-storage facility. Somatic compatibility data suggest that *F. acuminatum* isolates recovered in our survey were associated with a single SCG and hence a single clonal lineage. Likewise, all four *F. avenaceum* isolates corresponded to one SCG. Whether or not pathogenicity to almond is restricted to the lineages identified for these two species is unclear, but seems unlikely given the wide host ranges attributed to both *F. acuminatum* and *F. avenaceum*.

v.*Fusarium solani*: The inoculum threshold approach was used, represented by inoculum density, a highly used measure for epidemiological studies, easily estimated in field-soil samples (Macedo R, Sales et al.,2017). Plant pathogens and their diseases will likely follow the climate-mediated distributional shifts on crops, thus creating a dynamic scenario in the management of plant disease epidemics in the face of climate change. Highlighting which regions are expected to be most at risk of disease is therefore crucial for disease management policies and therefore, maps that anticipate disease risks can potentially increase efficiency and reduce the costs of disease management strategies in the future.

Root rots are a constraint for staple food crops and a long-lasting food security problem worldwide. In common beans, yield losses originating from root damage are frequently attributed to dry root rot, a disease caused by the *Fusarium solani* species complex. The aim of this study was to model the current potential distribution of common bean dry root rot on a global scale and to project changes based on future expectations of climate change. Macedo R, Sales et al., (2017) used a spatial proxy of the field disease occurrence, instead of solely the pathogen distribution; modeled the pathogen environmental requirements in locations where *in-situ* inoculum density seems ideal for disease manifestation. A dataset of 2,311 soil samples from commercial farms assessed from 2002 to 2015 allowed us to evaluate the environmental conditions associated with the pathogen's optimum inoculum density for disease occurrence, using a lower threshold as a spatial proxy. They encompassed not only the optimal conditions for disease occurrence but also the optimal pathogen's density required for host infection.

An intermediate inoculum density of the pathogen was the best disease proxy, suggesting density-dependent mechanisms on host infection. A strong convergence was found on the environmental requirements of both the host and the disease development in tropical areas, mostly in Brazil, Central America, and African countries. Precipitation and temperature variables were important for explaining the disease occurrence (from 17.63% to 43.84%). Climate change will probably move the disease toward cooler regions, which in Brazil are more representative of small-scale farming, although an overall shrink in total area (from 48% to 49% in 2050 and 26% to 41% in 2070) was also predicted. Understanding pathogen distribution and disease risks in an evolutionary context will therefore support breeding for resistance programs and strategies for dry root rot management in common beans.

An intermediate inoculum density (3700 propagules per soil gram) was the best proxy of disease occurrence, perhaps a more representative estimate of infestation by the FSSC in Brazilian common bean fields. The potential distribution of the disease in both the present and future times was also remarkably convergent to the predicted tropical areas (Central and South American and African continent) as highly suitable for common bean crops. The dry root rot distribution in the common bean was linked to all inoculum densities, but the intermediate threshold was the best spatial proxy for disease occurrence. The correlation between the inoculum threshold and disease occurrence therefore exhibited an intermediate saturation baseline. The density-dependent relationships of the population growth of soilborne pathogens may be the cause of such patterns in epidemiology as well as interspecific relationships in the soil community. Soilborne pathogens may be endemic and cause diseases in natural ecosystems, but their main nutrient resource, the host, is obviously a crucial driver of their distribution (Macedo R, Sales et al.,2017). However, pathogen populations are also dependent on a certain within-host density that does not overcome the host's carrying capacity, to avoid its complete depletion. This is the case of FSSC × common bean, which results in stunted, low-yield plants and rarely in plant death. A host supportability therefore exists so that the mechanisms of intra and inter competition might regulate the abundance of soil borne pathogens on a small scale.

5.20 Evaluation of Inoculum Potential of Pathogens in Seeds

Given what is already known in regard to seed health and the availability of molecular methods for detection of the pathogens *Stenocarpella maydis* and *Stenocarpella macrospora* in maize seeds, *Colletotrichum gossypii* var. *cephalosporioides* in cotton seeds, and *Corynespora cassiicola* in soybean seeds, the aim of this study was to evaluate seed vigor according to different inoculum potentials (Marina de Resende Faria Guimarães et.aal., 2017). The fungus isolates were inoculated on seeds by the technique of water restriction, through which different inoculum potentials are obtained, corresponding to times of seed exposure of 0, 24, 48, and 96 hours for maize and cotton seeds, and 0, 36, 108, and 144 hours for soybean seeds. The seeds were subjected to germination, electrical conductivity, health, and qPCR tests. Results of the blotter test showed that in most pathosystems, there was a higher incidence of the fungi with an increase in inoculum potential. A decrease in germination percentage was observed in all species as inoculum potential increased, as well as further degradation of seed membranes. The qPCR test confirmed that the most damaged seeds in the tests had higher presence of the pathogens.

In all the pathosystems studied, there was a sharp fall in seed germination percentage; the highest percentages occurred at the lowest inoculum potential and the lowest percentages at the highest inoculum potential, due to the death or, in some cases, poor formation of the seeds. These results reinforce what was reported in other pathosystems, such as *Sclerotinia sclerotiorum* and *Colletotrichum truncatum* in soybean and *Fusarium oxysporum* f. sp. *phaseoli* in dry edible bean seeds. The relationship between moisture content, the level of organization of seed cell membranes, and the amount of leachates in the imbibition solution allows association of electrical conductivity to seed vigor, in which high values of electrical conductivity indicate low vigor, and low values indicate high physiological quality of the seeds and, therefore, high vigor. Results in regard to electrical conductivity showed that the increase in the period of seed exposure to the pathogen led to reduction in seed vigor at higher inoculum potentials, considering that under this condition there is greater degradation of the cell membrane system, leading to leaching of solutes, which, consequently, increases seed deterioration. The lower values of conductivity correspond to lower release of exudates and indicate high seed vigor, as well as greater organization of the cell membrane systems. Upon comparing the results of the physiological and seed health test applied in this study and the concentrations of DNA from inoculated fungi at the different inoculum potentials, there is clearly proportionality between the values of the tests and the quantities of DNA extracted and evaluated by the qPCR technique. This study also confirms the efficiency of the primers in detection of pathogens in the seeds at different inoculum potentials.

The presence of the pathogens *Stenocarpella maydis* and *S. macrospora* in maize seeds, *Colletotrichum gossypii* var. *cephalosporioides* in cotton seeds, and *Corynespora cassiicola* in soybean seeds through inoculation by the osmotic conditioning method can lead to gradual and varied reductions in the physiological quality of the seeds at a proportion opposite to the inoculum potentials of these organisms quantified through qPCR Marina de Resende Faria Guimarães et al., 2017). The effects of *S. macrospora* in maize seeds are less than those observed for *S. maydis*. The effects of *C. cassiicola* in association with soybean seeds were less intense, regardless of the inoculum potential used. For all the pathosystems studied, the primers used in detection and quantification of the target DNA were specific, with linearity in the standard curve at each level of DNA dilution. Results showed proportionality between the fungal DNA extracted from the seeds, the inoculum potentials of each pathogen, the effects represented by germination percentages, and seed vigor.

Plant growth-promoting bacteria (PGPR), while generally considered to aid plant growth with the provision of nutrients, can also be used as biocontrol agents for plant pathogens. The study assessed the protective potential of inoculums and metabolites of plant growth promoting rhizobacterial strains against bacterial and fungal pathogens on soybean seedlings (Ajinde et al., 2024). In the setup infected with *Alternaria* sp., final germination values of seeds in the presence of the respective inoculums showed no significant variation between the treated only and the infected then treated setup. In the case of seeds infected with *Aspergillus niger,* higher germination and vigor index values were observed in the treated-only seeds when compared with the infected-then-treated seeds. For seeds infected with *Corynespora* sp., significantly lower germination and vigor index values were observed in the infected then-treated seeds than the treated-only seeds in the presence of the respective inoculums. With regards to setup infected with *Fusarium oxysporum*, significantly higher final germination and vigor index values were recorded for the treated only seeds when compared with the infected then treated setups. For the *Xanthomonas campestris* infected seeds, the majority of the infected then metabolite-treated seeds showed significantly lower final germination values when compared with the treated-only seeds. The study findings were able to establish the efficacy of some bacteria agents against economically important species of plant pathogens.

5.21 Effect of Inoculum Levels on Disease Development and Survival Potential

Being a homothallic fungus, a single ascospore of *Sclerotinia sclerotiorum* can completes the life cycle. Pathogen passes part of its life cycle as sclerotia, a resting structure, in the soil. These sclerotia mixed into the soil during harvesting and threshing, and act as a source of inoculum for future years (Khan et al., 2022). Sclerotia are made of three layers namely rind, cortex and medulla. These layers are consisting of hyphal aggregates. Melanin is a principal compound of cell wall of outer rind and it made up of several phenolic or indolic substances, which protect sclerotia from adverse environmental threats.Number of Sclerotia in the soil play significant role in disease epidemic incited by *S. sclerotiorum.* A positive correlation was documented between density of sclerotia and disease incidence.

Sclerotinia stem rot is constantly appearing from moderate to severe form in major mustard growing tracts. Most of life *S. sclerotiorum* passes in form of sclerotiain the soil. Present study was carried out in order to determine effect of inoculum level on disease development and survival of *S. sclerotiorum*. Different inoculum levels were prepared from culture of *S. sclerotiorum* and10 sclerotia/pot were buried at various depths ranging from 1 to 20 cm for 60 days. The maximum disease incidence (93.67%) was observed with 10 per cent inoculum level; however disease incidence was least (6.34%) in 1 per cent inoculum level. Disease incidence increased with increasing in inoculum levels. In case of plant mortality, maximum plant mortality (43.00%) was recorded at 10 per cent inoculum level followed by 32.67 and 29.34 per cent @ 9 and 8 per cent, respectively. No plant mortality was noticed at 1 and 2 per cent inoculum level. Maximum amount of sclerotia germination (83.33%) was observed at 1 cm soil depth followed by 75.65, 73.34, 70.34 per cent germination @ 2, 3 and 4 cm depth of soil, respectively. It was observed that sclerotial germination decreased in tune with increase in soil depth. Lowest sclerotial germination (10%) was observed in 20 cm vertical soil depth. Findings of this study can be harnessed in varietal screening and formulation of disease management modules.

Disease incidence was increased with increasing in inoculum levels. In case of plant mortality, maximum plant mortality (43.00%) was observed at 10 per cent inoculum level; however no plant mortality was noticed at 1 and 2 per cent inoculum levels. Disease frequency was increased from 10 to 80 per cent with increasing inoculums density by 7 to 250 sclerotia/100gm soil. Positive correlation had confirmed between infection of plant parts and inoculum density of *S. sclerotiorum* (Khan et al., 2022). The results of sclerotial survival in soil depth depicted that maximum (83.33%) germination of sclerotia was observed at 1 cm soil depth, while minimum sclerotial germination (10%) at 20 cm soil depth. It was observed that sclerotial germination decreased as the soil depth increased. Germination ability of sclerotia of *S. sclerotiorum* ranged from 22 to 39 % on the soil surface or placed at 5 cm depth in dry soil. Microbial activity of soil is the key factor for reduction of germination ability of sclerotia. Increase in vertical soil depth was negatively associated with viability of sclerotia that might be due to antagonistic action of soil dwelling microbes.

It has been evident that all the tested inoculum levels were responsible for causation of disease. Per cent disease incidence and plant mortality increase was with increasing inoculum levels. However, sclerotial germination decreased as the vertical soil depth increased. It is concluded, from the results, that inoculum level studies can be exploited in screening of verities in resistance breeding programme. On the other hand, findings on sclerotia survival in soil have the potential to assist in disease prediction and management strategy adoption.

5.22 Relationship between Infective Inoculum and Pathogenic Potential

The ability of a microbe, virus, or parasite to cause disease is dependent on multiple factors, virulence factors, host immune defenses, the infective inoculum, and the type of immune response. For many microbes their capacity for causing disease is highly dependent on the inoculum (Smith and Casadevall, 2022). The mathematical formula for Pathogenic Potential (PP) is a way to compare the ability of an organism to have a pathogenic effect, as measured by Fraction with signs or symptoms (Fs), mortality (M), and inoculum (I), and can include the median survival time of the host (T). Increasing inoculum of the fungus *Cryptococcus neoformans* for a moth host resulted in exponentially smaller pathogenic potential, and the Fs/T versus inoculum plot showed a logarithmic relationship. Together, these relationships show diminishing returns with increasing cryptococcal inoculum, in which each individual fungus plays a smaller role in pathogenicity. Literature data shows that other microbes, mostly bacteria, had linear Fs/T versus inoculum relationships, which indicate that each bacterium contributed an equal amount to pathogenicity. These differences in relationships can point to differences in host-microbe interactions and suggest new ways in which the organism causes disease.

Pathogenic Potential (PP) is a mathematical description of an individual microbe, virus, or parasite's ability to cause disease in a host, given the variables of inoculum, signs of disease, mortality, and in some instances, median survival time of the host. Smith and Casadevall (2022) investigated the relationship between pathogenic potential (PP) and infective inoculum (I) using two pathogenic fungi in the wax moth *Galleria mellonella* with mortality as the relevant outcome. Our analysis for *C. neoformans* infection revealed negative exponential relationship between PP and I. Plotting the log(I) versus the Fraction of animals with signs or symptoms (Fs) over median host survival time (T) revealed a linear relationship, with a slope that varied between the different fungi studied and a y-intercept corresponding to the inoculum that produced no signs of disease. The I vs Fs/T slope provided a measure of the pathogenicity of each microbial species, which we call the pathogenicity constant or kPath. The kPath provides a new parameter to quantitatively compare the relative virulence and pathogenicity of microbial species for a given host. In addition, we investigated the PP and Fs/T from values found in preexisting literature. Overall, the relationship between Fs/T and PP versus inoculum varied among microbial species and extrapolation to zero signs of disease allowed the calculation of the lowest pathogenic inoculum (LPI) of a microbe. Microbes tended to fall into two groups: those with positive linear relationships between PP and Fs/T vs I, and those that had a negative exponential PP vs I relationship with a positive logarithmic Fs/T vs I relationship. The microbes with linear relationships tended to be bacteria, whereas the exponential based relationships tended to be fungi or higher order eukaryotes. Differences in the type and sign of the PP vs I and Fs/T vs I relationships for pathogenic microbes suggest fundamental differences in host-microbe interactions leading to disease.

The PP and PPT analysis revealed the importance of comparing results from experiments performed using the same inoculum, especially when comparing the difference in pathogenicity of different strains of the same microbial species, or when comparing a mutant strain to the wild type (Smith and Casadevall 2022). Comparing different PP and PPT derived from experiments using different inoculum could cause the ΔPP to be off by orders of magnitude depending on the nature of the curve. However, it was also demonstrated how pathogenicity data collected using different inocula can be compared by fitting Fs/T versus I plots thus providing new options for comparative analysis. These results provide support for the view the capacity for virulence is relative, such that labelling a microbe a pathogen under one set of circumstances does not mean the microbe is equally as pathogenic under a separate set of circumstances. PP and PPT themselves are not intrinsic and immoveable statements on the absolute pathogenicity of a microbe, but rather provide a way to holistically and situationally evaluate pathogenicity given specific factors and variables. The PP would also change in the setting of an infection treated with an effective antimicrobial agent, where the expected Fs and M values decrease, T increases, and I (at the beginning of therapy) remains constant, and as such, changes in PP and PP T following treatment could be used to measure therapeutic efficacy. Conversely, immunosuppressive treatments or conditions that broadly enhance host susceptibility to infection would lead to increased PP and PP T, which can then be used to identify infection-related the risks involved in certain treatments.

The relationships between parameters of pathogenicity developed here (PP, PPT, Fs/T, kPath) provide new potential insights into how the organism cause disease and death within the host. If the microbe has a positive linear relationship in the PP vs I, PPT vs I, or Fs/T vs I plots, it is consistent with the notion that disease and death primarily result from increasing microbial burden, such that each additional microbial cell causes a proportional increase in host damage that when cumulative would result in the death of the host. This could be a pathogen that damages the host directly through the production of toxic substances or indirectly by eliciting a tissue-damaging inflammatory response that kills the host in a dose-dependent manner or that the host mounts a tissue-damaging inflammatory response that is dependent on microbial burden or a combination of both.

5.23 Effect of Inoculum Load on Disease Initiation, Threshold Point and Population Build up

Studies were undertaken by Saini et al., (2019) to find out the effect of inoculum load of post harvest pathogen (*Penicillium islandicum*) on disease initiation, threshold point and population build up of blue mould rot in Indian gooseberry. Inoculations were made on the fruits of Indian gooseberry at five concentrations (10, 50, 100, 200 and 500 spores/ml). The disease initiation commenced after 54 hrs. of inoculation with inoculum load of 500 spores/ml followed by 200 spores/ml where the initiation of disease was observed after 72 hrs. At lower spore load (10 spores/ml), there was no disease initiation even after 108 hrs of inoculation. Minimum number of spores required to initiate infection after 102 hrs was 50 spores /ml which can be considered as the threshold point. Maximum population build up (2 x 10^6 and 2 x 10^8) of the pathogen occurred at 500 spores/ml after 5 and 10 days of inoculation, respectively.

The time required for appearance of visible symptoms decreased with the increase in amount of inoculum. Concentration of 5 spores /ml failed to cause infection indicating that the threshold point for the infection is more than 5 spores /ml. No symptoms were visible up to 84 hours with a spore load of 50 spores/ml and symptoms could be noticed only after 90 hours. The disease was visualized within 48 hours when 250 spores were used. The rate of development was also faster with increased inoculum load. The inoculum load has influenced the disease development as well as the rate of multiplication of the fungus as observed from the number of spores produced from variously inoculated fruits. Perusal of data indicated that the number of propagules produced increased proportionately with the number of spores used for inoculating the fruits. The number of spores produced on the fruits varied from 1.9×10^5 to 2.1×10^8 for 25 to 250 spores/ml inoculum. Investigations on the inoculum potential of *P. islandicum* on aonla at optimum temperature and humidity have revealed that the time required for the establishment of infection is positively correlated with the inoculum load. The subsequent rate was also faster with the increasing amount of inoculum. The rate of multiplication has shown positive correlation with amount of inoculum. A minimum number of 100 spores /ml was necessary for initiating the disease which indicated that threshold point for this fungus is 50 spores per inoculation. A minimum number of 50 spores/ml was necessary for disease initiation which indicates that inoculum threshold point of this fungus to cause infection is 25 spores/inoculation (Saini et al., 2019). Maximum population build up of *P. islandicum* was observed at the concentration of 500 spores/ml after 5 and 10 days of inoculation on the fruits.

Aerobiology is the science that studies airborne microorganisms and their distribution, especially as agents of infection. Although this discipline has existed for decades, the development of new molecular technologies is contributing to an increase in the use of aerobiological data for several purposes, from day-to-day monitoring to improving our understanding of pathosystems. Although the importance of knowing the size and composition of plant pathogen populations present in the air is recognized, technical constraints hinder the development of agricultural aerobiology. Van der Heyden et al., (2021) reviewed the application of spore sampling systems in agriculture and discuss the main considerations underlying the implementation of airborne inoculum monitoring. The results of this literature review confirm that the use of aerobiological data to study the escape of inoculum from a source and its role in the development of diseases is well mastered, but point at a lack of knowledge to proceed with the deployment of these systems at the landscape scale. It was concluded that airborne inoculum surveillance networks are still in their early stages and although more and more initiatives are emerging, research must be conducted primarily to integrate evolving technologies and improve the access, analysis, interpretation and sharing of data. These tools are needed to estimate short- and medium-term risks, identify the most appropriate control measures with the lowest environmental risk, develop indicators to document the effects of climate change, and monitor the evolution of new genotypes at multiple scales.

Many factors have been shown to influence bacterial transfer between surfaces, including surface type, bacterial species, moisture level, pressure, and friction, but the effect of inoculum size on bacterial transfer has not yet been established (Montville and Schaffner Donald, 2003). Bacterial cross contamination rates during performance of common food service tasks were previously determined in the laboratory using nalidixic acid-resistant *Enterobacter aerogenes*. Eight different transfer rates were determined, each involving a minimum of 30 volunteers. The influence of source inoculum level on the percentage of bacteria transferred (percent transfer rates) and log10 CFU per recipient surface was determined using statistical analysis. The effect of inoculum size on transfer rate was highly statistically significant ($P < 0.0001$) for all transfer rate data combined (352 observations) and for each individual cross contamination rate, except for data on contamination via transfer from chicken to hand through a glove barrier ($P = 0.1643$). Where inoculum size on the source was greater, transfer rates were lower, and where inoculum size on the source was less, transfer rates were higher. The negative linear trend was more obvious for activities that had a larger range of inoculum sizes on the source surface. This phenomenon has serious implications for research seeking to determine bacterial cross contamination rates, since the different transfer efficiencies that were previously shown to be associated with certain activities may actually be the result of differing initial inoculum levels. The initial inoculum size on the source and the amount of bacteria transferred must both be considered to accurately determine bacterial transfer rates.

5.24 Effects of Inoculum Density and Substrate Type on Saprophytic Survival of Fungi

Sampling was carrid out on soils around pistachio trees in various regions of Rafsanjan, Iran. Following isolation and identification of *Phytophthora* isolates, the predominant species was found to be *P. drechsleri* and used for further investigation Saberi-Riseh et al., (2004). For studying saprophytic survival of the fungus, soils collected from different areas were combined and autoclaved. Sterile soil was divided into 10 parts and mixed with fungal inoculum at various concentrations of 0-9% (w/w) separately. Each soil part (100g) was placed in 15cm diameter plastic pot. Some soils in pots were supplemented with sterile wheat straw whereas others were mixed with pistachio leaves surface sterilized with 5% (v/v) sodium hypochlorite. After 3, 6 and 9 weeks of incubation, five leaves or straws samples were taken from each replicate and cultured on CMA-PARPH medium and the fungal colony formation was monitored. The experiment was performed using completely randomized design with factorial experiments including three factors (substrate type, inoculum density and time), 10 treatments (0-9 g inoculum levels) and nine replicates. The results showed that the type of substrate (wheat straw and pistachio leaf) was very important for the fungal saprophytic survival in that this was significantly greater for the pistachio leaves. Time was also considered another critical factor for the fungal survival. With passing incubation time, saprophytic survival of the fungus declined. Further, it was demonstrated that increasing inoculum density would result in longer survivability of *P. drechsleri* and maximum fungal survival on substrate was obtained when inoculum density was at 9% (w/w).

In field experiments, the density of *Macrophomina phaseolina* microsclerotia in root tissues of naturally colonized soybean cultivars was quantified (Reis et al., 2014). The density of free sclerotia on the soil was determined for plots of crop rotation (soybean-corn) and soybean monoculture soon after soybean harvest. *M. phaseolina* natural infection was also determined for the roots of weeds grown in the experimental area. To verify the ability of *M. phaseolina* to colonize dead substrates, senesced stem segments from the main plant species representing the agricultural system of southern Brazil were exposed on naturally infested soil for 30 and 60 days. Sclerotium density, assessed based on colony forming units (CFU), ranged from 156 to 1,108/g root tissue. Sclerotium longevity, also assessed according to CFU, was 157 days for the rotation and 163 days for the monoculture system. *M. phaseolina* did not colonize saprophytically any dead stem segment of *Avena strigosa, Avena sativa, Hordeum vulgare, Brassica napus, Gossypium hirsutum, Secale cereale, Helianthus annus, Triticosecale rimpaui* and *Triticum aestivum*. Mp was isolated from infected root tissues of *Amaranthus viridis, Bidens pilosa, Cardiospermum halicacabum, Euphorbia heterophylla, Ipomoea* sp., and *Richardia brasiliensis*. The survival mechanisms of *M. phaseolina* studied in this paper met the microsclerotium longevity in soybean root tissues, free on the soil, as well as asymptomatic colonization of weeds.

Stenocarpella species (*S.maydis* and *S.macrospora*) overwinters in maize stubble and no-till monoculture maize is related to ear and stalk rot outbreaks, but little is known about the factors that contribute to its survival (Lavras, 2016). Therefore this work aimed at determining the contribution of soybean crop rotation and fallow land on pathogen survival and dynamics of maize stubble breakdown, and determining factors that contribute to the pathogens survival. In the first article maize fields with history of severe white ear rot were used to determine the stubble sampling size to analyze the pathogen sporulation and viability, as well the maize decomposition. The maize monoculture was compared to one round of soybean rotation and fallow. In the second article, maize fields were selected in the maize growing areas in the state of Minas Gerais for sampling maize stubbles. During the offseason, maize stalks, grains, cobs, fallen maize leaves, straw and dead weed stubble were collected. During the 2015/16 season stalks and cobs were collected in areas under conventional tillage, no-tillage, either with crop rotation or maize monoculture. Samples were processed, freeze-dried, powdered, had their DNA extracted and the *Stenocarpella* sp inoculum quantified through qPCR. Soils from the sampled fields were checked for suppressivity to ear and stalk rot-causing fungi. Heat maps were generated with the obtained data. A sub-sample of one kilogram was the minimum necessary to estimate the pathogen survival. The pathogen sporulation was similar among treatments but viability was reduced (27.8- 33.3%) in the considered regions. The stubble decomposition was higher in only one location. The sample size, classical and molecular methods allowed the survey of the pathogen in maize stubble and pointed out the role of soybean crop rotation in the reduction of *Stenocarpella* sp viability. In cultivated areas adopting the conventional tillage system, the pathogen was not detected. However, under no-tillage managed areas, *Stenocarpella* sp was detected to a lesser extent than in areas adopting no-tillage system with maize monoculture. The results endorse the role of maize stubbles in the survival of *Stenocarpella* sp. and crop rotation as a useful method to manage the diseases caused by *Stenocarpella maydis*.

5.25 Inoculum Density and Virulence-dependent Interactions

Novel strains of *Plasmodiophora brassicae* were inoculated onto two susceptible canola cultivars, one resistant line and six CR cultivars. Although all cultivars/lines showed a susceptible response to inoculation with the new strains of *P. brassicae*, the severity of disease reaction, root hair infection rates and the amount of *P. brassicae* DNA present in each canola genotype varied depending on the strain (Hwang et al., 2017). In addition, the effect of inoculum density on disease severity and gall formation was recorded for one of these new strains on a universally susceptible Chinese cabbage cultivar and one susceptible and 10 resistant canola genotypes. Although root galls were observed at an inoculum density of 10^3 spores per mL of soil, clear differentiation of susceptible and resistant reactions among canola cultivars/lines was not observed until the inoculum density reached 10^5 spores mL^{-1}. At a spore density of 10^6 spores mL^{-1} and above, all cultivars/lines developed susceptible reactions, although there was some differentiation in the degree of reaction. This study showed the potential to develop a unique disease profile for emergent club root pathotypes and shows a useful range of spore densities at which to study new *P. brassicae* strains.

Strains of the pathogen capable of overcoming host resistance are probably present in some fields at very low levels, having arisen either through genetic recombination or spontaneous mutation. With the repeated cultivation of the CR canola, these compatible strains are the only ones that reproduce on the host, and hence increase quickly in number relative to other (avirulent) strains present in the same field. *P.brassicae* releases enormous quantities of resting spores into the soil. For example, one crop of susceptible canola contributed 1.4 × 10^8 spores per mL soil in mini-plot experiments, and 10^{10} spores per g gall under field conditions. As such, when there is a sufficiently high level of the new strains present in the soil, severe symptoms of club root can develop on what previously was considered to be a resistant cultivar. Information on the reactions of CR canola cultivars to new strains of *P. brassicae* and on the effect of inoculum density on the performance of CR host genotypes, is important for the effective evaluation of genetic resistance resources.

ID decreased with reductions in inoculum concentration on all of the canola cultivars. Differential resistance reactions among these canola cultivars were observed at inoculum density levels between 10^5 and 2 × 10^6 spores mL^{-1}. Gall formation was observed on the resistant cultivars at inoculum concentrations as low as 10^3 resting spores per mL of soil. Gall weight rose with increasing inoculum concentration (Hwang et al., 2017). At inoculum concentrations lower than 2 × 10^6 spores mL^{-1}, gall weight on all of the cultivars was similar to or greater than gall weight on at least one of the susceptible controls. The club root pathogen was able to infect all of the canola cultivars at inoculum concentrations as low as 5 × 10^4 resting spores per mL of soil, and infections in some hosts were noted at spore concentrations as low as 10^3 spores per mL of soil. Infection rates in resistant canola cultivars increased with exposure to higher inoculum concentrations, thereby selecting for virulent strains as occurs with repeated cultivation. Depending on moisture levels, symptoms begin to appear in infested commercial fields at 10^5 spores per g of soil, and every plant is usually infected where spore concentrations exceed 10^7 spores per g of soil. Spore concentrations exceeding 10^8 spores per g of soil have been observed in commercial fields. Pathogen populations rather than single-spore isolates were used in this study, and it is possible that as the spore concentrations increased, the proportion of virulent spores within these populations also increased. Populations of *P. brassicae*, which represent all resting spores initially isolated from a single gall, are often heterogeneous and may represent a mixture of pathotypes.

To properly assess the virulence spectrum of a particular pathogen isolate or population, meaningful comparisons must be made between that isolate and the susceptible controls. This research found that soilless mix inoculated with spore concentrations between 10^5 and 2×10^6 spores mL^{-1} showed clear differentiation between cultivars with some measure of resistance and the susceptible controls. As more sources of club root resistance are identified, this information may become useful for the identification of differential levels of resistance and for distinguishing *P. brassicae* strains.

5.26 Effects of Inoculum Density on Disease Severity Caused by Pythiaceous Fungi

The nature and quantification of the infective propagules of certain *Pythium* and *Phytophthora* in field conditions, the variation of spatial and temporal patterns of propagules in relation to pathogenicity and saprobic survival of essentially pathogenic species, the limits of saprophytism or parasitism in natural conditions including their role as minor pathogens, are only some of these questions (Yannis Raftoyannis and Michael W. Dick, 2002). The main reason for the fragmentary information is the biodiversity of the fungal species in these genera. An understanding of the influence of inoculum density (ID) on plant infection is important before attempting to elucidate quantitatively the overall interactions of the host, pathogen and the environment because of the complexity of factors influencing infection and then disease development. A range of inoculum densities should be selected that will characterize better the relationship between inoculum density and root infection. For some *Phytophthora* and *Pythium* species, infection and disease may result from very low zoospore numbers. The use of high ID may break down host barriers that otherwise would remain intact.

Plant age affects the pathogen- host relationship and physiological plant age has a greater influence on the expression of host susceptibility than chronological age. Older plants may show symptoms upon inoculation but later recover and grow to maturity. Diseases caused by pythiaceous pathogens occur under conditions unfavorable both to other pathogens and to the host. Temperature is a critical parameter for the development of infective propagules and for a successful infection and although most species grow well at a wide range of temperatures, the optimum conditions are found in a restricted spectrum. Temperature of the substratum affects the speed and severity of root symptom development. With few exceptions root-rot symptoms increase with rising temperatures to 30^0 C, while the symptoms may be suppressed at very low and high temperatures. Damping-off and other *Pythium* diseases occur over a wide range of temperature, which determines which species of *Pythium* may be active.

Alfalfa, maize, sorghum and sugar beet plants were inoculated with zoospores of *Phytophthora* and *Pythium* species in order to assess the effects of inoculum density, plant age and temperature on disease severity. Seedlings were grown axenically in test tubes and inoculated with zoospore suspensions. Disease severity was assessed by measuring the root growth and discoloration of treated and control seedlings. The incremental root length of all plants decreased and root discoloration increased as inoculum concentration of the pathogen increased. Changes were more intensive among low levels of zoospore concentrations and no significant differences in disease severity were found for inoculum densities higher than 10 zoospores per ml. Disease severity was negatively related to plant age. Disease development on sugar beet seedlings infected with *Pythium* and *Phytophthora* species was affected by temperature, but the pattern of response was determined by the pathogen's temperature preferences. The incremental root length decreased as temperature increased up to 25⁰ C. The effect of *Pythium dissimile* and *Phytophthora cactorum* on root length was significantly lower at 35⁰ C than at 25⁰ C, whereas *Pythium aphanidermatum* and *Phytophthora nicotianae* caused a significant damage to roots even at 35⁰ C.

Disease development on sugar beet seedlings infected with *Pythium* and *Phytophthora* species was affected by temperature, but the pattern of response was determined by the pathogen's temperature preferences (Yannis Raftoyannis and Michael W. Dick, 2002). Zoospores respond to chemicals released by the roots, and every effect of temperature on zoospore vigor or plant physiological condition, will affect the plant– pathogen relationship. Temperature can be seen as an environmental factor, which affects the zoospore responses toward roots, especially at sub-optimal levels for plant roots and zoospores. At low temperatures the zoospores accumulate in only small numbers, perhaps because of the low exudations from roots and the low rate of movement of zoospores. As the temperature increases, high amounts of chemicals are exuded from the roots and the speed of zoospores increases. At median temperatures there would be a positive correlation between rise of temperature, root exudation and zoospore motility. At higher temperatures zoospore motility would become limited, with a resultant decrease in zoospore encystment density on the roots. The relationship between zoospore encystment and temperature was determined by the responses toward temperature of plants and pathogens. The adaptation of *P. aphanidermatum* and *P. nicotianae* to relatively high temperatures might have resulted in satisfactory levels of zoospore vigor and high levels of encystment at such temperatures.

The differential virulence between warm and cool fungal species at temperatures outside their optima is well documented. In root rot of spinach there was a transitional temperature (23⁰C), above or below which *P. aphanidermatum* and *P. dissotocum* respectively, predominated. Temperatures higher than 30⁰C could greatly increase infection severity of *P. nicotianae* on Hibiscus plants. *Pythium* species are not especially temperature-specific for infection of adventitious roots of *Agrostis palustris*, but pathogenicity of some isolates is temperature dependent following infection. One *Pythium* isolate which inhibited growth of *Agrostis* at high temperature, stimulated growth at low temperature.

5.27 Inoculum Density Relationships for Infection by *Phytophthora ramorum, Rhizoctonia solani and Xanthomonas cucurbitae*

Inoculum density can also influence the accurate assessment of resistance to various plant pathogens. Numerous workers have shown that levels of disease development are directly related to concentrations of spores used for inoculation of host plant species and that resistance can be overcome by using high enough inoculum levels (Paul W. Tooley et al., 2013). It is likely that inoculum densities necessary to cause infection on different forest species may vary, as Eastern oak species were shown to differ in susceptibility to *P. ramorum. P.ramorum* sporangia can be disseminated by varied means; sporulation tends to be seasonal and correlated with rainy periods and warm temperatures. Although understory species such as California bay laurel (*Umbellularia californica*) are believed to be the greatest contributors of *P. ramorum* spore loads which drive epidemics, coast live oak leaves were recently shown to support sporulation by *P. ramorum* as well. Sporangia production has been quantified on common understory hosts, as well as on Northern red oak (*Quercus rubra*) and chestnut oak (*Q. prinus*), showing that under artificial conditions, numerous sporangia may be produced on whole (intact) plants. The rate at which plant disease epidemics develop is determined in large part by the initial levels of inoculum present. Knowledge of the amount of inoculum necessary to cause a certain level of disease on a given host can thus be used to predict when and where epidemics may occur. Measurement of spore loads obtained under natural conditions, such as those present in rain water, could thus be used to determine the likelihood and severity of future infection in a given forest ecosystem.

Paul W. Tooley et al., (2013) established inoculum density relationships between *P. ramorum* and selected hosts using detached leaf and whole-plant inoculations. Young plants and detached leaves of *Quercus prinus* (Chestnut oak), *Q. rubra* (Northern red oak), *Acer rubrum* (red maple), *Kalmia latifolia* (mountain laurel) and Rhododendron 'Cunningham's White' were dip inoculated with varying numbers of *P. ramorum* sporangia, and the total number of diseased and healthy leaves recorded following incubation at 20°C and 100% relative humidity. Calibration threshold estimates for obtaining 50% infected leaves based on linear analysis ranged from 36 to 750 sporangia /ml for the five hosts. Half-life (LD50) estimates (the number of spores for which the per cent of diseased leaves reaches 50% of its total) from asymptotic regression analysis ranged from 94 to 319 sporangia/ml. Statistically significant differences ($P = 0.0076$) were observed among hosts in per cent infection in response to increased inoculum density. Inoculum threshold estimates based on studies with detached leaves were comparable to those obtained using whole plants. The results provided estimates of inoculum levels necessary to cause disease on these five *P. ramorum* hosts and will be useful in disease prediction and for development of pest risk assessments.

Calibration threshold estimates for 50% disease we observed based on linear regression analysis using data from whole-plant studies ranged from 36 sporangia / ml for *Quercus rubra* to 750 sporangia/ml for *Acer rubrum.* Half-life estimates based on asymptotic regression analysis ranged from 94 sporangia /ml for Rhododendron 'Cunningham's White' to 319 sporangia/ml for *A. rubrum*. Based on both analyses, more sporangia were required for *A. rubrum* to reach 50% disease compared with the other hosts, which is in agreement with earlier findings that *A. rubrum* shows a relatively low level of susceptibility to *P. ramorum*. However, for detached leaf studies, threshold estimates for 50% disease based on linear regression ranged from 30 sporangia/ml for *A. rubrum* to 319 sporangia /ml for *Kalmia latifolia* and thus, the above relationship was not observed. The linear regression and asymptotic regression analyses were useful in different ways as they contained parameters intended to measure different features of the growth curves. The linear regression analysis was intended to measure rates of disease increase over differing concentrations of sporangia. The asymptotic regression analysis targeted the half-life, the amount of sporangia at which the per cent of diseased leaves reached 50% of its total.

Levels of sporangia required to obtain 50% disease as determined in experiments described here would likely be easily obtained in nature based on the sporulation capacity of a number of Eastern US forest species determined experimentally (Paul W. Tooley et al., 2013). Sporulation capacity for 25 understory plant species ranged from 36.3 to 2001.4 sporangia/cm^2 lesion tissue and was even higher for northern red oak.Thus, rainwater suspensions of sporangia produced in forests containing these species could be predicted to result in concentrations exceeding those required for 50% disease. *P. ramorum* propagules trapped in rainwater in naturally infested areas of California were found to be in the range of 437–530 colony-forming units (CFU)/l, but these could have consisted of zoospores, sporangia and perhaps hyphal fragments. These results will be useful to researchers attempting to determine spore loads necessary for causing epidemics of sudden oak death in diverse ecosystems. It was found in detached leaf studies that infection was obtained at concentrations as low as 10 sporangia/ml, thus, showing that very few spores are required to cause infection of various species. Spore concentrations occurring in nature have rarely been determined experimentally, so it is not known whether this level of spores is a common occurrence in native ecosystems.

Effects of Inoculum density of *Rhizoctona solani* on Sugar beet root rot: Understanding the effect of inoculum density and age of plants is essential to elucidate the overall interaction of the host and pathogen for better management decisions to protect sugar beet crops. The aim of the present study was to elucidate how inoculum density and growth stages of sugar beet plants affect varietal response to root rot severity in both greenhouse and field conditions (Bhuiyan et al., 2023). Root rot of sugar beet caused by *Rhizoctonia solani* anastomosis group AG 2-2 IIIB is responsible for significant crop losses in North Dakota and Minnesota. Understanding the association between plant age and inoculum density with disease severity of sugar beet cultivars is critical to properly screen for varietal resistance. Therefore, investigations were conducted for the responses of 4-, 6-, and 8-weeks old plants in seven commercial sugar beet cultivars to inoculum densities of one, two, and three seeds of *R. solani*-colonized barley in the greenhouse and with the corresponding three levels of colonized barley, mycelial plugs, and sclerotia in field experiments. In greenhouse conditions, disease severity was greatest before plants reached the 6-weeks of age (P=0.05). There was a positive linear relationship between the density of the inoculum and disease severity. All seven cultivars were equally susceptible (P>0.05) to *R. solani*. Interactions between cultivars and plant age and between plant age and intensity of inoculum were not significant (P>0.05). Field experiments showed that the density of inoculum was significant (P<0.001) and the disease severity was found to be highest in plants inoculated with three colonized barley seeds per plant compared to doses of other inoculum types.

Three colonized grains of barley per seedling demonstrated higher inoculum potentials than corresponding one or two grains in the greenhouse. When compared in the field with the three inoculum doses of mycelial plugs and sclerotia in two cultivation seasons (2020 and 2021), the aforesaid dose also showed the highest inoculum potential. Inoculum density influences the reaction of germplasm materials to various plant pathogens. Perhaps its clearest influence is on the severity of the diseases they cause. The knowledge and determination of level of pathogen inoculum in soil are thus essential for developing control strategies. The nature and quantification of the infective inoculum in field conditions and the variation of spatial and temporal patterns of propagules play a significant role in disease development. Previously, it was reported that an exponential reduction in the incubation period was due to increasing Fusarium oxysporum inoculum intensity in chickpeas.

A positive linear relationship was observed between the amount of inoculum (i.e., the number of colonized barley seeds per plant) and disease severity. This observation agrees with that of other researchers who reported a positive correlation was found between disease incidence of *Rhizoctonia* root rot in sugar beet and the number of viable sclerotia. Inoculum density was negatively associated with sucrose yield in sugar beet. Reaction of sugar beet to *R. solani* may be influenced by plant age and inoculum concentration. It was showed that the relationship between varietal response resistance, inoculum density, and growth stage is critical for determining the resistance response of sugar beet cultivars against *R. solani* in field conditions. Consequently, it is logical to assume the inoculation methods that cause maximum disease pressure may be better suited to identify resistance germplasm as well as to properly evaluate disease management practices for *R. solani* in sugar beet. It was also predicted that early application of fungicides would be more effective in protecting the most susceptible seedling stage of sugar beet plants against damping off caused by *R. solani* AG 2-2IIIB.

Effect of Inoculum density on the development of *Xanthomonas cucurbitae:* An understanding of impact of inoculum concentration on any plant infection is essential to understand any host pathogen interaction. A suitable seedling inoculation method for disease development would be highly desirable for defining conditions that allow optimal infection (Sujata Kumari et al., 2020). Bacterial leaf spot caused by *Xanthomonas cucurbitae* is one of the important diseases of bottle gourd and pumpkin and is gaining momentum in the sub-tropical zone of Himachal Pradesh. Studies were conducted under artificial inoculation conditions to standardize the inoculum concentration and method of inoculation for the development of disease. The bacterial concentration was standardized with the help of a growth curve of both bottle gourd and pumpkin isolates of *X. cucurbitae* with time and a sigmoid curve was obtained in both the isolates in which the bacterial concentration reached its maximum (107 cfu/ml) after 96 days of inoculation and then declined. Studies on effect of different inoculum densities of causal bacterium on disease development revealed that an inoculum concentration of 108 cfu /ml of both the isolates resulted in minimum incubation period (2 and 4 days) and maximum mean disease severity (38.89 and 42.00%) in both bottle gourd and pumpkin plants, respectively. Among various methods of inoculation with both isolates, syringe method of inoculation resulted in minimum incubation period (2.33 and 3.33 days) and maximum disease severity (46.33 and 37.97 %) in both bottle gourd and pumpkin, respectively followed by pin prick method of inoculation.

As far as the apparent infection rate was concerned, it was significantly maximum (0.28 per unit per day) in the seedlings inoculated with inoculum density 1.15×10^8 and 0.19×10^9 cfu/ml which was found to be statistically at par with those inoculated with inoculum density of 1.02×10^7 (0.27 per unit per day) and 1.48×10^6 cfu/ml (0.27 per unit per day). However, significantly minimum apparent infection rate (0.22 per unit per day) was recorded in the seedlings inoculated with inoculum density of 2.2×10^4 cfu/ml. As far as the AUDPC was concerned, it was found to be significantly minimum (2.34) in the seedlings inoculated with an inoculum density of 2.2×10^4 cfu/ml followed by those inoculated with 4.3×10^5 cfu/ml (2.68) which was statistically at par with seedlings inoculating of 1.48×10^6 cfu/ml (2.72). However, maximum (3.86) AUDPC were recorded in the seedlings inoculated.

All inoculum densities tested were able to induce the disease in test plants but, an inoculum density of range 10^7 cfu /ml and above was found to incite significantly higher levels of disease severity and rate of its spread as compared to lower concentrations of bacterial suspension in both bottle gourd and pumpkin isolates. An inoculum density of the range 10^7 cfu/ml is the optimum range of inoculum concentration to incite significant amount of disease in two cucurbits under study (Sujata Kumari et al., 2020). The results were in accordance with findings of many workers who have reported that a bacterial concentration of 10^5 to 10^8 cfu /ml is able to incite pathogenic reaction in different cucurbit hosts. Syringe inoculation method proved best for the initiation and development of the disease, followed by pin prick and leaf margin methods of inoculation in both bottle gourd and pumpkin isolates of the pathogen. It may be attributed to the fact that the bacterium moves systematically in the plant and thus, the syringe inoculation might have lead to faster spread of the pathogen in the inoculated seedlings resulting in early initiation and faster development of the disease. In case of pin prick inoculation, wounds were created by pin pricking resulting in easy entry of bacteria and further development of the disease. At leaf margin, hydathodes might have opened the avenues for the entry of bacteria, thus leading to development of disease. In case of swab and spray inoculations the bacteria could enter only through the stomatal openings, thus comparatively lowering the chances of entry of pathogen.

5.28 Nematode Inoculum Density on Host Growth and Nematode Reproduction

The minimal density that causes a measurable reduction in plant growth or yield varies with nematode species, host plant, cultivar and environment. The severity of plant damage by *Meloidogyne* spp. influenced by inoculum densities has been published for a number of crops. Similarly, the effects of initial population densities of *M. javanica* on cucumber have been reported but there is paucity of information on the relationships between initial density of *Meloidogyne* spp. and growth and yield of cucumber as influenced by plant age at the time of exposure to nematodes (Muhammad Zameer Kayani et al., 2018). It is generally assumed that young plants are more severely damaged by nematodes than are older plants. However, there are limited experimental data for supporting this notion. Therefore, the present studies were planned to investigate the relationship between initial population density of *M.*

incognita and damage to cucumber. Also, the effects of plant age at the time of exposure to nematodes on reductions in growth and yield of cucumber and reproduction of *M. incognita* were studied under a range of inoculum densities.

As the damaging effects of *M. incognita* have not been assessed on cucumber, the present studies report the effects of five initial population densities of *M. incognita* on growth and yield of cucumber and nematode reproduction. All inoculum densities caused significant reductions in growth and yield of cucumber and were found to be positively correlated with the latter (Muhammad Zameer Kayani et al., 2018). On the other hand, ages of plants at inoculation had negative correlations with reductions in these parameters at each inoculum density. The production of galls was found to be positively correlated with the inoculum densities and plant ages. However, rate of nematode build up decreased with an increase in inoculum density and appeared to be negatively correlated with inoculum densities and, on the contrary, was found to be positively correlated with plant ages. The results demonstrated that *M. incognita* has the potential to severely impair the growth of cucumber and by delaying early exposure of the latter to nematodes can significantly abate yield losses.

i. Effect of inoculum densities on growth parameters and yield of cucumber: All inoculum densities of *M. incognita* caused significant percent reductions in shoot and root lengths and fruit yield of cucumber at three plant ages. Minimum reductions in these parameters were observed with the lowest inoculum density while the reductions were found to be the maximum with the highest density (Figs 33-35). Inoculum densities were found to be positively correlated with these parameters indicating that with increase in inoculum densities, there were corresponding increases in percent reductions in these parameters. On the other hand, plant ages at inoculation had negative correlations with reductions in shoot and root lengths and fruit yield showing that the magnitude of damaging effects of nematode decreased as the age of plants increased at the time of inoculation. Linear models provided a good fit to the relationships between inoculum densities and reductions in shoot and root lengths and yield which have been shown by regression equations and trend lines in figures 19, 20 and 21, respectively.

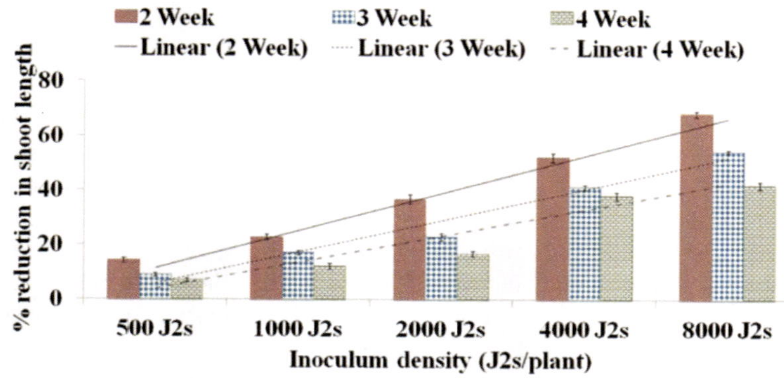

Fig. 43: Effect of *M. incognita* inoculum densities on reduction in shoot length. Data are means of ten replicates. Vertical error bars represent standard errors of differences of means

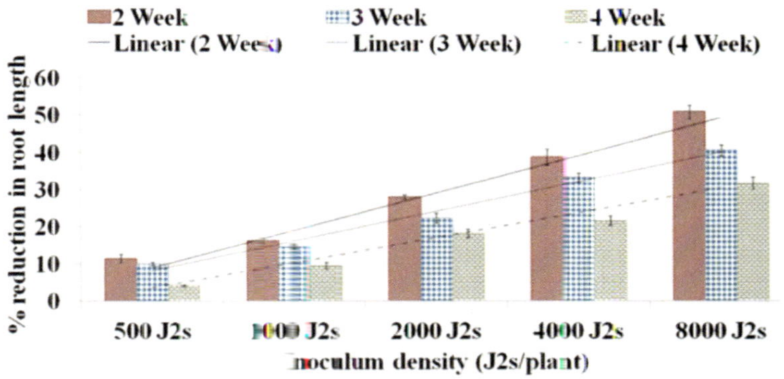

Fig. 34: Effect of *M. incognita* inoculum densities on reduction in root length. Data are means of ten replicates. Vertical error bars represent standard errors of differences of means

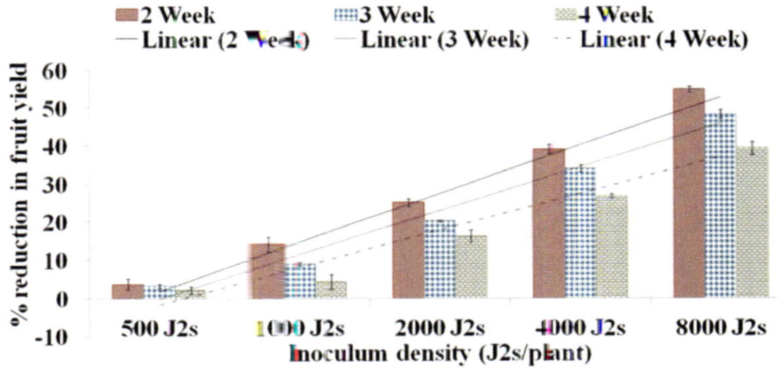

Fig. 35: Effect of *M. incognita* inoculum densities on reduction in fruit yield. Data are means of ten replicates. Vertical error bars represent standard errors of differences of means

ii. Effect of inoculum densities on nematode reproduction: Inoculum densities of *M. incognita*, plant ages and their interactions also had significant effects on number of galls and rate of nematode build up. Minimum galls were observed in plants with the lowest inoculum density. As the density of *M. incognita* increased, the number of galls also increased significantly and reached to the maximum at the highest inoculum density. In other words, production of galls was found to be positively correlated with the inoculum densities and plant ages (Fig. 36) and the relationships have been shown by regression equations. The numbers of galls on the roots inoculated after 2 and 3 weeks were not statistically different from each other at each inoculum density. Rate of nematode build up decreased with an increase in the inoculum density and appeared to be negatively correlated with inoculum densities and on the contrary was found to be positively correlated with plant ages (Fig. 37) (Muhammad Zameer Kayani et al., 2018). The relationship has been shown by regression equations.

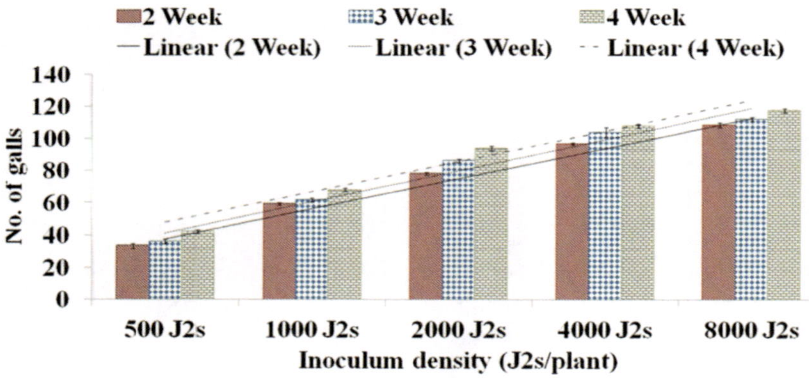

Fig. 36: Effect of *M. incognita* inoculum densities on number of galls. Data are means of ten replicates. Vertical error bars represent standard errors of differences of means.

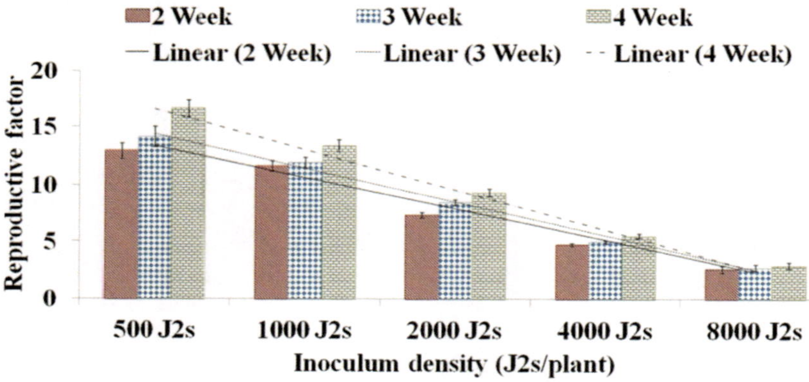

Fig. 37: Effect of *M. incognita* inoculum densities on reproductive factor. Data are means of ten replicates. Vertical error bars represent standard errors of differences of means.

Reduction in root length was due to extensive damage of root system as a result of nematode feeding on giant cells, causing root growth to stop. This damage and shortening of roots affected growth and development of cucumber plants by limiting their ability to absorb water and nutrients from the soil, eventually resulting in stunting and reduction in fruit yield. As the age of plant at the time of inoculation increased, the number of galls also increased (Muhammad Zameer Kayani et al., 2018). On the other hand, reproductive factor decreased with an increase in inoculum density and increased with an increase in plant age. The positive correlation between increase in plant age and number of galls was probably due to the higher number of sites available for nematode penetration and feeding on the older plants. The lower rate of nematode build up at higher inoculum densities may be due to overcrowding of larvae which compete for food.

5.29 Effect of Inoculum Densities on Seed Germination, Seedling Growth Parameters and Disease Incidence

Stem rot is a persistent soil borne disease throughout India and its incidence is increasing even at maturity stage of the groundnut crop. Though *Sclerotium rolfsii* survives both on seed and in soil, a greater threat is posed by soil borne inoculum (Raghavendra et al., 2018). *S.rolfsii* forms brownish sclerotia that can survive for long periods in the soil and retain their viability by tolerating biological and chemical degradation due to the presence of melanin in the outer membrane. *S. rolfsii* is a necrotrophic soil borne fungal pathogen, which has a very wide host range and distributed worldwide in warm climates. Effect of different inoculum densities of *S. rolfsii* on the development of stem rot of groundnut was assessed using soil infestation method in pot culture under green house and highest germination, root length and shoot length was observed at 1 per cent inoculum level. Germination percentage, root length and shoot length of groundnut was gradually decreased with increase of inoculum levels from 1 to 5 per cent. Lowest germination, root length, shoot length was observed in 5 per cent inoculum levels. Since the germination of groundnut seeds was reduced to more than 50 per cent at one per cent inoculums level it was followed throughout the investigations.

Germination of groundnut seeds gradually decreased with increase in inoculum densities from 1 to 5 per cent and was significantly different from each other. The germination percentages recorded at different inoculum densities were in the order, Control > 1% > 2% > 3% > 4% > 5%.The root length decreased with increase in the inoculums density however, the reduction in the root length was not significant at different densities from 2 per cent to 5 per cent. The root length recorded at different inoculums densities were in the order, Control > 1% > 2% > 3% > 4% > 5%. It is clear that even at one percent inoculum density the germination percentage of groundnut seeds was affected by more than 50 per cent. Hence throughout the experimentation one percent inoculum was applied and also sick soil was maintained by adding one percent inoculum. A positive correlation between seedling mortality and inoculum density has been reported in coniferous species, whereas plant height and weight of the surviving seedlings were negatively correlated with sclerotial density of *Macrophomina phaseolina*. Germination percentage, root length and shoot length of groundnut was gradually decreased with increase of inoculum levels from 1 to 5 per cent. Lowest germination, root length, shoot length was observed in 5 per cent inoculum levels. Since the germination of groundnut seeds was reduced to more than 50 per cent at one per cent inoculums level, it was followed throughout the investigation.

Rhizoctonia solani causes seedling blight and root rot in lentil, which reduces plant populations and the vigour and yield of surviving plants. Factors in the seedling environment, such as inoculum density, temperature, seeding depth, seeding date, and fungicidal seed treatment were studied to determine the degree to which they affectt he impact of *R. solani* on lentil seedlings (Chang Kan-Fa et al., 2008). Survival of lentil plants was evaluated after planting into soil artificially inoculated with various concentrations of a highly aggressive isolate of *R. solani* (AG-4). Emergence, seedling survival and shoot dry matter production decreased with increasing inoculum density, but these declines varied with temperature. Low soil temperatures delayed the emergence of lentil seedlings in non-inoculated soil, but in inoculated soils, emergence was inhibited with increasing temperatures. Depth of seeding did not affect seedling establishment, but root rot severity increased with depth of seeding in a growth cabinet trial. Root nodulation was reduced as root rot severity increased. In field experiments carried out over 3 station years, seeding date had a substantial effect on seedling emergence and yield of inoculated treatments, but the trends were not consistent between sites. In field assessments of fungicide efficacy, treatment of seed with thiabendazole plus carbathiin (Crown) and carbathiin plus thiram (Vitaflo 280) improved seedling establishment relative to the inoculated control.

In controlled environments, *Phomopsis convolvulus*, a fungus being examined as a potential bioherbicide for field bindweed, required a minimum of 18 h of dew to produce severe disease on seedlings inoculated with 109 conidia/m^2 (Morin et al., 1990). Under these conditions, excellent control of the weed (95% reduction in foliage biomass and up to 55% mortality) was achieved. Less disease was observed with shorter dew periods. Addition of gelatin, SORBO, or BOND to the inoculum did not increase infection under su b-optimal moisture conditions. A continuous dew period of 18 h was superior to the effect of three 6-h dew periods interrupted by humid periods. A high relative humidity (97 to 100%) during the humid periods favored infection compared to lower relative humidity (80 to 83%). Pycnidia and conidia were produced on diseased plants placed under humid conditions for 48 h or more. These results indicate that sub-optimal moisture conditions represent a possible constraint that may reduce the weed control efficacy of *P. convolvulus* on field bindweed.

Effect of inoculum density of *Stromatinia cepivora* on the amount of Garlic white rot: White rot, a garlic disease caused by the soil-borne fungus *Stromatinia cepivora* (Berk.) Whetzel, is a serious problem of garlic productions in Egypt. This study examines the potential of controlling the disease biologically by using three *Trichoderma* species, i.e., *Trichoderma harzianum*, *Trichoderma koningii*, and *Trichoderma virens* employed either alone or in combination (Elshahawy et al., 2019).

In in vitro assays, three *Trichoderma* species, i.e., *Trichoderma harzianum*, *Trichoderma koningii*, and *Trichoderma virens* and tebuconazole, were compared for their ability to suppress *S. cepivora* isolate (Sc8). In greenhouse experiments, the chemical treatment was the most effective, with the lowest incidence of garlic white rot compared with the control. The antagonistic fungi tested either individually or in combination significantly reduced the incidence of white rot on garlic. In general, dual and triple combinations of the fungal isolates were more effective than these isolates used individually. The combination of the three *Trichoderma* species was the most effective treatment, decreasing disease incidence by 50.0% in 2016/2017 season and 40.0% in 2017/2018 season, respectively. The three *Trichoderma* species employed alone or in combinations and tebuconazole were evaluated under low and high disease pressures in field trials to determine which situation (s) provided the best control of garlic white rot. Under low (40 sclerotia/kg of soil) and high (600 sclerotia/kg of soil) inoculum density, the standard fungicide programme (dipping of garlic cloves in tebuconazole (1 ml of Folicur 25% l^{-1} of water) plus spraying garlic stem bases with the same concentration of tebuconazole) gave statistically significant disease control, decreasing disease incidence by 67.7 and 29.4% in 2016/2017 season and 72.6 and 31.1% in 2017/2018 season, respectively. Under low disease pressure, significant control, equal to the fungicide treatment, was achieved with the trip combination of three *Trichoderma* species.

However, *Trichoderma* species employed alone gave insignificant control of garlic white rot under high disease pressure. The triple combination of three *Trichoderma* species decreasing disease incidence by 65.6 and 15.5% in 2016/2017 season and 74.2 and 18.6% during 2017/2018 season, under low and high inoculum density, respectively. The activities of defense enzymes, i.e., peroxidase, polyphenoloxidase, and chitinase due to application of *Trichoderma* species, were enhanced in garlic plants either grown under low or high disease pressures. Reduction of white rot disease incidence was accompanied by increased growth parameters and bulbs yield of garlic plants grown under field conditions (Elshahawy et al., 2019).These results indicated that the performance of three *Trichoderma* species may be influenced as much by the absolute disease pressure. It was concluded that, at the low disease pressure site, the low level of inoculum and disease incidence enabled three *Trichoderma* species to bring about disease control

5.30 Effect of Inoculum Density on the Early Reproduction and Damage Potential of the Burrowing Nematode, *Radopholus similis*

Breeding for nematode-resistant banana cultivars is an important component of an integrated nematode management system. Identification of resistant Musa germplasm involves early screening procedures in pot experiments. Yet these procedures are not standardized. Differences in growing medium, inoculum density, exposure time, pot volume and origin of the nematode population may influence the final results and conclusions (Moens et al., 2003). Goal of this series of experiments with pot plants was to optimize some steps of the method for resistance screening of Musa cultivars to *R. similis*. *Radopholus similis* per 100 g of roots was significantly higher in river sand than in local banana soil, however it followed the same tendency in both growing media. Reproductive index and final root weight decreased with increasing initial inoculation density. In a subsequent experiment with a broader initial inoculation range and two different *R. similis* populations, reproductive index, root and shoot weight decreased with increasing initial inoculation densities for both populations. When eight exposure times were evaluated, *R. similis* per 100 g of roots showed a lag phase of about five weeks, then started to grow during six weeks, and stabilized afterwards. Inoculation of plants in a variable pot volume experiment resulted in a lower reproductive index in higher pot volumes. Finally, *R. similis* per 100 g of roots was not different among Musa AAA cvs. Grand Naine, Gros Michel, and Musa cv. AA Pisang Mas ($P \leq 0.25$), but varied significantly between the first group and both Musa AAAA cv. FHIA-23 and Musa AAA cv. Yangambi Km5 ($P \leq 0.0001$). Too sum up, the use of 1.8 liter pots, filled with sterilized local banana soil and initially inoculated with 0.28 *R. similis* /ml of substrate and exposed to nematodes during 8 to 12 weeks, gave the most consistent results.

When inoculating in vitro propagated plants, nematode inoculum densities ranged between 100 and 10,000 per pot, resulting in a density between 0.125 and 5 nematodes /ml substrate. Exposure time (period between nematode inoculation and plant evaluation) for in vitro propagated plants ranged between 3 and 23 weeks. Pot volumes used in banana pot screening experiments varied between 0.4 and 10 litres. The higher reproduction rate observed in river sand is attributable to the growing medium, because plants grown in river sand or local banana soil received the same nutrition and management. Probably the larger pore diameter in river sand favored aeration and nematode growth. These physical characteristics could also improve nematode movement towards and penetration of the banana roots. The lowest inoculum density of 0.28 nematodes/ml of substrate resulted in the highest reproductive index in both growing media. For a specific pot volume an optimal inoculum density and time of inoculation after transplanting should be defined. In future research lower inoculum densities might be tested to check if a higher reproduction index can be obtained in pots of 1.8 litres. However, an equilibrium between a high reproductive index and certainty to infect roots, multiply and cause damage must exist to effectively identify resistant germplasm. In the bigger pot volumes, inoculum numbers were too high and not in relation with the almost equal root system of the plantlets in the four pot volumes at the moment of inoculation. An important focus of future research is thus the moment at which a certain pot volume turns into a limiting factor for root growth. Therefore, when running a resistance screening experiment, it is advisable to incorporate uninoculated plants for each tested Musa cultivar to distinguish root damage caused by lack of space from damage caused by nematodes.

In all experiments a clear correlation between *R. similis* per 100 g of roots and root suppression (pathogenicity) was observed. Lower root weight was found in the plants cultivated in river sand, more likely because the proportion of thick, thin and fine roots was different compared with local banana soil. With higher initial inoculations, *R. similis* per 100 g of roots increased and reproductive index and root weight decreased. The regression lines of both populations for initial inoculum and root weight did not cross. Therefore, differences in pathogenicity between both populations can only be explained by differences in final nematode numbers present in the roots.

5.31 The Effect of Different Initial Densities of *Meloidogyne javanica* on the Build-up of *Pasteuria penetrans* Population

Pasteuria penetrans will build-up faster where there is a high initial nematode density and can suppress root-knot nematode populations in the roots of tomato plants. The effect of different initial densities of nematode (*Meloidogyne javanica*) (150, 750, 1500, 3000) and *P. penetrans* infected females (F1, F3) densities (F0=control and AC=absolute control without nematode or *P. penetrans* inoculum) on the build-up of *Pasteuria* population was investigated over four crop cycles (Darban et al., 2005). Two major points of interest were highlighted. First, that within a confined soil volume, densities of *P. penetrans* can increase >100 times within 2 or 3 crop cycles. Second, from a relatively small amount of spore inoculum, infection of the host is very high. There were more infected females in the higher *P. penetrans* doses. The root growth data confirms the greater number of females in the controls particularly at the higher inoculum densities in the third and fourth crops. *P. penetrans* generally caused the fresh root weights to be higher than those in the control. *P. penetrans* has shown greater reduction of egg masses per plant at most densities. The effects of different initial densities of *M. javanica* and *P. penetrans* on the development of the pest and parasite populations were monitored. And no attempt was made to return the *P. penetrans* spores to the pots after each crop so the build-up in actual numbers of infected females and spores under natural conditions may be underestimated.

There were more infected females in the higher *P. penetrans* doses. The number of females/plant was lower in the 150 nematodes treatment as compared to the other densities so that the actual number of infected females was low. Often, but not always, there were significantly fewer females per root system in the *P. penetrans* treatments particularly at the higher nematode inoculum densities. To take into consideration the root growth, data are also presented which confirms the significantly greater number of females in the controls particularly at the higher inoculum densities in the third and fourth crops. The effects of *P. penetrans* on the reduction of egg masses per plant were highly significant (P<0.001) at most densities. After the completion of the 3rd and 4th crops the greatest reduction in the number of egg masses was generally found in the higher initial inoculation densities. *P. penetrans* treatments generally caused the fresh root weights to be higher than those in the control treatment. However, in the first crop, root weights were significantly (P<0.001) higher than those in the absolute control in the nematode treated plants at the two lower inoculum densities but significantly lower at the highest density. In the second crop all treatments except the lowest (150) showed heavier roots than in the absolute control. And in the final two crops the nematode control (without *Pasteuria*) had the lowest root mass.

Within a confined soil volume densities of *P. penetrans* can increase >100 times within 2 or 3 crop cycles. From a relatively small amount of spore inoculum, infection of the host is very high. Because of the reproductive potential of *Meloidogyne* species, only a few individuals escaping infection are sufficient to produce eggs that could result in crop damage (Darban et al., 2005). A major "failing" of biological control is that absolute control is unlikely to be achieved and certainly not at the required level of >99%. At the completion of the 4th crop, up to 70% of females in the root systems were infected with *P. penetrans*. There was a dose effect of *P. penetrans* in that the numbers of infected females were greater in the higher does *Pasteuria* treatment. No attempt was made to return the *P. penetrans* spores to the pots after each crop so the build-up in actual numbers of infected females and spores under natural conditions may be underestimated. Also the numbers of uninfected nematodes would have been higher if the root systems with egg masses had been returned to the pots. But, if they had, there would have been severe effects on the host plant growth.

Both higher initial densities of *P. penetrans* showed a suppressive effect on root galling and egg mass production over the successive crop cycles. However, with the lower initial nematode density both of the *Pasteuria* treatments suppressed the nematode populations less, presumably because there were fewer nematodes in the system to become infected and increase the density of the parasite. The number of egg masses was greater in the one-female treatments of all densities as compared to the three-female treatments. The number of egg masses was higher where there had been a lower initial nematode density. In the first crop untreated control plants of the higher nematode inoculum densities senesced prematurely due to the debilitating effects of the heavy nematode burden and were harvested after 8 weeks. However, the duration of the 3rd and 4th crop cycles was prolonged until 15 weeks instead of 12 weeks (in the second crop) which favoured the build-up of *P. penetrans*. There was no sign of senescence on the *Pasteuria* treated plants, however, control plants were dying, which helps to explain the lower fresh root weights in the controls. This is an indication of the benefit of *Pasteuria* by prolonging the duration of the crop (Darban et al., 2005). The duration of 15 weeks supported the build-up of *P. penetrans* and on the other hand it also provided enough root space to support the low numbers of uninfected nematodes which were then able to produce egg masses with high numbers of eggs which was reflected in the production of egg masses in lower density treatments at the 1st and the 4th crop harvests.

5.32 Synergies Between Inoculum Size & Heat Disturbance and Primer Inoculum Sources

The role of heat wave-induced disturbance in the invasiveness of soil-borne pathogens to healthy soils remains poorly understood. In addition, increasing empirical evidence has suggested that the size of the introduced inocula is vital in determining the invasion success of alien species (Gong et al., 2022). Alien pathogens with larger inoculum sizes are more likely to invade successfully than species with smaller inoculum sizes, as the population growth of many introduced species with small inoculum sizes tends to be negative, making these species vulnerable to extirpation. Inoculum size contributes to the invasion potential of pathogens in the soil. However, the role of inoculum size in determining the fate of pathogens in disturbed soils remains unclear. Gong et al., (2022) investigated the survival rates of a bacterial pathogen, *Ralstonia solanacearum*, in soils subjected to heat as a simulated disturbance. The results revealed that heating increased soil resource availability but reduced resource differentiation between *R. solanacearum* and indigenous bacterial communities. In both non-heated and heated soils, invader abundances increased with inoculum size, with a greater magnitude in heated soils. Inoculum size and heat-induced increases soil-available carbon and nitrogen best predicted invasion success. Altogether, these findings suggested that the invasion by soil pathogens could be predicted by synergies between heat perturbation and inoculum size.

The larger inoculum size of *R. solanacearum* survived better in heated than in non-heated soils on both Days 3 and 42 after being introduced into the paddy soils. *R. solanacearum* is a pathogen capable of infecting a variety of plant species, and its abundance can predict the severity of bacterial wilt disease outbreaks. The greater survival rates of *R. solanacearum* in heated than in non-heated soils suggest that the risk of disease incidence is elevated in heated soils such as those found under extreme hot weather conditions. Heating would increase the supply of available resources for

R. solanacearum, which might support the multiplication of *R. solanacearum* at low inoculum sizes.

Inoculum size and heat disturbance influenced the available resources (DOC and ammonium) and were the dominant drivers of *R. solanacearum* survival in the early and late stages, respectively. In the early incubation stages, inoculum size contributed to the survival rates of *R. solanacearum*. Allee effects, which indicate positive density dependence, suggest that fitness increases with increasing population density due to the benefits associated with the presence of conspecifics. The benefits for invaders from the addition of successive inoculum sizes outweigh the costs, such as competition among conspecifics for resources, which are abundant in the early incubation stages. Thus, a net gain in fitness for successful invasion of paddy soil contributed to the survival of *R. solanacearum*. In the late stages, the amount of *R. solanacearum* reached or nearly reached the carrying capacity of the habitats in both the non-heated and heated soils, making available resources more important than inoculum size in determining the survival of *R. solanacearum*. Inoculum size thus interacts with other factors, such as the amount of available resources and the community composition of the invaded habitat. Reducing the inoculum size is thus an effective approach to lowering the invasiveness of alien species (Stringham and Lockwood, 2021), yet the development of effective management to stop new invasions by *R. solanacearum* seldom considers inoculum size (Nion and Toyota, 2015). Hence, this study provides a unique dataset that is both essential and practical. It can be used to monitor the relationship between inoculum size and indigenous habitats, and to manage soil-borne pathogen invasions in the soil. Future studies should pay more attention to detecting the interactions between *R. solanacearum* and indigenous microbes (e.g., *Bacillus*) as an example of controlling pathogen invasions and the biocontrol of wilt disease.

Heat disturbances enhanced the effects of inoculum size in predicting *R. solanacearum* invasions in soil, and this was attributed to the increasing available niches of resource dimensions in disturbed habitats. Furthermore, the equilibrium dose of *R. solanacearum* was determined to be 10^7 CFU g^{-1} dry soil in paddy soils, but the value increased due to heating disturbance. Moreover, the survival of alien species was characterized by Allee effects, therefore, the invasion risk of *R. solanacearum* can be controlled if the inoculum size is under 10^3 CFU g^{-1} dry soil. Heating increased the competition for resources instead of niches between *R. solanacearum* and indigenous bacteria. This result can be used to formulate control strategies under global climate change, where heat waves may increase the release of available resources in the soil and simultaneously increase the competition for resources between *R. solanacearum* and the indigenous bacterial community.

Primer inoculum sources of *Erwinia* species: Etiology and epidemiology of soft rot pathogens on potato have been studied extensively because of their economic importance. But, the strains associated with soft rot of tomato have not been studied in great detail. There is no literature information available about the primary inoculum source of the disease (Aysan, et al., 2005). The disease has occurred the first time in some greenhouse tomatoes in 1994. Observations confirmed the symptom development start from root or crown region of seedlings, which may suggest that the pathogenic bacterium is either soil-borne or seed-borne. Bacterial stem rot caused by *Erwinia carotovora* subsp. *carotovora* (Ecc) and *E. chrysanthemi* (Ech) was observed in the eastern Mediterranean region of Turkey. The pathogens cause wilting of the whole plant, water-soaking areas on the stem, hollowing of the pith, browning of the vascular tissue and maceration in stem and fruits in plastic-covered greenhouses. Based on external disease symptoms, the prevalence of the disease ranged from 13 and 24 % in 1999 and 2000, respectively. Incidences of the disease were 12 and 20 % in 1999 and 2000 in the commercial greenhouses surveyed in the eastern Mediterranean region of Turkey. This is the first report of *Erwinia* species causing stem rot on tomato in Turkey. Primary inoculum sources of the disease were investigated with rifampicin-resistant strains in/on plant debris and seeds. Soft rot Erwinias were recovered from plant debris in summer period in the region for 2 years. The pathogenic bacteria were able to survive on the dried seeds for at least 7 months, but not more than 11 months. Tomato seedlings from naturally contaminated seeds grown in sterile soil under mist had leaf lesions, irregular, black-brown spots, no halo. This is the first study showing that both soft rot Erwinias were able to survive in/on the diseased tomato plant debris and tomato seeds, which may be potential primary inoculum of the disease. A polyclonal antiserum (MEOZ) has been developed in the present study, and successfully used in indirect ELISA for rapid diagnosis of stem rot disease on tomato. Furthermore, our data confirmed FAME analysis is a rapid and reliable method for identification and differentiation of Ecc and Ech strains isolated from soft rot samples of tomato.

Ecc and Ech can survive on seeds and in plant debris in summer periods in the region. There are some differences in surviving periods of the strains in plant debris and seeds tested in the present study. The differences in surviving periods of the pathogens may be explained by cultural and environmental factors effecting the decomposition rate of plant debris and microbial activities. Soft rot Erwinias have been reported to survive on pepper seeds and seed tuber. Ecc has been isolated from commercial pepper seed lots in Israel and contaminated pepper seeds can be a primary and important source of inoculum. But, there is no previous report on the transmission of the soft rot Erwinias by tomato seeds. This is the first study showing that both of Ecc and Ech were able to survive in/on the diseased plant debris and seeds, which may be potential primary inoculum of the disease.

5.33 Bacterial Inoculum and Fungal Disease Development

Winter wheat (*Triticum aestivum*) and winter barley (*Hordeum vulgare*) were grown on soil contaminated with different levels of a marked bacterial potato pathogen *Erwinia carotovora* ssp. *atroseptica* (*Eca*) from a previous blackleg-infected potato crop, to determine whether the presence of this pathogen could affect disease development on wheat and barley (Adrian C. Newton et al., 2004). The diseases 'septoria leaf blotch' and 'powdery mildew' on wheat (caused by *Septoria tritici* and *Blumeria graminis* f. sp. *tritici*, respectively) and 'scald' or 'rhynchosporium' on barley (caused by *Rhynchosporium secalis*) were assessed over two growing seasons, and the presence of the marked *Eca* strain monitored. Wheat diseases were greater on the areas with high levels of *Eca* contamination than on areas with low-level contamination. *Rhynchosporium* on barley was not affected overall, although it decreased on the high-level contamination areas early in the season. The *Eca* strain was detected on the upper foliage of both wheat and barley. Increased *S. tritici* symptoms in the field may be due to 'synergistic' interactions between the *Eca* and the fungal mycelium, as reported in previously published laboratory experiments. The potential importance of such fungal–bacterial interactions was discussed.

In the first trial, symptoms of *Rhynchosporium secalis* infection on barley, causing rhynchosporium or scald, were the first to show, and septoria caused by *Septoria tritici* (teleomorph *Mycosphaerella graminicola*) was observed 2 wk later on the wheat. Late in the season the 'powdery mildew' pathogen *Blumeria graminis* f. sp. *tritici* established on wheat, and was scored twice before it died out. All infections resulted from natural inoculum. There was no difference between the treatment areas that had the two different potato cultivars and there was no statistical difference between the two wheat cultivars or between the two barley cultivars in their disease score response to treatments overall. In the second trial, disease caused by *S. tritici* and *R. secalis* was again scored, but *B. graminis* was observed only at very low levels and was not scored. In both trials, initial disease caused by *R. secalis* (i.e. the first three scores) was reduced by the 10^5*Eca* inoculum treatment compared with the no-potato control. The data sets also showed the same trend, as the disease increased from the 10^5 to 10^3 to 10^1*Eca* treatments. In the first trial, over the full season (i.e. all eight scores) there was no *Eca* treatment effect on *R. secalis* infection (i.e. the early season differences were obscured).

While Koch's postulates for causal organisms of plant diseases are important, when trying to understand the aetiology and epidemiology of a pathogen, full account must be taken of the other organisms in the principal host–pathogen infection court as they may exhibit considerable modifying influences. Most research on this has hitherto been carried out in the context of biological control. Other organisms found on infected plants are generally thought to be secondary pathogens or saprophytes. However, where necrotrophs are concerned and the fungal spore inoculum is splash-dispersed from the soil, the presence of bacterial species is the norm and therefore one might expect interaction to the benefit of some of the organisms involved. Mutual benefit may be derived with the combination of certain bacteria and fungal pathogens. If such beneficial fungal pathogen–bacteria interactions are common, then it brings into question whether appropriate methodologies for assessment of pathogen resistance or fungicide efficacy are being used, and may help explain some discrepancies in efficacy data from field and controlled environment experiments. Documentation and characterization of such interactions may facilitate the development of better high-throughput screening tests.

The combination of different microorganisms should be considered, as a community of bacteria can have effects on other microorganisms that cannot be demonstrated by any of the components separately. These data highlight the fact that host–pathogen interactions can be highly complex. Molecular biology models of host–pathogen interactions indicate many complex interactions and normally exclude the third component of 'the environment', especially microbial components. As many host interactions with nonhost microorganisms nevertheless induce recognition responses, these must be included in a full epidemiological understanding of crop diseases.

5.34 Effect of Temperature on Wheat Streak Mosaic Disease Development in Winter Wheat

Temperature is one of the key factors that influence viral disease development in plants. Temperature effect on Wheat streak mosaic virus (WSMV) replication and in planta movement was determined using a green fluorescent protein (GFP)-tagged virus in two winter wheat cultivars (Wosula Everlyne et al., 2017). Virus-inoculated plants were first incubated at 10, 15, 20, and 25°C for 21 days, followed by 27°C for 14 days; and, in a second experiment, virus-inoculated plants were initially incubated at 27°C for 3 days, followed by 10, 15, 20, and 25°C for 21 days. In the first experiment, WSMV-GFP in susceptible 'Tomahawk' wheat at 10°C was restricted at the point of inoculation whereas, at 15°C, the virus moved systemically, accompanied with mild symptoms, and, at 20 and 25°C, WSMV elicited severe WSMV symptoms. In resistant 'Mace' wheat (PI 651043),WSMV-GFP was restricted at the point of inoculation at 10 and 15°C but, at 20 and 25°C, the virus infected systemically with no visual symptoms. Some plants that were not systemically infected at low temperatures expressed WSMV-GFP in regrowth shoots when later held at 27°C. In the second experiment, Tomahawk plants (100%) expressed systemic WSMV-GFP after 21 days at all four temperature levels; however, systemic WSMV expression in Mace was delayed at the lower temperatures. These results indicated that temperature played an important role in WSMV replication, movement, and symptom development in resistant and susceptible wheat cultivars. It also demonstrated that suboptimal temperatures impair WSMV movement but the virus rapidly begins to replicate and spread in planta under optimal temperatures.

Other studies that have evaluated the effect of temperature on WSMV severity in wheat cultivars used a minimum temperature of 18°C. In almost all cases, they report typical WSMV symptoms in the susceptible check within 14 days and no symptoms in resistant cultivars or germplasm carrying either the Wsm1 or the Wsm2 gene. A study showed that, when exposed to high temperatures (28°C) for 4 weeks after inoculation with WSMV, Mace and RonL wheat bearing the Wsm1 and Wsm2 genes, respectively, expressed severe WSMV symptoms and high virus titer. These plants failed to recover and even failed to produce heads after exposure to winter conditions with temperatures below the resistance threshold. This suggests that Mace can remain symptom-free, with little impact on yield under field conditions with cooler temperatures below 25°C, but it will suffer severe symptoms and yield loss under prolonged temperatures above 25°C. TMV has been reported to move passively with photo assimilates without replication and spread under low temperatures; however, when plants are moved to higher temperatures, lesions appear in a shorter duration than normally required from time of inoculation to symptom expression, suggesting systemic infection but no replication or detection. Prevailing climatic conditions, especially temperature, influence the impact of WSMV in both susceptible and resistant wheat cultivars.

Prevailing climatic conditions, especially temperature, influence the impact of WSMV in both susceptible and resistant wheat cultivars (Wosula Everlyne et al., 2017). This study demonstrates a rapid increase in virus activity with increasing temperatures beginning about 15°C for a WSMV-susceptible variety (Tomahawk); however, this response to temperature is delayed considerably until temperatures increase to about 25°C in a WSMV resistant (Wsm1) cultivar (Mace). In planta movement of WSMV at lower temperatures appears to be an important component of infection dynamics in winter wheat but this dynamic is significantly altered in plants with the Wsm1 resistance gene. Therefore, plants infected during the fall under cool temperatures that lose inoculated leaves during the winter may maintain the virus because of rapid translocation of WSMV to the crown. In susceptible plants, this will result in symptom development when warm conditions occur during the spring.

5.35 Inoculum Potential of *Polymyxa betae*, Beet Necrotic Yellow Vein Furovirus and Narcissus Mosaic Virus

The widespread occurrence of *P. betae* n the sugar beet area in Belgium is confirmed by this quantitative study. In the small number of surveyed fields where BNYVV was detected, the inoculum potential of the vector appeared not to be significantly higher than in the other fields (Goffart, et al., 1989). The range of inoculum potential of *P. betae* in Belgian soils was also similar to that in soils from other European countries, where rhizomania leads to decreasing yield in sugar production. Sugar beet plants reacting positively in the BNYW ELISA test were on the other hand found in several locations of the sugar beet growing areas. Yield losses were, however, since then noted only in the Prosper Polders at Doel-Kieldrecht. Several hypotheses can be put forward to explain these differences in evolution, viz., a lower inoculum potential of *P. betae* and BNYW in Belgium compared to other countries; the occurrence in Belgium of other strains; limited expression of inoculum potential due to unfavourable climatic or soil conditions, to crop management or possibly to microbial antagonism. In order to contribute to the analysis of these hypotheses, inoculum potential values were determined for *P. betae* and BNYW by a recently developed quantitative method and compared with those occurring in soils from other countries with or without rhizomania.

The inoculum potential of *P. betae* and BNYVV was studied from 52 random samples of Belgian soils and 10 samples from other European countries, by culture of bait plants in tubes under controlled conditions on serial dilutions of the soils in sterile sand. *P. betae* was detected in all samples within the range of0.0l to 27.1 infection units per g of soil. BNYW was detected by ELISA on root extracts of bait plants grown on three Belgian soil samples. All the tested samples from rhizomania-infested areas in France, FRG, the Netherlands, Switzerland and Austria, were found to be infested by BNYVV by this technique. For BNYVV survey, the plant bait technique appears more reliable than the analysis of rootlets collected in the field and observation of external symptoms in case of low BNYW infestations or non expression because of unfavourable environmental conditions. *P. betae* isolates from various origins heavily infected *Beta* spp. but only moderately spinach. *Chenopodium album* was slightly infected by 2 of the 7 isolates, *C. murale* by 4 of them.

Radial density distribution and symmetry of a Potexvirus, Narcissus Mosaic Virus: Narcissus mosaic virus (NMV) is a Potexvirus, a member of the Flexiviridae family of filamentous plant viruses. Fiber diffraction patterns from oriented sols of narcissus mosaic virus have been used to determine the symmetry and structural parameters of the viral helix (Amy Kendalla et al., 2007). The virions have a radius of 55±5 Å. The viral helix has a pitch of 34.45±0.5 Å, with 7.8 subunits per turn of the helix. We conclude that all members of the Potex virus genus have close to 8 subunits per helical turn.

Radial density distribution was obtained by assigning alternating positive and negative signs to the first five equatorial peaks of the diffraction pattern and calculating a Fourier–Bessel transform of the signed equator. Applying an artificial temperature factor, even as high as 1000, did not significantly affect this distribution. The distribution was significantly altered by assigning a negative sign to peak #5. By the minimum wavelength principle, such an assignment is only possible if one of peaks 4 or 5 is not present, so each of these peaks was omitted in separate calculations. In these calculations, the origin peak was taken to be positive and the second peak negative as before; the fifth peak was assigned a negative sign, and the remaining peak (3 or 4) was positive. The resulting distributions (discussed below) were similar to each other, but were completely dominated by a peak at 40 Å radius. Again, an artificial temperature factor had little effect. The radial density distribution confirms the viralouter radius of about 55 Å, with an uncertainty of no more than5 Å, and suggests the presence of a hole of about 20 Å radius running through the middle of the virion. By analogy with TMV(Caspar, 1956; Franklin, 1956) one might suggest that the RNA is at the radius of the highest peak, a little less than 30 Å.

For symmetry determination, the most useful features of the NMV diffraction pattern are the clear separation and definition of layer lines 19 and 20, together with consistent positions of intensity maxima on numerous other layer lines. It was concluded that NMV has 7.8 subunits per turn of the viral helix, and that potex viruses in general have approximately 8 subunits per turn, with precise value ranging from 7.7 to 7.9.

Bibliography

Aad Termorshuizen and Jeger, M.J., 2014. Assessing inoculum of soil borne plant pathogens: Theory and practice in decision-making for soil disinfestations. *Acta Hort.*, **1044:** 75-80

Abigail J. Stack, Stephen M. Marek, Thomas R. Gordon and Richard M. Bostock, 2022. Genetic Diversity and Potential Inoculum Sources of *Fusarium* Species Causing Cankers in Bareroot-Propagated Almond Trees in California Nurseries. *Pl. Dis.,* **106:** https://doi.org/10.1094/PDIS-08-21-1637-RE Adrian C. Newton, Ian K. Toth, Paul Neave and Lizbeth J. Hyman, 2004. Bacterial inoculum from a previous crop affects fungal disease development on subsequent non host crops. *New Phytologist*, **163:** 133-138

Ajinde, O. A., Ogunnusi, A. T. and Akpor, B. O., 2024. Assessment of the Protective Potential of Inoculums and Metabolites of Rhizobacteria on Soybean (*Glycine max*) Seedlings against Bacterial and Fungal Pathogens. *Recent Adv. Food Nutr. Agric.,* doi: **10:**2174/012772574X282130231206103404.

Amy Kendalla, Wen Bian, Justin Junna, Ian McCullougha, David Goreb and Gerald Stubbs, 2007. Radial density distribution and symmetry of a Potex virus, narcissus mosaic virus. *Virology,* **357**: 158–164

Aysan, Y., Sahin , F., Cetinkaya-Yildiz, R., Mirik, M. and Yucel, F., 2005. Occurrence and primer inoculum sources of bacterial stem rot caused by *Erwinia* Species on tomato in the eastern Mediterranean region of Turkey. *J. Pl. Dis. & Protn.*, **112**: 42–51

Berbegal, M., Ortega, A., García-Jiménez, J. and Armengol, J., 2007. Inoculum density-disease development relationship in Verticillium wilt of artichoke caused by *Verticillium dahliae*. *Plant Dis.*, **91**:1131-1136.

Bergamin Filho, A., Inoue-Nagata, A. K., Bassanezi, R. B., Belasque Jr, J., Amorim, L.,. Macedo, M. A. , Barbosa, J. C., Willocquet, L. and Savary, S., 2016. The importance of primary inoculum and area-wide disease management to crop health and food security. *Food Sec.*, **8**:221–238

Bhuiyan, M.Z.R., Mendoza, Luis E. Del Río, Lakshman Dilip, K., Qi Aiming and Khan Mohamed, 2023. Effects of Inoculum Density of *R. solani* AG 2-2IIIb and Age of Plant on Root Rot Severity in Sugar Beet. Available at SSRN: https://ssrn.com/abstract=4428900 or http://dx.doi.org/10.2139/ssrn.4428900

Bithell, S., Moore, K., Herdina, McKay, A., Harden, S. and Simpfendorfer, S., 2021. Phytophthora root rot of chickpea: inoculum concentration and seasonally dependent success for qPCR based predictions of disease and yield loss. *Australas. Plant Pathol.,* **50:** 91–103.

Botero-Ramirez, A., Hwang, S.F. and Strelkov, S.E., 2021. *Plasmodiophora brassicae* Inoculum Density and Spatial Patterns at the Field Level and Relation to Soil Characteristics. *Pathogens,* **10**:499.

Chang Kan-Fa, Hwang, S., Gossen Bruce, Turnbull, G.D., Wang, H. and Howard Ronald, 2008. Effects of inoculum density, temperature, seeding depth, seeding date and fungicidal seed treatment on the impact of *Rhizoctonia solani* on lentil. *Canadian J. Pl. Sci.*, **88:** 799-809.

Chatterton, S., Schwinghamer, T.D., Page,´ A., Davidson, R.B., Harding, M.W. and Banniza, S., 2023. Inoculum dose–disease response relationships for the pea root rot pathogen, *Aphanomyces euteiches*, are dependent on soil type and other pathogens. *Front. Plant Sci.,* **14**:1115420.

Claudio Cocozza, Emad Abdelhameed Abdeldaym, Gennaro Brunetti, Franco Nigro and Andreina Traversa. 2021. Synergistic efect of organic and inorganic fertilization on the soil inoculum density of the soilborne pathogens *Verticillium dahliae* and *Phytophthora* spp. under open-field conditions. *Chem. Biol. Technol. Agric*., **8**:24

Czislowski, E., Zeil-Rolfe, I. and Aitken, E.A.B., 2021. Effector profiles of endophytic *Fusarium* associated with asymptomatic banana (*Musa sp.*) hosts. *Int. J. Mol. Sci.,.* **22:** 2508.

Darban, D.A., Pathan, M.A., Bhatti, A.G. and Maitelo, S.A., 2005. The effect of different initial densities of nematode (*Meloidogyne javanica*) on the build-up of *Pasteuria penetrans* population. *J. Zhejiang Univ. Sci. B.,* **6**:113-118.

de la Lastra, E., Marín-Guirao, J.I., López-Moreno, F.J., Soriano, T., de Cara-García, M. and Capote, N., 2021. Potential inoculum sources of Fusarium species involved in asparagus decline syndrome and evaluation of soil disinfestation methods by qPCR protocols. *Pest Manag. Sci.,***77**: 4749-4757

Elshahawy, I.E., Saied, N., Abd-El-Kareem, F. *et al.,* 2019. Effect of inoculum density of *Stromatinia cepivora* on the amount of white rot reduced by *Trichoderma* species in garlic. *Bull. Natl. Res. Cent.,* **43:** 27

Gilligan, C. A. and Simons, S. A., 1987. Inoculum efficiency and pathozone width for two host-parasite systems. *New Phytol.*, **107**: 549-566

Goffart, J. P., V. Horta and H. Marait. Inoculum potential and host range of Polymyxa betae and beet necrotic yellow vein furovirus. Bulletin oEPP/EPPO Bulletin 19, 517- 525 ,1989

Gong, X., Zhang, Z., Wang, H., Li, H., Hu, F., Liu, M., Jiang, L., Chen, X. and Ma, C., 2022. Synergies between Heat Disturbance and Inoculum Size Promote the Invasion Potential of a Bacterial Pathogen in Soil. *Microorganisms,* **10**: 630.

Hofgaard, I.S., Seehusen, T., Aamot, H.U., Riley, H., Razzaghian, J., Le, V.H., Hjelkrem, A.G., Dill-Macky, R. and Brodal, G., 2016. Inoculum Potential of *Fusarium* spp. Relates to Tillage and Straw Management in Norwegian Fields of Spring Oats. *Front Microbiol.,* **7**:556

Hwang, S. F., Strelkov, S. E., Ahmed, H. U., Manolii, V. P., Zhou, Q., Fu, H., Turnbull, G., Fredua-Agyeman, R. and Feindel, D., 2017. Virulence and inoculum density-dependent interactions between club root resistant canola (*Brassica napus*) and *Plasmodiophora brassicae. Pl.Pathol.,* **66**: 1318-1328

Jason R. Brantner and Ashok K. Chanda, 2021. Effects of Inoculum Density and Cultivar Susceptibility on *Rhizoctonia* Damping-Off and Crown and Root Rot in Sugar Beet. *Pl. Disease*, **105:** https://doi.org/10.1094/PDIS-02-20-0336-RE

José da Cruz Machado, Andréia Quixabeira Machado, Edson Ampélio Pozza, Cibele Ferreira Machado and Willian Luis Antonio Zancan, 2013. Inoculum potential of *Fusarium verticillioides* and performance of maize seeds. *Trop. Pl. Pathol.*, **38:** 213-217

Juan A. Navas-Cortes, Antonio R. Alcala-Jimenez, Bernhard Hau and Rafael M. Jimenez-Dıaz, 2000. Influence of inoculum density of races 0 and 5 of *Fusarium oxysporum* f. sp. *ciceris* on development of Fusarium wilt in chickpea cultivars. *European J. Pl. Pathol.*, **106:** 135–146

Kessel, G.J.T., Kohl, J., Powell, J.A., Rabbinge, R. and van der Werf, W., 2005. Modeling spatial characteristics in the biological control of fungi at leaf scale: competitive substrate colonization by *Botrytis cinerea* and the saprophytic antagonist *Ulocladium atrum*. *Phytopathology*, **95:**439–48.

Khan, I., Rawal, P., Singh, R., Jatwa, T. kumar and Kumar, S., 2022. Effect of Inoculum Levels on Disease Development and Survival Potential of *Sclerotinia sclerotiorum* Sclerotia. *J. Exp. Agri. Intl.,* **44:** 32–36.

Kirsty Harris, Iain M. Young , Christopher A. Gilligan, Wilfred Otten and Karl Ritz, 2003. Effect of bulk density on the spatial organisation of the fungus *Rhizoctonia solani* in soil. *FEMS Microbiol.Ecol.*, **44**:45-56

Lavras, M.G., 2016. *Stenocarpella* spp. Inoculum quantification in tropical maize stubble. *Felipe augusto moretti ferreira pinto.* Universidade Federal De Larvas

Liu,Y., Qi, A. and Khan, M.F.R., 2019. Age dependent resistance to *Rhizoctonia solani* in sugar beet. *Plant Dis.*, **103**: 2322-2329

Macedo, R., Sales, L.P., Yoshida, F., Silva-Abud, L.L. and Lobo, M. Jr., 2017. Potential worldwide distribution of Fusarium dry root rot in common beans based on the optimal environment for disease occurrence. *PLoS ONE*, **12** (11): e0187770.

Maitlo,S.A., Syed, R.N., Rustamani, M.A., Khuhro, R.D. and Lodhi, A.M., 2016. Influence of inoculation methods and inoculum levels on the aggressiveness of *Fusarium oxysporum* f.sp. *ciceris* on chickpea and plant growth. *Int. J. Agric. Biol.,* **18**: 31–36

Marina de Resende Faria Guimarães, Carolina da Silva Siqueira, José da Cruz Machado, Sueny Kelly Santos de França and Gabriel Castanheira Guimarães, 2017. Evaluation of inoculum potential of pathogens in seeds: relation to physiological quality and DNA quantification by qPCR1. *J. Seed Sci.,* **39:** 233, 2017.

Maryam Khoshkdaman, Sedigheh Mousanejad, Seyed Ali Elahinia, Ali-Akbar Ebadi and Fereidoun Padasht-Dehkaei, 2020. Impact of soil-borne inoculum on sheath blight disease development in rice. *J. Crop Prot.,* **9**: 625–635

Matthews, A., Muthukumar, S.P.T., Hamill, S., Aitken, E.A.B and Chen, A., 2023. Impact of inoculum density of *Fusarium oxysporum* f. sp. *zingiberi* on symptomatic appearances and yield of ginger (*Zingiber officinale* Roscoe). *Access Microbiol.,***5:**000605.v3.

Moens, T., Araya, M., Swennen, R., De Waele, D. and Sandoval, J., 2003. Effect of growing medium, inoculum density, exposure time and pot volume on the early reproduction and damage potential of *Radopholus similis* in banana (*Musa* spp.). *Nematropica,* **33**:9-26.

Montville, R. and Schaffner Donald, 2003. Inoculum Size Influences Bacterial Cross Contamination between Surfaces. *Appl.& Environ. Microbiol.*, **69:** 7188-7193.

Morin, L., Watson, A.K. and Reeleder, R.D., 1990. Effect of dew, inoculum density, and spray additives on infection of field bindweed by *Phomopsis convolvulus. Canadian J.Pl. Pathol.*, **12**:48-56

Muhammad Zameer Kayani, Tariq Mukhtar and Muhammad Arshad Hussain, 2018. Interaction between Nematode Inoculum Density and Plant Age on Growth and Yield of Cucumber and Reproduction of *Meloidogyne incognita. Pakistan J. Zool.*, **50**: 897-902

Navas-Cortés, J. A., Alcalá-Jiménez, A. R., Hau, B. and Jiménez-Díaz, R. M., 2000. Influence of inoculum density of races 0 and 5 of *Fusarium oxysporum* f. sp. *ciceris* on development of Fusarium wilt in chickpea cultivars. *Eur. J. Plant Pathol.*, **106**: 135-146

Nehl David, Allen Stephen and Kochman Joseph, 2003. Fusarium wilt of cotton: A fatal fungal affliction?. *Microbiol.Australia*, **24:** 8. 10.1071/MA03308.

Nion, Y.A. and Toyota, K 2015. Recent trends in control methods for bacterial wilt diseases caused by *Ralstonia solanacearum. Microbes Environ.*, **30:** 1–11.

Nitzan, N., Cummings, T. F. and Johnson, D. A., 2008. Disease potential of soil- and tuberborne inocula of *Colletotrichum coccodes* and black dot severity on potato. *Plant Dis.*, **92:**1497-1502.

Nongkhlaw, F.M.W. and Joshi, S.R., 2014. Distribution pattern analysis of epiphytic bacteria on ethnomedicinal plant surfaces: A micrographical and molecular approach. *J. Microsc. Ultrastruct.*,**2**: 34–40.

Pandey, S., Ram Jilal, Srivastava, S.N. and Tripathi R.C., 1999. Role of inoculum density and incubation period on saprophytic behaviour of *Pythium ultimum* and *Sclerotium rolfsii. Indian Phytopath.*, **52:** 294-296

Pandey, S., Srivastava, S.N., Ram Ji Lal and Tripathi, R.C., 1999. Role of inoculum density and incubation period on saprophytic behaviour of *Rhizoctonia solani* and *Rhizoctonia bataticola. Indian Phytopath.*, **52** 56-58

Paul W. Tooley, Marsha Browning and Robert M. Leighty, 2013. Inoculum Density Relationships for Infection of Some Eastern US Forest Species by *Phytophthora ramorum. J. Phytopathol.*, **161**:595–603

Praveen Kumar Singh, Ramesh Chandra Vishwakarma, Jaswinder Kaur and Ravikant Singh, 2023. Growth and Survival of Plant Pathogens. P.3-18. *In: Principles of Plant Pathology*. R. C. Mishra & R Singh (Eds.).

Prussin, A. J., Li, Q., Malla, R., Ross, S. D. and Schmale, D. G., 2014. Monitoring the long-distance transport of *Fusarium graminearum* from field-scale sources of inoculum. *Plant Dis.,* **98**: 504–511

Raghavendra, B., Srinivas, T and Padmodaya, B., 2018. Effect of Different Inoculum Densities of *S. rolfsii* on Seed Germination, Seedling Growth Parameters and Disease Incidence. *Int.J.Curr. Microbiol. App.Sci.,* **7:** 84-91

Rao, A. and Rao, M., 1963. Inoculum Potential and the Fusarial Wilt of Cotton. *Nature,* **200:** 598–599

Reis, E. M., Boaretto, C. and Danelli, A. L. D., 2014. *Macrophomina phaseolina*: Density and longevity of microsclerotia in soybean root tissues and free on the soil, and competitive saprophytic ability. *Summa Phytopathologica,* **40**:128-133

Saberi-Riseh, R., Hajieghrari, B., Rouhani, H and Sharifi-Tehrani, A., 2004. Effects of inoculum density and substrate type on saprophytic survival of *Phytophthora drechsleri*, the causal agent of gummosis (crown and root rot) on pistachio in Rafsanjan, Iran. *Commun. Agric. Appl. Biol. Sci.,* **69:** 653-556.

Safarieskandari, S., Chatterton, S. and Hall, L. M., 2021. Pathogenicity and host range of Fusarium species associated with pea root rot in Alberta, Canada. *Can. J. Plant Pathol.,* **43:** 162–171.

Saini, A. K., Chauhan, R. S., Ashwani Kumar, Narender Singh and Satish Kumar, 2019. Effect of Inoculum Load on Disease Initiation, Threshold Point and Population Build Up of Post Harvest Pathogen (*Penicillium islandicum* Sopp.) of Indian Gooseberry (*Emblica officinalis* Goerth). *Int.J.Curr.Microbiol.App.Sci.,* **8**: 2082-2085

Smiley, R. W., Machado, S., Rhinhart, K. E. L., Reardon, C. L. and Wuest, S. B., 2016. Rapid quantification of soil borne pathogen communities in wheat-based long-term field experiments. *Plant Dis.,* **100**:1692-1708

Smith, D.F.Q. and Casadevall, A., 2022. On the relationship between Pathogenic Potential and Infective Inoculum. *PLoS Pathog.,* **18**(6): e1010484

Soum Sanogo, 2019. Genetic Soil Disinfestation, A Conceptual Framework to Reduce Inoculum Potential of Soil borne Plant Pathogens. *J. Agri. Hort. Res.,* **2**:1- 6

Stringham, O.C. and Lockwood, J.L., 2021. Managing propagule pressure to prevent invasive species establishments: Propagule size, number, and risk-release curve. *Ecol. Appl.,* **31:** e02314

Sujata Kumari, Kumud Jarial and R. S. Jarial, 2022. Effect of Inoculum Density and Methods of Inoculation on the Development of Bacterial Leaf Spot of Bottle Gourd and Pumpkin caused by *Xanthomonas cucurbitae. Int.J.Curr.Microbiol.App.Sci.*, **9**: 1358-1371

Tatiana Botelho Fantazzini, Renato Mendes Guimarães, Aline da Consolação Sampaio Clemente, Everson Reis Carvalho and José da Cruz Machado, 2016. *Fusarium verticillioides* Inoculum Potential and its relation with the physiological stored corn seeds quality. *Biosci. J.*, **32**: 1254-1262

Taylor, A., Coventry, E., Handy, C., West, J. S., Young, C. S., and Clarkson. J. P., 2018. Inoculum potential of *Sclerotinia sclerotiorum* sclerotia depends on isolate and host plant. *Pl.Pathol.,* **67**: 1286-1295

Tooley, P.W., Browning, M.and Leighty, R.M., 2013. Inoculum density relationships for infection of some Eastern US forest species by *Phytophthora ramorum*. *J. Phytopathol.,* **161**: 595–603

Van der Heyden, H., Dutilleul, P., Charron, JB. *et al.,* 2021. Monitoring airborne inoculum for improved plant disease management. A review. *Agron. Sustain. Dev.,* **41:** 40.

Wallenhammar, A.C, Omer, Z.S., Edin, E. and Jonsson, A., 2021. Influence of Soil-Borne Inoculum of *Plasmodiophora brassicae* Measured by qPCR on Disease Severity of Club root-Resistant Cultivars of Winter Oilseed Rape (*Brassica napus* L.). *Pathogens,* **10**: 433.

Wosula Everlyne, N., Tatineni Satyanarayana, Wegulo Stephen, N. and Hein Gary, L., 2017. Effect of Temperature on Wheat Streak Mosaic Disease Development in Winter Wheat. Papers in Plant Pathology. 464.

Xu, X. and Hu, X., 2020. The effect of aggregation of pathogen and biocontrol microbe propagules on biocontrol potential: A simple modelling study. *Phytopathol Res,* **2:** 5

Yannis Raftoyannis and Michael W. Dick, 2002. Effects of Inoculum Density, Plant Age and Temperature on Disease Severity caused by Pythiaceous Fungi on Several Plants. *Phytoparasitica,* **30**(1).

Young, C., Clarkson, J., West, J., Bouda Ahmadi, B. and Burnett, F., 2014. Reducing the Impact of Sclerotinia Disease on Arable Rotations, Vegetable Crops and Land Use. Kenilworth, UK: Agriculture and Horticulture Development Board: AHDB Final Project Report PR538

Zamani-Noor, N., Krohne, I. and Koopmann, B., 2021. Greenhouse Evaluation of Clubroot Resistant-*Brassica napus* cv. Mendel and Its Efficacy Concerning Virulence and Soil Inoculum Levels of *Plasmodiophora brassicae*. *Pathogens,* **10:** 151

Zhar, K., Sarkes, A., Yang, Y., Zhou, Q., Feindel, D. and Harding, M.W., 2021. *Plasmodiophora brassicae* in its environment- effects of temperature and light on resting spore survival in soil. *Phytopathology,* **111**:1743-1750

6

Competition, Predation Antibiosis and Fungistasis

6.1 Competition

Soil microorganisms in particular are perceived as active competitors that have evolved diverse strategies to facilitate resource acquisition. Activities including motility, coordinated behavior, and antibiotic production can tip the competitive balance, resulting in numerical dominance or local extinction of competing populations. Functional consequences of microbial social behaviors have been studied extensively and demonstrate the importance of factors such as relatedness, kin discrimination, enforcement of cooperation, and competition among relatives to microbial survival and fitness in the environment (Adil Essarioui et al., 2017). However, most investigations have focused on interactions among closely related taxa, often among species or genera. Yet, in natural environments, coexistence of individuals from diverse microbial phyla and kingdoms is common. In particular, bacteria and fungi commonly coexist in soil, but there has been relatively little study of the role of species interactions in mediating the dynamics of specific bacterial and fungal populations in soil. Moreover, we have little understanding of the extent to which the ecological context, and specifically plant host or plant community diversity, may influence bacterialfungal interactions in soil. Such studies are needed to begin to establish a more complete understanding of the ecological and coevolutionary dynamics of soil microbial populations.

Microbes have coevolved to engage in complex networks of interactions with coexisting kin and non-kin individuals in the soil and in association with plants. Microbial antagonism and resource competition within complex communities have been the focus of many studies that have explored the effects of diverse factors on the ecological outcomes of these interactions. However, the concurrent impacts on these processes of plant species richness and plant host, which jointly structure the environment within which rhizosphere microbes interact, have received far less attention. This is despite the fact that both plant community richness and plant species have been shown to impact microbial community composition and function in rhizosphere soil, with potential implications for microbial interactions and pathogen suppression in natural and agricultural habitats.

The ecology of competition is an old subject, developed through the efforts of numerous macrobiologists. Most of the key principles of competition theory arose through the observational data generated by field ecologists directly studying environmental systems (Hibbing et al., 2010). A limitation of these studies, however, is the difficulty in linking competitive behavior or processes to underlying genetic determinants. This is where bacteria present a wonderful opportunity. Their rapid generation time, genetic malleability and ability to engage in group behaviors make them excellent candidates to test some of the basic theories thought to underpin competition, and to examine how group level processes influence competition between species. For example, experiments with microbial populations have already helped refine and validate several predictions regarding the conditions that favor the maintenance of cooperation in the face of the potential for social cheating. The use of many active interference mechanisms of competition by bacteria species should provide particularly useful experimental systems for testing emerging theories regarding the conditions that favor interference over exploitative competition strategies. Finally, studies of microbial competition in the context of group level processes may also generate insight into the evolutionary transition to multicellularity, which was undoubtedly driven by competition.

One limitation of many studies of microbial competition conducted to date has been that predictions generated from *in vitro* studies have not been tested in more natural settings. The emergence of technologies such as high-throughput genomic sequencing, expression profiling, and sophisticated microscopy are now allowing us to ask questions regarding microbial distribution and physiology in complex systems that would not have been possible in the past, and which should facilitate testing of many of these predictions. Incorporation of additional complexity to mimic more natural settings will also enable comparisons between the competitive strategies favored in different settings. For example, antimicrobial production may be more important in densely colonized, nutrient-rich enviroments such as the human oral cavity, while motility could be particularly important in an oligotrophic environment like the open ocean.

Competition is the active demand in excess of immediate. Given the fact that soils generally have limited and inaccessible forms of nutrients for microbial utilization; competition for resources such as nutrients, oxygen and suitable sites at the root surfaces is a commonly occurring phenomenon among soil inhabiting organisms and is a fundamental mechanism for the control of soil borne pathogens. Starvation is an important and common cause of microbial death, so competition for limiting nutrients, may lead to biological control of fungal phytopathogens (Prashar Pratibha et al., 2013). However competition will take place only if the introduced microbial strain and the pathogen have similar demands in terms of space and resources. Since microbes generally have common requirements in terms of essential growth factors, lack of critical nutrients like carbon and mineral elements, like iron, along with suitable sites in the rhizosphere, is assumed to be responsible for inhibition of fungal spore germination in soil. Competition among biocontrol agent and target pathogen is illustrated in figure 5. It occurs for suitable niches and available resources in the rhizosphere. Possession of traits listed in the figure help the biocontrol agent to out-compete the pathogen thereby controlling the corresponding disease.

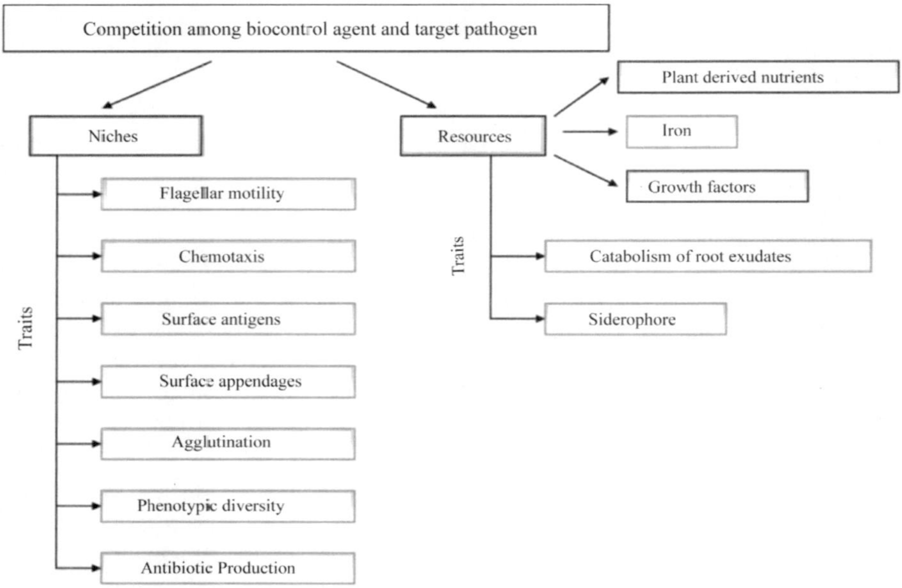

Fig. 38: Competition among biocontrol agent and target pathogen. It occurs for suitable niches and available resources in the rhizosphere. Possession of traits listed in the figure help the biocontrol agent to out-compete the pathogen thereby controlling the corresponding disease

Competition for Root Niches: Availability of carbon is the limiting factor for microbial activity in soil and thus affects a number of processes mediated by soil micro flora. Since the root surface itself and its immediate surroundings have high carbon concentrations (equivalent to 15–60% of the total photosynthetic production of the plant) owing to rhizodeposition, there is an acute competition among rhizospheric microorganism to occupy suitable nutrient rich sites along the root surface. Traits like flagellar motility of PGPR strains such as, chemotaxis toward root exudates and surface antigen (O-antigen of lipopolysaccharide) mediated attachment to the root surface, play crucial role in successfully establishing the introduced strain in the rhizosphere (de Weert and Bloomberg, 2006). Various studies involving the use of mutant strains lacking in above mentioned capabilities have further con fi rmed their significant role in the root colonization. For example nonmotile mutants of *Pseudomonas fluorescens* WCS365 have been proved to have reduced root colonization in tomato, wheat, and radish crop. The non-motile mutants of a well-known biocontrol strain *P. fluorescens* F113 was examined for competitive root colonization. It was observed that all the mutants were completely displaced from the root tip when competing with the wild-type F113 strain. Moreover one of the mutants which had longer flagella compared to the parent strain, when grown in rich medium showed a higher motility, suggesting the possibility of improving competitive root colonization by manipulating the motility processes. The independent role of chemotaxis along with motility was established. They evaluated cheA mutants of *P. fluorescens* that were defective in flagella-driven chemotaxis but still having motility for competitive tomato root-tip colonization in a gnotobiotic sand system and in nonsterile potting soil. Mutants were found to have a strongly reduced competitive root colonizing ability after seven days of plant growth, in both the cases. Substances that are known to create chemotactic response toward root exudates include organic acids, amino acids, sugars etc.

The mutant studies have proved that the presence of O-antigen lipopolysaccharide and ability to synthesize amino acids and vitamin B1 are important root colonization traits. However the role of O-antigen lipopolysaccharide may vary with strain. In mutants of strains WCS358 and WCS374 which lacked the 0-antigenic side chain of the lipopolysaccharide, the adhesion to potato roots was not effected, but the mutant of *P. fluorescens* PCL1205 which had shorter O-antigen chain was defective in colonization ability. Other cell surface exopolysaccharides, like cellulose fi brils, also play a role in the initial adherence of rhizobacteria like *A. tumefaciens* and *R. legunisprum* to plant surface and root hairs. Extracellular fi brils produced by *R. leguminosarum* were found to have a positive correlation between the percentage of fibrilated cells and the ability of the bacteria to form caps and to adhere to pea root hair tips.

Bacterial surface appendages like fimbriae also mediate the attachment of the introduced strain to the root surfaces. These proteinaceous, fi brillar structures are generally found on the surfaces of many gramnegative bacteria and play a role in attachment of the bacterium to different surfaces. It was observed that there was correlation between the total number cells that colonized the roots and the cells possessing pili. Role of fimbriae has also been established in case of *Pseudomonas.* It was found that the nonmucoidal strain 2–79 was highly piliated whereas the mucoidal type was nearly devoid of pili and the presence of pili was quantitatively correlated with hydrophobic attachment to roots. One more factor that has been shown to be instrumental in early steps of root colonization i.e. attachment of the bacterial cells to the root surface, is agglutination. Bacterial cells may be agglutinated by root surface mucilaginous substances like pectic compounds, other acidic polysaccharides and certain unknown compounds present in the root exudates. It has been shown that the agglutination and adherence to root surface leading to the long term colonization is rapid and frequent in rhizosphere bacteria like *P. fluorescens* than that of the non-rhizospheric bacteria in various plants like radish and pea. Another determinant of root colonization potential is the ability of introduced strain to generate phenotypically diverse population by site-specific recombinases of the one integrase family which control the phase variation (Weller 2007). Gene sss and xerD have been found to encode two different types of site-specific recombinases that determine phenotypic variation during rhizosphere colonization.

Production of antibiotics also influences the ecological competence of the producing strain in terms of root colonization and hence their persistence in the rhizosphere leading to bene fi cial interaction with plants by stimulating host defense mechanisms (Prashar Pratibha et al., 2013). For example phenazines produced by two strains of *Pseudomonas* which are known to suppress take-all of wheat caused by *Gaeumannomyces graminis* var. *tritici*, i.e. *P. fluorescens* 2–79 and *P. aureofaciens*

30–84, have been also shown to contribute to the ecological competence of these strains in the rhizosphere of wheat. Mutants defective in phenazine production were found to have signi fi cantly reduced population size compared to the parental strains. Similarly lipopeptides antibiotics: surfactins, iturins and fengycins produced by Bacillus species enhance their competitive potential. The ability to utilize nutrients like vitamins biotin; thiamine and amino acids like aspartic acid; glutamic acid; iso-leucine; leucine and lysine also plays a crucial role in root colonization by the PGPR.

ii. Competition for Iron: Iron is an essential nutrient for all living organisms. Though present abundantly on earth (fourth largest mineral), it is not readily available to microbes and plants as it get easily oxidized and exist predominantly in the form of sparingly soluble ferric ions which cannot be utilized directly (Prashar Pratibha et al., 2013). The concentration of iron in the soil may be as low as 10 −18 M which is signi fi cantly lesser than the required concentrations of approximately 10 −6 M for the growth of microorganisms. Soil microorganisms are known to secrete a class of low molecular weight molecules (400–1,000 Da) , known as siderophores, which act as ligands for binding ferric ions due to their high iron af fi nity (Kd = 10 −20 to 10 −50). Thus siderophore producing strains have a selective advantage over the non producing strains, including pathogens, as they may competitively scavenge the limited amount of ferric ions available in the rhizosphere resulting in inhibition of growth of the pathogens in their immediate vicinity due to iron limitation. A wide range of PGPR are known to produce siderphores including *Pseudomonas, Bacillus, Acinetobacter, Serratia* etc. and hence this serves as an important phenomenon contributing to suppression of soil borne pathogens.

Certain fungal phytopathogens may also synthesize siderophores, however their affinity for iron is much lower compared to those of antagonistic PGPR and as a result PGPR out-compete fungal phytopathogens for the available iron. Overall, soil borne pathogens whose route of infection is through mycelial contact, such as species of *Fusarium* and *Pythium*, are more susceptible to face competition from other soil and plant-associated microbes than those pathogens that germinate directly on plant surfaces and infect through appressoria and infection pegs. Siderophores are generally classified on the basis of the ligand that they use for chelation of ferric ions, as either catecholates (phenolates) or hydroxamates or carboxylates. While phenolate siderophores are produced exclusively by bacteria, hydroxamates are produced by both bacteria and fungi. *Pseudomonas* generally produce two kinds of siderophores: pyoverdins which are yellow-green siderophores produced by fluorescent *Pseudomonas* in iron-limited conditions and pyochelin which is a salicylate derivative and structurally unique phenolate siderophore produced by *P. aeruginosa*. Role of siderophores in control of pathogens either singly or synergistically with other metabolites like antibiotics has been well established, particularly by analyzing the mutants deficient in siderophore production. Numerous studies have reported the role of siderophore in suppression of plant diseases caused by *Pythium* spp., *Fusarium oxysporum*, *Magnaporthe grisea*, *Rhizoctonia solani, Pyricularia oryzae, Alternaria, Sclerotium* sp. and many other fungal pathogens in plants like wheat, tomato, rice, chick-pea, green gram by *Pseudomonas, Streptomyces* sp., *Alcaligenes* strain MFA1.

iii. Competition for Plant Derived Nutrients: Many studies have reported competition for root exudates and other nutrients among rhizosphere bacteria as the mechanism utilized by PGPR for biocontrol of pathogens. *P. putida* or *P. cepacia* were evaluated for the control of *Pythium aphanidermatum* in wheat, tomato, cucumber, melon, bean, pepper, tomato and cotton plants. It was found that competition for nutrients between germinating oospores of *P. aphanidermatum* and bacteria, was responsible for suppression of oospore germination and hence disease control up to 60–90%. The population of *P. putida* or *P. cepacia* was much higher compared to other bacteria and it retarded the establishment of *Pythium* along the roots, which are the potential sites for its attack. Further, the production of inhibitory substances, antibiotics and lytic enzymes was not observed and role of siderophores was excluded as the pH of the soil was 6.0 which ensured the availability of iron in high concentration. In a similar study competition for plant-derived unsaturated long-chain fatty acids was reported to be the mechanism of biocontrol of the seed-rotting oomycete, *Pythium ultimum* in cotton seedling, by *Enterobacter cloacae*, as fatty acids from seeds and roots are essential for germination responses of *P. ultimum* (van Dijk and Nelson 2000). As in case of siderophores, the use of mutants has con fi rmed that competition for nutrients is a major contributing factor in the control of soil borne pathogens. Another essential requirement for the successful establishment of the introduced microbial strain in the rhizosphere is the effective catabolism and utilization of the root exudates so that they are not available for the pathogen utilization.

Competition becomes the most important tool of action in cases where the biocontrol agent and target pathogen share high degree of relative proximity because of the similar requirements in terms of nutrients and growth factors (Prashar Pratibha et al., 2013). Role of siderphores is also irreplaceable as they contribute in many ways ranging from indirect disease control by promoting plant growth, inducing competition among rhizobacteria and pathogens to inducing systemic resistance in plants. However, at present certain issues need attention before moving ahead in this direction. Two most critical points are economic feasibility of the product and consistency in terms of performance under actual fi eld conditions. Though significant market growth for microbial/biological tools has been achieved in last few years, challenge is to attain a competitive level alongside the entrenched and effective chemical counterparts. Improvement in fi eld performance of these products is the key to success in this direction. A deeper and unambiguous understanding of the mode of action of the antagonistic organisms and detailed information about the signaling molecules involved in the same will help in optimizing their colonization and establishment in the rhizosphere thereby enhancing the effectiveness.

The phenomenon of competition has been characterized in liquid medium and sterile soil systems using a variety of soil bacteria and *Fusarium oxysporum* f.*cubense* as test organisms. For most of the bacteria, suppression of the fungus was the result of a biologically induced nitrogen deficiency, this effect being reversed by the addition of excess inorganic nitrogen. High populations of competitors were found in two soils of neutral pH, but no isolates competed in the acid San Alejo loam. *Agrobacterium radiobacter* was able to compete when San Alejo loam was limed to about pH 6.6

(Marshall and Alexander, 1960). Inhibition of the fungus by a number of gram-positive, spore-forming rods could not be accounted for in terms of competition for nutrients or by antibiotic production in artificial media.The competitive ability of *A. radiobacter* when tested in twelve Central American soils was found to be related to pH in acid and neutral environments but was correlated with texture, organic-matter content and total nitrogen in soils of intermediate pH. In all soils where inhibition occurred, the competitive effect was overcome by additions of inorganic nitrogen. Excluding the group of *Bacillus* spp., the competitive ability of soil bacteria was related to the ability to develop in the absence of amino acids and growth factors but could not be correlated with growth rates of the bacteria in soil or liquid medium. It is suggested that competition for nutrients is a significant means of ecological control among members of the soil microflora and, together with competitive interactions for space and oxygen, may be the major factors governing the biological control of soil-borne fungi.

6.2 Bacterial Competition

Most microorganisms face a constant battle for resources. Vast numbers of microbes are present in all but the most rarified environments. Tremendous microbial diversity has been revealed by new molecular methodologies such as metagenomic sequencing and deep microbial tag sequencing (Hibbing et al., 2010). These approaches and others have begun to reveal that underlying the numerically dominant microbial populations is a highly diverse, low-abundance population. Members of the rare biosphere that are amplified under favorable conditions to which they are pre-adapted can give rise to discrete, abundant populations. The potential pool of microbial competitors is therefore vast, and a wide range of mechanisms can be responsible for the emergence and radiation of dominant microbial populations. Most natural environments harbor a stunningly diverse collection of microbial species. Within these communities, bacteria compete with their neighbors for space and resources. Laboratory experiments with pure and mixed cultures have revealed many active mechanisms by which bacteria can impair or kill other microbes. Additionally, a growing body of theoretical and experimental population studies indicate that the interactions within and between bacterial species can profoundly impact the outcome of competition in nature. The next challenge is to integrate the findings of these laboratory and theoretical studies, and to evaluate the predictions they generate in more natural settings.

Nutritional resources are a focal point of microbial competition. Jacques Monod, a pioneer in the study of bacterial growth kinetics, first demonstrated the relationship between limiting nutrient concentrations and bacterial growth. In defined medium, in which all but one isolated nutrient was provided in excess, he demonstrated that "total growth", or bacterial growth yield, is linearly dependent on the initial concentration of the limiting nutrient. He then mathematically incorporated this relationship into the equation for exponential bacterial growth, thereby providing a model for the relationship between growth rate and the concentration of a limiting nutrient, an equation similar to the Michaelis-Menton representation of enzyme kinetics (Hibbing et al., 2010). Monod's equations were derived from extrapolations of data obtained from bacterial grown in batch cultures, but he proposed the method of continuous growth which was later used to verify many of his predictions.

Tilman later employed Monod kinetics for examining the competition between two different types of algal populations, as a function of limiting resource ratios. This resource ratio competition model proposed that the availability and individual demand for and rate of consumption of nutrients will determine the predominance of different taxa. Under certain ratios of nutrient concentrations, competing microbes can stably coexist, while under other conditions, specific taxa can be outcompeted due to acute nutrient limitation. Over time, consumption of limiting nutrients will shape the course of competition. These same principles have been applied to plant and animal communities, and they clearly explain some of the basic dynamics between competing organisms. More recently, the impact of limiting resources on bacterial competition has been studied for a wider range of growth substrates and population structures.

Collectively this "resource ratio" model of competitive interactions views all of the vacillations of microbial lineages through a nutritional framework. Although this provides a conceptual foundation for predicting the outcomes of microbial competition, it cannot, on its own, encapsulate the diversity of active mechanisms by which certain microbes compete (Hibbing et al., 2010). Microorganisms cannot be viewed as passive nutritional sinks, but rather have evolved numerous strategies to augment their acquisition of resources. Activities including motility, antibiotic production, and coordinated behavior can tip the competitive balance, resulting in outcomes that significantly differ from those predicted by resource abundance alone.

Competition and diversification within bacterial populations: In a well-mixed environment to which the input of new nutrients is minimal, such as a shaking liquid bacterial culture, individuals with similar nutritional requirements, such as members of the same population, will be in competition for acquisition of these nutrients as they become depleted by the growing population (Hibbing et al., 2010). In an environment providing multiple ecological niches, such as a static liquid culture or a biofilm, competition can lead to selection for variants that are better suited to colonize these alternative niches. For many bacterial species, the combination of rapid growth rates and large population sizes results in the introduction of many unique mutations, even if they occur at low frequencies. Some mutations give rise to variants that are adapted to particular niches and are maintained by negative frequency-dependent selection. For example, static cultures of *Pseudomonas fluorescens* generate several niche-specialized variants. One kind of variant overproduces extracellular polysaccharide (EPS), enabling the variant to float on the surface of the cultures, thus improving access to oxygen. However, this variant suffers if it becomes too dominant; the mats can become too thick to float, and then sink to the bottom of the culture.

Co-evolution and interspecies competition: Although it is well established that single-species populations of bacteria can evolve over time, relatively few studies have examined the potential for co-evolution in mixed-species environments (Hibbing et al., 2010). In the macroecological world, co-evolution between competitors, between pathogens and hosts, and between mutualists from different species has been repeatedly observed. For example, resistance to disease in plants is mediated in part by genes known as R-genes that recognize particular pathogen proteins; pathogens can subvert disease resistance if they no longer produce these proteins, but R-gene specificity can evolve in response to exposure to new or altered pathogen proteins, resulting in an evolutionary "arms race". Given that competition is a powerful selective pressure, similar arms races could potentially develop between bacterial species, as each responds to new competitive determinants deployed by the other.

Indirect evidence for coevolution of competitive interactions between bacterial species can be found among organisms colonizing the human oral cavity. For example, clinical studies have found that patients colonized by *Streptococcus oligofermentans* have a reduced incidence of dental caries, caused by *S. mutans*, which prompted an investigation of the interactions between these species *in vitro*. *S. mutans* is known to inhibit the growth of many other oral species by producing lactic acid from fermentable carbohydrates present in the host diet. Interestingly, *S. oligofermentans* has developed the counter-offensive strategy of using the *S. mutans*-produced lactic acid to generate hydrogen peroxide, which is in turn inhibitory to *S. mutans*.

Mechanisms of bacterial competition: Nicholson loosely categorized competition for a limiting resource into two broad groups, scramble and contest (Hibbing et al., 2010). Scramble competition, also called exploitation competition, involves rapid utilization of the limiting resource(s) without direct interaction between competitors. Contest competition (or interference competition) involves direct, antagonistic interactions between competitors, with the "winner" appropriating the resource(s). E.O. Wilson likened scramble competition to a group of young boys scrambling for pennies dropped on a floor and contest competition to a fight between the boys, with the winner taking all the pennies. Both strategies are likely employed in the microbial world.

Most research into competition between bacterial species has focused on elucidating the biochemical mechanisms underlying different interactions, and generally assumes that competition occurs between individual cells (Hibbing et al., 2010). However, the competitive mechanisms available to a single cell may differ from those available to an individual surrounded by a population of its near kin, because of the potential for competitive strategies to evolve requiring cooperative behavior. Below, we review several active competitive strategies described for bacteria, and highlight examples where population-level processes have been shown to be important.

i. **Protecting the supply lines:** As shown by the Monod experiments mentioned above, bacterial competition can often be framed in the context of nutrition - access to and protection of growth substrates. Antimicrobial production, space competition, predation and even a rapid growth rate can all be interpreted as a drive to maximize nutrient uptake by one organism at the expense of another. However, several mechanisms of competition function directly to actively restrict or remove a nutrient from one organism and supply it to another. For example, carbon and phosphorus sequestration by polyphosphate accumulating organisms in certain wastewater treatment configurations facilitates their dominance over other species. An additional example of this effect can be observed in the struggle to acquire iron.

Siderophore production is an important example of a competitive mechanism that involves a cooperative behavior. Because siderophores are secreted molecules **(public goods)** that are costly to produce, siderophore producing populations are vulnerable to social cheating by individuals that lose the ability to make these iron-binding products but maintain the capacity to take them up. Generally, the long-term evolutionary dynamics of siderophore-mediated interspecies interactions remain to be explored. One recent study found that competition for iron between *Staphylococcus aureus* and *P. aeruginosa* influenced the evolutionary stability of siderophore production by *P. aeruginosa*. In mixed cultures, *P. aeruginosa* can lyse *S. aureus*, liberating free iron. However viable *S. aureus* also compete with *P. aeruginosa* for free iron. In iron-limited mixed cultures with *S. aureus*, siderophore cheaters of *P. aeruginosa* arose more frequently than when free iron was provided, and also more frequently than in iron-limited pure cultures of *P. aeruginosa*. Thus, interspecies competition influenced intraspecies competition by increasing the selection pressure leading to the accumulation of social cheaters.

ii. **Taking and holding the High Ground:** A critical aspect of many competitive interactions is stable positioning at favorable sites in the environment. Obtaining access to favorable locations requires either colonizing new niches as they become available (scrambling), or actively displacing existing colonizers (contest). A number of species enhance their chances of winning the scramble to colonize newly available spaces by producing adhesins or receptors that bind to specific surface features. For example, in the human oral cavity, some bacteria specifically bind and colonize the host-derived pellicle coating tooth surfaces, while other species become established by expressing surface-exposed receptors that specifically recognize carbohydrates presented on the exterior of the primary surface colonizers. Clearing a space to colonize by eliminating prior residents can be accomplished by production of antimicrobials (discussed further below in "Calling out the artillery") or by production of molecules that facilitate competitors' dispersal without actually killing them. *P. aeruginosa*, for example, produces at least two molecules shown to stimulate dispersal of other species from established biofilms; rhamnolipid, shown to be active against biofilms of *Bordetella bronchiseptica* and the fatty acid *cis*-2-decenoic acid, which stimulated dispersal by a number of other species.

iii. **To fight or flee: motility in microbial competition:** Motility has been shown to affect the ability of some bacteria to compete, while other bacteria apparently use active locomotion to avoid competition. For example, in co-inoculated biofilms, *P. aeruginosa* uses motility, among other traits, to blanket *Agrobacterium tumefaciens* (Hibbing et al., 2010). Interestingly, at initial stages of colonization in the presence of *P. aeruginosa*, a non-motile *A. tumefaciens* mutant accumulated greater adherent biomass than the motile wild-type strain, possibly indicating that wild type *A. tumefaciens* actively evades contact with

P. aeruginosa. Wild type *P. aeruginosa* also uses motility to out compete its own non-motile variants for more suitable regions in biofilms. The motile *P. aeruginosa* migrate to the top of non-motile microcolonies, forming the caps of tall, mushroom-like structures. The motile cells thus access more oxygenated and nutrient rich regions of the culture. Motility is also critical for efficient predation by *Bdellovibrio bacteriovorus* and *Myxococcus xanthus*. In *B. bacteriovorus* a functional flagellar motor is necessary for efficient release from the prey bacteria. In *M. xanthus*, the adventurous motility system is necessary for predation.

Another important aspect of bacterial motility is its contribution to dispersal. Highly motile organisms that rapidly disperse into the surrounding environment may be less likely to interact with others of the same species, either via competition or cooperation. As a result, highly motile organisms will be more likely to encounter potential competitors as individuals, rather than in the context of a population of closely related organisms, thereby restricting their options for competitive strategies. However, some species may circumvent this problem by traveling in groups; swarming motility in numerous species and social motility in *M. xanthus* both involve concerted movements of multiple individuals across surfaces.

iii. **Calling out the artillery:** The most intensively studied mechanism of bacterial competition is the production of small antimicrobial compounds. Although some recent literature calls into question the role of antibiotics as bacterial growth inhibitors in the environment extensive *in situ* studies on a number of antibiotics have verified their importance in mediating competition in an antimicrobial capacity (Hibbing et al., 2010). Antimicrobial compounds can mediate competition between different species, between related strains of the same species, and also between genetically identical individuals in a population. Additionally, the outcome of antimicrobial production will clearly be affected by the context in which the compound(s) is produced. To effectively inhibit competitors, the antibiotic(s) must be produced in sufficient quantity, and this may require the concerted effort of a population. Accordingly, antibiotic production often is regulated by a quorum sensing mechanism. Finally, the habitat and lifestyle of an antibiotic producer may influence the target specificity of antibiotics produced. A generalist species occupying a broad spectrum of environments would be more likely to benefit from producing broad spectrum antimicrobials or a cocktail of toxins targeting different potential competitors, while those organisms highly specialized for a given habitat may produce antimicrobials with a narrower range, targeting specific competitors.

6.3 Native Soil Borne Pathogens Equalize Differences in Competitive Ability

Very little is known about the ecological role of native soilborne pathogens that have co-evolved with plant species in a given region, although a potential role of pathogens to the maintenance of local plant species diversity is receiving increasing attention in recent years. Plant pathogens can contribute to plant species coexistence and thus promote local plant diversity through different mechanisms (Felipe E. Albornoz et al., 2016). For example, this can occur through conspecific negative density dependence or by reducing differences in competitive ability between cooccurring plant species. Negative density dependence (i.e. Janzen–Connell effect) occurs when an increasing density of conspecific individuals leads to the local accumulation of host-specific pathogens, reducing conspecific seedling survival and growth. On the other hand, pathogens with low host specificity can still promote plant species diversity by being more detrimental to (or building up larger populations around) plant species showing higher competitive ability, thus enabling less competitive plant species to persist. These different effects of pathogens are not mutually exclusive and can both contribute to the maintenance of local plant diversity. Determining the ecological role of soilborne pathogens for plant species coexistence should help us understand how highly diverse plant communities are maintained.

Soil borne pathogens can contribute to the maintenance of local plant diversity by reducing differences in competitive ability between co-occurring plant species. It has been hypothesized that efficient phosphorus (P) acquisition by plants in P-impoverished ecosystems might trade off against resistance to root pathogens. This could help explain high plant diversity in severely nutrient-impoverished ecosystems. However, empirical evidence of such a trade-off remains scarce. In hyperdiverse shrublands in south-western Australia, non-mycorrhizal cluster-rooted Proteaceae are very efficient at acquiring P. However, Proteaceae co-occur with many other plant species using other P-acquisition strategies, such as ectomycorrhizal (ECM) associations. In a glasshouse experiment, Felipe E. Albornoz et al., (2016) grew Proteaceae and ECM plant species from hyperdiverse shrublands alone and in competition with each other, and in the presence or absence of native soil borne pathogens (*Phytophthora* spp.). It was hypothesized that native *Phytophthora* species are more detrimental to Proteaceae than co-occurring ECM plants, due to a trade-off between highly efficient P-acquisition and pathogen defence, and that this equalizes differences in competitive ability between these two plant groups. When seedlings were grown alone, biomass of non-mycorrhizal plants was reduced in the presence of *Phytopthora*, while ECM species were unaffected by this pathogen. When non-mycorrhizal and ECM species were planted together, ECM plants grew better in the presence of *Phytophthora* than in its absence, because *Phytophthora* reduced the growth of the non-mycorrhizal competitors. Growth of ECM plants was positively correlated with per cent root colonization by ECM fungi, but this was only significant when ECM plants were grown in the presence of *Phytophthora*.

This study showed that native soil borne pathogens equalized differences in competitive ability between seedlings of contrasting nutrient-acquisition strategies, thus supporting the hypothesis proposing a trade-off between highly efficient P-acquisition and resistance against root pathogens. It was found that non-mycorrhizal cluster-rooted species may be the most efficient at acquiring the growth-limiting resource, but that co-occurring ECM species are better defended against root pathogens. These results suggested that native soil borne pathogens and ECM contribute to the maintenance of the plant hyperdiversity in severely P-impoverished ecosystems

Non-mycorrhizal Proteaceae were more susceptible to native soilborne pathogens than ECM plant species, and this translated into a relaxation of competition between species with these two nutrient-acquisition strategies, presumably because non-mycorrhizal Proteaceae species are most effective in acquiring the growth limiting resource in these soils, P. Biomass gain of Proteaceae was reduced by c. 26% in the presence of native *Phytophthora* species, while the growth of ECM species was not affected. This supports the contention that non-mycorrhizal cluster-rooted species are more susceptible to soil borne pathogens than ECM species. Furthermore, when competing with Proteaceae, ECM species showed higher biomass gain in the presence of native Phytophthora species than in their absence, suggesting that the presence of native soil borne pathogens can modulate competitive interactions between ECM and Proteaceae species. Additionally, this increase in ECM plant biomass in the presence of *Phytophthora* was positively correlated with ECM root colonization, suggesting an important role in pathogen defence by ECM fungi. This study suggested that soil borne pathogens may contribute to the maintenance of highly diverse ecosystems by reducing differences in competitive ability among plant species of contrasting nutrient-acquisition strategies.

However, the experimental design does not allow to quantify how pathogens modulate density-dependent competition among Proteaceae and Myrtaceae. Quantifying how pathogen-mediated negative density dependence varies among co-occurring plant species of contrasting nutrient-acquisition strategies is an important avenue for future research that will help us better understand mechanisms of plant species coexistence in hyperdiverse vegetation (Lalibertet al. 2015). For the Proteaceae tested here, while not killed by the native *Phytophthora* species, there was a reduction in overall growth. On the other hand, growth of ECM plants was not negatively affected by the presence of these native *Phytophthora* species. The inoculation of chestnut seedlings by *P. cinamomi* or *P. cambivora* reduces leaf and root size of non-mycorrhizal chestnut seedlings by 43–48%, while not affecting growth of chestnut seedlings previously inoculated with ECM fungi. The results showed how plants with different nutrient acquisition can have contrasting responses to the same native pathogen.

It was showed how native soil borne pathogens can equalize plant competition among seedlings of contrasting nutrient-acquisition strategies. Root pathogens may play a key role in co-existence of plants with different nutrient-acquisition strategies in these hyper diverse shrub lands (Felipe E. Albornoz et al., 2016). Moreover, further evidence was provided for the hypothesis that there is a trade-off between P-acquisition efficiency and pathogen defence. It was proposed that in old, strongly weathered and severely Pimpoverished soils, ECM fungi are important for pathogen defence and potentially the persistence of their hosts. The results highlighted the need for considering soil microbiota in studies on plant interactions as well as plant diversity and ecosystem functioning, since pathogens and mycorrhizal fungi may strongly affect the outcome of plant competition.

6.4 Microbial Interactions Promote Rhizosphere Colonization

i. Colonization in the Plant Rhizosphere: Efficient colonization of the rhizosphere is the first and fundamental step to protect plants from soil-borne pathogens by microbes (Biocontrol agents-BCAs). Insufficient rhizosphere colonization can impair the beneficial effects of the biocontrol microbial strains, resulting in reduction or failure of disease control (Niu et al., 2020). Inoculation with MSBCAs may enhance the colonization of the rhizosphere by biocontrol microbes. The rhizoplane colonization ability of a fivestrain bacterial consortium suppressing a sudden wilt disease of *Nicotiana attenuata* was enhanced compared to that of each single community member. Survival of *Pseudomonas* species communities inhibiting bacterial wilt disease of tomato increased with increasing diversity In addition, the total bacterial abundance on bean root tips rose when a two-membered biocontrol *Pseudomonas* species consortium for anthracnose was added. Thus, using disease-controlling microorganisms as multi-strain consortia can indeed promote rhizosphere colonization by BCAs. Such positive effects on microbial colonization may be due to positive regulation of some colonizationrelated biological processes, such as biofilm formation, growth and migration, by the interactions among microorganisms within consortia.

Germination of fungal conidiospores and *Bacillus* endospores, is a critical step in successful colonization of BCAs. The contact of *Trichoderma* conidia with Bacillus biofilms did not impair the ability of fungal spores to germinate and establish the fungus in soil (Izquierdo-García et al., 2020). The mycelia of *Trichoderma* can serve as a supporting layer for formation of bacterial biofilms and can aid bacterial migration in the soil. Growth of bacteria was supported by nutrients present in the fungal exudates.

ii. Enhanced Biofilm Formation: Microbial colonization of plant roots can be promoted by the formation of biofilms. Biofilms are communities of surface-associated microorganisms encased in a self-produced extracellular matrix composed of exopolysaccharides, proteins and sometimes DNA. Beside the well-studied single-species biofilms, rhizosphere microorganisms belonging to multiple taxa are able to form multi-species biofilms, of which the formation might be elevated by the microbe–microbe interactions within biofilm communities. In a previous study, a consortium of five native bacterial isolates was found to be able to colonize the roots of *N. attenuata* by forming multiple-taxa biofilms on the root surfaces. Furthermore, under both in vitro and in vivo conditions, the amount of biofilm produced by each individual strain was significantly less than the biofilms formed by the five-membered bacterial community, which indicating a synergistic biofilm formation by the consortium (Santhanam et al., 2019). Similarly, a three-species biocontrol community composed of *Xanthomonas* sp. WCS2014-23, *Stenotrophomonas* sp. WCS2014-113 and *Microbacterium* sp. WCS2014-259 showed synergy, as the combination of three formed more biofilm than the single strains. Moreover, colonization of host roots by this community was stimulated by enhanced biofilm formation. Although the mechanisms of such positive effects on biofilm production are unclear, the improved efficacy can be attributed to the cooperative microbial interactions in consortia, triggering increased extracellular matrix deposition and cell-to cell signalling.

iii. Syntrophic Microbial Growth Promotion: Colonization of the rhizosphere requires robust microbial growth, which can be greatly improved by syntrophy, a nutritional situation in which multiple microorganisms combine their metabolic abilities to catabolize a substrate that cannot be degraded by either one of them alone (Mee et al., 2014). For example, *Azospirillum brasilense*, a well-known plant growth-promoting rhizobacterium (PGPR), is not able to use certain sugars and polysaccharides as carbon sources for growth in vitro. However, it is capable of associating with sugaror polysaccharide-degrading bacteria, establishing a metabolic association where the sugar- and/or polysaccharide-degrading bacteria degrade the metabolites to products, which can be utilized as carbon source by *A. brasilense*. In turn *A. brasilense* provides the sugar- or polysaccharide-degrading bacteria with nitrogen by fixing the atmospheric N2. Such a synergistic catabolic effect on sugars and polysaccharides may boost the survival of *A. brasilense* in the rhizosphere, because plant roots release 5–21% of their photosynthetically fixed carbon as root exudates, including sugars and polysaccharides, which are used by the rhizosphere microbial communities. Beside nutrients, there are always microbial growth-inhibiting materials in the rhizosphere. While methanol can suppress the growth of methanotrophs in the rhizosphere, the methanotrophs are able to survive by coexisting with *Hyphomicrobium* spp. to build a rhizospheric microbial association, in which the *Hyphomicrobium* spp. is capable of removing methanol (Liechty et al., 2020). Therefore, the promotion of the growth of rhizosphere microbes can be achieved by syntrophic interactions leading to effective nutrient utilization and removal of harmful substances.

iv. Facilitated Migration: Another crucial microbial trait for rhizosphere colonization is motility, defined as the ability of microorganisms to move or to perform mechanical work at the expense of metabolic energy. There are six different categories of surface motility including swimming, swarming, gliding, twitching, sliding and darting. Motility is required for early root colonization by BCAs. The migration of microorganisms can be enhanced by the interactions among community members. For example, fungal hyphae are capable of serving as vectors for the dispersion of bacteria in the rhizosphere, which is known as a "fungal highway". Extraradical mycelium formed by the mycorrhiza fungus *Glomus formosanum* CNPAB020 can facilitate the translocation of *Bradyrhizobium diazoefficiens* USDA 110 in the rhizosphere (de Novais et al., 2020) in addition to its main activity in nutrient transfer.

Microbe–microbe interactions can play a positive role in promoting rhizosphere colonization by beneficial microorganisms through boosting biofilm formation, microbial growth, migration inside of the microbiome, and interacting with plant roots. Thus, utilization of MSBCAs performing active interactions among their members may improve survival of disease-suppressing microbes, and their adaption to complex and changeable environmental conditions. In consequence, they may be able to stabilize their beneficial effects for the inhibition of soil-borne diseases.

Microbial interactions enhance growth suppression of soil-borne pathogens: Multi-strain biological control agents are able to exhibit stronger suppressive efficacy on the growth of soil-borne pathogens than SSBCAs (Niu et al., 2020). For instance, a bacterial strain mixture involving *Bacillus subtilis* S2BC-1 and GIBC-Jamog showed greater anti-fungal activity against the tomato vascular wilt pathogen, *Fusarium oxysporum* f. sp. *lycopersici,* than each of the individual strains. Similarly, *Pseudomonas fluorescens* T5 showed no inhibition against *Rhizoctonia solani* in vitro. However, when it was applied together with four non-antagonistic bacterial strains isolated from the rhizosphere of Tamarindus, this five-species bacterial community exhibited strong suppression of growth of *R. solani*. Although the understanding of the enhanced pathogen-inhibiting effect of biocontrol consortia is limited, changes in resource competition and secretion of antimicrobial compounds triggered by microbial interactions may contribute to the enhanced suppression.

i. Boosted Competition for Resources: Resource competition is a basic mechanism by which BCAs may protect plants from pathogens, implying that the beneficial microorganisms are able to rapidly and efficiently utilize the limited resources in the vicinity of the plant hosts to restrict or suppress the growth of phytopathogens. Plant exudates on root surfaces and in their surrounding rhizosphere, are the primary sources of nutrients for the rhizosphere microbiome. Successful suppression depends on the competition for nutrients in root exudates by biocontrol microbes and soil-borne pathogens. This contest can be elevated by the microbial interplay inside MSBCAs. Two biocontrol consortia for tomato bacterial wilt caused by *R. solanacearum*, consisting of eight Pseudomonas and five non-virulent Ralstonia strains, exhibited much stronger inhibiting effects on the population density of *R. solanacearum* than each individual strain. The enhanced inhibition is caused by an increase of niche overlaps exerted by these consortia with *R. solanacearum*. Niche overlaps may be defined as 'likeness' between the communities and *R. solanacearum* in the catabolism of 48 different single-carbon resources found in tomato root exudates. The more diverse soil bacterial communities are, the better they are able to acquire many of the 31 individual carbon sources typical for soil, than the pathogen *E. coli* O157:H7.

Limited assimilatable iron resources remain in the rhizosphere, following the competition between disease-suppressing microorganisms and soil-borne pathogens (Gu et al., 2020). Many soil microbes scavenge iron by secreting siderophores, a chemically diverse group of secondary metabolites with a high affinity for iron, because iron predominantly occurs in soil in its insoluble ferric Fe (III) form. The siderophores can both to facilitate and suppress competitors, depending on whether the competitors possess the transporters or channels for siderophore uptake. The production of siderophores can be positively regulated by interspecies interactions among soil microbes. The interplay of *Streptomyces coelicolor* with five other soil actinobacteria increased the diversity of siderophores. Production of desferrioxamines by *S. coelicolor*, was triggered by siderophores from neighboring strains. Therefore, the disease-inhibiting microorganisms in the rhizosphere may acquire elevated capability to utilize resources through microbial associations.

ii. Stimulated synthesis of Antimicrobial Compounds: Microorganisms are able to synthesize a multitude of compounds with antimicrobial activity, which is an important mode of action for direct inhibition or lethality on the microbial opponents in environments. So far, there have been a large number of reports of the antimicrobials produced by BCAs exhibiting suppressing effects on the growth of phytopathogens. These studies mainly focus on the biocontrol strains from the genera *Bacillus, Pseudomonas,* and *Trichoderma*, well known for the production of antibiotics including lipopeptides, polyketides, bacteriocins, phenazines, 2,4-diacetylphloroglucinol (DAPG) and chitinase. Some metabolites with inhibitory functions are found in low concentration or are not expressed in pure culture but may be upregulated in a community (Pishchany et al., 2018). Therefore, the modulating effect of microbial interactions on resource competition and production of antimicrobial compounds may contribute to strengthening the inhibition of growth of pathogens. Thus, applying BCAs as multi-strain mixtures can elevate the ability of biocontrol microorganisms to compete for the resources needed for rhizosphere survival with soil-borne pathogens and to stimulate the production of compounds toxic to specific pathogens. The increased niche overlaps and biosynthesis of novel antimicrobial compounds induced by microbe–microbe interactions may facilitate the BCAs to suppress a broader range of phytopathogens. The positive impact of interactions within MSBCAs may result in more efficient growth suppression of soil-borne pathogens, and improve the efficiency of soil-borne disease control by disease-inhibiting microbes.

Interactions of Microbial Communities with Plants and Soil

Plants rely on rhizosphere microbiota to facilitate nutrient acquisition, in exchange for carbon-rich root exudates for bacterial nutrition. In addition, the rhizosphere microbiome is important for plant health and fitness (Niu et al., 2020). The plant root microbiome consists of prokaryotic bacteria, eukaryotic filamentous fungi, and oomycetes. Besides a core microbiome ubiquitous in a multitude of hosts and geographical regions, a variable part of the microbiome is shaped by secretion of species-dependent plant secondary metabolites, which belong to diverse classes, such as coumarins, benzoxazinoids, phytoalexins and triterpenes (Jacoby et al., 2020). Consequently, diversity of species along the bulk-soil to root microbiota was found gradually decreasing. Positive correlations dominate within each of the three kingdoms. Reconstitution experiments performed with synthetic mono- or multi-kingdom microbial consortia and germ-free Arabidopsis plantlets revealed that the bacterial microbiota protects plants against potentially pathogenic fungi and oomycetes by mainly negative factors exerted against filamentous fungi. Widely distributed members of the core microbiota such as *Variovorax,* a gram-negative beta-proteobacterium, and *Pseudomonas* appeared to be important for plant protection but individual members of other bacterial taxa could overtake their function in biocontrol. Therefore, addition of an SSBCA or MSBCA might have positive effects in complex systems of agriculture and forestry.

The plant immune system also affects the composition of the microbiota in the vicinity of plant roots. The root-microbiome may expand plant immunity and acts as an additional layer of defense against plant pathogens (Teixeira et al., 2019). Interaction of beneficial microbes with plant roots can result in systemic host resistance to pathogens, which may be due to the activation of induced systemic resistance (ISR). The interactions between microbes and soil have always drawn the attention of microbiologists and ecologists. It has been widely accepted that microbial communities inhabiting soil are capable of alternating its physicochemical properties by organic litter deposition and metabolic activities, for example, by improving water retention, increasing carbon storage and mineral nutrition contents. Vice versa, the variability in soil traits may impact the composition and function of soil microbial communities.

Microbial interactions and Construction of MSBCAS: Application of BCAs in the community context as MSBCAs can increase the ability to control soil-borne diseases of crops through interaction-mediated promotion of rhizosphere colonization outcompeting soil-borne pathogens (Niu et al., 2020). Thus, construction and utilization of MSBCAs could augment soil-borne disease control in sustainable agriculture and forestry. So far, there are at least two strategies for preparing effective MSBCAs, (i) mixing the existing SSBCAs according to empirical criteria, and (ii) assembling MSBCAs by applying the reductionist SynCom approach, also named RSC. Using either one of the two strategies, microbe-microbe interactions need to be taken into account.

6.4.1 Molecular interaction between *Trichoderma* species and plants against soil-borne plant pathogens: *Trichoderma* is known as an aggressive colonizer of plant roots that competes for space, nutrients, water, or oxygen by mobilizing immobile soil nutrients, thereby eliminating other micro-organisms that inhabit their niche due to the diversified composition of root exudates secreted by plants (Dutta et al., 2023). Competition among micro-organisms is a strategy to utilize the nutrient hotspots present in the rhizosphere by eliminating other competitors (Guzmán-Guzmán et al., 2023). Therefore, to be an effective colonizer of plant roots, those organisms must have metabolic versatility and the competitive capacity to occupy the nutrient hotspots. In this regard, *Trichoderma* can be considered as an aggressive competitor because it has the capacity to secrete a plethora of chemically diverse secondary metabolites that have an antagonistic effect on other micro-organisms (i.e., competitive capacities) and it also exhibits rapid growth and colonization strategies (indicating metabolic versatility) that enable it to occupy space in rhizosphere, enhance plant growth, and restrict further growth of potentially pathogenic micro-organisms.

The interactions of *Trichoderma* spp. with host plants and pathogens at a molecular level are important in understanding the various mechanisms adopted by the fungus to attain a close relationship with their plant host through superior antifungal/antimicrobial activity. When working in synchrony, mycoparasitism, antibiosis, competition, and the induction of a system acquired resistance (SAR)-like response are considered key factors in deciding the biocontrol potential of *Trichoderma.* Sucrose-rich root exudates of the host plant attract *Trichoderma*. The soluble secretome of Trichoderma plays a significant role in attachment to and penetration and colonization of plant roots, as well as modulating the mycoparasitic and antibiosis activity of *Trichoderma*. This review aimed to gather information on how *Trichoderma* interacts with host plants and its role as a biocontrol agent of soil-borne phytopathogens, and to give a comprehensive account of the diverse molecular aspects of this interaction.

The presence of ATP-binding cassettes transporters (ABC transporters) in *Trichoderma* ensures enhanced competitive ability by conferring resistance to toxic metabolites secreted by other micro-organisms. Moreover, *Trichoderma* is compatible with sublethal doses of chemical fertilizers such as urea and muriate of potash, and many chemical pesticides such as thiamethoxam, methomyl, imidacloprid, and methyl bromide, which is attributed to the presence of ABC transporters in *Trichoderma. Trichoderma* releases certain iron chelators, i.e., siderophores, which become bound to iron present in soil (Dutta et al., 2023). Iron is a key micronutrient for the viability of fungi, and therefore the release of iron-chelating siderophores by *Trichoderma* is detrimental to the growth of other fungi. This is one of the main reasons for the biocontrol potential of *Trichoderma* against soil-borne phytopathogens such as *Pythium, Fusarium*, and *Botrytis,* which is inversely proportional to the concentration of nutrients in soil. The discovery of Gtt1 (highaffinity glucose transporter) in *T. harzianum* CECT 2413 raised questions about the probable role of glucose transporters during competition by *Trichoderma.* Gtt gene expression is upregulated when *T. harzianum* CECT 2413 is subjected to growth in nutrient-deficient media. Moreover, a mutant of *Trichoderma* with an additional copy of the glucose transporter gene performed strongly, with a two- to threefold increase in glucose uptake. Glucose metabolism is essential in the assimilation of enzymes and permeases, as well as proteins involved in membrane and cell wall modifications.

The interactions of *Trichoderma* spp. with host plants and pathogens at a molecular level will provide insights on the mechanisms that make *Trichoderma* a superior biocontrol agent. Mycoparasitism by *Trichoderma* is a complex process; therefore, a comprehensive study at the gene level is important to understand how the BCA safeguards itself from the defense strategies adopted by the parasitized fungi. Moreover, knowledge of secondary metabolites secreted by Trichoderma during their interaction with either the plant host or fungal host may be helpful in formulating effective bioactive molecule-based formulations that can provide enhanced protection to plants for a longer time. An understanding of the molecular dialogues between the host plant/fungus and Trichoderma is important to realizing the full potential of *Trichoderma* as a biocontrol agent.

6.4.2 Inhibitory interactions and Resource Competition between *Streptomyces* and *Fusarium* Populations in the Rhizosphere: Plant community characteristics impact rhizosphere *Streptomyces* nutrient competition and antagonistic capacities. However, the effects of *Streptomyces* on, and their responses to, coexisting microorganisms as a function of plant host or plant species richness have received little attention. In this work, we characterized antagonistic activities and nutrient use among *Streptomyces* and *Fusarium* from the rhizosphere of *Andropogon gerardii* (Ag) and *Lespedeza capitata* (Lc) plants growing in communities of 1 (monoculture) or 16 (polyculture) plant species (Adil Essarioui et al., 2017). *Streptomyces* from monoculture were more antagonistic against *Fusarium* than those from polyculture. In contrast, *Fusarium* isolates from polyculture had greater inhibitory capacities against *Streptomyces* than isolates from monoculture. Although *Fusarium* isolates had on average greater niche widths, the collection of *Streptomyces* isolates in total used a greater diversity of nutrients for growth. Plant richness, but not plant host, influenced the potential for resource competition between the two taxa. *Fusarium* isolates had greater niche overlap with *Streptomyces* in monoculture than polyculture, suggesting greater potential for *Fusarium* to competitively challenge *Streptomyces* in monoculture plant communities.

In contrast, *Streptomyces* had greater niche overlap with *Fusarium* in polyculture than monoculture, suggesting that *Fusarium* experiences greater resource competition with *Streptomyces* in polyculture than monoculture. These patterns of competitive and inhibitory phenotypes among *Streptomyces* and *Fusarium* populations are consistent with selection for *Fusarium*-antagonistic *Streptomyces* populations in the presence of strong *Fusarium* resource competition in plant monocultures. Similarly, these results suggested selection for *Streptomyces*-inhibitory *Fusarium* populations in the presence of strong *Streptomyces* resource competition in more diverse plant communities. Thus, landscape-scale variation in plant species richness may be critical to mediating the coevolutionary dynamics and selective trajectories for inhibitory and nutrient use phenotypes among *Streptomyces* and *Fusarium* populations in soil, with significant implications for microbial community functional characteristics.

Plant host and plant community richness have been shown to significantly influence *Streptomyces* competitive phenotypes. *Streptomyces* in monoculture were less niche-differentiated than *Streptomyces* in polyculture. Additionally, plant monocultures have more antagonistic *Streptomyces* than polycultures. Together, these findings are consistent with the hypotheses that polyculture plant communities, producing a greater diversity of soil nutrients, select for niche-differentiated populations, while monoculture plant communities induce a coevolutionary antagonistic arms race, resulting in highly antagonistic soil populations (Adil Essarioui et al., 2017). Although this framework provides a conceptual foundation for predicting the effects of resource competition among *Streptomyces* on their competitive phenotypes, and specifically how plant community richness may alter selective trajectories among competing populations, existing data do not address the potential for other microbial taxa to play significant roles in the selection for diverse resource use and inhibitory phenotypes in complex microbial communities.

Although *Fusarium* and *Streptomyces* are common symbionts of plants, little attention has been given to studying *Fusarium-Streptomyces* antagonistic and competitive interactions in relation to plant host or plant community characteristics. The effects of plant community richness and plant species were explored on the inhibitory interactions and nutrient niche overlap between *Fusarium* and *Streptomyces* populations from the rhizosphere of *Andropogon gerardii* (Ag) and *Lespedeza capitata* (Lc) plants, each growing in communities of 1 (monoculture) or 16 (polyculture) plant species. The specific objectives of this work were to characterize inhibitory interactions between *Fusarium* and *Streptomyces* isolates from the rhizosphere of Ag and Lc in monoculture and polyculture and determine the effects of plant host and plant community richness on the potential for nutrient competition between *Fusarium* and *Streptomyces*.

This investigation provides valuable insight into variation in resource use among *Streptomyces* and *Fusarium* and sheds light on the effects of plant community characteristics, as determinants of soil nutrient characteristics on the potential for nutrient competition and antagonistic interactions between these two common soil microbial taxa.

The increase in *Streptomyces* antagonistic capacity in monoculture versus polyculture plant communities was greater than increases in *Fusarium* inhibition in polyculture versus monoculture. Indeed, the proportions of antagonistic *Streptomyces* from monoculture were 685.8% higher than those from polyculture, whereas the proportions of inhibitory Fusarium from polyculture were only 53.2% greater than those from monoculture. Likewise, inhibition intensities of *Streptomyces* from monoculture exceeded by 667.5% those from polyculture, while the intensities of inhibition by *Fusarium* from polyculture were only 41.6% greater than those from monoculture (Adil Essarioui et al., 2017). Hence, in terms of inhibitory capacity, *Streptomyces* were more responsive than *Fusarium* to change in plant richness. If resource competition were explicitly reciprocal, it would be logical to predict that *Fusarium* antagonistic phenotypes against Streptomyces would be similarly most frequent or of greater intensity among *Fusarium* populations in monoculture than polyculture plant communities. However, *Fusarium* isolates from monoculture were less potent in their inhibitory activities toward *Streptomyces* than *Fusarium* isolates from polyculture. Antibacterial properties of chemically diverse compounds from members of the genus *Fusarium* have been long reported, but, the ecological forces influencing the production of such biologically active substances have not been addressed. These results suggested that plant diversity fundamentally alters the selective trajectories for antagonistic and nutrient use phenotypes for *Fusarium* and *Streptomyces* populations, but in distinct ways. Thus, though the niche overlap of *Fusarium* with *Streptomyces* was greater in monoculture than polyculture, the niche overlap of *Streptomyces* with *Fusarium* was greater in polyculture than monoculture.

Potential effects of plant communities and litter chemistry on resulting microbial functional diversity have been reported previously, and competition among microbes for space and nutrients in the soil has been suggested to lead to the evolution of distinct competitive strategies among microbes. The results suggested that soil carbon resources, which are a function of plant community richness, are a focal point of competition between *Fusarium* and *Streptomyces* and are central to mediating selection for inhibitory phenotypes among both *Fusarium* and *Streptomyces* populations in soil.

6.4.3 The improvement of competitive saprophytic capabilities of *Trichoderma* species: The extent to which an antagonist can survive in the presence of soil-borne pathogen depends on its competitive saprophytic ability, which may be assessed by the competitive saprophytic colonization of substrate units buried in graded series of inoculum dilutions of the pathogen (Singh Rashmi et al., 2016). The soil moisture content plays an important role in the activity and survival of *Trichoderma*. The higher the amount of inoculum of *T. harzianum* strain No. 94-016, the higher the percent colonization of rice straw in the soil. The CSA of selected *Trichoderma* strains was studied in the present investigation. *Th*-m1 possessed a significantly higher percent colonization than the other mutant strains at all population densities on all days of incubation, whereas *T. atroviride-m1* was recovered only after 4 days of incubation. The parent strains *Th*-m1 and *T. atroviride* were recovered from root segments less frequently than the mutant strains and were very slow to colonize the root segments at higher population densities. *Th*-1432 possesses higher CSA, which may be directly correlated with its higher cellulolytic activity.

The antagonistic potential of *Trichoderma* strains was assayed by studying the effect of their culture filtrate on the radial growth of *Sclerotium rolfsii*, the causal agent of chickpea collar rot. *Trichoderma harzianum*-1432 (42.2%) and *Trichoderma atroviride* (40.3%) were found to be strong antagonists. To enhance their antagonistic potential, mutagenesis of these two selected strains was performed. Two mutants, *Th*-m1 and *T. atroviride* m1, were found to be more effective than their parent strains. The enzymatic activities of the selected parent and mutant strains were assayed, and although both mutants were found to have enhanced enzymatic activities compared to their respective parent strains, *Th*-m1 possessed the maximum cellulase (5.69U/mL) and β-1,3-glucanase activity (61.9U/mL). *Th*-m1 also showed high competitive saprophytic ability (CSA) among all of the selected parent and mutant strains, and during field experiments, *Th*-m1 was found to successfully possess enhanced disease control (82.9%).

It has been shown that not all *Trichoderma* species possess the mechanisms and characteristics deemed necessary for optimum biocontrol. Very often, those strains that have the capacity to produce enzymes and antibiotics that are associated with biocontrol are not the ones that have good storage qualities or function well at temperatures and moisture levels where pathogen flourish. Additionally, effective pathogen control may require 10^5–10^6 propagules to be present in one gram of soil in the case of *Trichoderma*. The presence of such high amounts of conidia is undesirable from an ecological, economic, and public health perspective. To solve this problem, one of the important aims of the present investigation was to develop a biocontrol strain that had improved biocontrol potential and increased tolerance to carbandazim.

The data from the present study demonstrate that *Th*-m1 produces higher amount of enzymes with higher cellulase and β-1,3-glucanase activity than that of the other selected strain. These results strongly support the contention that *Th*-m1, as it released cellulase and β-1,3-glucanases in significant amounts, possessed stronger mycoparasitic capabilities than the other selected parent and mutant *Trichoderma* strains. The coordinated activity of these enzymes released by *Th*-m1 makes it capable of degrading different polymers, which leads to the effective breakdown of pathogen cell walls. The results obtained after field application of the antagonist revealed that the disease incidence was significantly reduced because of the high efficacy of the antagonist mutant strain against the pathogen by the enhanced production of extracellular enzymes (Szekeres et al., 2002). The disease control potential of *Th*-m1 may also be correlated with its high CSA. The results of biological control studies in pot and field conditions indicate that *Th*-m1 has the potential to serve as a biocontrol agent of chickpea collar rot.

6.5 Predation

Predation is the ecological process by which energy is transferred from living animal to living animal based on the behavior of a predator that captures and kills a prey before eating it. Predators occupy the upper levels of food chains. This top position helps explain their generally low density, frequently spaced distribution, and usually small total biomass. In terms of conservation, this is sometimes critical for the survival of local predator populations (Minelli, 2008). Effective predation requires structural, functional, and behavioral adaptations whose nature depends on the general bodily organization and size of the predator, the kind of predatory strategy adopted, the general environmental conditions where predation is practiced, and the defensive strategies of the available preys. The structural complexity of the environment at landscape, habitat, and microhabitat scale, and its seasonality affect the predator's behavior and efficiency. Mathematical modeling of population dynamics in predator–prey systems began with the pioneering work of Lotka (1925) and Volterra (1926) to define equilibrium conditions and lead to predictions about temporal changes, whether cyclical or not, in populations of interacting species. Removing top predators from a community may seriously destabilize its structure and eventually cause local extinctions of other species, with cascading effect.

In predation, one organism kills and consumes another. Predation provides energy to prolong the life and promote the reproduction of the organism that does the killing, the predator, to the detriment of the organism being consumed, the prey (Stevens, 2010). Predation influences organisms at two ecological levels. At the level of the individual, the prey organism has an abrupt decline in fitness, as measured by its lifetime reproductive success, because it will never reproduce again. At the level of the community, predation reduces the number of individuals in the prey population.

Ecological Consequences: Grazing affects bacterial populations. There is plenty of evidence that protozoan grazing significantly reduces bacterial abundance in systems with high microbial activity (Regin Rønn et al., 2012). Regular fluctuations in bacterial numbers were followed by regular but delayed fluctuations in protozoan numbers. Similarly, several microcosm experiments have demonstrated reduced bacterial abundance or microbial biomass in the presence of bacterial-feeding nematodes. Grazing-induced reduction of bacterial abundance is often accompanied by increased bacterial activity; in particular nitrification is also often stimulated by grazing. Protozoan and nematode carbon (C) to nitrogen (N) ratios are comparable to C to N ratios in the bacterial biomass. Hence, due to respiratory carbon loss the grazers ingest excess nitrogen when they ingest bacteria, which they excrete as ammonia (NH_3). Therefore, part of the grazing-induced stimulation of bacterial activity can be explained simply by N in excess from the grazers. Additionally, it has been suggested that bacterial populations that are grazed will be more metabolically active, because senescent cells are removed from the population and because fast-growing bacterial strains have a selective advantage over slower-growing strains. Finally, protozoan production of bacterial-stimulating metabolites may likewise stimulate bacterial activity.

Bacterial feeding nematodes are unable to discriminate between individual bacterial cells as many protozoa can; hence, the two groups will affect bacterial populations differently. In the presence of the flagellate *Cercomonas*, the bacterium *Pseudomonas fluorescens* CHA0 increased its abundance as compared to two other *Pseudomonas* strains, which were both eaten by the flagellate. In contrast, the number of *P. fluorescens* CHA0 declined relatively when the system was subjected to nematode predation pressure. *Arthrobacter* sp. (an Actinobacterium) thrived better when grazed by protozoa than when grazed by nematodes. As the bacterium is harmful to plants, this caused reduced plant performance in the experimental systems.

6.5.1 The functional significance of Bacterial Predators and Interaction among Bacteria, Protozoa and Nematodes in Soil: Although protists, rotifers, nematodes, and phages are thought to function as the dominant predators in microbiomes, predatory bacteria are common in both soil and aquatic systems. But beyond their common occurrence in these habitats, we know little of their activity in the wild, how rapidly they grow, their functional significance in food webs, and how they respond to enrichment at the base of the food web through substrate additions (Hungate et al., 2021). DNA sequencing and other 'omics techniques can provide detailed information on the composition and functional potential of the microbiome, but most measurements of *in situ* bacterial growth rates lack taxonomic resolution and are conducted at the scale of the entire microbial assemblage. Such aggregate measurements mask the contributions of genetically and functionally distinct populations. Even in macroscopic assemblages, taxa are known to vary in their influences on ecosystem processes. Techniques that combine isotopes and genetic sequencing hold promise for parsing the contributions of individual microbial taxa to interactions within microbial assemblages and to biogeochemical processes.

Most of the predatory bacteria detected were facultative, with 64.7% from the order *Myxococcales*, 16% from the class *Cytophagia*, and 9.2% from the order *Streptomycetales*; 8% were obligate predatory bacteria, with 7.0% from the order *Bdellovibrionales* and 1.0% from the order *Vampirovibrionales*.

Predation structures food webs, influences energy flow, and alters rates and pathways of nutrient cycling through ecosystems, effects that are well documented for macroscopic predators. In the microbial world, predatory bacteria are common, yet little is known about their rates of growth and roles in energy flows through microbial food webs, in part because these are difficult to quantify. Hungate et al., (2021) showed that growth and carbon uptake were higher in predatory bacteria compared to nonpredatory bacteria, a finding across 15 sites, synthesizing 82 experiments and over 100,000 taxon-specific measurements of element flow into newly synthesized bacterial DNA. Obligate predatory bacteria grew 36% faster and assimilated carbon at rates 211% higher than nonpredatory bacteria. These differences were less pronounced for facultative predators (6% higher growth rates, 17% higher carbon assimilation rates), though high growth and carbon assimilation rates were observed for some facultative predators, such as members of the genera *Lysobacter* and *Cytophaga*, both capable of gliding motility and wolf-pack hunting behavior. Added carbon substrates disproportionately stimulated growth of obligate predators, with responses 63% higher than those of nonpredators for the *Bdellovibrionales* and 81% higher for the *Vampirovibrionales*, whereas responses of facultative predators to substrate addition were no different from those of nonpredators. This finding supports the ecological theory that higher productivity increases predator control of lower trophic levels. These findings also indicated that the functional significance of bacterial predators increases with energy flow and that predatory bacteria influence element flow through microbial food webs.

This study quantified the growth of predatory and nonpredatory bacteria in soils (and one stream) by tracking isotopically labeled substrates into newly synthesized DNA. Predatory bacteria were more active than nonpredators, and obligate predators, such as *Bdellovibrionales* and *Vampirovibrionales*, increased in growth rate in response to added substrates at the base of the food chain, strong evidence of trophic control. This work provides quantitative measures of predator activity and suggests that predatory bacteria—along with protists, nematodes, and phages—are active and important in microbial food webs (Hungate et al., 2021)

Predatory bacteria are highly active in microbial food webs, synthesizing DNA with elements derived from added isotope tracers at rates higher than nonpredatory bacteria, consistent with evidence from experimental microcosms. These results suggest that bacteria should be considered alongside eukaryotes and viruses as important predators in microbial food webs. Similarly, a recent metagenomic qSIP analysis using a ^{13}C-CO_2 tracer introduced via plant root exudates found that ^{13}C recovery in metagenomes associated with putative predator bacteria was comparable to the recovery in viruses and substantially higher than that in predatory eukaryotes (Starr et al., 2020). Slower growth might be expected if bacterial predators were inactive or dormant, as are many soil microorganisms. The results presented here indicate that bacterial predators grow, metabolize, and feed at higher rates than most bacteria in the soil food web and those predatory bacteria may exert top-down effects in microbial food chains. Though our analysis focused on predation, techniques that combine isotopes and gene sequencing can also quantify evidence of other ecological interactions in microbiomes and how they shape carbon flow and nutrient cycling in microbiomes. Multiple signatures of interactions among bacteria have now been identified, informing the use of qSIP, metagenomics, and traits to evaluate the functional significance of interactions in diverse microbiomes.

Interaction among Bacteria, Protozoa and Nematodes in Soil: Bacteria, protozoa and nematodes interact closely in soil ecosystems. Protozoa and nematodes eat bacteria (and occasionally each other), while bacteria defend themselves using chemical substances, resistant cell walls, irregular shapes and motility. Protozoa and nematodes are very different types of organisms, and hence apply very different feeding mechanisms; thus many protozoa can pick and choose individual bacterial cells, whereas nematodes ingest bacterial patches more uncritically (Regin Rønn, et al., 2012). Protozoa and nematodes are both aquatic organisms whose activity depends on available soil water, but differences in size, motility, resting stages and reproductive strategies mean that the soil physico-chemical environment influences the activity of protozoa and nematodes differently. For example, the relative importance of protozoa compared to nematodes may shift towards protozoa in very clay-rich soils. The interactions between the three organism groups have major ecological consequences such as modification of the bacterial communities and increased nitrogen mineralisation, both of which affect plant growth. Increased nitrogen mineralisation will usually be beneficial for plant growth, whereas the grazing-induced changes in the bacterial communities can be both beneficial and detrimental to plants. Selective protozoan grazing can favour plant inhibiting bacteria. This may be a problem in clay rich soils where protozoa have better life conditions than nematodes. Mutually, the three groups affect each other; both positively and negatively. Both nematodes and protozoa stimulate nitrogen mineralisation and plant growth through their effect on the soil bacteria, but because the two groups of organisms have different ways of food-uptake, movement and colonisation, they affect bacteria differently. This means that they shape bacterial communities differently, and that they also can affect plant growth differently.

6.5.2 Mutations affecting predation ability of *Myxococcus xanthus*: Numerous types of evidence exist for predatory relationships among micro-organisms, suggesting that microbial competition for limiting resources is marked by food-web relationships analogous to those seen in the macroscopic world. Since the lysis of *Micrococcus* sp. by the myxobacterium *Chrondromyces crocatus* was first reported, myxobacteria have been recognized as important micropredators that possess the capacity for lysing diverse prey species (Vinh et al., 2005). *Myxococcus xanthus* is a ubiquitous, soil-residing predatory bacterium, which moves in colonial swarms during the vegetative part of its life cycle. It is thought that the collective secretion of biocidal factors by the swarm enhances the colony's predation of a wide range of susceptible microbial prey, including firmicutes, proteobacteria like *Serratia marcescens,* and both bacterial and fungal phytopathogens. Although *M. xanthus* growth can also be sustained by adequate concentrations of certain amino acids and central metabolic intermediates, predation appears to be an integral part of the *M. xanthus* survival strategy when such nutrients become limiting. The major characteristics of *M. xanthus* predation – mobility and collective group action – set these organisms apart from other bacterial predators with tractable genetic systems, and hence establish them as a suitable model for studying this particular mode of predatory behaviour.

Myxococcus xanthus genetic mutants with characterized phenotypes were analysed for the ability to prey on susceptible bacteria. Quantification of predatory ability was scored by a newly developed method under conditions in which prey bacteria provided the only source of nutrients (Vinh et al., 2005). These results were corroborated by data derived using a previously published protocol that measures predation in the presence of limited external nutrients. First, early developmental regulatory mutants were examined, because their likely functions in assessing the local nutrient status were predicted to be also important for predation. The results showed that predation efficiency is reduced by 64–80 % for mutants of three A-signalling components, AsgA, AsgC and AsgE, but not for AsgB. This suggests that an Asg regulon function that is separate from A-signal production is needed for predation. Besides the Asg components, mutations in the early developmental genes sdeK and csgA were also consistently observed to reduce predatory efficacy by 36 and 33 %, respectively. In contrast, later developmental components, such as DevRS, 4406 and PhoP4, did not appear to play significant roles in predation. The predatory abilities of mutants defective for motility were also tested. The data showed that adventurous, but not social, motility is required for predation in the assay. Also, mutants for components in the chemotaxis-like Frz system were found to be reduced in predation efficiency by between 62 and 85 %. In sum, it was demonstrated here that defects in development and development-related processes affect the ability of *M. xanthus* to prey on other bacteria.

Predation can be thought of as a behaviour that, like development, is dependent on a series of interlinked processes required for myxobacterial nutrient scavenging. Among these are systems for hunger alert (signalling), predator-to-predator communication (quorum sensing), prey detection (chemotaxis or chemokinesis), prey tracking (motility), production of killing implements (biolytic compounds) and feeding (lysis of prey and nutrient absorption). While some of these processes have been studied individually, the overall contribution of relevant genetic components to predation has received scant attention in the past. Under the conditions of the streak assay, variations in sensitivity to predation by *M. xanthus* were detected among different prey bacteria. The fact that *S. marcescens* was the species least sensitive to lysis in the streak assay suggested that using *S. marcescens* to score predation ability in the lawn assay would give the fewest false positive results. However, it is equally possible that the gradient of prey susceptibility observed is specific for the conditions of the streak assay, perhaps because the presence of biotic nutrients limits predation of certain prey bacteria. Regardless of this, the crucial point is that two distinct assays, measuring predation under different conditions, are available to monitor and quantify changes in the predation ability of a variety of *M. xanthus* strains. Since predation is complex and results may vary depending on the method used, both assays were employed in the study reported here. Predation in *M. xanthus* is dependent upon genes required for predationassociated processes, such as signalling and motility. However, the components involved may not play uniformly significant roles for predation of all susceptible prey species, and many more additional predation factors almost certainly remain to be identified. It is likely that these will include strain-specific biolytic factors useful for predation of different subsets of prey, as well as conserved global regulators instructing predation pathways common to all strains. The discovery of all these components is a prerequisite for a more thorough understanding of the genetics of *M. xanthus* predation.

6.5.3 Microbiome Predators and the chemical ecology of predatory soil bacteria

Microbiome Predators: Microbiome predators are increasingly studied in soils as more evidence points to their roles in shaping the soil microbiome. Through microbiome modifications, microbiome predators have a major impact on soil functioning, such as catalysing litter decomposition and enhancing plant performance (Shunran Hu et al., 2023). Microbiome predators include all organisms that feed on members of the microbiome (viruses, bacteria, archaea, fungi). The most abundant and important groups of microbiome predators are protists and nematodes as the major consumers of bacteria. Protists and nematodes also feed on fungi but have a weaker role in this energy channel than Collembola and mites. Protists and nematodes were focussed, as they are the most abundant microbiome predators that affect both bacterial and fungal communities.

Microbiome predators prey on diverse soil organisms. As predators of bacteria or fungi, they are key members of bacterial or fungal energy channels. Many microbiome predator species are generalists like other soil organisms and feed omnivorously on many bacterial and fungal species in both channels. Viruses are also preyed on by protists and nematodes, representing a link beyond classical energy channels. Even more, intraguild predation among microbiome predators seems to be common; already classical food webs propose that small flagellated protists are major prey of larger amoebae and nematodes. More recently, inverted food webs were shown with smaller protists (and other small biotas, such as myxobacteria, not covered here) preying on larger organisms, including nematodes. This complexity of diverse predator–prey links hampers a reliable placement of distinct microbiome predator species in soil food webs and, consequently, the understanding of their precise functions in soils.

Microbiome predators shape the soil microbiome and thereby soil functions. However, this knowledge has been obtained from small-scale observations in fundamental rather than applied settings and has focused on a few species under ambient conditions. Therefore, there are several unaddressed questions on soil microbiome predators, viz., What is the role of microbiome predators in soil functioning? ; How does global change affect microbiome predators and their functions? ; How can microbiome predators be applied in agriculture? Shunran Hu et al., (2023) showed that there is sufficient evidence for the vital role of microbiome predators in soils and stress that global changes impact their functions, something that urgently needs to be addressed to better understand soil functioning as a whole. There is a potential for the application of microbiome predators in agricultural settings, as they may help to sustainably increase plant growth.

Fundamental studies on the functioning of microbiome predators currently are focused on a few microbiome predator species, while field studies mostly investigate patterns of microbiome predator communities without testing their role in soil functioning. These approaches conflict with the fact that potentially millions of functionally diverse microbiome predator species exist. It is now known that microbiome predators have species-specific feeding differences that can change microbiome composition and functioning, which can explain the major impact of microbiome predators on the microbiome. Yet, we are missing an understanding of the functional importance of microbiome predator communities for key ecosystem functions that soils provide, such as elemental cycling and plant performance (including growth, health, and yield). Even less is known about microbiome predators functioning in soils exposed to ongoing global changes, including those in climate and land use. Several studies are showing that abiotic conditions are major determinants of protist and nematode communities, particularly soil moisture, structure, pH, and carbon availability (Erktan et al., 2020). All these factors are subject to change as drought, warming, and other climate change events are anthropogenically catalysed, while the need for land to meet human needs is rapidly rising. These global change drivers (GCDs) all affect biodiversity and soil functions, but how they affect microbiome predators and their impact on microbiome functioning remains unknown.

The extensive number of fundamental knowledge gaps also hampers the evaluation of the functional role of microbiome predators in agricultural systems. Protists and nematodes have long been claimed to serve as sensitive indicators of soil health (Du Preez et al., 2022) and were repetitively shown to improve crop health and production. As such, there is clear evidence of the potential of microbiome predators for application, but this remains a largely untouched field. To provide the basis for some next steps in increasing our understanding of microbiome predators in soil, three major questions are posed that should be focused on in future research, viz., What is the importance of microbiome predators in soil functioning? ; How do global changes impact microbiome predators? ; Are microbiome predators applicable such as in agriculture and if so, how? (Shunran Hu et al., 2023). While the first two are the basis to better understand the functioning of soil microbiome predators in current and future soils, the third is the needed link between fundamental science and potential application.

The study of natural products is entering a renaissance, driven by the discovery that the majority of bacterial secondary metabolites are not produced under standard laboratory conditions. Understanding the ecological role of natural products is key to efficiently directing our screening efforts, and to ensuring that each screen efficiently captures the full biosynthetic repertoire of the producing organisms (Brandon L. Findlay, 2016). Myxobacteria represent one of the most common and diverse groups of bacteria, with roughly 2500 strains publically available. Fed largely through predation, the myxobacteria have developed a large repertoire of natural products that target other microorganisms, including bacteria and fungi. Many of these interactions can be observed in predation assays, providing direct evidence for environmental interactions. With a focus on *Myxococcus xanthus*, this review will highlight how recent advances in myxobacteria are revealing the chemical ecology of bacterial natural products.

Myxobacteria are uniquely situated for research into the chemical ecology of soil natural products. As obligate predators and occasional prey, the effects of their secondary metabolism can be readily evaluated through a combination of genetic knockouts and simple predation assays. At the same time, their social lifestyle and altruistic behavior drives rapid speciation, leading to a plethora of strains and natural product classes. While the ecological role of many antibiotics is largely unknown, *M. xanthus* research suggests that many act as chemical defenses against predation. Production of bacillaene by *B. subtilis* significantly reduces grazing by *M. xanthus*, though the antibiotic is bacteriostatic and has no direct killing effect. Production of actinorhodin by *S. coelicolor* has a similar effect, while the antieukaryotic natural products of *Sorangium* sp. suggest this firewall approach is common to myxobacteria as well (Oßwald, et al., 2014). Long-term resistance to predation appears to require physical barriers, as both *Bacillus* and *Streptomyces* construct spore-filled structures when challenged with *M. xanthus*. Biofilms appear to form a similar role, providing resistance to grazing by both *M. xanthus* and *C. elegans*.

Predation by *M. xanthus* represents a significant challenge to the growth of other soil microorganisms, imposing selective pressure on prey species. Biofilms appear to act as a general physical barrier to predation, rendering *E. coli* less susceptible to killing by both *M. xanthus* and *Caenorhabditis elegans* (Brandon L. Findlay, 2016). Strains that are not prodigious biofilm producers may evade predation in multispecies biofilms, while others display more elaborate adaptations. Exposing cultures of the Grampositive soil bacterium *Bacillus subtilis* to *M. xanthus* leads to "mega-structures," large spore-filled cell aggregates that rise up to 200 µm above the agar plate. Fruiting bodies are found in close proximity to these mega-structures, indicating that they are largely resistant to *M. xanthus* predation.

M. xanthus does use at least one antibiotic as a molecular weapon, as myxovirescin expression improves predation against susceptible *E. coli* strains. While pure myxovirescin can inhibit the growth of these strains in vitro, its activity can also be replaced by subinhibitory concentrations of spectinomycin. This implies that full growth inhibition is not required for effective *M. xanthus* predation. The role of natural products is tied to the local environment, and several myxobacterial compounds appear to fill multiple roles. The simple amino acids that form A-signal both act as a quorum sensor during fruiting body formation and serve as a lure to draw in prey species like *E. coli*. The principle siderophore of *M. xanthus*, myxochelin, also appears to serve a dual purpose as a low micromolar inhibitor of eukaryotic 5- lipoxygenase and may contribute to the antagonistic interactions between myxobacteria and fungi. Myxochelin may also find a role in cancer therapy. From prey to protists to other myxobacteria, *M. xanthus* exhibits broad antagonistic activity in laboratory and soil environments. Key to many of these interactions is the production of natural products, which allows researchers to directly visualize complex interspecies interactions with genetic knockouts and simple in vitro competition assays. At present, these experiments suggest that many microorganisms use natural products as molecular firewalls that guard against competition or predation, while myxobacteria may find similar compounds useful for incapacitating prey. However, to date less than a third of the biosynthetic repertoire of *M. xanthus* has been well characterized. How the remainder will alter our understanding of bacterial chemical ecology remains to be seen.

6.5.4 The Role of Microbiome Predators in Soil Functioning

The positive roles of microbiome predators in ecosystem functioning are particularly linked to plant performance, for example, growth, health, and yield. Through predation, microbiome predators release the nutrients inside their prey (nitrogen, phosphorus, or silica) or enhance prey activity to mobilize these nutrients, which then become available and increase plant growth (Fig. 39) (Thakur & Geisen, 2019). Microbiome predators might also promote pathogen-suppressive microorganisms via changes in the community composition and feed on plant pests and pathogens, thereby reducing plant diseases (Schwelm et al., 2023). Additionally, microbiome predators play a role in carbon cycling, by enhancing litter decomposition. For instance, the addition of a single protist species increased CO_2 release and catalysed litter decomposition by 35%, through changing microbiome composition and increasing bacterial and fungal activities. Theoretically, and not reported yet, microbiome predators could also have negative effects on plant performance by preying on plant growth-promoting organisms, such as arbuscular mycorrhizal fungi, which contribute to litter decomposition and protect their host plants against pathogen infection.

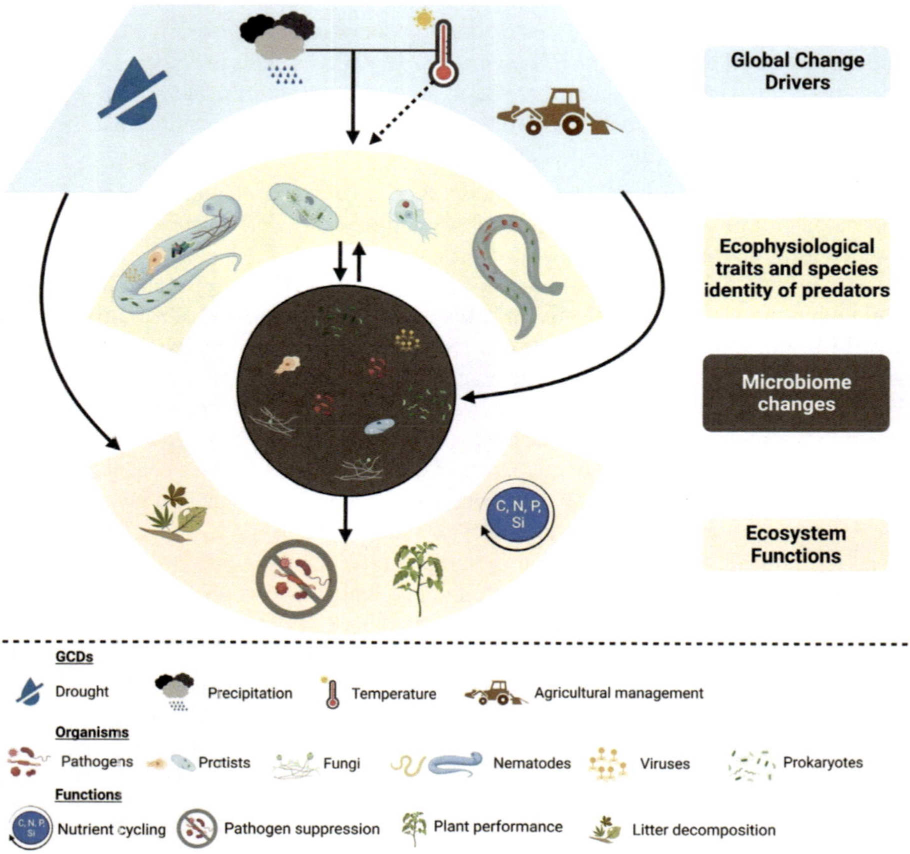

Fig. 39: Conceptual representation of the intimate link between soil microbiome predators and the microbiome leading to changes in ecosystem functions. Individual (dashed line, here using temperature as an example) or interactive (perpendicular lines, here connecting precipitation and temperature as an example) global change drivers affect these links. Arrows indicate a direct impact on microbiome predators, microbiome, or ecosystem functions. Symbols are explained at the bottom after the horizontal dashed line. Plant performance includes for example, growth, health, and yield.

Knowledge of those ecosystem functions driven by microbiome predators is limited and often based on a few microbiome predator taxa, plant species, and ecosystem functions. We thus still lack a global understanding of how microbiome predators affect plant performance, such as growth, health, and yield, and even more so on other ecosystem functions, such as the cycling of carbon, phosphorus, or silica.

The relationships between the taxonomic or ecophysiological trait diversity of microbiome predators and their impacts on soil functioning are not yet known. Following the biodiversity-ecosystem functioning (BEF) concept described, for example, aboveground plant communities, in which increasing plant diversity often positively and linearly relates to ecosystem functioning (e.g., increased plant biomass), we could expect a similar pattern for the relationship between microbiome predator diversity and soil functioning. However, there are major differences between aboveground and belowground diversity, suggesting that soil BEF (sBEF) relationships might be different from the ones described aboveground. For instance, microbial predators are in several ways more diverse than plants (e.g., phylogenetically), but many species may still have similar ecological roles in the ecosystem, quickly resulting in functional similarity. Keystone taxa, which are defined as species having a disproportionally influential role for a given function, might also exist among microbiome predators (Guo et al., 2021). These taxa could exert major predatory pressure on the microbiome compared with other microbiome predator taxa, which might be determined by their ecophysiological traits and prey specialization. The varying degree in strength and impact of predator organisms on the microbiome might change the sBEF relationship into more variable rather than linear patterns. As keystone taxa among microbiome predators are hardly known, information regarding how differences in diversity and taxonomic compositions of microbiome predator communities affect ecosystem functions is still lacking.

A framework linking the vast diversity of microbiome predator species and their functional traits to their impacts on soil functioning is currently missing. This leads us to the following questions for future research, viz., What are the main ecosystem functions performed by microbiome predators? ; Can the impacts of microbiome predators on the microbiome, and therefore on ecosystem functioning, be predicted by their ecophysiological traits? . Is the magnitude of ecosystem functioning determined by microbiome predator diversity and what is the contribution of microbiome predator keystone taxa in sBEF relationships?

6.5.5 Microbiome Predators in Agriculture and Global Changes

Microbiome Predators in Agriculture: As microbiome predators perform essential functions in soils, they are excellent candidates to be used as biological control agents (BCA) and biostimulants (BS) in agricultural systems (Nguyen et al., 2023). As such, microbiome predators might be part of the solution to reduce the agricultural application of pesticides and artificial fertilizers. Microbiome predators enhance plant performance of many plant species, including crops such as rice and wheat, through increasing plant phosphorus and nitrogen uptake. If these greenhouse and laboratory results are applicable in the field, fertilization needs for crop production would dramatically decrease. Similarly, pesticide usage could be lowered by the direct or indirect reduction of soil-borne pathogens by microbiome predators (see Introduction section). The wide variety of soil microbiome predator species and traits harbours the opportunity to produce specialized agricultural products for diverse tasks and environments. As microbiome predators are part of a more complex soil microbiome, research on links between the microbiome and microbiome predators can contribute to identifying microbiome predator species of potential use in agriculture (Figure 40) (Shunran Hu et al., 2023). Specific pathogen-preying predators and those that stimulate a pathogen-suppressive microbiome are likely to be found in various systems, but particularly in pathogen-suppressive soils. Identification of these main microbiome predator taxa that favourably regulate the microbiome could enhance crop performance and therefore warrants further investigation.

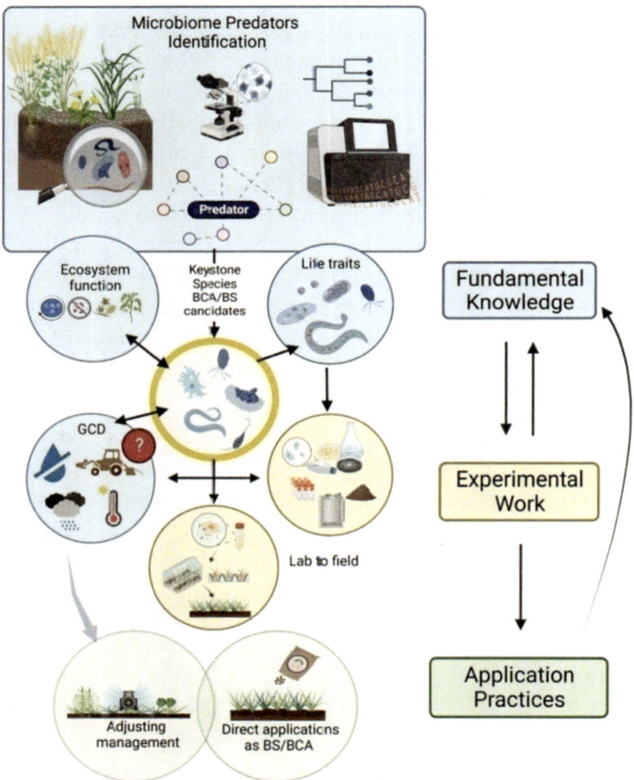

Fig. 40: The road to the application of soil microbiome predators. To apply predators, it will be essential to increase the fundamental knowledge, about the distribution, occurrence, ecosystem functions, life traits of soil microbiome predators, and how they are impacted by global change drivers (GCDs; blue). With this, keystone taxa can be identified, and their effect on crop performance, as well as culturing and storage characteristics, determined in experimental work (yellow). This experimental work will add to the knowledge about microbiome predator functions as well as how to practically apply them in the field (green). By understanding the impact of GCDs on those keystone taxa, crop performance could also be enhanced by adjusting

land management that favours beneficial microbiome predators. BCA, biological control agents; BS, biostimulants.

The advantage of microbiome predators as potential stimulators of crop growth compared with currently used BCA/BS in agricultural applications is their overarching top-down control of the entire soil microbiome, rather than providing distinct functions. This offers further strategies for application, such as co-inoculation of microbiome predators with non-prey plant-beneficial microorganisms, as spatial and resource competition of beneficial microorganisms is reduced. Indeed, soil microbiome predators might amplify the positive effects of BCA/BS organisms, such as *Trichoderma* and *Bacillus* species, by increasing their dominance and activity in the microbiome, or by phoresy of biocontrol agents by microbiome predators towards host roots (Hawxhurst et al., 2023). Using microbiome predators in synthetic communities with other microorganisms might therefore represent a boost to already existing BCA and BS products.

In addition to the interaction between microbiome predators and other microorganisms, the interaction between agricultural management and microbiome predator fitness is equally important. Indeed, the effectiveness of inoculum on crop performance is dependent on environmental conditions (Shunran Hu et al., 2023). Agricultural practices and various GCDs alter soil environmental characteristics, such as pH, temperature, and water availability, which can influence soil microorganisms and their predators. Thus, agronomic practices directly affect and stimulate native or applied microbiome predator communities. Fundamental knowledge, for example, on keystone species and environmental impacts on soil microbiome predators is needed to elucidate the effect of soil management on the native and/or applied soil microbiome predator community, to enhance crop performance through microbiome predator BCA/BS organisms and communities.

Despite their theoretical potential as both BCA and BS products in agricultural systems, the commercial development and agricultural application of microbiome predators are still in their infancy. The use of protists as microbial inoculum has not been widely adopted in the agricultural market. Nevertheless, the first protist-based BS and BCA products have been or are in the process of being established in the market (e.g., Ecostyle ProtoPlus®, Amoéba W. magna C2c Maky; Croze et al., 2021). Currently available nematode-based BCA products, however, focus entirely on entomopathogenic nematodes that control insect pests independent of the soil microbiome (Shunran Hu et al., 2023). These existing products can nevertheless considerably advance the development of new microbiome predatory protist and nematode products. This development depends on the selection of applicable microbiome predator species for mass production, field application, and commercial storage, taking into consideration traits like feeding preference, reproduction cycle, and replication rate, as well as their temperature and desiccation tolerance. As these factors are taxon specific, targeted functional research on the life history traits of microbiome predator species is a prerequisite for their application (Nguyen et al., 2023; Figure 2). the following questions are posed to be addressed to successfully apply microbiome predators in agricultural systems, viz., Which microbiome predator species determine a crop-beneficial soil microbiome and where to find them? ; How to formulate predator-based products for optimal application? ; How to adjust agricultural management practices enhance the beneficial effects of microbiome predators for crop production?

All these fields are in their infancy and need substantial work to resolve the more specific questions we defined for each of them. To address these questions and to increase our scientific understanding of soil microbiome predators and their application, a number of steps to be taken in the near future were proposed, which include trait-based studies of multiple microbiome predators, impact analyses of GCDs on microbiome predators, and applied studies on microbiome predators. It was concluded that microbiome predators should be more regularly studied in soil ecology as their impacts on the soil microbiome might be key for soil functioning, both in natural soils and agricultural ecosystems, now and in the future.

Predation regulates Microbial Structure: The importance of top-down control in regulating communities has long been argued in community ecology, which is gaining recognition also in macroscale ecology for predicting community patterns. It is believed that increasing interest in understanding the soil microbiome structure and functions at both the microscale and macroscale should adequately consider the importance of the predators (Madhav P. Thakur and Stefan Geisen, 2019). It has been argued that microbiome predators interact with other factors such as soil abiotic properties and plant-derived resources in influencing the soil microbiome. However, the overwhelming evidence of preferential feeding, intraguild predation, and omnivory obscures predicting the soil microbiome via trophic approaches. It was, thus advocated for experimental approaches that can simultaneously integrate predation and abiotic factors in understanding the soil microbiome structure and function at multiple scales.

The soil microbiome regulates vital ecosystem functions ranging from primary production to soil carbon sequestration. Yet, we have only begun to understand the factors regulating the soil microbiome. While the importance of abiotic factors is increasingly recognized, the roles of trophic regulations in driving the structure and function of the soil microbiome remain less explored. Madhav P. Thakur and Stefan Geisen (2019) reviewed the current understanding of how and when microbial and top predators of the soil shape the community structure and function of the soil microbiome via both direct and indirect effects. It was highlighted that the structure and function of the soil microbiome depend on the interactive effects among predation, plant inputs, and abiotic variables present in the soil.

Anthropogenic GCDs can strongly, and often negatively, affect the community composition, abundance, and diversity of microbiome predators (Shunran Hu et al., 2023). These GCDs include climate change-related factors (Ex. rising soil temperatures, droughts), land use change (e.g., fertilization and mechanical soil disturbance) and environmental pollution (i.e., agrochemicals, microplastics, antibiotics). GCDs may affect microbiome predator communities directly, as well as indirectly through bottom-up effects caused by alterations of the microbiome. Such bottom-up effects are, for example, driven by eutrophication, which stimulates fast-growing microorganisms, including most bacteria, more than slow-growing microorganisms, including most fungi. These differential microbiome changes result in increased abundances of bacterivorous compared with fungivorous protists and nematodes. However, the consequences of such microbiome-mediated changes in microbiome predator abundances and their resulting impacts on ecosystem functioning remain unexplored.

GCDs may also directly affect the performance and abundance of microbiome predators. In contrast to microbiome-driven changes in microbiome predator communities, direct effects of GCDs on microbiome predator communities may alter their top-down control of soil microorganisms, with consequences for ecosystem functioning. Indeed, such disruptions are likely to take place, as, for example, soil protist communities are more strongly affected by distinct types of environmental pollution and agricultural practices than bacterial and fungal communities (Zhu et al., 2021). Likewise, many nematode taxa show a particularly strong sensitivity to certain GCDs. Especially, drought is a major GCD that alters nematode community composition, with important implications for nematode top-down control impacts on the microbiome. More knowledge on the direct and indirect impacts of GCDs on microbiome predators and, especially, their consequences for soil functioning must be obtained.

GCDs can affect microbiome predator communities in various ways, and that microbiome predator traits may help to understand their responses to global change. Research on how GCDs affect microbiome predator communities has so far largely focused on examinations of single GCDs. However, soil communities are exposed to multiple simultaneously acting GCDs that together might induce different impacts on GCDs acting in isolation (Speißer et al., 2022). Moreover, while it has become evident that microbiome predators may be more sensitive to GCDs than their prey, the ecological consequences of this higher microbiome predator sensitivity to GCDs have remained unknown. Finally, while traits have been used to some extent to understand microbiome predator responses to GCDs, these responses have so far not been linked to subsequent impacts on the microbiome and soil functioning. Future research on how GCDs affect soil microbiome predator communities should therefore focus on the following questions, viz., How do GCDs affect microbiome predators and their impact on ecosystem functioning? ; Can microbiome predator traits be used to predict the impacts of GCDs on microbiome predator communities? ; How do microbiome predator communities and their impacts on the microbiome change ecosystem functioning in response to multiple, simultaneously acting GCDs?

6.6 Key Predatory Groups and their Direct Effects on the Soil Microbiome

This review refers to soil fungi and soil bacteria collectively as the soil microbiome. The other components of the soil microbiome, such as protists and viruses, are treated as the consumers within the soil microbiome. Broadly, there are four groups that consume the soil microbiome, viz., (i) protists (within the soil microbiome) ; (ii) free-living soil nematodes; (iii) soil microarthropods; and (iv) saprophagous soil animals (Madhav P. Thakur and Stefan Geisen, 2019). Among the four groups, protists, nematodes, and microarthropods are the direct predators of the soil microbiome, whereas saprophagous soil animals, such as earthworms and isopods, ingest bacteria and fungi via feeding on soil and soil organic matter. Importantly, these organisms that feed on microbes represent a large gradient in their body size, ranging from a few millimeters to a few centimeters. Moreover, as the body size varies among the consumers of soil microorganisms, their net consumption also varies; this imposes differential effects on the soil microbiome. Free-living soil nematodes are one of the major consumers of the soil microbiome with groups that feed on bacteria and fungi. Bacterial-feeding nematodes, representing half of all soil nematodes, are generalists that randomly ingest any bacteria by their filter-feeding habit. Fungal-feeding nematodes, by contrast, penetrate fungal hyphae by their stylet or spear; they have been shown to limit hyphal growth in soils. Soil microarthropods, particularly Collembola and mites, are the other key direct consumers of the soil microbiome.

Although collembolan species feed on soil bacteria and plant materials, the majority of Collembola species feed mainly on soil fungi. Among soil mites, oribatid mites are the major group that feed on a wide range of fungal species, such as on dark-pigmented fungal groups and ectomycorrhizal fungi in forest ecosystems. Some of the major consumers of soil microorganisms are in fact the ones that do not predate on them but rather consume them via the consumption of soil, litter, and soil organic matter, such as earthworms, isopods, millipedes, and enchytraeids. Among them, earthworms are the most dominant consumer (in terms of amount consumed) of the soil microbiome in most systems (Jackson et al., 2017). The earthworm's ingestion rate and assimilation efficiency of organic matter are the key determinants of how they influence the structure and function of the soil microbiome. In general, higher assimilation efficiency in some of the earthworms, such as those dwelling on the soil surface, promotes microbial growth in the soil. However, those earthworms with greater assimilation efficiency usually have the lower ingestion rates of organic matter. These differences among earthworms relate to their selective versus predominantly nonselective feeding on soil organic matter. An earthworm's gut passage can also act as an environmental filter for certain soil microorganisms, for example, by suppressing the colonies of bacteria and inhibiting the germination of spores and radial growth of fungal species.

Most soil microbiome predators exhibit preferential feeding on some prey over others. For instance, both bacterial- and fungal-feeding nematodes exhibit preference for some prey based on prey traits and the soil environment. Further, fungal palatability might determine the diet preference by collembolan species. Oribatid mites often prefer dark-pigmented fungi, which are crucial biotic drivers of litter decomposition. The preferential feeding in earthworms, based on the quality of the litter, has also been reported by several studies (Ferlian et al., 2018). Importantly, the vast physical structure present in the soil can constrain or facilitate preferential feeding on the soil microbiome. Such predominant preferential feeding among the microbiome predators can have important implications for the soil microbiome structure and function, such as competitive interactions between soil bacteria and fungi.

i. Predation within the Soil Microbiome: The soil microbiome consists not only of the commonly studied bacteria and fungi, but also a major group of microbiome predators - protists. Protists feed mainly on soil bacteria, reducing both abundance and changing the community composition of bacteria through specific prey selection (Madhav P. Thakur and Stefan Geisen, 2019). Generally, protists avoid too large or toxin-producing bacteria, such as pseudomonads, as well as increase the ratio of Gram-positive to Gram-negative bacteria through preferential feeding on easier-to-digest Gram-negative bacteria. These taxonomic changes in microbial communities have been found to result in functional changes that can result in enhanced plant performance (Geisen et al., 2018). A majority of protists also feed on fungi, whereas larger protists are often omnivores, feeding simultaneously on different microbial groups, such as bacteria, fungi, and smaller protists. Viruses are another important regulator of the soil microbiome. Despite the arguments whether viruses are truly a living form, viruses are often considered as part of the microbiome. Viruses, as key agents (bacteriophages) of bacterial mortality, have been widely studied in aquatic systems, and their presence in high abundance with high diversities has been confirmed for soils. Considering that many bacteriophages are often highly host-specific, viruses likely change the composition of bacterial communities. Viruses infecting fungi (mycoviruses) can further alter the structure of fungal communities.

Diverse groups of bacteria and fungi also kill other microorganisms through the production of antimicrobial compounds in the soil. This often results in increased growth rates of the antimicrobial compound-producing microorganisms as competitors for limiting nutrients are eliminated, while also additional nutrients from the killed competitors become available. Indeed, some bacterial groups, such as myxobacteria, have developed to obtain their nutrients through lysis of other bacteria, while other bacterial groups, such *Collimonas* spp., have specialized to use fungi as their main nutrient source. While our focus in this review is on the soil microbiome predators, we need to highlight the importance of parasites in indirectly structuring soil microbiomes by parasitizing predators of microorganisms. Among them are viruses, bacteria, fungi, and protists. So far, we have only begun to understand the diversity of some of these parasites in soils and soil animals. The role of these animal parasites in shaping animal communities in soils, and their effect on the soil microbiome, remains to be elucidated.

ii. Predator-Mediated Competition between Soil Bacteria and Fungi: Soil microbial ecology has long investigated the competitive interactions between bacteria and fungi. The general notion is that they compete for simple plant-derived substrates with important implications on microbe-mediated processes in the soil (Madhav P. Thakur and Stefan Geisen, 2019). A recent study also revealed that bacterial and fungal competition is strongly present in the top soil across the globe and is mediated by antibiotic-resistance genes. Negative interactions between fungi and soil bacteria can reduce the turnover of soil organic matter. Microbiome predators can further influence the competition between soil bacteria and fungi in numerous ways. For instance, fungi can have greater competitive advantage over bacteria when bacterial predators become dominant or more active (the same applies to competitive advantage to bacteria when fungal predators are dominant). However, these pathways of bacterial and fungal trophic regulation by their predators can also be highly connected; that is, many bacterial predators do not exclusively forage on bacteria but only preferentially do so (the same holds for fungal predators). Shared predators thus can also lead to apparent competition between soil bacteria and fungi.

Saprophagous consumers and top predators can also indirectly influence the competitive interaction between fungal and bacterial groups, such as via altering their habitat and trophic cascades, respectively. Besides, these predator-mediated competitive interactions in the soil microbiome can also occur within microbial groups, such as within fungi and within bacterial groups. Empirical evidence for predator-induced competitive interaction between soil fungi and bacteria is still scarce. Moreover, as shown by both theoretical and empirical studies, the interaction between predation and competition is one of the key determinants of variations in community structures and has rarely been tested in understanding the soil microbiome community structures. It is believed that experimental approaches can help us to understand the importance of predator-induced competitive interaction between and within the bacterial and fungal groups.

All microbial-feeding groups influence microbe-mediated ecosystem processes in the soil by affecting microbial population and traits. These processes range from increasing nutrient turnover to increasing decomposition rates to influencing plant performance. For instance, bacteria feeding predator groups have been shown to enhance the turnover rates of soil bacterial communities, particularly in the environment where resources for soil bacterial communities are sufficient, such as greater amounts of organic matter potentially in proximity to plant roots (Jiang et al., 2017). In such cases, predators of bacteria increase the rate of nutrient mineralization and the decomposition of organic matter in the soil, thereby benefitting plants. Fungus-feeding predator groups also affect plant performance and the decomposition process in a number of ways by feeding on soil fungi. For instance, they negatively affect the performance of mycorrhizal host plants when they feed on mycorrhizal fungi, but also could enhance the nutrient mineralization process via their stimulation of decomposer fungal communities with potential benefits to plants.

iii. Indirect effects of top Predators on the Soil Microbiome: Direct predators of the soil microbiome are further consumed by other carnivores in the soil, and they thereby exert indirect effects on the soil microbiome (Madhav P. Thakur and Stefan Geisen, 2019). Such indirect effects of predators of the soil microbiome predators (hereafter top predator for brevity) on the soil microbiome operate via two interdependent mechanisms: density- or biomass-mediated indirect effects and trait-mediated indirect effects. Density-mediated indirect effects are often studied as trophic cascades. Theoretically, the greater the prey suppression by the predator the larger is the strength of trophic cascades. Trophic cascades on the soil microbiome have mainly been demonstrated via predator-removal experiments. The removal of predatory mites which feed on microarthropods and nematodes suppressed bacterial population size, which, in turn, reduced decomposition rates in a desert soil. In another removal experiment, but with spiders, which are larger predatory species of microarthropods, a similar cascading effect was observed on litter decomposition, but only in drier conditions.

The three key groups of microbial consumers discussed above (nematodes, microarthropods, and saprophagous animals) have their own group of predators and it is thus likely that these top predators exert distinct cascading effects on the soil microbiome. In fact, some of these top predators are large enough to prey upon the predators of other groups, which typically results into intraguild predation (Eitzinger et al., 2018). Intraguild predation can potentially dampen the strength of trophic cascades in food webs. Owing to a greater probability of intraguild predation among the predators of microbe-feeding species, the size and structure of the soil microbiome are lesser affected by top predators. The same has been argued for widely present omnivory in the soil, which potentially reduces the prevalence of keystone predators in structuring prey communities as observed in other systems. In fact, several top predators in their juvenile stage feed on soil fungi and bacteria but change their diet to direct predators of the soil microbiome as they develop into adults. However, the effects of removal of top predators can manifest in soil processes carried out by the soil microbiome, such as decomposition and nutrient mineralization, even without a noticeable change in microbial community size.

Trait-mediated indirect effects in food webs often depend on how top predators influence the behaviour of their prey without affecting their population size, which then influences both the population and traits of the resources of the prey. When applying this to the soil microbiome, top predators should cause behavioural shifts (and potentially other trait changes, such as morphological and physiological changes) in microbial consumers to the extent that noticeable change occurs in microbial population, trait, and/or function. However, the role of traitmediated indirect effects on the soil microbiome is far less understood as the traits of microbiome predators have often been ignored in soil food-web studies (Buchkowski, 2016). Evidence of behavioural shifts in the consumers of soil microorganisms owing to top predators has shown that soil microbialmediated processes, such as soil respiration and N_2O emissions, can substantially change. It is thus likely that the structure of the soil microbiome is prone to change via the nonconsumptive effect of top predators on microbial consumers, and this merits further exploration.

iv. Climate change effects on Trophic Regulation of the Soil Microbiome: A global rise in the Earth's surface temperature and more erratic rainfall patterns posit key threats to biodiversity. Both soil microorganisms and their predators are vulnerable to such changes in soil temperatures and soil water availability, although the vulnerability may vary among the study groups (Madhav P. Thakur and Stefan Geisen, 2019). Importantly, the rate of predation by ectothermic organisms (all soil microbiome predators discussed in this review are ectotherms) often becomes higher at warmer temperatures due to elevated metabolic demands. However, this is only true until a certain temperature threshold beyond which predators die due to physiological constraints at high temperatures. The basic notion of metabolic ecology is that ectotherms with larger body sizes have lower temperature thresholds than smaller ones, which accordingly makes larger microbiome predators more sensitive to warming. In an experimental study, increase in temperature altered the physiology of microbiome predators by reducing their body size, which only occurred in a microbiome predator with a larger body size. Thus, while temperature enhances the foraging demand of microbiome predators, it also triggers phenotypic responses in microbiome predators that could balance their foraging demand.

The implications of greater foraging pressure by predators and their physiological adaptations can vary in time, and thus each of these stages (the adaptation process) may have distinct impacts on their microbial prey. This will further interact with how microorganisms themselves respond to increases in temperature, including their physiological adaptation. Soil fungal and bacterial responses to warming are often dependent on soil water and resource availability. For instance, drier and warmer soil seems to harm bacteria more than fungi, although this may again differ among the soil types. If both fungal and bacterial predators become active proportionally at higher temperature, it is likely that warming could favour fungal communities in the soil. Drier soil conditions, in particular, can lower the foraging efficiency of several soil microbiome predators in warmer soils (Thakur, M.P. et al., 2018). The implications of reduced foraging of soil microorganisms can slow down the stimulation of their population which often takes place via their consumption, such as for soil bacterial communities. Although soil bacterial communities can overcome such a competitive inferiority via their physiological adaptation to higher temperatures, a recent meta-analysis showed that the physiological adaptation of soil bacteria and fungi did not systematically vary over time in warming experiments. It is therefore encouraged to consider both the physiological adjustments in microorganisms and their predators to capture a better understanding of their dynamics in a changing world.

6.6.1 A predatory Myxobacterium controls Cucumber Fusarium wilt: Myxobacteria are micropredators in the soil ecosystem with the capacity to move and feed cooperatively. Some myxobacterial strains have been used to control soil-borne fungal phytopathogens (Ye et al., 2020). However, interactions among myxobacteria, plant pathogens, and the soil microbiome are largely unexplored. Myxobacteria are Gram-negative soil bacteria that display complex multicellular morphogenesis and social behaviors over their life cycles. Most myxobacteria can prey on various microorganisms, including bacteria and fungi. Predatory bacteria have been proposed as potential biological agents and have shown effective biocontrol activity against various plant soil-borne pathogens. Their abilities to slide on solid surfaces, significant capacity to produce antibiotics, formation of myxospores, and predatory behaviors make myxobacteria excellent candidates for BCAs.

The behaviors of the myxobacterium *Corallococcus* sp. strain EGB were studied in the soil and its effect on the soil microbiome after inoculation for controlling cucumber Fusarium wilt caused by *Fusarium oxysporum* f. sp. *cucumerinum* (FOC). A greenhouse and a 2-year field experiment demonstrated that the solid-state fermented strain EGB significantly reduced the cucumber Fusarium wilt by 79.6% (greenhouse), 66.0% (2015, field), and 53.9% (2016, field). Strain EGB adapted to the soil environment well and decreased the abundance of soil-borne FOC efficiently. Spatiotemporal analysis of the soil microbial community showed that strain EGB migrated towards the roots and root exudates of the cucumber plants via chemotaxis. Cooccurrence network analysis of the soil microbiome indicated a decreased modularity and community number but an increased connection number per node after the application of strain EGB. Several predatory bacteria, such as *Lysobacter*, *Microvirga*, and *Cupriavidus*, appearing as hubs or indicators, showed intensive connections with other bacteria.

The predatory myxobacterium *Corallococcus* sp. strain EGB controlled cucumber Fusarium wilt by migrating to the plant root and regulating the soil microbial community. This strain has the potential to be developed as a novel biological control agent of soil-borne Fusarium wilt.

Effective colonization and formation of biofilm-like structures on plant roots are crucial for the biocontrol of plant diseases by plant growth-promoting rhizobacteria (PGPR). The enhanced biofilm formation in *Pseudomonas fluorescens* CHA0 significantly enhanced the colonization of carrot roots and led to more stable interactions between mixed inocula and the rhizosphere. Strain EGB colonized the root elongation zone and displayed visible biofilm-like structures in hydroponic experiments. Myxobacteria secrete abundant extracellular matrices, which can facilitate biofilm formation and root colonization. The ability of myxobacteria to feed on other soil microbes leads to the propagation of these bacteria and reduction in the quantity of their prey. However, traditional BCAs have to compete for resources with the native microbial community to survive in the soil, which reduces their adaptive abilities (Ye et al., 2020).

Movement of bacteria in soil strongly affects their colonization efficiency, microbial fitness, and bacteria-mediated biocontrol abilities (Pu and Elsas, 2018). Active movements can be driven by flagella in the form of chemotaxis or swarming. Intact chemotaxis and swarming motility machineries were found to be important for PGPR to reach and colonize plant roots. Hypermotility mutations improved the colonization of alfalfa root surfaces by *P. fluorescens* F113 and improved its biocontrol activity. Swarming played an important role in root colonization by *B. subtilis*. However, movement driven by flagella was limited in wet soil. Vigorous movement of strain EGB towards cucumber roots was revealed by spatiotemporal high-throughput sequencing. A movement of 8 cm from the inoculation site to the roots was observed for strain EGB. Although myxobacteria lack flagella and are unable to swim in liquid environments, these organisms glide on solid surfaces in coordinated cell groups or as isolated individuals. This gliding motility driven by a focal adhesion complex allows myxobacterial cells to glide over relatively dry and hard soil matrices.

Biocontrol by predation offers a great opportunity for the management of soil-borne plant pathogens. Direct predation of phytopathogens and competing soil bacteria is also desirable for the adaptability of BCAs in the soil environment (Ye et al., 2020). Myxobacteria are slow-growing bacteria with the ability to produce heat-tolerant spores, and slow-growing, heat-tolerant bacterial families are proposed to have potential roles in plant protection. Notwithstanding, the nontargeted effect of strain EGB on soil microbiota should be given more attention in future research due to its broad prey spectrum.

6.6.2 Bacterial and Fungal predator - Prey Interactions modulate Soil Aggregation: The formation and stabilisation of soil macro-aggregates protects soils from erosion, a major worldwide threat on soils. While the role of bacteria and fungi in soil aggregation is well established, how predators feeding on microbes modify soil aggregation has hardly been tested. Amandine Erktan et al., (2020) studied how predators modulate the effect of microbial prey on soil aggregation. They focused on two predator - prey interactions: bacterial-based interactions comprising amoebae (*Acanthamoeba castellanii*) grazing on free-living bacteria (*Pseudomonas fluorescens*), and fungal-based interactions comprising collembolans (*Heteromurus nitidus*) grazing on saprotrophic fungi (*Chaetomium globosum*). They conducted a microcosm experiment lasting six weeks and assessed changes in soil aggregate formation and stabilisation, together with modifications in soil microbial communities (PLFAs). The food resource consumed by microbes was further traced using δ 13C isotopic tracing. The protest *A. castellanii* increased the formation of soil aggregates but decreased their stability, without affecting bacterial abundance and community composition, suggesting that the changes were due to amoebae-mediated changes in the production of bacterial mucilage. Saprotrophic fungi showed the highest positive effect on soil aggregate formation and stabilisation, associated with a more efficient use of particulate organic carbon (chopped litter) added to the microcosms. Adding collembolans decreased the abundance of fungi and their ability to capture carbon of litter origin, with negative consequences on soil aggregation. It has been demonstrated that trophic interactions are important for achieving a mechanistic understanding of biological contributions to soil aggregation.

Collembolans reduced the positive effect of *C. globosum* on both aggregate formation and stability. This negative effect was associated with reduced fungal biomass (as indicated by the fungal PLFA marker), which likely was due to the consumption of fungal hyphae by *H. nitidus*. Collembolans beneficially affected soil aggregate stability, which was due to the enhancement of the growth of AMF. These opposite effects presumably are due to the use of different fungi, AMF. Supporting this conclusion, it is known that collembolans preferentially feed on saprotrophic rather than AMF fungi. Preferential feeding on saprotrophic fungi rather than AMF contributed to enhanced growth of AMF and thereby to AMF-mediated increase in soil aggregate stability. In our study, only non-AMF was growing in the microcosms and collembolans fed on their hyphae, which led to a negative effect on soil aggregation. Moreover, our study captured the short-term effects of collembolans on soil aggregation, after 4 weeks of incubation, while a study was run over 14 weeks. Whether the negative short term effect observed here may turn positive in the long term remains to be tested.

Even though not related to soil aggregation, presence of collembolans as well as the "Collembola wash" altered bacterial community composition suggesting that bacteria associated with collembolans were responsible for these changes (Amandine Erktan et al., 2020). Interestingly, the effect was more pronounced when collembolans were added and not only their wash. One reason for that may be that washing selected only a fraction of bacteria associated with collembolans. Further, dispersion of bacteria through movement of collembolans may have contributed to the more pronounced changes in presence of collembolans. Finally, as collembolans also graze on bacteria, this may have contributed to changes in bacterial community composition. The positive effect of *C. globosum* on soil aggregation is consistent with previous studies highlighting the positive and transient effect of non-AMF on soil aggregate stability, with a peak at 4 weeks after incubation. As expected, the proportion of fungal PLFA markers was highest in the treatment with *C. globosum* only, indicating that our inoculation was successful and triggered fungal development. When *C. globosum* was added, alone or with *H. nitidus*, the proportion of bacterial PLFAs was reduced, and in parallel the proportion of fungal PLFAs was increased. Also, total microbial biomass was increased slightly presumably reflecting fungal growth. Finally, when incubated alone, *C. globosum* consumed higher proportions of C from the added litter compared to treatments with fungi and Collembolan. The ability to capture litter C likely is linked to the growth of fungal hyphae through air-filled pores, allowing *C. globosum* to reach a wide range of litter resources.

Bacteria can mobilize Nematode-Trapping Fungi to kill nematodes: In their natural habitat, bacteria are consumed by bacterivorous nematodes; however, they are not simply passive preys. Wang et al., (2014) reported a defensive mechanism used by certain bacteria to mobilize nematode-trapping fungi to kill nematodes. These bacteria release urea, which triggers a lifestyle switch in the fungus *Arthrobotrys oligospora* from saprophytic to nematode–predatory form; this predacious form is characterized by formation of specialized cellular structures or 'traps'. The bacteria significantly promote the elimination of nematodes by *A. oligospora*. Disruption of genes involved in urea transport and metabolism in *A. oligospora* abolishes the urea-induced trap formation. Furthermore, the urea metabolite ammonia functions as a signal molecule in the fungus to initiate the lifestyle switch to form trap structures. These findings highlighted the importance of multiple predator–prey interactions in prey defense mechanisms.

In natural ecosystems, bacterial communities play an important role in nutrient cycling, primary production and decomposition. Bacteria are also heavily consumed by bacterivorous nematodes that are present in compost, soil and aquatic systems. Thus, aside from competing for nutrients among themselves, bacteria experience significant predation pressure from nematodes. To counter predatory attacks from the metazoan enemies, bacteria have evolved a variety of defense mechanisms. First, many bacteria can produce and secrete toxins. For instance, the bacterium *Pseudomonas fluorescens* secretes secondary metabolites to repel nematodes, whereas *P. aeruginosa* produces phenazine compounds and cyanide to kill nematodes. Second, the secretion of extracellular proteases contributes to the nematode toxicity of the bacteria *Vibrio cholerae* and *Bacillus nematocida* against nematodes. Third, morphological adaptations by bacteria are also a common defense mechanism. For example, biofilm formation by the bacterium *Yersinia pestis* can prevent or reduce ingestion by nematodes (and other predators). These specific strategies provide certain bacteria an advantage against their nematode predators.

These bacterial defense mechanisms in turn create selective pressure on nematodes, resulting in predator–prey arms race. For example, nematodes have evolved to avoid the toxic/pathogenic bacteria and forage for nontoxic bacteria to support their growth and reproduction. Currently, the most detailed data on nematode behavioral responses to bacteria have come from the model nematode *Caenorhabditis elegans*. Although *C. elegans* has a relatively simple nervous system, its sensory neurons can detect many environmental cues. When encountering bacterial pathogens, *C. elegans* exhibits an avoidance behaviour and escapes. In the presence of nontoxic bacteria, *C. elegans* can recognize and hunt for them as food sources. However, it remains unclear whether counter response strategies exist in these nontoxic bacteria.

In natural environments, the nematode-trapping fungi can trap and kill nematodes. These fungi have two lifestyle phases: saprophytic and predacious. In the absence of nematodes, these fungi live as saprophytes. Upon encountering nematodes, these fungi enter the predacious stage by producing specialized predatory devices to capture and kill nematodes. Earlier studies have demonstrated that nematodes produce signal substances to promote trap formation in these fungi, but the chemical compositions of these substances remain largely unknown. A recent study has reported that ascarosides, a conserved family of signalling molecules that are widely biosynthesized among nematodes, induce trap formation in nematode-trapping fungi (Hsueh et. al, 2013). Aside from ascarosides, diffusible compounds from cow dung have also shown to induce lifestyle transitions in the model nematode-trapping fungus *Arthrobotrys oligospora*, which produces three-dimensional (3D) adhesive networks.

As major components of the soil ecosystem, bacterivorous nematodes and bacteria interact with each other and have evolved a diversity of mechanisms to counter balance each other (Jousset,, 2012). However, little is known about the mechanisms by which the nonpathogenic bacteria protect themselves against the nematode grazers. Theseresults demonstrated that certain bacteria can mobilize nematode-trapping fungi to reduce predation pressure by producing and releasing urea. The urea metabolite ammonia functions as a signal molecule to accelerate the switch in nematode-trapping fungi from saprophytic to predacious lifestyle, which is crucial for capturing nematodes.

6.6.3 Predatory Protists and Pathogen-suppressive bacterial & fungal impact on plant health: Microbial secondary metabolites have many important ecological roles. For instance, they fight against other microbial competitors and act as signaling molecules in inter- and intraspecies interactions (Guo et al.,, 2022). Nonribosomal peptides are among the microbial secondary metabolites produced by the keystone species mentioned above. Antimicrobial nonribosomal peptides, such as the antibiotics vancomycin, gramicidin and lipopeptides (surfactin, iturin A and bacillomycin), can be used as microbicides to inhibit plant pathogens. Soil protists might indirectly affect pathogens and plant health by altering the soil microbiome composition and thereby secondary metabolite production fungi. Predatory protists in the plant rhizosphere have also been reported to positively link with plant-beneficial microbes that improve plant performance. Through the production of secondary metabolites, these microbes not only suppress soil-borne pathogens, but also defend themselves against predation by protists. However, since plant-beneficial bacteria have been shown to produce metabolites that, in addition to suppress soil-borne pathogens, also defend themselves against protistan predation, it is worth disentangling the interactions between plant-beneficial bacteria and their protistan predators.

Little is known regarding the impact of microbiome predators on plant health-influencing microbes and the interactive links to plant health. It was aimed to decipher the importance of predator-prey interactions in influencing plant health. To achieve this goal, Guo et al., (2022) investigated soil and root-associated microbiomes (bacteria, fungi and protists) over nine years of banana planting under conventional and organic fertilization regimes differing in *Fusarium* wilt disease incidence. It was found that the reduced disease incidence and improved yield associated with organic fertilization could be best explained by higher abundances of protists and pathogen-suppressive bacteria (e.g. *Bacillus* spp.). The pathogen-suppressive actions of predatory protists and *Bacillus* spp. were mainly determined by their interactions that increased the relative abundance of secondary metabolite Q genes (e.g. nonribosomal peptide synthetase genes) within the microbiome. In a subsequent microcosm assay, the interactions between predatory protists and pathogen-suppressive *Bacillus* spp. was tested that showed strong improvements in plant defense. The study showed how protistan predators stimulate disease-suppressive bacteria in the plant microbiome, ultimately enhancing plant health and yield. A new biological model useful for improving sustainable agricultural practices was suggested that is based on complex interactions between different domains of life.

It was demonstrated that predator-prey interactions stimulated by organic fertilizer inputs change microbiome functioning and improve plant health. Organic fertilization supports a disease suppressive and consequently plant health-increasing soil microbiome, which supports former findings (Guo et al.,, 2022). The increase of particularly predatory protists is also in line with previous studies and can be explained by increased prey availability and physicochemical niche space. The observed positive effects of organic fertilization were not equally important for all soil microbes in all plant and soil habitats. It was found that especially protists and bacteria in the rhizosphere, rather than bulk or endosphere microbes determine disease incidence and plant yield. There is a common notion that the rhizosphere is the main area where soil-borne pathogens compete with beneficial microbiota, but interestingly the tightly plant-associated root endosphere, where many mutualists and pathogens directly invade the plant, seems of less direct importance in controlling plant health. This suggests that the determination of the composition of mutualistic and pathogenic microbes and their function is mainly determined in the rhizosphere, which we here show is determined by protists. Whether protists are also of key importance to shape the plant-beneficial rhizosphere microbiome beyond the system of banana with *Fusarium* wilt disease represents a key field for future research. Yet, previous studies showing that the importance of protistan communities in predicting tomato health in presence of pathogenic *Ralstonia solanacearum* supports this hypothesis.

Precise protistan predatory taxa and their importance in regulating bacteria with potential to suppress pathogens were identified. These findings confirm the pivotal role of rhizosphere Cercozoa, a major group of protists in soil and key microbiome links in agricultural systems, for plant health. It was shown that the disease-suppressive function of Cercozoa is likely linked to increases of disease-suppressive bacteria, especially *Bacillus*. In greenhouse study it was shown that Cercozoa enhanced

Bacillus likely by preying preferentially on other bacteria which led to increased production of secondary compounds that reduced pathogen abundances. This adds to the recent finding that cercozoan taxa increased cucumber yield effectively via their specific enhancement of plant-beneficial microorganisms (in that study *Trichoderma*) (Guo et al., 2021). Also the idea that protists selectively prey and commonly enhance secondary metabolite-producing bacteria is supported in the study. This suggests that effects of predatory protists on disease suppression might be a general phenomenon in soils, making protists keystone species that are worth exploring as biocontrol agents to increase sustainable soil management.

The pathogen population density was negatively affected by the secondary metabolite biosynthesis gene abundance, particularly non-ribosomal peptide syntheses (NRPS) genes (Guo et al.,, 2022). Microbial secondary metabolite biosynthesis function has been widely shown to be closely linked with plant pathogen suppression (Xiong et al., 2020). Previously, plenty of *Bacillus* species (e.g. *B. amyloliquefaciens*, *B. subtilis* and *B. velezensis*) were reported to produce antimicrobial compounds (e.g., iturin and bacillomycin), which provide them with a competing advantage against plant pathogens. It was proposed that the NRPS gene-induced pathogen suppression represents a plant-beneficial side effect of bacterial anti-predator defence.

Predatory Protists and Fungal Pathogens: Soil-borne fungal pathogens pose a major threat to global agricultural production and food security. Pathogen-suppressive bacteria and plant beneficial protists are important components of soil microbiomes and essential to plant health and performance, but it remains largely unknown regarding how agricultural management practices influence the relative importance of protists and bacteria in plant disease suppression. Ren et al., (2023) characterized soil microbiomes (including fungi, protists, and bacteria) in bulk and sorghum rhizosphere soils with various long-term inorganic and organic fertilization regimes, and linked the changes in fungal plant pathogens with the protistan and bacterial communities. It was found that the relative abundances of fungal pathogens were significantly decreased by organic fertilization regimes, and there was a significant difference in the community composition of fungal pathogens between inorganic and organic fertilization regimes. Organic fertilization significantly enhanced predatory protists but reduced the proportions of protistan phototrophs. Co-occurrence network analysis revealed more intensive connections between fungal plant pathogens with protists, especially predatory protists, than with bacterial taxa, which was further supported by stronger associations between the community structure of fungal pathogens and predatory protists. More protist consumer taxa were identified than bacterial taxa as predictors of fungal plant pathogens, and structural equation modelling revealed a more important impact of protist consumers than bacteria on fungal pathogens. Altogether, new evidence was provided that the disease inhibitory effects of long-term organic fertilization regimes could be best explained by the potential predation pressure of protists. The findings advanced the mechanistic understanding of the role of predator-prey interactions in controlling fungal diseases, and have implications for novel biocontrol strategies to mitigate the consequences of fungal infections for plant performance.

Soil organisms are affected by the presence of predatory protists. However, it remains poorly understood how predatory protists can affect plant disease incidence and how fertilization regimes can affect these interactions. Guo et al., (2024) characterised the rhizosphere bacteria, fungi and protists over eleven growing seasons of tomato planting under three fertilization regimes, i.e conventional, organic and bioorganic, and with different bacterial wilt disease incidence levels. It was found that predatory protists are negatively associated with disease incidence, especially two ciliophoran *Colpoda* OTUs, and that bioorganic fertilization enhances the abundance of predatory protists. In glasshouse experiments it was observed that the predatory protist *Colpoda* influences disease incidence by directly consuming pathogens and indirectly increasing the presence of pathogen-suppressive microorganisms in the soil. It was demonstrated that predatory protists reduce bacterial wilt disease incidence in tomato plants via direct and indirect reductions of pathogens.

6.6.4 Microbial root mutualists affect the Predators and Plant pathogens: Recent years have seen a surge in studies of the interactions between below- and aboveground biota, demonstrating that belowground organisms have major effects on aboveground ecological processes, including plant physical and chemical traits, plant performance, herbivore and pollinator performance, and their recruitment (Tao et al., 2017). However, our understanding of the role of belowground interactions between plants and other organisms on aboveground tri-trophic interactions remains in its infancy. Most of the limited work to date has focused on microbial root mutualists, their impacts on plant traits, and how these traits generate tri-trophic interactions between plants, arthropod herbivores, and arthropod natural enemies.

Tri-trophic interactions among plants, herbivores, and natural enemies of herbivores are common in nature, and are crucial components of trophic cascades and the dynamics of community composition. Plant traits are key determinants of the interactions between herbivores and their natural enemies aboveground, which in turn are affected by soil organisms. Recent years have seen a surge in studies of the interactions between below- and aboveground biota, including descriptions of how microbial root mutualists influence plant traits and herbivore performance. However, concomitant effects on the natural enemies of herbivores remain relatively poorly understood. Tao et al., (2017) reviewed the currently available literature to assess how and when mutualistic root microbes impose significant indirect effects on the performance of predators and pathogens of insect herbivores. Tao et al., (2017) focussed on how root microbes influence predator attraction, on-plant foraging efficiency, and the quality of prey tissues. The underappreciated effects of microbial root mutualists on the growth, transmission, and virulence of insect pathogens were also considered.

The interactions between herbivores and their parasitoids and predators are separated into three stages: a prey localization stage, during which enemies actively forage for herbivores over long distances; an on-plant foraging stage, during which plant physical traits mediate enemy foraging efficiency; and an after-contact stage, when enemies consume and/or reproduce in herbivore tissues, the quality of which determines enemy performance. Critically, the mechanisms underlying all of these interactions are based on variable plant traits (morphology, chemistry, physiology), that are subject to modifications by microbial root mutualists.

Prey Localization: Predators and parasitoids must locate their herbivorous prey before attacking those prey items. Prey location by invertebrate predators mainly occurs through visual and olfactory cues, which are strongly affected by plant morphology and chemical traits (the mechanisms). Plant size and architectural complexity are key determinants of foraging efficiencies of parasitoids, with increases in size and complexity reducing per capita foraging efficiency, while increasing the diversity of alternative prey (Tao et al., 2017). Plants recruit predators and parasitoids through indirect defense mechanisms, such as food rewards (e.g., extrafloral nectars), shelters (domatia), and volatile organic compounds (VOCs) that can attract predators and parasitoids over long distances. Because allocation to domatia and extra-floral nectaries depend on both nutrient status and plant size, microbial root mutualists are likely to influence the quality of such food and shelter rewards for predators and parasitoids. Notably, microbial root mutualists receive sugar from their plant hosts, in exchange for mineral nutrients and water. The carbon costs associated with hosting microbial mutualists may explain why some plants reduce their rewards to enemies aboveground when associated with root microbes. For example, in Vicia faba, association with AMF reduces extrafloral nectaries, which may result in reduced protection by ants against herbivores.

Belowground mutualists may affect the composition of plant VOCs aboveground through multiple mechanisms. First, mutualistic microbes may alter the production and emission of plant VOCs by modifying nutrient availability. For example, higher nutrient (nitrogen, phosphorus and potassium) concentrations in Eucalyptus tereticornis increase emission of the volatile terpene 1,8-cineole, which subsequently attracts more predators and parasitoids. Microbial root mutualists alter the nutrient and toxin concentrations of herbivore tissues, providing an additional mechanistic pathway by which root mutualists mediate tri-trophic interactions aboveground. Prey quality is important in determining the fitness of predators and parasitoids. Compared to herbivorous insects, predatory arthropods and parasitoids have higher body nitrogen (N) and P contents, so increases in plant nutrition can lead to greater performance of predators and parasitoids (Maure et al., 2016). Therefore, by affecting plant nutritional status, root microbes belowground can indirectly affect predators and parasitoids aboveground. Root associates may also alter the quality of prey for natural enemies by their impacts on plant secondary chemicals, which occur both passively in the hemolymph and midgut of herbivores, or may be sequestered in herbivore tissues. For example, plant glucosinolates occurring within herbivore prey negatively affect a wide range of parasitoids.

Soil organisms affect Plant Pathogens: Many of the same mechanistic pathways (chemical and physical traits) by which microbial root mutualists impact the efficacyof predators and parasitoids may also affect the pathogens of herbivores (Tao et al., 2017). Herbivorous insects are host to a wide diversity of disease agents, including protozoans, bacteria, and viruses. As with parasitoids and predators, the performance of herbivore pathogens is affected by both plant nutritional and secondary chemicals, and therefore influenced by belowground root mutualists. However, the effects of increased concentrations of nutritional chemicals on pathogens are not as readily predicted as they are for other types of natural enemy. This is because increases in plant nutritional quality can result in increased resources for pathogens, but can also result in improved host immunity. Therefore, when associations with soil mutualists result in higher plant nutrient concentrations, any subsequent increases in rates of pathogen replication may be counteracted by concomitant increases in host immunity.

Soil mutualists affect multiple plant traits simultaneously, generating multiple mechanistic pathways by which root microbes influence tri-trophic interactions above ground. As with other enemies, the overall impact of root mutualists on disease dynamics will depend on the relative strength of each mechanistic pathway. Following the example of the monarch butterfly and its protozoan parasite described above, while cardenolides (secondary chemicals) reduce parasite numbers and increase the lifespan of infected butterflies, macronutrients such as N and P also increase monarch performance. Since associations with AMF change macronutrients and cardenolides simultaneously in milkweed leaves (Tao et al., 2016), the net effects of AMF on butterflyparasite interactions are best explained by the combined changes in milkweed P and cardenolide concentrations. Net effects on monarchs vary from positive, through neutral, to negative, depending on how particular milkweed species respond phenotypically to their root microbial mutualists. Belowground biota have important effects on aboveground tri-trophic interactions. However, this topic remains in its infancy and many questions remain unresolved.

6.7 Antibiosis

Antibiosis is a process of biological interaction between two or more organisms that is detrimental to at least one of them; it can also be an antagonistic association between an organism and the metabolic substances produced by another. The term antibiosis, where anti means against and bio means life, as a whole it means against life and this term was introduced as a descriptive name for the phenomenon shown by antibacterial drugs by bacteriologist Vuillemin. It was the first time when antibiosis was studied in 1877 when Louis Pasteur and Robert Koch observed that an airborne bacillus had inhibited the growth of Bacillus anthracis. Various examples of antibiosis observed, but one of the most popular examples is the relationship between antibiotics and bacteria. A wide variety of chemical compounds could be involved in this process, including antimicrobial substances, bacteriocines, fungistatic compounds, surfactants, enzymes (Ex: chitinolytic etc.). Bacteria and fungi are the main actors responsible for the production of antibiotic and antagonistic compounds, and this process is mainly active against other microorganisms. Examples of antibiosis include the relationship between antibiotics and bacteria or animals and disease-causing pathogens.

Through antibiosis, microbially produced antibiotics are ultimately able to control plant pathogens and pests, and the damage they might cause to plants. Some antibiotics and secondary metabolites produced by microorganisms are used against other microorganisms and can also activate induced resistance in the plants, and/or inhibit other non-target organisms. Antibiosis can be estimated by measuring microbial antibiotic activity with lab-based functional antibiosis bioassays (Dundore-Arias et al., 2020) or with the agar well diffusion assay. Antibiosis can also be assessed through the measurement of enzymatic activities such as chitinolytic activity. Other methods rely on molecular quantification of functional genes and transcripts involved in antibiotic production with qPCR and RT-PCR, respectively (Imperiali et al., 2019).

There is abundant prior published information on antibiosis, one of the most studied biocontrol mechanisms for plant pathogens. Depending on their concentration, antibiotics can have various effects on target organisms, a phenomenon known as hormesis (Arseneault Tanya and Filion Martin, 2017). Under complex soil conditions where subinhibitory concentrations of antibiotics are thought to prevail, the mechanism of action responsible for disease reduction through antibiosis is often overlooked, where it is generally assumed that biocontrol occurs through mortality of the pathogen. This concept of dose-dependent response must be taken into account to better understand antibiosis and how it can contribute to biocontrol in various ways. This review aimed to focus on how antibiotics can operate and persist in soil, act as signalling molecules and enable interactions between soil microbial communities. It also aimed to pinpoint specific examples where low, subinhibitory concentrations of antibiotics, which are widespread under natural soil conditions, can reduce disease symptoms by modulating the pathogen's transcriptome, rather than by toxicity and death. This highlighted the need to better understand and characterize as much as possible the mode of action of antibiosis under various complex environmental conditions, in order to anticipate future development of resistance and loss of efficiency through changes in environmental conditions.

Until now, knowledge on the mechanisms of action in biocontrol agents includes antibiosis, production of lytic enzymes, parasitism, competition for nutrients and space and induction of resistance (Hernández et al., 2006). According to studies conducted to elucidate the mechanisms of action of antagonistic microorganisms, the production of antibiotics and competition for nutrients are the most prominent.

6.7.1 Secondary metabolites and Antibiosis potential of *Bacillus* sp. against *Sclerotium rolfsii*:Antibiosis is an antagonistic interaction incorporating low molecular weight diffusible secondary metabolites. In this, the antibiotics get released by microorganisms that are detrimental to the growth of pathogens. *Trichoderma* spp. produce a variety of antibiotics (Fig. 41), such as gliovirin, gliotoxin, viridin, viridol, koninginins, pyrones, and peptaibols, against fungal phytopathogens (Narendra Kumar and Paul Khurana, 2021). Trichokonins VI, a type of peptaibol from *Trichoderma pseudokoningii* SMF2, exhibited antibiotic activities by inducing extensive apoptotic programmed cell death in fungal pathogens. Gliovirin and Gliotoxin are the two most important *Trichoderma* secondary metabolites belonging to Q and P group strains, respectively. "P group strains of *Trichoderma* (*Gliocladium*) *virens* is active against *P. ultimum*, but not against *R. solani* and Q group is very active against *R. solani*, but less against *P. ultimum*". "The *T. virens veA* ortholog *vel*1 (VELVET protein Vel1) is involved in regulation of gliotoxin biosynthesis, biocontrol activity and many other secondary metabolism-related genes". "Koninginin D also inhibited the growth of soil-borne plant pathogens, such as *Rhizoctonia solani, Phytophthora cinnamomi, Pythium* *middletonii, Fusarium oxysporum* and *Bipolaris sorokiniana*". Viridins obtained by diverse *Trichoderma* spp. (*T. viride, T. koningii, T. virens*) inhibit spore germination **of** *Penicillium expansum, Botrytis allii, Fusarium caeruleum, Colletotrichum* ***lini,*** *Stachybotrys atra*, and *Aspergillus niger*. Harziaric acid obtained through the *T. harzianum* strain depicted in vitro antibiotic action against *Sclerotinia sclerotiorum, R. solani*, and *Pythium irregulare*. The isolation of trichodiene synthase (*tri5*) and its overexpression in the *T. brevicompactum Tb41tri5* transformant have enhanced the trichodermin released and thereby increased antifungal action against *Fusarium* spp. and *Aspergillus fumigatus*. This *Trichoderma* genus covers a group of filamentous fungi which is living freely. These species act as BCAs (biological control agents) against many pathogenic fungi. This power is the outcome of several mechanisms acting synergistically in these organisms, which in turn involves the release of secondary metabolites. They produce peptaibols, which belongs to the antibiotic peptide group. They have present nonproteinogenic amino acids such as Aib (α-aminoisobutyrate) along with N-terminal modifications and amino alcohols in the C-terminal region. The strains of *T. asperellum* resulted in production of five trichotoxins (T5D2, T5E, T5F, T5G, and 1717) and two asperelines (A and E) which are antibiosis in nature.

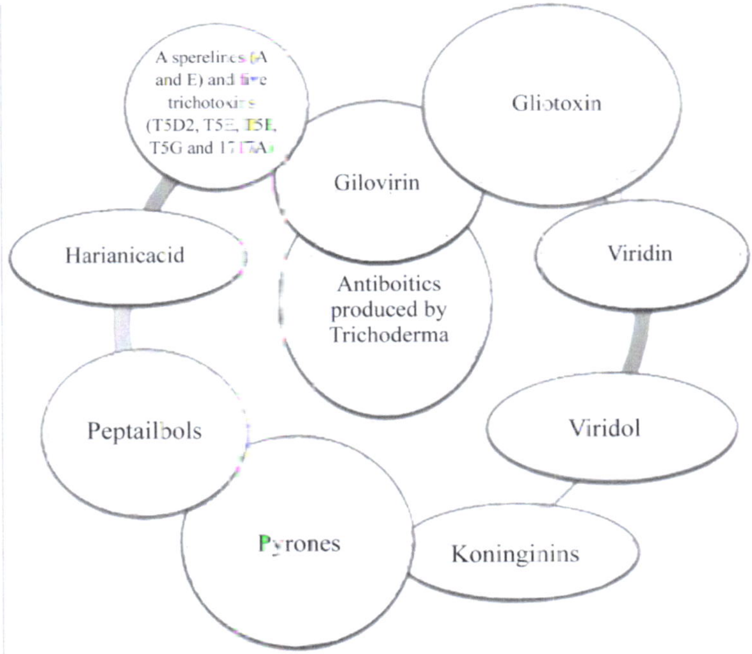

Fig. 41: Antibiotics produced by *Trichoderma* species

Antibiosis potential of ***Bacillus*** sp. against ***Sclerotium rolfsii*:** This study by Kumari, et al., (2021) aimed to investigate the rhizosphere bacterial isolates' antagonistic property against the soil-borne fungal phytopathogen, *Sclerotium rolfsii*. Biological control treatment of plant diseases utilizing antagonistic plant growth-promoting rhizobacteria (PGPR) gives a moderately effective, inexpensive, and environmentally secure alternative over the conventional chemical pesticides. One added advantage of these microbial control agents is that besides their antagonistic effect against fungal pathogens, they often harbor plant growth-promoting activities (siderophore production, ammonia production, HCN production, etc.), making the agents of choice to the growers. Among these, the bacterial species belonging to *Pseudomonas*, *Azospirillum*, *Klebsiella*, *Enterobacter*, *Alcaligenes*, *Arthrobacter*, *Burkholderia*, *Bacillus*, and *Serratia* genera are harboring plant growth-promoting activities (PGPA) (Kour et al. 2019). The species belonging to *Bacillus*, *Streptomyces*, *Pseudomonas*, *Burkholderia*, and *Agrobacterium* are a few of the extensively studied ones as far as the PGPR activities are concerned. Among these, the *Bacillus* and *Pseudomonas* genera are particularly crucial for their antibiosis capability against major plant fungal pathogens.

Various *Bacillus* species have been extensively studied, and a variety of antimicrobial cyclic lipopeptides such as iturins, fengycins, and surfactants playing an essential role in the fungal antagonism have been reported to be synthesized by them. The production of various hydrolytic enzymes viz. amylase, cellulase, protease, lipase, and chitinase by the *Bacillus* strains is also a part of this antagonistic activity, apart from the production of endospores and antibiotics (Jangir et al. 2019). Certain *Bacillus* species have also been reported to secrete volatile and non-volatile antimicrobial agents, which, in effect, inhibit the development of a variety of fungal infections. As part of the plant defense mutualism, *Bacillus* species secrete phenolic compounds, which are the secondary metabolites responsible for inhibiting mycelial growth. In the context of this background, the present study was focused on the isolation and screening of rice rhizospheric bacteria for potential antagonistic activity against *S. rolfsii* and their eventual evaluation for the synthesis of different antifungal metabolites, hydrolytic enzymes, and phenol producing capability.

Five rhizospheric *Bacillus* species viz. *B. subtilis* subsp. *Subtilis* str.168 (accession no. MH283878), *B. siamensis* strain PDA 10 (accession no. MH283879), *B. amyloliquefaciens* strain 1034 (accession no. MH283880), *B. velezensis* strain FZB42165 (accession no. MH283881), and *B. atrophaeus* strain NBRC 15539 (accession no. MH283882) were assessed for their antagonistic potential against *S. rolfsii* based on 3 different screening methods. Among these, 100% fungal growth inhibition by all 5 *Bacillus* spp. was observed in the novel ring method, whereas in the dual culture method, the maximum growth inhibition was (58%) exhibited by the strain NBRC 15539 of *B. atrophaeus*. The antagonistic activity showed by the modified dual culture method was also relatively high, and the highest activity (93.7%) was shown by the strain NBRC 15539 of *B. atrophaeus*. Besides, the *Bacillus* sp. was also evaluated for their plant growth-promoting attributes and other properties such as the production of siderophore, HCN, amylase, protease, lipase, and ammonia, including their assessment for chitinase and cellulase activity.

The study demonstrated the antifungal properties of *Bacillus* spp. isolated from rhizospheric soil of healthy rice crop plants against the pathogenic fungi S. *rolfsii*. All the five strains (CBK 2007 (*B. subtilis* subsp. *Subtilis* str.168), CBK 3005 (*B. siamensis* strain PDA 10) CBK 5001 (*B. amyloliquefaciens* strain 1034), CBK6004 (*B. velezensis* strain FZB42165), CBK 17001 (*B. atrophaeus* strain. NBRC 15539)) were found to exhibit antagonistic effects against fungal pathogen (Kumari, et al., 2021). Among all the five *Bacillus* spp., *B. atrophaeus* strain NBRC 15539 (accession no. MH283882) showed maximum inhibition of *S. rolfsii* by all three methods. Evidences suggest that the bacterial isolates identified in the present study can be used as an effective biocontrol agent solely or in combination for control of diseases caused by *S. rolfsii*. These strains can substitute certain chemical fungicides and inorganic fertilizers, making the agricultural production systems economically and environmentally sustainable.

6.7.2 Antibiotic production as mechanism of control: The antagonistic microorganisms (bacteria, yeasts and fungi) are able to provide biologic control against plant different pathogens, and are used to control several fruit and vegetables infections. To select antagonistic microorganisms, the general characteristics to be considered include, the ability to colonize the surface of vegetables and persist in the site effectively; better ability than the pathogen to obtain nutrients ; and the ability to survive under different environmental conditions (Ulloa-Ogaz et al., 2015). There are many other specific characteristics of the antagonist organism, including its genetic stability, that can be effective at low concentrations, should not require special nutritional elements; it has to be effective for a wide range of pathogenic microorganisms in a variety of fruits and vegetables. It is important also that the microorganism can be produced in large quantities in an economical growth medium, that can remain viable in the formulation for a long period of time, it needs to be easy to apply, should not produce secondary metabolites that cause damage to human health. It is also important that the microorganism is resistant to fungicides and can support commercial and nonpathogenic processes on the host plant. To properly use biocontrol microorganisms, it is important to understand the mechanisms of action involved in biocontrol activity, to develop safe application processes; this is also an important background to select new and efficient strains.

Basic information must be generated at both, the biochemical and the molecular level, contributing in this way in the elucidation of effects such as antibiosis, competition for nutrients and induction of resistance. It is desirable to propose the combined effect of various antagonists in order to assure and adequate disease control, with a reduction in dose and using the least amount of synthetic products; the use of additives to enhance the antagonistic effect of biocontrol microorganisms can also be considered. Until now, knowledge on the mechanisms of action in biocontrol agents includes antibiosis, production of lytic enzymes, parasitism, competition for nutrients and space and induction of resistance According to studies conducted to elucidate the mechanisms of action of antagonistic microorganisms, the production of antibiotics and competition for nutrients are the most prominent.

Antibiotic production: Multiple interactions between the diverse groups of soil organisms are common, including predation and competition for resources, where chemical substances are used (Ulloa-Ogaz et al., 2015). Some of these chemicals found in the soil and that have effect on other organisms, can be used for medicinal purposes as antibiotics, including Penicillin, isolated from the penicillin fungus which is found in soils ; Aminoglycosides, such as streptomycin and kanamycin, as well as tetracyclines were isolated from soil dwelling actinomycetes ; Lipopeptides such as daptomycin have also been derived from Steptomyces, a type of actinomycete. These bacterial antagonists enforce suppression of plant pathogens by the secretion of extracellular metabolites that are inhibitory at low concentration. Antibiotics produced by plant growth promoting rhizobacteria (PGPR) include 2,4 Diacetyl phloroglucinol, phenazine-1-carboxyclic acid, phenazine-1-carboxamide, pyoluteorin, pyrrolnitrin, oomycinA, butyrolactones, kanosamine, zwittermycin-A, aerugine, rhamnolipids, cepaciamide A, pseudomonic acid, azomycin, antitumor antibiotics FR901463, cepafungins and antiviral antibiotic karalicin. All these antibiotics have antiviral, antimicrobial, insecticidal, antihelminthic, phytotoxic, antioxidant, and cytotoxic effects, and can also produce a plant growth promoting effect.

Each of these antibiotics have a different mode of action, some attack the cellular membranes; others have inhibitory effects on the ribosome or other cellular constituents. This is the reason why some organisms are susceptible to some antibiotics but not others, depending on the specific form of cellular constituent the antibiotic attacks. One well-known example of production of antibiotics is pyrrolnitrin, a natural product produced by some *Pseudomonas* spp.; from this compound, the fungicide fludioxonil can be derived. This fungicide is used for seed treatment, foliar spray, or soil drench. A variety of pathogenic and non-pathogenic microorganisms can induce plant defenses and may be useful as biocontrol agents. Two other common antibiotics, streptomycin and oxytetracycline, are registered in the United States for control of fire blight, a disease of pear and apple caused by *Erwinia amylovora.* Streptomycin was an effective chemical compound used for the management of fire blight until some pathogenic strains resistant to the antibiotic appears in several growing regions. Oxytetracycline is registered for use only on pear and is considered less effective than streptomycin for suppression of antibiotic-sensitive populations of *E. amylovora*.

Two of the mechanisms usually cited for this microbial biocontrol are,

i. Antibiosis action, through the production of specific or non-specific metabolites with antibacterial, antifungal and antinematode activity. Several antibiotic substances produced by the pseudomonads have been particularly well characterized. The ones identified with biocontrol properties include the phloroglucinols, phenazine derivatives, pyoluteorin, pyrrolnitrin, cyclic lipopeptides and hydrogen cyanide. Among the other antibiotics characterized are agrocin 84 (*Agrobacterium* sp.), herbicolin A *(Erwinia* sp.), iturin A, surfactin, and zwittermicin A (*Bacillus* sp.) and xanthobacin (*Stenotrophomonas* sp.)
ii. Lytic enzyme action, is a feature of several bacteria with biocontrol ability, and involves the direct degradation of the pathogen cell wall material, or the

disruption of a particular developmental stage. For example, chitinase production by *Serratia plymuthica* has been reported that inhibits spore germination and germ-tube elongation in *Botrytis cinerea*, while ß- 1,3-glucanase synthesized by *Paenibacillus* sp. and *Streptomyces* sp. can lyse fungal cell walls of *Fusarium oxysporum* f. sp. *cucumerinum*. Other enzymes produced by bacteria with biocontrol activity include hydrolase, laminarinase and protease. Antibiotics can cause intense physiological effects on organisms at subinhibitory concentrations. Quinolone and macrolide antibiotics have been reported to block cell signaling, and the production of virulence factors in *P. aeruginosa*. Subinhibitory concentration of antibiotics can suppress adherence mechanisms and the production of extracellular virulence factors in bacteria. Secondary metabolites can impact soil microbial ecosystems in a variety of ways, and at a variety of levels. The production of antibiotics has been demonstrated to be a widespread mechanism exerted by microorganisms to control a wide variety of phytopathogens. Because of all these properties, advanced molecular techniques are now being used to characterize the diversity, abundance, and activities of microbes that live in and around plants, including those that significantly impact plant health.

6.7.3 Antibiotic Resistance: Some microorganisms that are not susceptible to a particular antibiotic, can be the initial source of a resistant strain over time (Ulloa-Ogaz et al., 2015). This represents a public health problem in human pathogens, but also antibiotic resistance seen in the clinical environment can often be found in the soil environment. The reason of this phenomenon is that soil microorganisms are often exposed to a wide range of compounds in their local environment, some of which may be harmful such as antibiotics, and this causes that these organisms develops resistance or go extinct (Beneduzi et al., 2012). For example, antibiotic producers must contain some antibiotic resistance mechanisms, to prevent them committing suicide through production of their own toxin. Soil environment is very important for research into the mechanisms of antibiotic resistance, including possible mechanisms which are not yet seen in clinical microbiology. The biochemical processes occurring in antibiotic resistance is just starting to gain much interest. Evolution has even taken antibiotic resistance one step further; it has been demonstrated that out of 18 antibiotics tested from 8 major classes of antibiotic of both natural and synthetic origin, 13 to 17 supported the growth of bacteria when the antibiotic was available as sole carbon source. Microorganisms are highly adaptable, in ways which we are only recently coming to understand. Antibiotic resistance occurs when antibiotics are a constant pressure on a given population; those organisms with natural resistance can survive and reproduce whereas those organisms which do not have the resistance factor die. Once a resistance factor has developed, it can be spread rapidly within a population where DNA is transferred from one bacterium to another. Transfer of DNA containing antibiotic resistance genes can occur through three processes, viz., Transformation, Transduction and conjugation.

There are numerous reports of the production of antifungal metabolites produced by bacteria in vitro that may also have activity in vivo (Ulloa-Ogaz et al., 2015).

These include ammonia, butyrolactones, 2,4- diacetylphloroglucinol (Ph1), HCN, kanosamine, Oligomycin A, Oomycin A, phenazine -1- carboxylic acid (PCA), pyoluterin (Plt), pyrrolnitrin (Pln), viscosinamide, xanthobaccin, and zwittermycin A as well as several other uncharacterized compounds. Mutants lacking production of antibiotics or over-producing mutants have been used to demonstrate the role of antibiotics in biocontrol. Alternatively, the use of reporter genes or probes to demonstrate production of antibiotics in the rhizosphere is becoming more commonly used. Antibiotic production by bacteria, particularly pseudomonads, seems to be closely regulated by a two-component system involving an environmental sensor (a membrane protein) and a cytoplasmic response factor. Mutation in either gene has similar multiple effects on antibiotic production.

Antibiotic resistance most commonly evolves in bacteria either through mutation of a targetsite protein, through the acquisition of an antibiotic-resistance gene (ARG) that confers resistance through efflux or inactivation of the antibiotic, or through synthesis of a new target protein that is insensitive to the antibiotic (George W. Sundin and Nian Wang, 2018). An extensive body of knowledge has been gained from studies of antibiotic resistance in human pathogens and in animal agriculture. The ability of bacterial pathogens to acquire ARGs and to assemble them into blocks of transferable DNA encoding multiple ARGs has resulted in significant issues that affect successful treatment interventions targeting some specific human infections. The current global antibiotic resistance crisis in bacterial populations has been fueled by basic processes in microbial ecology and population dynamics, engendering a rapid evolutionary response to the global deployment of antibiotics by humans in the millions of kilograms per year. What was not anticipated when antibiotics were discovered and introduced into clinical medicine is that ARGs preexisted in bacterial populations. Furthermore, the extent to which ARGs could be transferred between bacteria, and even between phylogenetically distinct bacteria, was not understood 70 years ago but is becoming more apparent through a number of elegant studies identifying the microbial antibiotic resistome. The collection of all known ARGs in the full-microbial pan-genome is defined as the antibiotic resistome. What is most important conceptually about the antibiotic resistome is the potential accessibility of individual ARGs to all bacteria.

Antibiotic resistance in plant-pathogenic bacteria has become a problem in pathosystems where these antibiotics have been used for many years. Where the genetic basis for resistance has been examined, antibiotic resistance in plant pathogens has most often evolved through the acquisition of a resistance determinant via horizontal gene transfer (George W. Sundin and Nian Wang, 2018). For example, the strAB streptomycin-resistance genes occur in *Erwinia amylovora, Pseudomonas syringae,* and *Xanthomonas campestris*, and these genes have presumably been acquired from nonpathogenic epiphytic bacteria colocated on plant hosts under antibiotic selection. We currently lack knowledge of the effect of the microbiome of commensal organisms on the potential of plant pathogens to evolve antibiotic resistance. Such knowledge is critical to the development of robust resistance management strategies to ensure the safe and effective continued use of antibiotics in the management of critically important diseases.

Native soil microbes hinder the soil enrichment with Antibiotic Resistance Genes: Organic wastes such as municipal sewage sludge and animal manure, containing high levels of antibiotic-resistant bacteria (ARB) and antibiotic resistance genes (ARG), are commonly used for agricultural fertilization (Pérez-Valera et al., 2019). Manure application to land represents one of the main routes for spreading ARB along with ARG to the environment. Although the survival of ARB in manured soils can be limited, it has been shown that ARG are able to persist in manured soils in the long term. There has been an increasing concern about the accumulation of ARGs in soils as they may be exchanged between soil bacteria and human pathogens. ARG are hence nowadays recognized as an emerging environmental pollutant. However, the factors contributing to the ARG dissemination and persistence in soil are not well understood yet.

Bacterial genes responsible for resistance to antibiotic agents (ARG) are spread from livestock to soil through application of manure, threatening environmental and human health. Pérez-Valera et al. (2019) investigated the mechanisms of ARG dissemination and persistence to disentangle i) the influence of nutrients and microorganisms on the soil tetracycline (TET) resistome, and ii) the role of indigenous soil microbiota in preventing ARG spread. short-term (7 days) and persistent (84 days) effects of manure were analysed on the resistome of three antibiotic-free pasture soils. Four microcosm treatments were evaluated: control, mineral nutrient fertilization, and deposition of a layer of fresh manure onto soil or γ-irradiated soil. Then quantified five TET-resistance genes, isolated 135 TET-resistant bacteria and sequenced both culturable TET-resistant and whole bacterial communities. Manure amendments, but not nutrient addition, increased the abundance of TET-r genes such as *tet* (Y). Such changes persisted with time, in contrast with the TET-resistant bacterial composition, which partially recovered after manure amendments. Manured γ-irradiated soils showed significantly lower nutrient content and higher TET-r gene abundance than non-irradiated soils, suggesting that native soil bacteria are essential for the fertilization effect of manure on soil as well as control the dissemination of potentially risky TET-r genes.

The impact of manure on total bacterial communities at the manure-soil interface was negligible in the case of native soils, compared to that in γ-irradiated soils. In addition, the identified TET-resistant subcommunity represented only 18% of total bacterial community of the native soil at T7, estimated as the relative abundance of OTUs shared between the two types of communities, while it accounted for as much as 69% of the total community in the case of γ-irradiated soil. This data suggests that the indigenous soil bacteria in the upper soil layer may act as a barrier against TET-resistant species from manure. This applies also on TET-r genes as the content of *tet*(Y) and *tet*(Q) at the manure-soil interface of γ-irradiated soil was higher, compared to native soils. Soil microorganisms, therefore, may hinder the entrance of certain TET-r genes from manure to the upper soil layer, most likely by preventing the establishment of their bacterial hosts. Comparable results were found for mixtures of manure or biogas digestates with soil in microcosm experiments suggesting that increased competition for nutrients, antibiosis or smaller niche availability prevent the establishment of TET-resistant or pathogenic bacteria from the applied organic fertilizers (Chen, et al., 2017).

Although the bacterial hosts of *tet*(Y) in fresh manure could not be identified, several strains of *Acinetobacter*, *Escherichia*, *Pelosinus* and *Rhizobium* were confirmed as hosts of *tet*(Y) at the manure-soil interface at 7 days. According to our sequencing data, at this time *Acinetobacter* was rare ($<0.1\%$) within the TET-resistant subcommunity but became relevant (ca. 3.4%) in the total community in γ-irradiated soils. *tet*(Y) showed a similar trend, suggesting that *Acinetobacter* could disseminate TET-r genes in soil (Pérez-Valera et al., 2019). The results showed that manure, but not the mineral nutrients in a similar concentration, increased the soil levels of TET-r genes over 3 months. Higher levels of TET-r genes and lower fertility in soils with reduced biological viability indicate that indigenous soil microorganisms are important for both improving the fertility and limiting the spreading of TET-r genes in soils amended with manure.

Exploring the role played by subinhibitory concentrations of Antibiotics in soil and their impact on plant pathogens: The quest to understand the roles and mechanisms of action of antibiotics under natural soil conditions, particularly with regards to the control of crop plant diseases, remains a challenge (Tanya Arseneault and Martin Filion, 2017). Access is available to ever-evolving analytical and molecular tools which will enable us to better characterize dose-dependent responses to antibiotics in plant pathogens and identify the associated global transcriptional patterns, providing clues as to the metabolic pathways targeted. This will be essential in order to better anticipate the future effects of a biocontrol treatment reliant upon antibiotic production, in particular with regards to the development of resistance and the looming threat of climate change which will modulate soil and environmental conditions. Gaining knowledge on how specific antibiotics can modulate pathogen virulence under agricultural soil conditions is the first step that will enable the filling in of pieces of the larger picture that constitutes soil health and suppressiveness to plant diseases in agroecosystems.

There is abundant prior published information on antibiosis, one of the most studied biocontrol mechanisms for plant pathogens.Depending on their concentration, antibiotics can have various effects on target organisms, a phenomenon known as hormesis. Under complexsoil conditions where subinhibitory concentrations of antibiotics are thought to prevail, the mechanism of action responsible for diseasereduction through antibiosis is often overlooked, where it is generally assumed that biocontrol occurs through mortality of the pathogen (Tanya Arseneault and Martin Filion, 2017). This concept of dose-dependent response must be taken into account to better understand antibiosis and how it can contribute to biocontrol invarious ways. This review aims to focus on how antibiotics can operate and persist in soil, act as signalling molecules and enable interactionsbetween soil microbial communities. It also aims to pinpoint specific examples where low, subinhibitory concentrations of antibiotics, whichare widespread under natural soil conditions, can reduce disease symptoms by modulating the pathogen's transcriptome, rather than by toxicityand death. This highlights the need to better understand and characterize as much as possible the mode of action of antibiosis under variouscomplex environmental conditions, in order to anticipate future development of resistance and loss of efficiency through changes inenvironmental conditions

Antibiotics are low-molecular weight compounds that caninclude volatile compounds, lytic agents, enzymes and othertoxic compounds. Antibiotics are 'inhibiting the growth or the metabolic activ-ities of bacteria and other microorganisms by a chemical substance of microbial origin'. A vital part of understanding the effects of antibiotics on a target microorganism is accounting for the concentration to which it is exposed. It has been shown that antibiotics that are applied at a concentration below the deadly minimal inhibitory concentration (MIC) can increase or decrease transcriptional gene expression of up to 5% of bacterial promoters and can specifically target certain processes such as colonization capacity and virulence.

6.7.4 Sources of Antibiotics, their fate and functions in soil: The production of antibiotics requires adequate and sufficient sources of carbon, nitrogen and other nutrients. The relatively poor nutrient content of bulk soils results in lowmicrobial metabolic activity, whereas the higher availability of substrates in the richer regions of the rhizoplane and the rhizosphere make them likely areas for antibiotic production, in addition to other metabolic activities (Thomashow et al., 2008). The rhizosphere is also one of the most dynamic and complex environments hosting several multitrophic interactions. In this environment, plants themselves produce rhizodeposits which can contain fungicidal and bactericidal com-pounds, such as phytoalexins, to select particular microbial communities which are able to thrive despite the presence of these molecules. Other sources of antibiotics in the environment include pharmaceutical antibiotics from animal or human waste. However, the major producers of antibiotics in the rhizosphere are the microorganisms present, whether they are pathogenic or beneficial.

Antibiotic production is part of the secondary metabolism of microorganisms, which produce metabolites that are not essential for their growth, but can have implications in survival, fitness and competitiveness. It occurs during late growth phases, and as such is dependent on environmental conditions and requires a critical mass of signal-producing cells(quorum-sensing) for its initiation. This suggests that even in an environ-ment favouring antibiotic production, such as the rhizosphere, production is limited to specific areas or pockets where all the conditions and the threshold populations needed are attained. This means that the concentration of a particular antibiotic can vary greatly, even within a distance of a few millimetres of roots, depending on the microorganisms that are present in a specific area (Tanya Arseneault and Martin Filion, 2017).

In the context of disease control, it has been shown that soils which are suppressive to disease often owe their activity to the presence of an array of plant-beneficial microorganisms with diverse microbial activities, including nutrient cycling, degradation of harmful compounds, but also and most importantly, production of antibiotics. As stated above, specific suppressiveness, or biological control, of particular diseases, has also been associated with the production of antibiotics by beneficial microorganisms. The identity of these antibiotics has become clearer with time, and include a diverse arrayof molecules and producing microorganisms, such as phenazines (phenazine-1-carboxylic acid (PCA),phenazine-1-carboxamide (PCN), pyocyanin (PYO)), 2,4-Diacetylphloroglucinol (DAPG), iturin A, bacillomycins, gliotoxins, pyoluteorins, pyrrolnitrins, etc. Several genera of microorganisms have also been recognized as multiple producers of antibiotics, particularly bacteria in the genus Pseudomonas.

The fate of antibiotics in soil: A major influence on the lifespan of antibiotics in soil is their degradation by soil microorganisms, which in some cases can even utilize them as nutrients. Inactivation might also occur through adsorption to soil particles depending on pH, water content, solubility and the composition of soil in clay, sand, silt, minerals and organic matter (Wegst-Uhrich et al., 2014). This results in relatively low half-lives (a few days) for some known antibiotics of biocontrol interest, highlighting the importance of consistent production to ensure their sufficient presence in soil to be effective against plant pathogens. While several studies following the fate of clinical antibiotics have been done to estimate their persistence in the environment, few have analysed the fate of antibiotics produced by soil microorganisms involved in plant disease control. The persistence of antibiotics in soil is evidently dependent on their chemical structure and physicochemical properties, as it affects their mobility and sorption in soil and water. Furthermore, adsorption to soil particles does not always inactivate antibiotics, as has been shown for two antibiotics used in the animal production industry (tetracycline and tylosin), which remained active despite being strongly adsorbed to clay particles. This highlight show information about antibiotic persistence in soil has important implications with regards to its role in the soil ecosystem.

Functions of antibiotics in soil: Beyond the toxic effects of antibiotics observed at high concentrations, it has become clear that at subinhibitory concentrations, these molecules can operate in diverse manners. Under soil conditions, where the concentration of antibiotics is very likely below the MIC, antibiotics are thought to have natural functions which are mostly involved in signalling and microbial community interactions as well as communication with plants (Venturi and Keel, 2016). For many antibiotics, these exact signalling roles remain a mystery, but there are some for which some information has surfaced. Fluorescent *Pseudomonas* spp. offer many examples of these signalling functions. Studies have shown that pyoluteorin and DAPG act as signals for their own auto-induction. Furthermore, a recent study has discovered that an intermediate in the production of DAPG, phloroglucinol, has a broad effect on the transcriptome of *P. protegens Pf5* (over 200 genes) depending on its concentration, and also regulates the production of pyoluteorin. In addition to modulating microbial responses, DAPG (along with Fusarium-produced zearalenone) has also been shown to affect plant roots, significantly altering the production (both in quantity and composition) of root exudates in several crops, which has ecological implications, particularly for soil microbial community composition. Other well studied Pseudomonas-produced antibiotics are the phenazines. These redox-active antibiotics have been implicated in the regulation of biofilm formation and one type of phenazine, pyocyanin, has been found able to bind DNA, perhaps suggesting a directrole in transcriptional modulation (Tanya Arseneault and Martin Filion, 2017).

6.7.5 Effects of Sulfonamide and Tetracycline antibiotics on soil microbial activity and microbial biomass Residual concentrations of pharmaceutical antibiotics exert a temporary selective pressure on soil microorganisms, which is clearly influenced by soil adsorption. To test time and concentration related effects, the fact that numerous antibiotics are biostatic and not biotoxic must be considered (Soren Thiele-Bruhn and Iris-Constanze Beck, 2005). Thus, long-term tests with a nutrient addition to stimulate microbial growth, such as the Fe (III) reduction test are suitable routine methods to test dose related effects of pharmaceutical antibiotics on soil microorganisms. The different effects of antibiotics on the soil microbial and fungal biomass show that further experiments are required, investigating the biodiversity of pharmaceutical affected soil microorganisms.

The determination of the microbial biomass C (EC) and ergosterol as indicators of the total microbial and fungal biomass, respectively, revealed different effects of SPY and OTC in the sandy Cambisol. After the addition of a nutrient substrate, the ergosterol concentration in the control sample increased during a 14-d incubation on average from 0.47 to 0.96 lg g1 , while the EC declined from 1190 to 450 lg g1. The addition of straw and glucose led especially to the growth of fungi. It has to be stated again that in this study, due to the high variability of the fungal ergosterol concentration, ergosterol was used as an indicator but not a measure of fungal biomass. Depending on the spiking concentration, SPY led to a reduction of the EC but did not affect the ergosterol concentration. In contrast, in the presence of OTC the ergosterol concentration strongly increased during the incubation time, whereas the EC did not significantly differ between amended samples and the unamended control.

Increasingly often soil residual concentrations of pharmaceutical antibiotics are detected, while their ecotoxic relevance is scarcely known. Thus, dose related effects of two antibiotics, sulfapyridine and oxytetracycline, on microorganisms of two different top soils were investigated. The fumigation-extracted microbial C (EC) and ergosterol were determined to indicate soil microbial and fungal biomass, respectively. Microbial activity was tested as basal respiration (BR), dehydrogenase activity (DHA), substrate-induced respiration (SIR), and Fe(III) reduction. The BR and DHA were uninfluenced even at antibiotic concentrations of 1000 lg g1. This revealed that an activation of microbial growth through nutrient substrate addition is required to test possible effects of the bacteriostatic antibiotics. In addition, the effects of both antibiotics were time dependent, showing that short-term tests were not suitable. Clear dose–response relations were determined with SIR when the short-term incubation of 4 h was extended into the growth phase of the microorganisms (24 and 48 h). The Fe(III) reduction test, with a 7-d incubation, was also found to be suitable for toxicity testing of antibiotics in soils. Effective doses inhibiting the microbial activity by 10% (ED10) ranged from total antibiotic concentrations of 0.003–7.35 lg g1 , depending on the antibiotic compound and its soil adsorption (Soren Thiele-Bruhn and Iris-Constanze Beck, 2005). Effective solution concentrations (EC10), calculated from distribution coefficients, ranged from 0.2 to 160 ng g1. The antibiotics significantly ($p < 0.05$) reduced numbers of soil bacteria, resulting in dose related shifts in the fungal: bacterial ratio, which increased during 14 d, as determined from analysis of ergosterol and EC. It was concluded that pharmaceutical antibiotics can exert a temporary selective pressure on soil microorganisms even at environmentally relevant concentrations.

6.7.6 Antibiosis of Siderophore producing bacterial isolates and the role of Antibiosis in the antagonism of different bacteria: Rhizobacteria have the capability to produce low molecular weight (500-1000 dt) metal chelating compound including iron, called as Siderophore. Siderophore chelate iron from mineral phases by formation of soluble Fe^{3+} complexes and make it available to the plant and enhance the plant (Pahari and Mishra, 2017). Generally siderophores are phnolic compound or citric acid derivatives and depending on the oxygen ligands for Fe^{3+} coordination, siderophores can be classified into three main categories i.e. hydroxamates, catecholates, and carboxylates growth. More over among the PGPR, siderophoreproducing rhizobacteria is the most promising biofertilizer which not only improve the plant iron nutrition but also protect the plant from different types of phytopathogens by chelating iron. Microbiologists have developed techniques for introduction of siderophore producing PGPR in soil system through seed, soil or root applications and enhance plant growth via suppression of phytopathogens. Hence, the use of siderophore producing rhizobacteria as a biofertilizer is one of the best modern tools for agriculture and it is a gift of our modern agricultural science.

Iron is one of the most essential micronutrient for any living cells but the availability of iron is limited due to very low solubility of the dominant ferric iron (Fe^{3+}) in soil. Bacteria can produce low molecular weight iron chelating compound called as siderophore. On account of that, an attempt was made in the present investigation to study the effect of siderophore producing rhizobacteria on growth of Okra and evaluation of biocontrol efficacy of the potential isolates against different phyto-pathogens in-vitro. It was observed that the siderophore producing isolates significantly increase the growth parameters like root length, shoot length and biomass of Okra which was also statistically significant and showed antagonistic effect against different phytopathogens including *Rahizoctonia solani* (ITCC-186) and *Fusarium oxysporum* (ITCC-578).

It was found that all the siderophore producing rhizobacterial isolates significantly increase the root length, shoot length and biomass of the plant and among them BGBA-1 showed highest activity. Highest root and shoot elongation was found in Okra when the Okra seeds were treated with BGBA-1. In case of biomass, maximum root biomass and shoot biomass gm was observed in BGBA-1 treatment in respect to control. Actually in the soil, plant roots normally coexist with bacteria that may produce siderophores capable of sequestering the available soluble iron, which could interfere with plant growth and function. However, plant roots are sometimes capable of taking up ferric complexes of siderophores and using these as sources of iron.

Siderophore producing bacteria also improve the chlorophyll content in plant because iron was absorbed in ferrous form (Fe2+), which was necessary for the formation of chlorophyll and functions in some enzymes of the plant's respiratory system. It was found that all the bacterial isolates increase the chlorophyll content of leaf in respect to control. Maximum total chlorophyll of fresh leaf was found in BRBA-1 treatment compared to the control.

The role of Antibiosis in the antagonism of different bacteria: Bacterial antagonists of *Helminthosporium solani* were submitted to different tests in order to determine the role of antibiosis in their antagonistic interaction. Among the bacterial strains tested, seven (*Alcaligenes piechaudii, Aquaspirillum autotrophicum, Cellulomonas fimi, Pseudomonas chlororaphis, Pseudomonas putida* (strains 94-19 and E-30) and *Streptomyces griseus*) were shown to produce agar diffusible metabolites inhibiting *H. solani* mycelial growth and/or conidial germination (Martinez et al., 2006). Differential activity was revealed when diffusible metabolites were extracted from either pure cultures of these antagonists or dual culture in the presence of *H. solani*. The results indicated that the methodology employed could be a decisive factor in whether or not antibiosis can be identified as a mode of action of biocontrol agents.

This study employed different methodologies to evaluate antibiosis in previously identified bacterial antagonists of *H. solani*. As such, solid or liquid cultures of the antagonists were carried out with or without the presence of the pathogen, and semipurified or crude bacterial culture extracts were tested for inhibition of both mycelial growth and conidial germination of *H. solani*. With this experimental design, antibiosis was assessed with regards to specific conditions governing the differential production of antifungal compounds from selected bacteria. In confrontation assays on solid media, including when precultures on cellophane membranes were used, *A. piechaudii, P. chlororaphis* and *S. griseus* inhibited both mycelial growth and conidial germination of *H. solani*, suggesting that antibiosis is a contributor to the antagonistic effect of these bacteria through the diffusion of antifungal compounds into the medium. Alcaligenes piechaudii, *P. chlororaphis* and *S. griseus* had previously been shown to produce antibiotics, though not against *H. solani*. *Arthrobacter oxydans, B. mycoides, K. varians, R. erythropolis* and *R. globerulus* did not inhibit mycelial growth or conidial germination of *H. solani*, suggesting that antibiosis is not a mechanism by which these bacteria antagonize the pathogen. Although *B. mycoides, K. varians* and *Rhodococcus* sp. had previously been shown to produce antibiotics, no antibiosis could be revealed with these species against *H. solani* in this study. In accordance with results presented herein, *A. oxydans* had not been previously reported to produce antibiotics.

It is possible that metabolites different from those implicated in mycelial growth inhibition would be involved in the inhibition of conidial germination. Indeed, *A. autotrophicum, C. fimi* and *P. putida* inhibited mycelial growth without affecting conidial germination while *B. cereus, K. rosea* and *P. fluorescens* demonstrated antibiosis against conidial germination of *H. solani*, but did not inhibit its mycelial growth. However, the possibility that *A. autotrophicum*, *C. fimi* and *P. putida* produced metabolites inhibiting conidial germination only in the presence of *H. solani* may not be excluded. Also not to be ruled out is the possibility that metabolic detoxification by *H. solani* occurred when the antifungal compounds from *B. cereus, K. rosea* and *P. fluorescens* were gradually produced in dual culture, whereas the high initial concentrations of antifungal metabolites (in the semi-purified fractions) to which conidia were exposed may have overwhelmed the detoxification mechanisms, thus revealing the antifungal effect of the compound(s) for these bacteria. Similar results were obtained with dual cultures of competing *Armillaria* species, where metabolites from pure cultures possessed inhibitory activity but were detoxified in dual culture.

The production of the antibiotic phenazine was six times higher in solid culture when compared with liquid culture in *Pseudomonas aureofaciens* (Martinez et al., 2006). It was observed that the production of antibiotic compounds may be influenced by the type of medium used (liquid or solid). Indeed, filtrates from pure liquid cultures of *A. piechaudii* and *P. chlororaphis* did not exhibit the antifungal activity on conidial germination they had displayed on solid media. This would indicate that liquid culture of these strains does not induce the production of antifungal compounds, or at least not in amounts that significantly reduce conidial germination.

Streptomyces hygroscopicus var. *geldanus* controlled rhizoctonia root rot of pea in previously sterilized soil if incubated for 2 or more days prior to infesting soil with *Rhizoctonia solani* and planting (Rothrock Craig and Gottlieb, David, 2011). *S.hygroscopicus* also reduced saprophytic growth and the population of *R. solani* in soil. Growth of *R. solani* was inhibited by geldanamycin, an antibiotic produced by *S. hygroscopicus,* on nutrient media. Methanol extracts of soils in which the antagonist was incubated for 2 or more days inhibited growth of *R. solani*. Geldanamycin concentration was 88 μg per gram of soil after 7 days of incubation. Bioautography of soil extracts indicated that the inhibitory compounds were geldanamycin and two other compounds, also found in the geldanamycin standard. The period of incubation necessary for antibiotic production and disease control was similar, with no disease control occurring where no antibiotic was detected. Amending soil with geldanamycin, in amounts equivalent to that produced after 2 or 7 days of incubation, controlled disease and reduced saprophytic growth of the pathogen. Lesser amounts of the antibiotic did neither. No evidence for antagonism owing to competition (nitrogen, carbon) or parasitism was found. *S.hygroscopicus and* geldanamycin also affected plant growth.

6.8 Fungistasis

Fungistasis refers to the inhibition of the growth and reproduction of fungi without destroying them. Fungal propagules are inhibited by most soils, a phenomenon first described as 'fungistasis' by Dobbs and Hinson (1953). Many factors, that affect the intensity of fungistasis including soil physical and chemical traits, environmental changes, fungal characteristics, the community composition, and the metabolic activities of other soil microbes, have been studied extensively over the last 50 years, and there are many explanations for the mechanism of soil fungistasis. The nutrient-deprivation hypothesis and germination inhibitor hypothesis are the two most popular. The former hypothesis states that exogenous energy sources, e.g. fixed carbon, is relatively unavailable to germinating spores due to strong competition from other soil microorganisms. The latter hypothesis states that fungistasis can be caused by antifungal compounds produced by other microorganisms. But until now, no definitive answers have emerged (Xu Chuankun et al., 2004). In recent years, significant data supporting microbial activity in soil fungistasis process has become available. For example, soil fungistasis can be reduced by soil sterilization or the addition of exogenous antibiotics. Soil microbial community composition was an important factor determining soil fungistasis, especially the pseudomonads, and is consistent with the germination inhibitor hypothesis of fungistasis.

Fungistasis in soil is a widespread phenomenon affecting most fungal propagules, though some are insensitive. In most instances, it is coexistent with the presence of living microorganisms, and is annulled by energy-yielding nutrients. Fungistasis with characteristics similar to that in soil may also occur on leaves of plants. Germination and growth of bacteria and actinomycetes is also restricted in soils. The characteristics of their inhibition appear to be the same as those for fungi. Therefore, the concept of a widespread microbial inhibition in soil can be applied to all three groups of microorganisms. Fungistasis can be detected by various direct methods, or indirectly by methods involving the use of porous or permeable carriers. It may be expressed as a restriction on the final amount of germination (the usual parameter), germination rate (with time), and rate of germ-tube or hyphal growth. Since the expression of fungistasis is often complete in soil, titration with nutrients may be required to distinguish between the sensitivities of different fungi. Fungistasis generally is expressed most strongly at soil moisture contents somewhat less than saturation. Its expression usually is maximal in neutral or slightly alkaline soils. In acidic conditions fungistasis may be lessened because of suppression of bacterial and actinomycete activity. Increased sensitivity of some fungi in soils of pH > 7.0 may be caused by a directly unfavourable effect of pH on the fungus.Fungal species with small spores tend to be highly sensitive to fungistasis (Lockwood, 1977). These spores tend to germinate slowly and to require exogenous nutrients for germination. By contrast, species with larger spores and sclerotia often do not require exogenous nutrients for germination. The larger spores tend to germinate rapidly and to exhibit low sensitivity, as compared with small spores. A few nutrient-independent spores are insensitive to fungistasis. At least a part of the difference in sensitivity is related to germination time; spores which germinate slowly compete poorly with the soil micro-flora for their nutrients. Fungistasis is often temporarily annulled by enriching the soil with energy-yielding nutrients. Usually, complex materials such as plant residues are most effective. A few weeks after such treatment, the level of fungistasis may, however, be increased. Annulment of fungistasis by compounds not utilized as energy sources has not yet been demonstrated. Several soils naturally suppressive to *Fusarium* wilt diseases were more fungistatic to *Fusarium* than soils conducive to wilt. Potential means by which fungistasis may be manipulated to control root-infecting fungi are (*a*) through stimulation of germination with nutrients, thus exposing the germ tube to lysis, and (*b*) by increasing the fungistatic level of soil through appropriate amendments.

Volatile substances identified in soils, some of which are potentially inhibitory to fungi include (*a*) ammonia, which apparently is evolved from ammonium salts in some arid soils of high pH, (*b*) ethylene, which has been identified in some soils of pH < 7.0 (though high levels of this gas seem to be tolerated by most fungi), (c) allyl alcohol, and (*d*) other unidentified substances. Non-volatile inhibitors include high molecular weight substances revealed by molecular sieve chromatography of soil extracts. Microbial metabolites such as those present in staled fungal cultures also have been proposed to account for fungistasis. In a few soils fungistasis persists after sterilization because of the presence of inhibitory concentrations of calcium carbonate, iron or aluminium. Inherent in the proposition that inhibitory substances provide the primary mechanism of fungistasis is the concept of a highly complex phenomenon, involving various highly specific inhibitory and counteracting stimulatory substances, with the outcome for the fungus depending on the kinds and relative amounts of each present. By the nutrient-deficiency hypothesis, the level of available nutrients in soil is insufficient to support germination of nutrient-dependent propagules, except in nutrient-rich microsites. Inhibition of nutrient-independent propagules is explained by loss of endogenous nutrients required for germination, through microbial nutrient competition. Evidence for this hypothesis is (*a*) the imposition of fungistasis on numerous nutrient-independent propagules during incubation on leaching model systems designed to simulate microbial nutrient competition in soil, (*b*) similar losses of endogenous nutrients occurring on soil and the leaching system, and (*c*) the fact that soils are chronically deficient in energy in relation to the microbial populations present, with the consequence that enforced inactivity is imposed upon most of the population at any given time for this reason alone, regardless of the presence or absence of fungistatic substances.

In most soils, fungal propagules are restricted to a certain extent in their ability to grow or germinate. This phenomenon, known as soil fungistasis, has received considerable attention for more than five decades, mostly due to its association with the general suppression of soil-borne fungal diseases. Paolina Garbeva et al., (2011) reviewed major breakthroughs in understanding the mechanisms of fungistasis. Integration of older fungistasis research and more recent findings from different biological and chemical disciplines has lead to the consensus opinion that fungistasis is most likely caused by a combination of microbial activities, namely withdrawal of nutrients from fungal propagules and production of fungistatic compounds. In addition, recent findings indicate that there are mechanistic links between these activities leading towards an integrated theory of fungistasis. Among the potentially fungistatic compounds volatiles have received particular attention. Whereas it has long been assumed that fungistasis is the result of the metabolic activity of the total soil microbial biomass, more recent research points at the importance of activities of specific components of the microbial community. These insights into fungistasis have also formed the basis for strategies to increase general soil suppression.

Fungistasis can be mediated by soil microorganisms and volatile organic compounds (VOCs). The relationship between soil microorganisms and VOCs is still unclear, however, and many fungistatic compounds remain to be identified. We assessed the effects of soils (soil direct fungistasis) and VOCs produced by natural soils (soil volatile fungistasis) on the spore germination of several fungi (Xu Chuankun et al., 2004). Both strong soil direct fungistasis and soil volatile fungistasis were observed in a wide range of soils. Soil fungistasis and VOC fungistasis were significantly correlated ($P<0.001$). The volatile fungistatic activity of soils stopped after autoclaving. Some VOCs were identified by using solid-phase microextraction-gas chromatography/mass spectrum. VOC composition and in vitro antagonism of relatively pure commercial compounds also were measured. Some VOCs, trimethylamine, 3-methyl-2-pentanone, dimethyl disulfide, methyl pyrazine, 2,5-dimethyl-pyrazine, benzaldehyde, *N,N*-dimethyloctylamine and nonadecane, were produced by various fungistatic soils. Moreover, antifungal activity test of above VOCs showed that trimethylamine, benzaldehyde, and *N,N*-dimethyloctylamine have strong antifungal activity even at low levels (4–12 mg l^{-1}).

6.8.1 The Nature of the widespread soil fungistasis: Spores of fungi which are able to germinate readily in distilled water or nutrient substrates fail to germinate in natural soils. Many authors have demonstrated fungistasis in different soils by using spores of numerous fungi. This apparently widespread fungistasis has been attributed to diffusible toxic factors in soil, and appears to coexist with biological activity since soils sterilized by heat or fumigants are not inhibitory. Toxicity to fungi has been restored to autoclaved soil by inoculating with a specific bacterium or with various actinomycetes. Fungistasis has been decreased by addition of plant residues or organic compounds such as glucose to soil. Many attempts have been made to elucidate the precise nature of soil fungistasis. The inhibitors are likely to be a simple mixture of a few components, probably of microbial origin. Some workers have reported unidentified fungistatic substances in soil extracts and diffusates (Lingappa and John L. Lockwood, 1961). Except for extraction of fungitoxic lignin-like material from soil these results have usually been inconclusive, and the factors responsible for fungitoxicity have not been isolated and identified.

Special kinds of soil fungitoxicity such as those of the highly calcareous Newborough sands and the very acid (pH 2.8) Charnwood forest soils seem to differ from the general fungistasis expressed by almost all soils. There is need to explain the nature of the widespread soil fungistasis. In the present investigation, the methods used by previous workers were repeated, and work was extended to include additional controls and new experiments. Attempts to determine whether the results from indirect assay methods in common use might lead to erroneous interpretation of soil fungistasis were of particular interest.

Fungistasis of Conover loam, muck, and hardwood forest soils to conidia of *Glomerella cingulata, Fusarium oxysporum* f. *lycopersici,* and *Penicillium* frequentans was demonstrated by indirect methods such as agar disks, double agar layer plates, and cellophan folds, as well as by direct placement of spores on or in soils. All attempts to extract toxic substances from soil with water or organic solvents failed. Fungistatic volatile substances could not be demonstrated in soil. Redox, pH, and osmotic conditions were not responsible for soil fungistasis. Various lines of evidence led to the conclusion that the so-called widespread soil fungistasis as observed by indirect methods is the result of production of antibiotics by soil microbes growing on the surface of the assay media, and is not due to a reserve of toxic substances in soils. The possibility is suggested that individual fungus spores serve as nutrient microsubstrates in soil and stimulate the rapid growth of soil microbes on their surface or in their immediate vicinity, and that this results in the production of sufficient fungistatic substances to prevent spore germination. Preliminary evidence in support of this suggestion are results with concentrated 50 % ethanol washings from teliospores of *Ustilago xeae* which markedly stimulated growth of mixed soil microbes and of pure cultures of *Streptomyces* sp. and *Pseudomonas* sp., known antibiotic producers, in agar media.

A possible group of suspected fungistatic substances in soil includes the antibiotics. On the basis of researches to date, the production and accumulation of detectable quantities of antibiotics from natural unamended and uninoculated soil is considered to be unlikely. The present workers carried out very extensive and intensive extractions of natural soils and failed to demonstrate any antifungal antibiotics. While the accumulation of antibiotics in natural soil has not been demonstrated, evidence generally favours the production of antibiotics on fresh organic substrates in micro-environments of the soil. As the amounts produced are likely to be extremely small, when the bulk of the soil is considered, their demonstration has been by necessity indirect. The addition of a fresh organic substrate, such as agar or cellophan, might supply a source of nutrients and support microbial growth and antibiotic production. Such a possibility is supported by the short generation time of many bacteria and the very large number of micro-organisms in soil. An average soil containing 10 million individual microorganisms/g. would provide an initial and enormous inoculum of up to 100,000 microbes in contact with a 1 cm. square agar block placed on it. Present information permits us to assume that 40-50 yo of the soil micro-organisms produce antibiotics. The present work and that of others shows that antibiotics are produced by actinomycetes growing on deficient media such as water agar, and that fungistatic substances are produced as well from mixed cultures as from single cultures of antibiotic-producing micro-organisms.

This explanation of the inhibition of fungal spore germination on assay media in contact with soil was supported by five lines of evidence , viz., Peptone agar became fungistatic more rapidly than water agar in contact with natural loam soil ; When agar media were pre-incubated on soil at lo, a temperature which permits diffusion of fungistatic substances but largely prevents microbial growth, no fungistasis occurred ; The same degree of fungistasis was obtained with agar disks preincubated with thick layers of soil, minute amounts of soil, mixed cultures of soil microbes, or pure cultures of *Streptornyces* sp. or *Pseudomonus* sp. ; Decrease of fungistasis was observed when high concentrations of antibacterial antibiotics were added to agar disks used for assaying soil ; Microscopic observations showed dense growth of micro-organisms within short periods on the contact surfaces of agar and cellophan with soil.

6.8.2 Microbial community composition and soil fungistasis: The vast majority of natural soils suppress the germination and growth of fungi to a certain extent. This phenomenon, referred to as widespread soil fungistasis, intensity of which, is dependent on the physical and chemical soil properties as well as soil microbial activity (de Boer et al., 2003). The importance of the last factor was demonstrated by the relief of fungistasis by (partial) sterilization treatments and addition of antibiotics.

Most soils inhibit fungal germination and growth to a certain extent, a phenomenon known as soil fungistasis. Previous observations have implicated microorganisms as the causal agents of fungistasis, with their action mediated either by available carbon limitation (nutrient deprivation hypothesis) or production of antifungal compounds (antibiosis hypothesis). To obtain evidence for either of these hypotheses, soil respiration and microbial numbers (as indicators of nutrient stress) and bacterial community composition (as an indicator of potential differences in the composition of antifungal components) was measured during the development of fungistasis. This was done for two fungistatic dune soils in which fungistasis was initially fully or partly relieved by partial sterilization treatment or nutrient addition. Fungistasis development was measured as restriction of the ability of the fungi *Chaetomium globosum*, *Fusarium culmorum*, *Fusarium oxysporum*, and *Trichoderma harzianum* to colonize soils. Fungistasis did not always reappear after soil treatments despite intense competition for carbon, suggesting that microbial community composition is important in the development of fungistasis. Both microbial community analysis and in vitro antagonism tests indicated that the presence of pseudomonads might be essential for the development of fungistasis. Overall, the results lend support to the antibiosis hypothesis.

Fungi differ strongly in their sensitivity to fungistasis. In general, plant pathogenic fungi appear to be more sensitive than saprophytic fungi. The sensitivity of plant pathogenic fungi to fungistasis is thought to protect them from germinating and initiating growth under unfavorable conditions. However, this protection comes at a cost, since the viability of resting structures decreases during prolonged incubation in soils. The negative effects of fungistasis on the inoculum density of plant pathogenic fungi have been suggested as a mechanism to explain the commonly found correlation between fungistasis and disease suppressiveness. Therefore, manipulation of fungistasis has often been mentioned as a potential measure to control plant pathogenic fungi. The idea is to stimulate germination, and subsequent dying, of plant pathogens when no host plant is present or to suppress their germination and growth in the presence of a host plant. So far, no consistent success has been obtained with fungistasis-manipulating measures and a major hurdle in the successful manipulation of fungistasis is our limited knowledge of its mechanisms.

The most popular explanation for the microbial cause of fungistasis is that strong competition by soil microorganisms limits carbon availability to the germinating spores or invading hyphae (de Boer et al., 2003). Limited carbon availability in soils is the rule rather than the exception, and not only fungi but also actinomycetes (actinostasis) and nonfilamentous bacteria (bacteriostasis) are restricted in their growth in soils. The importance of nutrients is further indicated by the partial or full alleviation of fungistasis after addition of simple energy substrates, like sugars or amino acids, to soil. Other studies have, however, suggested that fungistasis can be attributed to the presence of antifungal compounds of microbiological origin and the production of fungus-inhibiting compounds has been described for a wide range of soil microorganisms. Unfortunately, it is difficult to distinguish between these two proposed mechanisms of fungistasis, as nutrient status may be involved in both. Yet, the involvement of antibiosis in fungistasis may imply that the microbial community composition of a soil is critical because not all microorganisms produce antifungal compounds and the spectrum of antifungal compounds produced is different for different microbial species. If carbon limitation is the major causative factor of fungistasis, (potential) microbial activity should be more important than microbial composition.

Several results in this study indicate that soil microbial community composition can be a major determinant of soil fungistasis. The most pronounced result in this respect was the prolonged relief of suppression of hyphal growth of *C. globosum* in the soil from Oostvoorne (experiment 1) after partial sterilization or tryptic soy broth addition. The CO_2 production in these treatments dropped to the level of the untreated soil within 3 weeks, whereas the number of bacteria remained an order of magnitude higher than in the untreated soil. This indicates that intense competition for nutrients, i.e., low availability of carbon together with high carbon demand, must have occurred during the incubation of the microwave- and tryptic soy broth-treated soils but that this was not sufficient to induce fungistasis. These results strongly suggest that the changes in the bacterial community which were induced by these treatments resulted in a relative dominance of bacterial populations that produced smaller amounts of substances inhibiting *C. globosum*.

In contrast to observations for the soil from Oostvoorne, hyphal growth of *C. globosum* as well as that of all other fungi tested was significantly correlated with CO_2 production in the microwave-treated soil from Terschelling. The same was seen for the autoclaved-inoculated treatment. Although these results might suggest that hyphal growth was limited by carbon availability, several observations in experiment 2 indicate that carbon availability alone was not sufficient to explain the development of fungistasis in the soil from Terschelling (de Boer et al., 2003).The fact that fungistasis in the autoclaved-inoculated treatment returned very rapidly to the level of the untreated control may indicate that important microbial components were introduced that were no longer present after the microwaved, autoclaved-open, and autoclaved-closed treatments. For instance, actinomycetes, which are known for their production of antifungal compounds (Behal, 2000), may have been eliminated from these treatments. However, actinomycetes were not essential to the development of fungistasis, as prolonged incubation of microwaved and autoclaved-open treated soils also resulted in strong fungistasis despite no detection of actinomycetes in these soils.

Pseudomonads and stenotrophomonads/xanthomonads were always present when fungistasis reappeared in the different treatments and were absent from the nonfungistatic autoclaved-closed treated soil. These groups are known for their production of antifungal compounds (Minkwitz and Berg. 2001). The strong antifungal activity of *Pseudomonas* isolates against both *F. culmorum* and *T. harzianum* on water-agar indicated that they did produce inhibiting compounds under nutrient-limiting conditions. On the contrary, none of the stenotrophomonad/xanthomonad isolates inhibited *T. harzianum* on water-agar, and only a few strains inhibited *F. culmorum*. Therefore, the results suggest that the presence and activity of pseudomonads could be essential for the development of fungistasis in these soils.In most studies, fungistasis has been assayed by measuring the repression of spore germination and not by the restriction of hyphal extension, as was done here. In addition to hyphal extension measurements, spore germination tests (spores on nitrocellulose filters on top of soil) were done for a number of fungi as part of the main experiment. The patterns observed for the relief and reappearance of inhibition of spore germination mirrored the results obtained for hyphal extension. The same mechanism mediates both repression of spore germination and inhibition of hyphal extension, but that spore germination is more sensitive. Sterilization did not fully remove the fungistasis of some soils. This phenomenon was named residual mycostasis and attributed it to abiotic factors mostly linked to alkalinity. The sensitivity to this residual mycostasis was shown to be species dependent. a strong residual mycostasis for *F. culmorum* was found in the soil from Oostvoorne.

6.8.3 Soil fungistasis against *Fusarium graminearum* under different tillage systems: It is well known that extensive tillage has certain benefits over the conventional, such as reduced costs (especially through fuel use), soil erosion, nitrate leaching, increased soil organic matter and activity of soil organisms, improved soil structure and preserved soil moisture (Supronienė et al., 2023). Nevertheless, extensive tillage systems often face problems, including soil compaction, weed and disease management. Consequently, noninversion tillage is considered one of the major factors contributing to fusarium head blight (FHB) development, since it causes the increased inoculum resulting due to the presence of a high quantity of host plant residues on the soil surface.

The establishment of the harmful pathogen *Fusarium graminearum* in different agroecosystems may strongly depend on the ability of the soils to suppress its development and survival This study by Supronienė et al., (2023) aimed to evaluate the influence of different soil tillage systems (i.e., conventional tillage, reduced tillage and no-tillage) on soil fungistasis against *F. graminearum*. Soil samples were collected three times during the plant growing season in 2016 and 2017 from a long-term, 20-year soil tillage experiment. The *F. graminearum* in the soil samples was quantified by real-time qPCR. The soil fungistasis was evaluated by the reduction in the radial growth of *F. graminearum* in an in vitro assay. The antagonistic activity of the soil bacteria was tested using the dual culture method. The *F. graminearum* DNA contents in the soils were negatively correlated with soil fungistasis ($r = -0.649$ *). *F. graminearum* growth on the unfumigated soil was reduced by 70–87% compared to the chloroform fumigated soil. After the plant vegetation renewal, the soil fungistasis intensity was higher in the conventionally tilled fields than in the no-tillage. However, no significant differences were obtained among the tillage treatments at the mid-plant growth stage and after harvesting. 23 out of 104 bacteria isolated from the soil had a moderate effect, and only 1 had a strong inhibitory effect on the growth of *F. graminearum*. This bacterium was assigned 100% similarity to the *Bacillus amyloliquefaciens* Hy7 strain (gene bank no: JN382250) according to the sequence of the 16S ribosome subunit coding gene. The results suggested that the presence of *F. graminearum* in soil is suppressed by soil fungistasis; however, the role of tillage is influenced by other factors, such as soil biological activity, type and quantity of plant residues and environmental conditions

The development of *F. graminearum* in soil might be suppressed by fungistasis caused by the participation of soil microorganisms (Ossowicki, et al., 2020). The microbial community composition is essential for the development of fungistasis through the production of antifungal compounds. The soil fungistasis effect on *F. graminearum* mycelium growth might be detected using an in vitro assay. Using this method, Lisboa et al. (2015) found that the growth of *F. graminearum* was 88.1% greater (mean growth was 5.2 cm greater) for the soil samples fumigated with chloroform compared to unfumigated. This study also showed that the growth of *F. graminearum* mycelium was significantly increased (on average 80%) in chloroform fumigated soil compared to natural unfumigated. This suggests that inactivated native soil microbiota due to the fact of chloroform fumigation leads to the development of this fungus in the soil. In addition, during the soil fumigation, some of the nutrients can be released from dead microbial cells, which may also enhance fungus colonization.

Conventional tillage showed an advantage over reduced tillage and, especially, no-till practices in terms of the presence of *F. graminearum* in the soil, as significantly higher contents of *F. graminearum* DNA (on average 31.5 pg *F. graminearum* DNA/ng of total DNA) were detected in the no-tilled soils compared to the conventional (on average 6.9 pg *F. graminearum* DNA/ng of total DNA). *F. graminearum* growth on the chloroform fumigated soil was significantly increased (on average 80%) compared to the unfumigated soil, which shows the natural capacity of the tested loamy textured *Endocalcari-Epihypogleyic Cambisol* soil to suppress pathogen survival under specific biotic and abiotic conditions. Based on two-way ANOVA and Tukey's mean separation tests, the soil fungistasis against *F. graminearum* was significantly affected by the tillage practices or soil sampling time. Nevertheless, the overall results of the soil fungistasis suggest that only in spring after the renewal of the plant vegetation was the soil fungistasis intensity higher in the conventionally tilled fields than in the no-till, as no significant differences were obtained among the tillage treatments at the mid- and last soil sampling times. In the different soil tillage treatments, the *F. graminearum* DNA contents were negatively correlated with the soil fungistasis ($r = -0.649$ *), showing that the relationship between the fungistasis intensity and DNA content decreased. Further evaluation of fungistasis dynamics after soil amendment with different plant residues is needed (Suproniené et al., 2023).

6.8.4 Role of the Microbial nutrient sink and Fungistatic substances: Soil is maintained in a fungistatic state by continuous microbial competition for the minute qualities of nutrients that may become available, and which are required for germination by many spores (Bristow and Lockwood 1975). However, spores and sclerotia of certain fungi will germinate in water alone, but only sparsely in soil. Germination of such propagules is also suppressed when incubated on sand undergoing aqueous leaching in a system designed to impose upon the propagules a diffusion stress similar to that imposed in soil through microbial activity. There is involvement of a volatile fungistatic factor in addition to nutrient deprivation. Ammonia has been identified as the volatile fungistatic factor in two alkaline soils and ethylene was reported to be the active substance in several Australian soils. This research was undertaken to evaluate the role of the microbial nutrient sink in soil fungistasis in the light of recent reports of volatile inhibitory materials.

Sensitivity of conidia of *Cochliobolus victoriae* to fungistasis decreased markedly following incubation on moist sand for at least 1 h. Germination was greater on Conover loam or on sand being leached with water than on an alkaline clay loam soil known to produce a volatile fungistatic substance. Evolution of $^{14}CO_2$ began within 3 min after [^{14}C] glucose was applied to the soils, the rate of $^{14}CO_2$ evolution was faster with Conover loam. *Germinationot Thielaviopsis basicola* conidia per unit of glucose remaining in agar discs initially containing 0·1% glucose, was lower for discs incubated on the clay loam soil than on Conover loam, and was greatest on a bed of sand undergoing aqueous leaching. Germination of ascospores of *Neurospora tetrasperma* and conidia of *C. victoriae* was suppressed on discs of washed, Purified Agar or polyacrylamide gel incubated on or over the clay loam soil, but no suppression resulted when discs were incubated on Conover loam. Extensive aeration of either soil did not remove its fungistatic effect. Fungistasis in Conover loam appears to be caused primarily by nutrient deprivation, whereas volatile fungistatic substances may play a major role in the clay loam soil.

Correlations between decreasing fungal spore germination and decreasing nutrients because of microbial utilization have been reported. Results with glucose-impregnated agar discs indicated that nutrient deprivation may not be the only cause of fungistasis. However, in Conover loam the results were equivocal, since the unavoidable addition of glucose could have provided the energy needed for microbial synthesis of fungistatic substances. Failure to detect volatile or non-volatile inhibitory substances in Conover loam, or to remove the inhibition by aeration, suggests that the additional fungistatic factor observed is due to elevated local glucose concentrations in the soil. Since radioactive metabolites of glucose were not detected in the assay discs, glucose may have provided the energy required for their synthesis from organic matter in resistant soil. Certain decomposition products of lignin are capable of preventing spore germination.

Several lines of evidence from this or other research indicate that the microbial nutrient sink alone can account for the inhibition in Conover loam. Among these are the near instantaneous establishment of the microbial energy sink, and several correlations of results obtained with soil and the leaching system, namely: (i) the similarity of the fungistatic spectrum, as it affects nutrient-independent propagules, and the ability of ascospores of *N. tetrasperma* to germinate on either system; (ii) similarly increased germination of *C. victoriae* conidia on both systems after brief exposure to water; (iii) the similarity in the expression of fungistasis using agar or polyacrylamide gel discs; (iv) the similarity in degree of annulment of fungi-stasis by nutrients; (v) the very similar rates at which conidia of C. *victoriae* and *Cochliobolus sativus* were converted to nutrient-dependency; (vi) the very similar rates of reversal in the progress towards germination made by conidia of *Penicillium frequentans,* when subsequently incubated in either system. That enforced inactivity as a result of intense competition for scarce nutrients is the rule for micro-organisms in soil is also indicated from calculations of the yearly input of energy substrate to soil in relation to the biomass that must be supported. The energy available after cell maintenance requirements are met, would allow cells to divide only a few times per year. The presence of inhibitory volatile materials emanating from the alkaline clay loam soil was readily shown. Moreover, in contrast with Conover loam, the clay loam showed persistent suppression of conidia previously incubated on moist sand, utilized energy sources more slowly, showed only partial annulment of fungistasis by glucose or soil sterilization, and inhibited germination of ascospores of *N. tetrasperma.*

In a study by Wu et al., (2008), the bacterial community composition and diversity of a series of soils treated by autoclaving, which coming from the same original soil sample and showing gradient fungistasis to the target soil-borne pathogen fungi Fusarium graminearum, was investigated by soil bacterial 16S rDNA-PCR (polymerase chain reaction) cloning, restriction fragment length polymorphism (RFLP), and sequencing. The results showed that the soil fungistasis capacity was closely correlated with soil bacterial community composition and diversity, such as soil fungistasis declined with the decrease of soil bacterial diversity. Meanwhile, the bacterial community composition and structure were significantly different along the gradient of soil fungistasis tested. Alpha-Proteobacteria, beta-Proteobacteria, Flexibacter, and some uncultured soil bacteria were contributed to soil fungistasis in combination with some other special bacteria (*Pseudomonas* and *Acidobacteria*) which were know to be key species in suppression of fungal growth.

6.8.5 Microbial community composition and links between soil Bacteriobiomes & Fungistasis: Most soils inhibit fungal germination and growth to a certain extent, a phenomenon known as soil fungistasis. Previous observations have implicated microorganisms as the causal agents of fungistasis, with their action mediated either by available carbon limitation (nutrient deprivation hypothesis) or production of antifungal compounds (antibiosis hypothesis) (De Boer et al., 2003). To obtain evidence for either of these hypotheses, soil respiration and microbial numbers (as indicators of nutrient stress) and bacterial community composition (as an indicator of potential differences in the composition of antifungal components) were measured during the development of fungistasis. This was done for two fungistatic dune soils in which fungistasis were initially fully or partly relieved by partial sterilization treatment or nutrient addition. Fungistasis development was measured as restriction of the ability of the fungi *Chaetomium globosum, Fusarium culmorum, Fusarium oxysporum* and *Trichoderma harzianum* to colonize soils. Fungistasis did not always reappear after soil treatments despite intense competition for carbon, suggesting that microbial community composition is important in the development of fungistasis. Both microbial community analysis and in vitro antagonism tests indicated that the presence of pseudomonads might be essential for the development of fungistasis. Overall, the results lend support to the antibiosis hypothesis.

Links between Soil Bacteriobiomes and Fungistasis: The soil microbiota includes bacteria, fungi, protozoa, viruses, and archaea, which interact with each other. Soil microbial biomass is dominated by bacteria and fungi (Chertkova et al., 2023). The intensity of fungistasis depends on many factors, including the soil's physical and chemical properties, its microbial community, metabolic activity of the soil microbiota, and fungal characteristics. In particular, the sensitivity to fungistasis differs among various ecological groups of fungi.

Chertkova et al., (2023) investigated the level of fungistasis toward *Metarhizium robertsii* and *Beauveria bassiana* in soils of conventional potato fields and kitchen potato gardens. Agar diffusion methods, 16S rDNA metabarcoding, bacterial DNA quantification, and assays of *Leptinotarsa decemlineata* survival in soils inoculated with fungal conidia were used. Soils of kitchen gardens showed stronger fungistasis toward *M. robertsii* and *B. bassiana* and at the same time the highest density of the fungi compared to soils of conventional fields. The fungistasis level depended on the quantity of bacterial DNA and relative abundance of *Bacillus, Streptomyces*, and some *Proteobacteria,* whose abundance levels were the highest in kitchen garden soils. Cultivable isolates of bacilli exhibited antagonism to both fungi in vitro. Assays involving inoculation of nonsterile soils with *B. bassiana* conidia showed trends toward elevated mortality of *L. decemlineata* in highly fungistatic soils compared to low-fungistasis ones. Introduction of antagonistic bacilli into sterile soil did not significantly change infectivity of *B. bassiana* toward the insect. The results supported the idea that entomopathogenic fungi can infect insects within a hypogean habitat despite high abundance and diversity of soil antagonistic bacteria.

In soils of conventional fields, in addition to the low quantity of total bacterial DNA, the lowest Simpson 1-d and Berger–Parker 1/d indices were registered as compared with the kitchen garden soils. This finding is obviously due to greater soil degradation under intensive farming. It is well known that conventional field soils are characterized by reduced diversity and density of bacteria in comparison with organic field soils. One of the main distinctive features of bacterial composition of soils with high fungistasis (kitchen gardens) is an elevated relative abundance of Bacilli, specifically *Bacillus* and *Paenibacillus*. There is evidence of antagonistic properties of *Bacillus* species residing in the soil and rhizosphere (e.g., *B. pumilus, B. subtilis, B. cereus,* and *B. amyloliquefaciens*) toward phytopathogenic fungi, such as *Botrytis, Alternaria, Fusarium, Colletotrichum, Doratomyces, Penicillium, Pyricularia*, and *Sclerotiria* (Karim et al., 2022). Another important factor responsible for the differences in fungistasis among the studied soils is bacteria from the genus *Streptomyces*. Highly fungistatic soils (Karasuk location) were found to have the highest relative abundance of these bacteria. It is known that *Streptomyces* spp. have an antagonistic effect on fungi of various taxa. These actinobacteria are producers of a large set of volatiles, among which there are fungal inhibitors 2-phenylethanol, phellandrene, benzaldehyde, and dimethyldisulfide. In addition, *Streptomyces* spp.

produce many antifungal factors, including amphotericin, nystatin, phoslactomycins, kanchanamycins, phthoxazolins, dorrigocins, faeriefungin, polyenes, stendomycin, sultriecin, dunaimycins, phosmidosine, polyoxin, and others.

Highly fungistatic soils had higher relative abundance of Alphaproteobacteria, in particular, unc. *Rhizobiales, Bradyrhizobium*, and *Sphingomonas* (Chertkova et al., 2023). Rhizobiales and, in particular, *Bradyrhizobium* are known to produce various antifungal metabolites active against soil-born mycelial fungi, such as *Fusarium, Rhizoctonia*, and *Macrophomina,* although studies on entomopathogenic fungi are not known. *Sphingomonas paucimobilis* (formerly known as *Pseudomonas paucimobilis*) has been shown to exert antagonism toward the phytopathogenic fungus *Verticillium dahliae,* but research on entomopathogenic fungi is also lacking. After the joint introduction of *B. bassiana* and its antagonists *B. frigoritolerans* and *B. pumilus*, there were no significant effects on the CPB mortality. Nevertheless, at low concentrations of conidia, the combination of *B. bassiana* and *B. frigoritolerans* increased mortality as an additive effect when compared with the inoculation with *B. bassiana* alone ($p = 0.07$). It was believed that this outcome is due to an enhancement of fungistasis, and as a result, longer survival of *B. bassiana* conidia; certain pathogenicity of *B. frigoritolerans* to the CPB during the development of mycoses. It is known that conidia of *B. bassiana* can germinate in sterile but not in nonsterile soils, while fungistasis of sterile soils can be restored by the introduction of soil bacteria. Therefore, the introduction of bacteria may contribute to higher infectivity of *B. Bassiana*. Further research can be aimed at clarifying the impact of the fungal microbiota on fungistatic properties of soils toward entomopathogenic fungi and on the development of mycosis in insects in soils.

6.8.6 Biotic, Abiotic factors and Mechanisms of Glucose-Mediated soil fungistasis: Soil suppressiveness or receptivity can manifest itself through various forms including fungistasis, defined as the soil ability to restrict fungal germination and growth (Fabienne Legrand et al., 2019). Microbial activity has been recognized as a major component in determining soil fungistasis, notably through competition for nutrients and space but also production of antagonistic compounds. Soil physicochemical properties may also contribute to soil fungistasis, mostly by indirectly influencing microbial community diversity and structure. Crop management regimes, such as soil management, also inevitably alter the biotic and abiotic properties of soil, and ultimately result in changes of soil microbial communities and therefore in soil fungistasis.

The current study determined the levels of soil fungistasis against a soil-borne pathogen inoculum, *Fusarium graminearum* (Fg, a major causal agent of Fusarium Head Blight (FHB)), in 31 wheat fields by quantifying Fg growth after a 15-day incubation period using qPCR in autoclaved versus non-autoclaved soils. The results were used to define the six most Fg-resistant and the six most Fg-conducive soils. By using a metabarcoding approach, the diversity of the bacterial communities was significantly higher in Fg-resistant soils than in Fg-conducive soils. Microbial taxa potentially contributing to Fg-fungistasis of soil were selected if they were significantly more prevalent and/or abundant in Fg-resistant soils than in Fg-conducive soils. Some of these candidate indicators, e.g. *Pseudomonas* spp. and *Bacillus* spp., have been reported previously as effective biocontrol agents against plant pathogens. Correlation-based network analysis further showed that the members of the bacterial communities in Fg-resistant soils were more connected than in Fg-conducive soils. Moreover, network modules was found significantly correlated with certain edaphic abiotics factors (such as the soil manganese and nitrogen content) and Fg-fungistasis. Such observations may suggest and emphasize, although conceptual, the importance of synergistic rather than individual effects of network members, and the nutrient use efficiency in contributing to Fg-resistance of soils in wheat fields in France.

Besides characterizing the changes in species richness or community compositional structure, understanding the ecological associations between community members can further reveal ecosystem functionality, especially in complex systems such as soils. In fact, microbial diversity may not be susceptible enough to detect changes in the ecosystem functioning in response to perturbations (Karimi et al. 2016). It was found that the bacterial communities of the Fg-resistant soils exhibited a much more densely connected co-occurrence network than those of the Fg-conducive soils did, which may suggest that the microbiota members in Fg-resistant soils are more sensitive and responsive to disturbance such as the introduction of pathogens, and therefore more resilient. Among the nodes of the biological network of the Fg-resistant soils, one Bradyrhizobiaceae OTU had the highest centrality and betweenness, while OTUs of Rhodoplanes and Devosia were recognized as connectors (linking modules of biological networks). These nitrogen-fixing bacteria in *Rhizobiales,* usually forms symbiosis with legume plants, may be, therefore, considered as 'keystone taxa' in the ecological networks of Fg-resistant soils. Keystone taxa and bioindicators associated with Fg-resistant soils were also identified, many of which had been described previously as biocontrol agents against fungal plant pathogens or displayed plant growth-promoting activities. Correlation network analysis further indicated that Fg-resistant soils possessed a highly connected and dense network of the bacterial communities, which could be associated to higher resilience of such soils to perturbations, such as Fg incidence.

Mechanisms of Glucose-Mediated Soil Fungistasis: Soil is a complicated ecological environment, and soil fungistasis (mycostasis) strongly represses the germination and growth of fungal bio-control agents, rendering biocontrol agents inefficient in soil. Soil fungistasis was identified in 1953 as a phenomenon that caused the inhibition of conidial germination and hyphal growth in soils (Liu et al., 2020). Reaching a sufficient population density is a prerequisite for fungal agents to function effectively in soils. Soil fungistasis is prevalent in most soils, making the relief of soil fungistasis necessary for fungal agents to reach a sufficient population density.

Past decades studies argued that the most likely explanation for fungistasis is probably a combination of the nutrientdeficiency and fungistatic factor-mediated inhibition. Previous studies have also shown that soil supplementation with nutrients, including simple sugars, amino acids, corncobs, tapioca, and soybean powder, can relieve fungistasis. Nutrient combinations and complex materials such as plant residues are typically more effective than simple nutrients. Fungistasis relief by amino acids is greater in acidic soils, but the efficacy of glucose is unaffected by changes in soil pH. Palmitic acid, octadecanoic acid, and citric acid can also promote the germination of endospores and chlamydospores. Based on these studies, biocontrol agents have been developed by combining nutrients with fungal agents, and the spore germination rate of bio-control agents is improved in soil. However, the fungistasis relief effect of nutrients was shown to be temporary. For example, the spore germination rate of *Thielaviopsis basicola* was improved for the first 4 days following the addition of alfalfa hay in soil, but greater fungistasis was observed after the fifth day. Although many compounds can relieve soil fungistasis, the underlying mechanisms remain poorly understood. It was speculated that fatty acids are broken down by acid-metabolizing microorganisms to simple compounds that have stimulatory activity on fungal germination. Some plant-pathogenic fungi can prevent accumulation of toxins in the cell to toxic concentrations through the function of ABC transporter which is an energy-driven (ATP or proton motive force) efflux pump. Energy-driven efflux pumps might help fungi to resist soil fungistasis, and the exogenous nutrients in soil would provide energy for efflux pumps, this speculation remains to be proved by further research. Except for these reports, there is little report about the mechanisms behind soil fungistasis relief by nutrients. Further researches are very necessary to elucidate the underlying mechanisms (Liu et al., 2020).

Arthrobotrys oligospora is a nematode-trapping fungus that is widespread in soils. It is used to control plant-parasitic nematodes, especially *Meloidogyne* spp. *A. oligospora* conidial germination is repressed by soil fungistasis, making it a suitable model organism for researching the mechanisms behind soil fungistasis relief by nutrients. In recent years, omics techniques have been widely used to study scientific topics of interest in fungi. For example, proteomics was used to understand the molecular mechanisms of conidial germination (Deng et al., 2015). Transcriptomics was used to elucidate the divergent lifestyle features of the nematode endoparasitic fungus *Hirsutella minnesotensis*. In addition, comparative transcriptomics revealed the different strategies of *Trichoderma* mycoparasitism. These studies suggest that omics techniques can be used to elucidate the mechanisms behind soil fungistasis relief by nutrients. In this study, the transcriptomic features and possible mechanisms of glucoseinduced soil fungistasis relief were revealed based on comparative transcriptomics of *A. oligospora* conidia.

Nutrients can relieve soil fungistasis, but the mechanisms behind this process remain poorly understood. Liu et al., (2020) determined and quantified the transcriptomes of *A.oligospora*, a nematodetrapping fungus, derived from samples of fresh conidia, germinated conidia, soil fungistatic conidia, and glucose-relieved conidia. The transcriptomes of fungistatic and glucose-relieved conidia were significantly different from those of the other two conidia samples. KEGG pathway analyses showed that those genes upregulated in fungistatic and glucose-relieved conidia were mainly involved in translation and substance metabolism, and the down regulated genes were mainly involved in MAPK pathway, autophagy, mitophagy, and endocytosis. As being different from the transcriptome of fungistatic conidia, upregulated genes in the transcriptome of glucose-relieved conidia are also related to replication and repair, spliceosome, oxidative phosphorylation, autophagy, and degradation pathway (lysosome, proteasome, and RNA degradation). And the upregulated genes resulted from comparison of glucose-relieved conidia and fungistatic conidia were enriched in metabolic pathways, cycle, DNA replication, and repair. The differentially splicing events in the transcriptome of glucose-relieved conidia are far more than that of other two transcriptomes, and genes regulated by differentially splicing were analyzed through KEGG pathway analysis. Furthermore, autophagy genes were proved to play important role in resisting soil fungistasis and glucose-mediated soil fungistasis relief. These data indicate that, in addition to being a carbon and energy source for conidia germination, glucose may also help to relieve soil fungistasis by activating many cellular processes, including autophagy, DNA replication and repair, RNA alternative splicing, and degradation pathways.

Several genes involved in oxidative phosphorylation were upregulated after glucose addition. This suggested that a small portion of added glucose might provide energy through the activated oxidative phosphorylation pathway, although most glucose must be diffused to outside of dialysis bag (Liu et al., 2020). Besides, glucose also activated the transcription of 21 transporter genes, including two ABC transporter genes, three MFS transporter genes, and ion channel genes. It remains to be explored whether energy-driven efflux pumps help in the resistance of soil fungistasis. The expression of genes involved in membrane trafficking, autophagy, and mitophagy was also activated in glucoserelieved conidia. Membrane trafficking is involved in endocytosis, exocytosis, autophagy and vesicle formation and fusion and is, therefore, vital for fungal growth and differentiation. In this study, conidial germination rates of autophagy pathway mutants 1ATG1 and 1ATG5 decreased significantly in soil suspension. This result confirmed the important role of autophagy in resisting soil fungistasis, and thus glucose activation of the autophagy pathway would promote the conidial germination of glucoserelieved conidia. The transcription of 15 cytoskeleton genes was also activated by glucose. The actin cytoskeleton proteins play important roles in various fundamental cellular processes, including the maintenance of cell shape, polarity, cell division, cell migration, endocytosis, and vesicular trafficking (Mishra et al., 2014). The tubulin cytoskeleton proteins play important roles in mitosis and meiosis. Thus, cytoskeleton proteins are involved in both substance transport and the cell cycle.

Spore germination is a process in which non-dividing haploid spores re-enter the mitotic cell cycle and resume vegetative growth. Thus, the upregulation of genes in the above mentioned categories should promote the cell cycle and conidial germination process in *A. oligospora*. These observations are consistent with other reports. For example, glucose can modify cell proliferation in maize during germination (Lara-Núñez et al., 2017). Glucose in plant tissues triggers cell division and cell cycle-related markers can be regulated at different levels by glucose in plants. In addition to being a carbon and energy source, glucose is also a signaling molecule that helps to relieve soil fungistasis by activating many cellular processes, including autophagy, DNA replication and repair, RNA alternative splicing, and degradation pathways.

6.8.7 Fungistasis of essential oil extracted from *Tagetes lucida*: In the case of some species of *Tagetes,* the plant extracts have antifungal effects, for example, aqueous extracts, powder of leaves, flowers and roots of *T. erecta* have fungicidal effect against *Drechslera oryzae, Pyricularia orizae* and *Uromyces phaseoli*) and extracts of leaves and flowers of *T. patula* are fungistatic in *Monilia* spp.; however, there are few toxicological studies, particularly bioassays with essential oils (Edgar López López et al., 2018). The methanolic extract of this plant inhibits the growth of bacteria such as *Escherichia coli, Salmonella* sp. and *Shigella* sp., as well as in the species of the *Fusarium* genus due to the action of the dimethoxyphenolic components present in the extract, such as the scorpion (6,7-dimethoxycoumarin) responsible for presenting a high inhibition on the mycelial growth of *Fusarium* in doses of up to 250 μg mL-1. The characterization of the essential oil profile of populations of *T. lucida* and its toxicological evaluation against phytopathogenic fungi of economic importance such as *F. oxysporum, P. janthinellum, A. niger* and *R. solani* is a necessary step for the utilization of local natural resources in organic control strategies for agricultural production.

The *T. lucida* species known as 'pericón', is a natural resource of Mexico whose essential oil has antifungal properties, although not all the phytopathogenic fungi have been explored the effect that this vegetable substance can cause. From a population collected in Atotonilco the Grande, Hidalgo, Mexico, essential oil was obtained by hydrodistillation with 0.2% yield; using the CG-EM technique, eight different components were identified in the oil, but the abundant ones were estragole (48%) and anethole (35%). The in vitro antifungal activity of the essential oil against *Aspergillus niger, Fusarium oxysporum, Penicillium janthinellum* and *Rhizoctonia solani* was also evaluated by means of the agar diffusion technique, performing two experiments: the first testing concentrations of 0, 0.1 and 1% and in the second 0, 2 and 3%. The concentration of 1% reduced the mycelial growth 46% for *F. oxysporum,* 39% in *R. solani*, 21% in A. niger and 16% in *P. janthinellum*; however, in high concentrations of oil, such as 3%, only in *R. solani* was the greatest reduction in mycelial growth observed (72%).

In general, phenylpropanoids such as estragole and anethole, which are found in the essential oil of the Atotonilco population, are secondary metabolites that are also present in other plant species such as fennel (*Foeniculum* spp.) or anise (*Pimpinela anisum*) (Edgar López López et al., 2018). Oil has fungistatic or fungicidal activity. In the case of the biological activity of Benomil®, the mechanism is given by the inhibitory action of ergosterol synthesis, which is a structural component of the fungal membrane, causing deformation and abnormal proliferation of germinative tubes. The application of synthetic or natural substances against phytopathogenic fungi should evaluate the implications to the environment and the human, as well as its technical-economic viability, in this last sense, the extrapolation of the biomass and oil yield of the population of *T. lucida* in dry conditions in Atotonilco, Hidalgo, as well as the outstanding fungistatic effect evaluated against the fungi *F. oxysporum* and *R. solani,* allow estimating that with a yield of 2 000 kg of fresh biomass per hectare obtained by harvesting in the natural areas of Atotonilco and with an essential oil yield of 0.28%, approximately 5.6 L of essential oil could be obtained, which could be used to formulate more than 1000 L of biopesticide at 1% and more than 10 000 L at 0.1%. Hence, this natural resource of Atotonilco, Hidalgo can have good prospects in ecological agriculture against such fungi; its evaluation in the field will be convenient, at the same time as promoting the agroecological management of the 'pericón' population to increase the number of plants per unit area.

Bibliography

Adil Essarioui, Nicholas LeBlanc, Harold C. Kistler, and Linda L. Kinkel, 2017. Plant Community Richness Mediates Inhibitory Interactions and Resource Competition between *Streptomyces* and *Fusarium* Populations in the Rhizosphere. *Microb. Ecol.*, DOI 10.1007/s00248-016-0907-5 2017.

Amandine Erktan, Matthias C. Rillig, Andrea Carminati, Alexandre Jousset and Stefan Scheu, 2020. Protists and collembolans alter microbial community composition, C dynamics and soil aggregation in simplified consumer–prey systems. *Biogeosci.*, **17**: 4961–4980

Arseneault Tanya and Filion Martin, 2017. Biocontrol through antibiosis: Exploring the role played by sub inhibitory concentrations of antibiotics in soil and their impact on plant pathogens. *Canadian J. Pl. Pathol.*, **39**: 10.1080/07060661.2017.1354335.

Behal, V., 2000. Bioactive products from *Streptomyces*. *Adv. Appl. Microbiol.*, **47**:113-156

Beneduzi, A., Ambrosini, A. and Passaglia, M.P., 2012. Plant growth-promoting rhizobacteria (PGPR): Their potential as antagonists and biocontrol agents. *Genetics & Molecular Biol.*, **35**:1044-1051.

Brandon L. Findlay. 2016. The Chemical Ecology of Predatory Soil Bacteria. ACS Chem. Biol.

Bristow, P. R., Lockwood, J.L.,.1975. Soil Fungistasis: Role of the Microbial Nutrient Sink and of Fungistatic Substances in Two Soils. *Microbiol.*, **90:** https://doi.org/10.1099/00221287-90-1-147

Buchkowski, R.W., 2016. Top-down consumptive and traitmediated control do affect soil food webs: It's time for a new model. *Soil Biol. Biochem.*, **102:** 29–32

Chen, Q. L. *et al.*, 2017. Do manure-borne or indigenous soil microorganisms influence the spread of antibiotic resistance genes in manured soil? *Soil Biol. Biochem.*, **114**: 229–237.

Chertkova, E., Kabilov, M.R., Yaroslavtseva, O., Polenogova, O., Kosman, E., Sidorenko, D.. Alikina, T., Noskov, Y., Krivopalov, A. and Glupov. V.V., 2023. Links between Soil Bacteriobiomes and Fungistasis toward Fungi Infecting the Colorado Potato Beetle. *Microorganisms*, **11**: 943.

Croze, A., Carlino, A., Quélard, B., Saha, A., Convert, T., Eberst, J.B. et al., 2021. Intracellular behaviour of legionella non-pneumophila strains within three *Amoeba* strains, including *Willaertia magna* C2c Maky. *Pathogens*, **10:** 1350.

de Boer, W., Verheggen. P., Klein Gunnewiek, P.J., Kowalchuk, G.A. and van Veen, J.A., 2003. Microbial community composition affects soil fungistasis. *Appl. Environ. Microbiol.*, **69:** 835-844

de Novais, C. B., Sbrana, C., da Conceicao Jesus, E., Rouws, L. F. M., Giovannetti, M., Avio, L., et al.. 2020. Mycorrhizal networks facilitate the colonization of legume roots by a symbiotic nitrogen-fixing bacterium. *Mycorrhiza*, **30**: 389–396.

de Weert, S. and Bloomberg. G.V., 2006. Rhizosphere competence and role of root colonization in biocontrol. *In*. *Plant-associated bacteria*. Gnanamanickam, S.S. (Ed.) Springer, Dordrecht, p 317

Deng, G. M., Yang, Q. S., He, W. D., Li, C. Y., Yang, J., Zuo, C. W., et al., 2015. Proteomic analysis of conidia germination in *Fusarium oxysporum* f. sp. *cubense* tropical race 4 reveals new targets in ergosterol biosynthesis pathway for controlling Fusarium wilt of banana. *Appl. Microbiol. Biotechnol.,* **99:** 7189–7207

Dobbs, C. and Hinson, W. A., 1953. Widespread *Fungistasis* in Soils. *Nature,* **172:** 197–199

Du Preez, G., Daneel, M., De Goede, R., Du Toit, M.J., Ferris, H., Fourie, H. et al., 2022. Nematode-based indices in soil ecology: Application, Utility, and Future directions. *Soil Biol Biochem,* **169**: 108640

Dundore-Arias, J.P. et al., 2020. Carbon Amendments Influence Composition and Functional Capacities of Indigenous Soil Microbiomes. *Front. Mol. Bioscie.,* **6.** DOI=10.3389/fmolb.2019.00151

Dutta, P., Mahanta, M., Singh, S.B., Thakuria, D., Deb, L., Kumari, A., Upamanya, G.K., Boruah, S., Dey, U., Mishra, A.K., Vanlaltani, L., Vijay Reddy, D., Heisnam, P. and Pandey, A.K., 2023. Molecular interaction between plants and *Trichoderma* species against soil-borne plant pathogens. *Front. Plant Sci.,* **14:**1145715

Edgar López López, Margarita Gisela Peña Ortega, María Teresa Beryl Colinas León, Francisco Díaz Cedillo and Miguel Ángel Serrato Cruz, 2018. Fungistasis of essential oil extracted from a *Tagetes lucida* population of Hidalgo, México. *Rev. Mex. Cienc. Agríc.,*[online]. **9:** 329-341

Eitzinger, B. et al., 2018. Testing the validity of functional response models using molecular gut content analysis for prey choice in soil predators. *Oikos*, **127:** 915–926

Erktan, A., Or, D. and Scheu, S., 2020. The physical structure of soil: determinant and consequence of trophic interactions. *Soil Biol. Biochem.*, **148**: 107876.

Fabienne Legrand, Wen Chen, Jose Francisco Cobo-Dıaz, Adeline Picot and Gaetan Le Floch, 2019. Co-occurrence analysis reveal that biotic and abiotic factors influence soil fungistasis against *Fusarium graminearum. FEMS Microbiol.Ecol.,* **95**: fiz056

Felipe E. Albornoz, Treena I. Burgess, Hans Lambers, Hannah Etchells and Etienne Laliberte, 2016. Native soil borne pathogens equalize differences in competitive ability between plants of contrasting nutrient-acquisition strategies. *J. Ecol.,* doi: **10.**1111/1365-2745.12638

Ferlian, O. et al., 2018. Invasive earthworms erode soil biodiversity: A meta-analysis. *J. Anim. Ecol.,* **87:** 162–172

Geisen, S. et al., 2018. Soil protists: A fertile frontier in soil biology research. *FEMS Microbiol. Rev.,* **42:** 293–323

George W. Sundin and Nian Wang, 2018. Antibiotic Resistance in Plant-Pathogenic Bacteria. *Ann. Rev. Phytopathol.,* **56**: 8.1–8.20

Gu, S., Wei, Z., Shao, Z., Friman, V.-P., Cao, K., Yang, T., et al., 2020. Competition for iron drives phytopathogen control by natural rhizosphere microbiomes. *Nat. Microbiol.,* **5:** 1002–1010

Guo, S., Jiao, Z., Yan, Z. *et al.,* 2024. Predatory protists reduce bacteria wilt disease incidence in tomato plants. *Nat. Commun.,* **15:** 829.

Guo, S., Tao, C., Jousset, A., Xiong, W., Wang, Z., Shen, Z., Wang, B., Xu, Z., Gao, Z., Liu, S., Li, R., Ruan, Y., Shen, Q., Kowalchuk, G.A. and Geisen, S., 2022. Trophic interactions between predatory protists and pathogen-suppressive bacteria impact plant health. *ISME J.*, **16**:1932-1943.

Guo, S., Xiong, W., Hang, X., Gao, Z., Jiao, Z., Liu, H., et al., 2021. Protists as main indicators and determinants of plant performance. *Microbiome,* **9:** 64.

Guzmán-Guzmán, P., Kumar, A., de los Santos-Villalobos, S., Parra-Cota, F. I., Orozco-Mosqueda, M., Fadiji, A. E., et al., 2023. *Trichoderma* species: our best fungal allies in the biocontrol of plant diseases–a review. *Plants,* **12:** 432

Hawxhurst, C.J., Micciulla, J.L., Bridges, C.M., Shor, M., Gage, D.J. and Shor, L.M., 2023. Soil protists can actively redistribute beneficial bacteria along *Medicago truncatula* roots. *Appl. Environ. Microbiol.,* **89:** e0181922.

Hernández, A., Heydrich, P.M., Velazquez, M.G. and Hernandez, L.A., 2006. Perspectivas del empleo de rizobacterias como agentes de control biológico en cultivos de importancia económica. *Revista Mexicana de Fitopatología*, **24**: 42-49.

Hibbing, M.E., Fuqua, C., Parsek, M.R. and Peterson, S.B., 2010. Bacterial competition: surviving and thriving in the microbial jungle. *Nat. Rev. Microbiol.,* **8**:15-25

Hsueh, Y. P., Mahanti, P., Schroeder, F. C. and Sternberg, P. W., 2013. Nematode-trapping fungi eavesdrop on nematode pheromones. *Curr. Biol.,* **23:** 83–86

Hungate, B.A., Marks, J.C., Power, M.E., Schwartz, E., van Groenigen, K.J, et al., 2021. The functional significance of bacterial predators. *mBio,* **12**:e00466-21.

Imperiali, N. et al., 2019. Relationships between Root Pathogen Resistance, Abundance and Expression of *Pseudomonas* Antimicrobial Genes, and Soil Properties in Representative Swiss Agricultural Soils. *Front. Plant Sci.,* **8:** 427

Izquierdo-García, L. F., González-Almario, A., Cotes, A. M., and MorenoVelandia, C. A., 2020. *Trichoderma virens* Gl006 and *Bacillus velezensis* Bs006: a compatible interaction controlling Fusarium wilt of cape gooseberry. *Sci. Rep.,* **10:**6857.

Jackson, R.B. et al., 2017. The ecology of soil carbon: pools, vulnerabilities, and biotic and abiotic controls. *Ann. Rev. Ecol. Evol. Syst.,* **48:** 419–445

Jacoby, R., Peukert, M., Succurro, A., Koprivova, A., and Kopriva, S., 2017. The role of soil microorganisms in plant mineral nutrition-current knowledge and future directions. *Front. Plant Sci.,* **8:** 1617

Jangir, M., Pathak, R., Sharma, A., Sharma, S. and Sharma, S., 2019. Volatiles as strong markers for antifungal activity against *Fusarium oxysporum* f. sp. *lycopersici*. *Indian Phytopathol.,* **72:**681–687

Jiang, Y. et al., 2017. Nematode grazing promotes bacterial community dynamics in soil at the aggregate level. *ISME J.,* **11:** 2705–2717

Jousset, A., 2012. Ecological and evolutive implications of bacterial defences against predators. *Environ. Microbiol.,* **14**: 1830–1843

Karim, H., Azis, A.A. and Jumadi, O., 2022. Antagonistic activity and characterization of indigenous soil isolates of bacteria and fungi against onion wilt incited by *Fusarium* sp. *Arch. Microbiol.,* **204:** 68.

Karimi, B., Meyer, C., Gilbert, D. et al., 2016. Air pollution below WHO levels decreases by 40% the links of terrestrial microbial networks. *Environ. Chem. Lett.,* **14**:467–475.

Kour, D., Rana, K.L., Yadav, N., Yadav, A.N., Kumar, A., Meena, V.S., Singh, B., Chauhan, V.S., Dhaliwal, H.S. and Saxena, A.K., 2019. Rhizospheric microbiomes: biodiversity, mechanisms of plant growth promotion, and biotechnological applications for sustainable agriculture. In: *Plant Growth Promoting Rhizobacteria for Agricultural Sustainability*. Springer, Singapore, pp 19–65.

Kumari, P., Bishnoi, S.K. and Chandra, S., 2021. Assessment of antibiosis potential of *Bacillus* sp. against the soil-borne fungal pathogen *Sclerotium rolfsii* Sacc. (*Athelia rolfsii* (Curzi) Tu & Kimbrough). *Egypt. J. Biol. Pest Control,* **31:** 17

Laliberte, E., Lambers, H., Burgess, T.I. and Wright, S.J., 2015. Phosphorus limitation, soil-borne pathogens and the coexistence of plant species in hyper diverse forests and shrub lands. *New Phytologist,* **206:** 507–521

Lara-Núñez, A., García-Ayala, B. B., Garza-Aguilar, S. M., Flores-Sánchez, J., Sánchez-Camargo, V. A., Bravo-Alberto, C. E., et al., 2017. Glucose and sucrose differentially modify cell proliferation in maize during germination. *Plant Physiol. Biochem.,* **113:** 20–31

Liechty, Z., Santos-Medellin, C., Edwards, J., Nguyen, B., Mikhail, D., Eason, S., et al., 2020. Comparative analysis of root microbiomes of rice cultivars with high and low methane emissions reveals differences in abundance of methanogenic archaea and putative upstream fermenters. *mSystems,* **5**:e00897- 19

Lingappa, B. T. and John L. Lockwood, 1961. The Nature of the Widespread Soil Fungistasis. *J. Gen. Microbiol.,* **26**: 473-485

Lisboa, B.B., Bayer, C., Passaglia, L.M.P., de Oliveira Camargo, F.A., Beneduzi, A., Ambrosini, A. and Vargas, L.K., 2015. Soil fungistasis against *Fusarium graminearum* under different crop management systems. *Rev. Bras. Cisnc. Solo,* **39**: 69–77.

Liu, T., Huang, Y., Chen, X.X., Long, X., Yang, Y.H., Zhu, M.L., Mo, M.H. and Zhang, K.Q., 2020. Comparative Transcriptomics Reveals Features and Possible Mechanisms of Glucose-Mediated Soil Fungistasis Relief in *Arthrobotrys oligospora. Front. Microbiol.,* **10:**3143

Lockwood. J. L., 1977. Fungistasis in Soils. Journal series article no. 7747 from the Michigan Agricultural Experiment Station.

Madhav P. Thakur and Stefan Geisen, 2019. Trophic Regulations of the Soil Microbiome. *Tre. Microbiol.,* **27**: 771-780

Marshall, K.C. and Alexander, M., 1960. Competition between soil bacteria and fusarium. *Plant Soil,* **12:** 143–153

Martinez, C., Avis, T. J., Simard, J. N., Labonté, J., Bélanger, R. R. and Tweddell, R. J., 2006. The role of antibiosis in the antagonism of different bacteria towards *Helminthosporium solani,* the causal agent of potato silver scurf / Le rôle de l'antibiose dans l'antagonisme de différentes bactéries contre *Helminthosporium solani*, agent responsable de la gale argentée de la pomme de terre. *Phytoprotection,* **87**: 69–75.

Mee, M. T., Collins, J. J., Church, G. M. and Wang, H. H., 2014. Syntrophic exchange in synthetic microbial communities. *Proc. Natl. Acad. Sci. U.S.A.,* **111:** E2149–E2156.

Minelli, A., 2008. Predation. *Encyclo. Ecol.*, 2923-2929

Minkwitz, A. and Berg, G., 2001. Comparison of antifungal activities and 16S ribosomal DNA sequences of clinical and environmental isolates of *Stenotrophomonas maltophilia. J. Clin. Microbiol.,* **39:**139-145.

Mishra, M., Huang, J. and Balasubramanian, M. K., 2014. The yeast actin cytoskeleton. *FEMS Microbiol. Rev.*, **38:** 213–227

Narendra Kumar, S.M. Paul Khurana, 2021. 3 - Trichoderma-plant-pathogen interactions for benefit of agriculture and environment. In: *Biocontrol Agents and Secondary Metabolites*, Sudisha Jogaiah (Ed.), Woodhead Publishing, P. 41-63.

Nguyen, B.T., Dumack, K., Trivedi, P., Islam, Z. and Hu, H.W., 2023. Plant associated protists-untapped promising candidates for agrifood tools. *Environ. Microbiol.,* **25**: 229–240.

Niu, B., Wang, W., Yuan, Z., Sederoff, R.R., Sederoff, H., Chiang, V.L. and Borriss, R., 2020. Microbial Interactions within Multiple-Strain Biological Control Agents Impact Soil-Borne Plant Disease. *Front. Microbiol.,* **11**:585404

Ossowicki, A., Tracanna, V., Petrus, M.L.C., van Wezel, G., Raaijmakers, J.M., Medema, M.H. and Garbeva, P., 2020. Microbial and volatile profiling of soils suppressive to *Fusarium culmorum* of wheat. *Proc. Royal Soc.,* **287:** 20192527

Oßwald, C., Zipf, G., Schmidt, G., Maier, J., Bernauer, H. S., Müller, R. and Wenzel, S. C., 2014. Modular Construction of a Functional Artificial Epothilone Polyketide Pathway. *ACS Synth. Biol.*, **3:** 759–772.

Pahari A. and B.B. Mishra, 2017. Antibiosis of Siderophore Producing Bacterial Isolates against Phythopathogens and Their Effect on Growth of Okra. *Int.J.Curr.Microbiol.App.Sci.*, **6:** 1925-1929.

Paolina Garbeva, W.H. Gera Hol, Aad J. Termorshuizen , George A. Kowalchuk and Wietse de Boer, Fungistasis and general soil biostasis – A new synthesis. *Soil Biol. & Biochem.*,**43**: 469-477

Pérez-Valera, E., Kyselková, M., Ahmed, E. *et al.,* 2019. Native soil microorganisms hinder the soil enrichment with antibiotic resistance genes following manure applications. *Sci. Rep*., **9:** 6760

Pishchany, G., Mevers, E., Ndousse-Fetter, S., Horvath, D. J. Jr., Paludo, C. R., Silva-Junior, E. A., et al., 2018. Amycomicin is a potent and specific antibiotic discovered with a targeted interaction screen. *Proc. Natl. Acad. Sci. U.S.A.,* **115:** 10124–10129.

Prashar Pratibha, Kapoor Neera and Sachdeva Sarita, 2013. Biocontrol of Plant Pathogens Using Plant Growth Promoting Bacteria. **10.**1007/978-94-007-5961-9_10.

Pu, Y. and Elsas, J.D.V., 2018. Mechanisms and ecological implications of the movement of bacteria in soil. *Appl. Soil Ecol.,* **129**:112–120.

Regin Rønn, Mette Vestergård and Flemming Ekelund, 2012. Interactions between Bacteria, Protozoa and Nematodes in Soil. *Acta Protozool.,* **51**: 223–235

Ren, P., Sun, A., Jiao, X., Shen, J.P., Yu, D.T., Li, F., Wu, B., He, J.Z. and Hu, H.W., 2023. Predatory protists play predominant roles in suppressing soil-borne fungal pathogens under organic fertilization regimes. *Sci. Total Environ.,* **863**:160986

Rothrock Craig and Gottlieb David, 2011. Role of antibiosis in antagonism of *Streptomyces hygroscopicus* var. *geldanus* to *Rhizoctonia solani* in soil. *Canadian J. Microbiol.,* **30**:1440-1447.

Santhanam, R., Menezes, R. C., Grabe, V., Li, D., Baldwin, I. T. and Groten, K., 2019. A suite of complementary biocontrol traits allows a native consortium of root-associated bacteria to protect their host plant from a fungal sudden-wilt disease. *Mol. Ecol.,* **28:** 1154–1169

Schwelm, A., Brennan, F. and Geisen, S., 2023. No rest for resting spores: can predators mitigate club root disease? *J. Sustain Agric. Environ.*, **2:** 131–139.

Shunran Hu, Guixin Li, Alejandro Berlinches de Gea, Joliese Teunissen, Stefan Geisen, Rutger A. Wilschut, Arne Schwelm and Yuxin Wang, 2023. Microbiome predators in changing soils. *Environ. Microbiol.*, **25**: 2057-2067

Singh Rashmi, Sudarshan Maurya and Ram Sanmukh Upadhyay, 2016. The improvement of competitive saprophytic capabilities of *Trichoderma* species through the use of chemical mutagens. *Brazilian J. Microbiol.,* **47**: 10-17

Soren Thiele-Bruhn and Iris-Constanze Beck, 2005. Effects of sulfonamide and tetracycline antibiotics on soil microbial activity and microbial biomass. *Chemosphere*, **59**: 457–465

Speißer, B., Wilschut, R.A. and van Kleunen, M., 2022. Number of simultaneously acting global change factors affects composition, diversity and productivity of grassland plant communities. *Nat. Commun.*, **13:** 7811.

Starr, E.P., Shi, S., Blazewicz, S.J., Koch, B.J., Probst, A.J., Hungate, B.A., Pett-Ridge, J., Firestone, M.K. and Banfield, J.F., 2020. Stable isotope informed genome-resolved metagenomics uncovers potential trophic interactions in rhizosphere soil. *bioRxiv:* 2008.2021.262063

Stevens, A., 2010. Predation, Herbivory, and Parasitism. *Nature Edu. Knowledge,* **3**:36

Supronienė, S., Kadžienė, G., Shamshitov, A., Veršulienė, A., Šneideris, D., Ivanauskas, A. and Žvirdauskienė, R., 2023. Soil Fungistasis against *Fusarium Graminearum* under Different tillage Systems. *Plants,* **12**:966.

Szekeres, A., Manczinger, L., Kredics, L., Antal, Z. and Kevei, F., 2002. Acta Microbiol Immunol Hung, **49:** 405-406

Tanya Arseneault and Martin Filion, 2017. Biocontrol through antibiosis: Exploring the role played by sub inhibitory concentrations of antibiotics in soil and their impact on plant pathogens. *Can. J. Plant Pathol.,* **39:** 267–274

Tao, L., Ahmad, A., de Roode, J. C., and Hunter, M. D., 2016. Arbuscular mycorrhizal fungi affect plant tolerance and chemical defences to herbivory through different mechanisms. *J. Ecol.,* **104:** 561–571.

Tao, L., Hunter, M.D. and de Roode, J.C., 2017. Microbial Root Mutualists Affect the Predators and Pathogens of Herbivores above Ground: Mechanisms, Magnitudes, and Missing Links. *Front. Ecol. Evol.,* **5:**160.

Teixeira, P. J. P., Colaianni, N. R., Fitzpatrick, C. R. and Dangl, J. L., 2019. Beyond pathogens: Microbiota interactions with the plant immune system. *Curr. Opin. Microbiol.,* **49:** 7–17

Thakur, M.P. et al., 2018. Reduced feeding activity of soil detritivores under warmer and drier conditions. *Nat. Clim. Chang.,* **8:** 75–78

Thakur, M.P. and Geisen, S., 2019. Trophic regulations of the soil microbiome. *Tre. Microbiol.,* **27:** 771–780.

Thomashow, L.S., Bonsall, R.F. and Weller, D.M., 2008. Detection of antibiotics produced by soil and rhizosphere microbes in situ. *In: Secondary metabolites in soil ecology.* Karlovsky, P., (Ed.), Berlin, Heidelberg: Springer Berlin Heidelberg; p. 23–36.

Ulloa-Ogaz, A.L., Muñoz-Castellanos, L.N. and Nevárez-Moorillón, G.V., 2015. Biocontrol of phytopathogens: Antibiotic production as mechanism of control. In: *The Battle Against Microbial Pathogens: Basic Science, Technological Advances and Educational Programs.* A. Méndez-Vilas (Ed.). 305-309

van Dijk, K and Nelson, E.B., 2000. Fatty acid competition as a mechanism by which *Enterobacter cloacae* suppresses *Pythium ultimum* sporangium germination and damping-off. *Appl. Environ. Microbiol.,* **66**:5340–5347

Venturi, V. and Keel, C., 2016. Signaling in the rhizosphere. *Tre. Plant Sci.,* **21**:187–198.

Vinh D. Pham, Conrad W. Shebelut, Michelle E. Diodati, Carolee T. Bull and Mitchell Singer, 2005. Mutations affecting predation ability of the soil bacterium *Myxococcus xanthus. Microbiology*, **151**: 1865–1874

Wang, X., Li, GH., Zou, CG. *et al.,* 2014. Bacteria can mobilize nematode-trapping fungi to kill nematodes. *Nat. Commun.,* **5:** 5776

Wegst-Uhrich, S.R., Navarro, D.A., Zimmerman, L. and Aga, D.S., 2014. Assessing antibiotic sorption in soil: a literature review and new case studies on sulfonamides and macrolides. *Chem. Cent. J.,* **8:**5

Weller, D.M., 2007. Pseudomonas biocontrol agents of soil borne pathogens: looking back over 30 years. *Phytopathology,* **97**: 250–256

Wu, M., Zhang, H., Li, X., Zhang, Y., Su, Z. and Zhang, C., 2008. Soil fungistasis and its relations to soil microbial composition and diversity: A case study of a series of soils with different fungistasis. *J. Environ. Sci., (China),* **20**:871-7

Xiong, W., Song, Y., Yang, K., Gu, Y., Wei, Z., Kowalchuk, G.A, et al.,2020. Rhizosphere protists are key determinants of plant health. *Microbiome,* **8**:27.

Xu Chuankun, Mo Minghe, Zhang Leming and Zhang Keqin, 2004. Soil volatile fungistasis and volatile fungistatic compounds. *Soil Biol. & Biochem*., **36:**1997-2004

Ye, X., Li, Z., Luo, X. *et al.,* 2020. A predatory myxobacterium controls cucumber *Fusarium* wilt by regulating the soil microbial community. *Microbiome,* **8**: 49.

Zhu, D., Li, G., Wang, H.T. and Duan, G.L., 2021. Effects of nano- or microplastic exposure combined with arsenic on soil bacterial, fungal and protistan communities. *Chemosphere,* **281:** 130998.

7

Conducive and Suppressive Soils

7.1 Conducive Soils

Without a conducive environment, plant disease does not occur. Soils that favour the disease/ and the pathogen are generally referred to as conducive soils. Inoculum of powdery scab is the either the spore ball (an aggregation of thick-walled, tough, resting spores, comprising an average of 700 spores) or the swimming spore (zoospore) that each spore of the spore ball releases. Spore balls of powdery scab have been shown to persist for long periods (Brierley et al., 2008). Exactly, how long they can survive has not been fully determined, but experimentally and anecdotally, it is much longer than a typical rotation. Inoculum may be seed-borne or soil-borne and whatever the source spore balls are able to cause infection and disease. Seed tubers may carry spore balls within powdery scab lesions or may carry spore balls symptomlessly where they have been picked up from adjacent infected tubers or contaminated machinery (e.g. on grading lines). In addition, recently, it has been found that tubers may be latently infected. That is where tubers are infected but symptoms are not expressed. Latent infections may be associated with conditions that are unfavourable for disease development, such as sub-optimal temperatures.

i. **Soil Moisture:** A high soil moisture content both encourages swimming spore release and facilitates movement to the host. It is generally accepted that swimming spores are unable to move through the soil without free water within the soil matrix (i.e. soil saturation). Many studies confirm the importance of free water in soil as a requirement for infection. The occurrence of wet soil conditions prior to tuber initiation, enabling inoculum multiplication, has been cited as important for high levels of powdery scab. Alternatively, prolonged wet soil conditions after tuber initiation can lead to high levels. In experiments investigating soil moisture, there is disagreement over the importance of constant versus fluctuating soil moisture conditions in powdery scab development. However, it is difficult to make clear comparisons because soil moisture levels have not always been measured and factors that may affect disease development (e.g. soil type, initial inoculum level) will have differed from experiment to experiment. Some studies have shown that constant high soil moisture has suppressed powdery scab development possibly due to low oxygen/high carbon dioxide levels rendering swimming spores inactive.

ii. **Temperature:** The relationships between temperature and infection are much better understood than soil water relationships. The release of swimming spores is more rapid at warmer temperatures (15-25° C) compared to cooler temperatures (3-10° C) and root infection is very limited at low temperatures (Brierley et al., 2008). Thus, at 9° C infection is more limited than at 12° C or 17 °C. The optimum temperature for root galling is around 17° C. In contrast, the optimum temperature for tuber infection is around 12° C. Around this optimum, infection can occur from around 9° C to 17° C. Outside this range the extent of tuber infection has proved very limited.

iii. **Weather and Powdery Scab:** It is clear from the studies on soil moisture and temperature summarised above that powdery scab is favoured by cool, wet conditions. In the UK, these occur in the north and west of the country predominantly. The occurrence and significance of powdery scab in the UK reported by industry confirms that northern Britain is most at risk, although in cool wet springs powdery scab can occur anywhere in the GB. In the British Potato Council Reference Crop project, funded between 1998 and 2000, the same stocks were planted in four locations and trials were carried out with and without irrigation. Seed infected with powdery scab (assessed visually) was planted in all three years. Whilst soil inoculum was not measured, in the two extreme sites (Cambridge and Dundee), tuber initiation temperatures at the two sites were c. 17.6° C and 14.2° C respectively. At Cambridge, very little powdery scab developed even in the irrigated plots. In the two years when rainfall was low at both sites at tuber initiation, high levels of powdery scab developed in the irrigated plots at Dundee. This suggests the lower temperatures at tuber initiation at Dundee were more conducive to infection.

iv. **Soil pH**:Despite a reasonable number of studies, there is little evidence to suggest *S. subterranea* infection is influenced by pH within the normal range found in arable soils. In other countries, lowered pH associated with the addition of sulphur has reduced powdery scab (although the high levels of sulphur applied may have affected the pathogen) and increasing pH by use of lime or nitro chalk has indicated that the risk of powdery scab is increased (although the use of lime prior to planting a potato crop will affect incidence of common scab).

v. **Soil Type**:There is no clear evidence that powdery scab development is influenced by soil type. However, anecdotally and supported by some studies, lighter sandy soils are often reported to have the worst infection. There is no evidence of 'suppressive' or 'conducive' soil types although development of powdery scab in clay soils (Ex: in Essex) has been limited despite infected seed of susceptible varieties being planted. There is some evidence that powdery scab development in soils with high levels of zinc may be less (Brierley et al., 2008).

vi. **Host Resistance**:Trials have clearly demonstrated that changing from a susceptible variety to a more resistant one gives the most consistent reduction in incidence and severity of powdery scab in tubers of all control measures available. However, in GB, 75% of all varieties grown are susceptible or moderately susceptible to powdery scab. Published resistance ratings refer to tuber resistance but infection of roots and development of galls on roots, stems and stolons may still occur. Thus, whilst growing a (tuber) resistant variety will probably result in a lower incidence and severity of powdery scab on harvested tubers, effect on soil inoculum is not known. With some varieties, the occurrence of root galls may increase the level of soil inoculum. There are no published ratings on relative susceptibility in roots and it has been established that there is no correlation between root and tuber infection.

7.1.1 Soil Properties conducive to the formation of *Tuber aestivum* fruiting bodies: Summer truffle (*Tuber aestivum*), also known as Burgundy truffle, is getting interest in Poland in terms of cultivation as a promising incentive for rural areas. Yet the occurrence of the fungus in wider scale in our country has been confirmed in the last decade. Ecological factors that determine the occurrence of *T. aestivum* are rather well known in the Mediterranean region, whereas such knowledge is limited in northern Europe (Dorota Hilszczańska et al., 2019). The aim of this work was to find the correlations between essential nutrients in surface horizons of soils typical of truffle occurrence. The study area is situated in the Nida Basin in southern Poland. Principal component analysis (PCA) showed that active carbonate content is the variable that accounts for the greatest percentage of occupancy in the *T. aestivum* habitat. It was proposed that active carbonate is a major factor in the fruiting of summer truffle. The obtained results could have applications in natural harvesting and truffle culture.

Burgundy truffle *Tuber aestivum,* one of the species among black truffles, is a prized fungus due to its taste and aroma. This hypogeous species is found throughout Europe and belongs to the Pezizales, a large group of ectomycorrhizal fungi that form mutualistic symbioses with the roots of angiosperms and gymnosperms. The importance of soil chemical and physical characteristics for truffle (e.g. *T. aestivum*) ectomycorrhiza development and fruiting bodies (ascocarps) formation in plantation and natural stands is well documented. Data on *T. aestivum* ecology are still rather sparse due to the fact that the species is less valued (economically and culinary) than, for example, *T. melanosporum*. Although *T. aestivum* is widely distributed throughout Europe , researchers have given it less attention. This data showed that the development of Burgundy truffle fruiting bodies is influenced by content of carbonates. Both forms of the carbonates, namely $CaCO_3$ and exchangeable Ca_2+, regulate pH, the latter form being more relevant for truffle development. An increase of $CaCO_3$ content positively affected a number of *T. aestivum* fruiting bodies. This corresponds to the broad opinion that truffles in general and Burgundy truffle in particular, require high calcium availability in the soils. High bioavailability of calcium stimulates the development of truffle mycelium. The importance of carbonates for proper development of three truffle species, *T. melanosporum, T. aestivum* and *T. mesentericum* has been highlighted (Dorota Hilszczańska et al., 2019).

Sandy textures are less conducive to truffle development. The result is in accordance with findings by Chevalier and Sourzat. At *T. aestivum* sites in natural forests in Poland, the soil texture is highly variable, ranging from silty-clay to clayey-silt and more rarely silty, silty-sandy, or clayey. The analysed soils varied from "heavy" (up to 55.4% clay) to "light" (up to 46% sand). The "light" soils are generally thought to be less favourable for truffle cultivation. However, even excessively sandy soils can support truffle development if they are sufficiently rich in calcium (Chevalier and Sourzat, 2012). Thus, *T. aestivum* appears to have a wide tolerance for soil texture as it develops in sedimentary terrain of various geological ages, from Palaeozoic to Quaternary and recent alluvia. Moreover, *T. aestivum* fruit bodies are able to grow on alkaline volcanic substrates.

The ecological and pedo-climatic requirements of the Burgundy truffle allows the fungus to adaptat to different environments typical of the Mediterranean region and in northern and eastern Europe as well. The main challenge of research about *T. aestivum* soils is to estimate very precisely the optimal conditions favourable for fruiting bodies' growth. It is also important to characterize conditions that facilitate the colonization and persistence of *T. aestivum* mycorrhizae on root systems. This knowledge is a key factor for establishing best management practices in truffle orchards, such as irrigation, soil tillage, supplies of organic matter, and nutrients. These results are a new input of knowledge in this aspect and are important for truffle growers, especially in new areas outside the Mediterranean region.

7.1.2 Characterization of soils conducive and non-conducive to *Prunus* replant disease, *Phelipanche ramose* and Tobacco black root rot: Successive orchard plantings of almond and other *Prunus* species exhibit reduced growth and yield in many California soils. This phenomenon, known as *Prunus* replant disease (PRD), can be prevented by pre plant soil fumigation or anaerobic soil disinfestation, but its etiology is poorly understood and its incidence and severity are hard to predict. Khan Abdur et al., (2021) reported the relationships among physicochemical variables, microbial community structure, and PRD induction in 25 diverse replant soils from California. In a greenhouse bioassay, soil was considered to be "PRD-inducing" when growth of peach seedlings in it was significantly increased by pre plant fumigation and pasteurization, compared to an untreated control. PRD was induced in 18 of the 25 soils, and PRD severity correlated positively with soil exchangeable-K, pH, %clay, total %N, and electrical conductivity. The structure of bacterial, fungal, and oomycete communities differed significantly between the PRD-inducing and non-inducing soils, based on PERMANOVA of Bray Curtis dissimilarities. Bacterial class MB-A2-108 of phylum Actinobacteria had high relative abundances among PRD-inducing soils, while Bacteroidia were relatively abundant among non-inducing soils. Among fungi, many ASVs classified only to kingdom level were relatively abundant among PRD-inducing soils whereas ASVs of *Trichoderma* were relatively abundant among non-inducing soils. Random forest classification effectively discriminated between PRD-inducing and non-inducing soils, revealing many bacterial ASVs with high explanatory values. Random forest regression effectively accounted for PRD severity, with soil exchangeable-K and pH having high predictive value. This work revealed several biotic and abiotic variables worthy of further examination in PRD etiology.

This report identified multiple soil physicochemical and microbial variables in 25 diverse orchard and vineyard replant soils that collectively discriminated between soils that did or did not induce PRD in a greenhouse bioassay. It is logical to consider whether or not the greenhouse bioassay reflected PRD induction capacities under orchard conditions. Although the bioassay has not been rigorously validated against experimental field responses, there is evidence for useful correspondence between greenhouse bioassay and orchard trial responses. For example, the orchard supplying soil no. 6 did not respond significantly to preplant soil fumigation in the field, nor did peach seedlings respond significantly to fumigation or pasteurization of that soil in the bioassay. Conversely, orchards planted where soils 8 and 13 were collected responded positively to preplant soil fumigation, as did plants in the bioassay (note: in soil 8, PRD was induced according to plant total weights but not top weights). Bioassay-field correspondences were also evident in soil nos. 9 and 14: these soils were collected from zones (plots) that had been field fumigated, and accordingly, PRD was not induced by them in the bioassay. Also, grape vineyard soils do not typically induce immediate growth suppression in subsequent orchard plantings of *Prunus* (Browne, *unpublished*), nor did they induce PRD in the bioassays of soil nos. 11 and 12 from grape vineyards. Interestingly, however, PRD induction from soil no. 15 in the bioassay was not consistent with corresponding orchard data. Soils no. 13, 14, and 15 were collected, respectively, from the control, fumigated, and ASD-treated plots of a field trial. In the orchard, both the ASD and the fumigation treatments significantly and equally increased orchard tree growth and nut yields (Browne et al., 2018), which is contrary to the PRD induction shown by soil no. 15 in the bioassay. Although reasons for this bioassay-orchard discrepancy are unknown, it is possible that mechanisms of fumigation and ASD differ in ways not accounted for adequately in the bioassay. Clearly, further orchard validation of the PRD bioassay is advisable.

Comparative analysis of two neighboring conducive and suppressive soils towards plant parasitism caused by *Phelipanche ramosa* on *Brassica napus*: Background In western France, rapeseed farmers face significant yield losses due to root parasitism caused by *Phelipanche ramosa*, a holoparasite plant of the Orobanchaceae family. Recently, a reduction of parasitic plant development has been observed in fields with history of severe infestation (Martinez Lisa, et al., 2023). In a same given pedoclimatic environment, this so-called soil-suppression of parasitism might presumably result from microbial mechanisms targeting *P. ramosa*. This hypothesis was tested on two soils from neighboring rapeseed fields, with similar physicochemical properties but contrasted parasitism, characterized as suppressive and conducive. These soils were assessed in a hydroponic co-cultivation system of *P. ramosa* and *B. napus*, and simultaneously sampled rhizosphere exudates weekly, as well as rhizosphere and rapeseed roots at three time points along parasite development. Comparisons were thus drawn between conducive and suppressive soils, both in untreated or gamma-sterilized conditions, regarding the effects of soil derived signaling metabolites on broomrape key early parasitic stages (i.e. germination and pre-haustorium development), late parasite development as well as soil bacterial and fungal structures.

It was demonstrated that the suppressive soil mitigates broomrape parasitism by reducing both parasite attachments and development, while causing tubercle necrosis. Activity assays on initial soils as well as co-cultivation rhizosphere exudates reveal that pre-attachment stages of broomrape are not affected by suppression. Root exudates, either they were collected from plants cultivated on suppressive or conducive soils, exerted similar contribution on the induction of broomrape seed germination and on haustoria formation. Thus, attachment and post-attachment stages are more likely to be suppressed by soil microbiota. Given the observed dissimilarities in microbial composition between the soils, it was suggested that rather than bacteria, fungi play a greater part in parasitism suppression. Because suppression is not correlated with fungal diversity during co-cultivation, it was suspected that in addition to general microbial interactions, specific groups of fungi participate in parasitism reduction. Further correlation network analyses (WGCNA) propose that three clusters of ASVs are indeed correlated with suppressive parasitic traits by their abundance. In these clusters, 7 ASVs are inversely correlated with parasitic attachment and one is positively correlated with the occurrence of necrotic tubercles, while being more abundant in the suppressive soil (DEseq2). These 7 ASVs are from Nectriaceae, Niessliaceae, Bartaliniaceae, Holtermanniaceae families as well as from Helotiales and Pleosporales orders, while the last ASV is identified as a Berkeleyomyces sp., a necrotrophic fungal genus known as a causal agent of black root rot. This study explains the observed contrasted parasite development on two physiochemically similar soils by linking parasitic infestation to fungal dynamics in the rhizosphere. This finding extends our knowledge of disease suppressive soils to plant parasitism and proposes biological leads for potential biocontrol of broomrape.

Prevalence of fluorescent pseudomonads in soils naturally suppressive or conducive to tobacco black root rot: Certain soils from Morens, Switzerland, are naturally suppressive to *Thielaviopsis basicola*-mediated black root rot of tobacco, and fluorescent pseudomonads are involved in this suppressiveness. Alban Ramette et al., (2003) compared two conducive, one moderately suppressive and one suppressive soil from Morens. Disease levels on tobacco after heavy *T. basicola* inoculation varied from 29% to 85% for the two conducive soils, 10% to 78% for the moderately suppressive soil and 11% to 42% for the suppressive soil, depending on time of the year. In the absence of *T. basicola* inoculation, disease levels were between 0% and 40% and varied also in time. Fluorescent pseudomonads were isolated from the rhizosphere and roots of tobacco subjected to *T. basicola* inoculation and characterized for production of the biocontrol metabolites 2,4-diacetylphloroglucinol (Phl) and HCN. No difference in population size was found between the suppressive and the conducive soils for total, Phl^+ and HCN^+ fluorescent pseudomonads colonizing the rhizosphere or roots of tobacco. Yet, the percentage of Phl^+ isolates was significantly higher (30–32% vs. 6–11%) in the rhizosphere and roots for plants grown in the suppressive soil compared with the moderately suppressive and conducive soils. Different restriction profiles for *phlD*, one of the Phl biosynthetic genes, were often found when analyzing Phl^+ isolates colonizing the same plant. Most *phlD* alleles were recovered from both suppressive and conducive soils, except one allele found only in root isolates from the suppressive soil.

7.2 Suppressive Soils

Achieving the best possible results for you crop begins below the ground, and specifically in the top few centimetres of healthy topsoil. Topsoil is so important because it contains all the nutrients that plants need to survive. It is also the city in which the all-important soil organisms live – where the Bacteria, fungi, archaea, nematodes and nameless other organisms quietly feed and defend our crops against pests and pathogens. It is also in this zone that the crop protection battle begins. Crop protection is a challenge for every grower but boosting soil life can give added Armor for this battle. Within the myriad of organisms which make up the soil food web countless species have demonstrated defence traits against pathogens. By fostering these beneficial microorganisms, we can switch on these defence mechanisms to suppress these diseases and outcompete pathogens.

Suppressive soils are referred to as "soils in which certain diseases are suppressed due to the presence in the soil of microorganisms antagonistic to the pathogen or pathogens." Another definition states "soils in which the pathogen does not establish or persist, establishes but causes little or no damage, or establishes and causes disease for a while but thereafter the disease is less important, although the pathogen may persist in the soil" (David M. Weller et al., 2002). In contrast, conducive (non suppressive) soils are soils in which disease readily occurs. The terms disease-suppressive soil and pathogen suppressive soil often are used interchangeably. However, some consider pathogen suppression as the suppression of saprophytic growth or survival of the pathogen in the soil, and disease suppression as the suppression of the pathogen growing parasitically. Disease-suppressive soils are soils in which specific soil-borne plant pathogens cause only limited disease although the pathogen and susceptible host plants are both present. Suppressiveness is in most cases of microbial origin.

Suppressive soils represent an underutilized resource for the control of soil borne pathogens of food, fiber, and ornamental crops. Early research identified the characteristics of soil suppressiveness and the major groups of microorganisms involved, but only during the past decade has the coupling of molecular and traditional approaches made it possible to begin to dissect the genetic and functional determinants underlying the activity of some biologically suppressive soils. These studies provide a foundation for more widespread integration of suppressive soils into organic and conventional cropping systems.Natural disease-suppressive soils probably are the best examples in which the indigenous micro flora effectively protects plants against soil borne pathogens. Suppressive soils initially become apparent because the incidence or severity of disease is lower than expected for the prevailing environment or as compared to that in surrounding soil. Suppressive soils have been described for many soil borne pathogens, including *Gaeumannomyces graminis* var. *tritici, Fusarium oxysporum, Aphanomyces euteiches , Heterodera avenae, H. schachtii, Meloidogyne* spp., *Criconemella xenoplax, Thielaviopsis basicola, Phytophthora cinnamomi, P.infestans, Pythium splendens, P.ultimum, Rhizoctonia solani, Streptomyces scabies, Plasmodiophora brassicae* and *Ralstonia solanacearum.*

Agricultural soils suppressive to soilborne plant pathogens occur worldwide, and for several of these soils the biological basis of suppressiveness has been described. Two classical types of suppressiveness are known. General suppression owes its activity to the total microbial biomass in soil and is not transferable between soils. Specific suppression owes its activity to the effects of individual or select groups of microorganisms and is transferable. The microbial basis of specific suppression to four diseases, Fusarium wilts, potato scab, apple replant disease, and take-all, is discussed. One of the best-described examples occurs in take-all decline soils. In Washington State, take-all decline results from the build-up of fluorescent *Pseudomonas* spp. that produce the antifungal metabolite 2,4-diacetylphloroglucinol. Producers of this metabolite may have a broader role in disease-suppressive soils worldwide. By coupling molecular technologies with traditional approaches used in plant pathology and microbiology, it is possible to dissect the microbial composition and complex interactions in suppressive soils.

Disease suppressive soils function to prevent the establishment of pathogens in the rhizosphere of plants. These soils develop through the establishment of beneficial microbes, known as plant growth-promoting rhizobacteria (PGPR) in the rhizosphere of plant roots. These mutualistic microbes function to increase plant health by fighting against harmful soil microbes either directly or indirectly. As beneficial bacteria occupy space around plant roots they outcompete harmful pathogens by releasing pathogenic suppressive metabolites.

PGPR and harmful microbes: Plant growth promoting rhizobacteria are bacteria that promote plant growth, through fixing nitrogen, producing growth hormones, or even suppressing pathogens. A variety of PGPR genera provide a wide array of functions that directly improve plant health. Bacteria genera *Rhizobium* and *Mesorhizobium* work to fix nitrogen into a usable form (Terpolilli et al., 2012). *Actinomyces* and *Azospirillum* produce growth hormones that increase root growth and uptake of nutrients.

While many of these genera increase plant health directly some PGPR promote plant health indirectly through pathogen suppression. Pathogenic fungi create a complex of hyphae that migrate through the soil. When these fungi reach the rhizosphere they release enzymes that degrade the cell wall of plant root cells. This allows them to enter and infect the host plant and prevent the uptake of nutrients. Beneficial *Pseudomonas* and *Bacillus* produce fungal suppressing metabolites that break up this migrating fungal hyphae. Over time, plants are able to create disease-suppressive soils in response to these pathogens through the increasing the abundance of these PGPR in their rhizosphere area.

Establishment of suppressive soils: Plants respond to pathogens by recruiting PGPR to their root rhizosphere from the bulk soil to fill in and prevent pathogen establishment. This ultimately leads to the development of disease-suppressive soils. PGPR are recruited through the natural release plant exudates from root cells as they push through the soil. Different species of plants release different exudates and therefore recruit different microbial communities to their rhizosphere from the microbes already present in the surrounding bulk soil. If a beneficial microbe genera is not present in a soil plants are not able to recruit it as a defense. Therefore, suppressive soils are a function of the microbes already present in a soil and able to be recruited. The creation of these soils must be done by increasing PGPR in the overall bulk soil. Specific plants are able to recruit beneficial bacteria through the secretion of different root exudates. A greater diversity of plants in a soil leads to a greater diversity of microbes in the rhizosphere and furthermore can lead to greater suppression of soil diseases (Latz, et al., 2016). Management, such as informed crop rotation and soil solarization, can create suppressive soils that naturally suppress pathogens.

7.2.1 Disease Suppressive Soils- A Critical Review: Disease-suppressive soil (DSS), as the name suggests, does not refer to the complete elimination of the soil-borne pathogens from the soil environment. It refers to those soils in which disease development is at the minimum level even in the presence of the soil-borne pathogen and the susceptible host plant (Jayaraman et al., 2021). Although it is of prime importance to understand the functioning of the DSS, relatively few soils with suppressive properties have been described in the global literature to date. Disease suppressiveness not only refers to the healthy soils with a stable soil microbial community but also consists of advantageous physical and chemical soil properties that enhance crop protection. Soil disease suppression can be regarded as the soil biological property conferred by the microbial community (microbiome) because sterilization of DSS destroys the disease controlling capacity of the suppressive soils and, in addition, the transfer of disease suppressiveness could be possible when conducive soil receives an amount of suppressive soil. Plant species tend to develop their defence strategy against soil-borne diseases/pathogen through selective stimulation and support of antagonistic microorganisms also, indicates the potential of DSS in disease control.

Among the soil-borne pathogens, fungi and Oomycetes are one of the major pathogens which limit the productivity of the agroecosystems and are relatively difficult to control with the introduction of resistant host cultivars and use of synthetic fungicides. Few soil-borne diseases, for example, *Aphanomyces euteiches* Dress. that causes root rot of pea are difficult to control with fungicides or developing plant resistance. The only practical method to reduce the losses is to completely avoid infection of the crops with pathogens. Therefore, in such cases, developing soil-related disease suppressiveness could potentially work in reducing the loss of productivity due to the soil-borne pathogens. Therefore, the development of DSS could be one of the effective examples in which the indigenous soil microbiota effectively protects the host plant against the infections from soil-borne pathogens. Deciphering the rhizosphere microbiome to screen and identify those beneficial microbes that contribute to disease suppression is the first step in unravelling the direct and indirect mechanisms of DSS.

Disease-suppressive soils (DSS) have been identified and studied around the world. Soil disease suppression is the reduction in the incidence of soil-borne diseases even in the presence of a host plant and inoculum in the soil. The disease-suppressive capacity is mainly attributed to diverse microbial communities present in the soil that could act against soil-borne pathogens in multifaceted ways (Jayaraman et al., 2021). The beneficial microorganisms employ some specific functions such as antibiosis, parasitism, competition for resources, and predation. However, there has been increasing evidence on the role of soil abiotic factors that largely influence the disease suppression. The intricate interactions of the soil, plant, and environmental components in a disease triangle make this process complex yet crucial to study to reduce disease incidence. Increasing resistance of the pathogen to presently available chemicals has led to the shift from culturable microbes to unexplored and unculturable microbes. Agricultural management practices such as tillage, fertilization, manures, irrigation, and amendment applications significantly alter the soil physicochemical environment and influence the growth and behaviour of antagonistic microbes. Plant factors such as age, type of crop, and root behaviour of the plant could stimulate or limit the diversity and structure of soil microorganisms in the rhizosphere. Further, identification and in-depth of disease-suppressive soils could lead to the discovery of more beneficial microorganisms with novel anti-microbial and plant promoting traits. To date, several microbial species have been isolated and proposed as key contributors in disease suppression, but the complexities as well as the mechanisms of the microbial and abiotic interactions remain elusive for most of the disease-suppressive soils. This review critically explores disease-suppressive attributes in soils, mechanisms involved, and biotic and abiotic factors affecting DSS and also briefly reviewing soil microbiome for anti-microbial drugs, in fact, a consequence of DSS phenomenon.

The DSS has been observed for several soil-borne pathogens such as *Streptomyces scabies*, *Pythium splendens*, *Pythium ultimum*, *Thielaviopsis basicola*, *Heterodera avenae*, *Phytophthora cinnamomi*, *Criconemella xenoplax*, *Phytophthora infestans*, *Fusarium oxysporum*, *Rhizoctonia solani*, *Meloidogyne* spp., *Gaeumannomyces graminis* var. *tritici*, *H. schachtii*, *Ralstonia solanacearum*, *Aphanomyces euteiches* and *Plasmodiophora brassicae*. The so-called disease-suppressive soils have the ability to suppress the various plant pathogens by improving the plant health, induce natural plant defence, produce antibiotics, compete against pathogens, modulate plant immune system, or hyperparasite the pathogen (Liu et al., 2020). These DSS have been identified for almost 60 years before; however, recent advances in next-generation sequencing methods have open the new arena in the understanding of DSS Moreover, comprehensive information on DSS is lacking. Thus, an attempt has been made in this review to understand different aspects of DSS such as types of disease suppressiveness, mechanisms, biotic and abiotic factors on DSS, method of study, and soil attributes on disease suppressiveness. This article also deliberates DSS under different management practices and indicates undiscovered soil microbiome for antibiotics and drugs and complexities in understanding disease suppressiveness (Jayaraman et al., 2021).

7.2.2 Types of Soil-related disease suppression and Disease Triangle Model: The basic reason for the high incidence of soil-borne diseases in croplands is the deterioration of the soil micro-ecological environment that often disrupts the soil microbial community balance (Jayaraman et al., 2021). Therefore, attempts have been made to differentiate the microbial community structure of DSS from those of disease-conducive soil (DCS). Maintaining a dynamic microbial balance between species, high soil microbial biomass, and high microbial diversity are key factors which facilitate the development of DSS. High microbial diversity in the soil allows fewer pathogens to survive, and may also prevent the invasion of exogenous pathogens. Several soil microorganisms confer benefits in nutrient acquisition and protect host plants by preventing colonization from soil-borne pathogens.

There are two distinct types of disease suppressiveness in soil, viz., general and specific mechanisms. General suppression mechanism (GSM) refers to the widespread but limited potential of the soil in suppressing the disease-causing activity of soil-borne pathogens. It is also referred to as non-specific antagonism or biological buffering. The GSM category refers to the disease suppression through the competition between the resident soil microbes and pathogens for a common resource such as nutrients and/or space and the release of antibiotic compounds and toxins from an active soil microbial community that hampers the growth and development of the specific pathogen. On the other hand, specific suppression mechanism (SSM) is another type of disease suppression but superimposed over the main concept of GSM. The SSM refers to the effects of an individual or specific groups of soil microorganisms (or during specific stages of pathogen life cycle) on disease suppression. The distinctive characteristic of SSM is its transferability factor; i.e. it can be transferred from suppressive soil to a conducive soil. The specific mechanism includes *parasitism* and *predation* of the pathogens by the soil-dwelling microbes. However, in general, the mechanism of soil suppression includes antibiosis, competition, parasitism, and predation (Chandrashekara et al., 2012). They have their activity in a blend of "*general*" and "*specific*" suppression.

General suppression is the capacity of soils to suppress the growth and activity of soil-borne pathogens to a certain level due to the combined competitive and antagonistic activity of the total soil microbiome fighting with the pathogens. Though some argue for limiting the use of the term disease suppressiveness to situations involving a clear biological component, there is plentiful evidence for the role of both biotic and abiotic elements of the soil having a role in disease suppression (Mousa and Raizada, 2016). Chemical and physical attributes of soil, including pH, soil organic matter (SOM), and clay content, can operate in the suppression of plant diseases directly or indirectly through their impact on soil microbial activity. These soil attributes are largely influenced under different management practices, and thereby control the microbial population and diversity in soil rhizosphere. Long-standing type of disease suppression occurs in soils with specific disease suppression, in which the disease suppression is naturally associated with the soil without the presence of plants, although its origins are unknown. The Fusarium wilt suppressive soils from Chateaurenard region of France are one such example of long-standing disease suppression. On the other hand, "induced" type of specific disease suppression is associated with sustainance of disease suppression through crop monoculture or inoculation of the soil with pathogens or by growing susceptible crops (Weller 2002). A prominent example of "induced" specific suppression is decline of take-all disease through wheat monoculture.

The Disease Triangle Model: As the soil microbes and the pathogens share a common space in the rhizosphere, their interactions during pre- and post-planting have a great influence on plant productivity (Jayaraman et al., 2021). The whole disease suppression concept can be visualized with a disease triangle concept that consists of three major determinants: host plant, soil-borne pathogen, and environmental factors.

Since plants are the main providers of soil C and energy sources, plant diversity affects the composition and structure of microbial communities. The physical and chemical properties of soil such as pH, electrical conductivity, soil nutrients, and soil organic carbon (SOC) also determine the microbial activities as soil provides the optimum habitat for the growth and development of the soil microbes. In addition to these, crop management practices such as tillage, fertilization, weeding, and irrigation manipulate the soil environment and affect the population of soil microbiota (Fig. 42). It is nearly an impossible task to study the roles of these factors independently in disease suppression, and researchers need to address them simultaneously. Understanding the disease triangle model (i.e. a model based on the interactions between the host plant, pathogen, and the environment conducive for disease development) is one of the key steps in studying the complex system of the DSS. Diseases are developed when all the components of this model are in favourable conditions for the pathogen. The "environment" component needs to be manipulated and specifically tailored to develop DSS or reduce the conduciveness of disease development even in the presence of the pathogen and the host. Until and unless soil properties are completely modified or pathogens are mutated, the persistence period of disease suppression is quite long even with the repeated introduction of the pathogen in soil.

Commonly available amendments such as vermicompost, rice straw, animal manure, and green waste are potential amendments which have disease-suppressive qualities through their influence on soil microbial communities. However, despite showing satisfactory properties of the biocontrol amended soils in either laboratory or controlled conditions, it is a major challenge to achieve the same results under field conditions. Such a differential response is attributed to the complex and specific interactions between these components in the disease triangle model, and better mixing of the bio-inoculant enriched compost with the soil in controlled conditions than in the fields

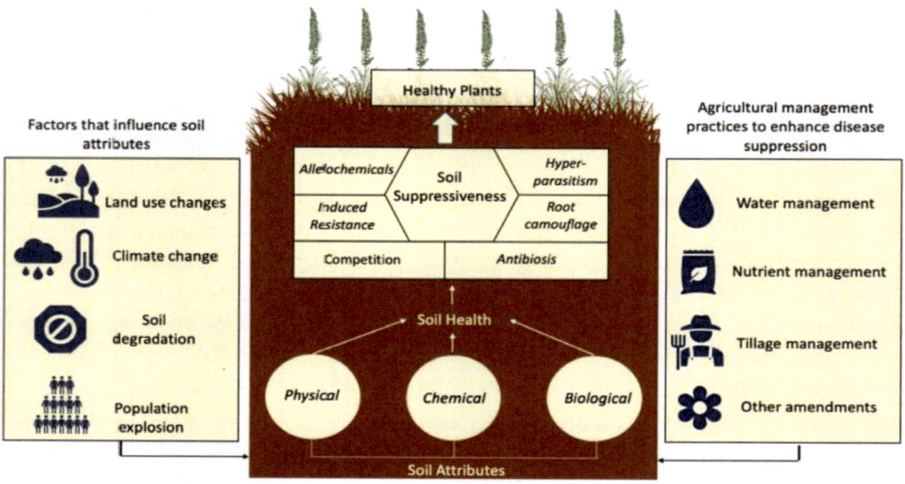

Fig. 42: The interplay of the soil inherent factors and environmental changes in the development of disease-suppressive soils

The degree of suppressiveness is linked to soil physical conditions, fertility level, biodiversity and populations of soil organisms, and soil management practices. For example, use of manure application modifies soil physical, chemical, and biological parameters that directly affect crop infection and the survival of the pathogen. *Pythium* sp. suppression was linked to volatilization of ammonia from manure amendments. The application of liquid swine manure reduced the occurrence of wilt, as well as a common scab in potato fields. It also lowered the number of plant-parasitic nematodes for a period of 3 years following a single-use. A significant reduction in root disease of the red stele strawberry was also observed in fields treated with steer/poultry and dairy manure compost, comparative to control.

7.2.3 Biotic and Abiotic Contributors to Disease Suppressive Soil

i. Soil Bacterial and Archaeal Community: The composition of the microbial community is one of the major factors in the mechanism of disease suppression (Shi et al., 2019). In microarray analysis, higher signals in suppressive soil could be observed from non-pathogenic *Streptomyces* (*Actinobacteria*), *Bradyrhizobium*, *Burkholderia* whereas the conducive soil exhibits higher signals from *Acidobacteria*, *Pseudomonas*, *Agrobacterium*, and *Janthobacterium* (*Proteobacteria*) (Jayaraman et al., 2021). Actinobacteria (multicellular bacteria) are known for their inherent ability to produce a large number of bioactive secondary metabolites such as 2,4 –diacetylphloroglucinol (DAPG) that can inhibit the growth of the soil-borne pathogens. Similarly, the archeal community is also neglected in biocontrol and disease suppressiveness even though it is also a part of the rhizosphere microbiome. A suppressive soil was differentiated from a conducive soil by lower quantities of actinobacteria and a specific composition in archaea and micro-eukaryotes. More disease-suppressive characteristics were detected in *Proteobacteria*, *Firmicutes*, and *Actinobacteria.* Other pathogens suppressing members often detected are *Xylariaceae*, *Lactobacillaceae* and *Bacillus* , *Enterobacter* spp., *Flavobacterium aluminium* Harrison, *Pseudomonas* spp., *Streptomyces* spp., *Penicillium* spp., and *Trichoderma* spp. The suppression of *Gaeumannomyces graminis* var. *tritici* (the fungal pathogen responsible for development of take-all disease in wheat) is mostly mediated through bacterial endophytes rather than rhizospheric microoorganisms and emphasized that mere reduction of pathogen biomass is not key factor in disease suppression.

ii. Soil Fungal Community: Most of the soil suppressive assessment studies are focused on soil bacteria (Shi et al., 2019). Although fungi and micro-eukaryotes are closely associated with disease suppressiveness and are crucial for crop protection, most of them are comparatively neglected. Micro fauna and meso fauna can feed on pathogens, help in nutrient cycling/turnover, and maintain species diversity by consuming dominant bacterial taxa. Soils with high disease suppressiveness are also associated with higher fungal diversity. As there exists a large diversity of uncultured fungi in the soil, culture-independent approaches must be developed to describe the fungal community and screen the fungi with high disease suppressiveness. The fungal composition revealed differences between suppressive and non-suppressive soils using culture-independent terminal restriction fragment length polymorphism (T-RFLP) method.

Arbuscular mycorrhizal (AM) fungi could help in disease suppression in several ways. The AM plants could recruit more pathogen–antagonistic actinomycetes than the non-mycorrhizal plants. AM fungi do not compete with other growth-promoting rhizobacteria; rather they interact for the mutual establishment to increase plant health. AM fungi are often known to increase the nutritional status of the host plant and help indirectly in the suppression of plant diseases. AM increases available phosphorus and is found to increase tolerance of the plants to pathogen damage. Other non-nutritional mechanisms of AM fungi are changes in exudation patterns, activation of plant defence systems and increased lignifications of cell walls, and competition for space for colonization and infection sites. AM fungi also release a variety of antibiotics and other toxins that act against pathogenic organisms. The feeding preferences of the biocontrol agent are one of the major determinants in developing specific rhizosphere microbiome. The interactions between different trophic levels (an organism's position it occupies in a food web) can modify the nutrient cycling and influence soil nutrient status and plant nutrition that further affects the soil suppressiveness.

Abiotic contributors to DSS

i. Temperature: Soil physical and chemical properties are also useful indicators in predicting the suppressive level of the soil (Löbmann et al., 2016) (Fig. 43). Soil abiotic factors could influence the disease suppression in soil by limiting the strength and altering the direction of plant-soil feedbacks (Jayaraman et al., 2021). The sensitivity of the disease suppressiveness to temperature depicts the contribution of several temperature-sensitive and/or resistant microbes in disease suppression. The reduction of pathogen suppressiveness was observed at 50 °C and complete destruction of soil disease suppressiveness at 80° C, suggesting that total suppressiveness could have been controlled by a mixed group of biocontrol agents: a group of temperature-sensitive Gram-negative bacteria that worked till the temperature of the soil reached 50° C and another group of less sensitive spore-forming micro-organisms that are killed beyond 80° C. This temperature sensitivity clearly indicates the contribution of more complex microbial communities in the suppression of diseases in soil.

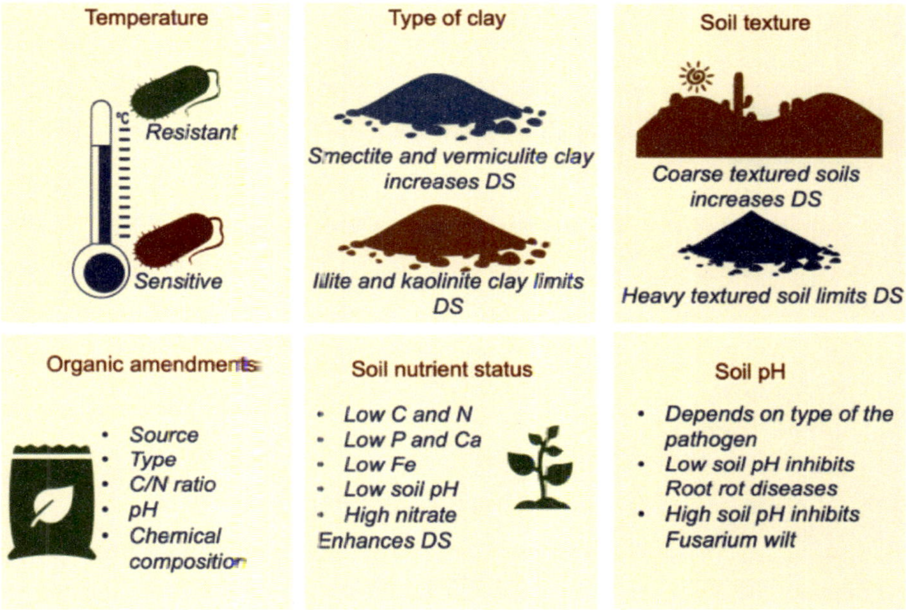

Fig. 43: Effects of soil abiotic factors on disease suppressiveness

ii. Clay Content and Cation Exchange Capacity: Abiotic factors such as high clay content in soil and high cation exchange capacity (CEC) can suppress the infection rate of *Pythium ultimum* (Jayaraman et al., 2021). Several investigations have revealed the impact of clay minerals in disease suppression. Presence of smectite type of clay has been shown to enhance bacterial respiration and increase disease suppression. A strong positive relationship was found between disease suppression of *Aphanomyces* root rot and ratio of vermiculite–smectite to illite–kaolinite ratio, increase in clay content, soil calcium content and soil pH. Smectite and vermiculite have high CEC, and the resulting high buffering capacity could relate to the abundance of charged components such as organic matter and thus have higher suppressiveness to diseases such as Fusarium wilt (*Fusarium oxysporum f.* sp. *cubense* (Foc)) of banana. Few corpora of evidence presented contrasting results where smectite- and illite-rich soils were conducive to diseases, and vermiculite rich soils were suppressive to black rot (*Thielaviopsis basicola* Berk & Broome) of tobacco. Therefore, it appears that it is not the specific type of clay minerals that affects the disease suppressiveness; rather it is the ratio of clay minerals with high and low CEC that determines the disease suppressiveness of a particular soil (Persson and Olsson, 2000) (Fig. 44). However, the sole presence of clay minerals with high surface area and CEC do not favour disease suppression because sterilization or pasteurization of vermiculite rich soils could not impede the infection of soil-borne pathogens such as *Thielaviopsis basicola* in tobacco.

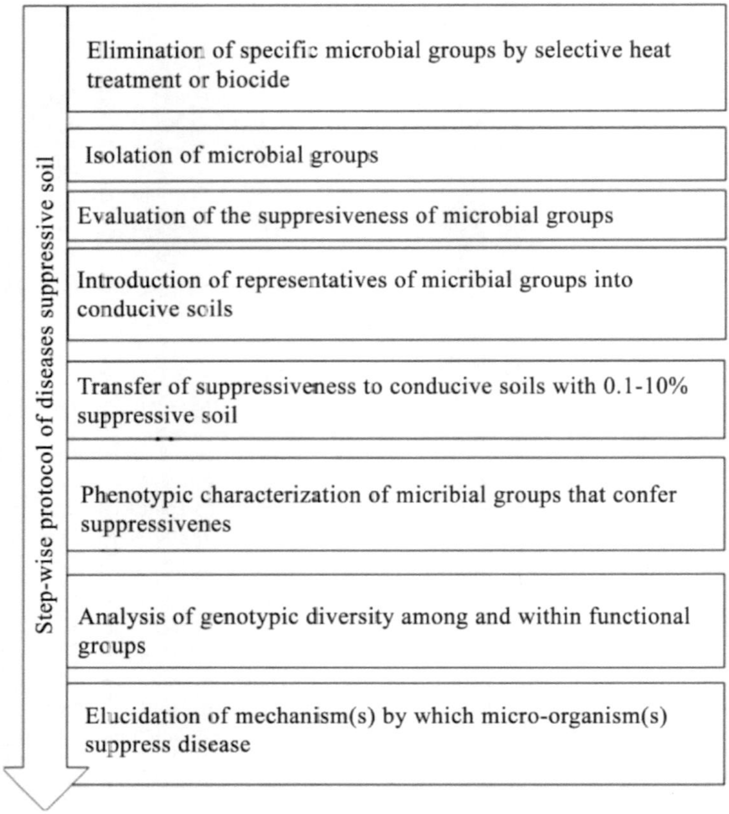

Fig. 44: Flowchart depicting the general process to study specific disease-suppressive soils

7.2.4 Mechanisms of soil related disease suppression: Maintaining high functional soil microbial biodiversity increases ecosystem resilience making the soil less vulnerable to short-term changes in the environment (Jayaraman et al., 2021). Therefore, soil biodiversity can significantly contribute to disease suppressiveness through direct (i.e. supporting higher trophic level organisms that feed on pathogens) and indirect mode (i.e. improving the plant health and resistance). In general, the biological mechanisms through which the disease-suppressive soils control the disease development are as follows (Hoitink and Boehm, 1999) (Fig. 45):

- Parasitism against pathogens by a group of beneficial microbes/microbial communities
- Production of metabolites, toxins, and antibiotics
- Competition for nutrients/resources/substrates
- Activation of disease-resistant genes in the host plant by the beneficial microbes
- Improved plant nutrition and soil health

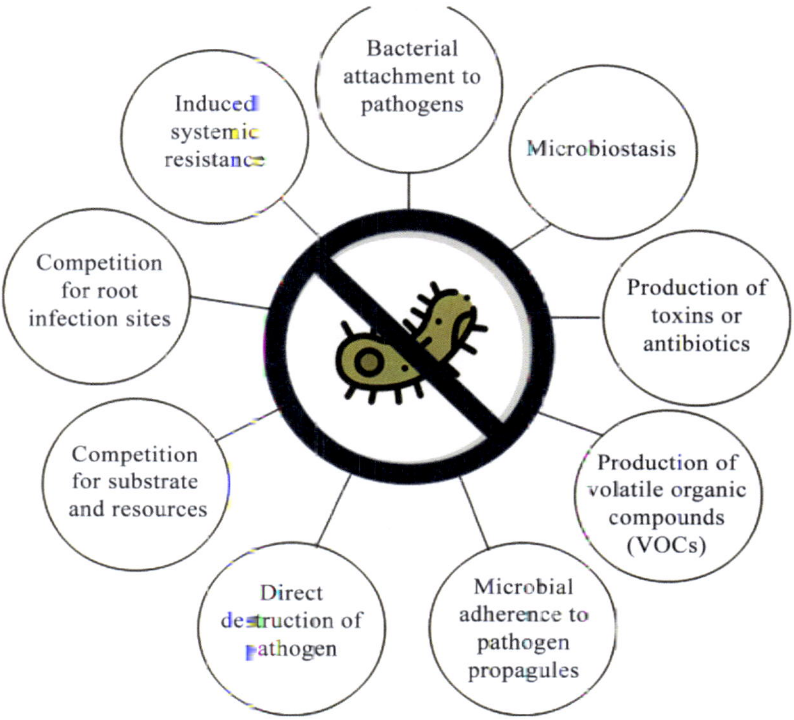

Fig. 45: Possible mechanisms of disease-suppressive soils—a close view

Disease suppressiveness could be attributed to combinations of antagonistic microorganisms which support each other forming a microbial consortium and acting on a specific soil-borne pathogen. For example, the combined actions of mutually compatible non-pathogenic *Fusarium* strains and fluorescent *Pseudomonas* could stimulate the soil suppressiveness in Châteaurenard soil in France. The relative abundance of disease-suppressive functional genes can be assessed by targeting the *prn*D gene (Peralta et al., 2018). The *prn*D gene is responsible for producing antifungal compounds such as pyrrolnitrin (PRN). The natural development of disease-suppressive soil is a very slow process and can take several years during which outbreaks of disease incidence may be at the higher rate. The process of natural disease suppression is a time-consuming phenomenon as soil microorganisms need to stabilize the soil with a multitude of physical and chemical processes so that diverse soil microorganisms are likely to dominate. Farmers might be reluctant to promote naturally occurring disease suppression. Therefore, the speed of developing DSS can be accelerated with the artificial management strategies in mainstream practice.

i. **Bacterial attachment to pathogens:** Attachment of beneficial soil microbes to the cuticles of parasitic nematodes, such as root-knot nematode (*Meloidogyne hapla*), is an interesting mode of action for disease suppression (Jayaraman et al., 2021). There is increasing evidence of microbial attachment of antagonists to the cuticles and surface coats of disease-causing parasites. The highly glycosylated mucins of the surface coat of the nematode play an important role in microbial specificity (Davies et al. 2011). The form of bacterial adhesion changes with the modification of the complex carbohydrates of the nematode surface coat. A fungus of the genus *Rhizophydium* was reported attaching to the surface coat of *Criconemoides* sp. A fungus related to *Cylindrocarpon olidum* could be found on the surface coat of *M. hapla* and reduce the number of galls in tomato roots or inhibit the egg hatch of the nematode through the production of metabolites.

ii. **Microbiostasis:** Manipulating the soil nutrient content to provide nutrient stress to the soil microbial community and to repress the spore germination and growth of the pathogen is termed as "microbiostasis" or "fungistasis" (Jambhulkar et al., 2015). It results from the loss of energy of the propagule and finally die due to the nutrient stress. For example, insufficient nutrients in soil restrict the germination of fungal conidia and the chlamydospores (thick-walled spores produced asexually from mycial cells) of *Fusarium* spp.

iii. **Production of Toxins or Antibiotics:** Common scab of potatoes is caused by a soil-borne pathogen *Streptomyces* that releases phytotoxins (*Thaxtomin*) (Jayaraman et al., 2021). Disease control of common scab of potato is carried out through biological interactions (antibiotic production and enzyme activities) between beneficial microbiota and the pathogens. Another example of antibiosis is bacterial production of pyrrolnitrin (PRN) and 2,4-diacetylphloro-glucinol (DAPG) that are known to suppress the fungal pathogens in the soil. Presence of 2,4-diacetylphloroglucinol (DAPG)–producing bacteria such as *Burkholderia cepacia* and *Peanibacillus azotofixans* largely depends on the age of the host plant. In maize, the abundance of DAPG producers was relatively low in the first stage of plant growth than at the advanced stage. Other studies confirmed that young and immature roots recruit more r-strategists (microbes typically living in unstable environments) and mature roots stimulate more abundance of k-strategists (microbes that occupy more stable environments). Therefore, disease suppression depends not only on plant type but also on phenological stage of plant development.

iv. **Production of Volatile Organic Compounds:** The genus *Streptomyces* also accounts for 80% of the currently available antibiotics (Jayaraman et al., 2021). *Streptomyces* produces several volatile organic compounds (VOCs) that reduce the severity of plant diseases by causing morphological abnormalities in different fungal pathogens. Members of the genus *Streptomyces* that are found antagonistic to *R. solani* could produce more than 10,000 secondary metabolites including VOCs. The chemical composition of VOCs is highly diverse, complex, and unique to each microorganism. These microbial VOCs can exhibit versatile functions as they can inhibit pathogen growth, enhance plant growth, and stimulate plant resistance. The chemical composition of VOCs is the main factor for their specificity in their target pathogens. For example, *Pseudomonas* can produce VOCs such as cyclohexanal, decanal, 2-ethyl 1-hexanol, nonanal, benzothiazole, and dimethyl trisulfide that can inhibit the fungal growth and germination of *Sclerotinia sclerotiorum*. *Streptomyces* can produce butanone (methyl vinyl ketone) and dimethyl disulphide that inhibit the spore germination of *Cladosporium cladosporioides*. *Streptomyces albus* produces anisole that acts against *S. sclerotiorum* and *F. oxysporum*. These VOCs also display their role in plant growth promotion such as the production of indole, 1- hexanol, pentadecane, 13-tetradecadien-1-ol, 2-butanone, and 2-methyl-n-1-tridecene and indirectly influence the disease suppression. In addition, VOCs also function as signalling molecules in intraspecific interactions and indirectly help in disease suppression but its primary modes of action are not known. Thus, further studies must be conducted to provide a more conclusive proof of the role of VOCs in disease suppression using specific soil bioassays where VOC producers and the pathogen are physically separated.

v. **Adherence and Microbial colonization of pathogen propagules:** The propagules of soil borne pathogens are typically colonized by higher populations of bacteria, fungi and protozoa in suppressive soils (Jayaraman et al., 2021). The colonized pathogen spores are difficult to germinate and lyse readily than non-colonized spores. For example, bacterial colonization of *Cochliobolus* spp. (causal organism of root rots of grasses) can decrease the virulence of the pathogen.

vi. **Destruction of pathogen propagules:** The antagonists stimulate lysis of pathogens, degradation of chlamydospores, conidia and zoospores (Jambhulkar et al., 2015). For example, *Trichoderma* spp. (biocontrol agent) act on *Phytophthora* spp. (pathogen) by stimulating hyphal lysis of the pathogen.

vii. **Competition for substrate and Resources:** As most of the plant pathogens are weak saprophytes, there exists a strong competition in the soil environment for the organic substrates (Jayaraman et al., 2021). For example, *Pythium nunn* wins over *Pythium ultimum* for colonization of added organic compounds, resulting in the death of *P. ultimum*.

viii. **Competition for Root infection sites:** As rhizosphere is a rich source of organic substrates for the soil microbes including soil-borne pathogens, the pathogen and the biocontrol agents compete for root colonization and lead to the suppression of the disease (Jambhulkar et al., 2015). For example, non-pathogenic *Fusarium equiseti* suppresses *Verticillium* wilt by competing with the causal pathogen for root site.

ix. **Improving plant health through Induced Systemic Resistance:** One indirect way to suppress disease incidence is to increase plant resistance to soil-borne infections (Jambhulkar et al., 2015). For example, non-pathogenic *F. oxysporum* soil isolates stimulate induced systemic resistance to *Fusarium* wilt in watermelon (*Citrulus lanatus*). The resistance is increased with better plant health or modification of certain parts of the plant to reduce infections. In some cases, disease suppression is associated with an increase in physical barriers in plant root such as callus-rich, multi-layered wall appositions.

x. **Organic matter mediated soil suppression:** High SOM content or incorporation of organic manure (OM) into the soils is usually associated with a lower incidence of root diseases. Many scientific reports have addressed the positive effects of compost addition on disease suppressiveness (Jayaraman et al., 2021). Application of organic amendments not only manipulates the biological factors but also influences the physical and chemical aspects, thus creating several pathways in disease suppression. Addition of disease-suppressive amendments such as composts enhances the plant defence system through induced system resistance and has an edge over the application of a single bio-inoculant. Further, well-prepared compost also contains a multifaceted microbial consortium with diverse mechanisms which cause disease suppression. In addition, the efficiency of composts in disease suppression can be enhanced with the inoculation of specific biocontrol agents such as *Trichoderma hamatum* or *Bacillus subtilis*. Several treatments commonly applied to comprehend the mechanisms of disease suppression through the application of composts include sterilization or pasteurization or heat treatment of the soil–compost mixture.

   Such treatments lead to the observations of reduction in the disease-suppressive capacity of the composts, indicating the biological nature of disease suppressiveness. Water extracts collected from several composts are also reported to be suppressive to several soil-borne pathogens although no significant amount of antibiotics and siderophores have been detected in the extracts. Such observations provide hints/clues on the major contribution of biological aspects than the physical and chemical perspectives of disease suppression. Another mechanism of biocontrol through composts is the release of toxic or stimulatory volatile compounds that lead to changes in physical and chemical properties of the soil and further affect the development of the pathogens. Such organic amendments enhance the growth and development of antagonistic microbes such as *Lysobacter* spp. (*L. antibioticus* and *L.gummosus*) which strongly inhibit important soil-borne pathogens such as *Rhizoctonia solani.*

xi. **Biopolymers in enhancement of disease suppression:** Biopolymers have also been suspected to have the potential to enhance disease suppressiveness in soil (Jayaraman et al., 2021). Application of chitin and/or chitosan extracted from animal waste could temporarily increase root growth and reduce the incidence of nematode infection on plant roots. Though most of the underlying mechanisms of the disease suppression with biopolymer treatment are still unknown, one of the possible mechanisms could be the change in the diversity and/or activity of soil microbiota that confers the benefits of disease suppression. The application of chitin stimulates chitinolytic soil microorganisms, which are capable of hydrolysing chitinous fungal hyphae of the pathogens, and the hydrolysed chitin attracts secondary responders and enhances disease suppression. In this context, several studies have postulated that the addition of chitinous biopolymers to soil stimulates more bacterial communities such as members of the genus *Streptomyces* than the fungal communities. The ubiquitous actinobacteria have been widely studied for their primary role in degrading complex organic molecules such as chitin while some have reported its secondary role in disease suppression.

Application of biopolymers can be recommended for those soil-borne pathogens that are currently controlled only with the use of some environmentally harmful chemicals. For example, *Verticillium* wilt, caused mainly by *V. dahlia*, is controlled by the use of volatile chemicals. Application of green manure and chitin has been reported to increase disease suppression of *V. dahliae* in the green-house and field studies. Despite these reports, there are still unexplored areas in disease suppression through the application of biopolymers (Giotis et al., 2012). Effects of chitin in relation to crop rotation, soil properties, and nutrient management must be further studied to understand its complex behaviour in disease suppression.

7.3 Agricultural Management Regime and Disease Suppression

i. **Crop Selection- An important decision in disease suppression:** Among several ways to manage the soil microbiome, crop selection is one of the strategies that can alter the physical, chemical, and biological properties of soil in their rhizosphere (Jayaraman et al., 2021). The resistance of disease differs from one cultivar to another due to their differences in the ecophysiological properties that influence the type and diversity of microbial activity in their rhizosphere. The resistant cultivars differ from the susceptible ones by their higher microbial diversity, the higher number of putative bacterial interactions, and specific microbial community structure. The composition of the microbial community also changes with the stages of growth of the host plant because of the presence of different types of rhizodeposition during different plant growth stages. However the dominant mode of biocontrol might not differ along the growth stages.

   Plants can also help in attracting specific antagonistic microorganisms in disease suppressiveness. Depending on the most dominant and active soil-borne pathogen infecting the host species, beneficial *Pseudomonads* are promising options. Plants manipulate the soil microbial biota through its specific root exudates. The chemical structure of the root exudates directs the type of microbial colonization in the rhizosphere. A mono-cropped cotton exudes more amino acids but less sugar and phenolic acids in its rhizosphere than fallow agricultural soil (control) with the increasing growth phase of the crop. The physiological shift in root exudation systematically affects the plant–microbial interactions. An increase of amino acids (e.g. Glu, Ala, Gly) in the cotton rhizosphere leads to the decrease of beneficial bacteria such as *Xanthomonadaceae*, *Comamonadaceae*, and *Oxalobacteraceae* in the rhizosphere, thereby reducing the disease-suppressive levels of the soil under mono-cropped cotton.

ii. **Land Use and its effects on disease suppression:** Natural soil ecosystems are generally more disease suppressive than are arable lands, and such a differential response is attributed mainly to differences in microbial community and structure (Garbeva et al., 2006). The type of plants used as crops in arable land and the naturally grown plants in wild ecosystems differ in their impacts on the characteristics of resident soil microbial communities and, therefore, play different roles in disease suppression. Higher above-ground biodiversity richness can maintain higher soil microbial diversity. Therefore, land-use changes are often associated with the change in soil microbiota and thus affect disease suppressiveness. This concept of grassland as "preserver of microbial diversity" can be explored to identify more microbial taxa capable of disease suppression. The abundance of DAPG and PRN producers has been reported to increase with plant diversity but with greater spatial diversity. Therefore, a higher abundance of DAPG and PRN producers is detected in grassland soils than in cultivable soils. Grasses tend to increase the prnD gene abundance whereas legumes decrease DAPG and PRN producer abundance.

iii. **The Potential effects of Conservation Agriculture on disease suppression:** Agricultural management regime plays a complex role in developing disease suppression as the outcome may vary and the results are indirect. Effects of long-term use of agricultural management practices such as minimum tillage, crop rotation, crop residue retention and organic farming have been assessed on disease suppressiveness (Jayaraman et al., 2021). The conventional intensive farming is often associated with increased destruction of soil structure and deterioration of soil health, which lead to decrease in soil biodiversity including the natural enemies and antagonistic microbial community. Therefore, it is expected that disease suppression is higher with the use of conservation agriculture (CA) than that with conventional farming practices. The adoption of CA/NT and stubble retention practices can, in some cases, increase soil-borne plant diseases in the short-term. However, long-term adoption of crop management practices that supply higher levels of biologically available C inputs, either through crop residues or the addition of composts and organic manures, can lead to higher levels of disease suppression. This occurs through changes to the composition and activity of soil microbial community.

iv. **Disease Suppression in Organic Farms:** Löbmann et al., (2016) found no difference in disease suppression levels in conventional and organic farms. Similar or equal distribution of disease suppressiveness in conventional and organic sites was also reported. In contrast, some studies have also reported higher suppressiveness with conventional farming systems than that in organic ones. In general, organic systems have higher disease suppressiveness than conventional systems as the application of organic manure can increase soil biological health.

v. **Rotation effect and Crop diversification:** The significance of crop rotation in developing disease-suppressive soil is colloquially termed as "rotation effect", and it has a long history in agronomic research (Jayaraman et al.,

2021). It has been proved that crop rotation can increase agronomic yield. Crop rotation provides some of the essential ecosystem functions such as SOC addition/storage, nutrient cycling, and pest control. The type of crop used in the rotation has strong implications in developing soil microbial structure. For example, maize is often considered a strong root exuder where 30% of the total photosynthate is released in the rhizosphere and it supports more diverse groups of soil microbes. However, to elucidate the specific micro-organisms from a complex and diverse soil microflora that predominantly contributes to disease suppression is inherently a difficult task.

Some studies have reported better disease control in monoculture than that in diverse crop rotations. For example, there is a control of take-all disease in wheat monoculture due to the increased abundances of antagonistic *Pseudomonas* spp. in monoculture than in rotation (Kwak and Weller, 2013). The mixed benefits of crop rotations on disease control are related to crop diversity that promotes different pathogen suppressing microbiome. Despite decreased microbial diversity, the disease suppression capacity can be increased with crop diversity due to enrichment of a specific group of soil microorganisms with disease suppression traits. This trend suggests the importance of the composition of soil microbial community rather than diversity in disease suppression.

vi. **Stimulation of disease suppression with Intercropping and Cover Crops:** Some large-scale field experiments have highlighted the importance of intercropping in increasing total crop yields and disease reduction (Jayaraman et al., 2021). Intercropping could be beneficial in disease suppression against soil-borne bacteria, fungi and nematode. For example, pepper (*Piper nigrum*) monoculture could lead to the outbreak of Phytophthora blight. Intercropping of maize with pepper could restrict the spread of *Phytophthora* blight which is attributed to the formation of a "root wall" in maize roots that act as a physical barrier against *Phytophthora*. Moreover, maize secretes a significant quantity of antimicrobial compounds that inhibit the growth and spread of the pathogen.

Several studies have correlated the expression of antimicrobial genes and soil factors such as soil texture and nutrient availability. In chemical aspects, cover crops enhance SOC content through the decomposition of crop residues and the release of exudates from plant roots. Soil microbial activities are stimulated as a result of increased C flow from root exudates released from cover crops. Therefore, with the difference in quantity and quality of the root exudates from cover crops, there is a shift in microbial community composition, either from a diverse to specific or vice-versa.

vii. **Reductive Soil Disinfestation:** Another form of soil management practice is reductive soil disinfestation (RSD), a pre-plant soil disinfestation practice in which organic matter is incorporated into the soil before planting followed by irrigation to maximum field capacity and covering the soil with plastic film (Liu 2019). RSD has been reported to increase tolerance to soil-borne pathogens in upland-paddy rotation. RSD treatment enhances the production of antagonistic compounds, manganese, ferrous cations, and ammonia that help in suppressing a wide range of soil-borne pathogens. RSD indirectly influences disease suppression by improving soil properties such as pH, electrical conductivity, microbial population, and SOC content. Thus, RSD is considered an alternative to chemical fumigation in several countries including Japan, the USA, and China.

7.4 Antibiotics and Drugs from Soil Microbiome

Majority of the research on DSS is restricted to easily cultural bacteria with fast growth rate in vitro (Jayaraman et al., 2021). Culture-independent metagenomics studies have shown that the top 10% of most dynamic taxa found in suppressive soil belong to actinobacteria. Existing conventional microbial cultivation approaches could exploit only a minor fraction of soil microbiota. Therefore, molecular-based techniques, which derive the results from a direct exploration of soil extracted DNA, are increasingly used in disease suppression studies. Although a number of medicines have been developed from soil biota to control diseases, infectious diseases such as COVID-19 remain one of the prominent causes of human death worldwide (Chakraborty and Maity, 2020). By the year 2050, it is predicted that mortality rates due to the infectious diseases will increase more than tenfold. Therefore, to come up with new effective antibiotics, the search for new bacteria that could produce new bioactive bacterial metabolites has led to the global search in natural ecosystems such as soil system that harbours around 10,000 unique bacterial species. Although 1,500,000 fungi species have been estimated, it is rather difficult to grow them by standard methods. Another class of soil microorganisms mined for antibiotics is actinomycetes (diverse family of soil bacteria). Actinomycetes have numerous uses in the pharmaceutical industry as they are a potential source of bioactive compounds. As much as 80% of the world antibiotics are derived from the genera *Micromonospora* and *Streptomyces.*

The discovery of new antibiotics from soil microbial natural product (NP) starts with the collection and isolation of required soil microorganisms. They are grown in conditions required for their growth and development and certain growth factors are added to induce production of desired metabolite (Table 13) (Pela´ez, 2006). The natural metabolites are then extracted and tested. The specific compound needs to be isolated and purified from other mixtures before it is formulated as new drugs. The list of the natural products isolated from soil microorganisms that are found to be effective towards antibiotic-resistant pathogens is shown in Table 13.

Table 13: List of the bioactive compounds isolated from soil microbiome

Soil microorganism	Bioactive compounds	Bioactivity	Mechanisms
β-proteobacteria provisionally named *Eleftheria terrae*	Teixobactin	Kill Staphylococcus aureus or Mycobacterium tuberculosis resistant	Inhibits cell wall synthesis by binding to a highly conserved motif of lipid II and lipid III
Streptomyces strain PM0324667	MFAT-133	Antidiabetic	Induced glucose uptake in L6 skeletal muscle cells
Clostridium cellulolyticum	Closthioamide	Antibiotic	*Staphylococci* multiresistant inhibition
Uncultured soil microbiome	Malacidins	Calcium-dependent antibiotics	Kills multidrug-resistant Gram-positive pathogens
Gordonia sputi DSM 43896	G43 JF905613 Compound	Antimicrobial	*C. albicans, S. aureus inhibition*
Unspecified	Kanglemycins/ Kangs	Rifamycin antibiotics	Target rifamycin-resistant pathogens such as *Mycobacterium tuberculosis*
Actinomycetes	Ea3 Compound	Antibacterial	*E. amylovora, P. viridiflova, A. tumefaciens, B. subtilis* ATCC 663, *E. coli* ATCC 29998 3 inhibition
Micromonospora sp.	Diazepinomicin/ ECO-4601	Antimicrobial	Unspecific

7.5 Complexities & Climate Change and Disease Suppressing Soils

Knowing the potential of disease-suppressive soils, studies attempting to control soil-borne diseases such as *Verticillium* wilt, blight, root rot, and bacterial wilt through artificial construction of diverse microbial communities have been accelerating (Gao et al., 2019). The natural soil disease suppressiveness involves far more than the extensively discussed biocontrol agents or antagonists several other microbial taxa could have played an important role in disease suppression either by releasing pathogen inhibitors or by enhancing plant growth and resistance. Plants are often considered the main influencer of soil microbial community composition and diversity. However, the underlying mechanisms of the influence of crop management on the microbial community are still limited. Knowing the complexity of the biochemical and ecological interactions between the host plant and the pathogens, it is unlikely for developing DSS to succeed (Jayaraman et al., 2021). The existence of heterogeneous soil ecosystems and a large number of farming practices make it difficult to evaluate the specific conditions needed for developing and maintaining DSS in cultivable lands. Although disease suppression largely depends on soil and weather conditions, soil disease suppressiveness at the same site may vary with the type of long-term management practices. Therefore, manipulation of soil management practices has a large potential in suppressing soil-borne diseases.

Despite the fact that disease suppressiveness is regarded as an important component of soil quality, it is not well understood even though a corpus of documentation has been released. The issue in studying the mechanisms of disease suppression is the inability to accurately identify the actual causal agent of the disease suppression. The positive or negative correlations between the pathogen abundance and the antagonistic population only suggest the possible role of the antagonistic bacteria in disease suppression. Therefore, unravelling the complex mechanisms of soil-related disease suppressiveness after the application of organic materials remains a researchable issue. Although the actual extent of soil microbial diversity is not yet known, the soil system is the largest reservoir of biological diversity on Earth (Guerra et al., 2020). The infection of a host plant by a pathogen can trigger and manipulate the rhizosphere of the conducive soil in such a way that the other pathogens are developed in the rhizosphere. For example, in a mono-cropped cotton soil, specific populations of *Flavobacteriaceae*, *Streptomycetaceae*, and *Sphingomonadaceae* are significantly increased in the rhizosphere after the infection of cotton with *Fusarium oxysporum*. Higher abundance of *Flavobacteriaceae*, *Streptomycetaceae*, and *Sphingomonadaceae* are found in the conducive soil after infection as they can actively produce cellulose, pectin, and chitin decomposing enzymes that feed on the root leakage constituents. Therefore, the concomitant occurrence of these taxa with the specific pathogens in the rhizosphere of the host plants causes greater damage in a conducive soil and working on a specific group of antagonists might not work in these cases. The nature of the soil-related disease suppressiveness is highly unpredictable because of the complex interactions of different pathogenic and beneficial microbes and the heterogeneities of soil and weather at the field, landscape, and regional level.

The organic amendments prepared from the same base material may differ in disease-suppressive potential. For example, compost made from the middle part of hardwood bark is often less suppressive to disease development than when it is produced from the thermophilic region with lesser microbial activity and higher reducing sugars. However, the conducive material when incubated at room temperature can become suppressive with the increase of microbial activity and decline in reducing sugar levels.

Climate change and Disease suppressing soils: Plant diseases could threaten up to 82% of attainable yield in cotton and more than 50% for most of the major crops. These yield losses due to plant diseases are a huge concern, especially in resource-poor countries (Döring et al., 2020). Several authors highlight the "fertilization effect" of increasing atmospheric CO_2 that might help in increasing food production. Even though there is a possibility of the fertilization effect, it may be outweighed due to climate change–mediated increase in plant diseases. The impacts of plant diseases under increasing climate change are mostly ignored in several global food security assessments. In order to secure sufficient and nutritious food for the expanding human population, an increase in the incidence of plant disease will demand more pesticide usage. This also highlights the future need for DSS as a cost-effective and environment-friendly disease management technique that could be pursued by growers. Therefore, continuous exploration and research on the impacts of climate change on DSS must be carried out to illustrate the changes in pathogen–soil–plant–environment complex and secure a sustainable future for an ever-expanding population.

While we are worried about the impacts of rising temperature stress on crop production, there is complete lack of sufficient quantitative evidences on use of DSS in climate change conditions (Jayaraman et al., 2021). The possible effects of climate change on plant diseases include acceleration of pathogen evolution, shorter incubation periods, early attack of the disease, higher susceptibility of the host plants to soil-borne pathogens, and geographical expansion of plant diseases. While few soil-borne pathogens, such as *Pythium cinnamon*, showed a null response to climate change, others have demonstrated a reduction of disease suppressiveness against *Rhizoctonia solani* after heat treatment of the soil (50° C). The combined heat and drought stress could reduce the disease suppressiveness of the test soils (Döring et al.,2020), although the change varies in different soils due to differences in soil microbial community. Soils from dry and arid areas or located in high altitude with summer temperatures, and that has been already experienced heat stress are more vulnerable to decrease its disease suppressiveness due to predicted temperature rise in future. Such experiments could also show recovery of the soil microbial communities after the removal of temperature stress. Soil physical and chemical properties, through their effects on soil microbial community could govern the resistance and resilience of DSS. Soil abiotic characteristics are closely linked with the soil biological and physical resilience. Higher organic matter and clay contents positively affect the resilience of DSS. Therefore, keeping these views in mind, future studies must be conducted to understand the relationships between soil abiotic factors and soil resilience in DSS.

Future Perspectives on DSS: Although it has been a primary goal of plant breeders to develop high-yielding plant cultivars, tools to develop cultivars which develop selective microbial communities in the rhizosphere, are still lacking. In-depth studies are needed to understand the complexities of soil microbial community interactions in DSS which can enhance the melding of this information with plant breeding strategies for improving disease suppression (Jayaraman et al., 2021). However, several questions need to be answered through future studies on DSS. And important among these include: are there any genotype differences of 2, 4–DAPG producers in disease-suppressive capacities? Although soil disease suppressiveness is closely associated with the microbial community, a few questions remain in respect to the compatibility of the soil microbes among each other and their associated effects on disease suppression. Why the crop monoculture can enrich the population of 2, 4–DAPG producers remains to be answered. Future biological studies of soil suppressiveness need to generate new insights into the microbial soil ecology, which can serve as a foundation for the development of cost-effective and environmentally friendly management strategies for disease suppression. It is still a challenge to further develop molecular assessment tools (including specialized DNA microarrays and metagenomics) to unravel the antagonistic behaviour of soil towards soil-borne pathogens. In addition, the complex interactions between two main drivers of DSS (i.e. abiotic and biotic factors) and their fluctuations must be further studied. The intricate knowledge of DSS can be broadened with the promotion of collaborations between soil microbiologists, soil microbial ecologists, plant pathologists, physiologists, soil scientists, and agronomists. This multi-disciplinary approach is needed to understand the complexity of the DSS system and simplify the painstaking process of identifying, decoding the mechanisms of DSS, and widen its practical applications.

7.6 Current Insights into the Role of Rhizosphere Bacteria in Disease Suppressive Soils

In the early days of research on disease suppressive soils, several valuable insights were obtained for the role of individual microbial genera (Gómez Expósito et al., 2017). In most disease suppressive soils, however, suppressiveness appears to be due to the concerted activities of multiple microbial genera working together at specific sites or operating at different stages of the infection process of the pathogen. Understanding the temporal and spatial microbial dynamics of disease suppressive soils as well as the corresponding modes of action will be needed to facilitate the development of effective, consistent and durable disease management tools. A model predicting *Fusarium* wilt suppressiveness, including several soil factors combined with the abundance of three keystone microbial taxa, was designed to support the choice of crops or cultivars referred to as "Know before you Sow." Suppressive soils constitute a valuable source of biocontrol agents. Isolation of these microorganisms follows a "taxonomybased" approach and their activities are typically tested in in vitro assays that do not mimic field conditions. Additionally, the reintroduction of these microorganisms in non-suppressive soils often lead to inconsistent protective activities, mostly driven by a lack of a sufficient root colonization and/or inhibition of their modes of action under field conditions.

Relevant functions involved in disease suppressiveness can be executed by multiple microbial taxa, but metatranscriptome, metaproteome and metabolome studies of disease suppressive soils are still underrepresented. The combination of the "taxonomy-based approaches with "trait-based" approaches would be preferred to unravel the complexity of the specific microorganisms and mechanisms underlying disease suppressiveness. Thus, rather than introducing beneficial microorganisms, agricultural research should focus on identifying the factors that influence key microorganisms or traits responsible for suppressiveness, meanwhile eliminating the practical and legislative difficulties of introducing microorganisms in the environment. Hence, research on management practices aiming to select or stimulate resident microbial communities or activities that enhance suppressiveness is emerging. Examples are the use of specific soil amendments including chitin, chitosan or fish emulsion, the introduction of agricultural practices such as crop rotation or minimum tillage or the use of cover crops or by host-mediated microbiome engineering, where the protective microbiome is artificially selected over multiple generations.

Disease suppressive soils offer effective protection to plants against infection by soil borne pathogens, including fungi, oomycetes, bacteria, and nematodes. The specific disease suppression that operates in these soils is, in most cases, microbial in origin. Therefore, suppressive soils are considered as a rich resource for the discovery of beneficial microorganisms with novel antimicrobial and other plant protective traits. To date, several microbial genera have been proposed as key players in disease suppressiveness of soils, but the complexity of the microbial interactions as well as the underlying mechanisms and microbial traits remain elusive for most disease suppressive soils. Recent developments in next generation sequencing and other 'omics' technologies have provided new insights into the microbial ecology of disease suppressive soils and the identification of microbial consortia and traits involved in disease suppressiveness. Gómez Expósito et al., (2017) reviewed the results of recent 'omics'-based studies on the microbial basis of disease suppressive soils, with specific emphasis on the role of rhizosphere bacteria in this intriguing microbiological phenomenon.

Several 'omics'-based studies have been conducted recently to compare the microbial (mainly bacterial) community composition of soils suppressive or conducive for specific plant pathogens, including *F. oxysporum, G. graminis var. tritici, T. basicola, R. solani, S. scabies,* or *M. hapla*. A wide range of bacterial taxa were found in higher abundance in suppressive soils (Table 14). With regard to fungi, differences associated with disease suppressiveness of soils to *R. solani* on wheat were attributed to less than 40 fungal genera, including a number of endophytic species and mycoparasites. Among the fungi most frequently associated with disease suppressive soils to other pathogens are *Mortierella, Trichoderma, Fusarium*, and *Malasezzia*. Poudel et al. (2016) highlighted the importance of constructing microbial networks to determine microbial community structure and assemblage for disease management. Documenting differences in relative abundance between bacterial and/or fungal communities in suppressive and conducive soils by network analyses can be highly instrumental to zoom in on specific microbial consortia. However, these descriptive analyses need to be combined with other techniques to pinpoint the specific microbial traits involved in suppressiveness and to distinguish between cause and effect. Mechanistically, recent studies pointed to antimicrobial volatiles, including sesquiterpenes, methyl 2-methylpentanoate and 1,3,5-trichloro-2-methoxy benzene, 2-methylfuran, 2-furaldehyde, 2-(methylthio)benzo thiazole and murolool for their potential role in disease suppressive soils. In these studies, however, these volatile compounds were detected under in vitro conditions and their production in vivo should be validated to provide more conclusive proof of the role of antimicrobial volatiles in disease suppressiveness of soils.

Table 14: Summary cf microbia taxa associated with disease suppressive soils and identified by different cultivation-independent techniques

Pathogan	Crop	Location	Technique	Microbial taxa'
Gaetxnannomycos gramns var. tntro	Wheat	New Zealand	16SDGGE	Pseudomonas putida. Pseudomonas flourescens, Nocardodes olevorans, Streptomyces bingchengensis. Tenabacter
			ITSDGGE	Gibberella zeae, Penicillium, echinulatum Penicillium allii, Fusarium lateritium, Mortierella elongata, Microdochium bolleyi
Gaeixnannomyces gramnsvar. trbd	Barley	Germany	16S • Mcroarray	Plactomycetes, Chloroflexi Nitrospira, Acidobacteria Proteobacteria
Gaeixnannomyces gramnsvar. trbo	Wheat	France	16S • Mcroarray	Ptanctomyoetes. Ntrospra. Acdobactena. Ooroflen. Proteobactena CAzospraum. (Thermoanaerobacter. LactobacAus). Cyanobactena iLyngbya)
Fusanun oxysponxn	Vanaa	Chna	16S- Ampfccon	Acidobacteria Gp1 Gp13 Verrucomicrobia Actinobacteria Firmicutes (groups Gp2. Gpi. Gp3. Gpi 3). > Verrucomcrotxa. Actnobactena Wedonobacter). F rmcutes Ktedenobacter
			ITS- Ampkcon	Zygomycota (Morterda). Basidiomycota Staphylotrichum Actinobacteria Cylindrocladium Gliocladiopsis

Fusanm oxysponjm	Strawberry	Korea		Actinobacteria Proteobacteria Acidobacteria Gemmatimonadetes. Urosprs, Chloroflexi
Fusamxn oxysponjm	Varia	Chna	16S-Ampkcon	
			ITS • Ampkcon	
Fusanm oxysporjm f. sp. *cubonse*	Banana	Chna	16S-Ampfccn	
Fusanm oxysporjm f. sp. *cubonse*	Banana	Ch^a	16S-Ampkcon	

Nevertheless, its validation in situ has technical challenges since volatile-producing microorganisms should be positioned in their ecological context (the rhizosphere) but also physically separated from the pathogen to exclude the role of compounds other than volatiles. Among the antimicrobial peptides, specific emphasis has been given in recent studies to the role of lipopeptides in disease suppressive soils. In independent studies, the two structurally similar, chlorinated lipopeptides thanamycin and nunapeptin were shown to contribute to suppressiveness of soils against the fungal root pathogen R. solani. Using a combination of different techniques, Cha et al. (2016) elegantly revealed that the production of the thiopeptide conprimycin by *Streptomyces* played a role in a soil suppressive to *Fusarium* wilt of strawberry. Next-generation sequencing analyses revealed an increase of Actinobacteria in this suppressive soil leading to the isolation and genomic characterization of *Streptomyces* isolate S4-7. Genome mining of *Streptomyces* S4-7 pointed at the production of conprimycin as a metabolite involved in suppressing *Fusarium*. A chemogenomic approach further suggested that conprimycin acts by interfering with fungal cell wall biosynthesis (Cha et al., 2016).

To further target the active microbial communities and to identify other microbial traits involved in disease suppressive soils, DNA-SIP or metatranscriptomics should be applied and/or combined. For example, by using metagenomic approaches a clone was obtained from a soil suppressive to club-root disease on cabbage producing the antifungal chitinase Chi18H8, and a clone was obtained producing the salt-tolerant chitinase 53D1 from a soil suppressive to *V. dahliae*. Within the METACONTROL project, several novel polyketide antibiotics were identified. Chapelle et al. (2015) combined metagenomic and metatranscriptomic analyses to resolve the transcriptional changes in the rhizobacterial community of sugar beet plants in a *Rhizoctonia*-suppressive soil challenged with the fungal pathogen. They found that upon pathogen exposure, stress-related genes were upregulated in rhizobacteria belonging to the Oxalobacteraceae, Sphingobacteriaceae, Burkholderiaceae, Alcaligenaceae, Cystobacteraceae, Sphingomonadaceae, Cytophagaceae, Comamonadaceae, and Verrucomicrobia. Based on these results they proposed a model in which the fungal pathogen secretes oxalate and phenylacetic acid during colonization of the root system, thereby exerting oxidative stress in the rhizobacterial community as well as in the plant. This stress response in turn leads to the activation of survival strategies of the rhizobacterial community leading to enhanced motility, biofilm formation and the production of yet unknown secondary metabolites.

7.7 Disease Suppressive Soils- New Insights from the Soil Microbiome

Suppressive soils provide the best examples of natural microbe-based plant defense, whereby via rhizodeposition plant roots stimulate, enrich, and support soil microorganisms as the first line of defense against soil borne pathogens (Daniel Schlatte et al., 2017). They are soils in which, because of their microbial makeup and activity, a pathogen does not establish or persist, establishes but causes little or no disease, or establishes and causes disease at first but then the disease declines with successive cropping of a susceptible host even though the pathogens may still persist in the soil. Suppressive soils owe their activity to a combination of "general" and "specific" suppression. General suppression is the ability of soils to inhibit the growth and activity of soil borne pathogens to some extent, owing to the collective competitive and antagonistic activity of the total soil microbiome competing with the pathogen(s). The microbiome refers to a community of microbes in a particular environment (in this case the soil and rhizosphere), which includes not only their genes, but their transcripts, metabolites and proteins. General suppression is a natural and preexisting characteristic of soil; is often effective against a broad spectrum of soil borne diseases; is not transferrable from field to field or soil to soil with very small amounts of microbial inoculum or soil; is reduced by steaming and eliminated by sterilizing the soil; and can be enhanced by management practices that increase the population size, diversity and/or activity of the soil microbiome.

It is important to remember that most soil borne pathogens exist primarily as dormant structures, waiting for root or seed exudates or added nutrients to stimulate and fuel germination, growth and root infection. General suppression is equated to a microbial fire that is "burning" root exudates and other nutrients and limiting their availability to soil borne pathogens. Thus, general suppression is manifested as a continuum of suppressiveness, and the faster the burn rate from increased microbial activity, the greater the general suppression limiting pathogen growth and infection. General suppression is easily visualized and quantified when inoculum of a soil borne pathogen is added to an unamended or raw field soil and to the same soil that has been sterilized. Less disease develops on the host plant in the raw compared with the sterilized soil and the amount of suppressiveness can be quantified by calculating the difference in disease incidence or severity between the two soils. Soils that express only this "basal level" of general disease inhibition are considered "conducive" or "nonsuppressive" soils when compared with soils with specific suppression.

Some soils with specific disease suppression are characterized as "long standing" because the suppressiveness is naturally associated with the soil, its origins are not known, and suppressiveness persists in the absence of plants. One of the best examples of long-standing suppression occurs in certain soils from the Chateaurenard region of France, which have long been known to be highly suppressive to Fusarium wilts of several crops (Daniel Schlatte et al., 2017). Other suppressive soils are "induced," meaning that specific suppressiveness is initiated and sustained by crop monoculture, by growing crops susceptible to the disease, or by adding inoculum of the pathogen into the soil. The best described example of induced specific suppression is take-all decline, which is "induced" by wheat or barley monoculture following a severe outbreak of take-all. Soils with specific suppressiveness against fungi, oomycetes, bacteria, and nematodes occur worldwide. The mechanisms of specific (and general) suppression have not been fully defined for most suppressive soils. There are numerous examples in which individual antagonistic strains of bacteria or fungi with an affinity for a host crop or pathogen propagule reach threshold population sizes (10^5 g^{-1} root tissue) and initiate sustained disease suppression. A classic example is the natural suppression of crown gall, caused by *Agrobacterium tumefaciens*. The observation of over 40 years in South Australia that the incidence of crown gall on almond correlated with the ratio of pathogenic to nonpathogenic agrobacteria suggested the ability of a single bacterial species, *A. radiobacter* (nonpathogenic), to suppress a disease. This resulted in the development of one of the most successful biocontrol agents, strain K84. This phenomenon has since been seen in other orchards. Specific suppressive microorganisms often directly attack the pathogen (i.e., *Dactylella oviparasitica* against the beet-cyst nematode, *Heterodera schachtii*) and/or show intimate niche overlap with the pathogen especially in infection courts (i.e., *A. radiobacter* and *A. tumefaciens*; *P. brassicacearum* and *Gaeumannomyces graminis* var. *tritici*; pathogenic and nonpathogenic *F. oxysporum*).

Selective feedbacks are negative in specific suppressive soils and positive in general suppressive soils: If different species interactions impose selection for specific versus general suppression, the ecological and evolutionary dynamics of suppression in the two systems will have fundamentally different dynamics. Because antagonists in specific suppression are selected in the presence of a particular pathogen which is sensitive to the antagonist, specific suppression will exhibit a *negative feedback* dynamic with disease (Fig. 46) (Daniel Schlatte et al., 2017). Thus, for example, within the take-all system: (i) disease imposes selection for specific pathogen-suppressive population; (ii) pathogen-suppressive population reduces disease frequency and intensity; (iii) there is a loss of positive selection for the disease-suppressive population; (iv) pathogen-suppressive populations decline; and (v) disease increases. This negative feedback requires continued engagement of the pathogen in selection for the antagonist(s), and results in a continually-cycling selection landscape.

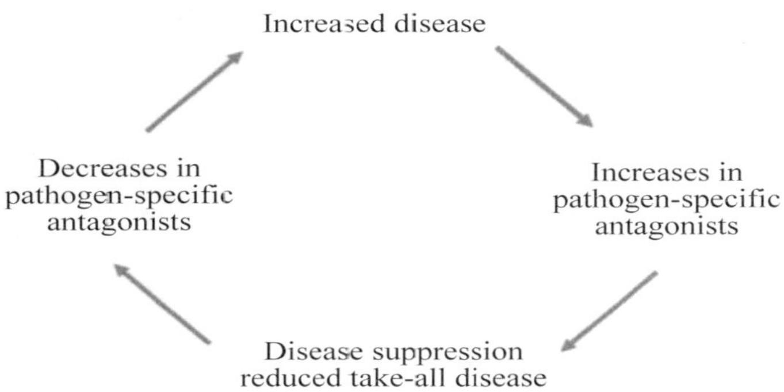

Fig. 46: Negative feedbacks illustrated by take-all specific suppression

In contrast, in general suppression, with no explicit pathogen role in selection, directional, positive feedbacks in the development and maintenance of antagonistic populations are possible (Fig.47) (Daniel Schlatte et al., 2017). Specifically, antagonistic and competitive interactions within the highly diverse network of interacting soil populations may be likely to be self-reinforcing. Competition begets selection for ever-better competitors, so that positive feedbacks may serve to reinforce suppressive potential. In general suppression, pathogen populations suffer apparent collateral damage stemming from competitive interactions within the saprophytic soil microbiome. Note that the competitive dynamics within the community may conceivably induce either arms race or niche differentiation coevolutionary dynamics (or perhaps a combination of both occurring simultaneously). Though these coevolutionary trajectories will yield communities with substantially different functional characteristics (highly antagonistic versus very niche-differentiated), it may be that either one could result in general pathogen suppression through different mechanisms. However, the pathogens that may be inhibited by highly-antagonistic populations may be quite different from those that are sensitive to strong resource competition. In either case, positive feedbacks offer the potential for sustained directional selection to maintain and enhance selection for general suppression, in contrast to specific suppressive soils in which intermittent disease/pathogen pressure is required to maintain suppression.

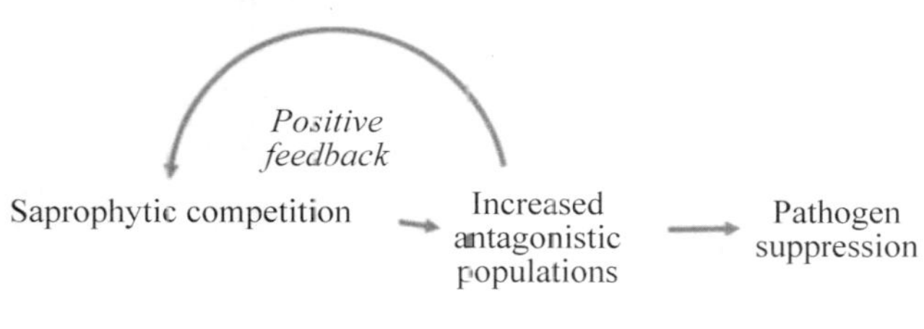

Fig. 47: Hypothesized positive feedbacks in the development and maintenance of general pathogen suppression

The distinct feedback relationships within specific versus general suppressive populations have significant implications for the development of practical management approaches for optimizing disease suppressive capacities within soil microbiomes. For example, with specific suppression, crop rotation (growing a nonhost) can reduce or eliminate the disease, so acts as a negative feedback to reduce the suppressive capacity of the soil community. This suggests a potential challenge in management to both minimize disease and maximize disease suppressive potential. In the case of general suppression, consideration of pathogen dynamics is less critical to management. Instead, factors that influence antagonist competitive interactions are likely to be much more important to disease suppression. Thus, for example, cover crops or green manures may offer a significant food source that generates positive selection for microbial competitive interactions among microbes able to quickly utilize the substrate. Ultimately, management to enhance suppression will require a clear focus on the distinct interactions that select for disease suppressive populations or communities, and an understanding of the feedback dynamics associated with those selection events.

7.8 New Approaches for Understanding Disease-Suppressive Microbiomes

New technologies have made it possible to characterize the composition, functional potential, and activities of microbial communities and their relationships with plants and pathogens at an unprecedented scope and level of detail (Daniel Schlatte et al., 2017). Recent DNA-based methods, enabled by high-throughput (HT) sequencing (HTS) of PCR-amplified platforms (e.g., Illumina MiSeq/HiSeq, IonTorrent, PacBio) have revolutionized microbial ecology. Most commonly, profiling of PCR-amplified marker genes (e.g., 16S ribosomal RNA, ITS [internal transcribed spacer], COI [cytochrome oxidase I]) from environmental samples at depths of thousands to hundreds-of-thousands of sequences per sample now allows researchers to efficiently characterize the diversity of microbial communities. As read length, accuracy, and accessibility continue to improve and costs decline, these tools are expected to become routine in the lab and in the field for understanding microbial communities. Network analyses are another tool to provide a means for characterizing and contrasting the structure and connectivity of microbial co-associations among communities having diverse functional characteristics (Poudel et al. 2016).

Despite its power, HT amplicon sequencing has substantial limitations. For example, sampling strategy, DNA extraction, and PCR amplification can introduce significant bias in observed communities. Variation in gene copy number among taxa (especially 16S rRNA genes) introduces further bias in observed sequence counts, rendering HT amplicon sequencing only semiquantitative. Further, including any relevant metadata, such as soil or environmental characteristics, plant age or health status, or field history, is invaluable in determining the potential drivers of microbial communities and will aid in the development of predictive models for microbial community functional dynamics. Finally, best practices should be followed in every study using HT amplicon sequencing. These include the use of negative and extraction controls free of source material to detect contaminant sequences, sequencing mock communities alongside samples to quantify sequencing error rates and calibrate clustering thresholds and replicating a subset of samples within and between sequencing runs to examine intra- and interrun variability and bias. Even after generating high quality data, many analytical and statistical challenges remain and care should be taken to use proper normalization procedures and statistical tests for sparse, nonnormal, compositional count data in downstream analyses.

Because HTS data are noisy, sequences are typically clustered into OTUs at a particular similarity threshold (generally 97%). Numerous approaches for OTU clustering have been developed that fall into three distinct categories based on their reliance on reference databases (Daniel Schlatte et al., 2017). Closed-reference OTU clustering simply finds a best match of a query sequence to a reference sequence in a database, but sequences without a match (e.g., uncharacterized taxa) are discarded. De-novo clustering approaches don't rely on a database to define OTUs, and instead cluster all sequences in a dataset relative to one other to generate OTU bins; thus, taxa without representatives in a database will be retained. Open-reference approaches are a hybrid of closed-reference and de novo clustering, first relying on hits to a database and then reclustering those sequences failing to hit the database using a de novo approach. Regardless of the algorithm used, OTU clustering at arbitrary similarity thresholds does not necessarily represent biologically coherent units. Due to variation in primer specificity, amplicon length, and evolutionary rates, different markers (e.g., variable regions of the 16S, ITS1, ITS2, 18S, and 28S) have different capacities to detect and delineate taxonomic groups and may resolve some groups while binning others into a single OTU (Nguyen et al., 2016). However, recently developed "denoising" methods may offer robust alternatives to OTU clustering and improved sensitivity to resolve single base pair polymorphisms. Still, because functional characteristics of microbes such as antibiotic production or resource use can be highly variable within species-groups with identical marker gene sequences, the functional characteristics of an OTU cannot be confidently inferred even with a genus- or species-level taxonomic identification. For example, a *Pseudomonas* or *Streptomyces* OTU found in a sample will not necessarily be an antibiotic producer involved in disease suppression. Thus, care must be taken not to over-interpret community sequence data, and patterns found among individual OTUs with hypothetical functions should be confirmed by using complementary approaches for quantifying taxon abundances or functional characteristics.

Identifying members of a soil microbial community to the genus or species level does not precisely characterize the functional capabilities or activities of the community. Metagenomics, in which all DNA extracted from a sample is sequenced in parallel, can offer additional insight into the diversity and functional potential of microbial communities. Because this technique does not rely on PCR, it lacks primer and amplification biases and can capture a much greater diversity of organisms and genes. Marker genes and functional genes from the same community can be analyzed together to generate a snapshot of community composition and functional potential. Similarly, meta-transcriptomics, or the sequencing of RNA transcripts (RNA-seq), provides a snapshot of only the portion of the metagenome that is actively transcribed. However, there are unique analytical challenges for both shotgun metagenomics and meta-transcriptomic data including metagenome assembly, binning, gene prediction/ annotation, and assignment of taxonomy to metagenomics reads (Daniel Schlatte et al., 2017). These methods require a much greater sequencing depth to effectively characterize a complex community, and thus they are much more costly to generate and computationally intensive to analyze than amplicon-based community profiles.

Further, as with amplicon profiling, the value of a metagenomic/transcriptomic "snapshot" and the linkages between meta-genomic and transcriptomic information and meaningful soil processes can be unclear. For example, treatments that induce subtle changes in community activity may only be apparent in transcriptomes, while not affecting the gene content of a metagenome. In contrast, treatments that impose selection on some taxa will be apparent in marker gene profiling, but combinations of metagenomics and metatranscriptomics will further allow for the identification of not only who is responding, but in what way. Single-cell-omics or fine-scale genomic reconstruction may provide insight into evolutionary responses of populations. Like other methods, the use of omics tools should be carefully considered in the context of their strengths and limitations to effectively answer a scientific question or hypothesis of interest.DNA-based methods are only one of many emerging technologies that will be invaluable for dissecting microbiomes of suppressive soils. Recent advances in metabolomics, proteomics, stable-isotopes, and microscopy/imaging not detailed here are proving to be powerful tools for investigating the microbes, genes, and molecules involved in microbial interactions among community members and with plants and pathogens. Combinations of complementary approaches will be especially useful for understanding disease suppressive systems. For example, the metabolism of plant-produced compounds may be tracked through soil food webs using stable isotopes, centrifugation, and HT sequencing (DNA-SIP) (Coyotzi et al. 2016). Integrating new and emerging technologies with traditional approaches will prove powerful for advancing our understanding of suppressive soils. As new technologies continue to improve the efficiency and scale of sample preparation and the read length and quality of DNA sequencing (e.g., Boreal Genomics Aurora system, 10× genomics linked reads), we will be able to gain a more detailed and complete characterization of plant and soil microbiomes.

7.9 Disease Suppression as a Tool For Disease Management

Over the last 3 years there has been increasing interest in the concept of "soil health." The year 2015 was the Year of Soil, and 180 papers with the term "soil health" were published in 2016. Companies are marketing tests of soil health and growers have become increasingly interested in this concept (Daniel Schlatte et al., 2017). These tests are based on organic C and N, respiration, and biological activity. Most of the measures of soil health have emphasized soil properties, organic matter and general microbial activity. Is this concept of soil health missing the most obvious connection-a healthy soil will be buffered against pathogen activity? Is disease suppression really a more specific manifestation of this concept? Regrettably, as often happens, different disciplines and research communities are not aware of developments and theories from other disciplines. The real question is: how we can capitalize on these soil health concepts to develop management approaches to achieve sustainable production? How can the take-all/*Rhizoctonia* / *Streptomyces* suppression examples help us think through novel approaches or hypotheses?

In one respect, specific suppression to take-all is already operating across over 800,000 hectares of wheat in the Pacific Northwest. *Rhizoctonia* bare patch is primarily seen in this region in early conversion to direct-seed no-till, and there are thousands of acres that have been in long-term no-till without any acute *Rhizoctonia* problem. But what about growers that have severe disease-how can they enhance and accelerate the development of suppression? How can they reduce the period of conversion from a conducive soil to one harboring microbial populations in a suppressive equilibrium, thus reducing economic losses during the transition period? Can we describe the starting microbial communities and identify those fields that will develop suppression and those that will not? The technology to describe and characterize the microbiome is becoming increasingly less expensive to the point that microbial profiles on a field-by-field basis will be available as a commercial service in the near future. In the same way that personalized genomics and microbiomics are now possible in human medicine, will this be the future of agriculture? Or are microbial communities too complex to develop simple predictive models? Will this become the new "snake oil" of agriculture, where over simplistic prescriptions will be offered, based on technology that may be feasible, resulting finally in a solution in search of a problem?

A more direct benefit to growers may be knowledge about how cultural practices can enhance the suppressive capacity of beneficial indigenous microbial populations. For example, how does crop rotation affect the development and sustainability of suppressive populations? With take-all, rotation to a nonhost crop can break the suppression, but the identity of that crop can influence how long it will take to revive suppressiveness when returning to monoculture. In the case of *Rhizoctonia*, crop rotation may not be as detrimental. In the long-term rotation study near Ritzville, WA, *Rhizoctonia* suppression of *R. solani* occurred in rotation strips with safflower, mustard, and winter pea (all hosts for the pathogen), raising questions about whether particular cropping sequences might favor the development of suppressiveness. Tillage is another variable that growers might exploit. For example, no-till seems to be important to develop suppression to *Rhizoctonia* bare patch, possibly because it initially favors disease. Residue management is another cultural practice that can be exploited. Growers use different types of machinery to manage residue and increase its decomposition including flail mowers, choppers, harrows, cultivators, rakes, and sickle bars and chaff spreaders on combines. Does crop residue play a role in suppression? We know that herbicides may also play a role in the development of suppression because of the "greenbridge" effect on soil borne pathogens. Although many root pathogens persist in soil by colonizing dying roots of crops and weeds, what about the microbial communities that eventually displace these pathogens on the root? Can they serve as a source of suppression? Recent work by D. C. Schlatter, C. Yin, S. Hulbert, I. Burke, and T. C. Paulitz (unpublished data) has described bacterial communities associated with roots killed by glyphosate. They showed that specific bacterial OTUs are enriched on roots in pots treated with multiple cycles of glyphosate, and more

OTUs are increased by glyphosate than are reduced. Can herbicide timing be part of establishing suppressive populations?

Cover crops and green manures promoted by the National Resource Conservation Service of the U.S. Department of Agriculture are touted as an important part of soil health. They are being used increasingly, especially in wetter areas of the U.S. Midwest and South than in the more arid dryland regions. These practices are designed to maintain a cover year-round to prevent soil erosion, increase organic matter, and improve soil structure by fostering deeper taproots that can break up tillage pans. Can these crops, as drivers of soil microbial communities, influence the development of soil suppressiveness? The use of substrate amendments to enhance general suppression in perennial and other high-value crops has been explored recently by Mazzola and Freilich (2017). For example, the addition of brassica seed meals has been shown to be a biologically based and effective management practice for apple replant disease. Similarly, the use of a substrate to shift microbial communities is employed in anaerobic soil disinfestation (ASD), in which organic substrates are added to soil and then flooded to produce oxygen stress and stimulate anaerobic communities that produce organic acids and alcohols suppressive to pathogens.

Breeding crop varieties that promote beneficial microbial communities is the approach with perhaps the greatest potential impact. There is increasing evidence that the cultivar can influence the composition of the rhizosphere microbial community. Mahoney et al. (2017) showed that the cultivars differentially selected soil bacteria, some of which provided potential benefits to the host such as pathogen suppression or nutrient mobilization. The development of cultivars selective for suppressive or biocontrol communities in the rhizosphere has long been a dream for plant breeding, but until now we have lacked the tools to attempt it. The more we understand about the complex microbial interactions involved in disease suppression, the greater the chances of melding this information with host plant breeding strategies and specific grower practices and recommendations supportive of greater crop productivity.

Beneficial Microbes are the key to Stopping Pathogens: There's a lot of microbiology in your soil including beneficial microbes, and pathogenic microbes. Beneficial microbes improve soil and crop health; whereas, pathogenic microbes can be devastating for your crops.The good news is you don't have to get rid of pathogens to mitigate soil-borne diseases. Even with the presence of disease-causing pathogens, healthy soils can keep their attacks on your crop to a minimum. This is because the beneficial soil microbes have a symbiotic relationship with your crops. These microbes use the exudates and nutrients created and released by your crops. This enables their survival and blocks pathogenic microbes from accessing and utilizing the resources needed to survive. When beneficial microbes outnumber pathogenic microbes, there's an increase in microbial competition. Additional crop-supporting microbes reduce the establishment and growth of pathogenic fungi and bacteria in your microbiome. Occasionally, beneficial microbes display hyperparasitism, secreting antibiotic-like enzymes, and release exudates.

- When soil microbes display hyperparasitism, they infect pathogenic microbes that pose a threat to the crops around them.
- Antibiotic-like enzymes and toxins called antibiosis kill pathogenic microbes and increase beneficial microbes' ability to defend themselves and the host plant from other pathogens.
- Exudates can trigger disease-resistant responses from host plants.

It's in the beneficial microbes' best interest to keep your crops alive, so they have multiple ways of fighting off dangerous pathogens.

Some soils have been observed to suppress diseases in crops grown upon them. Soils are a rich source of microbes that are thought to help plants suppress pathogens by improving the health of the plant, induce natural plant defense, produce antibiotics, compete against pathogens, or hyperparasitize the pathogen (Mousa and Raizada, 2016). Soil that suppresses crop disease due to the specific structure of its microbial community is known as disease-suppressive soil. Suppressive soil is an attractive method of biocontrol, because it has the potential to be sustainable over many seasons under favorable conditions. Suppressive soil is divided into two categories, general and specific.

7.10 General Suppressive Soil vs. Specific Suppressive Soil and Characteristics

Healthy, active, beneficial microbes are key to creating DSS. There are two kinds of Disease-Suppressive Soils (DSS) worth noting, including general and specific suppressive soil. DSS naturally defend against pathogenic diseases before the disease attacks a crop. Or, if a pathogen attacks, DSS causes the disease to decline with successive cropping. To understand DSS, you must first understand beneficial microbes.

General Suppressive Soil: General suppressive soil provides protection against multiple pathogenic microbes. To do this, beneficial microbes must out-compete pathogenic microbes to create general suppression. Thus, to achieve general suppressive soil, your soil microbiome needs to have an abundance (total number) and diversity (different types) of beneficial microbes. General suppressive soil doesn't necessarily protect against every pathogen in all conditions, but it does protect against many pathogens in many conditions. Think of general suppression as a way to have extra players on the field during a football game – it's a great advantage for your crops.

Specific Suppressive Soil: Specific suppressive soil is associated with a specific species of beneficial microbes or group of microbes. These beneficial microbes provide complete suppression of one pathogen. Think of specific suppression as creating a twenty-foot wall around your team's end zone. While pathogens are still present in your soil, it's almost impossible for pathogens to strike your crops. Even when they do strike, you will see a major reduction in incidents of crops affected by pathogens. Specific suppressive soil can occur naturally or through induction. When it occurs naturally, the suppressive soil is long-standing and is often found even without the presence of plants. Induced specific suppression can be produced by implementing monoculture of a crop (or monocropping) and the use of green manure. Due to the disadvantages of monocropping, it may be in a grower's best interest to take steps towards creating general suppressive soil over specific suppressive soil.

Characteristics of Suppressive Soils: The widespread but limited ability of soils to suppress the growth or activity of soilborne pathogens has been referred to as "general suppression", "general" or "nonspecific antagonism" or "biological buffering" (David M. Weller et al., 2002). General suppression is related to the total microbial biomass in soil, which competes with the pathogen for resources or causes inhibition through more direct forms of antagonism. General suppression often is enhanced by the addition of organic matter, certain agronomic practices, or the build up of soil fertility, all of which can increase soil microbial activity. No one microorganism is responsible for general suppression and the suppressiveness is not transferable between soils. When inoculum of a pathogen is added to pairs of raw and sterilized soil samples, the effect of general suppression is apparent by the greater severity of disease on a host grown in the sterilized as compared to the raw soil. "Specific suppression" is superimposed over the background of general suppression and is due, at least in part, to the effects of individual or select groups of microorganisms during some stage in the life cycle of a pathogen. Transferability is the key characteristic of specific suppression and the term "transferable suppression" has been used synonymously with specific suppression. Suppressive soils undoubtedly owe their activity to a combination of general and specific suppression. The two function as a continuum in the soil, although they may be affected differently by edaphic, climatic, and agronomic conditions (Rondon et al., 2000). Suppressive soils also have been differentiated according to their longevity. Hornby (1998) divided suppressive soils into the categories of "long-standing suppression" and "induced suppression." Long-standing suppression is a biological condition naturally associated with the soil, its origin is not known and it appears to survive in the absence of plants. In contrast, induced suppressiveness is initiated and sustained by crop monoculture or by the addition of inoculum of the target pathogen.

Most suppressive soils maintain their activity when brought into the greenhouse or laboratory, which facilitates assessment of their properties and mechanisms of suppression under more controlled and reproducible conditions. The strategy used to determine the microbiological basis of suppressive soils and to identify the contribution of specific microbial groups varies. The first step is to determine whether suppressiveness can be destroyed by pasteurization (moist heat, 60°C for 30 min) or selective biocides (e.g., novobiocin or chloropicrin), or whether harsher treatments (e.g., steam, methyl bromide, autoclaving, or gamma radiation) are required. Both general and specific suppression are eliminated by autoclaving and gamma radiation. General suppression is reduced but not eliminated by soil fumigation, and usually survives 70°C moist heat. In many examples of specific suppression, suppressiveness was eliminated by pasteurization; however, this characteristic is not a prerequisite for specific suppression. For example, the suppressiveness of a root-knot nematode–suppressive soil in Florida, which apparently resulted from the spore-forming bacterium *Pasteuria penetrans*, would not be eliminated by pasteurization. A second step, which allows confirmation of the biological basis of suppression, involves transfer of suppressiveness to a raw conducive, fumigated, or sterilized soil by addition of 0.1% to 10% or less (w/w) of the suppressive soil. The impact of soil edaphic factors on disease development in soil transfer studies is minimized when suppressive and conducive soils are diluted into a common background soil, allowing a direct comparison of the introduced microbiological components. The research strategies used historically to elucidate the microbiological basis of suppressive soils are in some ways analogous to those currently applied to studies of microbial gene function even though the two areas of research are completely different (David M. Weller et al., 2002).

7.11 Examples and Case History Studies Of Suppressive Soils

i. *Fusarium* wilt–Suppressive soils: Fusarium wilts are caused by pathogenic *F. oxysporum*, a soil borne fungus found worldwide. Wilts cause significant yield losses in numerous crops, and fungicides and host resistance often do not give adequate and sustainable control (David M. Weller et al., 2002). Natural suppressiveness of soils to Fusarium wilt was first recognized in the nineteenth century and was later described for other soils around the globe. Wilt–suppressive soils limit the incidence or severity of wilts of many plant species. The suppressiveness is specific to Fusarium wilts and not effective against diseases caused by nonvascular *Fusarium* species including *F. roseum* and *F. solani*, or other soil borne pathogens. Long-standing suppression operates in most *Fusarium* wilt–suppressive soils, although there are a few examples of induced suppression. For example, suppressiveness to *F. oxysporum* f. sp. *melonis* and *F. oxysporum* f. sp. *niveum* was induced following continuous cropping of melon and watermelon, respectively. Interestingly, the induction of suppressiveness in these cases was associated with continuous cropping of partially resistant cultivars, whereas induction of suppressiveness against other soil borne pathogens normally involves monoculture of susceptible cultivars.

The microbiological nature of *Fusarium* wilt-suppressiveness was demonstrated by the strategies described previously. Suppressiveness was eliminated upon treatment with moist heat, methyl bromide, or gamma radiation and was transferred by mixing small amounts of suppressive soil into a heattreated conducive soil. Among the bacterial and fungal genera proposed to contribute to *Fusarium* wilt–suppressiveness are *Alcaligenes* sp., *Bacillus, Trichoderma*, *Pseudomonas* spp., Actinomycetes and nonpathogenic *F. oxysporum*. Although introduction of representative strains of each of these genera increased the level of soil suppressiveness in most cases, the introduction of large populations is unlikely to reproduce the microbial community structure and interactions that occur naturally in suppressive soils. The introduction of microorganisms isolated from suppressive soils into conducive soils, therefore, does not necessarily provide conclusive information about their contribution to soil suppressiveness. The extensive studies of Fusarium wilt–suppressive soils from Chateaurenard (France) and the Salinas Valley in California (USA), however, have provided substantial insight into specific microorganisms and mechanisms involved in suppression. In these soils, natural suppressiveness is associated with a reduction in the saprophytic growth and inhibition of chlamydospore germination of pathogenic

F. oxysporum. This suppressiveness has been attributed mainly to the activity of nonpathogenic *F. oxysporum* and fluorescent *Pseudomonas* spp., and for both microbial groups, similar mechanisms including competition and induced systemic resistance were shown to be active. Particularly interesting from the work of Lemanceau and co-workers is the intimate and complementary association between these two groups of microorganisms, which in combination, provided enhanced disease suppression mediated by competition for iron via siderophores produced by the pseudomonads and for carbon by nonpathogenic *F. oxysporum* strain Fo47 (David M. Weller et al., 2002).

In contrast to the suppressiveness of the Chateurenard and Salinas Valley soils, the monoculture-induced suppressiveness to *Fusarium* wilt of watermelon does not result in a reduction in saprophytic growth of the pathogen nor in inhibition of chlamydospore germination. Among a large collection of bacteria, fungi, and actinomycetes isolated from this suppressive soil, only nonpathogenic *F. oxysporum* isolates consistently suppressed the disease in both microwave-treated and natural soil. Induced systemic resistance was the primary mode of action for several of these isolates, but it is not yet clear if the mechanism is similar to that described for induced systemic resistance by rhizobacterial. Strains of nonpathogenic *F. oxysporum* differ considerably in their efficacy against *Fusarium* wilt. For example, strain Fo20 was the least effective of eight strains tested, whereas Fo47 proved to be the most effective in controlling *Fusarium* wilt. Furthermore, Larkin & Fravel showed that nonpathogenic strains differed not only in their efficacy, but also in the mechanism(s) and dose required to suppress disease. For example, strain CS-20 required doses as low as 100 chlamydospores per gram of soil to reduce disease incidence significantly, whereas strain Fo47 was effective only at doses of 104 to 105 chlamydospores per gram of soil. These studies clearly illustrate the need for knowledge of the diversity within a group of antagonistic microorganisms when studying disease-suppressive soils. The degree of intraspecific diversity varied widely among isolates from different French soils, but not as much among isolates from the same soils collected over time.

ii. Potato scab decline: Common scab is an important disease of potato caused by *Streptomyces scabies* and other *Streptomyces* species (David M. Weller et al., 2002). Pathogenic strains produce thaxtomins, phytotoxins that induce symptoms of scab when applied to tubers in the absence of the pathogen. Production of thaxtomin A, the major phytotoxin produced by *S. scabies*, is positively related to the ability to cause disease and the amount produced by a strain in vitro is correlated with its aggressiveness in vivo. Thaxtomin nonproducers are nonpathogenic. Potatoes grown in "old" irrigated fields in central Washington that had undergone many years of potato production were almost scab free. In contrast, potatoes grown in "new" fields that had been brought into production only in the previous 15 years suffered from scab. Where monoculture potato production was attempted, scab occurred uniformly on potatoes from new fields but did not appear on potatoes grown in the old fields. Scab has declined with potato monoculture in other potato growing regions. For example, a plot for screening scab-resistant germplasm at Grand Rapids, MN, was maintained in potato monoculture from 1943 to 1971, and in 1965 scab decline was first observed. Susceptible cultivars grown in the plot from 1985 to 1987 failed to develop scab and addition of the Grand

Rapids soil to a conducive plot at Becker, MN, transferred suppressiveness. A diverse collection of *Streptomyces* isolates from scab-free potatoes grown in the suppressive soil produced antibiotics inhibitory to *S. scabies* in vitro and the pathogenic strains were much less inhibitory than the suppressive strains against other isolates, whether pathogenic or not. DNA fingerprinting by repetitive DNA sequence analysis (rep-PCR) did not distinguish between pathogenic and suppressive strains; cellular fatty acid analysis was more useful, but also did not distinguish perfectly between pathogens and nonpathogens.

Suppressive strains introduced into scab-infested soil reduced the severity of scab. PonSSII and PonR were introduced in-furrow and potatoes were grown for two years. PonSSII and PonR reduced pathogen populations at harvest by 93% and 85%, respectively, in the first year and 36% and 44%, respectively, in the second year. Scab lesions on tubers were reduced in both years. These two studies are especially notable because, as is the case with the transfer of a suppressive soil, the suppressive strains were introduced only once at the beginning of the experiment in order to initiate suppression in subsequent years. Suppressive strains with more vigorous growth and antibiotic production provided better scab control than strains with less aggressive growth and antibiotic production. However, more recent studies indicate that suppressive strains function through a combination of resource competition and production of inhibitory compounds. The picture emerging from these studies is that suppressive *Streptomyces* strains play an important role in scab decline, but the contribution of other microorganisms needs further study. In general, *Streptomyces* spp. have received less attention than other microorganisms as agents of specific suppressiveness. Expanded studies of *Streptomyces* spp. in suppressive soils is warranted given their abundance in soil, ability to produce broad-spectrum antibiotics, and well-documented biocontrol abilities. *Streptomyces* spp. appear to have a major role in the disease suppressiveness of lightcolored Sphagnum peat that commonly has been used as a growth medium in glasshouse cultivation in Finland. The biofungicide MYCOSTOP® is based on a strain of *Streptomyces griseoviridis* isolated from this peat (David M. Weller et al., 2002).

iii. Induction of suppressiveness to Apple Replant Disease: The poor growth of apple trees that occurs after replanting on a site previously cropped to apples is known as replant disease. Symptoms include stunting, shortened internodes, rosetted leaves, and stunted and decayed or discolored roots (David M. Weller et al., 2002). In Washington State, the dominant cause of replant disease is a complex of fungi including *Cylindrocarpon destructans*, *Phytophthora cactorum, Pythium* spp., and *Rhizoctonia solani.* Disease control depends largely on treatment of preplant soils with broad-spectrum pesticides such as methyl bromide that are being phased out of agricultural production. Soils that have not undergone apple cultivation are suppressive to replant disease. However, in contrast to take-all and potato scab-suppressive soils that are induced by monoculture, orchard soils become progressively more conducive to replant disease the longer the orchard is in production. This phenomenon was demonstrated by introducing inoculum of *R.solani* AG-5 (a member of the replant pathogen complex) into soils collected from orchard blocks in their first to fifth years of growth, and from nearby noncultivated areas. Apple seedling growth was significantly reduced in soils from the third-, fourth-, and fifth-year blocks as compared to growth in noncultivated soil or in soil from firstand second-year blocks. Concomitant with diminished soil suppressiveness was an increase in the populations of decline pathogens isolated from seedling roots and a decrease in populations of *Burkholderia cepacia* and *Pseudomonas putida*. *B. cepacia* produces multiple antibiotics and has biocontrol activity against soil borne pathogens including *R. solani* and *Pythium* spp. Cultivation of old orchard soils in the greenhouse with three cycles of wheat prior to planting apple seedlings induced suppressiveness, resulting in increased seedling growth, increased rhizosphere populations of *P. putida*, and decreased populations of the replant pathogens and *P. fluorescens* biovar C. These findings suggest that changes in fluorescent pseudomonad community structure play a role in suppression of replant disease. Interestingly, although the wheat cultivars Penewawa, Eltan, and Rely all enhanced apple seedling growth, Penewawa induced larger populations of *P. putida* and better apple seedling growth than did Eltan or Rely. These results suggest a cultivar x *Pseudomonas* interaction possibly related to differences in root exudates among the cultivars resulting in differences in growth of the antagonistic strains in the rhizosphere. Of particular interest now is whether specific genotypes or mixtures of subspecies of *B. cepacia* and *P. putida* contribute to suppressiveness.

iv. Take-all decline: Take-all, caused by the fungus *Gaeumannomyces graminis* var. *tritici,* is an important root disease of wheat worldwide. Wheat is particularly susceptible to the take-all fungus, and other *Gramineae* such as barley, rye, and triticale also can be infected. Breeding for resistance has been unsuccessful and methods of chemical control are limited (David M. Weller et al., 2002). Take-all can be controlled by a combination of crop rotation and tillage, practices that reduce the inoculum potential of the pathogen, but the current trend in cereal production is toward less tillage and two or three consecutive wheat crops before a break. *G. g.* var. *tritici* is vulnerable to a number of types of soil suppressiveness, which have been reviewed extensively over the past two decades and categorized on the basis of the requirements for the presence of the host, the pathogen, and/or severe disease in order to develop suppressiveness. Take-all decline (TAD), the most thoroughly studied of the various types of take-all suppressiveness, requires three components: monoculture of a susceptible host, *G. g.* var. *tritici,* and at least one severe outbreak of take-all. TAD is defined as the spontaneous decrease in the incidence and severity of take-all that occurs with monoculture of wheat or other susceptible host crops after one or more severe outbreaks of the disease.

Suppressiveness can be reduced or eliminated by breaking monoculture with a non host crop, but a field with a long history of TAD can regain suppressiveness once wheat or barley is again grown. Suppressiveness was induced by repeatedly adding mycelium of *G. g.* var. *tritici* to soil. Whether this and other types of take-all suppressiveness that develop without all of the classical components of TAD share a common microbial basis with TAD remains unknown. Most of the mechanisms reported to be responsible for TAD involve microbiological changes in the bulk soil, the rhizosphere soil, and/or the rhizoplane, resulting in antagonism of the pathogen.

Several lines of evidence support the widely held opinion that different microbial antagonists and mechanisms are responsible for TAD worldwide. First, TAD develops in multiple agroecosystems. Second, the time needed for TAD to develop differs among fields and crops grown in monoculture. Third, the site of suppression (i.e., bulk soil, rhizosphere, root surface) and the stage in the life cycle of *G. g.* var. *tritici* when suppression is thought to occur (i.e., parasitic, saprophytic, or both) appear to differ among TAD soils. Fourth, microorganisms with biocontrol ability from TAD soils represent many different taxonomic groups. Fifth, the take-all fungus is sensitive to a variety of forms of antagonism including destruction of hyphae by amoebae; cross protection by *G. graminis* var. *graminis* or *Phialophora graminicola*; hyphal lysis by a sterile red fungus and antibiosis by actinomycetes, *Trichoderma* spp. , *Bacillus* spp. and *Pseudomonas* spp. Of the many microorganisms implicated in TAD, however, several exhibit biological properties not consistent with a role in suppressiveness. For example, the sensitivity of TAD to pasteurization rules out the involvement of heat-resistant bacteria like Bacillus spp. and probably also many actinomycetes.

Antagonistic *Pseudomonas* spp. have been implicated in TAD throughout the world and have characteristics consistent with their involvement in suppressiveness. They are well adapted to the rhizosphere environment; they utilize many organic substrates, they synthesize a variety of antifungal metabolites inhibitory to *G. g.* var. *tritici,* their populations increase dramatically on roots with take-all lesions, and they are eliminated by soil pasteurization. Efforts of Weller and co-workers to understand the role of pseudomonads in TAD have focused on TAD soils from three irrigated fields near the cities of Lind, Quincy, and Moses Lake, and a nonirrigated field near Pullman, Washington, USA. They have been compared to conducive virgin (covered by native vegetation) and nonsuppressive agricultural soils collected from sites near the TAD fields. Transferability of the suppressive factor in these TAD soils has been demonstrated in field and greenhouse studies over the past 30 years.

7.12 Characterizing the Microbial Components Responsible for Soil Suppressiveness

Biocontrol of soil borne pathogens traditionally has been studied as a three-way interaction among the pathogen, the host, and the biocontrol agent, with little attention to other interspecies interactions in the rhizosphere (David M. Weller et al., 2002). To fully understand the nature of soil suppressiveness, however, the ecological context within which the major players act also must be considered. With specific suppressiveness, one or a few groups of microorganisms may be responsible for biocontrol activity, but interactions with other members of the rhizosphere community can significantly modulate the degree of suppressiveness observed. Indeed, there are probably hundreds, if not thousands, of genotypically distinct microbial species inhabiting the rhizosphere of individual plants and biotic as well as abiotic variables influence both the structure and activities of rhizosphere populations including those of pathogens and their antagonists (Simon et al., 2001). A clear understanding of the biological factors responsible for soil suppressiveness may require broader knowledge of the identity, relative abundance, and biological activity of the phylogenetically diverse microbial populations that inhabit the rhizosphere.

Several approaches are now available for detailed, high-throughput analysis of the distribution and importance of microbial antagonists and root pathogens in suppressive soils. The first involves direct separation of PCR-amplified DNA products based on electrophoretic mobility. Differences in primary sequence composition can be detected under partially denaturing conditions as in denaturingor temperature-gradient gel electrophoresis [DGGE and TGGE] or by single-strand conformation polymorphism [SSCP]. Alternatively, length polymorphisms can be assayed directly, as in amplified ribosomal intergenic spacer analysis [ARISA or by digesting amplified products with selected restriction enzymes and then characterizing the resulting fragment length polymorphisms (RFLPs). Differences in community structure can be detected by analyzing complete mixtures of restriction fragments, but it usually is advantageous to visualize only the terminal restriction fragments (TRFs) by fluorescently labeling one of the two PCR primers. Because only a single band is visualized from each distinct amplification product, the number of TRFs in a T-RFLP profile indicates the species richness of the amplified sequences.

T-RFLP analyses can be used to study diversity at various taxonomic levels (Braker et al., 2001). When applied to ribosomal gene sequences, RFLP analyses are more properly termed amplified ribosomal DNA (rDNA) restriction analysis [ARDRA] or fluorescently tagged ARDRA [FT-ARDRA] , depending on whether all or only the terminal restriction fragments are visualized. The choice of approach depends on the degree of sequence diversity in the amplified products. FT-ARDRA is particularly useful to distinguish among larger taxonomic groups such as genera within a domain because the amplified fragments are likely to differ at multiple restriction sites. DGGE and SSCP are appropriate for diversity studies on a more limited scale, e.g., encompassing species within a genus, because the sequence e differences that distinguish taxa may not correspond to most restriction sites. In the future, a third approach to characterizing microbial diversity likely will involve hybridization of amplified sequences to DNA arrays of known sequences. Early reports indicate the potential to identify several taxa within a single mixture or individual sequences by serial hybridization. All of these approaches also can be adapted to examine metabolically active populations within microbial communities by using extracted RNA to generate DNA templates by reverse-transcriptase mediated PCR (RT-PCR). Alternatively, newly synthesized template DNA, present in metabolically active microorganisms, can be isolated from cells grown in the presence of appropriate nucleotide analogues.

Profiles based on amplified ribosomal sequences have become the foundation upon which our descriptions of microbial communities are built. However, the precision with which taxonomic assignments can be made is limited by the number of nucleotides sequenced and the potential for the generation of sequence heterogeneity during PCR amplification. For community analyses involving PCR with domain-specific primers, FT-ARDRA and partial sequencing of cloned 16S sequences each provided for phylogenetic assignments roughly at the genus level. Other genes also have been targeted for analyses of microbial diversity including bacterial RNA polymerase and genes involved in nitrogen cycling (Poly et al., 2001). For dissection of community structure at the species or subspecies level, PCR primers also can be developed from clones of the targeted DNA fragments themselves, as was done to identify phlD+ strains, molecular markers derived from related isolates, or sequences linked to the original target sequence and identified by screening the soil metagenome. Future studies will incorporate an ever-increasing number of molecular probes to evaluate the diversity of genes and their expression in complex communities.

7.13 Functional Analysis of key Populations Involved in Soil Suppressiveness

All populations are characterized by some degree of variation among individuals, but not all variation is functionally significant. However, among phlD+ fluorescent *Pseudomonas* spp. antibiotic production, rhizosphere competence and disease suppression vary at the subspecies level. For example, ARDRA group 2 strains were more effective than those of group 1 against *Fusarium* crown and root rot of tomato and *Pythium* damping-off of cucumber. Raaijmakers & Weller (2001) while examining, found the relationship between genotype and rhizosphere colonization by indigenous populations of 2,4-DAPG producers on roots of wheat grown in Quincy TAD soil for eight successive cycles. Of the 16 RAPD groups identified among 101 isolates, one group comprised 50% of the phlD+ isolates and had an average population density of 2 × 105 CFU g−1 root. This genotype also was dominant on roots of wheat cycled in Moses Lake and Lind TAD soils, belongs to BOX-PCR group D, and is exemplified by *P. fluorescens* Q8r1-96. Short- and long-term colonization studies demonstrated that Q8r1-96 and other group D strains are much more competitive in the wheat rhizosphere than representatives of other phlD+ genotypes that have so far been tested. For example, Q8r1-96 required a much lower dose (only 10 to 100 CFU seed−1 or g−1 of soil) to establish population densities of up to 107 CFU g−1 of root on wheat grown in Quincy virgin soil than did strains Q2-87 (group B) or 1M1-96 (group L).

The ability of Q8r1-96 to suppress take-all after eight months in Quincy virgin soil reflected its superior rhizosphere competence, which allowed it to maintain threshold densities throughout the cycling process (David M. Weller et al., 2002). All three strains provide equivalent biocontrol if present at densities above the threshold needed for take-all control (Raaijmakers and Weller, 2001). The exceptional rhizosphere competence of Q8r1-96 is especially remarkable given that all three strains belong to biovar II of *P. fluorescens*, are nearly identical physiologically, and produce similar amounts of 2,4-DAPG in situ. Collectively, the results of these studies suggest that D genotype strains may be primarily responsible for TAD in Washington State soils, and raise the possibility of accelerating the process of TAD by introducing this specific genotype into the soil at the onset of a take-all outbreak. Field studies have verified that very low doses of D genotype strains introduced on wheat seed are sufficient to establish population densities required for root disease suppression. The populations are sustained throughout the growing season, survive between crops, and re-establish on the roots in successive years. Studies of the role of 2,4-DAPG producers in TAD fulfill four criteria that should be met to define the functional importance of a microbial population to soil suppressiveness: (a) the ability of members of the population to perform a suppressive function; (b) the presence in suppressive soils of the population at densities above the threshold required for disease control (and its absence from conducive soils) (c) duplication of the function in conducive soil by addition of populations obtained from a suppressive soil with similar physical and chemical properties; and (d) detection of a functional activity (e.g., antibiotic production) in situ. However, interactions among microbial populations and environmental variables are complex, and even when two soils share a common suppressive mechanism, the degree of suppressiveness may differ because local conditions affect on microbial community structure and function.

Interactions between microbial components and soil properties including pH, clay type, and specific ions have a determinative role in *Fusarium* wilt–suppressive soils (Dom´ınguez et al., 2001). Similarly, the type of clay greatly influenced the suppressive activity of Pseudomonas spp. in black root rot–suppressive soils and fertilization with NH+ 4 -N as compared to NO− 3 - N resulted in a greater percentage of antagonistic pseudomonads on roots of wheat grown in take-all suppressive soils. Production of 2,4-DAPG by biocontrol strains is influenced by their chemical and physical environment; notable is the finding that expression of the 2,4-DAPG biosynthesis gene phlA in strain CHA0 was significantly greater in the rhizosphere of monocots (maize and wheat) than in the rhizosphere of dicots (bean and cucumber). Interactions with other microorganisms also can affect the degree of root disease suppression observed. Interpopulation signaling among bacteria via N-acetylhomoserine lactones, which regulate the synthesis of phenazine antibiotics suppressive of take-all, has been detected in the wheat rhizosphere and several recent studies indicate that increases in the abundance of 2,4- DAPG producers (by inoculation) can influence the structure of rhizosphere microbial communities.

Conversely, the effects of microbial communities on 2,4-DAPG producers are only now being explored. Altered root health was associated with multiple changes in rhizosphere bacterial community structure and some of these changes might create an environment more favorable to the onset of take-all suppression. Production of fusaric acid by *F. oxysporum* f. sp. *radicis-lycopersici* repressed production of 2,4-DAPG by strain CHA0 and *Pythium ultimum* decreased the ecological fitness of the 2,4-DAPG producer *P. fluorescens* F113. These and numerous other observations substantiate the importance of the total rhizosphere microbial community as a modulator of the activity of suppressive microorganisms, and undoubtedly account for some of the variability in the degree of soil suppressiveness seen over time and space.

Understanding the diversity within populations of biocontrol agents holds the promise of pairing specific genotypes with their most supportive plant hosts or soil environments to maximize root colonization and disease suppression (David M. Weller et al., 2002). Such knowledge also can streamline the process of selecting new biocontrol agents because targeted molecular screens can be used to identify isolates with specific traits such as superior competitiveness or the production of 2,4-DAPG. Establishing the presence and functionality of individual populations within a particular soil is just one first step toward fully understanding the nature of suppressiveness within that soil. Ultimately, the parameters within which the activities of functionally important microbial populations combine to produce a suppressive soil also must be defined. To identify those parameters, new and more detailed studies will be required to characterize the soil structure and composition, the environmental conditions under which suppression occurs, the molecular interactions among functionally important populations under different conditions, and the biogeography and population dynamics of beneficial as well as pathogenic microbial populations in the field. Because of the complexity of field soils, high throughput methods will be required to adequately characterize these populations, but the pay-off will be worth the effort. Future studies of biologically based soil suppressiveness will present new insights into the microbial ecology of agricultural soils and lay the foundation for the development of creative management strategies for the suppression of soil borne diseases.

7.14 Microbial Community Composition in Take-all Suppressive Soils

Suppression is the ability of a natural soil to reduce or suppress the activity of plant pathogens, mostly due to the presence and activity of soil microorganisms (Paola Durán et al., 2020). Their presence increases the ecosystem resilience by creating redundancy in ecosystem services, making soil less vulnerable to short-term changes in the environment. Studies showed that conducive soils, where the incidence of take-all disease is elevated, become suppressive under certain characteristics: monoculture of susceptible host, Ggt presence, and take-all disease outbreak. Thus, suppressive soils are defined as soils where disease development is minimal despite the presence of an infective pathogen and a susceptible plant host. Despite suppressive soils have been known for over 100 years and have been studied for more than five decades, they remain poorly understood. Considering the great potential that this type of soil offers for sustainable pest management and plant biocontrol, understanding the mechanisms underlying suppression and soil properties should be priority line of research. Recently, we found six suppressive soils against take all disease, managed under ancestral agronomic practices as monoculture for more than 10 years and we demonstrated the essential role of soil microbial communities in Ggt suppression.

Gaeumannomyces graminis var. *tritici* (Ggt) is the main biotic factor that affects wheat production around the world. Paola Durán et al., (2020) reported the occurrence of six suppressive soils in monoculture areas from indigenous "Mapuche" communities, and evidenced that the suppression relied on the microbial component of those soils. They determined the occurrence of the microbial community (total bacteria, actinomycete, total fungi, and ascomycete) on the rhizosphere and endosphere of wheat plants grown in suppressive soils compared with a conducive soil. It was found that Ggt suppression is mediated mostly by endophytic, rather than rhizosphere microorganisms, especially by bacteria since the community structure was similar in all suppressive soils. Interestingly, despite the lower incidence of take-all disease in suppressive soils, the Ggt concentration quantified by qPCR was not different among plants grown in suppressive soils compared to conducive soil. Therefore, the fungal activity is not a determinant factor for suppressiveness classification. Due to the importance of endophytic bacteria, *Serratia* sp and *Enterobacter* sp were isolated from suppressive soils, which only diminish around 20% of fungal growth. Thus, induced resistance or tolerance mechanisms in plants are involved in suppression against take-all more than direct effect of microorganisms.

G. graminis can survive as saprophytically on infected or dead root and crown debris from previous crops causing primary infection through parasitism, thus this pathogen infects the next wheat crop. Roots come into contact with the ascospores and dark runner hyphae of *Ggt*, they colonize the surface, and then penetrate directly through hyaline hyphae beneath the hyphopodia into the roots cortex and across the endodermis into the stele, obtaining nutrients, carbon, and energy, hence triggering a secondary infection. Consequently, crop rotation is determinant cultural practice to diminish disease incidence due to the control of the pathogen is inefficient. The total endophytic bacteria communities were similar in all suppressive soils and were clearly different from conductive control soil. This is why it was hypothesized that endophytic microorganisms have an important role in take-all suppression. Additionally, it was found that the ascomycete group (group that belong Ggt) in suppressive soils were associated with endophytic bacteria in comparison with the rest of the microbial group analyzed, thus confirming the importance of these microorganisms in specific Ggt suppression as showed in other study where indicate that bacteria could be linked to a specific form of suppression (Paola Durán et al., 2020).

Studies have showed the importance of some strains of actinobacteria in reducing *Ggt* disease symptoms up to 70%, under steamed soil and field conditions. A similar role of actinomycetes groups was found in suppression of other soil borne pathogens as *Pythium, Phytophthora, Ustilago crameri* and *Ralstonia solani* In contrast in fungal structure community no differences were found. Early studies reported that fungal microorganisms have a general role in suppression, but no role in specific suppression. However, it was found that the ascomycete group, where *Ggt* belongs, has major dominance in suppressive soils endosphere. In contrast with the relevance of endophytic microorganisms in soil suppression found in this study, rhizosphere microorganism composition did not show significant differences in relation to microbial composition in suppressive soils as compared to the conductive control soil. On the other hand, microbial biodiversity was directly related to essential parameters of volcanic-ash-derived soils as pH, OM and Al Sat. Thus, biodiversity of the rhizosphere suppressive soils was higher when pH and SOM were high, and lower in relation to Al Sat, as we reported previously (Duran et al., 2017). Other authors through DGGE analyses of the microbial communities showed similar banding patterns in suppressive and conducive soils supporting that rhizosphere communities have no direct influence over soil suppressive effects.

Suppressive soils had low disease incidence despite *Ggt* DNA concentration. For example, g Ggt DNA g endosphere μ soils 2 and 4 showed around 30 samples and less than 3% blackening roots compared with the control with more than 30% blackening roots. Similarly, other study on suppressive where low disease severity coupled with high *Ggt* DNA concentrations in roots (Chng et al., 2015). On the other hand, it was found that *Ggt* presence in wheat plant endosphere with no associated symptoms, confirming our hypothesis that endophytic bacteria located inside the root control the infectivity of the pathogen. It was hypothesized that the suppressivenes could be related with the capacity of endophytic bacteria to induce resistance to plants due to *Serratia* sp. and other Enterobacteriacea as *Enterobacter* sp. are involved in the Nacyl-L-homoserine lactone (AHL) and salicilic acid production, hence increasing systemic resistance. Therefore, inducing hormone analysis and defense genes in planta and mechanisms related with plant tolerance are requested in order to determine the role of endophytic bacteria in take-all suppression.

Bacteria from wheat plants endosphere, grown in suppressive soils, have a direct effect on take-all suppression, whereas rhizosphere microorganisms are influenced by soil chemical parameters that are more relevant in volcanic soils such as pH, MO and Al sat (Paola Durán et al., 2020). Notwithstanding, the lower incidence of take-all disease in suppressive soils and the Ggt concentration in wheat roots are not key factors to determine supressiveness, possibly due to the fact that the fungus could remain infective in endosphere as a result of the action of microorganisms, mainly endophytic bacteria. However, *Serratia* sp and *Enterobacter* sp isolated from suppressive soils endosphere were only able to inhibit mycelia growth around 20% under in vitro conditions. Regarding this, we consider that the suppressing effect of bacteria is related with factors involved in induced resistance in plants. Therefore, in planta analyses with endophytic bacteria isolated from wheat grown in suppressive soils are needed.

7.15 Rhizosphere Suppression Hinders Antibiotic Resistance Gene (ARG) Spread under *Ralstonia solanacearum* invasion

Genes conferring resistance to antibiotics (antibiotic-resistance genes, ARGs) occur naturally. ARGs can be easily exchanged among organisms and surrounding environments. For instance, ARGs present in the soil can enter the food chain through rhizosphere dwelling microorganisms, the accumulation of which can dramatically increase the health risks of the exposed population (Li et al., 2023). In fact, the acquisition and transfer of ARGs imposes significant harm to human health by generating multi-resistance human pathogens. Therefore, strategies to reduce the amount of ARGs and hinder their spread in soils are crucial. The accumulation of ARGs in the environment is closely related to artificial interference. Many studies have reported that exogenous contamination, such as the discharge of urban wastewater, manure and sludge from agricultural animals increases ARGs in soil, and the invasion of ARG-rich microorganisms in these wastes is directly linked to the higher human risk. In agriculture, intensive cropping and conventional mineral fertilizer application usually increase soil pathogen density and ARG abundance. In the presence of mobile genetic elements (MGEs), these ARGs can be easily transferred between microbial community members. However, the subsequent interaction among the co-accumulation factors (pathogens, ARGs and MGEs) around crops remains unclear. After all, soil microbiome under biotic and abiotic stress seem to change their associated microenvironment and consequently leads to the enrichment of ARGs. In this context, it is crucial to understand the ARG dynamics of the rhizosphere microbiome of disease-induced soils, as pathogen invasion could promote the spread of ARGs.

The rhizosphere is an extremely important component of the "one health" scenario by linking the soil microbiome and plants, in which the potential enrichment of antibiotic resistance genes (ARGs) might ultimately flow into the human food chain. Despite the increased occurrence of soil-borne diseases, which can lead to increased use of pesticides and antibiotic-producing biocontrol agents, the understanding of the dynamics of ARG spread in the rhizosphere is largely overlooked. Here, tomato seedlings grown in soils conducive and suppressive to the pathogen *R. solanacearum* were selected as a model to investigate ARG spread in the rhizosphere with and without pathogen invasion. Metagenomics data revealed that *R. solanacearum* invasion increased the density of ARGs and mobile genetic elements (MGEs). Although ARGs were found originating from human pathogenic bacteria in both soils, the enrichment was alleviated in the suppressive soil (Li et al., 2023). The suppressive soil hindered ARG spread through pathogen suppression and had a lower number of taxa carrying antibiotic resistance.

Almost all ARG contributors in the conducive rhizosphere were more abundant than those in the suppressive rhizosphere following invasion. Similar to the abundance of ARGs in R. solanacearum, some biocontrol agents, such as *Streptomyces* and *Pseudomonas*, and many human conditional pathogens, such as *Staphylococcus aureus, Pseudomonas aeruginosa*, and *Escherichia coli*, which contain multidrug-type ARGs, were significantly enriched in the conducive rhizosphere (Li et al., 2023). To understand the patterns associated with MGEs following invasion, their associated ARGs, and the potential hosts of the ARGs were visualized as a network. The results showed that the pathogen mainly contributed to the multidrug resistance gene mexC, which was potentially connected with various MGEs, such as tnpA, IS91, istB, intl1, and istA9. These MGEs were also potentially linked with many ARGs belonging to other microbial hosts, including some of the human conditional pathogens. Co-occurrence networks were constructed to examine the interaction between the rhizosphere microbiome and ARGs. Importantly, *Ralstonia* was abundant following invasion and was significantly positively correlated with ARG density, which indicates that pathogen invasion affected ARG enrichment. More specifically, the network of interactions between bacteria and ARGs was simpler in the absence of invasion but increased in complexity in the presence of the pathogen. Most of these resistance genes are multidrugresistant, followed by aminoglycoside, *chloramphenicol, fosmidomycin*, and macrolide-resistant. Some bacteria, such as Sphingomonas, Pseudoxanthomonas, and *Stenotrophomonas*, were significantly enriched in the suppressive rhizosphere and negatively correlated with ARGs. In contrast, the bacteria enriched in the conducive rhizosphere showed the opposite trend. Finally, an in vitro test demonstrated that the antibiotic resistance of these taxa (*Pseudoxanthomonas* spp. and *Sphingomonas* spp.) was significantly less enriched in the suppressive rhizosphere than the potentially responsive ARG contributors (*Staphylococcus aureus* and *Pseudomonas* spp.) that were enriched in the conducive rhizosphere.

The results support the overlooked potential link between pathogen incidence and the increase in antibiotic resistance spread risk. In general, *R. solanacearum* invasion increased ARG and MGE abundances in both rhizospheres, indicating that the biotic pressure imposed by the pathogen boosted the abundance of ARG-rich taxa. However, ARG enrichment was relatively moderate in the suppressive rhizosphere soil, which occurs in soil following long-term (bio)-organic fertilization, confirming that the response depends on the health status of the soil. It is widely known that soil suppressiveness can be primed by organic amendments through microbiome manipulation (Deng et al., 2022), which can also decrease the soil ARG level. ARG-poor taxa such as *Sphingomonas* and *Pseudoxanthomonas*, belonging to Sphingomonadaceae and *Xanthomonadaceae*, respectively, were stimulated by pathogen invasion in suppressive soils. In suppressive rhizosphere soil, these species can control bacterial wilt, contributing to lower pathogen density while potentially alleviating the ARG horizontal transfer risk. In contrast, conducive soils, which are more susceptible to disturbances and prone to pathogen colonization, pose a higher risk of ARG. Therefore, we hypothesize that the high suppression induced by organic management is a promising agricultural mode in plant-associated disease control.

Antimicrobial resistomes in the soil can be absorbed by plants and then migrate into the food chain, serving as an additional risk to human health (Li et al., 2023). Therefore, reducing the amount of ARGs in agricultural soil, especially in plant-associated soil, is particularly important for human health. A conceptual model was proposed for ARG spread following biotic stress based on our results. First, the functional microorganisms in suppressive soil limit the migration and colonization of *R. solanacearum* and directly inhibit the horizontal and vertical transfer of its ARGs; second, the resident response of the ARG-poor microbiome reduces the space of microorganisms with a high ARG microbiome, thus indirectly limiting the abundance of ARGs in indigenous communities, and reducing the risk to human health.

7.16 Microbial Diversity in Soils Suppressive to *Fusarium* diseases

Disease-suppressiveness of soils is a useful model to understand microbial phytoprotection and develop sustainable plant protection strategies for soils devoid of this property. Fusarium species are cosmopolitan soil phytopathogens from the division Ascomycota, which produce mycotoxins and cause significant economic losses of crop plants. However, soils suppressive to Fusarium diseases are known to occur, and recent knowledge on microbial diversity in these soils has shed new lights on phytoprotection effects (Irena Todorović et al., 2023). Suppressive soils represent a reservoir of beneficial microorganisms, which may confer effective plant protection against various soil borne diseases. This biocontrol potential of suppressive soils is of great importance when considering phytopathogens like *Fusarium* spp., which are causing increasing damage to crops in the on-going climate change context. Insight into the time and space microbial dynamics of soils suppressive to Fusarium diseases, together with the understanding of microbial modes of action and agricultural practices applied, is needed in order to develop safe, effective, and stable tools for disease management.

Irena Todorović et al., (2023), in a review, synthesized current knowledge on soils suppressive to *Fusarium* diseases and the role of their rhizosphere microbiota in phytoprotection. This is an important issue, as disease does not develop significantly in suppressive soils even though pathogenic *Fusarium* and susceptible host plant are present, and weather conditions are suitable for disease. Soils suppressive to *Fusarium* diseases are documented in different regions of the world. They contain biocontrol microorganisms, which act by inducing plants' resistance to the pathogen, competing with or inhibiting the pathogen, or parasitizing the pathogen. In particular, some of the *Bacillus, Pseudomonas*, *Paenibacillus* and *Streptomyces* species are involved in plant protection from *Fusarium* diseases. Besides specific bacterial populations involved in disease suppression, next-generation sequencing and ecological networks have largely contributed to the understanding of microbial communities in soils suppressive or not to *Fusarium* diseases, revealing different microbial community patterns and differences for a notable number of taxa, according to the *Fusarium* pathosystem, the host plant and the origin of the soil. Agricultural practices can significantly influence soil suppressiveness to *Fusarium* diseases by influencing soil microbiota ecology. Research on microbial modes of action and diversity in suppressive soils should help guide the development of effective farming practices for *Fusarium* disease management in sustainable agriculture.

By selecting their rhizosphere microbiome, plants may contribute themselves to suppressiveness (Gómez Expósito et al., 2017). Soil represents the richest known reservoir of microbial biodiversity and displays several compartments, i.e. the bulk soil containing microorganisms that are not affected by the roots, the rhizosphere where soil microorganisms are under the influence of roots (and roots exudates), the rhizoplane with root-adhering microorganisms, and the endosphere for root tissues colonized by microorganisms. The rhizosphere and rhizoplane harbor an abundant community of bacteria, archaea, oomycetes and fungi, whose individual members can have beneficial, deleterious or neutral effects on the plant. The collective genome of this microbial community is larger than that of the plant itself, and is often referred to as the plant's second genome. Thus, this alliance of the plant and its associated microorganisms represents a holobiont, which has interdependent, fine-tuned and complex functioning. In this system, the plant is a key player, as nearly 40% of all photosynthates are released directly by roots into the rhizosphere, serving as a fuel for microbial communities, thus recruiting and shaping this microbiome. These photosynthates are conditioned by the plant genotype, developmental stage, metabolism, immune system and its ability to exudate (Sánchez-Cañizares et al., 2017). In this context, suppressiveness will depend on microbiome diversity and functioning.

On one hand, extensive information is available on environmental and microbial properties responsible for suppressiveness to *Fusarium* diseases (Irena Todorović et al., 2023). One prominent feature is the diversity of *Fusarium*-based pathosystems for which suppressive soils are documented, in terms of *Fusarium* species (often F. oxysporum, but not only), host plants (both monocots and dicots), types of disease (often wilt, but not only) geographic locations of soil and farming conditions, and types of suppressiveness (i.e. natural suppressiveness to *Fusarium* diseases, but also monocultureinduced suppressiveness as well as fungistasis towards *Fusarium* pathogens). This diversity is paralleled by differences in microbiota composition and diversity associated with disease control in the different cases of suppressiveness. On the other hand, despite the fact that soils suppressive to *Fusarium* diseases have been studied for decades, they are still poorly understood in terms of microbiota functioning, and knowledge remains fragmented.

Additional research is needed to integrate further the scientific approaches used to decipher suppressiveness to *Fusarium* diseases. First, by combining complementary assessment methodology with current next-generation sequencing and ecological networks research, and incorporating experimental strategies to manipulate and transplant rhizosphere microbiome (or single microorganisms) of plants grown in suppressive soils to those in conducive soils to go beyond correlative work, as started recently (Jiang et al., 2022). Second, by extending the range of soil conditions investigated, and develop meta-analyses to estimate key microbiota differences between suppressive and conducive soils. Third, by considering a wider range of biological actors, including beneficial fungi (often neglected), soil fauna (likely to influence microbial communities, *Fusarium vectorisation* and plant health). Fourth, by taking into account plant genetics, behavior and physiological responses to *Fusarium* pathogens. Therefore, there is a need for a more multidisciplinary approach to understand microbiota functioning in soils suppressive to *Fusarium* diseases.

7.17 Suppression of Banana *Fusarium* wilt Disease with Soil Microbial Mechanisms

The large outbreak of banana *Fusarium* wilt has become a bottleneck limiting the industry's development, and crop rotation is a cost-effective and essential measure to overcome the obstacles of banana crop monoculture. The present work by Yang et al., (2023) was carried out to explore the mechanisms of how changes in soil chemical properties and the reestablishment of soil microorganisms in high-incidence soils are affected by crop rotation and plant residue. In this study, pineapple–banana crop rotation and pineapple residue amendment were carried out to alleviate banana *Fusarium* wilt, and their effects on bacterial and fungal communities were analyzed using the MiSeq Illumina sequencing platform. Both pineapple–banana rotation and residue addition significantly reduced disease incidence. Moreover, pineapple rotation and residue amendment altered the bacterial and fungal community composition. The taxonomic and phylogenetic alpha diversity of bacteria and fungi significantly increased against disease suppression and nutrition competition. The relative abundances of the *Burkholderia*, *Pseudomonas*, *Elaphocordyceps*, *Penicillium*, and *Talaromyces genera* were higher, and the number of *Fusarium* was significantly lower in rotational soil than in banana monoculture soil. Finally, linear models (LM) showed that the *Burkholderia and Talaromyces* in crop rotation, and *Aspergillus* in residue amendment had a significantly negative relationship to disease incidence, which plays a key role in *Fusarium* reduction. To consider the economic benefits and protect the vitality of the soil, this study suggested that pineapple–banana rotation and pineapple residue amendment both could be considered for the sustainable management of banana wilt.

Different banana crop rotations can effectively reduce banana *Fusarium* wilt to varying degrees (Fan et al., 2022). In this study, pot experiments revealed that pineapple–banana rotation and residue pineapple amendment could significantly enhance the suppression of *Fusarium* wilt disease compared with banana monocropping and residue addition. The results were a further extension of our previous report. Notably, disease incidence reduction in crop rotation and residue amendment treatments could be attributed to a reduction in the number of *Fusarium*, as well as an increase in beneficial genera and changes in soil physicochemistry. Both the two steps in banana–pineapple rotation system, rotation, and residue amendment could inhibit fungal infections. Similar results have been shown for strawberry soil disease, likely due to the interruption of the host pathogen cycle in the root. Both high-throughput sequencing and qPCR showed that the *Fusarium* abundance was lower in both rotation and residue amendment treatments than in banana monoculture. Moreover, disease incidence has a significantly positive correlation with residue addition amount. Pathogen density and disease incidence have been decreased in each pineapple residue treatment (P_L, P_R, and P_F). This finding aligns with our previous studies, indicating that using a pineapple–banana crop rotation system can effectively minimize the incidence of this pathogen. Further, this study provides strong evidence that pineapple–banana rotation and residue amendment were both effective approaches for disease suppression. Pineapple rotation may result in a significant decline in pathogen number via increasing some beneficial microorganisms. The disease suppression was potentially caused by a decrease in nutrient availability of pathogens which resulted from antagonistic and nutritional competition, parasitism, predation, induction of plant resistance, and interference of pathogenic signals. In addition, it seems possible that pineapple crops can stimulate beneficial bacterial microbiomes through root exudates to suppress pathogens during the crop rotation season.

Consistent with previous findings, the major bacterial phyla were Proteobacteria, Acidobacteria, Bacteroidetes, Firmicutes, Gemmatimonadetes, Actinobacteria, Verrucomicrobia, Chloroflexi, Planctomycetes, Cyanobacteria/Chloroplast, and Thaumarchaeota and the fungal phyla were Ascomycota and Basidiomycetes in this study (Wang et al., 2022). The relative abundance of Firmicutes and Ascomycota increased after pineapple rotation and residue addition which also showed a strong negative correlation with *F. oxysporum* abundance. This finding suggests that Firmicutes and Ascomycota may be involved in disease suppression through the representatives with biocontrol potential.

Pineapple–banana rotation significantly stimulated the relative abundance of the *Burkholderia*, *Pseudomonas* and *Talaromyces* genera in bulk and rhizosphere soils. These microorganisms showed significantly negative correlations with the abundance of *F. oxysporum* and there is a large body of work demonstrating the stable suppression of disease by these beneficial microorganisms (Hong et al., 2022). For example, *Burkholderia* spp. has been shown to be abundant in golden pineapple–banana rotations and *Pseudomonas* spp. plays a key role in pathogen antagonism by stimulating the synthesis of a beneficial microbiome common to the native beneficial microorganisms, after the application of bioorganic fertilizers, to reduce the occurrence of *Fusarium* in banana. Thus, they are widely reported as biological control agents. These microorganisms are capable of producing antifungal metabolites and can colonize the roots of plants, making them targets for biocontrol. Additionally, studies have shown that some fungal microorganisms pre-empt the microecological environment with pathogens, such as the effect of the mass production of *Talaromyces* spp. on pathogens populations. Both pineapple crop rotation and residue addition significantly reduced the incidence of banana seedling wilt, closely related to the decrease in the number of *Fusarium oxysporum* in the soil after pineapple cultivation.

Plant and soil microflora acted as important roles in soil-borne diseases suppression. The species and number of beneficial microflorae in soil microorganisms are crucial for suppressing pathogens (Tian, et al., 2020). Numbers of *Burkholderia*, *Pseudomonas*, *Penicillium*, and *Talaromyces* were significantly increased in pineapple rotation treatment in the residue amendment, and *Pseudomonas*, *Elaphocordyceps*, and *Aspergillus* were significantly reduced in the residue amendment treatment. The numbers of culturable bacteria and *Burkholderia* increased and the number of *Fusarium oxysporum* decreased significantly in pineapple rotation with bioorganic fertilizer. In herb and "maize–potato" rotations, *Pseudomonas* genus was abundant and significantly suppressed *Fusarium oxysporum* populations. The incidence of banana replanted after pineapple residue addition was significantly positively correlated with *Fusarium oxysporum* and culturable *Fusarium oxysporum*. In addition, it was significantly negatively correlated with soil pH, *Pseudomonas*, *Aspergillus* and culturable bacteria. *Penicillium* and *Burkholderia* were abundantly present in the bulk and rhizosphere soils after pineapple residue addition, while *Talaromyces* and *Aspergillus* could pass through the pineapple rhizosphere to the soil and then survive in the next banana soil and *Talaromyces* could colonize the next banana soil. It was similar to the golden pineapple by Yuan et al., and the residual energy alleviated the occurrence of banana fusarium wilts, which verified that the abundance and colonization ability of *Talaromyces* and *Aspergillus* significantly increased after pineapple rotation (Yuan et al., 2021). Therefore, they are also potential key microorganisms in this case. In summary, these results provide the basis for isolating and screening key beneficial microorganisms and the effect of their root exudates on these beneficial microorganisms. It provides some directions for the study of banana–pineapple rotation and residue amendment mechanisms.

This study further indicated that the addition of pineapple plant residues in the soil can also trigger soil suppression of a serious soil-borne pathogen (*Fusarium* spp.). Yang et al., (2023) described that this state of suppression is modulated by changes in the soil bacterial and fungal communities and highlighted one bacterial taxa (*Burkholderia* spp.) and two fungal taxa (*Aspergillus* spp. and *Talaromyces* spp.) that may be directly involved in pathogen suppression. This research may provided new avenues for exploring agricultural practices with a focus on beneficial outcomes that directly affect soil health and crop productivity in a viable and sustainable manner.

7.18 Functional Amplicon Sequencing and 10× Metagenomics to Dissect Disease-Suppressive Rhizosphere Microbiomes

Disease-suppressive soils protect plants against soil borne fungal pathogens that would otherwise cause root infections. Soil suppressiveness is, in most cases, mediated by the antagonistic activity of the microbial community associated with the plant roots. Considering the enormous taxonomic and functional diversity of the root-associated microbiome, identification of the microbial genera and mechanisms underlying this phenotype is challenging. One approach to unravel the underlying mechanisms is to identify metabolic pathways enriched in the disease-suppressive microbial community, in particular, pathways that harbor natural products with antifungal properties. An important class of these natural products includes peptides produced by nonribosomal peptide synthetases (NRPSs). Tracanna et al., (2021) applied functional amplicon sequencing of NRPS-associated adenylation domains (A domains) to a collection of eight soils that are suppressive or nonsuppressive (i.e., conducive) to *Fusarium culmorum,* a fungal root pathogen of wheat. To identify functional elements in the root-associated bacterial community, we developed an open-source pipeline, referred to as dom2BGC, for amplicon annotation and putative gene cluster reconstruction through analyzing A domain co-occurrence across samples. This pipeline was applied to rhizosphere communities from four disease-suppressive and four conducive soils and found significant similarities in NRPS repertoires between suppressive soils. Specifically, several siderophore biosynthetic gene clusters were consistently associated with suppressive soils, hinting at competition for iron as a potential mechanism of suppression. Finally, to validate dom2BGC and to allow more unbiased functional metagenomics 10× metagenomic sequencing of one suppressive soil was performed, leading to the identification of multiple gene clusters potentially associated with the disease-suppressive phenotype.

Disease suppression is not associated with increased adenylation domain diversity but shows a distinct community structure. There is great need for diagnostic tools to assess the disease-suppressive potential of agricultural soils based on their microbial and functional compositions. Suppressive soils are enriched in cyclic peptide-associated A domains: Adenylation domains activate and incorporate specific amino acids in the growing nonribosomal peptide during synthesis by an NRPS assembly line (Tracanna et al., 2021). The substrate specificity for different A domains is determined by a restricted number of residues in their sequence. A domains incorporate a large variety of both proteogenic and nonproteogenic amino acids, which facilitate the structural diversity of the final peptide products. We reasoned that prediction of the substrate specificities of the domain amplicons detected in suppressive and conducive rhizosphere samples could provide new insights into the abundance and diversity of different products and a classifier was trained to predict these specificities (see Materials and Methods). Intriguingly, it was found predicted threonine-specific domains to be significantly more common in suppressive soils than in conducive soils (rank sum test P value <0.001).

This is particularly interesting as threonine is an amino acid commonly involved in lactone ring formation of cyclic and branched cyclic (lipo) peptides. Such peptides have a large variety of natural functions, which encompass, among others, the induction of systemic resistance in plants to fungal infection and direct antifungal activity. Rank sum statistics were calculated using relative counts of appearances of amplicons annotated to the monomer for calculating ranks. Since a domain functional amplicon data showed more distinctive patterns than 16S data between soils with conducive and suppressive phenotypes, it was decided to set out to explore if it might be feasible to use functional amplicon sequencing as a diagnostic tool of disease suppressiveness. To test the possible association of within-sample amplicon diversity measures with the suppressive phenotype, within-sample richness, evenness, and phylogenetic diversity (PD) were calculated for all samples based on observed unique amplicons, Simpson *e*, and Faith PD, respectively. Wilcoxon rank sum tests showed no significant association of alpha diversity measures with the presence of the pathogen or with the suppressive phenotype for any of these metrics.

This study provides novel insights into the NRPS AMP-binding domain diversity of agricultural rhizosphere samples. Remarkably, the diversity of the set of unique amplicons from this rhizosphere collection equals the level of diversity of adenylation domains found across all publicly available genomes. Annotation rates for nAMPs were generally low, which highlights the incredible potential of plant-associated microbiomes for discovering novel natural products. Significant community structure overlap among suppressive rhizobacterial adenylation domain profiles was reported and generated new hypotheses regarding possible roles for siderophores in disease suppression against *Fusarium culmorum*. A pipeline was also developed for taxonomic and functional annotation of NRPS amplicons without the requirement of a BAC clone library. The dom2BGC pipeline can be extended to and currently supports annotation of any natural product-associated domain that occurs multiple times within a BGC and, to some extent, for any BGC-associated domain. The amplicon clustering results were validated by reconstructing the delftibactin BGC, a siderophore associated with suppressive soils, using a combination of amplicon sequencing and novel 10× genomics shotgun metagenomics sequencing. It was concluded that combining functional amplicon sequencing and shotgun metagenomics represents a powerful approach to probe complex microbiome-associated plant phenotypes and to generate new hypotheses on the functional roles of microbial metabolites in microbe-microbe and microbe-host interactions.

7.19 28S LSU Gene Sequencing in Assessing Fungal Community Structure in Disease Suppressive Soils

Soil suppressiveness is the ability of a soil to prevent/suppress disease even in the presence of a pathogen, suitable host plant and favorable climatic conditions (Penton et al., 2014). The term 'non-suppressive' was used for soils that are unable to suppress disease incidence by the pathogen. Biological suppression of soil borne pathogens has been reported from a variety of cropping systems worldwide. In the case of wheat and barley crops, this suppression has been shown against a number of soil borne diseases including *Fusarium* wilt, Take-all and *Rhizoctonia* bare patch. In Australia, biologically-based disease suppression has been reported in long-term experimental plots and farmer fields. This suppression has been attributed to diverse microbial communities including bacteria, fungi and protozoa and is reported to affect pathogen survival, growth in bulk soil and rhizosphere and root infection. The adoption of no-till and stubble retention practices can, in some cases, increase soil borne plant diseases in the short-term. However, long-term adoption of crop management practices that supply higher levels of biologically-available carbon inputs either through crop residues or addition of composts and organic manures can support higher levels of suppression. This occurs through changes to the composition and activity of the soil microbial community.

Natural biological suppression of soil-borne diseases is a function of the activity and composition of soil microbial communities. Soil microbe and phytopathogen interactions can occur prior to crop sowing and/or in the rhizosphere, subsequently influencing both plant growth and productivity (Penton et al., 2014). Research on suppressive microbial communities has concentrated on bacteria although fungi can also influence soil-borne disease. Fungi were analyzed in co-located soils 'suppressive' or 'non-suppressive' for disease caused by *Rhizoctonia solani* AG 8 at two sites in South Australia using 454 pyrosequencing targeting the fungal 28S LSU rRNA gene. DNA was extracted from a minimum of 125 g of soil per replicate to reduce the micro-scale community variability, and from soil samples taken at sowing and from the rhizosphere at 7 weeks to cover the peak *Rhizoctonia* infection period. A total of ~994,000 reads were classified into 917 genera covering 54% of the RDP Fungal Classifier database, a high diversity for an alkaline, low organic matter soil. Statistical analyses and community ordinations revealed significant differences in fungal community composition between suppressive and non-suppressive soil and between soil type/location. The majority of differences associated with suppressive soils were attributed to less than 40 genera including a number of endophytic species with plant pathogen suppression potentials and mycoparasites such as *Xylaria* spp. Non-suppressive soils were dominated by *Alternaria*, *Gibberella* and *Penicillum*. Pyrosequencing generated a detailed description of fungal community structure and identified candidate taxa that may influence pathogen-plant interactions in stable disease suppression.

Research on microbial communities in disease suppressive soils has mainly focused on bacteria (Rosenzweig et al., 2012). A wide range of bacterial groups have been suggested as contributing to disease suppression through antibiosis, plant growth promotion or systemic induced resistance. The functional diversity of soil fungi and their capacity to colonize diverse microhabitats can influence pathogen levels and play a significant role in improving plant health, e.g. *Trichoderma* spp and mycorrhizal fungi. The genus *Trichoderma* has been studied extensively for its biocontrol potential and a number of fungi and oomycetes are registered as biocontrol agents. Soils with higher disease suppressive potential have been found to exhibit higher fungal diversity. In view of the large diversity of uncultured fungi in soil, culture-independent methods are required to describe their composition and to identify community differences between soils. Recently, based on high-throughput sequencing, soils from pea fields with different degrees of disease were discriminated on the basis of their fungal communities.

Disease suppression: The level of impact of soil-borne disease incidence on plant growth is influenced by the amount of plant pathogen present and its interactions with the plant, soil and environmental factors. A main criterion defining biological disease suppression is the ability of soil biotic communities to suppress the disease expression in the host plants even in the presence of adequate virulent inoculum and a susceptible host (Penton et al., 2014). *R. solani* AG 8 levels measured at sowing indicate the presence of sufficient inoculum to cause significant disease. Differences in pathogen inoculum levels between suppressive and non-suppressive soils were small, indicating similar pathogen pressure. Field disease assessments showing distinct differences in the disease levels in the SP and NSP fields validate the suitability of these field soils to investigate disease suppressive communities. These field results are supported by the differences in the DS potential of suppressive and non-suppressive soils. Grain yields of crops were higher (avg. >30%) in the suppressive fields compared to non-suppressive fields confirming better plant performance partly due to the lack of disease in the SP fields.

The biological disease suppression observed at these field sites is effective against multiple soil-borne pathogens and under multi-crop rotations. It was hypothesized that both the general and specific disease suppression may be acting simultaneously. Continued supply of crop residue carbon stimulates general microbial activity including fungal populations that can compete with *R. solani* for resources and also may increase the build-up of antibiotic producing/antagonistic microflora including fungi. TRF profiling and sequence data both show differentiation of fungal communities of SP and NSP soils from both field sites but the combined lack of consistent trends in the total fungal biomass (measured as DNA concentration) suggests that the change in community composition may play an important role in the observed suppression. Since a number of pathogenic fungi can become active saprophytes depending upon resource availability and host presence, total fungal biomass may not truly reflect the net ecological role of entire soil fungal community. The significant difference (PERMANOVA) in the fungal community structure in the soils from 7 weeks, representing the root microbiome suggests a relationship between the disease incidence (root health) and associated fungal community. While we found that the fungal component of soil communities was different in suppressive versus their paired non-suppressive soil, bacteria and their interactions with fungi and plants could also be important to biological disease suppression and is the subject of continuing work for these soils.

7.20 Involvement of Burkholderiaceae and Sulfurous Volatiles in Disease-Suppressive Soils

Suppressive soils have been described for plant pathogenic fungi, oomycetes, bacteria, nematodes, and parasitic weeds. While soil physicochemical properties modulate the onset and intensity of the disease-suppressive state, functionality of the response is in most cases (micro) biological in nature (Carrión et al., 2018). The specific suppression that operates in these soils can be eliminated by soil pasteurization or biocides and can be transferred to non-suppressive (conducive) soils via soil transplantations.

The characteristics of soils that are suppressive to specific fungal root pathogens are comparable to the character stics of innate and adaptive immunity in animals. Similar to the requirement of specialized cell types in adaptive immunity in animals, induced suppressiveness of soils to fungal root pathogens requires the enrichment and activation of select microbial genera and antagonistic traits that interfere at some stages of the pathogen infection cycle. For most disease-suppressive soils, however, the responsible microorganisms and underlying mechanisms involved in the soil suppressiveness are largely unknown.

Disease-suppressive soils are ecosystems in which plants suffer less from root infections due to the activities of specific microbial consortia. The characteristics of soils suppressive to specific fungal root pathogens are comparable to those of adaptive immunity in animals, but the mechanisms and microbial species involved in the soil suppressiveness are largely unknown (Carrión et al., 2018). Previous taxonomic and metatranscriptome analyses of a soil suppressive to the fungal root pathogen *Rhizoctonia solani* revealed that members of the Burkholderiaceae family were more abundant and more active in suppressive than in non-suppressive soils. Isolation, phylogeny, and soil bioassays revealed a significant disease-suppressive activity for representative isolates of *Burkholderia pyrrocinia*, *Paraburkholderia caledonica*, *P. graminis*, *P. hospita*, and *P. terricola*. In vitro antifungal activity was only observed for *P. graminis*. Comparative genomics and metabolite profiling further showed that the antifungal activity of *P. graminis* PHS1 was associated with the production of sulfurous volatile compounds encoded by genes not found in the other four genera. Site-directed mutagenesis of two of these genes, encoding a dimethyl sulfoxide reductase and a cysteine desulfurase, resulted in a loss of antifungal activity both in vitro and in situ. These results indicated that specific members of the Burkholderiaceae family contribute to soil suppressiveness via the production of sulfurous volatile compounds

Over the past 7 years, the role and identity of microbial consortia has been unravelled in disease suppressiveness of soils to the fungus *Rhizoctonia solani*, a devastating root pathogen of multiple crops including sugar beet, wheat, potato, and rice. Initial PhyloChip-based taxonomic analyses of the rhizosphere of sugar beet plants grown in *Rhizoctonia*-suppressive soil pinpointed several bacterial genera that were consistently associated with the disease-suppressive state, in particular bacterial genera belonging to the Gammaproteobacteria, Actinobacteria, and Betaproteobacteria. Subsequent isolations and functional analyses showed that, for the Gammaproteobacteria, *Pseudomonas* species contribute to disease suppressiveness via the production of the NRPS-encoded lipopeptide thanamycin. Work on the potential role of *Lysobacter* species (Gammaproteobacteria) demonstrated that, in spite of their extensive metabolic potential and antifungal activities in vitro, none of the species tested were able, either alone or in a consortium, to consistently suppress *Rhizoctonia* damping-off disease in situ (Gómez Expósito et al., 2017). Among the Actinobacteria, the *Streptomyces* species showed significant inhibition of hyphal growth of *R. solani* via the production of volatile organic compounds (VOCs), in particular the VOCs methyl 2-methylpentanoate and 1,3,5-trichloro-2-methoxybenzene.

For the Betaproteobacteria, taxonomic analyses combined with metatranscriptomics of the rhizosphere microbiome showed that members of the Burkholderiaceae family were significantly more abundant and more active in *Rhizoctonia*-suppressive soil. The Burkholderiaceae family is ubiquitous in diverse environments and associated with different plant species as well as with fungi (Stopnisek et al., 2022). They are metabolically versatile and can inhibit fungal growth through the production of various secondary metabolites, including VOCs.

Disease-suppressive soils provide intriguing examples of microbiome-mediated protection of plants against fungal infections (Carrión et al., 2018). It was showed that different *Burkholderia* and *Paraburkholderia* species contribute to this soil suppressiveness, with one species (*P. graminis* PHS1) operating via the production of sulfurous antifungal VOCs. This study exemplified that by integrating different "omics" technologies with classic microbiology and molecular biology, it is feasible to unravel highly complex ecosystem services provided by soil and plant-associated microorganisms. How the other four *Burkholderia / Paraburkholderia* species contribute to this soil suppressiveness remains to be investigated. Due to the lack of antifungal activity of these four species in our in vitro assays, other approaches need to be adopted to identify the underlying mechanisms and genes. These include screening of genome-wide random mutant libraries for reduced or loss of activity in situ, as recently demonstrated for identifying genes and mechanisms involved in plant growth promotion and induced systemic resistance by *Pseudomonas* (Cheng et al., 2017). Other approaches include targeted mutagenesis of a number of potential candidate gene clusters, in particular genes encoding PKS and nonribosomal peptide synthetases (NRPSs). Preliminary bioinformatic analyses of the genome sequences of these four other *Burkholderia / Paraburkholderia* species revealed various PKS and NRPS gene clusters of unknown function. Identifying the roles of these and other biosynthetic gene clusters, as well as their encoded metabolites in the specific soil suppressiveness will be subject of future studies.

7.21 The Microbiome of Suppressive Soils against Vascular wilt

Cape gooseberry (*Physalis peruviana*) is Colombia's second most exported fruit, with a market worth 37.8 million USD in 2021. *Fusarium oxysporum* f sp. *physalis* (*Foph*) is arguably the most devastating pathogen causing losses of up to 80%. Managing this disease is challenging due to pathogen resistance or the reduced efficacy of commercial fungicides and the production of resistant structures allowing pathogen survival in the soil for up to 30 years (Daniel Bautista et al., 2023). Thus, new methods of control are necessary. Two cape gooseberry farms (organic vs. conventional) were detected free from *Foph* in Nariño. It was hypothesized that the soil microbiome might have a suppressive effect against vascular wilt, caused by *Foph*. To test this, farm soils were propagated by adding 10% farm soil and 90% peat soil. Then, peat soil (control) and propagated soils were inoculated with *Foph*. A decrease of 65%–68% in disease incidence and a 70% in disease severity reduction was observed in seedlings grown in propagated soils compared to peat soil. Next-generation sequencing was used to study the soil microbiome to understand the possible mechanisms for disease suppression of propagated soils. It was concluded that despite the high diversity of soil microbiomes, the relative abundance of some taxa might be a more important indicator of disease suppression than the presence of specific taxa.

Fusarium wilt is an important disease in a vast number of economically important crops worldwide, such as tomato, banana and cape gooseberry. It is a soil-borne pathogen that is extremely difficult to manage because it can remain in the soil for long periods even when the host crop is absent. Suppressive soils protect plants against soil-borne pathogens in a microbiome-mediated manner. The present work studied microbial communities of *Fusarium* wilt suppressive and conducive soils. As the experimental propagated soils share 90% of conducive soil, it was expected to see a similar taxa composition among all soil types. Clear discrimination of bacterial and fungal communities based on soil types was not observed. Nevertheless, both propagated soils were suppressive against *Foph*, suggesting that the taxa that may induce such properties were still present, and represented low prevalence and abundance. For instance, propagated organic and conventional soils presented the phylum Planctomycetota, which has been associated with suppressive soils, in higher prevalence and abundance than conducive soil. However, there was no difference in bacterial abundance (Shannon index) among soils. These results suggest that the relative abundance of some bacterial taxa might be a more important indicator of disease suppression than the exclusive presence of specific taxa. For example, the Phyla Acidobacteriota, Actinobacteriota and Proteobacteria are present both in conducive and suppressive soils, but the ASVs for these taxa seem to be more prevalent in suppressive soils, as reported previously (Daniel Bautista et al., 2023).

The taxonomical composition at the Phylum rank for fungi was similar among the different soils. However, the Shannon index shows a significant difference ($p<0.05$) in ITS abundance for propagated organic soils, which could be a consequence of the organic management versus the conventional management with pesticide use that may reduce fungal diversity. Fungal abundance comparison between propagated conventional vs conducive soil showed that Aspergilliaceae, *Ascobolus, Cephaliophora, Fusarium, Chrysosporium* and *Rhizopus* taxa together with the protist Spirotrichea, were more abundant in conducive soils. These results are interesting since some of these taxa are plant pathogens such as *Fusarium, Cephaliophora* and *Rhizopus*, and some species are enriched in conducive soils (Wang et al., 2020).

Studies of soil microbiomes have uncovered the complex mechanisms governing suppressive soils and have shown that the complexity of soil microbiomes can be overwhelming. The advent of high-throughput sequencing has permitted a holistic study of soil microbiomes in a vast number of pathosystems and geographic distributions (Döring et al., 2020). These studies show that despite the high diversity of soil microbiomes, the consistent presence of some bacterial and fungal taxa in suppressive soils might indicate that soil microbiome composition is a crucial factor for plant health, suggesting that despite the complexity of microbial communities, some predictive microbial diversity models can be used as biomarkers for healthy plant microbiomes for biocontrol strategies. The next step will be to challenge this hypothesis by selecting some of the microbial markers to assess for plant health both under control conditions and in the field.

7.22 Multiple Parallel Mineralization Technique in generation of *Fusarium oxysporum*-suppressive Soil with Non-soil carriers

Hypotheses to explain the disease-suppression mechanisms have been based on differences in the soil minerals, antagonistic microorganisms, and microbial ecosystem, but the actual mechanism is unknown (Meeboon et al., 2022). Previous attempts to generate disease-suppressive soil through inoculation with antagonistic microbes that suppress or out-compete the pathogens have been ineffective, since the microbes could not become established within the soil5. To establish these antagonistic microorganisms, it's necessary to reconstruct or simulate natural soil microbial ecosystems, but the ability of natural soil to degrade organic matter into inorganic nutrients has not been artificially reproduced. Achieving this ability would be an important breakthrough because of problems such as eutrophication caused by inorganic fertilizers and their high cost, as well as the growing availability of organic wastes that should be recycled in agriculture rather than discarded. In the soil, degradation of organic nitrogen proceeds in two steps: ammonification and nitrification. Nitrogen-cycling bacteria (which produces nitrates) are inactivated by exposure to high concentrations of organic matter. When these bacteria are inactivated, degradation of organic matter proceeds only until ammonia is produced, which is problematic, because many plants prefer nitrates, and this results in poor growth.

Shinohara et al. (2011) developed a novel method to efficiently generate nitrate in non-soil media, such as in the water used in hydroponic systems. In summary, the method consists of three steps, i.e., add 10 g/L of soil to water as a source of microbial inoculum, add small amounts of organic fertilizer, and aerate the solution. The resulting system can degrade organic matter into inorganic nutrients such as nitrate by enabling simultaneous ammonification and nitrification. This was called the "multiple-parallel-mineralization" (MPM) method. Hydroponic conditions with no solid medium, cultivation with MPM solution suppressed root diseases such as bacterial wilt disease and *Fusarium* wilt (Fujiwara *et al,* 2016). When the MPM culture solution is inoculated into a non-soil carrier, the carrier acquires the ability of a natural soil to degrade organic matter into inorganic nutrients8. This non-soil medium may also suppress the growth of pathogenic organisms such as *Fusarium* species. To test this possibility, it was examined whether immobilizing the MPM microorganisms on a non-soil solid carrier could achieve the same effect as a disease-suppressing soil.

The disease-suppression mechanism is unknown, and it's unclear how to produce such soils. The microbiota that develop in a multiple-parallel-mineralization system (MPM) can increase nutrient production efficiency and decrease root disease in hydroponic systems (Meeboon et al., 2022). Artificial media inoculated with MPM microorganisms can degrade organic matter to produce inorganic nutrients similarly to natural soil, but it's unknown whether they can also suppress pathogen growth. Here, we produced an artificial medium that inhibited root disease similarly to disease-suppressive soils.

Microbial MPM culture solution was inoculated into non-soil carriers (rockwool, rice husk charcoal, and vermiculite) to test whether it could suppress growth of *Fusarium oxysporum* f. sp. *lactucae*. *F. oxysporum* f. sp. *conglutinans* strain Cong:11 and *F. oxysporum* f. sp. *lactucae* were inoculated into artificial media sown each with *Arabidopsis thaliana* and *Lactuca sativa* L. var. *capitata* supplemented with MPM culture microbes. The MPM microorganisms suppressed *F. oxysporum* f. sp. *lactucae* growth and prevented plant disease. Thus, MPM inoculated non-soil carriers that can generate inorganic nutrients from organic matter may also suppress disease in the absence of natural soil. This study showed novel creation of a disease-suppressive effect in non-soil media using the microbial community from MPM culture solution.

In soil-less culture, non-soil media such as rice husk charcoal, rockwool, and vermiculite are used instead of soil, but they do not contain a microbial ecosystem that performs the same function as microorganisms in soil (i.e., degrading organic matter into inorganic nutrients). Conventional hydroponics is limited by the inability to use organic fertilizer in the solution; if organic fertilizers are added, the solution becomes turbid and the plant roots become damaged. If solid media are used in hydroponics, they must be able to degrade organic fertilizers by immobilizing the microbial ecosystem. The hydroponic techniques developed by Shinohara et al. (2011) can use organic fertilizer (because the media are capable of degrading the organic fertilizer into inorganic forms) and can also lower the incidence of root rot.

The results indicated that the disease-suppressive effect can be reproduced in a non-soil carrier. The microbial ecosystem that developed here had abilities similar to that of soil to degrade organic fertilizer into inorganic nutrients and to suppress root disease (Meeboon et al., 2022). The inoculation of MPM microorganisms can therefore be an eco-friendly method to control *Fusarium* wilt and other root diseases in hydroponic systems, although it will be necessary to test it to determine its effectiveness against other pathogens. In addition, the MPM microbial ecosystem contains more than 50,000 microbes, making it difficult to identify the relationship between disease suppression and specific components of the microbial ecosystem. It will be necessary to see whether it is possible to reproduce the observed disease control in a microbial ecosystem composed of a smaller number of microorganisms, as this would facilitate its implementation on an operational scale.

7.23 Suppressive Soils in Plant Disease Management-Examples

The phenomena of disease suppressive soils have been documented for numerous plant-pathogen systems around the world. Among plant/fungal patho-systems, soils suppressive to *Fusarium* wilts caused by *F. oxysporum* or take-all of wheat caused by *G. graminis* var. *tritici* have been studied most extensively across a diversity of geographic regions (Peddi Durga et al., 2012).

Advantages of Suppressive Soils:· Reduces the legal, environmental and public issues; Safety hazards of chemicals ; Should achieve control equal to the best;· Currently available by other methods;· Can be used in organic or reduced pesticide ; Systems, thereby adding value to the product.

Disadvantages of Suppressive Soils·:Takes more intensive management and planning ; May restrict flexibility in early years ;· Requires new skills ; Disease control is not achieved immediately.

How long does it take for a soil to become Suppressive?: Suppression develops over a period of time. The duration will depend on the conditions and the return of organic residues. Pathogen-suppressive soils ca be defined into two different broad types of diseases suppressiveness viz., natural and induced. Natural suppressiveness is frequently associated with the physical properties of soils and is relatively independent of crop history (Peddi Durga et al.,2012). On the contrary, induced suppressiveness is wholly dependent on agricultural practices. The isolation, identification, and culture of the antagonistic microorganism(s) responsible for suppressiveness in soils opens up the exciting possibility for controlling plant diseases by adding these antagonists to previously conducive soils or substrates. The presence of a pathogenic micro-organism in a substrate is not alone sufficient to cause severe disease in a susceptible host. The nature of the substrate and the physiological state of the host must also be favourable to disease development. The composition of the substrate has been shown in recent years to have an overriding influence on disease intensity. The term "disease suppressiveness" is commonly used to designate agricultural soils or substrates, where certain specific soil-borne diseases are absent or occur only to a low degree, when the pathogen is present naturally or artificially introduced. The inhospitality of these "suppressive" soils to some plant pathogens is such that either the pathogen cannot establish or they establish but fail to produce disease or they establish and cause disease at first but diminish with continued culture of the crop.

i. **Fusarium wilt suppressive soils:** *Fusarium* wilt caused by *formae speciales* of *Fusarium oxysporum* infects wide variety of plant species. Soils that naturally suppress the development of *Fusarium* wilt diseases have been identified in numerous cropping systems and in several regions of the world. The existence of such soils has long been recognized and an attempt to elucidate the mechanisms involved in disease suppression has yielded information on numerous potential biological control agents (Peddi Durga et al.,2012). These soils consistently share many of the same biological and physical features and several abiotic factors including soil pH, organic matter content and clay content likely play a role in disease suppression. However, these abiotic elements appear to function in disease suppression predominantly, though not exclusively, through their impact on the composition and activity of the saprophytic soil microflora. Numerous studies have demonstrated that non-pathogenic *Fusarium* spp. are functional elements of the resident microflora in *Fusarium* wilt suppressive soils from diverse geographic regions.

   Suppression of *Fusarium* wilt diseases by non-pathogenic *Fusarium* spp. occurs via various mechanisms, including saprophytic competition for substrate, induction of systemic resistance in the plant host and parasitic competition for infection sites. In addition to non-pathogenic *Fusarium* spp., suppression of *Fusarium* wilt has also been attributed to certain elements of the fluorescent pseudomonad community resident to these suppressive soils via direct antagonism through competition for iron, direct inhibition of saprophytic fungal growth and induced systemic resistance. It is likely that these microbial elements and other elements of the saprophytic microflora act in concert rather than independently of each other in the suppression of *Fusarium* wilts. Enhanced competition for iron via the production of siderophores by fluorescent Pseudomonads resulted in an elevated level of carbon competition between nonpathogenic and pathogenic strains of *F. oxysporum*. Biological control of *Fusarium* wilt of flax was greater with nonpathogenic *F. oxysporum* Fo47 and *P. putida* WCS358 used in combination than with either organism applied individually. Effect of artificially induced suppressive soil on *Fusarium* wilt: The effectiveness of the artificially induced suppressive soil on *Fusarium* wilt was assessed based on disease incidence (%) and disease progression, estimated by using the area under the disease progress curve.

The mechanism of disease suppression was evaluated based on the effect of *T. harzianum* and Ca(NO3)2 on the host and on *Fusarium oxysporum* f. sp. *cubense* race 4 FocR4. The resistance of the host was assessed based on the enzymatic activity of peroxidase and polyphenoloxidase, and the phenol content was assayed. The effect of the treatments on plant growth and vigour was also estimated by measuring plant height, pseudostem diameter and root weight. Histological observations were also conducted to detect the presence of lignified cell walls and Ca-pectate. In addition, scanning electron microscopy was used to observe the effect of *T. harzianum* on FocR4. Disease suppression was mostly attributed to the role of Ca2+ in enhancing the resistance of the host. Plantlets treated with $Ca(NO_3)_2$ suppressed disease development by enhancing cell wall lignification. This mechanism prevented further penetration of FocR4 and slowed its spread. This increase in cell wall lignification was marked by an increase in peroxidase and polyphenoloxidase activities. Formation of Ca-pectate in the cell walls also strengthened the physical barrier by resisting the effects of FocR4 polygalcturonase enzymes. In addition, plant growth and vigour was also significantly enhanced in plantlets treated with $Ca(NO_3)_2$.

ii. **Take-all suppressive soils**: Soils suppressive to take-all of wheat have been observed worldwide under cropping conditions that employ continuous wheat monoculture for a period of years. The spontaneous decline in the severity of take all after a severe outbreak of the disease for one or two crops early in the monoculture sequence is termed take-all decline (Peddi Durga et al.,2012). First report of take-all decline was noted in early in the 20th century and the first measures of take-all decline are attributed to Glynne (1935) at Woburn, England. Take-all decline is a biological phenomenon that is transferable to a conducive soil is eliminated by steam pasteurization and develops only in the presence of the active pathogen, *G. graminis* var. *tritici*. Numerous biological and chemical elements have been put forth as the operative factors responsible for take-all suppressive soils. However, a large body of research indicates that the reduction in disease severity and induction of soil suppressiveness in response to wheat monoculture most commonly can be attributed to increased populations and activity of certain antibiotic-producing fluorescent pseudomonads or parasitism by *Trichoderma* spp. Production of antibiotics including phenazine- 1-carboxylic acid, 2,4 diacetylphloroglucinol (2,4-DAPG), pyoluteorin and pyrrolnitrin play an important role in the biological control of soil borne pathogens by certain strains of fluorescent Pseudomonas spp. that produce these antibiotics. McSpadden Gardener et al. (2000) reported the predominant role of 2,4-DAPG producing fluorescent pseudomonads in the take-all suppressive soils of the Pacific Northwest. High population of 2,4-DAPG producing fluorescent pseudomonads were recovered from the suppressive soils at populations of approximately 10^5 to 10^6 cfu g^{-1}. Populations of 2,4-DAPG-producers were no higher than 1.2×10^4 in conducive soils and more typically were not detected in these soils. These bacteria were also found to colonize the rhizosphere of wheat grown in suppressive soils at populations that were at or above the threshold population required for suppression of take-all. Although several studies demonstrated a role for phenazine production in the capacity of certain fluorescent pseudomonads to provide biocontrol of take-all.

iii. ***Rhizoctonia* suppressive soils:** *Rhizoctonia solani* has great importance among soil borne plant pathogens and damages a wide range of hosts worldwide. The contribution and relationships of abioticfactors (pH; electrical conductivity; organic matter content; N total; P; K; Ca; Mg; Al; H; S; Na; Fe; Mn; Cu; Zn; B; cation exchange capacity; sum of bases and base saturation) was evaluated and biotic factors (total microbial activity evaluated by the CO_2 evolution and fluorescein diacetate hydrolysis; culturable bacterial, fungal, actinomycetes, protozoa, fluorescent *Pseudomonas* and *Fusarium* spp. communities) to the suppressiveness of soils from the Córrego Taquara Branca microbasin to R. solani (Peddi Durga et al.,2012). Soil sampling areas included undisturbed forest, pasture and fallow ground areas, annual crops, perennial crops and ploughed soil. The contribution and relationships of these variables to suppression to *R. solani* were assessed by path analysis. When all samples were analyzed together) only abiotic variables correlated with suppression of *R. solani*, but the entire set of variables explained only 51% of the total variation. However, when samples were grouped and analyzed by vegetation cover, the set of evaluated variables in all cases accounted for more than 90% of the variation in suppression of the pathogen. In highly suppressive soils of forest and pasture/ fallow ground areas, several abiotic variables and fluorescein diacetate hydrolysis correlated with suppression of *R. solani* and the set of variables explained more than 98% of suppressiveness. Decline in the severity of disease caused by *R. solani* also has been documented in response to successive plantings of a given plant host and in general this has been attributed to increased parasitism by *Trichoderma* spp. The development of suppressiveness was observed after four to five successive plantings of radish, but observed no correlation between the onset of suppressiveness and the in vitro antagonism of *R. solani* by the resident soil microflora. However, suppressive soils possessed higher populations of

Trichoderma spp. than the corresponding conducive soil. *T. harzianum* was identified as the primary cause of the disease decline in these soils. Soil pH had an indirect influence on the phenomenon through its impact on *T. harzianum*; growth and spore germination of *T. harzianum* was stimulated at low pH and suppressiveness to *F. solani* was induced by crop monoculture only in acidified or naturally acidic soils.

iv. ***Phytophthora* suppressive soils**: Avocado root rot, caused by *Phytophthora cinnamomi* Rands, is considered the most important root disease of avocado (Peddi Durga et al.,2012). The suppressive nature of the soils was affirmed by greenhouse experiments and was in part ascribed to higher numbers of microorganisms as well as higher nitrogen and calcium content (Broadbent et al., 1974). Suppressive soils contain higher percentages of antagonistic micro-organisms than conducive soils, especially actinomycetes, aerobic spore forming bacteria, and fungi. Results obtained thus implicate the microflora of the soils for influencing the suppressiveness. The biological nature of suppression is confirmed by the fact that all suppressiveness is lost when the soils were first sterilized.

v. ***Pythium* suppressive soils**: Suppression of saprophytic increases in inoculum densities of *Pythium ultimum* was found in finely textured soils in the San Joaquin Valley of California (Peddi Durga et al.,2012). Suppressive soils had greater mean concentrations of sodium, sulfate, and chloride than conducive soils, only chloride was inhibitory to *P. ultimum*. When conducive soils were amended with chloride at concentrations found in suppressive soils, colonization of leaf debris by *P. ultimum* was partially suppressed. In suppressive soils, *P. oligandrum* was the most commonly isolated primary colonizing fungus and tended to be found at higher propagule densities than observed in conducive soils. When propagule densities of *P. oligandrum* were increased artificially in conducive soils, colonization and subsequent inoculum increases of *P. ultimum* were reduced. Suppressiveness was overcome by successive soil amendments with dried leaf debris, which resulted in progressive reductions in the frequencies of colonization by *P. oligandrum*. Apparently soils with elevated chloride concentrations allowed *P. oligandrum* to successfully compete with *P. ultimum* and thus increase its propagule density and further suppress the saprophytic activity of *P. ultimum*.

vi. **Bacterial wilt suppressive soils**: The state of microbial community at a given time or during its establishment has a significant impact on the growth and survival of unfavorable microorganisms (Peddi Durga et al.,2012). Thus it is possible to displace the plant pathogens through community evolution within the community evolution/suppressive soils. For instance, it has been known fro many years that there are two types of soil in terms of survival of *Ralstonia solanacearum*, the bacterial pathogen responsible for southern bacterial wilt in many solanaceaus plant: one in which bacterial wilt is a consistent problem (persistent soils) and another in which disease does not persist (non- persistent soils/ suppressive soils). It is no doubt that in such suppressive soils, microbial communities are regulated and maintained by a most inextricable web of 'signals' and 'interactions' among soil microorganisms, although the exact mechanisms of soil suppressiveness have not been elucidated in all cases.

7.24 Managing Soil Suppressiveness in Agro-Ecosystems

The study of suppressive soils has provided an extensive resource of agents for the biological control of soil borne plant diseases. Although a valuable tool, these agents in and of themselves characteristically provide a level of disease control that is inferior to that achieved in the source suppressive soil (Peddi Durga et al.,2012). This is not surprising as disease suppression commonly operates through communities of microorganisms and multiple genotypes of specific microbes. We cannot restore the -balance of organisms that was present under native, undisturbed circumstances, but we can build a new, stable balance of soil organisms that will be adapted to the altered soil conditions. This is a proactive plan that moves us toward the desired outcome of disease prevention. As such, the transfer or management of the biologically complex components that contribute to disease suppressive soils to agricultural production systems is often a prominent goal of research programs.

Soil Enrichment: Sustainable Approach to Biological Control of Soil borne Diseases: Soil enrichment with a specific nutrient is one approach to the study of community evolution in soil microorganisms and its impact on unfavorable microorganisms. In response to specific nutrients, a soil microbial community starts to proliferate and evolve into a specialized biological network that is most suitable for its survival and further growth (Peddi Durga et al.,2012). Many studies have indicated that soil amendment with green manure, stable manure, or compost is effective in mitigating disease problems caused by various soil borne fungal pathogens including *Pythium* spp., *Phytophthora* spp., and *Rhizoctonia solani*. For instance incorporation of green manure into sand before establishment of the lawn controlled damping-off and root rot of creeping bentgrass by *Pythium graminicola*. Soil amendment with cotton-gin trash was very effective in control of southern blight of tomato caused by *Sclerotium rofsii*. The mechanism of disease mitigation by organic amendment is due primarily to amplified competition for available nutrients and suitable ecological niches through enhancement of edaphic microorganisms or parasitism or antibiosis by antagonists that also proliferate with organic amendment. Management practices including crop rotation, input system (organic vs. conventional), tillage and fertilization, intensification in cropping, moderate to high levels of nutrients, management of weeds, pests and other diseases, stubble retention and limited grazing will influence ecological processes that impact microbial communities involved in suppression of soil borne plant pathogens.

Suppressive Soils Management

i. **Incorporation of root colonizing rhizosphere microorganisms:** These organisms can promote Phytostimulatory and biofertilising effects plant health by making the plant 'stronger'. Many rhizosphere microorganisms can induce a systemic response in the plant, resulting in the activation of plant defence mechanisms (Peddi Durga et al.,2012).

ii. **Better agronomic practices:** Adaptation of cultural practices has been proposed as a means to decrease the soil inoculum potential or increase the level of suppressiveness to diseases Indeed, disease suppressiveness has been obtained through crop rotation, biofumigation, intercropping, residue destruction, organic amendments, tillage management practices and a combination of those regimes (Peddi Durga et al.,2012).

- **Biofumigation:** This strategy better adapted to the cooler regions of the World, it involves fermentation of organic matter under plastic results in anaerobic conditions in soil and which leads to production of toxic metabolites. Both these processes contribute to the inactivation or destruction of pathogenic fungi. Many species of Brassicaceae (Cruciferae) produce glucosinolates, a class of organic molecules that may represent a source of allelopathic control of various soilborne plant pathogens. For example, soil amendment with *Brassica napus* seed meal controlled root infection by *Rhizoctonia* spp. and the nematode *Pratylenchus penetrans*. Similarly degradation of garlic, onion, and leek tissues releases sulfurous volatiles such as thiosulfinates and zwiebelanes which are converted into disulfides that have biocidal activities against fungi, nematodes and arthropods.
- **Soil solarization:** Solarization or solar heating is a method that uses the solar energy to enhance the soil temperature to levels at which many plant pathogens will be killed or sufficiently weakened to obtain significant control of the diseases.
- **Residue management:** Plant residues left on or near the soil surface may contribute to an increase of disease suppressiveness through the promotion of the general microbial activity. In some cases, however, the debris not only promotes the microbial activity but also helps to preserve the pathogens, preventing a decrease of the inoculum density. This is the case for *Macrophomina phaseolina* causing charcoal rot in soybean, *Fusarium* sp. causing root and crown rot on maize and *R. solani* causing crown and root rot on sugar beet.
- **Organic amendments**: The mechanism of disease mitigation by organic amendment is due primarily to amplified competition for available nutrients and suitable ecological niches through enhancement of edaphic microorganisms or parasitism or antibiosis by antagonists that also proliferate with organic amendment. This approach to disease control is important in sustainable agriculture, since it reuses natural resources in the process of disease control without using chemicals. In addition it does not require the mass culturing of biocontrol agents for applications to large fields. There are also interactions between organic matter decomposition level and soil physical conditions and those interactions also affect disease incidence and severity in some crops. The structure of the soil is improved by partially decomposed organic matter. This transformation results in better water retention under dry weather conditions and improved drainage during periods of high precipitation. The improved soil condition, in turn leads to

natural root rot suppression in wet soils and some degree of suppression of wilt diseases in dry soils. Examples are *Phytophthora* root rot, which is prominent in wet soils, and the early dying disease of potato, caused by a complex involving nematodes and *Verticillium* in dry soils.

The phenomena of disease suppressive soils have been documented for numerous plant-pathogen systems around the world (Peddi Durga et al.,2012). Harnessing the potential of these soils as a practical means to manage diseases in agro-ecosystems has long been a goal of plant pathologists. The findings reported here and those of other investigators demonstrate that the manipulation of microbial communities to induce a disease suppressive soil environment does possess potential as a tool in the management of soil borne plant diseases. The initial objective of biological control is to maximize soil suppressiveness through the manipulation of resident antagonists. These microorganisms probably resided in equilibrium with pathogens and plants before the intervention of agriculture. However, the residents may not be present at sufficient levels for effective disease control at specific vulnerable sites on the plant or at specific vulnerable stages of crop growth. In these circumstances, the population has to be enhanced with inoculant formulation, knowing that the site conditions are favourable for the control agents.

7.25 Chemical Characters of Disease Suppressive and Conducive Soils

A soil acidified by ammonium sulphate following annual application of the fertilizer for 9 years was suppressive of the saprophytic growth of *Gaeumannomyces graminis* var. *tritici* in soil (pathogen suppressive) (Simon and Sivasithamparam, 1988). The same soil amended with lime was pathogen conducive. In natural field soil microbial respiration and the 'total' number of aerobic microorganisms were greater in the conducive than in the suppressive soil. In a soil-sandwich bioassay of the transferable suppression of saprophytic growth of the pathogen there were higher numbers of 'total' aerobic microorganisms, fluorescent pseudomonads, and Gram-negative organisms, but lower numbers of filamentous fungi and yeasts in the conducive than in the suppressive soil. It was estimated that *Trichoderma* spp. made up 71 and 34% of the total numbers of fungi counted in the suppressive and conducive soils, respectively. It is proposed that *Trichoderma* spp. played a major role in the transferable pathogen suppression in the suppressive soil.

Three conducive and three suppresive soils were examined for differences in physical and chemical properties, as well as qualitative and quantitative content of phenolic compounds. The three suppressive soils were slightly alkaline, whereas the three conducive soils were slightly acid (Young, et al., 1991). The concentrations of Ca and Mg in suppressive soils were 3–15 times those in conducive soils. The total concentrations of phenolics and humic acids in conducive soils were 2–4 and 2–10 times those in suppressive soils, respectively, whereas the concentrations of gentisic (2,5-dihydroxybenzoic) acid in suppressive soils was higher than that in conducive soils. In bioassay, the radicle growth of *Brassica pekinensis* was inhibited more by extracts from the conducive soils than those from the suppressive soils. Clubroot disease was significantly suppressed by gentisic acid applied to conducive soil at the rate of 800 mg kg^{-1}. When seeds were soaked in of gentisic acid solution (800 mg kg^{-1}, pH 5.5) 24 h before planting, the incidence of the disease dropped from 78 to 41%.

Shallots are the leading commodity of lowland farmers in Brebes, Central Java, contributing the most prominent (30%) of shallot production in Indonesia. In recent years, moler disease (twisting disease) caused by *Fusarium oxysporum* f.sp. *cepae* (FOCe) has become an epidemic in the region. Cultivation practice factors play an important role in the epidemic (Poromarto et al., 2021). Lack of organic fertilizers and lack of crop rotation by farmers, on the one hand, are responsible for decreasing soil fertility, and on the other hand, potentially increasing moler disease intensity. However, the intensity of moler disease varies widely between fields, ranging from very low (less than 5%) to very high (more than 50%). There are both suppressive and conducive soils to the disease. This paper reports the phenomenon of variations in the intensity of moler disease in 30 shallot fields in Brebes, Central Java.

Analysis of soil chemical character in this area supports the relationship between soil suppressive and soil conducive to moler disease intensity. A disease epidemy of Moler caused by FOCe has occurred in Brebes Central Java. The disease intensity, however, varies between the shallot production lands. Some lands show suppressive with disease intensity lower than 5%, and others are conducive to disease intensity over 50%. It is interesting that in Brebes occur suppressive and conducive soil to moler disease. The suppressiveness may be determined by environmental conditions, including chemical soil characters. This paper reports the chemical character of suppressive and conducive soil to moler disease in Brebes. The evidence shows that the suppressive soil is more fertile than that conducive one. The suppressive soil is chemically characterized by significantly higher organic mineral, C-organic, P-available, K-exchangeable, and Cation Exchange Capacity than that conducive one.

The relatively high moler disease intensity may be related to overall soil chemical properties. The number (%) of C-organic and N-total are low. Even the C/N ratio is very low. Currently, farmers in Brebes rarely use organic fertilizers in shallot cultivation (Supriyadi, et al., 2020). However, the organic matter content in the suppressive soil was higher than in the conducive soil. Organic matter provides not only essential nutrients but also a complex of micronutrients, improves soil structure, and increases the activity of functional microorganisms. They can play a role in suppressing diseases caused by weak pathogens such as *Fusarium*. The P-availability and K-exchangeable in this study may be the most influential factor in the intensity of moler disease in suppressive and conducive soils. The P-available and K-exchangeable in the suppressive soil were significantly higher than in the conducive soil. K (potassium) and P (phosphate) are essential nutrients that play an important role in various physiological and structural processes of plants in responding and protecting themselves against biotic and biotic environmental stresses. Phosphate-solubilizing-fungi increasing P-available have been shown to suppress the development of moler disease. The application of 100 kg.ha-1 of KCl fertilizer on sandy soil can extend the disease's incubation period and suppress moler disease on shallots with a very good effectiveness value (89.23%). On the other hand, phosphorus and potassium deficiency in garlic has been reported to increase the moler intensity of garlic.

In this study, although the soil acidity (pH) in the suppressive soil (6.39) was significantly lower than the conducive soil (6.61), the difference was not much, only 0.22. Thus, the pH factor does not seem to play a role in the difference in moler disease intensity between suppressive and conducive soils. To be noted, *Fusarium* can grow well in acidic media. Media acidity (pH) affects the production of degradative enzymes involved in the pathogenesis of *Fusarium.* Using lime reduces pathogen persistence by increasing soil pH to a neutral range. The study shows that the suppressive soil is more fertile than that conducive one. The suppressive soil is chemically characterized by significantly higher organic mineral, C-organic, P-available, Kexchangeable, and Kation Exchange Capacity than that conducive one.

7.26 Responses of Microbial communities to Banana *Fusarium* wilt in suppressive and conducive soils: In general, the soils of banana plantations with a high *Fusarium* wilt disease rate (>60%) over five consecutive production years are defined as conducive soils, and the soils with a low wilt disease rate (<15%) are defined as suppressive soils (Shen et al., 2015). The conducive and suppressive soils can be located in adjacent banana orchards. The differences in the *Fusarium* wilt disease incidence between the geographically proximate fields that share the same climatic conditions and agronomic management may be explained by the distinct resident microbial communities in these soils (Wang et al., 2022). For sustainable agriculture and environmental protection, suppressive soils have recently attracted attention. Microbial identifications based on sequencing technology are now being applied to investigate banana soils infected by *Fusarium* wilt, but it is still unclear as to whether there is a difference in the assemblages of microbial communities between suppressive and conducive soils across soil particle-size fractions (PSFs).

Previous studies on bananas in the Canary Islands (Spain) have shown that soils are undoubtedly composed of a wide range of differently sized fractions. They act as microhabitats for diverse microbial communities. These contexts are considered to be important for determining the mineral–organic interactions of many soils' biochemical processes, which can affect the renewed assemblages of microbial communities in suppressive and conducive soils. Similarly, soil microorganisms and other binding agents (e.g., organic matter) play major roles in the formation of aggregates. For example, filamentous fungi are closely related to the formation and stabilization of macro-aggregates (>0.25 mm). In addition, the extracellular polymeric substances of bacteria and fungi usually contribute to the formation of micro-aggregates (<0.25 mm). In particular, the soil enzyme pattern has been proposed as an appropriate biological indicator. Soil enzyme activities have been widely used as indicators of soil quality and functional potency (Loeppmann et al., 2016). They could also be sensitive to both shifts in soil properties and microbial traits. Therefore, soil enzyme activities might offer new insights into the functional implications of soil condition shifts (such as suppressive versus conducive soils).

The microbiota plays a primary role in inhibiting plant pathogens in the soils. However, the correlation between soil particles and local microbial communities has not been fully confirmed. Wang et al., (2022) contrasted the different assemblages of microbial communities between suppressive and conducive soils via the differentiation of soil particle-size fractions (PSFs). The direct and indirect interactive associations among the soil biotic and abiotic factors were further extracted by using samples from two continuous banana cropping systems. Notable differences were shown in PSF composition, biological traits (microbial communities and enzyme patterns) and physiochemical parameters between suppressive and conducive soils among the different soil fractions. For example, compared with conducive soils, suppressive soils have higher nutrient contents, fungal abundance and diversity and enzyme activities, and the extent of these differences was explored for fractions of different sizes. Moreover, the microbial taxonomic composition strongly varied between disease-suppressive and disease-conducive soils. For instance, there were significant differences in the relative abundance among key microbiology communities, such as *Actinobacteria*, *Firmicutes*, *Bacteroidetes*, *Proteobacteria* and *Ascomycota*, especially for antagonistic microorganisms (i.e., *Streptomyces*, *Pseudomonas*, *Trichoderma*, etc.) across various soil fractions. In addition, structural equation modeling (SEM) showed that the complex associations among soil PSFs, physiochemical parameters and microbial communities were mediated by multiple pathways, which then drive the soil enzyme activities and may further influence the suppressiveness of the soil. These results demonstrate that the resident microbial communities in specific soil particles may play a crucial role in the development of soil suppressiveness against banana *Fusarium* wilt disease.

The microbial taxonomic composition strongly varied between the disease-suppressive and -conducive soils. The distribution of the composition, however, was different. For example, for bacteria, the relative abundances of *Actinobacteria*, *Firmicutes* and *Proteobacteria* in the 2–0.25 mm-sized fractions of the suppressive soils were all higher than those in the conducive soils. The importance of the phyla was associated with their rapid growth rate and suppression of soil-borne disease through the production of high levels of secondary metabolites (Palaniyandi, et al., 2013). Previous studies have also observed higher abundances of *Actinobacteria* and *Firmicutes* in *Rhizoctonia*-suppressive soils, which appear to have a key role in soils with specific suppressiveness against *Fusarium* wilt. At the genus level, the *Pseudomonas* in soils shows its potential contribution to the suppressive ability. *Pseudomonas* was reported to be related to the resistance to banana *Fusarium* wilt. In addition, several *Pseudomonas* spp. have been widely recognized as biocontrol agents due to their ability to suppress soil-borne disease and to promote plant growth, e.g., *Thielaviopsis basicola*, *Pythium* pathogens and *R. solani*. The *Streptomyces* in the suppressive soils were also higher than those in the conducive soils across all the soil fractions, and are usually considered as the important antagonistic microorganisms in soils.

For the fungal phyla, the relative abundances of *Ascomycota* in different PSFs of conducive soils were all higher than those in suppressive soils. Many destructive plant pathogens that cause diseases including strawberry root rot, wheat scab and Dutch elm disease are *Ascomycetes*. Thus, we speculated that suppressive soil formation was related to a lower relative abundance of *Ascomycetes*. *F. oxysporum* belongs to the *Ascomycetes*. Its abundance data show comparative enrichment in conducive soils, which suggests that the quantity of the pathogen in banana soils directly determines the suppressive ability against *Fusarium* wilt. Among other fungal genera, *Trichoderma* is a potential biocontrol agent against several phyto-pathogenic fungi, e.g., the relative abundances of *Trichoderma* in a macro-aggregate (>0.25 mm) of suppressive soil were significantly higher than those in the conducive soils. This fungus can be attached to the host hyphae via coiling and absorbing actions. Therefore, the occurrence of *Trichoderma* in the banana soils could contribute to the suppressive ability against *Fusarium* wilt, and it is primarily located in macro-aggregates. Notably, the *F. oxysporum* abundance has a significantly negative effect on the microbial abundance and community structure, and this effect probably resulted in the ultimate regulation of enzyme activities and also directly and negatively mediated the enzyme activities. Our results also show that the difference in *F. oxysporum* abundances in different aggregates between suppressive and conducive soils were not large, while the disease incidence in suppressive soils was lower (Xiong et al., 2017). This phenomenon may suggest that the manifestation of the disease is impeded by the regulation from the resident microbial communities. Moreover, the suppressive soils showed a higher number of positive co-occurrence relationships than the conducive soil for microbial networks which might suggest more cooperation in the complex microbial community ecosystems.

The microbial taxonomic composition strongly varied between disease-suppressive and -conducive soils (Wang et al., 2022). For instance, there were relative abundance differences in key microbiology communities, such as *Actinobacteria*, *Firmicutes*, *Bacteroidetes*, *Proteobacteria* and *Ascomycota*, especially antagonistic microorganisms (i.e., *Streptomyces*, *Pseudomonas*, *Trichoderma*, etc.) across various fractions. The SEM showed that the complex associations among the soil PSFs composition, availability of physiochemical parameters and microbial communities within different soil fractions were mediated by multiple pathways, which further drove the soil enzyme activities, influencing the suppressiveness of soil towards disease. Collectively, these results suggested that microbial communities may be important in the development of soil suppressiveness against banana *Fusarium* wilt disease.

7.27 The Metagenomics of Disease-Suppressive Soils

A European research project denoted Metacontrol, which ran between 2002 and 2007, aimed to unravel the antagonistic capacities locked up in the microbiota of disease-suppressive soils. The basic idea was to find clues as to the involvement of such capacities in the suppression of plant pathogens as well as to explore them for application purposes (Jan Dirk van Elsas, 2008). The project has yielded a wealth of methodological advances and has given glimpses of the antagonistic potential of the soils studied. However, a full understanding of the antagonism exerted against plant pathogens is still missing, and this is largely due to the enormous diversity of the soil microbiota. Living soils commonly contain a vast microbial diversity and are to be seen as the best reservoir of genetic novelty on Earth. Furthermore, the novel genetic information in the soil microbiota offers an excellent source that can be exploited to foster our understanding of soil functioning. Alternatively, the expected genetic novelty will spur biotechnological innovation. However, only a fraction of the soil microbiota is easily cultivable and therefore our ability to unlock the genetic treasures locked up in it has been strongly hampered. Recently-developed advanced DNA-based methods enable us to directly examine genomic DNA derived from the soil microbiota, obviating the need to cultivate the organisms. Thus, information on the collective soil metagenome is obtained and a particular aim is to explore genes that encode functions of key interest to soil ecology and functioning, as well as biotechnology. In particular, the metagenome of disease-suppressive soils is of interest given the expected prevalence of antibiotic biosynthetic clusters. Due to the complexity of the soil microbiota, deciphering the key genetic information locked up in it is, however, challenging. Jan Dirk van Elsas (2008) drew on the experience gained in the context of a trans-European project on the metagenomics of disease-suppressive soils denoted METACONTROL. They first described the progress achieved in developing the necessary tools for metagenomic exploration, later reported the progress and examined the critical issues and challenges that still impinge on the metagenomic exploration of soil.

In the light of the currently available DNA-based technologies, the potential for exploration of the genetic treasures of the soil microbiota is enormous. Thus, examination of the entire soil metagenome (here defined as the collective genomes of the microorganisms present in a soil sample) has been proposed as a means to address the issue. However, there are definite problems in this approach, most of them being technical.

Major advances of the Metaexplore project: Disease-suppressive soils can restrict the activity of plant-pathogenic microorganisms. The key to suppression often lies in the soil microbiota and in situ production of antibiotics may be involved. Great part of this antagonistic potential may be in the as-yet-uncultured microbiota (Jan Dirk van Elsas, 2008).Therefore, unlocking this microbiota via metagenomics is important. The METACONTROL project, which ran as from 2002, united seven European laboratories, mainly from academia but including the small enterprise LibraGen (Toulouse, France), with the aim of accessing and examining phytopathogen-suppressive soils. The assumption was that the microbiota of suppressive soils would show to be reservoirs of genetic loci involved in antibiosis, such as those involved in the production of polyketides and chitinases. During the project, four suppressive soils were identified in the Netherlands (W, *Rhizoctonia solani* AG3), Sweden (U, *Plasmodiophora brassicae*), France (C, *Fusarium*) and the UK (Wy, *Fusarium*). Metagenomic libraries were constructed from these soils plus one control soil (M) and screened for the occurrence of antibiotic functions. In addition, a range of methodologies were developed that facilitated the exploration of the suppressive soil libraries. Discussions were held on the technology developed and the choices that had to be made prior to each analytical step with respect to the soil DNA extraction methodology, the potential to "bias" the soil community or DNA, the suitability of the vector/host system for the objectives, the optimal screening procedure and the final analysis.

i. **Soil DNA extraction and processing**: For reliable library preparation, soil DNA that accurately represents the genetic make-up of the soil microbiota is required in a quantity that allows efficient cloning. In addition, the metagenomic DNA needs to be of sufficient quality with regard to purity, integrity and length of the fragments in order to be suitable for cloning into a relevant vector. A minimal size of 40 – 100 kb will increase the chance that the entire target pathways, e.g. those involved in the biosynthesis of polyketide antibiotics, can be cloned. The METACONTROL consortium developed and applied advanced methodology that allowed to produce pure high molecular weight (HMW) DNA from all soils. The most efficient approach consisted of the extraction of microbial cells from soil followed by gentle DNA extraction and purification using pulsed-field gel electrophoresis (PFGE) (Jan Dirk van Elsas, 2008). Cushion (Percoll and/or Nycodenz) pre-separation of cells from soil was required for subsequent isolation of the HMW metagenomic DNA. Moreover, the microbial growth status in soil was an important determinant of soil DNA quality. The quality could be boosted by incubation with growth substrates. Typically, this approach produced adequate HMW DNA, often with size > 60 -100 kb. It was found that high amounts of cells, i.e. minimally ca. 1011, were required to yield sufficient DNA for efficient library construction. As soils often contain on the order of 108 to 1010 cells per g, this finding set the rule for efficient construction of the soil metagenomic libraries.

ii. **Deliberate bias and preselection:** In the project, mainly unselective approaches were used to examine the four selected naturally disease-suppressive soils for their anti-phytopathogen functions (Jan Dirk van Elsas, 2008). It was thus disregarded the possibility of positive growth selection, as the chosen target genes/operons do not a priori imply a growth advantage in the host used for cloning, even in the presence of the target phytopathogens. However, targeted approaches were developed to assess specific members of the microbial community. For instance, the soil was pre-treated with chitin in order to enhance the levels of chitinase producers. A strong focus was also placed on the enrichment of metabolically-active bacterial cell fractions with the help of FACS analysis. This strategy was hoped to increase the prevalence of organisms carrying the target genes, in light of their presumed function in the suppressive soil. Metabolically-active cells were indeed successfully sorted from the soil. However, as a result of the limited flow rate of the cell sorters used, throughput was limited (106 cells/hour). This resulted in insufficient biomass for library construction. To meet the high sorting demands in order to establish a library, more sophisticated cell sorters with higher flow rates are needed.

iii. **Metagenomic library production**: The clone libraries obtained for the four disease-suppressive soils consisted of approximately 6,000 to 60,000 clones and were constructed in Escherichia coli. Both large insert size vectors, such as bacterial artificial chromosomes (BACs), that allow cloning of inserts up to 200 kb, and fosmids (that allow insertion of 35 - 45 kb fragments) were used. BAC vectors allow for the cloning of complex large operons and facilitate the analysis of a gene/operon within its original genomic context (Jan Dirk van Elsas, 2008). In contrast, fosmids are able to accommodate smaller inserts and thereby only allow the cloning of smaller operons. Using a fosmid vector (such as the Epicentre ccFos system) allows for the positive selection of vectors that have acquired inserts. Three libraries were based on fosmid vectors, the reason being the ease of obtaining appropriately-sized libraries in reasonable time. One library, for the M soil, was constructed in a BAC vector. The latter vector also contained a replicon compatible with a *Streptomyces* host, which allowed shuttling between *E. coli* and *Streptomyces* metagenomic hosts. Consequently, the probability of heterologous gene expression was enhanced for the clones obtained in this library.

iv. **Metagenomic Library Screening**: Soil metagenomic libraries may contain only few clones that carry a gene/operon of interest. This in addition to the potential constraints to gene expression in the metagenomic host strain (Jan Dirk van Elsas, 2008). Hence, screening of the libraries was a daunting task in all cases. In the METACONTROL project, both functional and molecular screenings were used to uncover antagonistic functions, at varying success (hit) rates. Overall, rather low numbers of phytopathogen-suppressive clones were found across the four libraries.

v. **Functional screening**: Functional screenings of the libraries were performed in dual-culture assays. These assays allow target phytopathogenic organisms to grow out over metagenomic library clones arrayed on Petri dishes (Jan Dirk van Elsas, 2008). Scoring during and following growth was for irregularities / inhibitions in growth of the target organism. This experimental set-up led to the detection of few positive clones (range up to 48 per library), amounting to < 0.05% of positives for all libraries. Such low numbers can be attributed either to a rare occurrence of target genes/operons in the vectors used, or to the required molecular machineries being significantly larger than the vector inserts, a fact reported for many polyketide production loci. Other factors that potentially impede the detection of function of the target genes could be a low expression rate from the signals present in the inserts or unsuitable expression conditions on the plates used in these assays. For instance, for systems controlled by quorum sensing, a cellular communication mechanism commonly found in bacteria, the appropriate system might have been lacking in the E. coli host used. For instance, shuttling from *E. coli* to *Streptomyces* as the metagenomic host facilitated the expression of an antibiotic (amphotericin) production locus, as indicated by high activity in resulting clones against target fungi. Considering the potential expression bottlenecks, the low numbers of functionally-positive clones did not come as a surprise. In support, a screening of forest soil libraries for antifungal traits also yielded only one positive signal among the 113,700 fosmid clones examined.

vi. **Molecular screening**:The libraries obtained in the METACONTROL project were also screened using molecular tools, such as hybridization and PCR amplification methods. The success in detecting novel operons, such as those involved in polyketide biosynthesis, was supposed to depend on the application of deliberate degeneracy in the probes and primers used (Jan Dirk van Elsas, 2008). The rationale was that, this way, the molecular screening is not restricted to previously-known exact sequence information and will allow a broad range of meaningful positive hits within the metagenomic library. The approach facilitated the identification of target genes that are similar to the query sequences. Thus, it constituted a rapid crude search step prior to a more resource-demanding investigation. Using the total soil community DNA as the target, we amplified the genes of the polyketide biosynthesis operon (PKS1) from soil DNA with degenerate primers. The amplicons obtained were used for the generation of probes that would thus detect the relevant PKS sequences in the library. This approach yielded seven positive clones in one (W) soil library that contained genes that were likely involved in PKS biosynthesis. This contention was subsequently confirmed by endsequencing of the clones. In the M soil, the roughly 60,000 clones from the soil metagenomic library were divided into pools, which were subsequently used as templates for PKS-based PCR screenings. This allowed the detection of over 100 positive pools (0.22% hit rate).

The amplicons were then sequenced to check for redundancies and for known PKS sequences. The redundancy level was low (about 2 %) and 39 unique PKS sequences were found from the pools. All of these represented promising novel PKS biosynthesis operons. Then, positive clones were identified using colony hybridization with relevant probes. These were subsequently tested, following shuttling into *Streptomyces,* for antagonistic activity against *Bacillus subtilis* 1A72, *Staphylococcus aureus* 21, *Enterococcus faecalis* 40, *Escherichia coli* 9, *Pseudomonas aeruginosa* 39, *Fusarium oxysporum* LNPV, *Aspergillus fumigatus* Gasp 4707 and *Neurospora crassa* HK. They exhibited 56% antimicrobial activity against at least *B. subtilis,* 13% against *S. aureus*, 4% against *E. faecalis* and less than 1 % partial inhibition of growth of *Neurospora crassa* mycelium (LibraGen). These screening procedures thus allowed quick access to novel PKS sequences that are typical for the soil community, and can be used for production.

It may be may concluded that major advances in the metagenomic exploration of suppressive soil are still required. Such advances will boost our capacity to finally come to grips with the astounding soil microbial diversity and possibly harness it to our advantage (Jan Dirk van Elsas, 2008). Guidance by GGM will be an asset in the progress. The improved or fine tuned soil metagenomics approaches will enable us to mine soil for genes / pathways of interest to biotechnological applications, decipher identity and function of as-yet- uncultured microorganisms and obtain a characterization of soil with regard to antagonistic function, diversity and genetic complement. However, soil metagenomics will continue to be labor-intensive as well as resource-demanding. On the positive side, the quickly-increasing throughput of sequencing technologies will assist in the rapid assessment of the prevalence of target genes, shedding increasing light on the soil genetic reservoir and potential for biotechnological exploration and application.

Bibliography

Alban Ramette, Yvan Moënne-Loccoz, Geneviève Défago, 2003. Prevalence of fluorescent pseudomonads producing antifungal phloroglucinols and/or hydrogen cyanide in soils naturally suppressive or conducive to tobacco black root rot. *FEMS Microbial. Ecol.*, **44:** 35-43

Braker, G., Ayala-del-Rio H.L., Devol, A.H., Fesefeldt, A. and Tiedje, J.M., 2001. Community structure of denitrifiers, Bacteria, and Arcahe along redox gradients in Pacific Northwest marine sediments by terminal restriction fragment length polymorphism analysis of amplified nitrite reductase (nirS) and 16S rRNA genes. *Appl. Environ. Microbiol.*, **67:** 1893–901

Brierley, J., Lees Scri, A. and Wale Sac, S., 2008. Powdery Scab - Strains and Conducive Conditions A review of literature and other sources of information. *Res.Rev.,* Potato Council, Oxford.

Browne, G.T., Ott, N.J., Poret-Peterson, A., Lampinen, B.D., Metcalf, S.G., Yaghmour, M.A., et al., 2018. Anaerobic soil disinfestation as an alternative to soil fumigation for management of *Prunus* replant disease: Effects on tree growth and yield. *Acta Hort.*, DOI:10.17660/ActaHortic.2018.1219.**44**: 287–294

Carrión, V.J., Cordovez, V., Tyc, O. *et al.,* 2018. Involvement of Burkholderiaceae and sulfurous volatiles in disease-suppressive soils. *ISME J.,* **12:** 2307–2321

Cha, J.Y., Han, S., Hong, H. J., Cho, H., Kim, D., Kwon, Y., et al., 2016. Microbial and biochemical basis of a Fusarium wilt-suppressive soil. *ISME J.,* **10:** 119–129.

Chakraborty, I. and Maity, P., 2020. COVID-19 outbreak: migration, effects on society, global environment and prevention. *Sci. Tot. Environ.,* **728:**138882.

Chandrashekara, C., Kumar, R., Bhatt, J.C. and Chandrashekara, K.N., 2012. Suppressive soils in plant disease management. *In: Eco-friendly innovative approaches in plant disease management.* Singh, V.K., Singh, Y. and Singh, A.,(Eds.) India: International Book Distributors; 2012. pp. 241–256

Chapelle, E., Mendes, R., Bakker, P. A., and Raaijmakers, J. M., 2015. Fungal invasion of the rhizosphere microbiome. *ISME J.,* **10:** 265–268.

Cheng, X., Etalo, D.W., Mortel, J.Evd, Dekkers, E., Nguyen, L., Medema, M.H. and Raaijmakers, J.M., 2017. Genome-wide analysis of bacterial determinants of plant growth promotion and induced systemic resistance by *Pseudomonas fluorescens*. *Environ. Microbiol.,* **19**:4638–4656.

Chevalier, G. and Sourzat, P., 2012. Soils and Techniques for Cultivating *Tuber melanosporum* and *Tuber aestivum* in Europe. *In: Edible Ectomycorrhizal Mushrooms: Current Knowledge and Future Prospects*. Zambonelli A., Bonito G.M., (Eds.), Springer Berlin Heidelberg, Berlin, Heidelberg, Germany, 163, 2012.

Chng, S., Cromey, M.G., Dodd, S.L., Stewart, A., Butler, R.C, et al., 2015. Take-all decline in New Zealand wheat soils and the microorganisms associated with the potential mechanisms of disease suppression. *Plant & Soil,* **397**: 239–259.

Coyotzi, S., Pratscher, J., Murrell, J. C. and Neufeld, J. D., 2016. Targeted metagenomics of active microbial populations with stable-isotope probing. *Curr. Opin. Biotechnol***.,** **41**:1-8.

Daniel Bautista, Diana Garcia, Lorena Dávila, Alejandro Caro-Quintero, Alba Marina Cotes, Adriana González, A. and Paola Zuluaga, 2023. Studying the microbiome of suppressive soils against vascular wilt, caused by *Fusarium oxysporum* in cape gooseberry (*Physalis peruviana*). *Environ. Microbiol.,* **15:** 757-768.

Daniel Schlatter, Linda Kinkel, Linda Thomashow, David Weller and Timothy Paulitz, 2017. Disease Suppressive Soils: New Insights from the Soil Microbiome. *Phytopatholgy,* **107:**

David M. Weller, Jos, M., Brian B. McSpadden and Linda S. Thomashow, 2002. Microbial populations responsible for specific soil suppressiveness to plant pathogens. *Ann. Rev. Phytopathol.,* **40**: 309–348

Deng, X., Zhang, N., Li, Y., Zhu, C., Qu, B., Liu, H., Li, R., Bai, Y., Shen, Q. And Falcao Salles, J., 2022. Bio-organic soil amendment promotes the suppression of *Ralstonia solanacearum* by inducing changes in the functionality and composition of rhizosphere bacterial communities. *New Phytol.,* **235**: 1558–1574

Dom´ınguez, J., Negrin, M.A and, Rodr´ıguez, C.M., 2001. Aggregate water-stability, particle size and soil solution properties in conducive and suppressive soils to Fusarium wilt of banana from Canary islands (Spain). *Soil Biol. Biochem.,* **33**:449–55

Döring, T.F., Rosslenbroich, D., Giese, C., Athmann, M., Watson, C., Vágó, I., Kátai, J., Tállai, M. and Bruns, C., 2020. Disease suppressive soils vary in resilience to stress. *Appl. Soil Ecol.,* **149**:103482

Dorota Hilszczańska, Hanna Szmidla, Katarzyna Sikora and Aleksandra Rosa-Gruszecka, 2019. Soil Properties Conducive to the Formation of *Tuber aestivum* Fruiting Bodies. *Pol. J. Environ. Stud.,* **28**: 1713-1718

Durán, P., Jorquera, M., Viscardi, S. and Carrion, V.J., 2017. Screening and Characterization of Potentially Suppressive Soils against *Gaeumannomyces graminis* under Extensive Wheat Cropping by Chilean Indigenous Communities. *Front. Microbiol.,* **8**: 1–16.

Fan, P.S., Lai, C.Y., Yang, J.M., Hong, S., Yang, Y., Wang, Q., Wang, B., Zhang, R., Jia, Z., Zhao, Y., et al., 2022. Crop rotation suppresses soil-borne *Fusarium* wilt of banana and alters microbial communities. *Arch. Agron. Soil Sci.,* **68:** 447–459

Fujiwara, K. *Et al.,* 2016. Emergence of antagonism against the pathogenic fungus *Fusarium* oxysporum by interplay among non-antagonistic bacteria in a hydroponics using multiple parallel mineralization. *J. Phytopathol.,* **164:** 853–862.

Gao, Z., Karlsson, I., Geisen, S., Kowalchuk, G. and Jousset, A., 2019. Protists: puppet masters of the rhizosphere microbiome. *Tre. Plant Sci.,* **24:** 165–176.

Garbeva, P., Postma, J., van Veen, J.A. and van Elsas, J.D., 2006. Effect of above- ground plant species on soil microbial community structure and its impact on suppression of *Rhizoctonia solani* AG3. *Environ. Microbiol.,* **8**:233–246.

Giotis, C., Theodoropoulou, A., Cooper, J., Hodgson, R., Shotton, P., Shiel, R., Cretoiu Eyre, M., Wilcockson, S., Markellou, E., Liopa-Tsakalidis, A., Volakakis, N. and Leifert, C., 2012. Effect of variety choice, resistant rootstocks and chitin soil amendments on soil-borne diseases in soil-based, protected tomato production systems. *Eur. J. Plant Pathol.,* **134**: 605–617

Gómez Expósito, R., de Bruijn, I., Postma, J. and Raaijmakers, J.M., 2017. Current Insights into the Role of Rhizosphere Bacteria in Disease Suppressive Soils. *Front. Microbiol.*, **8**:2529

Guerra, C.A., Heintz-Buschart, A., Sikorski, J., Chatzinotas, A., Guerrero-Ramírez, N., Cesarz, S., Beaumelle, L., Rillig, M.C., Maestre, F.T., Delgado-Baquerizo. M. and Buscot, F., 2020. Blind spots in global soil biodiversity and ecosystem function research. *Nature commun.,* **11**:1-13.

Hoitink, H.A.J. and Boehm, M.J., 1999. Biocontrol within the context of soil microbial communities: a substrate-dependent phenomenon. *Ann. Rev. Phytopathol.,* **37**:427e446.

Hong, S., Jv, H.L., Yuan, X.F., Geng, J.J., Wang, B.B., Zhao, Y., Wang, Q., Li, R., Jia, Z.J. and Ruan, Y.Z., 2022. Soil Organic Nitrogen Indirectly Enhances Pepper-Residue-Mediated Soil Disease Suppression through Manipulation of Soil Microbiome. *Agronomy,* **12:** 2077

Hornby, D., 1998. Take-All of Cereals: A Regional Perspective. Wallingford, UK: CAB Int. 384 pp

Irena Todorović, Yvan Moënne-Loccoz, Vera Raic, Jelena Jovičić-Petrović and Daniel Muller, 2023. Microbial diversity in soils suppressive to Fusarium diseases. *Front. Pl. Sci*., **14**: ff10.3389/fpls.2023.1228749ff.

Jambhulkar, P.P., Sharma, M., Lakshman, D. and Sharma, P., 2015. Natural mechanisms of soil suppressiveness against diseases caused by *Fusarium, Rhizoctonia, Pythium*, and *Phytophthora. In: Organic amendments and soil suppressiveness in plant disease management.* Meghvansi, M.K., Varma, A. (Eds.,), Switzerland: Springer International Publishing; pp. 95–123

Jan Dirk van Elsas, 2008 The Metagenomics of Disease-Suppressive Soils. W5-06. Department of Microbial Ecology, University of Groningen Kerklaan 30, 9750AA Haren, the Netherlands

Jayaraman, S., Naorem, A.K., Lal, R., Dalal, R.C., Sinha, N.K., Patra, A.K. and Chaudhari, S.K., 2021. Disease-Suppressive Soils-Beyond Food Production: a Critical Review. *J. Soil Sci. Plant Nutr*., **21**:1437-1465

Jiang, G., Zhang, Y., Gan, G., Li, W., Wan, W., Jiang, Y., et al., 2022. Exploring rhizo-microbiome transplants as a tool for protective plant-microbiome manipulation. *ISME Commun*., **2:** 1–10.

Khan Abdur, R., Wicaksono Wisnu, A., Ott Natalia, J., Poret-Peterson Amisha, T. and Browne Greg, T., 2021. Characterization of soils conducive and non-conducive to *Prunus* replant disease. *PLoS ONE*, **16**:1

Kwak, Y,S, and Weller, D.M., 2013. Take-all of wheat and natural disease suppression: a review. *Plant Pathol. J.,* **29:**125–135

Latz, E., Eisenhauer, N., Rall, B. C., Scheu, S. and Jousset, A., 2016. Unravelling linkages between plant community composition and the pathogen-suppressive potential of soils. *Sci.Rep.,* **6:** 23584.

Li, Y., Deng, X., Zhang, N., Shen, Z., Li, R., Shen, Q. and Salles, J. F., 2023. Rhizosphere suppression hinders antibiotic resistance gene (ARG) spread under bacterial invasion. *One Health,* **16**, Article 100481. https://doi.org/10.1016/j.onehlt.2023.10048

Liu, H., L,i J., Cavalhais, L.C., Percy, C., Verma, J.P., Schenk, P.M. and Singh, B.K.,2020. Evidence for the plant recruitment of beneficial microbes to suppress soil–borne pathogen. *New Phytologist,* **10.**1101/2020.07.31.231886

Löbmann, M.T., Vetukuri, R.R., de Zinger, L., Alsanius, B.W., Grenville-Briggs, L.J. and Walter, A.J., 2016. The occurrence of pathogen suppressive soils in Sweden in relation to soil biota, soil properties and farming practices. *Appl Soil Ecol.,* **107:** 57–65.

Loeppmann, S., Blagodatskaya, E., Pausch, J. and Kuzyakov, Y., 2016. Substrate quality affects kinetics and catalytic efficiency of exoenzymes in rhizosphere and detritusphere. *Soil Biol. Biochem.* **92:** 111–118.

Mahoney, A. K., Yin, C., and Hulbert, S. H., 2017. Community structure, species variation, and potential functions of rhizosphere-associated bacteria of different winter wheat (*Triticum aestivum*) cultivars. *Front. Plant Sci.,* **8**:132.

Martinez Lisa, Pouvreau Jean-Bernard, Jestin Christophe, Montiel Grégory, Gravot Antoine, Berardocco Solenne, Marnet Nathalie, Bouchereau Alain, Delage Erwan, Philippe Simier and Poulin Lucie, 2023. Comparative analysis of two neighboring conducive and suppressive soils towards plant parasitism caused by *Phelipanche ramosa* L. Pomel on *Brassica napus* L.. 10.21203/rs.3.rs-3059250/v1.

Mazzola, M. and Freilich, S., 2017. Prospects for biological soilborne disease control: Application of indigenous versus synthetic microbiomes. *Phytopathology,* **107**:256-263.

Meeboon, J., Ando, A., Ogawa, J. *et al.,* 2022. Generation of *Fusarium oxysporum*-suppressive soil with non-soil carriers using a multiple-parallel-mineralization technique. *Sci. Rep.,* **12**: 7968.

Mousa, W.K. and Raizada, M.N., 2016. *Natural Disease Control in Cereal Grains.* 2. Oxford: Academic Press

Nguyen, N. P., Warnow, T., Pop, M., and White, B., 2016. A perspective on 16S rRNA operational taxonomic unit clustering using sequence similarity. *NPJ Biofilms and Microbiomes*, **2**:16004.

Palaniyandi, S.A., Yang, S.H., Zhang, L. and Suh, J.W., 2013. Effects of actinobacteria on plant disease suppression and growth promotion. *Appl. Microbiol. Biotechnol.,* **97**: 9621–9636.

Paola Durán, Gonzalo Tortella, Sharon Viscardi, María de la Luz Mora and María José Pozo, 2020. Microbial Community Composition in Take-All Suppressive Soils. *In: Prime Archives in Microbiology.* Alexandre Morrot (Ed.). Hyderabad, India: Vide Leaf. 2020

Peddi Durga, Kumar Ravinder, Bhatt, J.C. and Kn, Chandrashekara, 2012. Suppressive Soils in Plant Disease Management. 10.13140/2.1.5173.7608.

Penton, C.R., Gupta, V.V.S.R., Tiedje, J.M., Neate, S.M., Ophel-Keller, K., Gillings, M, et al., 2014. Fungal Community Structure in Disease Suppressive Soils Assessed by 28S LSU Gene Sequencing. *PLoS ONE* **9**(4): e93893.

Peralta, A.L., Sun, Y., McDaniel, M.D. and Lennon, J.T., 2018. Crop rotational diversity increases disease suppressive capacity of soil microbiomes. *Ecosphere*, **9**(5). 10.1002/ecs2.2235

Persson, L. and Olsson, S., 2000. Abiotic characteristics of soils suppressive to *Aphanomyces* root rot. *Soil Biol. Biochem.,* **32**:1141–1150

Poly, F., Rnajard, L., Nazaret, S., Gourbiere, F. and Jocteur Nonrozier, L., 2001. Comparison of nifH gene pools in soils and soil microenvironments with contrasting properties. *Appl. Environ. Microbiol.,* **67**:2255–62

Poromarto, S, H., Supyani, Supriyadi and Hadiwiyono, 2021. Chemical characters of disease suppressive and conducive soil of Moler on shallot in Brebes Central Java. *IOP Conf. Ser.: Earth Environ. Sci*., **905:** 012057

Poudel, R., Jumpponen, A., Schlatter, D. C., Paulitz, T. C., Gardener, B. B., Kinkel, L. L., et al., 2016. Microbiome networks: A systems framework for identifying candidate microbial assemblages for disease management. *Phytopathology,* **106:** 1083–1096.

Raaijmakers, J.M. and Weller, D.M., 2001. Exploiting the genetic diversity of *Pseudomonas* spp: characterization of superior colonizing *P. fluorescens* strain Q8r1-96. *Appl. Environ. Microbiol.,* **67:** 2545–2554

Rondon, M.R., August, P.R., Bettermann, A.D., Brady, S.F., Grossman, T.H., et al., 2000. Cloning the soil metagenome: A strategy for accessing the genetic and functional diversity of uncultured microorganisms. *Appl. Environ. Microbiol.,* **66:**2541–47

Rosenzweig, N., Tiedje, J.M., Quensen, J.F., Meng, Q., Hao, J.J., 2012. Microbial communities associated with potato common scab-suppressive soil determined by pyrosequencing analyses. *Plant Dis.,* **96:** 718–725

Sánchez-Cañizares, C., Jorrín, B., Poole, P. S. and Tkacz, A., 2017. Understanding the holobiont: The interdependence of plants and their microbiome. *Curr. Opin. Microbiol.*, **38:** 188–196.

Shen, Z., Ruan, Y., Xue, C., Zhong, S., Li, R. and Shen, Q., 2015. Soils naturally suppressive to banana *Fusarium* wilt disease harbor unique bacterial communities. *Plant Soil,* 393: 21–33.

Shi W, et al., 2019. The occurrence of potato common scab correlates with the community composition and function of the geocaulosphere soil microbiome. *Microbiome,* **7:** 1–18

Shinohara, M. *et al.*, 2011. Microbial mineralization of organic nitrogen into nitrate to allow the use of organic fertilizer in hydroponics. *Soil Sci. Plant Nutr.,* **57:** 190–203

Simon, H., Smith, K., Dodsworth, J., Guenthner, B., Handelsman, J. and Goodman, R., 2001. Influence of tomato genotype on growth of inoculated and indigenous bacteria in the spermosphere. *Appl. Environ. Microbiol.,* **67:**514–20

Simon, A. and Sivasithamparam, K., 1988. Microbiological differences between soils suppressive and conducive of the saprophytic growth of *Gaeumannomyces graminis* var. *tritici*. *Canadian J. Microbiol.,* **34**: 860-864

Stopnisek, N., Zuhlke, D., Carlier, A., Barberan, A., Fierer, N., Becher, D., et al., 2016. Molecular mechanisms underlying the close association between soil *Burkholderia* and fungi. *ISME J.,* **10:**253–264.

Supriyadi, Supyani, Poromarto, S. H. and Hadiwiyono, 2020. The 1st International Seminar on Agriculture, Biodiversity, Food Security, and Health

Terpolilli, J. J., Hood, G. A. and Poole, P. S., 2012. What determines the efficiency of N2-fixing Rhizobium-legume symbioses?. *Adv.Microbial physiolo.,* **60**: 325-389.

Tian, B.N., Xie, J.T., Fu, Y.P., Cheng, J.S., Li, B., Chen, T., Zhao, Y., Gao, Z.X., Yang, P.Y., Barbetti, M.J., et al., 2020. A cosmopolitan fungal pathogen of dicots adopts an endophytic lifestyle on cereal crops and protects them from major fungal diseases. *ISME J.,* **14:** 3120–3135.

Tracanna, V., Ossowicki, A., Petrus, M.L.C., Overduin, S., Terlouw, B.R., Lund, G., Robinson, S.L., Warris, S., Schijlen, E.G.W.M., van Wezel, G.P., Raaijmakers, J.M., Garbeva, P. and Medema, M.H., 2021. Dissecting disease-suppressive rhizosphere microbiomes by functional amplicon sequencing and 10× metagenomics. *mSystems*, **6**:e01116-20.

Wang, B.B., Sun, M.Z., Yang, J.M., Shen, Z.Z., Ou, Y.N., Fu, L., Zhao, Y., Li, R., Ruan, Y.Z. and Shen, Q.R., 2022. Inducing banana Fusarium wilt disease suppression through soil microbiome reshaping by pineapple–banana rotation combined with biofertilizer application. *Soil,* **8:** 17–29

Wang, G., Ma, Y., Chenia, H.Y., Govinden, R., Luo, J. and Ren, G., 2020. Biochar-mediated control of Phytophthora blight of pepper is closely related to the improvement of the rhizosphere fungal community. *Front.Microbiol.,* **11:** 1427.

Wang, Q., Zhou, L., Jin, H., Cong, B., Yang, H. and Wang, S., 2022. Investigating the Responses of Microbial Communities to Banana *Fusarium* Wilt in Suppressive and Conducive Soils Based on Soil Particle-Size Differentiation. *Agronomy,* **12:**229.

Xiong, W., Li, R., Ren, Y., Liu, C., Zhao, Q., Wu, H., Jousset, A. and Shen, Q., 2017. Distinct roles for soil fungal and bacterial communities associated with the suppression of vanilla *Fusarium* wilt disease. *Soil Biol. Biochem.,* **107:** 198–207

Yang, J., Wu, Q., Wang, Y., Chen, X., Gao, W., Zhao, Y., Wang, B. and Ruan, Y., 2023. Suppression of Banana *Fusarium* Wilt Disease with Soil Microbial Mechanisms via Pineapple Rotation and Residue Amendment. *Agronomy* **13:**377.

Young, C.C., Cheng, K.T. and Waller, G.R., 1991. Phenolic compounds in conducive and suppressive soils on club root disease of crucifers. *Soil Biol. & Biochem.*, **23:** 1183-1189,

Yuan, X.F., Wang, B.B., Hong, S., Xiong, W., Shen, Z.Z., Ruan, Y.Z., Li, R., Shen, Q.R. and Dini-Andreote, F., 2021. Promoting soil microbial-mediated suppressiveness against *Fusarium* wilt disease by the enrichment of specific fungal taxa via crop rotation. *Biol. Fert. Soils,* **57:** 1137–1153

8

Biological Control: Concepts and Potentialities of Bioagents for Managing Soil Borne Pathogens

Soil borne pathogens cause significant economic losses in agricultural production all over the world. These species can survive for many years in the absence of a host plant by forming persistent structures such as microsclerotia, sclerotia, chlamydospores or oospores. Consequently, soilborne diseases are particularly difficult to predict, detect, diagnose and successfully control. Over the past 30 years, a fumigant, methyl bromide, has been widely used for their control in many crops (Milica Mihajlović et al., 2017). In 1992, methyl bromide was listed as an ozone-depleting substance under the Montreal Protocol - an international treaty to protect the ozone layer. During the phaseout of methyl bromide, problems generated in agricultural production made it clear that dependence on a single method or a single chemical should be avoided. Milica Mihajlović et al., (2017) summarized the current knowledge about different methods of soilborne disease control including: crop rotation, steam soil disinfection, soil amendments, hydroponics and soilless growing systems, soil solarization, grafting, biological control and use of natural compounds, and chemical control. Positive and negative aspects of all available methods were reviewed. Benefits, achieved by simultaneous application of several methods based on different mechanisms of actions, were discussed.

Effective use of biological control agents (BCAs) is a potentially important component of sustainable agriculture. The principal biocontrol mechanisms involved include mycoparasitism, antibiosis, competition, and induced resistance; additional mechanisms are hypovirulence mediated through fungal viruses and inhibition of enzymes involved in plant pathogenicity (Xu et al., 2011). Individual biocontrol mechanisms have been shown to be predominant for some BCAs but there are also many instances where more than one mechanism may operate in a given BCA isolate and prolonged study of any individual BCA will often reveal additional biocontrol mechanisms. The ecological processes determining the success or otherwise of biological control are complex. Thus, it is not surprising that, in addition to variable control efficacies, biocontrol success in field crops has been limited despite much research effort. Most biocontrol success has been achieved in greenhouse cultivation, where ecological parameters are less variable. Because of the inconsistent or limited biocontrol achieved in the field, BCAs have also been used in combination with fungicides or cultural practices.

Biological control of plant diseases, involving the use of (micro-) organisms or compounds of biological origin, is now considered one of the most promising alternatives to the use of chemical pesticides. Sales of so-called biocontrol agents are expected to reach circa 10% of the overall global crop protection market in the near future with an annual growth rate estimated between 15% and 20% (Adrien Anckaert et al., 2022). Biocontrol can provide plant protection against many diseases and is environment friendly. Since it often involves several modes of action to restrict the growth of microbial pathogens, biocontrol may limit the development of resistance mechanisms in pathogens but further studies are needed to assess their durability. BCAs may act against pathogens via a number of mechanisms: competition for space and nutrients, parasitism, antibiosis and/or by stimulating the host plant defences. Biocontrol organisms are natural enemies of plant pathogens including viruses, bacteria, fungi, insects and nematodes. Non-living agents of biological origin include semiochemicals (pheromones, plant volatiles) and biochemical products mainly secreted or extracted from plants and microorganisms. Microbial products currently dominate the BCA market, particularly bacteria such as *Bacillus thuringiensis* which is widely used as a bio-insecticide. Selected beneficial bacteria naturally living in close association with plants, and referred to as plant growth-promoting rhizobacteria (PGPR), are also used as BCAs against plant pathogens. PGPRs constitute a diverse group of bacteria isolated from the rhizosphere and belonging mainly to the genera *Pseudomonas, Streptomyces, Acetobacter, Azospirillum, Paenibacillus, Serratia, Burkholderia, Herbaspirillum, Rhodococcus, Rhizobium* and *Bacillus*.

Use of mixtures of cultivars or fungicides has been successfully adopted in many crops to increase and maintain disease control efficacy when individual cultivars or fungicides may not be able to control disease effectively. To improve biocontrol efficacies achieved through use of a single BCA, there has been increasing interest recently among researchers in using mixtures of BCAs. A range of biocontrol mechanisms may operate in mixed BCA populations, although their relative significance is unknown but may vary with circumstances. In mixed BCA populations, we may need to consider both direct and indirect interactions between different BCA populations. Compared with the more efficacious BCA, combined use of two BCAs may lead to increased, reduced or similar biocontrol efficacy. However, in many such studies, the hypothesis testing for synergistic, additive, or antagonistic interactions among BCAs in controlling diseases was not clearly formulated and, thus, not tested statistically.

Why need of bio agents?: Environmental safety is one of the leading drivers of bio agent usage because chemical pesticides also kill beneficial insects (bees, butterflies etc.) and microbes (*Rhizibium,* entomopathogenic fungus etc.). Non-toxic to non-target organisms, including beneficial insects and wildlife, many bio agents also are biodegradable (Narender Kumar et al., 2019). They decompose quickly and do not negatively impact surface water and groundwater. Bio agents typically are effective in small a quantity which eliminates pollution concerns sometimes associated with traditional chemicals. In addition, bio agents are manufactured from naturally occurring raw materials in an environmentally responsible and sustainable manner.

Growing educational opportunities in plant protection training over the past few decades have led to the widespread successful use of biological control, particularly in developing countries like Brazil, where research and implementation of both augmentative and classical biocontrol are gaining momentum (Jaiswal et al., 2022). This trend has been accelerated by the demands due to the awareness created by researchers, academicians, and non-governmental organizations that have all been hastening this shift. Growing education opportunities in plant protection training over the past few decades have resulted in the successful use of biological control on a large scale, especially in developing nations like India, China & Brazil, where research and implementation of both augmentative and classical biocontrol are gaining momentum. Realizing that synthetic pesticides and fertilizers have damaged ecosystems and exacerbated food security concerns, China and India have invested in biological control research, training, and adoption.

Soil borne pathogens affect crops from their initial stages in nurseries until harvest. Plant diseases caused by *Rhizoctonia* spp., *Fusarium* spp., *Verticillium* spp., *Sclerotinia* spp., *Pythium* spp., and *Phytophthora* spp. affect a number of important crops, including wheat, cotton, vegetables and temperate fruits. Economic losses caused by Fusarium wilt of column stock pathogens are estimated at 50-75% of the attainable yield for many crops. The symptoms of soilborne disease, caused by different pathogens, are very similar. They include root rot, root blackening, wilt, yellowing, stunting or seedling damping-off, bark cracking and twig or branch dieback. Soil borne species can survive for many years in the absence of a host plant by forming resistant structures such as microsclerotia, sclerotia, chlamydospore or oospores. Consequently, these pathogens are particularly difficult to predict, detect, diagnose and control. In an attempt to reduce the use of pesticides, there is an increasing interest in introducing biological agents and putting to use plant compounds as natural commercial products for managing soilborne pathogens.

8.1 Antagonistic Microorganisms

Numerous studies have shown suppression of disease incidence in different crops after supplementing soils with fungal or bacterial antagonists. Biocontrol of soilborne diseases, as an alternative to synthetic pesticides, is particularly complex because such diseases occur in dynamic environments at the interface of root and soil known as the rhizosphere. A rhizosphere is defined as the region surrounding a root that is affected by it. It is important to point out that bioagents can reduce harmful effects of some pathogens below a certain threshold with no substantial changes in the soil microbiological balance, something that does not occur when chemicals are applied (Neshev, 2008). Various mechanisms are involved in the biological control of fungal pathogens. These mechanisms include: the production of secondary metabolites (antibiotics, siderophores, hydrolytic enzymes, volatile extracellular metabolites, hydrogen cyanide), parasitism, competition for nutrients, promotion of plant growth and, finally, induced resistance within the plants. Biological control agents, as living organisms, are much more sensitive to various external conditions than synthetic pesticides. On the other side, soil is a complex ecosystem in which many factors, such as soil structure, pH and moisture can limit the success of introduced biocontrol agents. Therefore, many of these biological control agents can be effective but require specific environmental conditions to flourish (Milica Mihajlović et al., 2017).

Numerous studies have indicated that several established biocontrol agents, including strains from the genera *Bacillus, Pseudomonas, Sphingomonas, Stenotrophomonas* and *Serratia,* can suppress vascular or soilborne fungal pathogens. Fungi belonging to the genus *Trichoderma* and bacteria such as *Pseudomonas.*spp, or *Bacillus subtilis,* are the most promising biocontrol agents. On the one hand, they stimulate plant growth, while on the other they eliminate plant pathogens by their unique antimicrobial activities, including the production of antibiotics and toxins to compete with pathogenic organisms. *B. subtilis, Trichoderma harzianum*, and *T. virens* have been reported as biocontrol agents against soilborne potato diseases. *Trichoderma* spp. are well-studied, efficient mycoparasites that perform best in moist, slightly acidic soils. In a strawberry trial, the amendment of *Trichoderma* sp. to a soil treated with ozone gas decreased *Verticillium* wilt in the first year, but this was not repeated the following year. *Trichoderma harzianum* and *T. koningii* are effectively used for seed treatment against *R. solani, Pythium* spp., and *Sclerotia rolfsii*. *Coniothyrium minitans* and *Sporidesmium sclerotivorum* were effective in controlling Sclerotinia diseases in lettuce. *Pseudomonas fluorescens* is a commonly used biocontrol agent against soilborne diseases. The bacterium is a soildwelling antagonist that has several modes of action against pathogenic microorganisms. It favors moist soils with high organic matter and is compatible with mulching. Root pretreatment of olive plants with some isolates of *P. fluorescens* during nursery propagation can help in the biocontrol of *V. dahliae. Bacillus* and *Pseudomonas* species are very beneficial and effective in the biocontrol of Fusarium wilt of chickpea.

One possible approach to improving biological control may be the application of combinations of biocontrol agents. By combining microorganisms, multiple antifungal traits can be achieved and one may assume that at least one biocontrol mechanism will be functional under the conditions faced by the released biocontrol agents. Moreover, combinations of biocontrol strains are expected to result in a higher level of protection, stabilization of biocontrol efficacy and suppressiveness of multiple plant diseases. It has been demonstrated that natural suppressiveness of the Châteaurenard soil in France against Fusarium wilt was based on various mechanisms involving several microbial populations acting alone or together to limit the activity of the pathogen. Yet, the use of biological control agents against soil borne pathogens cannot fully replace methyl bromide fumigation, they could rather function together in an integrated pest management strategy.

The phenomenon of reducing the amount of inoculum or the disease causing activity of plant pathogens accomplished through the use of living organisms other than the man is called as biological control. According to Baker and Cook's (1974) definition "Biological control is the reduction of inoculum or disease producing activity of a pathogen accomplished by one or more organisms other than man." In other words Biological control refers to the purposeful utilization of introduced or resident living organisms, other than disease resistant host plants, to suppress the activities and populations of one or more plant pathogens (Pal and Gardener, 2006). Biological control was originally defined "the action of parasites, predators, or pathogens in maintaining another organism's population density at a lower average than would occur in their absence". Biological control provided by these living organisms is especially important for reducing the numbers of pest insects, mites and pathogens. It is an environmentally sound and effective means of reducing or mitigating pests and pest effects through the use of natural enemies (Zerihun Tsegaye et al., 2018). Interest in biological control has increased over recent decades for many reasons. First, a greater appreciation for environmental stewardship among regulators, government, politician, growers, and the public has promoted development of more sustainable farming practices. Second, highly economic, self propagating, no harmful effects on human, other organisms, environment, have great ability to search their prey, compatible with IPM components and permanents. Lastly, consumers increasingly demand products that are grown in a sustainable manner and are free of insecticide residue.

Biological control of plant diseases has been considered a viable alternative method to chemical control. These include Augmentation, Classical and Conservation biological control. There are various naturally occurring bio-control agents that aggressively attack on plant pathogens and suppress plant disease by different mode of actions (Zerihun Tsegaye et al., 2018). Bio-control agents comprise of multiple beneficial characters such as rhizosphere competence, antagonistic potential, and ability to produce antibiotics, lytic enzymes, computation for nutrient and niche. These biological control activities are exerted either directly through antagonism of soil-borne pathogens or indirectly by eliciting a plant-mediated resistance response. Understanding the mechanisms of biological control of plant diseases through the interactions between antagonists and pathogens may allow us to select and construct the more effective bio-control agents and to manipulate the soil environment to create conducive condition for successful bio-control. Some of the important factors that affect the efficacy of microbial bio-control agents in controlling plant diseases which should carefully be considered include biotic, a biotic, method of application, formulation of bio-control microorganisms and timing of application. Application and commercial use of biological control agents have been slow mainly due to their variable performances under different environmental conditions in the field. To overcome this problem and in order to take the bio-control technology to the field and improve the commercialization of bio-control, it is important to develop new formulations of bio-control microorganisms with higher degree of stability and survival. A significant improvement have been made in different aspects of biological control of plant diseases, but this area still need much more research for development of efficient bio-control, including development of novel formulations, understanding the impact of environmental factors on bio-control agents to improve its efficacy.

8.2 Strategy and Principles of Biological Control

There are 3 basics strategies in biological control of pests, viz; Classical Biological Control (Importation), Augmentation and Conservation. Classical biological control is defined by Eilenberg et al. (200 1) as: The intentional introduction of an exotic, usually co-evolved, biological control agent for permanent establishment and long-term pest control. It is the importation of pest natural enemies from other countries, to a new locale where they do not occur naturally. When an organism is introduced either intentionally or accidentally into an area in which it did not occur previously, it can sometimes increase to a high population density and become a serious pest. This population increase is mostly due to the fact that the pest was introduced without its natural enemies. The goal of classical biological control is to find useful natural enemies, introduce them into the area of the target pest, and permanently establish them so that they will provide continuing pest control with the result that the pest population decreases in population density, hopefully below the economic injury level of the pest. The time scale on figure 1 can be years (Zerihun Tsegaye et al., 2018).

The process of classical biological control involves Determining the origin of the introduced pest ; Collecting appropriate natural enemies associated with the pest or closely related species ; Selected natural enemies are then passed through a rigorous assessment, testing and quarantine process, to ensure that they will work and that no unwanted organisms (such as hyper parasitoids) are introduced ; Mass production and release of selected natural enemies ; Follow-up studies are conducted to determine if the natural enemy becomes successfully established at the site of release, and to assess the long-term benefit of its presence.

Augmentation is the other strategy of biological control is also defined as: The intentional release of a living organism asa biological control agent with the expectation that it will multiply and control the pest for an extended period, but not permanently. It includes the periodic release of a natural enemy that does not occur naturally in sufficient numbers to keep a pest below damaging levels. The practice of augmentation is based on the knowledge or assumption that in some situations there are not adequate numbers or species of natural enemies to provide optimal biological control, but that the numbers can be increased by releases. This relies on an ability to mass-produce large numbers of the natural enemy in a laboratory or by companies to produce and sell them. There are two general approaches to augmentation (Inoculative releases and Inundative releases). Inoculation biological control is a type of augmented biocontrol strategy and defined as: intentional release of a living organism as a biological control agent with the expectation that it will multiply and control the pest for an extended period, but not permanently. Its control involves releasing small numbers of natural enemies at prescribed intervals throughout the pest period, starting when the pest population is very low. The natural enemies are expected to reproduce themselves to provide more long-term control. However, the expected outcome of inoculative releases is to keep the pest at low numbers, never allowing it to approach an economic injury level; therefore, it is more of a preventive measure. A pest population increases in size but in due course, before this population density has reached the potential maximum, a biocontrol agent is inoculated in small to moderate amounts. Its goal is to increase population size of the natural enemy in order to control the pest over a period of time. The inoculated bio-control organisms will, however, not establish permanently at a sufficient high population density. The pest will therefore increase in population size after a period of time and a new inoculation would then be needed.

Inoculation biological control has much in common with other inoculation practices. The major factor is that the bio control organism is expected to proliferate, at least temporarily. Conceptually it is therefore comparable to classicalbiological control but with the main differences that inoculation biological control uses mostly organisms which already occur in the area of application and only temporary establishment is achieved. Typical examples are the releases of *Encarsia formosa* and other parasitoids in glasshouses and: the inoculation of soil with the insect pathogenic fungus-*Beauveria brongniartii.* It can also be postulated that inoculation biological control represents a reestablishment of a natural balance, temporarllydistorted by man. Soil for cropping is for example inoculated with other additives to enhance growth (mycorrhiza) and inoculation with a bio-control agent can be seen as a moderate helpto speed up a natural process. Inoculation bio control has always, however, the advantage of being closely linked to monitoring pest populations and thus understanding population interactions. In glasshouses, a successful inoculation of bio control agents requires proper diagnosis of the pests present and in due course, releaseof the correct agents at the optimum density and time.

Inundation biological control is other type of augmentation bio-control and defined as the use of living organisms to control the pests when control is achieved exclusively by the released organisms themselves (Zerihun Tsegaye et al., 2018). A pest population increases in size, but at a certain time , for example when the economic injury level has been passed) a bio-control organism is applied in large amounts (inundated). The pest is quickly controlled and the population density of both the pest and the biocontrol agent decrease over time. Inundation biological control with its strong resemblance to chemical control can be perceived as 'less natural' than the other biological control strategies, especially when using a microorganism for biocontrol. The amount of the control agent to be applied is often several magnitudes higher than would ever occur under so-called natural conditions. The presentation of the inundation biological control agent gives association to chemical pesticides; for microorganisms to be used in bio-control, the product label has the appearance of a chemical product with information about the concentration and application rate expressed per square unit. Conservation biological control is also other strategy of the biological control can be defined as Modification of the environment or existing practices to protect and enhance specific enemies or other organisms to reduce the effect of pests (Eilenberg et al. 2001). Habitat manipulation often involves increasing the species diversity and structural complexity of agro ecosystems. Habitat manipulation approaches provide natural enemies with resources such as nectar, pollen, physical refugia, alternative prey and alternative hosts and operate to reduce pest densities via an enhancement of natural enemies.

Conservation biological control is thus completely different from the three other biological control strategies, since no organisms are released. Only organisms, which are already present, are enhanced in order to avoid damage. It is important to keep in mind that the definition allows both passive and active conservation. An example of passive conservation is the avoidance of actions which disfavour the natural enemies, for example spraying with certain chemical pesticides. An active conservation could be the initiation of actions to support the natural enemies actively by establishing for example 'beetle banks'. Among the four biological control strategies, conservation biological control can be seen as the most tightly connected to the main principles of organic farming, which have the protection of the existing natural enemies as one of the main principles. There is a tight connection also to conservation biology, since conservation biological control to large extent builds on ecological theory about meta populations, spatial fragmentation, and fate of species in a habitat. Conservation biological control can thus be seen as an example of habitat restoration with the specific purpose of supporting natural enemies to control pests.

Biological control agents and Characteristics: Biological control agents are organisms that control pathogens and diseases through antagonists (Zerihun Tsegaye et al., 2018). Microbial antagonism of plant pathogens occurs in several ways, the most common mechanisms being parasitism and predation, competition for nutrients or space, production of antimicrobial substances, lytic enzymes and induced resistance. As our knowledge of these mechanisms has increased, it has become apparent that antagonism often involves the synergistic action of several mechanisms.

Characteristics of the successful biological agents

- Highly effective bio-control agents must be fulfill the criteria: Be able to compete and live longer in soil and host tissue; Be able to colonize and proliferate; Be non-pathogenic to host plant and environment.
- Inexpensive production and formulation of agent must be developed: Production must result in biomass with excellent shelf live; To be successful as agricultural agent must be inexpensive, Able to produce in large quantities and Maintain viability.
- Delivery and application must permit full expression of the agent: Must ensure agents will grow and achieve their purpose.

8.3 Mechanisms Employed by Bio-Control Agents for Plant Disease Management

The bio-control activity is exerted either directly through antagonism of soil-borne pathogens or indirectly by eliciting a plant-mediated resistance response (Zerihun Tsegaye et al., 2018). Direct antagonism results from physical contact and/or a high-degree of selectivity for the pathogen by the mechanism(s) expressed by the bio-control agents whereas indirect antagonisms result from activities that do not involve sensing or targeting a pathogen by the bio-control agents. Stimulation of plant host defense pathways by nonpathogenic bio-control agents is the most indirect form of antagonism. Understanding the mechanism of action of a bio control agent may improve the consistency of control either by improving the mechanism or by using the bio- control agents under conditions where it is predicted to be more successful.

i. **Competition for available resources:** Competition occurs when two or more organisms require the same resource for growth and survival. The use of this resource by one organism reduces the amount available to the other. A soil-borne pathogen which infects only certain parts of the root may therefore be limited by competition for suitable colonisable sites or space on the root surface. The rhzosphere is a region of intense microbial activity where there may also be competition for oxygen. On the leaf surface, where nutrients are in short supply, competition for nutrients is thought to play a significant role in disease suppression. Competition for the same site is sometimes called 'site exclusion' and frequently takes place amongst organisms which are closely related taxonomically (e.g. the fire blight bacterium *Erwinia amylovora* and the saprophytic species *E. herbicola*). In the case of competition for nutrients the situation is similar. Antagonists with the same nutrient requirements as the pathogen are its most effective competitors. Competition for the same carbon source between *Pythium ulttimum*, a common cause of seedling damping-off and rhizosphere bacteria has resulted in effective biological control of *P. uttimum* in several crops. Ethanol and acetaldehyde released from seeds of pea and soybeans following imbibition stimulate hyphal growth from sporangia of *P. uttimurn*. Treating seeds with a strain of *Pseudomonas putida* which utilises ethanol as its sole carbon source in culture reduces the concentration of volatiles released, lessens hyphal growth or sporangial germination of the pathogen and increases seedling emergence in *Pythium* infested soil. Similarly, ethanol metabolising strains of *Enterobacter cloacae* reduce the ability of cotton seed exudates and volatiles to stimulate sporangium germination of *P. ultimum.*

   One of the best documented examples of nutrient competition in biological control involves competition for iron between fluorescent pseudomonas and soil borne fungal pathogens such as *Fusarium oxysporum* (Zerihun Tsegaye et al., 2018). Strains of bacteria including *Pseudomonas fluorescent* and *P. putida* produce siderophores metabolic products of micro-organisms that bind iron and facilitate it transport from the environment into the microbial cell. The siderophores pyoverdine and pseudobactin have a high affinity for the soluble ferric iron (Fe*") and inhibit the growth of pathogens by limiting the availability of iron. Evidence for involvement of siderophores in biological control has been obtained from experiments with well-characterised and genetically-manipulated bacterial strains. Pyoverdine production is correlated with biological control while pyoverdine deficient strains have reduced capabilities in biological control. The usefulness of nutrient competition as a mechanism of biological control depends on the type of pathogen that is targeted. It may not be useful in suppressing biotrophs such as rusts and powdery mildews because they do not require exogenous nutrients to infect the host. On the other hand, a necrotrophic pathogen such as *Botrytis cinerea* is directly affected. Such pathogens require some exogenous nutrients during a definite saprophytic phase prior to infecting the host and are therefore vulnerable to nutrient competition. There are many examples of bacteria and yeasts effectively reducing spore germination or germ tube growth in necrotrophic pathogens by competing for nutrients. For instance, a *Pseudomonas* species inhibited the germination of conidia of *Botrytis cinerea* by competing for amino acids. Pollen stimulates infection of sugar beet leaves by *Phoma betae*. However, pink and white yeasts present on the leaves can reduce this effect, presumably by utilising the nutrient source before the pathogen. Active motility and chemotactic response towards chemical attractants present in root exudates include organic acids, amino acids, and specific sugars govern arrival of bio-control agent to the root surface (Welbaum et al., 2004).

ii. **Active metabolites mediated suppression of pathogens**: Production of active microbial metabolites including antibiotics (2,4 – Diacetylpholoroglucinol, Agrocin 84, Bacillomycin D, Xanthobacin A, Gliotoxin, Zwitermycin A, Mycostubilin, Herbicolin and Iturin), iron-chelating siderophores, biocidal volatiles, lytic enzymes, and toxins play a very significant role in determining the offensive bio-control activity.

iii. **Antibiotics**: Antibiotics can poison or kill other microorganisms at low concentrations. To be effective, antibiotics must be produced in sufficient quantities near the pathogen to result in a bio-control effect (Zerihun Tsegaye et al., 2018). Some examples of antibiotics reported to be involved in plant pathogen suppression include 2, 4-diacetyl phloroglucinol against *Pythium* spp., Agrocin 84 against *Agrobacterium tumefaciens*, Iturin A against *Botrytis cinerea* and *Rhizoctonia solani*, Phenazines against *Gaeumannomyces graminis* var. *tritici*. A native isolate of *P. fluorescens* was noticed producing an antibiotic compound, pyrollnitrin, which inhibited growth of *Macrophomiea phaseolina* by producing an inhibition zone of 12 mm. *Bacillus cereus* strain UW85 is known to produce both zwittermycin and kanosamine suppress plant pathogens. The ability to produce multiple antibiotics probably helps to suppress diverse microbial competitors, some of which are likely to be plant pathogens. The ability to produce multiple classes of antibiotics, that differentially inhibit different pathogens, is likely to enhance biological control. More recently, *Pseudomonas putida* WCS358r strains genetically engineered to produce phenazine and DAPG displayed improved capacities to suppress plant diseases in field-grown wheat.

iv. **Hydrogen Cyanide (HCN) production**: Considerable numbers of free-living rhizospheric bacterial communities, mainly *Pseudomonas* sp., are capable of generating HCN by oxidative decarboxylation from direct precursors such as glycine, glutamate, or methionine (Zerihun Tsegaye et al., 2018). Other rhizobacterial genera reported to produce HCN include Bacillus and Chromo bacterium. HCN secreted by *Pseudomonas* fluorescent strain CHAO has been demonstrated to stimulate root hair formation and suppress back root rot caused by Thielaviopsis basicola in tobacco plant. Cyanogenesis in bacteria accounts in part for the bio-control capacity of the strains that suppress fungal diseases of some economically important plants. For instance, for many Pseudomonades, production of metabolites such as HCN is the primary mechanism in the suppression of root fungal pathogens. HCN effectively blocks the cytochrome oxidase pathway and is highly toxic to all aerobic microorganisms at picomolar concentrations. However, producer microbes, mainly Pseudomonades, are reported to be resistant.

v. **Iron Chelating Siderophores**: Biocontrol agents are able to decrease the availability of particular substance/nutrients to the pathogens because of their efficient uptake or utilizing capacity. This phenomenon of competition for nutrients can limit the growth of pathogens (Zerihun Tsegaye et al., 2018). Iron is an essential growth element for all living organisms and scarcity of its bioavailability in soil habitats and on plant surfaces creates a furious competition. Under iron-limiting conditions, bio-control agents produce low-molecular-weight compounds called siderophores to competitively acquire ferric ion. Kloepper et al. (1980) were the first to demonstrate the importance of siderophore production as a mechanism of biological control of *Erwinia carotovora* by several plant growth promoting *Pseudomonas fluorescens* strains A1, BK1, TL3B1 and B10. Some bacteria, especially fluorescent pseudomonas, produce siderophores that have very high affinities for iron as compared to fungal siderophores and can sequester this limited resource from other micro flora thereby preventing their growth. Earlier reports have demonstrated the importance of *P. fluorescens* siderophores in disease suppression.

vi. **Lytic Enzymes**: Various extracellular hydrolytic enzymes produced by microbes play important role in suppression of plant pathogens. Chitinase and -1, 3-glucanase attack on chitin and -1, 3-glucan, major constituents of many fungal cell walls resulting in its degradation which further kills the pathogens (Zerihun Tsegaye et al., 2018). Chitinase produced by *Serratia plymuthica, Serratia marcescens*, *Paenibacillus* sp. and *Streptomyces* sp. was found to be inhibitory against *Botrytis cinerea, Sclerotium rolfsii, Fusarium oxysporum.* Similarly, laminarinase (beta-glucanase) produced by *Pseudomonas stutzeri* digest and lyse mycelia of F. solani. 1, 3-glucanase synthesized by *Paenibacillus, B. cepacia* destroys *F. oxysporum, R. solani, Sclerotium rolfsii* and *Pythium ultimum* cell walls. Recently, genetic evidence for the role of these enzymes in bio-control has been obtained where ChiA from *Serratia.* marcescens was inserted into the non biocontrol agent *Escherichia coli* and the resulting transgenic bacterium reduced disease incidence of Southern blight of bean caused by *S.rolfsii.* Similarly, transformed *Trichoderma harzianum* with ChiA from *S. marcescens* was more capable of suppressing *S.rolfsii* than the original strain. More recently, a trademark in the history of bio-control was established by generating transgenic plants containing the gene for endochitinase from *T. harzianum* with increased resistance against plant pathogenic fungi. These results indicate that these enzymes play an important role in bio-control and the bio-control ability of some microbes may be improved by transformation with chitinolytic enzymes.

vii. **Root colonization and Protection of infection sites**: Root colonization ability of bio-control agents and potential to survive and proliferate along growing plant roots over a considerable period, in the presence of the indigenous micro flora results in intimate associations that directly provide a selective adaptation to plants towards specific ecological niches (Zerihun Tsegaye et al., 2018). Also, the ability of bio-control agent to colonize specific substrates or sites, whether a seed, root, shoot area, stump or fruit surface, provides protection to infection site from pathogen attack. However, they are effective only when provided with an additional competitive advantage, such as high initial cell numbers, earlier establishment than the pest or pathogen, or the production of antibiotic substances. Therefore, rhizosphere competence is considered as a prerequisite of effective biological control. Understanding root-microbe communication, as affected by genetic and environmental determinants in spatial and temporal contexts, will significantly contribute to improve the efficacy of these bio-control agents. Once bio-control agents establish on the site, the mechanism of antagonism might be competition for nutrients, space, siderophore production, antibiosis, production of hydrolytic enzymes or other active substances.

viii. **Detoxification of Virulence factors**: The detoxification mechanisms involve production of a protein that binds with pathogen toxin reversibly or irreversibly which ultimately results in decrease in the virulence potential of pathogen toxin. For example, role of certain bio-control agents such as Alcaligenes denitrificans and *P. dispersa* in detoxifying albicidin toxin produced by *Xanthomonas albilinears* has been reported previously (Zerihun Tsegaye et al., 2018). Similarly, strains like *B. cepacia* and *Ralstonia solanacearum* hydrolyze fusaric acid, a phytotoxin produced by various *Fusarium* species.

ix. **Parasitism and Predation**: The parasitism of one fungus by another (hyperparasitism or mycoparasitism) is well documented and is manifested as morphological disturbance, direct penetration of hyphae and hyphal lysis. The degree of hyper parasitism is affected by environmental factors such as temperature, light and pH, but nutrient status, especially the C: N ratio is also important (Zerihun Tsegaye et al., 2018). The genus *Trichoderrna* contains some of the most studied mycoparasites. Formulations of some species are available commercially and are used to control fungal pathogens in the soil and on aerial plant surfaces. This mycoparasite penetrates resting structures such as sclerotia or may parasitise growing hyphae by coiling round them. *Trichoderma harzianum* degrades fungal cell walls by the lytic action of glucanases and chitinases, while other species also produce cellulose. *Trtchoderma* species are amenable to genetic manipulation and their usefulness in integrated disease management programs has been enhanced by the production of pesticide-tolerant strains with enhanced hyper parasitic and lytic capabilities. The attributes of the partially successful mycoparasite *Ampelomyces quisqualis*, which can control powdery mildew in glasshouse grown cucumber, are discussed later in this chapter. Other widely studied mycoparasites include

Coniothyrium minitans and *Sporidesmium sclerotiorum,* which are antagonists of sclerotial fungi, and *Gliocladium* spp which parasitize a range of soil-borne pathogens. Mycoparasitic *Pythium* spp appear to have some affinity for plant-parasitic members of the same genus. On the phylloplane, several fungi, including *Verticillium lecanii, Sphaerellopsis filum* and *Cladosporium* sp are known to attack rust fungi. Bacteria which occur on the phylloplane and in the rhizosphere are also known to parasitize plant pathogens.

x. **Induction of Systemic Resistance**: Certain biocontrol agents show indirect mode of antagonism against the pathogens by inducing a state of plant resistance (Zerihun Tsegaye et al., 2018). This induced resistance is of two types representing two distinct pathway responses, systemic acquired resistance (SAR) and induced systemic resistance (ISR). Typically, SAR is induced by pathogens while ISR is salicylic acid in dependent and is induced by non-pathogenic bacteria. SAR is mediated by a compound called salicylic acid which is frequently produced following pathogen infection that leads to expression of pathogenesis related (PR) proteins such as PR-1, PR- 2, chitinases, and some peroxidases. These PR proteins can cause lysis of invading cells, reinforcement of cell membranes to resist infections, or induce localized cell death. In contrary, certain biocontrol agents do not induce PR proteins but increase accumulation of peroxidase, phenylalanine ammonia lyase, phytoalexins, polyphenol oxidase, and/or chalcone synthase. A second pathway referred to as ISR is mediated by jasmonic acid (JA) and/or ethylene, which are produced following applications of some nonpathogenic rhizobacteria. ISR was first observed on carnation (Dianthus caryophillus) with reduced susceptibility to wilt caused by *Fusarium* sp. and on cucumber with reduced susceptibility to foliar disease caused by *Colletotrichum orbiculare*. Several bacterial traits (i.e., flagellation and production of siderophores and lipopolysaccharides) have been proposed to trigger ISR but there is no compelling evidence for an overall ISR signal produced by bacteria. ISR results in strengthening of plant cell wall and alteration of host plant physiology and metabolic responses, leading to an enhanced synthesis of plant defence chemicals upon challenge by pathogens and/or abiotic stress factors. It has recently been reported that volatile organic compounds may play a key role in this process. For example, volatiles secreted by *B. subtilis* GBO3 and *B. amyloquefaciens* IN937a were able to activate an ISR pathway in *Arabidopsis* seedlings challenged with the soft-rot pathogen *Erwinia carotovora* subsp. *carotovora* (Ryan et al., 2001).

8.4 Soil and Phytomicrobiome for Plant Disease Suppression and Management

The phytomicrobiome plays a crucial role in soil and ecosystem health, encompassing both beneficial members providing critical ecosystem goods and services and pathogens threatening food safety and security. The potential benefits of harnessing the power of the phytomicrobiome for plant disease suppression and management are indisputable and of interest in agriculture but also in forestry and landscaping. Indeed, plant diseases can be mitigated by in situ manipulations of resident microorganisms through agronomic practices (such as minimum tillage, crop rotation, cover cropping, organic mulching, etc.) as well as by applying microbial inoculants. However, numerous challenges, such as the lack of standardized methods for microbiome analysis and the difficulty in translating research findings into practical applications are at stake. Moreover, climate change is affecting the distribution, abundance, and virulence of many plant pathogens, while also altering the phytomicrobiome functioning, further compounding disease management strategies. Chen et al.,(2023) reviewed literature demonstrating how agricultural practices have been found effective in promoting soil health and enhancing disease suppressiveness and mitigation through a shift of the phytomicrobiome. Challenges and barriers to the identification and use of the phytomicrobiome for plant disease management will then be discussed before focusing on the potential impacts of climate change on the phytomicrobiome functioning and disease outcome.

The original definition, proposed by Whipps et al.(1988), referred to "microbiome" as "the microorganisms involved but also encompasses their theatre of activity", which has been considered the most comprehensive definition, according to an online survey and workshop discussions and was further detailed (Berg, et al., 2020). While "microbiota" is usually defined as the assemblage of living microorganisms in a specified habitat, the word "microbiome' also includes a wide array of metabolites (such as signaling molecules, toxins, etc.) and microbial structural elements, including proteins and their building blocks-peptides, lipids, polysaccharides, nucleic acids, or mobile genetic elements. In line with this definition, the phytomicrobiome can thus be defined as the microorganisms associated with any internal or external component of a plant, from the aboveground to the belowground, as well as the results of their activity. From an anthropogenic viewpoint, the phytomicrobiome includes beneficial members providing critical ecosystem goods and services and pathogens threatening food safety and security.

In terms of applications, two broad phytomicrobiome-based approaches that may offer alternative strategies for plant disease suppression and management can be defined, depending on the necessity to cultivate microorganisms before application in the field (Chen et al., 2023). Indeed, microorganisms can be directly introduced as inoculants, either alone or as a mixture, to the ecosystems. Globally, the agricultural biocontrol market, including microorganism-based products, is valued at ~USD 6.6 billion with 39.5% attributable to North America. It was also predicted to reach USD 13.7 billion by 2027, based on a report published in 2022. Alternatively, plant diseases may also be mitigated by in situ manipulations of resident microorganisms through agronomic practices without the need for industrial-scale production of microorganisms, such as those targeted in "conservation agriculture". It relies on three principles, (i) minimal soil disturbance or absence of tillage, (ii) continuous soil cover with crops, cover crops, or a mulch of crop residues, and (iii) crop rotation (FAO., 2023). While conservation agriculture primarily focuses on farming practices and not on breeding, it is worth noting that microbiome-assisted plant breeding, which involves selecting plants based on their enhanced capability to recruit beneficial microorganisms, can be considered within the second category from a broader perspective. This has notably become possible thanks to the identification of specific genetic loci controlling such traits (Morales Moreira, et al., 2023). While soil amendments with compost, biochar, or even manure are primarily used for fertilization, the modified soil microbiota may indirectly impact plant protection against pests and diseases. In addition, at the frontier of these two approaches, the use of rhizo-microbiome transplants (RMT), although still in its infant stages, has recently shown promising results. In this study, transplanting the rhizobiome of eggplant-resistant donors to tomato-susceptible recipients under greenhouse conditions resulted in a reduction of bacterial wilt disease incidence, with reductions of up to nearly 50%. It is crucial to consider the distinction between these two broad approaches based on the phytomicrobiome since their implications in terms of challenges and constraints may differ.

The phytomicrobiome is an essential component of soil and ecosystem health, and its potential to mitigate plant diseases and improve resilience to climate change is promising. However, identifying and utilizing the phytomicrobiome for disease management presents numerous challenges. Moreover, as climate change continues to affect global temperatures and precipitation patterns, the phytomicrobiome's functioning and its interactions with plant pathogens may be altered, further complicating disease management strategies. Nevertheless, the potential benefits of harnessing the power of the phytomicrobiome are significant, not just for agriculture but also for other fields such as forestry and urban landscaping. Further research is needed to fully understand the mechanisms underlying the phytomicrobiome's effects on plant disease and to develop practical management strategies that can be implemented on a larger scale.

8.5 Major Bioagents

8.5.1 *Trichoderma* spp.: The root rot of tomato, in particular, is commonly caused by soil-borne fungal pathogens such as *Rhizoctonia solani*, *Fusarium solani*, *Phytophora* species and *Sclerotium rolfsii* that delays growth, reduce harvest yield and quality, and subsequent death of severely infected plants. *Trichoderma* spp. are among the front-line microorganisms commonly employed in novel biotechnology applications. They have been well-proven as biopesticides, biofertilizers, and biostimulants for managing plants against biotic and abiotic stresses (Olumayowa Mary Olowe et al., 2022). They are instrumental in managing plant diseases of economic importance, such as tomato root rot. However, this group of fungi has not been well-exploited en-mass in developing countries, while the use of bioagents in-lieu of chemical pesticides is still not a common practice in many African countries. Africa contributes 11.8% to global tomato production. Unfortunately, more than half of the actual product is lost due to diseases. The root rot of tomatoes predominantly caused by soil-borne fungal pathogens are among significant problems of tomato cultivation in Africa. Olumayowa Mary Olowe et al., (2022) reviewed the constraints of tomato root rot in Africa and the roles of *Trichoderma* in repositioning the crop for optimum productivity. A comprehensive overview was given of the economic importance, root rot epidemiology, and how to circumvent it through gene pool to resistant tomato and employ Trichoderma's biological control potentials. Furthermore, this review gives an overview of the mechanisms of action of *Trichoderma*, gaps in the advocacy, adoption, commercialization, and regulation of *Trichoderma* as biocontrol agents of tomato rot diseases in Africa.

Mycoparasitic attributes of many *Trichoderma* strains have been established and have been well explored for the biological control of various plant pathogens. Many of such strains are now packaged under different trade names for the usage of farmers. *T. stromaticum* is a notable example patented as 'Tricovab' for the control of fungus *Moniliophthora perniciosa* causing the witches' broom disease of cacao which is the major problem of cacao in Neotropics. This product has been well distributed to farmers in southern Bahia in Brazil for many years and has now been employed as part of the integrated pest management strategy for controlling cacao pests. Another prominent example is *T. harzianum* which can colonize and degrade sclerotia of many fungal pathogens, this fungus has been widely distributed as bio-fungicide. Since *Trichoderma* can grow in different habitats, these abilities have made them very popular among the science community. As a result, several works have been published on the importance of *Trichoderma* in agriculture and plant disease control. Their mode of action against plant pathogens is shown in fig.48. Mechanisms of *Trichoderma* induced resistance and signaling pathways in plants against plant diseases has been illustrated in fig.49.

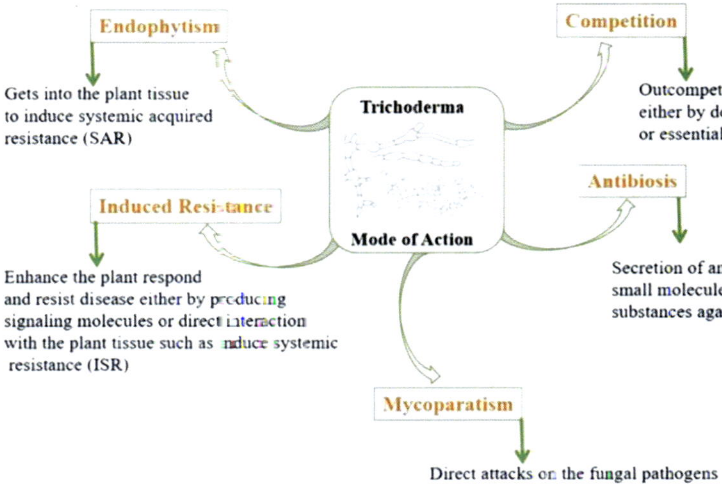

Fig. 48: Trichoderma modes of action as bioagent

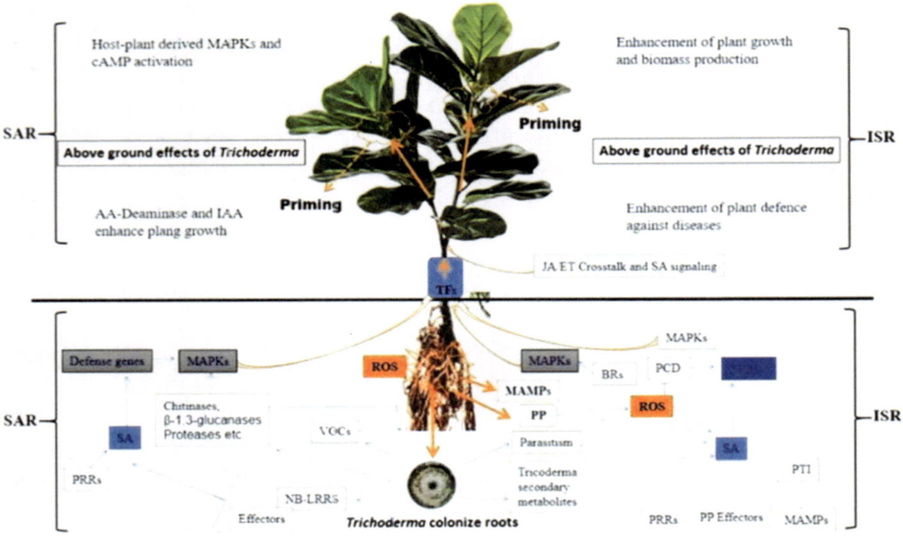

Fig. 49: Mechanisms of *Trichoderma* induced resistance and signaling pathways in plants against plant diseases. The Plant pathogens (PP) like bacteria, nematodes, viruses or fungi can secret various enzymes to degrade the plant's cell wall, and insight infection or develop feeding sites for their growth and multiplication. Some PP can also produce some effector proteins to suppress the plant triggered immunity (PTI). Trichoderma can inhibit PP by producing different kinds of toxic metabolites or directly parasitize on tissue, egg, hyphae, or cell of PP, they can as well produce some elicitor proteins or chemical signals like the salicylic acid (SA) or jasmonic acid (JA) to either enhance the plant's defense mechanism against the PP a process called The induce systemic resistance (ISR). Some PP-like nematodes can produce a high level of reactive oxygen species (ROS) to avoid cell death inside living cells or tissue, *Trichoderma* can help plants by triggering cell death around parts already colonized by the PP. In addition, another biocontrol mechanism of Trichoderma on PP is by induction of systemic acquired resistance (SAR) in plants. This they do by their interference with or triggering in plants, the production of the microbial-associated molecular pattern (MAMP), and volatile organic compounds (VOCs) which can also induce indole acetic acid (IAA) and 1-aminocyclopropane-1-carboxylate-deaminase (ACC-deaminase) secretion by plants. Others include the cell wall degrading enzymes (CWDEs) such as Chitinases, β-1,3 glucanases, and proteases, and induce mitogen-activated protein kinases (MAPKs). The cross-talk among the elements of SAR triggered pathogens and ISR signaling by *Trichoderma* can activate the plant defense response pathways of cyclo adenosine monophosphate (cAMP) and MAPKs from uninfected plant's part to the infected parts.

***Trichoderma* Advocacy, Commercialization and Regulations:** Biocontrol activity involves a variety of biological control agents such as bacteria and fungi, among which the fungal genus, *Trichoderma*, is significant in the management of plant diseases (Olumayowa Mary Olowe et al., 2022). This fungus produces a variety of enzymes that aid in the biocontrol activities such as biotic and abiotic stress tolerance, cell wall disintegration, hyphal development, and antagonistic action against plant diseases. Trichoderma species are pervasive and frequently dominating components of the mycoflora in natural and agricultural soils across all temperate zones. They are resident in both the terrestrial and marine environment and have adapted to almost every ecology, and contribute significantly to the health status of the ecosystem. Species of Trichoderma colonize the plant endosphere and rhizosphere and have also been recovered from soil organic matter, plant litter, and human tissues.

The capacity of *Trichoderma* fungi to function as biocontrol agents against plant diseases has been recognized as early as the 1920s. The ability to detect, invade, and destroy other fungi are the reason for its economic success and commercial application, as *Trichoderma*-based biopesticides constitute over 60% of all registered biopesticides. Apart from providing resistance against plant diseases, these fungi also increase plant growth and vigor, solubilize plant nutrients, enhance nutrient absorption in plants, and bioremediate heavy metals and environmental contaminants while producing clinically important secondary metabolites and enzymes to both human and plant safety (Sood et al., 2020). *Trichoderma* species are often employed as biological control agents against phytopathogenic microbes, including those which cause the tomato root rotting disease that is of interest in this study. They act through mechanisms such as antibiosis, mycoparasitism, and competition for space and resources, while their impacts on plants such as induced systemic or localized resistance are also of importance in the biological management processes. *Trichoderma* spp. are not only successful plant symbiotic organisms, but they also act as a low-cost, effective, and environmentally benign biocontrol agent. They may establish themselves in a variety of pathosystems, have a low influence on soil balance, and do not harm beneficial species that aid in pathogen management.

Trichoderma products are promoted as biofungicides, biopesticides, bioprotectants, bio-stimulants, bio-inoculants, bio-fertilizers, bio-decomposers, plant growth promoters, and so on. Some *Trichoderma* isolates such as *T. polysporum, T. harzianum, T. asperellum, T. atroviride, T. viride T. polysporum* are particularly known for their broad-spectrum biofungicidal potentials against a range of foliar, vascular, and soilborne diseases caused by *Rhizoctonia solani,* Botrytis cinerea, *Sclerotium* spp., *Sclerotinia sclerotiorum*, *Pythium ultimum*, *Armillaria* spp., *Phytophthora* spp., *Verticillium* spp., *Fusarium oxysporum* and *Gaeumannomyces graminis*. These properties broaden the range of potential applications for *Trichoderma*-based products in the agricultural market, not only as a phytopathogen biofungicide, but also as a general bio-inoculant that boosts plant resistance to biotic and abiotic stresses, improves plant growth and yield, and enhances the agro-ecosystem.

Despite its efficacy, *Trichoderma* as a biocontrol agent has its limitations. Although *Trichoderma* successfully colonizes the spermosphere, it does not thrive in the rhizosphere. Also, *Trichoderma* spp. are active only in certain types of soil and seasons, therefore, achieving only transitory localized dominance of the rhizosphere. Moreover, the inactive and dormant nature of *Trichoderma* spores in soil suggests that *Trichoderma* cannot be supplied as spores (Aamir et al., 2020). In addition, various mechanisms have been uncovered in vitro, but the validity of such has not been successfully validated in the field of biocontrol systems. More so, biological control agents are fragile and could be denatured in hostile environmental conditions. Some of the measures to curtail these limitations are through adequate formulation. Hence the development of cost-effective, stable, and easy to apply biological control formulations is essential for the enhanced efficiency of the deployed biocontrol formulations. The *Trichoderma*-based formulation in pelletized form is viable for over three months (Cuevas et al., 2001). It has been shown as effective against seedling diseases of vegetables caused by *S. rolfsii* when applied in pelletized form (Ram et al., 2020). In addition, *Trichoderma* formulations had also been commercialized as a wettable powder (WP), granular, powder preparation, liquid, and solids products. In contrast, others are emulsions, dry flowable, concentrated liquid suspensions, and talc, among others (Woo et al., 2014).

The European Commission and other agencies around the world have indicated that about 38 *Trichoderma* commercial products (single species and combination mixtures) are registered or in the process of being registered in various countries, while four products have received special registration or temporary approval only for use in field experiments (Topolovec-Pintaric, 2019). Furthermore, the species of *Trichoderma* mostly present on the label of the commercial products are *T. atroviride, T. asperellum, T. hamatum, T. gamsii, T. koningii, T. harzianum, T. lignorum, T. polysporum, T. parceanamosum, T. viride T. virens and Trichoderma* spp. However, following the trend in the usage of several *Trichoderma* spp., nearly half of these new products contain *T. harzianum*. More than 250 *Trichoderma*-based products have recently been registered and identified on the international market (Ram et al., 2020). Yet, there occurs a huge disparity in the actual number of biological control agents in use compared to those registered. However, out of the *Trichoderma*-based registered products, Asia made the most significant contributions of 39%, and Africa, with nine registered products, had the least contributions of 3.5% among the regions of the world (Fig. 50). On the other hand, South Africa, Kenya, Zambia, Morocco, and Tunisia had registered and produced at least a *Trichoderma-based* product among the African nations. While *Trichoderma* poses enormous promise in Africa, as evidenced in many scientific investigations into the genus conducted in the region, most African scientists and entrepreneurs seem not to be aware of the full potential of this product, especially in solving the peculiar cropping problem in Africa, as in the case of tomato farming. Hence, the smallholder farmers that constitute the bulk of agricultural productions in most African countries are stocked with chemical fertilizer in the face of non-viable and available organic alternatives.

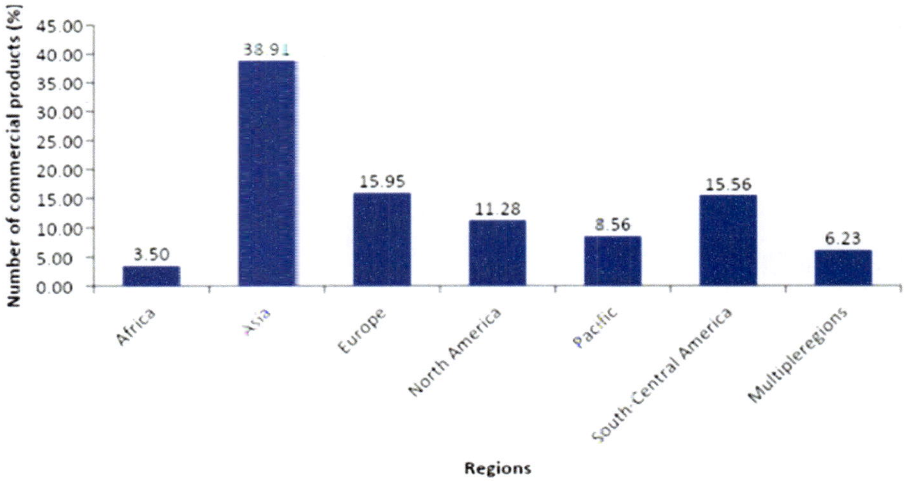

Fig. 50: Regional contribution to the registered *Trichoderma*-based biological control formulations

Governing authorities across the globe, notably in Europe, are currently enacting legislation to reduce agriculture's reliance on pesticides to improve consumer and environmental safety (Olumayowa Mary Olowe et al., 2022). Directives will promote low pesticide input by implementing integrated pest management (IPM) and will provide the means to establish the necessary conditions and measures to employ these practices, as well as to ensure the security of commercial products, to reduce the risks associated with pesticide applications and reduce dependency on their use. As an alternative to synthetic agrochemicals, one strategy is to employ biological control agents and their products (Masip et al., 2013). Hence, *Trichoderma* species are qualified candidates, having been extensively researched and one of the most widely employed microbial biological control agents (MCBAs) in Agriculture, and is presently marketed as biofertilizers, growth promoters, biopesticides, and natural resistance stimulants.

Genetic engineering has also been used to enhance the biocontrol abilities of *Trichoderma* fungi. In a study, *T. virens* strains that were transformed with *TvBgn2* and *TvBgn3,* which are genes that encode β-1,3 and β-1,6 glucanase, efficiently inhibited the growth of *R. solani, Rhizopus oryzae,* and *Pythium ultimum,* using pGEM-T vector, successfully cloned an antifungal glucan1,3-β-glucosidase gene called *gluc78* into *T. atroviridae* cells for effective biocontrol of several plant pathogens (Olumayowa Mary Olowe et al., 2022). This gene was in another study carried out was also cloned a glucose repressor gene from *T. harzianum* and cloned it into *E. coli* for expression of cellulase and xylanase from pathogens using pTZ57R/T plasmid vector. Several other works that have reported various genes in *Trichoderma* for the control of plant pathogens are presented in table 15.

Table 15: Biocontrol genes that have been isolated or expressed in *Trichoderma* strains and their functions

S/N	Biocontrol gene	*Trichoderma* strain	Function
1	*tga1* gene	*T. atroviride* strain P1ATCC 74058	Chitinase formation and production of antifungal metabolites
2	tvsp1 serine protease encoding gene	*T. virens* wildtype strain Gv29–8 and an arginine auxotrophic strain, Tv10.4	Involved in pathogenesis or biocontrol process of *R. solani.*
3	1,3- β-glucosidase gene, *gluc78*	*T. atroviride* strain P1(ATCC 74058)	Cell wall degradation of pathogens *Pythium* and *Phytophthora*
4	Beta tubulin gene	*T. harzianum* T88	Exhibits biocontrol activities like mycoparasitism, and antifungal activity.
5	Chitinase chit42 and proteinase prb1 gene	*T. hamatum* strain LU593	Mycoparasitic activity against *Sclerotinia sclerotiorum*
6	CRE1 gene	*T. harzianum* E58	Production of cellulase and hemicellulase enzymes.
7	Di/tri peptide transporter gene, *ThPTR2*	*T.harzianum* CECT 2413	Involves in mycoparasitic activity against *B. cinerea*
8	Endochitinase encoding gene, *ech42*	*T. harzianum* strain P1 74058	T*richoderma* expressing this gene show efficient inhibition of *Botrytis cinerea* and *R. solani*
9	Endopolygalacturonase ThPG1 gene	*T. harzianum* T34 CECT 2413	Secretion of plant cell wall degrading enzymes against *R. solani* and *P. ultimum*
10	Endo-β-1–6-glactanase gene	*T. viride* IFO31137	A type of arabinogalactan proteins that involves in cell-cell adhesion, expansion and cell death.
11	Enzymology Group collection,UFG-ICB, tag 3 gene	*T. asperellum*	Production of cell wall degrading enzyme glucanase
12	erg1 gene	*T.harzianum* CECT 2413	Silencing of the erg1 gene enhances resistance to terbinafine that shows antifungal activity
13	Genes encoding for NAGases (exc1 and exc2), chitinases (chit42 and chit33), proteases (prb1) and b-glucanases (bgn13.1)	*T. harzianum* CECT 2413	Mycoparasitic activity against *F. oxysporum*

S/N	Biocontrol gene	*Trichoderma* strain	Function
14	Mitogen Activated Protein kinase gene, *TmkA*	*T. virens* strain IMI 304061	Biocontrol activity against pathogens like *S. rolfsii* and *R. solani*
15	Monooxygenase gene	T. hamatum LU593	Antagonist activity against *S. sclerotiorum, S. minor* and *S. cepivorum*
16	Proteinase gene *prb1* and endochitinase gene ech42	*T harzianum* strain IMI206040	Regulation of hyrodrolytic enzymes to enhance parasitism of *Trichoderma* against *Sclerotium rolfsii* and *Rhizoctonia solani*
17	qid74 gene	*T harzianum* Rifai CECT 2413	Involved in cell protection and adherence to hydrophobic surfaces that helps in antagonism against *R.solani*
18	Serine protease gene SL41	*T harzianum*	Biocontrol activity against pathogens
19	Sm1 gene, cysteine-rich protein	*T virens* strain TvSMOE38	A small cysteine-rich protein that induces defense responses in dicot and monocot plants and in protecting crop diseases
20	T34 hsp70 gene	*T harzianum* CECT 2413	Increases fungal resistance to heat and abiotic stresses
21	Taabc2 gene	*T atroviride* P1 (ATCC 74058)	Plays key role in antagonism against *R.solani, P.ultimum*, and *B.cinerea*
22	*tac1*, adenylate cyclase gene	*T virens* IMI 304061	Mycoparasitism against *R. solani, S.rolfsii, Pythium* Spp. and production of secondary metabolism
23	TgaA, TgaB genes	*T virens* strain IMI 304061	Antagonism against *S. rolfsii* and *R. solani*
24	Th-Chit gene	*T harzianum*	Antifungal activity in transgenic tobacco
25	Thctf1 transcription factor gene	*T harzianum* CECT 2413	Antifungal activity against *R.solani*, *Fussarium oxysporum* and *B.cinerea* and production of 6-pentyl-2 H-pyran-2
26	Thke11 gene	*T harzianum* CECT 2413	Expression of this gene in *A. thaliana* modulates glucosidase activity, and enhances tolerance to salt and osmotic stresses

S/N	Biocontrol gene	*Trichoderma* strain	Function
27	tmkA gene	*T. virens* IMI 304061	Induction of plant systemic resistance and biocontrol activity against *R. solani.* (Tested in green house condition)
28	TrCCD1 gene	*T. reesei* QM9414 (ATCC 26921)	Helps in hyphal growth, conidiospore development and carotenoid pigment production.
29	tri5 gene	*T. brevicompactum* IBT40841	Production of trichodermin and antifungal activity against *C. albicans*, *C. glabrata* and *A. fumigatus*
30	Trichodiene synthase gene, *tri5*	*T. harzianum* strain ATCC 90237	Production of toxin for inhibiting DNA or protein synthesis, and enhances virulence against *Fusarium* spp.
31	TvBgn2 and TvBgn3 genes	*T. virens* Gv29–8	These genes help in encoding cell wall degrading enzymes
32	TvGST glutathione transferase gene	*T. virens* (strain IMI 304061)	Enhances cadmium tolerance in plants
33	Xylanase gene, *Xyl*	Trichoderma strain SY	Helps in breakdown of hemicellulose
34	β-1,4-endoglucanase gene, *egl1*	*T. longibrachiatum*strain CECT2606	Cell wall degrading enzyme for inhibition against *Pythium ultimum* on cucumber.

However, there is a need for adequate knowledge of the biocontrol mechanism of a specific species before its employment as a commercial product, although, reports from some earlier studies had recognized *Trichoderma harzianum* as the most effective biocontrol agent among the *Trichoderma* species.

Future Perspectives

New contributions of new valuable *Trichoderma* species are being made yearly thanks to molecular or genome analysis. Valuable genes, related to biological control potentials, their defense induction in plants, or stress tolerance, have been identified using these methods. These genes can hence be enhanced using the advancing biotechnological/ genetic tools. Their example of the gene that controls complex mechanisms used by *Trichoderma* against pathogens, for example, *T. atrobrunneum* (Olumayowa Mary Olowe et al., 2022). In other studies, it was revealed that the bioactivity of secondary metabolites produced by some *Trichoderma* strains helps attack plant pathogens, and the genes that control the secretion of these metabolites were cloned in bacterial cells. Likewise, the metabolites of *T. viridae* and *T. pseudoharzianum* (T113) were toxic against the eggs of *M. incognita*. More than twenty species of *Trichoderma* were reported to be able to secrete about 390 important non-volatile nematicidal metabolites.

In tomato plants, *T. harzianum* T22 and *T. asperellum* T-34 strains were employed for systemic resistance induction against *M. incognita*. This was reflected in the over-expression of *Mi-1.2* gene in the inoculated tomato plants (Pocurull et al., 2020).

Trichoderma genus is already a well-known group of fungi with known importance and applications for sustainable agriculture. It is one of the most studied groups of fungi for use as plant disease control. They can evolve quickly with diversified metabolic pathways leading to the production of several bioactive compounds, enzymes, and secondary metabolites that can, in addition, be explored for human benefits. They have vast importance with potential value commercially and industrially apart from their importance in agriculture. Their ability to produce commercially important enzymes like amylases cellulases, glucanases, and chitinases are reported every year with interesting results. Recently, they have been associated with the production of siderophores and silver nanoparticles. More than 200 species have been reported as biopesticides, while many have been well packaged commercially for use as biofungicides. The mass production of *Trichoderma* is of interest due to its valuable application in the biological control of plant disease pathogens. Commercial production of these bioagents has now come into reality, and several formulations have been reported. The potential of *Trichoderma* as a tool for managing the plant's pest is now widely recognized globally. However, this approach has virtually not been well implemented in the management of many tropical crops, especially in Africa. Most of the studies that investigated and reported the diversity of endophytic *Trichoderma* strains on different plant hosts were based on surveys restricted to the United States, Hawaii, and other countries other than the Africas.

However, Africa is rich in biodiversity of different plants, which could also be rich in novel species of *Trichoderma*. For example, del Carmen et al. (2021) investigated the endophytic fungi associated with coffee plants observing that the fungi from America and Hawaii were dominated by Colletotrichum, *Fusarium*, Penicillium, *Pestalotia* and *Xylaria* species which are in most cases existing as opportunistic endophytes of their host plants. The study from Africa in contrast, yielded more of *Trichoderma* as endophytic fungi; in detail, they reported endophytic *Trichoderma* from the aerial tissues of coffee plants, including four new species. The new taxa described by these authors may be geographically restricted to Africa or even to coffee plants. The Africans are rich in diversity of *Trichoderma*, especially in the Neotropics. Some studies have reported a close association between *Trichoderma* species and *C. arabia* from Ethiopia. These studies focused majorly on the rhizosphere and root tissue species. They reported several species of *Trichoderma* such as *T. asperelloides, T. atroviride, T. gamsii, T. hamatum, T. koningiopsis, T. longibrachiatum, T. spirale, and T. sensulato. T. hamatum, T. spirale,* and *T. koningiopsis* are cosmopolitan species commonly associated with stems, leaves, berries from tropical habitats and the soil. Some species, such as *T. koningiopsis* and *T. spirale* have been isolated from more than one plant tissue from the leaf and stem of *C. canephora*, *C. arabica,* and tomatoes. Other reports have also identified *T. koningiopsis, T. hamatum*, and *T. spirale* as endophytes of tropical woody plants, notably cacao and rubber. *T. atroviride, T. guizhouense, T. breve,* and *T. theobromicola* are very common in the tropical soils, mushroom compost, in leaf-cutting ant colonies decaying wood and bark, they have also been reported as endophytes of *Theobroma* sp. (Malvaceae) (Olumayowa Mary Olowe et al., 2022).

The genetic pool of major bioagents in this group are those from the above-ground habitat (López-Quintero et al., 2013). There is limited information about the ecology of new strains. *Trichoderma* species from Africa in general coupled with their potential application. The shelf-life of these bioagents under different formulations is a significant factor affecting their commercial distribution. *T. harzianum* and *T. viride* are well-known bioagents in different formulations. They have been grown on solid substrates like wheat straw, wheat bran, coffee husk, sorghum grains, and much other biomass to enhance their life span. Fermented molasses and brewer›s yeast were suggested to prepare *Trichoderma* formulations (del Carmen et al., 2021) while some organic matter farmyard manure, as well as seasoned press, have also been suggested.

Trichoderma spp.

Characteristics: *Trichoderma* is a genus of asexually reproducing fungi that are often the most frequently isolated soil fungi; nearly all temperate and tropical soils contain 101–103 cultivable propagules per gram. These fungi also colonize woody and herbaceous plant materials, in which the teleomorph (genus *Hypocrea*) has most often been found (Ilaria Pertot et al., 2020). However, many strains, including most biocontrol strains, have no known sexual stage. They show a high level of genetic diversity, and can be used to produce a wide range of products of commercial and ecological interest. Several strains belonging to the genus *Trichoderma* have been identified so far as biocontrol agents of plant diseases and nematodes and a few of them have been developed and registered as biofungicides. Several species have been found to be very effective as biocontrol agents, namely *T. atroviride, T. asperellum, T. harzianum, T. viride, T. gamsii, T. polysporum*, etc.

Mechanism of action: *Trichoderma* strains may have one or all mechanisms of action according to species and strain. In some strains prevails the direct, hyper-parasitic activity against pathogens, in others the induction of resistance mechanism prevails or they compete with the pathogens for space and nutrients (Ilaria Pertot et al., 2020). Trichoderma strains produce a great variety of lytic enzymes most of which play a great role in biocontrol (cell wall degrading enzymes, CWDEs). CWDEs from different Trichoderma strains showed antifungal activity towards a broad spectrum of fungal pathogens (i.e. species of *Rhizoctonia, Fusarium, Alternaria, Ustilago, Venturia* and *Colletotrichum*, as well as funguslike organisms such as the Oomycetes *Pythium* and *Phytophthora* which lack chitin in their cell walls). *Trichoderma* produces also many secondary metabolites, some of them specifically involved in the direct antibiosis, like i) volatile antibiotics, i.e. 6-pentyl-a-pyrone (6PP) and most of the isocyanide derivates; ii) water-soluble compounds, i.e. heptelidic acid or koningic acid; iii) peptaibols, which are linear oligopeptides of 12–22 amino acids rich in a-aminoisobutyric acid, N-acetylated at the N-terminus and containing an amino alcohol (Pheol or Trpol) at the Cterminus. The production of lytic enzymes and antibiotic metabolites may greatly vary among strains. Each strain shows specific characteristics and adaptation to the environment. Some strains are very good colonizers of soil other prefer wood material, others are found on dead plant tissue in the phyllosphere or colonizing the rhizosphere of plants and the root tissues. Several strains can also promote growth.

Use: Being soil inhabitants and good colonizers of organic matter the main use of *Trichoderma* strains is against soil-borne pathogens, wood disease and grey mould (*Botrytis cinerea*). In particular some strains represent the only possible solution against wood diseases of grape (no chemical substance have been shown to be effective so far), in fact they can colonize the pruning wounds on the plant for several months thus preventing the entrance of the pathogens. In infested soils they can significantly reduce several soil-borne pathogens and thus the related diseases. On grape, strawberry, tomatoes and other crops they can colonise the dead plant material as leaf residues, dead leaves, etc. preventing the establishment of *B. cinerea* on those materials and thus reducing the disease. Some strains have been used to induce resistance; however this mechanism seems to be very limited in term of efficacy (Ilaria Pertot et al., 2020).

Advantages, limitations and way to improve efficacy: Trichoderma can offer several advantages over synthetic chemical fumigants: they are not toxic for mammals and environment, they do not leave toxic residues, they have no adverse effect on soil microflora and microfauna and they do not negatively affect the natural biocontrol population. To be effective the *Trichoderma* strains should be carefully chosen according to their mechanism of action and adaptation to the specific environment (Ilaria Pertot et al., 2020). The intrinsic capability of the strain of producing lytic enzymes and secondary antimicrobial metabolites can make the difference. The use of poorly competitive strains or not adapted to soil characteristics may also impede the efficacy. Most of the failures in term of efficacy can be related to the use of product with poor viability of conidia (this often happens when *Trichoderma*-based fertilizers are applied), insufficient concentration in the soil, short time for activity (Trichodermas need a certain amount of time to kill the pathogen), too high disease pressure. In few weeks after its application the concentration of exogenous Trichoderma tends to decrease to the basal level of that specific soil. There are several possible way to improve its efficacy as: the increase of the applied quantity per unit of soil, which is commonly limited by the cost of the commercial product, the use of carriers or substrates which can guarantee a longer persistency (i.e. cellulose, or chitin rich substrates), to give sufficient time to *Trichoderma* to reduce inoculum before planting, to apply when the temperature of the soil is sufficiently high, to combine with other practices, for example as bioinoculum after solarisation or anaerobic soil disinfestation or biofumigation.

8.5.2 Arbuscular Mycorrhizal Fungi: The environmentally safe solutions have been expanded in recent years and researchers are working on the application of arbuscular mycorrhizal fungi (AMF) (belonging to subdivision Glomeromycotina) as a disease control agent for the crop plants. AMF are obligate biotrophs and form a mutualistic-symbioses with many crops plant roots. As a result of this relationship, plants take an advantage on the improvement of plant nutrient uptake and confer a bio-protective effect against soil-borne pathogens via direct and indirect mechanisms (Ruwanthika Kalamulla et al., 2022). However, the bio-protective effect of AMF symbiosis in rice is poorly studied because rice is primarily cultivated in wetland ecosystems and form lower percentage of colonization due to anaerobic conditions.

AMF are soil-borne fungi that form mutualistic symbiosis with plant roots, and can signifi cantly improve plant nutrient uptake and increase resistance to several biotic and abiotic stresses. Though most rice cultivation systems involved submerged growing conditions, researchers are exploring the potential of AMF colonization and its benefits for rice. AMF has a potential of serving as a biocontrol agent in rice plants via direct mechanisms viz. nutrient improvement, competition, integration, morphological, biochemical, and physiological alteration and indirectly via induction of systemic induced resistance (ISR) in plants. Furthermore, other biotic factors (host genotype, age of the plant, species of AMF that is involved, degree of mycorrhization, presence of weed plants) and abiotic factors (soil temperature, moisture, nutrient content, and atmospheric gas composition, crop management systems) can be affected the process of biocontrol by AMF. These different mechanisms cannot be considered completely independent from one another. The potential for biocontrol by AMF depends mainly on all these mechanisms in conjunction with all of the other biotic and abiotic factors that affect mycorrhization. By considering such mechanisms, the suitability of the application of AMF as a pathogen biocontrol agent in rice and the other crop plants, while reducing the application of chemical agents to control the diseases, was discussed. However, for considering AMF as a better alternative for the synthetic agrochemicals, more research should be conducted by applying AMF and synthetic chemicals to compare the levels of pathogen control.

In plants, AMF exhibit antagonistic behavior to soil-borne pathogens and act to prime pathogen resistance. It was reported that the effective biocontrol mechanisms of AMF required by the plants before the invasion of pathogens. an overview was presented of the direct mechanisms viz. nutrient improvement, competition, integration, morphological, biochemical, and physiological alteration, and induction of systemic induced resistance (ISR) as an indirect mechanism, which have been suggested to play a role in AMF-mediated biocontrol in rice plants. The biocontrol effect of AMF can be resulted with a combination of different mechanisms under the influence of several biotic and abiotic factors. The biotic factors are host genotype, age of the plant, species of AMF that is involved, degree of mycorrhization, and the presence of weed plants. Soil temperature, moisture, nutrient content, and atmospheric gas composition and crop management systems are considered as abiotic factors (Schouteden et al., 2015).

Mechanisms of AMF-mediated biocontrol of pathogens: The conventional pathogens in crops like rice depends on the use of agrochemicals. The application of these synthesized chemicals can cause adverse effects on human health and the overall environment (Ruwanthika Kalamulla et al., 2022). AMF facilitate plant resistance towards pathogens via improvement of plant nutrition, activation of plant defense mechanisms, altering of the rhizosphere microbial population, the morphological, biochemical, and physiological alteration of the plant rhizosphere and through direct competition with the pathogen in colonization. In contrary, it was reported that AMF can trigger the susceptibility of the plant towards the pathogen, called as mycorrhizal-induced susceptibility. Colonization of AMF in the Italian rice variety named Maratelli increased the blast susceptibility of the plant to Magnaporthe oryzae. Several studies have shown that pathogenic viral multiplication is higher in AMF colonized plants that belong to family Poaceae. When plants interact with AMF, the rice roots decrease their resistance by increasing of the infection area of roots for the pathogen Rhizoctonia solani that causes sheath blight.

i. **Defense through the activation of Induced Systemic Resistance**: Induced systemic resistance (ISR) is a resistance mechanism in the plant that is activated by virulence, avirulence, non-pathogenic microorganisms and alteration of environmental conditions, or chemicals. Researchers have suggested that AMF induce ISR in plants as a mechanism of biocontrol. Activation of ISR by AMF is difficult to identify because of the root localization of both the pathogens and AMF. By using the theory of the split-root experimental system, scientists have observed reduction of disease symptoms in the non-mycorrhizal part of the mycorrhizal root system (Schouteden et al., 2015). This phenomenon is called "mycorrhizal autoregulation", which means that once a plant is subjected to mycorrhization, it suppresses further colonization of roots by AMF or any other pathogenic fungi, in order to limit the energy cost to the plant. Scientists have also proposed the mechanism for activation of ISR by AMF. Initially, plants identify the beneficial microorganisms as putative pathogens using microbe-associated molecular patterns (MAMPs), because MAMPs are conserved between pathogenic and beneficial fungi. After recognition of MAMPs by pattern recognition receptors (PRR) in plants, an MAMP-triggered immunity response (MTI) is activated and forms the plant's first line of defense, limiting the pathogenic invasion. The phenomena of Induced resistance in AMF can happen via localized and systemic resistance. As a result of AMF-mediated systemic resistance, plants make changes to their gene expression levels. The symbiosis between rice roots and *G. intraradices* triggers transcriptional changes in leaves and induces the ISR via signal transduction genes. These genes are related to the calcium-mediated signaling process, and also play a regulatory role in host defense. In addition, the mycorrhizal rice plant forms a resistance to the pathogen *M. oryzae* via ISR. Non-mycorrhizal plants exhibit a low level of pathogen-related (PR) transcript accumulation when compared to mycorrhizal plants. Additionally, after *M. oryzae* infection of AMF colonized plants, the plants exhibit a resistance phenotype to the blast fungus by strongly increasing the expression of the pathogenesis-related (PR) gene in leaves.

ii. **Integration of AMF with Rhizosphere microorganisms:** A microbiome of rhizosphere-inhabiting antagonistic bio-agents facilitates disease resistance by providing multiple mechanisms of action, and establishing a more stable rhizosphere community in many crop plants. In the rhizosphere, these co-existing organisms occupy their realized niches or complementary niches by using different resources. Integration of AMF with other bio-agents causes a decreasing in disease development in the plant when compared to the individual effect of AMF on the pathogen (Saleh and El-Akshar, 2020). A combination of

AMF and *Trichoderma viride* showed 65.60% greater efficacy against the rice brown spot pathogen *Bipolaris oryzae*, when compared with the application of *T. viride* (40.12%) alone. Application of the combination of AMF, *T. viride* and *Pseudomonas fluorescens* has also displayed a higher percentage of efficacy in resistance to pathogens than when they are used without AMF. Associations between Rhizophagus irregularis and plant growth-promoting rhizobacteria (PGPR) improve the rice plant resistance against rice blast pathogen *Magnaporthe oryzae* by inducing both ISR by PGPR and mycorrhizae-induced resistance (MIR). Furthermore, a demonstration of the wild rice variety Cyclops, scientists concluded that depending on the host genotype, AMF suppresses the biocontrol effect of PGPR or vice versa because same defense-related genes were induced by both microorganisms in the rice plants. Fusion of the *G. intraradices* with the fluorescent bio-agent *Pseudomonas* sp. and *T. harzianum* escalated the resistance against the pathogen *Fusarium oxysporum* in tomato plants than the application of *G. intraradices* alone.

iii. **Competition with Pathogens**: Within the same ecological niche, AMF can directly compete with pathogens for space, nutrition, and infection sites on a particular host (Dar and Reshi, 2017). Physical competition arises between AMF and rhizospheric microorganisms to obtain habitable space in the host root. This competition for space can also be considered a biocontrol mechanism, as long as AMF colonize the plant before the pathogen encounters the host, because AMF have slower colonization rates than soilborne pathogens. In-vitro examination of *F. virguliforme* pathogenicity with the presence and absence of the AMF *R. irregularis*, resulted in a low degree of pathogen colonization with the presence of AMF, because of the competition for the root space. With that, plants show a decrease in symptoms and tissue infection. Furthermore, the majority of AMF belong to the families Paraglomeraceae and Glomeraceae, can suppress root-knot nematodes by colonizing the galls that are produced in the susceptible root. Both AMF and pathogenic fungi depend on the host plant photosynthate. If AMF colonized first, they consume more photosynthates from host and leaving less available carbon for the soilborne pathogens. With that, carbon availability will determine the level of competition. When a plant is colonized with AMF, 4-20% of net photosynthates are transferred from the host to the fungus. With that, the carbon demand for the rhizosphere-inhabiting soil-borne pathogens increases. For the first time, Lerat et al. (2003) explored the effect of three different strains of G. mosseae on variation in carbon partitioning in barley plants, and revealed that the biocontrol effect facilitates altering levels of carbon available for the pathogen.

Biotic and Abiotic factors affecting biocontrol through Mycorrhization: Bio-protection through mycorrhization is an inter-relational process among the host plant, pathogen, and AMF spp. within the rhizosphere (Ruwanthika Kalamulla et al., 2022). The effectiveness of AMF on pathogen control can depend on both biotic and abiotic factors, like host genotype, the age of the plant, the species of AMF that is involved, the degree of mycorrhization, the presence of weed plants, and abiotic factors such as soil temperature, soil moisture, soil nutrient content and atmospheric gas composition.

i. **Biotic Factors**: Different varieties of rice form different degrees of resistance to soil-borne pathogens when plants are inoculated with the same AMF species (Campo et al., 2020). Scientists have shown that colonization percentage also depends on the genetic expression within the host plant's genotype. Low levels of colonization conferred little to no bio-protection on the host. Varieties like Gigante Vercelli already have the basal level of resistance to the blast pathogen *Magnaporthe oryzae*. Thus, the effect of AMF on pathogen resistance can be varied. The evidence shows that the mycorrhizal wild rice ancestor (*Oryza rufipogon*) was more resistant to the pathogen *M. oryzae* than cultivated rice. Further, transcriptome analysis indicates that the genes that are related to the secondary metabolic and biotic stress-related pathways are more enriched with mycorrhizal wild rice than with mycorrhizal cultivated rice (Tian et al., 2019). When AMF inoculates the rice variety Maratelli, blast susceptibility is increased. The lowest degree of colonization of AMF during the seedling stage, and the highest colonization in the tillering and mature stages were recorded in rice plants. The susceptibility of the rice plant against the pathogen *Pyricularia oryzae* was gradually decreased upon plant maturation. Hypothetically, as in previous study, the resistance to the pathogen should become higher with the maturity, because of the increasing AMF colonization. In contrast, when rice seedlings were infected with the bacterial blight pathogen *Xanthomonas campestris* pv. *oryzae* formed higher resistance than plant in their maturity stage. Depending on the AMF species, the biocontrol effect can be different. As an example, *G. intraradices* does not protect the plant against nematode infection, while *G. mosseae* does protect the plant against nematode infection. Through formation of symbiotic relationships, agronomic weeds can increase the diversity and abundance of agronomically beneficial AMF taxa. The presence of weed plants like *Cyperus rotundus*, *Echinochloa colonum*, *Paspalum paspaloides* and *Marsilea minuta* in cultivated fields affects mycorrhizal colonization by stimulating mycelial, arbuscular and vesicle colonization in rice var. Basmati Super.

ii. **Abiotic Factors**: Changing of the environmental conditions affect AMF biocontrol. As an example, soil temperature will rise because of increasing of the atmospheric carbon dioxide (CO_2) levels, which will subsequently affect rainfall, which changes the soil moisture levels. Scientists have shown that this will affect the interaction between crops and-*Verticillium* spp.-AMF. With elevated CO_2 levels, photosynthesis of crops increases, structural resistance to pathogen attacks also increases, and jasmonate acid (JA) and salicylic acid (SA) dependent signaling pathways are activated. Elevation of CO_2 will favor the increasing abundance and the extension of the extraradical hyphae of AMF spp. which will increase the biocontrol ability of AMF. Some researchers have reported that the elevation of atmospheric CO_2 has negatively affected cotton plant defenses against the *Verticillium dahlia* pathogen. The abundance of the family Glomeraceae is increased in drought conditions, and activates the JA defense pathway in the plant to increase resistance against pathogens (Goicoechea, 2020).

Researchers are interested in environmentally safe solutions for agricultural problems, and they have reported that AMF was contributed in pathogen biocontrol in rice and the other agricultural crop as well (Ruwanthika Kalamulla et al., 2022). Further studies, including transcriptomic and metabolomics analysis, are needed to understand and manipulate the molecular mechanisms, and signal exchange pathways to gain maximum benefit from AMF biocontrol capability. More than 75% of rice is cultivated under wetland waterlogging conditions, and that limits AMF root colonization because they are aerobic in nature and further it can affect their functionality. The construction of genetically engineered AMF with the ability of tolerance to the anaerobic condition is much important. To consider AMF as a better alternative for the synthetic agrochemicals, more research should be conducted by applying AMF and synthetic chemicals to compare the levels of pathogen control.

8.5.3 *Bacillus* spp.

Characteristics: Members of the *Bacillus* genus are among the beneficial bacteria mostly exploited as biopesticides to control plant diseases but also nematodes. The most known species hosting biocontrol agents are *B. subtilis* and *B. amyloliquefasciens*, but also *B. firmus* and *B. Pumilus* (Ilaria Pertot et al., 2020). They are Gram positive rod-shaped bacteria that can form a protective endospore that can tolerate extreme environmental conditions.

Diversity of *Bacillus* species: The biocontrol activity of *Bacillus* species has mainly been linked to their ability to produce a wide range of chemically diverse compounds (Adrien Anckaert et al., 2022). Comparative genomics has revealed that species of the *B. subtilis* group, including plant-associated clades, are particularly rich in biosynthetic gene clusters (BGCs) encoding bioactive secondary metabolites (BSMs). Up to 12% of the genome is devoted to the synthesis of those compounds. Some BGCs are widespread across species in the *B. subtilis* group while others seem to be more species-specific. *B. velezensis* has the highest number of different genes coding for BSMs, making this species one of the most efficient and commonly used bacilli in biocontrol. Other species of the *B. subtilis* group produce fewer BSMs in terms of diversity, for example, *B. amyloliquefaciens* and *B. subtilis*. Many of the commercialized strains have been registered as *B. subtilis* or *B. amyloliquefaciens*. However, according to recent re-classification based on phylogenetic analyses at the whole genome level, most of these strains actually belong to *B. velezensis* species.. Some lipopeptides such as lichenysin or pumilacidin are specific to the producing species (*B. licheniformis* and *B. pumilus*, respectively). *B. subtilis* and *B. cereus* produce plipastatins to prevent phospholipase A_2 which plays an important role in different cytological process (inflammation and acute hypersensitivity) (Jamil Shafi et al., 2017).

Mechanism of action: *Bacillus* spp. can control fungal pathogens by competition, direct antibiosis and induced resistance. In the rhizosphere, competition takes place for space at the root surface and for nutrients, noticeably those released as seed or root exudates. Competitive colonisation of the rhizosphere and successful establishment in the root zone is a prerequisite for effective biocontrol, regardless of the mechanism(s) involved. Direct antagonism involves the production of several microbial metabolites among which lipopeptides play the major role. Surfactin in particular is involved in the mechanism of resistance induction. The bio-nematicide Bacillus firmus also colonizes the rhizosphere of the plant where it parasitizes the eggs and larvae of nematodes especially of the root knot nematodes.

Bacillus species are recognized as safe bacteria that produce substances that are beneficial for crops and the production of industrial compounds (Hashem et al., 2019). In addition, *Bacillus* spp. also produce endospores, which helps the bacteria to survive harsh environmental conditions, can allow for germination by different environmental cues, can allow for long-term storage of the biocontrol agent, and reduce the complexity of the formulation process (Collins and Jacobsen, 2003). Notably, *Bacillus* species that are used for rhizosphere applications can also function as plant endophytes that also protect plants from pathogens. *Bacillus* spp. produce antimicrobial metabolites that can be used as a substitute to the use of synthetic chemicals or as a supplement to the use of bio-pesticides, and biofertilizers, for controlling plant diseases. The success of biocontrol approaches depends on the proper selection of effective biocontrol agents and their ability to provide protection against specific target pathogens in specific crops.

B. subtilis is a gram-positive bacterium that forms biofilms on inert surfaces and possesses many transcriptional factors (Stanley et al., 2003). Different strains of *B. subtilis* synthesize a variety of hydrolytic enzymes, including i.e. cellulases, proteases, and β-glucanases. Since *B. subtilis* has the ability to secrete antibiotics and hydrolytic enzymes, it can modify its' environment in a self-beneficial manner and also produce resistant endospores to sustain itself under adverse conditions. The ability of *B. subtilis* to exhibit biocontrol activity is dependent upon three factors, viz., host vulnerability; pathogen virulence; and the environment. Potential biocontrol mechanisms of *B. subtilis* are presented in fig 57. Importantly, any molecular changes (changes in gene expression) can directly or indirectly influence the mechanisms illustrated in fig. 51 (Hashem et al., 2019). Additionally, genetically engineered enhancement of *B. subtilis* with known biocontrol traits may interact with existing mechanisms in a synergistic manner.

Bacteria also produce the cell-wall-degrading enzymes and various metabolites that can limit the growth or activity of other microorganisms. Notably, *B. subtilis* strains are known to synthesize antibiotic lipopeptides, including fengycin, surfactin, and iturin. Lipopeptides are low molecular weight compounds with amphiphilic features. Surfactants and antimicrobial compounds produced by *B. subtilis* are receiving more attention. Lipopeptide genes occur in many species and strains of biocontrol agents and some with enhanced capacity to produce antibiotics and limit fungal root pathogens have been commercialized. Lipopeptides provide protection to plants under both pre- and post-harvest conditions by directly suppressing pathogenic fungi or by inducing systemic resistance in host plants.

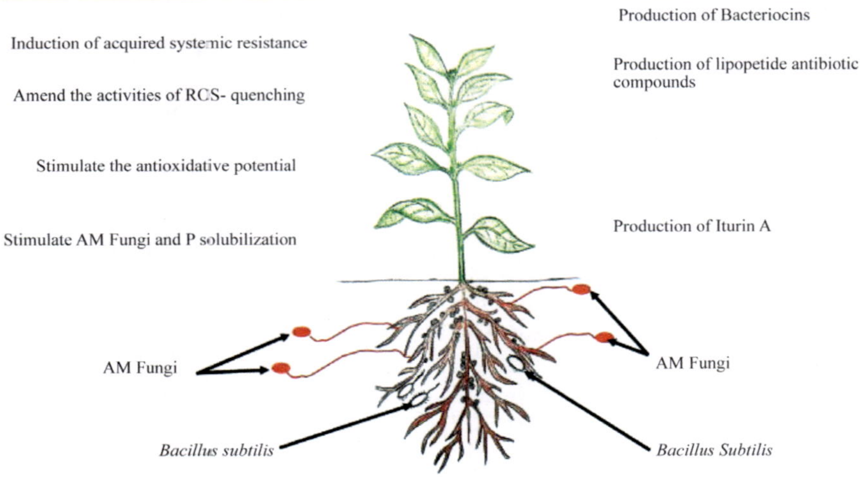

Fig. 51: Mechanisms of *Bacillus subtilis* in biological control of biotic stress

Iturins are categorized into A, C, D, and E iturins; mycosubtilin; D, F, and L bacillomycins; and bacillopeptin. Iturins exhibit antimicrobial activity against fungi and yeast and are considered as an excellent biopesticide (Hashem et al., 2019). Fengycins, a type of plipastatin, and A, B, or C fengycin are less hemolytic than surfactins and iturins but have strong antifungal activity and limit the growth of bacteria and fungi. *B. subtilis* also produces peptide antibiotics called bacteriocins that play an important role in innate host immunity. Bacteriocins are grouped into four classes based on their genetic and biochemical properties. Class 1 bacteriocins, called lantibiotics, are commonly used as an antibiotic. Lantibiotics synthesized by *B. subtilis* are categorized into A and B types based on their antimicrobial activity and chemical structure. The biocontrol activity exhibited by *B. subtilis* can also be attributed to indirect mechanisms. *B. subtilis* is a common soil microbe but is present freely except in soils where it has been applied in high doses. Evidence also indicates that *B. subtilis* occurs as an endophyte of plant roots. Indirect mechanisms associated with the biocontrol activity of *B. subtilis* against plant pathogens include, biofilm formation, plant growth promotion (PGP), competition for nutrients and colonization sites, ability to induce cell lysis, and induced systemic resistance (ISR). Antibiotic substances play an important role in disease control by microbes, including *B. subtilis*. More than 24 antibiotic substances have been reported to be produced by *B. subtilis*. The produced substances include peptides, proteins, and non-peptides based substances. Non-peptide antibiotics can be categorized as ribosomal and non-ribosomal peptide antibiotics. *B. subtilis* also forms biofilms on plant roots which help to produced lipopeptides and augment their antimicrobial activity in the soil.

Use: *Bacillus* strains are effective against a broad spectrum of plant pathogens and they can be used either as foliar application or root application before trasplanting. In case of soil or root application the ability of the specific strain to colonise and permanently establish on the roots of the specific crop is crucial. Sometime the colonization is not simply crop-specific, but cultivar-specific. Root application of Bacillus spp. should be preferably targeted to improve resistance of the plant or to protect the early stage of seed germination, rather to directly control soil-borne inoculum of pathogens (Ilaria Pertot et al., 2020).

Advantages, limitations and ways to improve efficacy: The advantage of *Bacillus* spp.-based products is their long shelf-life at room temperature, given by the fact that they produce endospores. There is also a lot of knowledge gained in the fermentation process and the ways of production are commonly optimised. It is a relatively easy microorganism to ferment with relatively low cost of production compared for example to fungal biocontrol agents. Main limitation is due to the fact that the *Bacillus* strains should colonise the roots in order to be active. Since no major direct effect on the pathogen's inoculum is expected, its efficacy may be low in case of high pathogen's inoculum in soil or against very aggressive soil-borne pathogens. The highest efficacy is seen on germinating seeds or young plantlets, because of the combination of direct antibiosis, induced resistance and grow promotion of several *Bacillus* strains. Possible ways to improve efficacy are: verification of compatibility between strains and cultivars, application in strategies where another biocontrol agents are targeted to reducing soil-borne inoculum, application in hydroponic culture or fertigation, selection of *Bacillus* strains with high production of antimicrobial metabolites.

8.5.4 *Coniothyrium minitans*

Characteristics: *Coniothyrium minitans* is a coelomycete with a worldwide distribution. In nature it occurs mainly in sclerotia of *Sclerotinia sclerotiorum* and *S. trifoliorum*, but can colonise other *Sclerotinia* species (Ilaria Pertot et al., 2020). Spore germination, mycelial extension and sclerotial infection take place at temperatures between 5 and 25 °C. Field control of *S. sclerotiorum* and *S. minor* can be achieved following soil incorporation of solid-substrate formulations before planting. *C. minitans* can survive and spread in soil for at least 2 years and continue to give some control of disease although limited.

Molecular characterization of *Coniothyrium* sp.: Agarose gel electrophoresis amplifed the genomic DNA of three *C. minitans* isolates and yielded an amplicon size of approximately 560 base pairs. Te rDNA homology sequences analysis using BLAST revealed that all the three isolates share more than 99% nucleotide sequence identity with the existing C. minitans isolates available in the NCBI database. A nucleotide sequence of three isolates was submitted to NCBI for getting accession number. Te accession numbers of three *Coniothyrium* isolates are listed in Table 16 (Sivagnanapazham et al., 2022).

Table 16: Molecular identification of *Coniothyrium minitans* isolates

S. No.	Isolates	Sources	Species identified	Accession number	Per cent identity	Query coverage (%)
1	KET Cm 1	Ketti	*C.minitans*	MW093134	100	100
2	PAL Cm 2	Palada	*C. minitans*	MW111282	99	100
3	TDK Cm 3	Thadiyankudusai	*C. minitans*	MW136938	99.95	100

C.minitans is a promising biocontrol agent of S. sclerotiorum both in glasshouse and in feld trials. Culturing of *C. minitans* for large-scale production requires standard culture media, with suitable pH and temperature. In the present study, eight diferent cultural media were tested for their support in the growth of *C. minitans* and its pycnidial production. Among the diferent cultural media tested, PDA recorded maximum mycelial growth than other cultural media.

Mechanism of action: *C. minitans* enters through the small pores or a lacerated surface of the target organism, and/or lysing by chitinase and $-1,3 glucanase for entry. Upon entry, *C. minitans* will penetrate the inter- and intra-cellular sub-cortex and medulla, producing fruiting bodies, through asexual reproduction. The infection process will cause the cell to shrink, due to high osmotic potential, and the membrane will be detached and/or degraded (Anonymous, 2020). *C.minitans* is highly specialized antifungal agent that targets sclerotia of Ascomycotina and Deuteromycotina (e.g., *Sclerotinia sclerotiorum* and *S.minor*). *C. minitans* requires a host to be in a vegetative stage.

Mycoparasitism illuminated by Genome and Transcriptome Sequencing of *C.minitans*: To further understand the parasitism of *C. minitans*, Huizhang Zhao et al., (2020) assembled and analysed its genome and performed transcriptome analyses. The genome of *C. minitans* strain ZS-1 was assembled into 350 scaffolds and had a size of 39.8 Mb. A total of 11 437 predicted genes and proteins were annotated, and 30.8% of the BLAST hits matched proteins encoded by another member of the Pleosporales, *Paraphaeosphaeria sporulosa*, a worldwide soilborne fungus with biocontrol ability. The transcriptome of strain ZS-1 during the early interaction with *S. sclerotiorum* at 0, 4 and 12 h was analysed. The detected expressed genes were involved in responses to host defenses, including cell-wall-degrading enzymes, transporters, secretory proteins and secondary metabolite productions. Seventeen differentially expressed genes (DEGs) of fungal cell-wall-degrading enzymes (FCWDs) were up-regulated during parasitism, with only one down-regulated. Most of the monocarboxylate transporter genes of the major facilitator superfamily and all the detected ABC transporters, especially the heavy metal transporters, were significantly up-regulated. Approximately 8% of the 11 437 proteins in *C. minitans* were predicted to be secretory proteins with catalytic activity. In the molecular function category, hydrolase activity, peptidase activity and serine hydrolase activity were enriched. Most genes involved in serine hydrolase activity were significantly up-regulated. This genomic analysis and genome-wide expression study demonstrates that the mycoparasitism process of *C. minitans* is complex and a broad range of proteins are deployed by *C. minitans* to successfully invade its host. This study provides insights into the mechanisms of the mycoparasitism between *C. minitans* and *S. sclerotiorum* and identifies potential secondary metabolites from *C. minitans* for application as a biocontrol agent.

Mycoparasitism of *C. minitans* is a two-way process: Previously, a model for the interaction between *C. minitans* and *S. sclerotiorum* was established. *C. minitans* senses oxalic acid produced by *S. sclerotiorum,* or the low pH caused by oxalic acid, and produces antifungal substances to inhibit the growth of *S. sclerotiorum* and degrade the oxalic acid with oxalate decarboxylase (Huizhang Zhao et al., (2020). After the ambient pH rises, *C. minitans* secretes fungal cell-wall-degrading enzymes and enters the parasitic life stage. However, in this interaction system, although *C. minitans* could produce antifungal substances to weaken *S. sclerotiorum* and destroy the cell wall of *S. sclerotiorum* using FCWDEs, counteraction by *S. sclerotiorum* should be considered. As shown in this study, many genes for MFS transporters, ABC transporters, effector-like proteins and secondary metabolites were significantly up-regulated during the early stage of interaction, suggesting that these substances play essential roles in parasitism of *S. sclerotiorum*. ABC transporters, especially HMTs, may be used by *C. minitans* to compete with *S. sclerotiorum* for heavy metal elements such as iron; MFS transporters may play roles in releasing antifungal substances and in the efflux of toxic compounds produced by *S. sclerotiorum*. There is a high possibility that the effector-like proteins enter the hyphae of *S. sclerotiorum* and specifically inhibit the counteraction system of *S. sclerotiorum*. Once it subdues *S. sclerotiorum*, *C. minitans* carries on a parasitic life in its host.

Use: The use is mainly in high value crops as lettuce, beans, peas, but also in sunflower, oilseed rape. Application is commonly by spraying onto soil and incorporating into upper soil layer either in pre-plant, by applying the product to the soil 3 to 4 months prior to the onset of disease or in postharvest, by applying product to harvest residues and incorporate into the soil.

Advantages, limitations and ways to improve efficacy: The main advantage is that *C. minitans* controls specifically *Sclerotinia* species without any interference with the soil microflora, the treatment is easy to apply and with no risk for the operators. Main limiting factors are the cost per hectare (the concentration should be sufficiently high per unit of soil and sufficiently deeply incorporated) and the time needed to reduce the inoculum (few months) which is incompatible with frequent growing cycles as for lettuce in some areas. If the time between the application of the biocontrol agent and the subsequent growing cycle is too short, they biocontrol activity and the reduction of inoculum can be insufficient. Another limiting factor is the habit of yearly renting of land by growers, which is common in some countries. In such context farmers do not invest in a soil treatment the year before, which can benefit the next farmer renting that land (Ilaria Pertot et al., 2020).

8.5.5 *Gliocladium catenulatum*: *Gliocladium* is a genus known for its production of bioactive secondary metabolites. Terpenes, sulfur-containing diketopiperazines and polyketides are just a few examples of the diverse structural and biological activity of these compounds (Bertinetti et al., 2010). Some *Gliocladium* strains are also employed commercially as biocontrol agents in the field against pests. *Gliocladium* sp. is a genus from Fungi that is often described as a counterpart of *Penicillium* sp. Most studies related to this genus have been focused on its powerful capacity as biocontrol agent, especially of fungal pathogens (Castillo et al., 2015). Nevertheless, its potential applications include numerous products derived from its metabolism capable of generate a variety of chemicals. It is remarkable as well that *Gliocladium* sp. is an important endophyte organism whom besides providing benefits to the wide range of host plants, can successfully mimic their chemical behavior. Among the latest applications stands out its potential for the production of compounds associated with diesel fuel and its ability as bio-absorbent organism of heavy metals. This research describes and illustrates these applications showing this genus of fungi as a promising biotechnological too.

Characteristics: *Gliocladium catenulatum* is a naturally-occurring saprophytic fungus which is widespread in the environment (Ilaria Pertot et al., 2020).

Mechanism of action:The mode of action is reported to be by an enzymatic mechanism. There are no reports indicating that *G. catenulatum* produces any toxins or antibiotics of concern. Use It can be used to control damping-off, seed- root- and stem-rots, and wilt diseases caused by *Rhizoctonia, Pythium, Phytophthora, Fusarium, Didymella, Botrytis, Verticillium, Alternaria, Cladosporium, Helminthosporium, Penicillium,* and *Plicaria.* It can be applied by soil incorporation, soil drench, foliar spray, and dipping of cuttings, bulbs, and tubers. Timing of application is at sowing, potting or transplanting and 2-6 weeks later for potted plants and 2-8 weeks later for other plants. Cuttings, bulbs, or tubers may be sprayed after planting or treated before planting or storage.

Hyphal interactions between the antagonist *G.catenulatum* and the soil-borne plant pathogenic fungi, *Pythium ultimum* and *Rhizoctonia solani,* were investigated in dual culture by scanning electron microscopy (Mcquilken et al., 2008). Gliocladium catenulatum hyphae grew along and sometimes coiled loosely around those of the pathogens. Appressorium-like structures were produced by the antagonist which aided in holding and penetrating the hosts' hyphae. Partial destruction of these was observed in the later stages of parasitism. A wettable powder formulation of the antagonist was then evaluated for biological control of damping-off in bedding plant seedlings grown in a peat-based growing medium, artificially infested with *P. ultimum* or *R. solani*. A growing medium incorporation or drench of the formulation were, in general, equally effective in reducing damping-off and, in most cases, as effective as fungicide drenches of propamocarb HCl or tolclofos-methyl. Populations of *G. catenulatum* survived well in the pathogen-infested growing medium with levels of > 106 colony forming units/cm3 being detected 28 days after application. Survival of *G. catenulatum* in the formulation and subsequent ability to reduce damping-off was unaffected by storage for 48 weeks at $\leq$ 5°C. Further, agar tests with fungicides showed that furalaxyl, propamocarb HCl and fosetyl-aluminium were minimally detrimental to the growth of the antagonist.

Advantages, limitations and ways to improve efficacy: Advantages, limitations and ways to improve efficacy are the same as reported for the *Trichoderma* spp.

8.5.6 *Purpureocillium lilacinum* (previously *Paecilomyces lilacinus*): *Purpureocillum lilacinum* is an endophytic, entomopathogenic microorganism with several important abilities that can be used for various purposes. It can be used for nematode and insect control, as well as a biofertilizer to make nutrients available to plants in the soil (Carlos Henrique Barbosa Santos et al., 2023). Additionally, it can be used for phytopathogenic control to reduce the presence of pathogens and subsequently decrease the incidence of diseases. By harnessing these diverse skills, *P. lilacinum* presents promising applications in biotechnology and agriculture. Its use can help reduce the dependency on chemical fertilizers, pesticides, and nematicides and contribute to sustainable production practices.

Characteristics: *P.lilacinum* is a naturally occurring fungus commonly found in soils. It has been isolated from a wide range of habitats, including cultivated and uncultivated soils, forests, grassland, deserts, estuarine sediments and sewage sludge, and insects. It has also been found in nematode eggs, and occasionally from females of root-knot and cyst nematodes. In addition, it has frequently been detected in the rhizosphere of many crops. The species can grow at a wide range of temperatures - from 8 to 38 °C (46 to 100 °F) for a few isolates, with optimal growth in the range 26 to 30 °C (79 to 86 °F). It also has a wide pH tolerance and can grow on a variety of substrates. *P. lilacinum* has shown promising results for use as a biocontrol agent to control the growth of destructive root-knot nematodes. Unlike many other *P. lilacinum* strains, the registered strain does not produce mycotoxins or paecilotoxins. The registered strain was isolated from infected nematode eggs (Ilaria Pertot et al., 2020).

P. lilacinum was first observed in association with nematode eggs in 1966 and the fungus was subsequently found parasitising the eggs of *Meloidogyne incognita* in Peru.It has now been isolated from many cyst and root-knot nematodes and from soil in many locations.Several successful field trials using *P. lilacinum* against pest nematodes were conducted in Peru.The Peruvian isolate was then sent to nematologists in 46 countries for testing, as part of the International Meloidogyne project, resulting in many more field trials on a range of crops in many soil types and climates. Field trials, glasshouse trials and in vitro testing of *P. lilacinum* continues and more isolates have been collected from soil, nematodes and occasionally from insects. Isolates vary in their pathogenicity to plant-parasitic nematodes. Some isolates are aggressive parasites while others, though morphologically indistinguishable, are less or non-pathogenic. Sometimes isolates that looked promising *in vitro* or in glasshouse trials have failed to provide control in the field.

Mechanism of action: *P. lilacinum* parasitizes and subsequently kills eggs, juveniles, and adult females of various plant parasitic nematodes. *P.lilacinum* is a fungus that produces a variety of secondary metabolites with potentially important applications in medicine, agriculture, and industry. *P. lilacinum* has shown great potential as an endophyte and entomopathogenic fungus with numerous applications in biotechnology and agriculture. Various studies have highlighted its abilities, including biocontrol, biofertilization, and bioremediation, achieved through different modes of action. In the case of biocontrol of nematodes, the fungus produces various enzymes, including chitinases and proteases, which digest the chitin and proteins in the nematode's body, leading to its death (Carlos Henrique Barbosa Santos et al., 2023). Moreover, this fungus is capable of reducing the incidence of phytopathogenic fungi by damaging their cell membranes and components, ultimately leading to their demise. Some of these secondary metabolites have been found to possess antimicrobial, anticancer, and antiviral properties, making them promising candidates for developing new drugs. Plants and microbes only create secondary metabolites during certain stages of development. Pigments, hormones, poisons, and antibiotics are all tiny molecules with complicated chemical structures that are not essential for normal development and reproduction. *P.lilacinum* has been identified as a great producer of leucinostatin D, acremoxanthone G, and acremonidin. Leucinostatin A, B, C, and D are considered biologically active secondary metabolites. They were found to be effective against a wide variety of yeasts, both pathogenic and nonpathogenic, as well as filamentous fungi, and to have moderate activity against gram-positive bacteria. In addition to secondary metabolites with potential use in biotechnology, the fungus also produces a substance of agricultural importance. The effects of two secondary metabolites produced by *P. lilacinum*, indole-3-carboxaldehyde and indole-3- carboxylic acid, on regulating tobacco mosaic virus have been evaluated. In vitro and in vivo, these secondary metabolites demonstrated potent antiviral activity. They split virus virions and caused the fragmented particles to clump together. Virus application also caused immunological reactions or resilience to virus infection in tobacco, including the expression of hypersensitive reactions, increased defense-related enzymes, and overexpression of pathogenesis-related proteins.

The increase in the salicylic acid biosynthesis of several genes showed that salicylic acid functioned as a defense-related signaling molecule in plants. As a result, indole compounds can activate tobacco defenses against mosaic viruses and other viruses and can be used for disease management. Various food-grade antioxidants and natural polyphenols are compatible with *P. lilacinum*. The most suitable compounds with the 21 genotypes of *P. lilacinum* were found to be 2(3)-tert-butyl-4 hydroxyanisole (BHA), 2,6- di(tert-butyl)-p-cresol (BHT), 3-phenyl-2-propenoic acid (CA), and the mixture of BHA and BHT. Furthermore, certain types develop better in the presence of certain therapies. The effect of these compounds and entomopathogenic fungus mixtures on *Aspergillus flavus* growth metrics and toxin generation in vitro was also investigated. The study found that combining entomopathogenic fungus with BHA (1 mM) enhanced the latency period and reduced the Aspergillus flavus growth rate and aflatoxin B1 generation at three water activities evaluated. These results imply that combining foodgrade antioxidants and natural polyphenols with entomopathogenic fungi could be a successful and long-term approach to managing aflatoxin-producing fungi.

Whole genome annotation and comparative genomic analyses of *P.lilacinum*: The study aims to decipher the putative genetic components of the fungus involved in nematode pathogenesis by performing comparative genomic analysis with nine closely related fungal species in Hypocreales. de novo genomic assembly was done and a total of 301 scaffolds were constructed for *P. lilacinum* genomic DNA. By employing structural genome prediction models, 13, 266 genes coding for proteins were predicted in the genome (Prasad et al., 2022). Approximately 73% of the predicted genes were functionally annotated using Blastp, InterProScan and Gene Ontology. A 14.7% fraction of the predicted genes shared significant homology with genes in the Pathogen Host Interactions (PHI) database. The phylogenomic analysis carried out using maximum likelihood RAxML algorithm provided insight into the evolutionary relationship of *P. lilacinum*. In congruence with other closely related species in the Hypocreales namely, *Metarhizium* spp., *Pochonia chlamydosporia, Cordyceps militaris*, *Trichoderma reesei* and *Fusarium* spp., *P. lilacinum* has large gene sets coding for G-protein coupled receptors (GPCRs), proteases, glycoside hydrolases and carbohydrate esterases that are required for degradation of nematode-egg shell components. Screening of the genome by Antibiotics & Secondary Metabolite Analysis Shell (AntiSMASH) pipeline indicated that the genome potentially codes for a variety of secondary metabolites, possibly required for adaptation to heterogeneous lifestyles reported for *P. lilacinum*. Significant up-regulation of subtilisin-like serine protease genes in presence of nematode eggs in quantitative real-time analyses suggested potential role of serine proteases in nematode pathogenesis. The data offer a better understanding of *P. lilacinum* genome and will enhance our understanding on the molecular mechanism involved in nematophagy.

Use: As a pesticide active ingredient, *P. lilacinum* can be used to control plant root nematodes on many food and non-food crops. It acts by infecting eggs, juveniles, and adult females of various plant pathogenic nematodes including *Meloidogyne* spp. (root knot nematodes), *Radopholus similis* (burrowing nematode); *Heterodera* spp. and *Globodera* spp. (cyst nematodes); *Pratylenchus* spp. (root lesion nematodes). It can be applied to agricultural soil through drip irrigation, or water in the suspension around base of each plant. If neither of these methods is possible, spray the soil surface around the base of each plant, and drench in afterwards using the irrigation system. The timing varies with crop. In general, 14 days pre-plant; just before planting, 6 weeks after planting, repeat every 6 weeks to 4 months.

Advantages, limitations and ways to improve efficacy: The main advantages are related to safety and the possibility to treat the soil immediately before planting or after. Main limitations are related to the short shelf-life, temperature and the cost per hectare.

8.5.7 *Pseudomonas* spp.: *Pseudomonas* bacteria are inhabitants of sea, freshwater and soil-related environments. Many species live in association with plants and animals, mostly as saprophytes but some are pathogenic for plants or animals. *Pseudomonas* bacteria have useful applications in biotechnology, plant growth promotion, bioremediation and biocontrol (Monica Höfte et al., 2022). The *P. fluorescens* group is composed of 10 subgroups since they distinguish the *P. protegens* subgroup from the *P. chlororaphis* subgroup and, in their phylogeny, *P. asplenii* is clustered within the *P. fluorescens* group. *P. chlororaphis* and *P. protegens* will be considered as two separate subgroups since their biocontrol traits are clearly different.

The *P. fluorescens*, *P. putida*, *P. syringae* and *P. aeruginosa* groups harbor biocontrol strains. Within the *P. fluorescens* group, *P. chlororaphis*, *P. protegens* and *P. corrugata* subgroups, the so-called CPC cluster are especially rich in biocontrol agents and biocontrol properties. The *P. fluorescens, P. mandelii, P. jessenii* and *P. koreensis* subgroups, the so-called FMJK cluster, have fewer biocontrol properties but are enriched in phytostimulatory properties such as plant hormone modulation and plant nutrition.

Characteristics: *Pseudomonas* spp. are aerobic, gram-negative bacteria, ubiquitous in agricultural soils, and are well adapted to growing in the rhizosphere. Pseudomonads possess many traits that make them well suited as biocontrol and growth-promoting agents (Ilaria Pertot et al., 2020).

Mechanism of action:The traits related to biocontrol include the ability to rapidly utilize seed and root exudates, colonize and multiply in the rhizosphere and spermosphere environments and in the interior of the plant, produce a wide spectrum of bioactive metabolites (i.e., antibiotics, siderophores, volatiles, and growth-promoting substances), compete aggressively with other microorganisms; and adapt to environmental stresses. Secondary metabolite production in *Pseudomonas* include hydrogen cyanide (HCN); 2,4-diacetylphloroglucinol (DAPG); phenazines; pyrrolnitrin; pyoluteorin; 2-hexyl-5-propyl-alkylresorcinol; siderophores; and (cyclic) lipopeptides. In addition, various strain-specific bio-active metabolites are only produced by a limited number of *Pseudomonas* bacteria. Gene clusters carrying the biosynthetic genes for these metabolites have most likely been obtained by horizontal gene transfer. Examples are rhizoxins; promysalin; sessilin; and L-furanomycin.

i. ***P. fluorescens***: *P. fluorescens* forms a large subgroup within the *P. fluorescens* group. At least 20 described type species cluster within this group but many genomes that fall within this subgroup are distinct from the type species. Various well-known and intensively studied biocontrol strains belong to this subgroup (Monica Höfte et al., 2022). These include two of the so-called Dutch strains that were isolated in the 1980s at the Dutch Phytopathological Laboratory 'Willy Commelin Scholten', WCS374 obtained from the rhizosphere of potato and WCS417 from the rhizosphere of wheat grown in a field suppressive to take-all disease. These strains have intensively been studied for their plantgrowth-promoting and biocontrol abilities. Whole-genome sequencing has clarified their taxonomic position. WCS417 belongs to the species P. simiae. WCS374 is closely related to the biocontrol strains Pseudomonas sp. A506 isolated from the pear phyllosphere in California, USA and Pseudomonas sp. SS101 from the rhizosphere of wheat in the Netherlands. It has been proposed to group these strains under the new species name P. defensor with WCS374 as the type strain.

   P. simiae WCS417 has intensively been studied for its capacity to trigger induced systemic resistance in Arabidopsis, grapevine, radish, banana, bean, carnation and tomato. *P. defensor* WCS374 does not trigger ISR in Arabidopsis or grapevine, but is able to do so in Eucalypt, radish and rice.

   Specific strains within the *P. fluorescens* subgroup are known to produce CLPs that belong to the Viscosin group (viscosin, massetolide, WLIP, viscosinamide).

Well-studied strains in this aspect include the massetolide producer *P. fluorescens* (lactis) SS101 and the viscosin-producer *P. fluorescens* SBW25. *P. fluorescens* (lactis) SS101 is a plant growth-promoting bacterium and has biocontrol activities against *Pythium intermedium* root rot of flower bulbs, tomato late blight caused by *P. infestans, P. syringae* pv. *tomato* and *Spodoptera exigua* on Arabidopsis. Massetolide production is involved in the control of *Pythium* root rot on flower bulbs and *P. infestans* on tomato. Various strains within the *P. fluorescens* subgroup produce the phenazine PCA. Taxonomically, they are associated with the species *P. synxantha, P. libanensis, P. orientalis* and *P. aridus. P. aridus* is a provisionally named species and has not been formally described (Monica Höfte et al., 2022)

ii. ***P. putida***: *P. putida* is ubiquitous and abundant in temperate soils and waters. Microorganisms belonging to this group are commonly found in polluted soil. *P. putida* is characterized by its physiological robustness, metabolic versatility and high tolerance to stress (Monica Höfte et al., 2022). Many *P. putida* strains are excellent colonizers of plant roots and promote their growth. Plant growth promotion is due to their ability to secrete phytohormones such as indole acetic acid that promote root elongation. Many isolates can solubilize phosphates and have the ability to induce systemic resistance or stress tolerance in plants. Inhibition of fungal growth is mainly through the production of siderophores. Isolates of the *P. putida* group do not usually produce the archetypical *Pseudomonas* biocontrol antibiotics such as phenazines, DAPG, pyoluteorin or pyrrolnitrin. But their genomes contain orphan gene clusters with uncharacterized NPRS or hybrid NRPS/PKS orthologs.

Production of CLPs is a common feature and include WLIP, xantholysin, entolysin and putisolvin. WLIP and entolysin can induce systemic resistance to *Pyricularia oryzae* in rice (Omoboye et al., 2019). Putisolvin-producing *P. putida* strains were enriched in the cocoyam rhizosphere from soils in Nigeria and Cameroon conducive to the cocoyam root rot disease caused by *P. myriotylum* and in the rhizosphere of black pepper in Vietnam. They appear to abound in harsh or contaminated environments where CLP diversity is low, but their role in biocontrol is marginal. Xantholysin can inhibit appressoria formation in *P. oryzae* and shows strong insecticidal activity against aphids.

Recently, a new bioactive metabolite has been discovered in *P. donguensis* SVBP6, an isolate from an agricultural plot in Argentina with broad-spectrum antifungal activity (Muzio et al., 2020). The compound was identified as 7-hydroxytropolone, a compound with a seven-membered nonbenzenoid aromatic ring. The biosynthetic gene cluster was identified by Tn5 mutagenesis and harbors 15 genes encoding metabolic enzyme functions, a putative efflux transport system and transcriptional regulators. Genome mining revealed that 15 strains among all *Pseudomonas* genomes available at the *Pseudomonas* Genome Database harbor this gene cluster. They are representatives of *P. donghuensis* (5 strains), *P. qingdaonensis* (7 strains) and *P. wadenswilerensis* (1 strain). These species were recently described are closely related and belong to the *P. putida* group. The 7-hydroxytropolone genomic region is associated with antibacterial activity against the black leg and soft rot pathogens Dickeya solani and *P. carotovorum* subsp. brasiliense in *P. donghuensis* P482.

iii. ***P. syringae***:The best-studied *P. syringae* biocontrol strains are ESC-10 and ESC-11 (also known as strain L-59-66), isolated from apple fruits and leaves, respectively. Both strains have been developed in commercial products Bio-Save 10 and BioSave 11 which have been registered in 1995 and are commercially available in the USA (Monica Höfte et al., 2022). The strains are effective biocontrol agents against various postharvest diseases on apple, pear and citrus. Strain ESC-10 produces a hypersensitive response on tobacco and is a weakly virulent pathogen of citrus. Strain ESC-11 is not a pathogen and does not produce a hypersensitive response. Strain ESC-10 is superior for the protection of citrus from green and blue molds, while strain ESC-11 is superior to control postharvest pathogens on apples and pears. In vitro, both strains produce the CLP syringomycin. Purified syringomycin E inhibited *P. digitatum* spore germination at about 5 µg/mL but about 100 µg syringomycin E per wound was needed to control green mold on lemon to an extent equal to that provided by the biocontrol strains ESC-10 and ESC-11. Strain ESC-10 reduced *Rhizopus* rot on peach caused by *Rhizopus stolonifera* to levels comparable with that of dichloran and iprodione.

iv. ***P. aeruginosa***: It is an ubiquitous organism with minimal requirements for survival that is commonly found in water, soil, animal hosts and associated with plants (Monica Höfte et al., 2022). *P. aeruginosa* has many secondary metabolites in common with biocontrol *Pseudomonas* strains and produces phenazine antibiotics, HCN, the siderophores pyoverdine and pyochelin, and rhamnolipid-type biosurfactants. Phenazines produced by *P. aeruginosa* are pyocyanin, a blue pigment, but some strains can also produce 1-OH-PHZ, PCA and PCN. *P. aeruginosa* 7NSK2 produces pyocanin and the siderophores, pyoverdine and pyochelin. Pyoverdine and pyochelin are involved in suppression of Pythium-damping-off of tomato. ISR against *B. cinerea* in tomato is mediated by pyocyanin and pyochelin, while pyocyanin triggers ISR against *P. oryzae* in rice, but enhances susceptibility to *R. solani. P. aeruginosa* PNA1 produces rhamnolipids, PCA and PCN. Phenazines contributed to the suppression of *Fusarium* wilt caused by *F. oxysporum* f. sp. *ciceris* in chickpea and *Fusarium udum* in pigeonpea. Phenazines and rhamnolipids acted synergistically in the control of *Pythium splendens* on bean and *P. myriotylum* on cocoyam. Ginger is also susceptible to *P. myriotylum* and an endophytic PCA-producing *P. aeruginosa* isolate obtained from the ginger rhizome gave protection against the pathogen. *P. aeruginosa* M18 can effectively control *Mycosphaerelia melonis* on sweat melon plants and has strong activity against various fungal, oomycete and bacterial pathogens. The strain predominantly produces PCA rather than pyocyanin especially at 28°C. In addition it produces pyoluteorin, an antibiotic that is commonly found in *P. protegens* isolates but not in *P. aeruginosa*.

Despite the overwhelming literature on *Pseudomonas* biocontrol, only few *Pseudomonas* strains are commercialized. This is mainly due to inconsistent field performance, which can be explained by poor adaptation to environmental conditions, suboptimal formulation, and host-specific responses to microorganisms. In addition, the expensive, tedious and time-consuming registration procedure, especially in Europe, hampers the introduction of new microbial bioprotectants.

Use: *Pseudomonas* strains are commonly uses to treat seeds or roots of plants before planting. They can also be used to treat tubers and bulbs. Advantages, limitations and ways to improve efficacy. One of the major advantages is that they grow rapidly in vitro and can be mass produced at relatively low cost. Other advantages are the good compatibility with several fungicides. The major weakness of pseudomonads as biocontrol agents is their inability to produce resting spores (as do many *Bacillus* spp.), which complicates formulation of the bacteria for commercial use. The identification of new formulations, which can prolong their shelf-life, can dramatically increase their use in practice. Other point to be implemented is the understanding of the best combination between the specific strain and plant species, soil type and environmental conditions.

8.5.8 *Pythium oligandrum*

Characteristics: *Pythium oligandrum* is a mycoparasite belonging to the *Pythium* family. This mycoparasite is widely distributed throughout the world. *P.oligandrum* is common in soil and in or on plants. It is a parasite of many fungi and other oomycetes including *Botrytis*, *Fusarium* and *Phytophthora*. It has been licensed as a biocontrol agent in the form of an oospore soil treatment, which reduces pathogen load and concomitant plant disease. *P. oligandrum* have been found to express several genes belonging to the CAZy-family when feeding on prey. *P. oligandrum* can grow within the roots of certain plants, including tomato and sugar beet. Production of auxin-like substances stimulate plant growth. Defense responses can be induced in the plant, which primes the plant from further infection by pathogenic fungi, oomycetes or bacteria.It helps in reducing the inflammation caused by harmful microbes via mycoparasitism and antagonism or by incorporating a defense mechanism (Belur Satyan Kumudini et al., 2023). The formation of pathogen-related molecular patterns *P. oligandrum* helps in the activation of jasmonic acid a growth-regulating substance, thus decreasing plant systemic resistance against harmful microbes.

Mechanism of action: *Pythium oligandrum* acts as a hyperparasite by colonizing other plant pathogenic fungi in and around seeds and the rhizosphere of treated plants, thereby suppressing the growth of several soil-borne pathogenic fungi, including *Alternaria*, *Botrytis, Fusarium, Gaeumannomyces, Ophiostoma, Phoma, Pseudocercosporella*, *Pythium, Sclerotinia* and *Sclerotium*. *P. oligandrum* produces the protein oligandrin and other compounds that stimulate plants' cell walls to fend off pathogen invasion, and also stimulates natural plant defense mechanisms called pathogenesis related (PR) proteins which help plants resist disease, without harming the plant (Ilaria Pertot et al., 2020). *P. oligandrum* during the first step of mycoparasitism transcripts the productivity of encoding proteases, glucanases, cellulases, protease inhibitors, putative effectors, and elicitors, after this attachment of antagonist to the host, occurs (Belur Satyan Kumudini et al., 2023). Hydrolytic enzyme deteriorates the host cell wall, therefore enabling entry to the cell and helping in the growth and development by providing carbon. After the multiplication of antagonistic cells, *P. oligandrum* develops many papilla-like structures at the entry point. As a result of this interaction, the cells of *P. oligandrum* can be seen in host hyphae. *P. oligandrum* helps in protecting plants from harmful pathogens by instigating resistance against several microbial and biotic responses. Its presence in the rhizospheric area of soil does not affect the native microflora but reduces the effect of harmful microbes on plants. The mode of action of *P. oligandrum* protecting plant roots in the rhizospheric area of soil from harmful microbes is the same as that of *Trichoderma* species. The synthesis of phytohormones and secondary metabolites by favorable microbes is a well-known fact for enhancing plant growth. The beneficial microbe *P. oligandrum* produces an ample amount of auxin and TNH_2 in the rhizospheric area that enhances plant growth. Therefore, *P. oligandrum* can be considered a plant growth promoting oomycete.

Fungal and oomycete populations and their dynamics were investigated following the introduction of the biocontrol agent *P.oligandrum* into the rhizosphere of tomato plants grown in soilless culture (Vallance et al., 2009). Three strains of *P. oligandrum* were selected on the basis of their ability to form oospores (resting structures) and to produce tryptamine (an auxin-like compound) and oligandrin (a glycoprotein elicitor). Real-time PCR and plate counting demonstrated the persistence of large amounts of the antagonistic oomycete in the rhizosphere throughout the cropping season (April to September). Inter-simple-sequence-repeat analysis of the *P.oligandrum* strains collected from root samples at the end of the cropping season showed that among the three strains used for inoculation, the one producing the smallest amount of oospores was detected at 90%. Single-strand conformational polymorphism analysis revealed increases in the number of members and the complexity of the fungal community over time. There were no significant differences between the microbial ecosystems inoculated with *P.oligandrum* and those that were not treated, except for a reduction of *Pythium dissotocum* (ubiquitous tomato root minor pathogen) populations in inoculated systems during the last 3 months of culture. These findings raise interesting issues concerning the use of *P. oligandrum* strains producing elicitor and auxin molecules for plant protection and the development of biocontrol.

A review by Kateřina Bělonožníková et al., (2022) primarily focussed on biochemical processes underlying their positive effects. For example, *P. oligandrum* degrades host cell wall polysaccharides using chitinases, cellulases,endo-β-1,3-glucanases, and various exoglycosidases. Proteases from various classes also participate in the cell wall hydrolysis. All these processes can modify cell surface structures and help *Pythium* spp. compete for space and nutrition. Accordingly, enzyme secretion most likely plays a key role in plant root colonisation. Plant-*P. oligandrum* interactions, nevertheless, do not involve tissue injury but instead activate plant defence mechanisms, thereby strengthening future plant responses to pathogen attacks. Priming induces the phenylpropanoid and terpenoid pathways and thus synthesis of secondary metabolites, including lignin, for cell wall fortification and other metabolic adjustments. Such metabolic changes are mediated by elicitins, cell wall glycoproteins and oligandrins produced by *P. oligandrum*. As homologous proteins of β-cinnamomin from *Phytophthora cinnamomi* with similar essential amino acids for sterol binding, oligandrins stand out for their structure, which they share with cell wall glycoproteins, albeit without the Ser-Thr-rich O-glycosylated domain for cell wall attachment. *P. oligandrum* also provides plant with tryptamine used for auxin synthesis, promoting plant growth. Overall, in addition to discussing plant metabolic and phytohormonal changes after *P. oligandrum* inoculation, we review data on *P. oligandrum* applications as researchers increasingly search for effective and environmentally friendly ways to protect crops. In this context, *P. oligandrum* emerges as a highly suitable biotechnological solution.

Nep1-Like Proteins from *P. oligandrum* enhance plant disease resistance independent of cell death and Reactive Oxygen Species: Microbial necrosis and ethylene-inducing peptide 1 (Nep1)-like proteins (NLPs) act as cytolytic toxins and immunogenic patterns in plants. Previous work showed that cytolytic NLPs (i.e., PyolNLP5 and PyolNLP7) from the biocontrol agent *P.oligandrum* enhance plant resistance against *Phytophthora* pathogens by inducing the expression of plant defensins. However the relevance between PyolNLP-induced necrosis and plant resistance activation is still unclear (Yang et al., 2022). It was found that the necrosisinducing activity of PyolNLP5 requires amino acid residues D127 and E129 within the conserved "GHRHDLE" motif. However, PyolNLP5-mediated plant disease resistance is irrelevant to its necrosis-inducing activity and the accumulation of reactive oxygen species (ROS). Furthermore, we reveal the positive role of non-cytotoxic PyolNLPs in enhancing plant resistance against *Phytophthora* pathogens and the fugal pathogen Sclerotinia sclerotiorum. Similarly, non-cytotoxic PyolNLPs also activate plant defense in a cell death-independent manner and induce defensin expression. The functions of non-cytotoxic PyolNLP13/14 rely on their conserved nlp24-like peptide pattern. Synthetic Pyolnlp24s derived from both cytotoxic and non-cytotoxic PyolNLPs can induce plant defensin expression. Unlike classic nlp24, Pyolnlp24s lack the ability of inducing ROS burst in plants with the presence of Arabidopsis nlp24 receptor RLP23. Taken together, this work demonstrated that PyolNLPs enhance plant resistance in an RLP23-independent manner, which requires the conserved nlp24-like peptide pattern but is uncoupled with ROS burst and cell death.

Necrosis and ethylene-inducing peptide 1-like proteins have been proposed to have dual functions in plant-pathogen interactions, acting as both toxin-like virulence factors and triggers of immune responses. However, it is unclear whether cytotoxic NLPs directly trigger immune responses or these responses are indirectly induced by cell death. Constitutive expression of a mutant NLP1 lacking cytotoxic activity in the hemibiotrophic pathogen Colletotrichum orbiculare can still block its infection in cucumber (Azmi et al., 2018). In our work, PyolNLP5 mutants with completely abolished necrosis-inducing activity retain the ability of suppressing *Phytophthora* infection in *N. benthamiana*. These consistent results suggest that the cytotoxin and immunity induction activity of NLPs are largely independent.

Use: It can be applied as seed dressing, pre-plant soak, overhead spray, soil drench, or through irrigation system application.

Advantages, limitations and ways to improve efficacy:The main advantage is the wide spectrum of activity and the possibility to be applied with irrigation systems. However to function effectively, the *P. oligandrum* requires a moist environment and a temperature range of 20-35°C during the infection period, which can last for three to four hours. Under dry conditions, the onset of infection may be delayed until increased moisture is available. In addition the quantity per hectare should be sufficiently high, which can make the treatment very expensive. The implementation of pre-planting treatments with the irrigation systems and the reduction of the cost per hectare can make this biocontrol agent more appealing.

8.5.9 *Streptomyces* spp.: To date, *Bacillus* spp., *Pseudomonas* spp., and *Streptomyces* spp. have been reported as effective biocontrol agents in controlling diverse phytopathogens (Le et al., 2022). Numerous recent studies have focused on the use of *Streptomyces* spp. as biocontrol agents to control various plant diseases caused by bacteria and fungi. By several mechanisms, bacteria of the genus *Streptomyces* can effectively suppress various plant fungal and bacterial diseases. The production of volatile and non-volatile antibiotics, cell wall-degrading enzymes, hyperparasitism on pathogenic organisms, promotion of plant growth, and induction of systemic resistance in the host plant have been reported as their action mechanisms. Since 1955, after its discovery and isolation from *Streptomyces griseus* subsp. *griseus*, the aminoglycoside antibiotic streptomycin has been applied in plant agriculture. It is commonly applied to control plant bacterial diseases such as fire blight caused by *Erwinia amylovora*, soft rot caused by *Pectobacterium* spp., leaf spot caused by *Xanthomonas campestris* pv. *vesicatoria*, and crown gall caused by *Agrobacterium tumefaciens*. However, the emergence of resistant bacterial strains has impeded the control of plant bacterial diseases using streptomycin. Since the discovery of the first resistant strain in Florida in 1960, streptomycin-resistant strains of several bacteria, including *E. amylovora*, *Pectobacterium carotovorum*, *Xanthomonas* spp., and *Pseudomonas* spp., have been detected. In this context, new antibiotics must be developed that can effectively suppress the development of plant diseases caused by streptomycin-resistant bacterial pathogens. Meanwhile, a secondary method to overcome streptomycin resistance is the use of streptomycin in combination with a material that produces a synergistic effect against antibiotic-resistant pathogens. A number of secondary metabolites have been identified in *Streptomyces* spp.

The *Streptomyces* genus of gram-positive bacteria is a potentially viable natural alternative that has been extensively researched due to its capacity to generate diverse antimicrobial compounds, such as metabolites and organic compounds (Khan et al.,

2023). Scientists globally use diverse approaches and methodologies to extract new bioactive compounds from these bacteria. The efficacy of bioactive compounds in mitigating various phytopathogens that pose a significant threat to crops and plants has been demonstrated. Hence, the *Streptomyces* genus exhibits potential as a biological control agent for combating plant pathogens. This review aimed to provide further insight into the *Streptomyces* genus as a source of antimicrobial compounds that can potentially be a biological control against plant pathogens. The investigation of various bioactive compounds synthesized by this genus can enhance our comprehension of their prospective utilization in agriculture (Fig. 52).

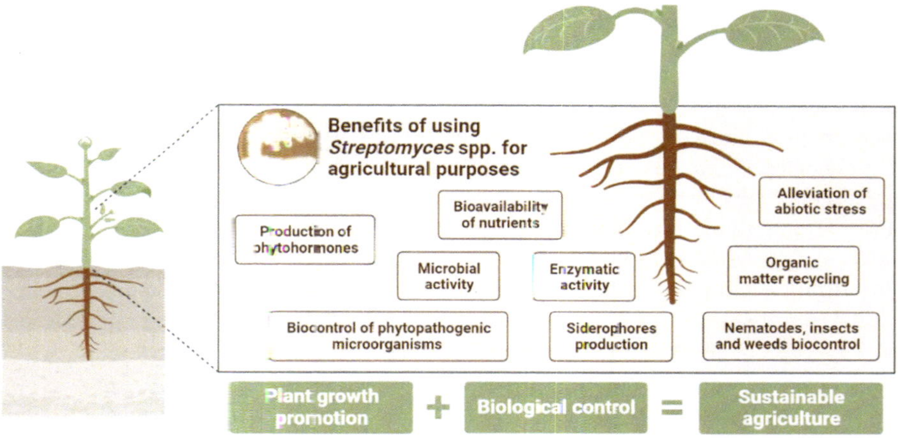

Fig. 52: Benefits of *Streptomyces* in Agriculture

Characteristics: The active ingredients, *Streptomyces* spp. and *S. lydicus*, are ubiquitous and naturally-occurring bacteria that are commonly found in soil environments. *Streptomyces* sp. AN090126, isolated from agricultural suppressive soil in Korea, showed broad-spectrum antagonistic activity against various phytopathogenic bacteria and fungi (Le et al., 2022). In the 96-well plate assay, the fermentation filtrate of *Streptomyces* sp. AN090126 exhibited antimicrobial activity, with a minimum inhibitory concentration (MIC) of 0.63-10% for bacteria and 0.63-3.3% for fungi. The MIC of the partially purified fraction was 20.82-250 μg/mL for bacteria and 15.6-83.33 μg/mL for fungi. Gas chromatography-mass spectrometry (GC-MS) analysis revealed that AN090126 produced various volatile organic compounds (VOCs), including dimethyl sulfide and trimethyl sulfide, which inhibited the growth of pathogenic bacteria and fungi in in vitro VOC assays. In pot experiments, the fermentation broth of *Streptomyces* sp. AN090126 reduced tomato bacterial wilt caused by *Ralstonia solanacearum*, red pepper leaf spot caused by *Xanthomonas euvesicatoria*, and creeping bentgrass dollar spot caused by *Sclerotinia homoeocarpa* in a dose-dependent manner. Moreover, the secondary metabolites derived from this strain showed a synergistic effect with streptomycin sulfate against streptomycin-resistant *Pectobacterium carotovorum* subsp. *carotovorum*, the causative agent of Kimchi cabbage soft rot, in both in vitro and in vivo experiments. Therefore, *Streptomyces* sp. AN090126 is a potential biocontrol agent in controlling plant diseases caused by pathogenic bacteria and fungi, specifically by the streptomycin-resistant strains.

Mechanism of action: *S. lydicus* colonizes growing root tips of plants and acts as a mycoparasite of fungal root pathogens which help protect the plants. Other possible mechanisms include the production and excretion of antifungal metabolites (e.g., antibiotics and/or low molecular weight antifungal compounds or lytic enzymes like chitinase) after colonization. *Streptomyces,* the largest genus of actinobacteria, produce numerous biological substances, which have important applications in human health and agriculture. These substances can suppress pathogen growth via multiple mechanisms, such as the action of antibiotics and wall-degrading enzymes or the induction of host resistance (Le et al., 2022). *Streptomyces* spp. are important biocontrol agents of plant diseases as their products represent a major portion of all antibiotics used in agriculture. Interestingly, when the FB of AN090126 strain with chloroform, ethyl acetate, or butanol was partitioned, antimicrobial activity was not detected in any organic solvent layers, but only the aqueous layer exhibited both antibacterial and antifungal activities. This indicated that the active substances are likely to be highly polar compounds, such as peptide or protein. For testing these hypotheses, an ultrafiltration process revealed that molecular sizes of active substances are smaller than 3 kDa, meaning that they do not seem to be proteins. Additionally, their antimicrobial activity was stable at room temperature, similar to that under UV light and low pH. Nevertheless, following exposure to high temperatures of 80 °C and 100 °C for 1 h, the activity decreased by 50% and 80%, respectively. Furthermore, the PPF interacted with ninhydrin in the bioautography assay, indicating that active substances may harbor amino groups. Taken together, the active compounds produced by AN090126 may be hydrophilic peptides that were photo- and acid-stable, but not thermostable. On the other hand, the AN090126 strain displayed the ability to produce hydrolytic enzymes such as cellulase, chitinase, protease, and gelatinase, some of which were reported to play a role in controlling plant disease. The antibacterial and antifungal activities of *Streptomyces* sp. AN090126 were quite similar to those of *Streptomyces* sp. JCK-6131, which also displayed the two biological activities. However, most *Streptomyces* strains have been reported to have either only antibacterial or antifungal activity. Thereby, the AN090126 strain is a potent candidate for developing a biocontrol agent in controlling plant diseases caused by both bacteria and fungi.

The production of antimicrobial VOCs is one of the biocontrol mechanisms of *Streptomyces* species against plant diseases. Until now, many studies have shown that the VOCs produced by *Streptomyces* can inhibit the growth of plant pathogens. *Streptomyces philanthi* RM-1-138 produced 36 VOCs which inhibit the growth of four plant pathogenic fungi. *S. coeruleoprunus* UAE1 produced antifungal VOCs against *F. solani* (Alblooshi, et al., 2022). Moreover, *Streptomyces alboflavus* TD-1 produced VOCs that suppressed the mycelial growth of *Fusarium moniloforme*, *Penicillum citrinum*, and three *Aspergillus* spp. Similarly, in the present study, the VOCs produced by *Streptomyces* sp. AN090126 could inhibit the growth of 9 fungi and 14 bacteria in vitro. Streptomycin inhibits the growth of bacteria by binding with the small ribosomal subunit during protein synthesis. Bacteria carrying specific mutations in the 16S rRNA gene can survive even at high concentrations of streptomycin. Various benefits of using Streptomyces in agriculture are illustrated in fig. (Mateus Torres Nazari et al., 2023).

Use: These strains can be used as an antifungal agent for greenhouse, nursery, turf grass, and agricultural use sites and control *Fusarium, Rhizoctonia, Pythium, Phytophthora, Phytomatotricum, Aphanomyces, Monosprascus, Armillaria, Sclerotinia, Postia, Verticillium, Geotrichum*, and other root decay fungi. They can be applied as a soil mix/drench to potted plants or turf grass or to plant foliage in greenhouses (Ilaria Pertot et al., 2020).

Advantages, limitations and ways to improve efficacy: The main advantage is related that once it has successfully colonised the tip of the roots, it can continuously protect the plant. On the other hand the efficacy is not extremely high and the rhizosphere competence is not validated in many crops. A possible way to improve the efficacy is to combine with other biocontrol agents more active in reducing the inoculum.

Future perspective: *Streptomyces* spp. have shown significant promise as biological control agents for plant pathogens. These filamentous, Grampositive bacteria are known for their ability to produce a wide range of secondary metabolites, including antibiotics and antifungal compounds, which can inhibit the growth of plant pathogens. Here are some future perspectives (Khan et al., 2023).

- **Understanding Diversity:** Researchers are likely to continue exploring the diversity within *Streptomyces* species. Different strains may possess varying abilities to produce bioactive compounds, and understanding this diversity can lead to the discovery of more potent biological control agents.
- **Metagenomics and Genomics:** Advances in metagenomics and genomics can aid in the identification and characterization of novel *Streptomyces* strains with enhanced biocontrol properties. These techniques can also help in understanding the genetic basis of antibiotic production in *Streptomyces*.
- **Synthetic Biology** Synthetic biology approaches may be employed to engineer *Streptomyces* strains for improved biocontrol capabilities. This could involve enhancing the production of specific antifungal or antibacterial compounds or optimizing their environmental survival and competitiveness.
- **Eco-friendly Formulations:** Researchers will likely work on developing eco-friendly formulations of *Streptomyces*-based biopesticides. This could involve creating stable formulations, improving shelf life, and ensuring ease of application.
- **Integrated Pest Management (IPM):** *Streptomyces*-based biocontrol agents may become an integral part of IPM strategies. Combining biological control with other pest management approaches like crop rotation, organic farming practices, and resistant crop varieties can provide a more comprehensive solution.
- **Regulatory Framework:** As *Streptomyces*-based products gain prominence, regulatory agencies may establish clearer guidelines and standards for their registration and use. Ensuring safety and efficacy will be crucial.

- **Field trials and Commercialization:** Large-scale field trials and commercialization efforts will be essential to bring *Streptomyces*-based biopesticides to the market. This may involve partnerships between research institutions, agribusinesses, and governmental agencies.
- **Environmental Impact Assessment:** Continual assessment of the environmental impact of using *Streptomyces*-based biocontrol agents will be necessary. This includes evaluating their effects on non-target organisms and assessing their long-term sustainability.
- **Education and Outreach**: Raising awareness among farmers and agricultural stakeholders about the benefits and proper use of *Streptomyces*-based biopesticides will be crucial for their adoption.
- **Global Collaboration:** Given the global nature of agriculture and the spread of plant pathogens, international collaboration in research, development, and sharing of best practices will be important.

8.5.10 *Penicillium* spp.: *Penicillium* is much more ubiquitous in the environment than other species of fungi and they also exist as endophytes inside plant tissues (Liang et al., 2021). Due to a paucity of morphological characteristics, overlap of morphs among different species and largely similar cultural characteristics, traditional morphological classification somehow been somehow unreliable to delineate species in this genus. With the development of modern molecular based phylogeny, the concept of multi-loci sequence typing (MLST) for phylogenetic species was proposed in the classification of *Penicillium*, thus making identification of strains more precise. MLST is a typing method initially used for distinguishing bacterial mutations by directly determining the nucleotide sequences of multiple housekeeping genes that are highly conservative and usually stably expressed in all cells. It was first applied in typing of the bacterial pathogen (*Neisseria meningitidis*), gradually implemented to other bacteria and fungi predominantly in epidemiology and taxonomy. Based on multiple gene locus sequence information to identify species, it came up with a short standardized gene region namely DNA barcoding which was used in the identification of *Penicillium.*

i. ***Penicillium adametzioides*, a potential biocontrol agent:** *Penicillium adametzioides* (S3) and *Penicillium expansum* (J1) (at certain stages) reduced the OTA (Ochratoxin A) production and mycelial growth. Ochratoxin-A (OTA) is a mycotoxin with nephrotoxic, teratogenic, and immunosuppressive properties. It has been classified as a possible human carcinogen (group 2 B) by the International Agency for Research on Cancer (IARC). P. expansum was excluded as a bio-control agent because of its mycotoxin production ability (Hoda Ahmed, et al., 2015). The higher challenge distance between certain strains of P. adametzioides (S3) and other Penicillium strains (as J1, J2, J3 and J4; at three and seven days) reduced the secretion of OTA by OTA-PF. This OTA production reduction could possibly prevent OTA contamination prevention in the case of epidemic favourable conditions by reducing the OTA produced in grape post-harvest products (i.e., juice). This could be accomplished by applying as the elicitor one of the tested fungi with an antagonistic effect on OTA production, such as *P. adametzioides* (at 10 days). Certain strains, such as *P. adametzioides* (S3) and J2 (*P.* spp.) should be further investigated to determine the details of the underlying mechanism of their OTA reduction and their ecosystem effects in cases of *in vivo* application.

ii. *P. frequentans:* To improve efficacies achieved through biocontrol agent application, there has been increasing interest in studying the efficacy of biocontrol agent with other existing technologies like chemicals or different biocontrol agent combinations. The compatibility of *Penicillium frequentans* (Pf 909) with modern inputs in plant protection like fungicides (chemicals and biologicals) and insecticides is a pre-requisite for developing integrated brown rot management strategies (Belen Guijarro et al., 2019). The combination of *P. frequentans* and *Bacillus amyloliquefaciens*, each with different mechanisms of action, competition and antibiosis, respectively, did not improve the biocontrol of brown rot when each treatment was applied individually. *P.frequent*ans and *B. amyloliquefaciens* could not be combined together in the same tank, because bacteria inhibited the germination and growth of *P. frequentans* tubes in suspension. Furthermore, *B. amyloliquefaciens* also competes with *P. frequentans* once applied on fruit surfaces. Additionally, bacteria reduced the growth of *P. frequentans* CFUs on media, at different 25 °C and 30 °C throughout the incubation periods. However, 76% of fungicides and 90% of insecticides commonly used against stone fruit pests are compatible with *P. frequentans*. Results indicated that the most commercial chemical pesticides could be simultaneously applied with Pf909 in integrated pest control, but not *B. Amyloliquefaciens.*

iii. *P. linzhiense:* A novel species of *Penicillium*, proposed as *P. linzhiense* sp.nov was isolated from soil collected in Linzhi Town, Linzhi County, Tibet Autonomous Region, China. DNA sequence analyses from eight different gene regions indicate that the isolate represents a novel species and most closely related to *P. jarczewskii* (Liang et al., 2021). The phylogenetic analysis based on a concatenated dataset of three genes, ITS, *CaM*, and *BenA*, also confirmed the placement of the novel species within the *Canescentia* section of the genus *Penicillium*. Differences in morphology among similar species are detailed and single gene phylogenies based on ITS, *CaM* and *BenA* genes as well as a multi-loci gene phylogeny are presented. Cultural studies were performed to study inhibitory activities on plant pathogens. The results reveal a notable antifungal activity against *Pyricularia oryzae* causing rice blast with an inhibition rate up to 77%, while for other three citrus pathogens, *Diaporthe citri*, *Phyllosticta citrichinaensis* and *Colletotrichum gloeosporioides*, inhibition rate was 40, 50, and 55% respectively. No noticeable effects were observed for *Fusarium graminearum*, *Botryosphaeria kuwatsukai*, and *Rhizoctonia solani*. Interestingly, unlike other reported members of *Canescentia*, *P. linzhiense* showed no antagonistic effect on root rotting fungi. The new taxon isolated here has the potential to be used as a biocontrol agent especially for economically important phytopathogens or emerging pathogens on diseases occurring on citrus or rice.

The multigene phylogeny reveals that *P. linzhiense* is close to *P. janczewskii*, *P. dunedinense*, *P. echinata*, *P. griseoazureum*, and *P. nigricans*. However, *P. linzhiense* is distinct from species mentioned above particularly in morphs with distinctly thinner mycelium on CZ medium and with single-whorled broom branches mainly accompanied by few double-whorled one. With respect to *P. janczewskii*, *P. linzhiense* mainly differs in the color of colony after two weeks' culture and the broom branches (at the start, there is gray-white and mainly monoverticillate, with fewer biverticillate, with solitary conidiophore, and with time turns grayish green to grayish black and mainly biverticillate, with terverticillate or few monoverticillate, with two to four metula per conidiophore).

iv. ***P. funiculosum:*** *Penicillium funiculosum* was evaluated in the greenhouse for its ability to suppress *Phytophthora* root rots of azalea (*Rhododendron* spp.) and sweet orange (*Citrus sinensis*) as measured by increased shoot and root growth (Fang and Tsao, 1995). The antagonist was grown in a wheat bran-peat moss (1:1) medium, which was then mixed into a peat-perlite (3:1) planting mix at a concentration of 0.7% (wt/vol). Rooted azalea cuttings or sweet orange seedlings were transplanted into the planting mix, which was inoculated with *Phytophthora* 5 to 7 days later. In most of the tests, *P. funiculosum* effectively increased the shoot growth and root weight of azalea in media infested with *P. parasitica* and those of sweet orange in media infested with *P. citrophthora.* It also reduced, to some degree, root rots caused by *P. cinnamomi* on azalea and by *P. parasitica* on sweet orange. Three subisolates of *P. funiculosum* (T327H, T327L, and T327S) exhibited different degrees of root rot suppression; T327S provided the best control. The population density of each of these subisolates increased in the planting mix during the first 4 weeks and decreased thereafter. Mixing the *P. funiculosum* bran-peat inoculum into the planting mix was a more effective method for delivering the biocontrol agent than dipping the plant root systems into a *P. funiculosum* spore suspension before transplantation. *P. funiculosum* was also effective in increasing plant growth when the bran-peat inoculum was used at concentrations lower than 0.7%. *P. funiculosum* inoculum applied at a concentration of 0.35% twice during a 12-week period was more effective in the biocontrol of azalea root rot caused by *P. cinnamomi* than when it was applied only once. *P. funiculosum* also reduced sweet orange root rot caused by *P. citrophthora* in some other planting mixes.

v. ***P. rubens* (*Penicillium oxalicum*)**: The soil-borne fungus, *Penicillium oxalicum* Currie and Thom strain 212 (PO212), is an effective biocontrol agent (BCA) against a large number of economically important fungal plant pathogens that infect different horticultural crops. *Penicillium oxalicum* strain 212 (PO212) is an effective biocontrol agent (BCA) against a large number of economically important fungal plant pathogens. For successful registration as a BCA in Europe, PO212 must be accurately identified. Maria Villarino et al., (2016) described the use of classical genetic and molecular markers to characterize and identify PO212 in order to understand its ecological role in the environment or host. We successfully generated pyrimidine (*pyr-*) auxotrophic mutants. In addition we also designed specific oligonucleotides for the *pyrF* gene at their untranslated regions for rapid and reliable identification and classification of strains of *P. oxalicum* and *P. rubens*, formerly *P. chrysogenum*. Using these DNA-based technologies, we found that PO212 is a strain of *P. rubens*, and is not a strain of *P. oxalicum*. This work presents PO212 as the unique *P. rubens* strain to be described as a BCA and the information contained here serves for its registration and commercialization in Europe.

Previous taxonomic classifications of the BCA PO212 and the soil-borne Spanish isolates identified these isolates as strains of *P. oxalicum*, and this identity was based on colour, size and shape of the colony and the various characteristics, such as conidial size and conidiophore morphology (Maria Villarino et al., 2016). Using these morphological and phenotypic characteristics to identify *P. oxalicum* strains can be challenging, and contradictory findings and grouping of more than one species often occur. Species in the *Penicillium* genus exhibit only few distinguishing features, which additionally may vary depending on the growth conditions and the culture medium. Taxonomic classification in the genus *Penicillium* is challenging because it is being continually reviewed. This ongoing review stems from the genetic diversity of this genus that has been unmasked by random amplified polymorphic DNA markers and nucleotide sequencing and restriction fragment length polymorphism markers of the ITS1-5.8S-ITS2 regions of their rDNA. For example, the reference fungus of the genus *Penicillum*, *P. chrysogenum*, has recently been reclassified following a phylogenetic analysis of numerous *P. chrysogenum* isolates. The results of this analysis revealed the presence of two highly supported clades that represent two species: *P. chrysogenum* and *P. rubens*. Even Fleming's original penicillin-producing strain and the full genome-sequenced strain of the filamentous fungus, *P. chrysogenum* Wisconsin 54–1255 (NRRL 1951), have been re-identified as *P. rubens*. Such reclassifications show that no strain of *P. chrysogenum sensu stricto* has had its entire genome sequenced. In this study, we found that well-defined clusters in the dendrograms were based on the ITS1-5.8S-ITS2 regions and the BOX and REP DNA fingerprints of PO212 and the 27 *Penicillium* isolates. These phylogenetic analyses also revealed that all the Spanish *Penicillium* isolates are grouped together, suggesting that they belong to the same *Penicillium* species. Thus, the identification of a strain needs to be based on the newest available methodologies and knowledge about the species and its genus. Progress and advances in molecular and *in silico* genetics have resulted in the invention of novel genetic and molecular tools for rapid identification and accurate taxonomic classification of fungal species.

This study generated auxotrophic mutants of PO212 in order to improve our current understanding of PO212 as a BCA. Specific oligonucleotides of the *pyrF* gene to be used were designed in a standard PCR that can be used to rapidly and reliably identify two species of the genus *Penicillium*, *P. rubens* and *P. oxalicum*, in field samples (Maria Villarino et al., (2016). Furthermore, nucleotide sequencing of the CDS of the *pyrG* and *pyrF* genes revealed that PO212 is a strain of *P. rubens*, and is not a strain of *P. oxalicum*. The availability of phylogenetic markers and nucleotide sequencing of the genome of fungi has created exciting opportunities for comparative genomics between fungal species. Additionally, their availability and genome sequencing will enable reliable and definitive classification of fungal strains at the species level within a genus. Furthermore, these technologies can be used to characterize the mode of action, the ecology and the fitness of PO212, which is also a required information for their registration and commercialization as a BCA in Europe. In the future, we plan to fully sequence the genome of PO212 and then compare its genome to full genome-sequenced strains of *Penicillium* species. Results of all these studies will improve the efficacy and practical application of PO212 when it is used as a BCA for plant disease control.

vi. ***P. citrinum:*** *Earlier, a* fungal isolate *Penicillium citrinum* VFI-51 and its secondary metabolite was reported to have antagonistic potential against *Botrytis cinerea*, the causative agent of *Botrytis* gray mold disease in chickpea. In the present investigation, *P. citrinum* VFI-51 was further evaluated for its antagonistic potential against *Macrophomina phaseolina*, the causative agent of charcoal rot in sorghum (Sreevidya and Gopalakrishnan, 2016). P. citrinum VFI-51 inhibited *M. phaseolina* in both dual culture as well as secondary metabolite production assays. In the in vivo blotter paper assay, under light chamber conditions, *P. citrinum* VFI-51 controlled 85% of the charcoal rot disease on the roots when compared to the positive control. Under greenhouse conditions, when *M. phaseolina* was inoculated by tooth pick method in to the stalk of sorghum plant, the charcoal rot disease was controlled by 75% in *P. citrinum* VFI-51 treatment over the positive control. This study demonstrates the biocontrol potential of *P. citrinum* VFI-51 against charcoal rot of sorghum.

The production of citrinin, a secondary metabolite, has been reported by *P. citrinum* VFI-51 which was responsible for controlling the *Botrytis* gray mold disease in chickpea. Production of citrinin was also reported by *Aspergillus* spp. and many species of *Penicillium,* including *P. citrinum* (Sreevidya and Gopalakrishnan, 2016). Citrinin is also reported for its antagonistic activity against soil and seed-borne plant pathogenic fungi such as *Sclerotium rolfsii, Rhizoctonia solani* and *Sclerotinia minor.* In the present investigation, the organic fraction of the culture free extract of *P. citrinum* VFI-51 was found to inhibit *M. phaseolina* while in our previous study, the citrinin was extracted from the organic fraction only. Hence, it can be concluded that citrinin may be also responsible for the inhibition of *M. phaseolina*. Though, citrinin is a reported mycotoxin, it is non-phytotoxic and not altering ATPase activity, respiration and photosynthetic rates when applied on sorghum leaves. The LD50 of citrinin on various animal models was also very high when compared with the concentrations used for the control of disease *Botrytis* gray mold. The control of charcoal rot disease in sorghum by *P. citrinum* VFI-51 could also be due to its capability to produce hydrolytic enzymes. In our previous investigation, *P. citrinum* VFI-51 was reported to produce siderophore, indole acetic acid (IAA), hydrocyanic acid (HCN), lipase, protease and β-1,3-glucanase. Siderophores help plants not only to acquire iron but also helps in disease suppression. IAA helps the host plants to stimulate seed germination, root formation and root elongation whereas HCN was also reported to help in disease suppression. Microorganisms producing lytic enzymes reported to play not only a role in nutrient mineralization and thus help the plants in growth promotion but also help in lysis of pathogenic fungal cell walls.

vii. ***P. multicolour*:** *Penicillium multicolor* a soil-borne fungus was evaluated for its activity against root pathogens *Fusarium oxysporum* and *Sclerotium rolfsii*. In vitro plate assay exhibited mycelia inhibition of the pathogens on PDA medium (Nair et al., 2020). Microscopic analysis revealed conidiophores and conidia deformations and mycelia piercing in the case of *F. oxysporum* where as mycelia coiling and cytoplasmic disintegration resulted for *S. rolfsii*. Rhizosphere competence of *P. multicolor* where sterile paddy straw segments placed at different centimetres up to 7cm depth to which 40ml of 2 x 106 conidial suspensions were poured and determined at 21 days of incubation exhibited mycelia growth up to 6 cm depth but observed very low colonisation frequency rate, 22%. In vitro root colonisation of *P. multicolor* and pathogens was studied for Teak and Mahogany seedlings. 2 x 105 and 2 x 106 conidial suspensions of pathogens and the bio-agent respectively at different concentrations v/v (1pathogen:1 biological agent and 1 pathogen: 5 biological agent) were tested and 1:5 ratio was found to be effective where complete suppression of pathogens were observed. The present study exerted on the further standardisation of biological agent for their success in field application.

P. multicolor against root pathogens *Fusarium oxysporum* and *Sclerotium rolfsii* exhibited parasitic activity inhibiting mycelium growth by lysing and coiling of hyphae and disintegrating cytoplasmic contents. Many bio-control agents against *Fusarium* wilt such as *T. harzianum*, *Penicillium oxalicum* and non-pathogenic *F. oxysporum* having ability to inhibit conidia production over 90% have been demonstrated. Ability of *P. oxalicum* inhibiting *F. oxysporum* mycelium and subsequent reduction in wilt incidence in Tomato plants are also reported by various researchers. There are limited studies on the activity of *Penicillium* sp. over *S. rolfsii*. The activity of *Penicillium* members performs different actions against various pathogens which involve hyphal interactions, production of various metabolites and in some cases inducing resistance on host plants. However, proper identification of *Penicillium* sp. and management strategies need to standardised for its effectiveness in field conditions. Hence, understanding antagonism mechanism and modifying development protocols need to be aimed before application in natural ecosystems.

viii. ***P. hordei* and *P.polonicum*:** To Strawberry anthracnose disease, caused by *Colletotrichum nymphaeae* is a major limiting factor in strawberry production in Kurdistan province of Iran. In this research, endophytic fungi were isolated from stem and crown organs of strawberry and their antifungal activity against *C. nymphaeae* was tested (Darya Zargar and Jahanshir amini, 2019). The selected isolates were identified as *Penicillium hordei* and *P. polonicum* based on morphology and molecular characteristics. Antagonistic effects of selected isolates were evaluated against *C. nymphaeae*, through dual culture, volatile and non-volatile compounds production mechanisms in in vitro and greenhouse conditions. In volatile compounds test, two isolates of *P. hordei* and *P. polonicum* were able to reduce *C. nymphaeae* growth (34.05 and 42.36%) compared with control. In addition, isolates of *P. hordei* and *P. polonicum* reduced the strawberry anthracnose disease severity significantly in greenhouse conditions (58.9 to 61.4%) compared to the control. According to the results both in vitro and greenhouse experiments, both isolates reduced the growth of pathogen. Therefore, the results of this research showed that it may be possible to manage strawberry anthracnose disease effectively by using *Penicillium* species as biocontrol fungus. This is the first research related to the effects of *Penicillium* species on strawberry anthracnose.

Penicillium spp. has not been previously reported as endophytic fungal isolated from strawberry plant and their beneficial effects i.e. strawberry anthracnose suppression under in vitro and greenhouse conditions. In conclusion, the results of this research indicated that selected endophytic fungi may be applied in the field to manage strawberry anthracnose and contribute to reduce chemical fungicide use and maintain the stability and durability of the resistance in strawberry cultivars against strawberry anthracnose disease. Therefore further research is needed to confirm the findings. Isolates of *Penicillium* spp. that reduced the disease severity of strawberry anthracnose should be studied in greater detail with the aim of increasing their positive effect on diseased plant in field conditions.

ix. ***P. chrysogenum* against Root-knot nematodes:** Root-knot nematode (*Meloidogyne incognita*) has become a serious risk for cucumber globally. Haphazard use of synthetic chemicals poses a serious threat to agricultural crops, eliminating predators and also polluting natural community. Thus these negative effects led towards development of ecofriendly approaches which are safe and effective for *M. incognita* management. The study has been planned to coat seeds with Snef1216 (*Penicillium chrysogenum*) and evaluate its ability to control *M. incognita* in cucumber (Sikandar et al., 2019). It reduced *M. incognita* invasion rate significantly in different inoculums 500J2 (second stage juveniles) 67.09% followed by 1000J2 and 2000J2 plant-1 60.44% and 36.02%, respectively. It inhibited development of nematodes 60.30%, 50.37% and 38.77% at 500, 1000 and 2000J2 plant-1 inoculation levels respectively compared to control. Snef1216 reduced significant reproduction rate at 500J2 (69.46%), 1000J2 (62.89%) and 2000J2 (63.62%) of *M. incognita*. It interfered in galls formation and nematodes g-1 in root mass. Snef1216 enhanced seed germination (22.09%) with germination index (53.77%) and germination rate

(64.49%). Additionally, seed dressing with Snef1216 exhibited additional biomass, reduced invasion of second-stage juveniles and also restrained development of nematode. The results suggested that Snef1216 can be introduced as a biomass enhancer and potential bio-control agent against *M. incognita* in cucumber.

Seed treatment used as a standard way for the application of *P. chrysogenum* that could concurrently confer resistance to root-knot nematodes in agriculture. The results also revealed that pathogenic level of root-knot nematode was also observed in cucumber that the number of galls and egg masses was increased with increase in inoculums level while, seeds treated with Snef1216 significantly reduced compared to control. It also inhibited the development and rate of reproduction of nematodes in the roots. Cucumber is susceptible and an excellent host for multiplication of *M. incognita* but, seeds treated with Snef1216 significantly reduced the infection of nematodes and galls at a different level of inoculums also enhanced biomass. It can be considered as a commercial biocontrol agent, however, before recommending this Snef1216 as commercial nematicides further study is needed to evaluate its active component screening and mechanisms of action. However, present findings may provide a theoretical foundation as a biocontrol agent for better control of root-knot nematodes.

8.5.11 Phages: Phages are viruses that attack specifc bacterial species, and within current years much consideration is received by them in controlling diferent diseases caused by bacteria. Phages can be used for controlling diferent crop-related diseases. Several phage-based products are accessible in the market. Compared to chemical control methods, phage biocontrol ofers several advantages (Nawaz et al., 2023). Mixtures of phages can be employed to target pathogenic bacteria. Unlike chemical control strategies, phage mixtures can be readily adjusted to counter any potential resistance. This review summarizes the use of phages as a biocontrol agent against phytopathogens.

Bacteriophages have been discovered in the twentieth century and are considered as most abundant and diverse in the natural environment and afect various ecological and biological processes having their role in bacterial mortality and genetic exchange (Chevallereau et al. 2022). They can be found in every domain or habitat, mainly in the soil and oceans. Tey infect the bacteria by using lysogenic or lytic cycles. In the case of the lysogenic cycle, a bacteriophage integrates its genome directly into bacterial cell chromosomes after which they replicate and form daughter cells. Tey initiate the infection by attaching themselves to specifc bacteria by adsorption and inserting their genome into the bacterial cytoplasm. In the case of the lytic cycle, phages are intended to utilize the ribosomes of host bacteria to make their proteins right after the injection of genetic material. At the end of this cycle, phage has produced some virion components that are lysin and holins. Lysins are also known as endolysin that cause lysis of bacterial cells. The virulent nature of lytic phages along with their exponential increase in numbers after each infection cycle makes them the best alternative of pesticides use. In the year 1920, phages were discovered, and right after a decade, they were subjected to be used as a potential therapeutic agent in the agriculture sector. Many plant diseases caused by bacterial pathogens like *Xanthomonas campestris, Pseudomonas syringae, Dickeya solani*, and *Erwinia amylovora* have been reported to be cured by bacteriophages. In this review by Nawaz et al., (2023), some major aspects regarding bacteriophages like history, life cycle, limitations, advantages as well as commercialization and control applications of bacteriophages as biocontrol agents have been summarized.

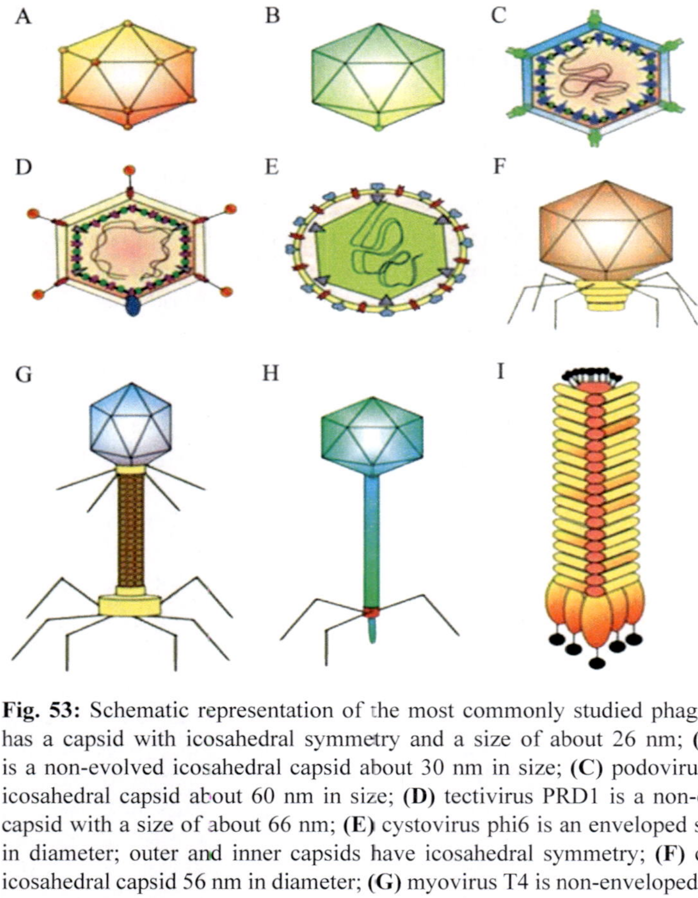

Fig. 53: Schematic representation of the most commonly studied phages: **(A)** levivirus MS2 has a capsid with icosahedral symmetry and a size of about 26 nm; **(B)** microvirus ϕX174 is a non-evolved icosahedral capsid about 30 nm in size; **(C)** podovirus T7 is a non-evolved icosahedral capsid about 60 nm in size; **(D)** tectivirus PRD1 is a non-enveloped icosahedral capsid with a size of about 66 nm; **(E)** cystovirus phi6 is an enveloped spherical virion 85 nm in diameter; outer and inner capsids have icosahedral symmetry; **(F)** corticovirusPM2 is an icosahedral capsid 56 nm in diameter; **(G)** myovirus T4 is non-enveloped, with a morphological head–tail structure about 110 nm in length; **(H)** siphovirus T5 is non-enveloped, with a head–tail structure; head is about 60 nm in diameter; and **(I)** inovirus M13 is non-enveloped, with rods of filaments 7 nm in diameter and 700 to 2000 nm in length.

The class *Caudoviricetes* is the most studied among the phage groups, as it is the order in which phages for biocontrol of agricultural crop diseases are found. Its members have an icosahedral capsid and a tail containing trimeric protein fibers in the lower part that serves as a receptor for interaction with the host cell, with stability in L morphology

(Villalpando-Aguilar et al., 2022). The class *Caudoviricetes* include phages according to the tail shape: the myovirus includes phages with long and contractile tails that stand out, the podovirus have short non-contractile tails, and the siphovirus have long and flexible tails (Fig. 39) (Subirats et al., 2016).

Proteobacteria are mainly gram-negative and cause many diseases in agricultural crops. From a structural point of view, some examples show how the structure of bacteriophage is related to bacterial species that cause damage to agricultural crops. For instance, *Xanthomonas* spp. are infected by phage Xaj2 and phage Xf2. Phage RSM3 infects *Ralstonia solanacearum*. *Pseudomonas syringae* is infected by like phage phobos or phage MR1-MR18, classified in podovirus or myovirus, and cystovirus as the phage phi6 or phi8. The most studied are myovirus phage T4 and levivirus as phage MS2, which infect *E. coli*.

i. Recent applications of bacteriophages for controlling phytopathogens: Almost every phage-based research and publication is confned to the isolation and characterization of bacteriophages, while some of them gave insight into the potential of phage therapy (Hatfull et al. 2022). Bacteriophages possess a wide range of biocontrol applications that are widely used in the feld of agriculture to protect plants from various bacterial diseases. Several transgenic plants express the specifc bacteriophage enzymes for their survival and protection from bacterial pathogens. In the last decade, several bacterial plant pathogens were described by numerous scientists and pathologists. Out of them, one of the most important plant pathogens is the *P. syringae* responsible to cause several plant diseases in monocots as well as in dicot plants from all over the world. To check the efcacy of phage in this case, two diferent feld trials were conducted on a phage cocktail of about six isolated bacteriophages. Tis trial was carried out by using a phage as a biocontrol against *P. syringae* that caused the bacterial blight disease in leek. So, it was evident from the experiment that the phage cocktail has the potential therapeutic cure for the bacterial blight disease and successfully eradicates the plant infection caused by the *P. syringe* (Nawaz et al., 2023).

Bacteriophages that are used for biocontrol in agriculture must be stable enough in the environment to tolerate (UV) radiations, temperature fuctuations, and chemical agents and must be lytic (Wang et al. 2022). Agriphage is a phage-based product, produced by the approved US company Omnilytics, which can control the bacterial spot diseases of peppers and tomatoes. Several bacteriophage enzymes such as ΦXo411 and Lys411 have been isolated showing lytic activity against the *Xanthomonas*. Phage should be applied directly while treating bacterial infections at early stages for better and more efcient results. Referring to experimental trials, *Xanthomonas campestris* pv. *pruni* is a bacterial pathogen that causes leaf spots. It was subjected to two diferent treatment approaches. In the initial phase, the treatment was administered one hour prior to the bacterial inoculations, while in the subsequent phase, it was applied 24 h in advance of the bacterial inoculation. Scientists have also demonstrated that the activity of phage concerning their time for a specifc application like dawn and dusk is the most convenient time for the better activity of phage. At those specifc times, the reduction of UV radiation is observed, which increases the efciency of phage activity. The selection of suitable bacteriophages in growing conditions is also an important element. RSL1 is a phage against bacterium *Ralstonia solanacearum*, which showed a signifcant resistance pattern when exposed to high temperatures, i.e., 37–50°C.

Streptomyces scabies is a Gram-positive bacterium responsible for causing infections on potatoes leading to the formation of corky lesions generally known as common scabs and results in reduced growth of seedlings. The pathogen can be efectively treated by using phage so the chances of infection in potato crop can be successfully eliminated. In summary, the above studies demonstrate phage potential in treating various plant diseases (Nawaz et al., 2023). Bacteriophages have also been used in feld conditions and greenhouses. To control *R. solanacearum*, phage was applied directly into the rhizosphere through soil drenching that was efective in the suppression of the development of wilting in the tomato plant. In the case of other soil-borne pathogens like *Xanthomonas euvesicatoria, X. campestris pv. campestris*, and *P. carotovorum* subsp. *carotovorum,* foliar spraying method was used to eradicate the disease incidence in plants caused by these pathogens. Furthermore, another application regarding the potential of flamentous phage has also been reported. Te flamentous ΦRSM-type phage enhances the eradication of *R. solanacearum* in tomato plants (Umrao et al. 2021).

Besides biocontrol applications, some signifcant factors afect the phage efectiveness used in the feld of agriculture. In a recent evident study, it was demonstrated that the pH of the fruit has a vital impact on the phage stability as well as its activity (Bastas and Baysal 2022). In the case of fresh-cut melon, application of the potential phage cocktail that is specifc to Salmonella was efective, as the bacterial population was signifcantly reduced by 3.5 logs at 10 °C and 2.5 logs at 20 °C. Conversely, in the case of apple, the same phage cocktail was applied and no reduction in bacterial population was observed. It was predicted that it may be possible due to the low pH of apples. To ensure the formulations of phage in such a way that it can possess a wide host range, multiple phages were combined into a mixture known as a bacteriophage cocktail. It is an efective way to recompense the host range limitation of a single bacteriophage. This technique can also eradicate the chances of the development of phage-resistant bacteria. Studies regarding the efectiveness of phage cocktail on *Xanthomonas* are also quite important to invective the impact of diferent environmental factors on phage therapy. Recently, a study reported regarding the biocontrol activity of phages against *P. syringae.* Aerosol of single phage containing 5% sucrose and 3% corn four sprayed on bean leaves before inoculation of *P. syringae*. Te disease severity was reduced about 60% when a single phage was used and about 70% reduction was observed by using a phage cocktail. Moreover, further research is needed in the feld of agriculture to implement the role of bacteriophages on a global scale. In crops propagation, the stability of phage cocktails mainly depends upon phage resistance to adverse environmental factors (Nawaz et al., 2023).

ii. Lytic proteins of bacteriophages for controlling phytopathogens: Bacteriophages have evolved numerous strategies to release hundreds of new progenies from the host to the external environment (Łoś et al. 2021). Most flamentous phages constantly release new virions from the host cell through extracellular vesicles without lysing the bacterium. While in some cases, bacteriophages release their progenies through lytic enzymes which are a quite benefcial in controlling the pathogenic bacteria as the phages possess a wide range of hosts. That's why they can easily eliminate the risk of the development of resistant strains by reproducing and infecting a wide variety of bacterial strains. Bacteriophages have evolved group of lytic enzymes that are responsible for the degradation of the peptidoglycan layer of the bacterial host. These enzymes are known as endolysin. During the late stages of replication, these hydrolases are activated with the help of holin proteins and destroy the peptidoglycan (PG) layer of the bacterial host which results in the release of progeny virions. Endolysins are classifed into fve diferent groups based on the bonds in peptidoglycan they target. Phage infecting Gram-negative bacteria produces endolysin with a single globular domain known as enzymatically active domain EAD that is responsible for the digestion of the PG layer (Nawaz et al., 2023).

Endolysins CN77 and CMP1 have also been reported for lysing *Clavibacter nebraskensis* and *C. michiganensis* respectively. Te endolysin can lyse *C. michiganensis* subspecies specifcally without afecting the soil bacteria (Nawaz et al., 2023). Moreover, endolysins have extended activity and they are capable to infect other opportunistic pathogens like *Stenotrophomonas maltophilla*, *Pseudomonas aeruginosa* as well as *Xanthomonas* species by simply degrading their peptidoglycan. Expression of genes of endolysin in several plants is a way to reduce resistance in pathogenic bacteria. *C michiganensis* is a species that causes bacterial infections such as wilt and canker in the tomato plant. This infection can be prevented by using a bacteriophage endolysin, namely CMP1, normally expressed by transgenic tomato plants. The application of endolysin was not only limited to bacteria as its efect has also been noticed in various cases of fungi. A remarkable resistance pattern to infection has been observed when T4 lysozymes bacteriophage was introduced into those plants that have been infected by several species of fungi like *Magnaporthe oryzae* and *Rhizoctonia solani*. Te endolysin can disrupt the cell wall of Gram-negative and Gram-positive bacteria when applied exogenously and have the potential to hydrolyze the peptidoglycan layer of Gram-positive bacteria more easily due to the absence of their outer membrane (Lee et al. 2022).

iii. Use of bacteriophages with other strategies for plant disease management: Researchers have started examining outcomes caused by a combination of phages with other disease control measures. The hypersensitive response of diferent phages in several infected plant models has been studied against *Xanthomonas* spp (Nawaz et al., 2023). Te combined technique of phage with other control agents and chemicals like copper hydroxide with mancozeb resulted in synergistic benefts in the control of pathogens of various plants. Copper and mancozeb can help in the control of plant disease as they can enhance the penetration of the plasma membrane. Other evidence regarding the combination of phage with other control strategies in plants has also been reported. *Pantoea agglomerans* has been considered as an excellent biocontrol agent as they can suppress fre blight in some plants caused by *Erwinia amylovora*. Besides, the enhanced biocontrol activity would be achieved if it is used with the combination of phage. A similar enhancement has also been observed in the case of treating tobacco bacterial wilt. When *Ralstonia solanacearum*, a bacteriocin-producing strain, was used with the combination of phage, remarkable results were obtained (Nakao et al. 2021). Besides, there is an organic compound, namely Acibenzolar-S-methyl (ASM), that is commonly used as a fungicide has shown increased control activity when it was combined with bacteriophage against bacterial disease in the tomato plant. Te combination of phage with antimicrobial agents has contributed well to the reduction of several plant diseases. Systemic acquired resistance (SAR) is a kind of resistance in plants acquired due to exposure to several virulent and pathogenic microbes.

Employing carrier bacteria with bacteriophages for managing phytopathogens: Besides several environmental factors, the abundance of host bacteria can also cause variations in phage count. To improve the activity and efciency of phage, the idea of the combination of phage with non-pathogenic bacteria has been raised (Nawaz et al., 2023). Non-pathogenic carrier bacterium should be used with the combination of phage, as it will not cause any harm to the plant nor phage properties. In this regard, several strategies have been investigated. Recently, a group of researchers have isolated *Erwinia amylovora* phages to characterize them and to use them with carrier bacteria. Initially, a temperate phage was used that was capable to infect *E. amylovora* species as well as a saprophyte. Currently, a carrier system of *P. agglomerans* has been combined with the phages that had reduced the risk of blossom blight diseases. Tis combination also reduced the risk of fre blight in the plant as well. In a nutshell, co-application of phage with carrier bacteria has reduced the risk of infection to a greater rate (Meile et al. 2022). Tis process is considered as economical as they do not need any purifcation which is an advantage too.

Challenges and issues associated with bacteriophages for controlling phytopathogens: Tough the use of bacteriophages to control bacterial plant pathogens was successful at early stages, phage therapy did not end up to its potential and applied antibacterial stratagem for the control of pathogenic plantbacteria because of issues with reliability and efcacy (Nawaz et al., 2023). In 1963, Okabe, a forerunner in the feld, concluded, in general, that phages appeared to be inefcient as a controlling agent. Furthermore, their narrow range of action against a particular bacterial strain puts them in a difcult spot against various antibacterial agents, i.e., antibiotics, which have had an extensive range of action (Majkowska□Skrobek et al. 2021).

Emergence of bacterial resistant strains against bacteriophages: While considering phages to be used as biological control agents, the real concern is the possibility of bacterial mutation that makes them resistant to specifc phages. As various studies have shown, when identifying an appropriate phage as a biocontrol agent, the major factor that could be drastic is genetic variability shown by bacterial strains related to specifc plants that might not have existed previously (Larrahondo-Rodríguez et al. 2022). For example, *Xanthomonas campestris* pv *vesicatoria*, a bacteria linked with the bacterial spot disease of pepper and tomato, now has about four diferent species. Diferent phages are required to control the disease for each of the four strains because phages are mostly specifc to a particular strain of a species of bacteria. Besides, there could be a signifcant change in the specifcity of phage to diferent strains in a bacterial species.

Advantages of using bacteriophages over other techniques: Without any harm, humans are exposed to phages daily as phages exist in the environment naturally, as compared to chemical biocides. If the target bacterial host is reachable to the phage, their numbers increase after application (Majdinasab et al. 2021). However, phages manage to multiply if the target host is available to them. Unlike metal-based pesticides, phage does not efciently accumulate in the soil. Generally, phages only target a specifc strain of bacterium within a species. Basit et al. (1992) isolated a phage that could kill the bacterium that did not help in nitrogen fxation, but was not able to infect the strain *Bradyrhizobium japonicum* that help in soyabean growth because of its nitrogen-fxing properties, ultimately leading to better crop growth. A signifcant factor in the virulence of plant pathogens is bioflm formation. It is involved in the resistance of bacterial plant pathogens to metal bactericides. But phages have adapted to cope this problem of bioflm, by the release of depolymerase enzymes that destroy the bioflm material, making the receptors on the surface of the target bacterium accessible to the phage. The increasing consumers demand for chemical-free preservatives, and biocide-free food products have led to limited use of chemicals on crops. Phages could be registered as biopesticides as they naturally exist in the environment, which makes them appropriate for organic consumer-friendly farming (Nawaz et al., 2023).

Methods for applications of bacteriophages for getting maximum activity: Poor persistence on the phyllosphere is one of the limiting factors for efcient biocontrol on crops by phage, though various methods have been studied to minimize this issue. With the availability of living hosts in the rhizosphere and phyllosphere, phage persistence could be improved (Nawaz et al., 2023). Phage-based control could be improved by applying the phage during dark hours. Certainly, longer phage persistence was observed in the phyllosphere when phage was applied in the evening, allowing phage to infect and kill their targeted bacteria. It shows that a vast range of substances, which absorb UV resulting in limiting its exposure to phage, can boost the performance of phage in the phyllosphere. Tese protective efects are also observed with the use of biodegradable polymers (Nga et al. 2021). Improved activity of phage was showed by applying the following combinations with phage: (1) 0.5% sucrose and 0.75% skim milk, (2) 0.5% sucrose and 0.5% (PCF) pregelatinized corn, and (3) 0.25% (PCF) pregelatinized corn, 0.5% sucrose and 0.5% Casecrete NH400. These experiments were performed with phages against *Xanthomonas campestris* pv. *vesicatoria* on tomato plants in feld and greenhouse trials. All formulations mostly showed improved disease control (Collinge et al. 2022).

Challenges: Challenges that may hinder successful phages application and how they can be overcome are provided in table 17 (Halawa, 2023).

Table 17: Challenges that may hinder successful phages application and how they can be overcome

Challenges that may hinder successful phages application	How they can be overcome
1. Inactivation of phages by UV irradiation	- Using a protective formulation that protects phages against UV - Applying phages after sunset
2. Poor persistence of phages in the rhizosphere and phyllosphere	- Accompanying phages by a viable host - Avoiding daylight during phages application
3. Possibility of lysogens or pseudo lysogen production	- Using only lytic phages
4. The difficulty of eliminating every member of a particular bacterial genus or species due to the narrow host range for phages	- Development of phage cocktails
5. Instability of phages survival under improper storage conditions	- Keeping phages refrigerated and protected from light - Production of stable bio formulation and conversion of phage Formulation from liquid to powder state
6. Possible development of phage resistance in the bacterial host	- Development of phage cocktails
7. The inability of phages to disperse or interact with their target bacteria when there is a lack of moisture on the leaf surface	- Applying phages when free moisture is expected to stay on the leaves (such as dew or rain)
8. The difficulty of applying phages evenly over large tree leaves	- Applying phages via tree vascular system

Commercialization of bacteriophage as a biocontrol agent: In the last decade, various biocontrol agents have been introduced into the market commercially. Omnilytics is a first-ever USA-based company that received registration regarding phage-based biocontrol products. Agriphage is a phage-based product that can control the bacterial spots of peppers and tomatoes (Nawaz et al., 2023). Another Hungarian-based company Enviroinvest received registration regarding their product, namely Erwiphage, designed to control apple tree disease. Besides, a Scottish company owns a product, namely APS biocontrol, specialized for the control of soft rot diseases caused by Enterobacteriace. However, in some countries, there are some gaps and delays in legislation to allow the biocontrol agent in controlling bacterial plant diseases. This problem was arisen due to the constant changing of newly emerging strains of target bacterium. The problem could be overcome by the adaptations of new emerging strains as well as constant updating. Moreover, the legislation related to phage biocontrol should be pliable for the best applications and implementation of biocontrol agents.

iv. Legal status of bacteriophages as biopesticides: Notably, microbial biocontrol agents (MBAs) are highly efective biopesticides made of living microorganisms. Tough viruses ought not to be categorized this way, the international regulatory framework categorizes viruses along with other true microorganisms, i.e., fungi and bacteria

(Nawaz et al., 2023). Concerning various pesticides, before marketing and registration, MBAs ought to go through very cautious hazard assessment to environmental fate and food safety. Te evaluation system of the EU was frst established in regulation 91/414/EEC, followed by directive 1107/2009 which annulled and succeeded the regulation. Along with its execution, just 26% of registered Plant Protection Products (PPP) and active substances qualifed the audit, compared to those under regulation 91/414/EEC. Hence, because of increased awareness about environmental and consumers safety, only 1/4th of the active substances have not been annulled in agriculture. In the USA, the regulatory framework is less complicated, because only two administrations are concerned, the Food and Drug Administration (FDA) and Environmental Protection Agency (EPA). On the other hand, in European Union EU has at least four main administrations involved: the committee on Plants, Animals, Food and Feed (PAFF Committee), the Directorate-General for Health and Food Safety (DG SANTE), the European Food Safety Agency (EFSA), that provides risk communication and risk assessment and the Rapporteur Member State, where the candidate presents a report that contains all related data on the MBA.

Approved patent calls are quite comprehensive covering application methodology. The increased patenting activity and publishing show there is great interest in bacteriophages. On the contrary, very few bacteriophages based legal biopesticides are available commercially. In a reported opinion, EFSA removed bacteriophages from the Qualifed Presumption of Safety (QPS). The QPS list enlists microorganisms deliberately introduced into the feed and food production chain. In 2007, it was frst established that in the framework of commercial approval, it is updated and revised every year by EFSA. Lest as, i) the last level of taxonomy (Caudovirales order) is too big, ii) shortage of detailed assessment at the genomic level to diferentiate between non-transducing and transducing bacteriophages or if they contain virulence factors.

8.5.12 Identification and Characterization of Triple Action Bioagents (TAB) and their potency against *Fusarium* Wilt of Lentil: The current work focused on isolating potentialbio-agents exhibiting copper oxychloride resistance and evaluating their efficacy in seed treatment for ecologically sustainable management of *Fusarium* wilt of lentil (Nisa et al., 2021). Seventy biocontrol agent isolates were isolated and tested for resistance by growing them on Potato Dextrose Agar medium (PDA) amended with copper oxychloride at the rate of 2500 ppm. Isolate-H10 and isolate-C9 showed more excellent compatibility with copper oxychloride fungicide with 69 mm and 65 mm radial growths, respectively. The isolates H10 and C9 had the highest inhibitory percentages of 84.30% and 83.94% against *Fusarium oxysporum* f. sp. *lentis*, respectively, and the highest phosphorus solubilization index (PSI). Primers (ITS 1 and ITS 4) identified these putative bioagents as *Trichoderma harzianum* isolate skua-tab-1 and *Penicillium crysogenum* strain Tab2. Sequences were submitted to the NCBI and assigned the accession numbers MK414603 and MK418066. In pot culture, these isolates also demonstrated their superiority in reducing the disease incidence and severity if seeds were treated with H10 and C9 alone or in combination with copper oxychloride fungicide. The two isolated bioagents exhibit three fundamental properties: compatibility with copper oxychloride, antagonistic activity toward the pathogen fall armyworm, and the ability to dissolve phosphorus minerals.

The present study found that the pathogen associated with wilt in lentils in the Himalayan state of Kashmir was found to be *F. oxysporum* f. sp. *lentis*. A study was conducted to isolate the copper oxychloride resistant bioagents and use them against *Fusarium* wilt of lentils. Bioagents H10 and C9 were found resistant to copper oxychloride at 2500 ppm (Nisa et al., 2021). This technique proved to be a reliable and effective way to manage the disease on an ecological level. Highly effective antagonistic (copper oxychloride resistant) fungi that were screened by the dual culture test were further evaluated for phosphorus solubility. It is noteworthy that the top two fungal isolates, H10 and C9, which proved to be highly antagonistic, also had the highest "P" solubilizing activity. Fungal isolates H10 and C9 were selected for further pot culture evaluation to determine if both effectively suppressed the disease alone or combined with the fungicide copper oxychloride. Due to their triple effect, i.e., resistance/compatibility with copper oxychloride fungicide, antagonistic activity, and superior solubilizing activity of "P", isolates H10 and C9 were renamed TAB-1 and TAB-2 (triple effect biological control). After morpho-cultural characteristics and DNA molecular approaches using ITS-1 and ITS-4, the bioagent TAB-1 (H10) was identified as *Trichoderma harzianum* isolate skua-tab-1 andbioagent TAB-2 (C9) as *Penicillium crysogenum* strain Tab-2. TAB-1 and TAB-2 were found to be more effective in controlling wilt disease in lentils in pot culture trials than untreated control and copper oxychloride fungicide alone. The control variant improved significantly when combined with a fungicide containing copper oxychloride. The current investigation demonstrated that the isolated and described bioagents are essential in integrated disease management and sustainable agriculture, preventing pathogen resistance to synthetic pesticides. Apart from that, they are economically significant due to their excellent field-level applicability compared to any other common antagonistic bioagent and their biofungicidal and bio-fertilizer potentialities (as evidenced by high "P" solubilization indices). As a result, their use will not only minimize losses but will also increase yield through "P" solubilization, as "P" fixation is a common concern in this region's soils. These identified potential bioagents could very well be helpful in a variety of soil-borne diseases (such as wilts and root rots) in a variety of other crops.

Furthermore, copper fungicide inhibits pathogen growth quickly, and the biocontrol agent establishes itself effectively by the time the copper fungicide's efficacy has expired. As a result, their combination can give long-term protection against diseases. Copper oxychloride and antagonistic/bio-fertilizer fungi are not only a successful IDM/IPM approach, but they are also an organic management strategy since copper oxychloride is accepted in organic farming systems.The compatibility of bioagents with chemicals and fertilizers has been studied extensively by various researchers. *Trichoderma harzianum* was found highly compatible with copper oxychloride, phorate, and carbofuran. *Penicillium* spp. EU0013 exhibited resistance against benomyl fungicide. In previous studies, bioagents successfully inhibited the *Fusarium* wilt and showed their ability to solubilize phosphorus by forming a wide zone of clearance on pikoviskya medium (PKA) (Nisa et al., 2021). *Trichoderma atroviride*, *T. hamatum*, *T. harzianum* and *T. koningii* successfully inhibit the growth of *F. oxysporum* f. sp.

ciceri and the inhibition mechanisms by these biocontrol agents include competition, mycoparasitism, antibiosis (Poveda et al., 2021). *T.harzianum* has ability to induce systemic resistance in plants and is considered a good agricultural biostimulant. *T. harzianum* and *T.parareesei* have the ability to promote stress tolerance to salinity and drought in plants. *Trichoderma parareesei* was previously reported that cumin seed treatment with the combination of carbendazim and *T.viride* resulted in only 10.6% disease incidence of cumin wilt caused by *F. oxysporum* f. sp. *cumini*.The combination of all three properties, i.e., copper tolerance, antagonism to soil pathogen(s), and highly efficient solubilization of "P", is a potential novelty explored and offered by this research that could be useful in a wide range of contexts other than the lens wilt disease studied in this case

8.5.13 Soil-borne and compost-borne *Penicillium* sp. and ***Gliocladium*** spp. against anthracnose-induced decay on tomato fruits: Anthracnose is a soil-borne tomato (*Solanum lycopersicum* L.) disease caused by *Colletotrichum coccodes* and associated with early dying syndrome and postharvest fruits decays leading to significant yield losses. A biocontrol trial was conducted using *Penicillium* sp. and *Gliocladium* spp. isolates recovered from soil and compost for the evaluation of their antifungal potential against the target pathogen (Hassine, et al., 2022). Cell-free culture filtrates of the microbial agents tested at different concentrations displayed an important antifungal activity based on pathogen mycelial growth inhibition and spore germination suppression. The highest antifungal potential was induced by *Penicillium* sp. CH6, *G. catenulatum* Gc1 and *G. virens* Gv1 cell-free filtrates applied at 3 concentrations (10, 15 and 20% v/v). *C. coccodes* mycelial growth was significantly reduced by 16 to 84% with chloroform and ethyl acetate extracts of *Penicillium* sp. CH6 and *G. catenulatum* Gc1 isolates applied at the concentrations 1, 2.5 and 5% v/v. These treatments had induced an important morphological alteration to pathogen mycelium expressed mainly by a decrease in colony melanization. Testing on pathogen-inoculated fruits, treatment based on cell-free culture filtrates and organic extracts had significantly reduced anthracnose severity as compared to control. Rot lesion diameter was significantly decreased about 22% following treatments with *Penicillium* sp. CH6 and *G. catenulatum* Gc1 cell-free filtrates. All chloroform and ethyl acetate extracts of isolates tested had suppressed disease severity by 11 to 59% and 14 to 85%, respectively, as compared to the untreated control. The results demonstrated the involvement of bioactive compounds in the cultural filtrates and organic extracts tested. These microbial agents may be explored as an alternative tool for the protection of tomato fruits from fungi-associated decays.

8.5.14 Microbial cell-free filtrates in suppressing anthracnose-induced fruit decay: The cell-free cultural filtrates of the microbial agents tested were treated on tomato fruits in order to check their ability to suppress anthracnose decay severity (Hassine, et al., 2022). ANOVA analysis revealed that the external diameter of the rot developed within treated and inoculated wounds significantly varied (at $P \leq 0.05$) depending on treatments tested. Disease severity was significantly lowered by 22 to 25% versus control following fruit treatments with cell-free filtrates from *Penicillium* sp. (MC1, CH6) and *G. catenulatum* (Gc1) cultures but the decreases induced by the other tested treatments were not significant as compared to control.

Inhibitory effect of biocontrol agents tested was observed using the cultural filtrate and the different organic extracts. Similar trend was recorded with the cultural filtrate from *T. harzianum* against *C. coccodes* growth. The cell-wall-degrading enzymes (CWDEs), such as chitinase and *β-1,3*-glucanase, released in the cell-free cultural filtrates were responsible of the inhibition of *C. gloeosporioides* (Ruangwong et al., 2021). The cultural filtrate of *T. viride* and *Pseudomonas fluorescens* were very effective in inhibiting the mycelial growth of *C. capsici*, the causal agent of anthracnose on chili. The antifungal potential of *T. viride* and *T. harzianum* cultural filtrates against *C. coccodes* growth were highlighted which ranged between 18 and 55% when used at 3.5 to 5% v/v, respectively. Also, some bacterial cell-free culture filtrates such as those of *Bacillus amylolique faciens* displayed antifungal activity against several phytopathogenic fungi including *C. dematium* where this effect was more attributed to antibiotics released in these filtrates. Similarly, the application of the cultural filtrate of the novel endophytic bacterium, *Stenotrophomonas maltophilia* at 1 ml has suppressed *C. nymphaeae* mycelial growth.

Several previous studies reported the effectiveness of microbial culture filtrates against postharvest pathogens, where *Streptomyces* filtrates had reduced anthracnose incidence and severity under controlled conditions by 85–88% over control. Nevertheless, some filtrates have reduced anthracnose severity and pathogen sporulation at the inoculation site showed the ability of *Penicillium* sp. isolates to reduce *Sclerotium* sp. mycelial development through the production of antibiotics. This reduction of fruit infection levels was also recorded using cell-free filtrates from *T. viridae*, *T. pseudokoningii* and *T. harzianum* cultures against *C. capsici* development (Hassine, et al., 2022). In addition, the application of *T. harzianum* culture filtrate under greenhouse conditions has reduced greatly disease severity associated to *C. sublineolum* infection and has also concomitantly promoted sorghum growth.

Ruangwong et al. (2021) demonstrated the effectiveness of *T. koningiopsis* culture filtrate in controlling *C. gloeosporioides* on chili pepper and recorded no disease symptoms following treatments. Similar decreases in disease lesion and severity on chili pepper (*C. gloeosporioides*) were also observed using the culture filtrate of the entomopathogenic fungus *Ophiocordyceps sobolifera*. Similar disease suppressive effects have been obtained using 50 µl of methanolic extracts of the actinomycete bacterium (*Streptomyces* sp.) for the control of *C. musae* on banana fruits. The n-butanol extracted of the chitinolytic bacterium *Paenibacillus ehimensis* has also completely inhibited the spore germination of the phytopathogen *C. gleosporioides* on pepper fruits. Similarly, the ethanol extracts of *O. sobolifera* applied at 50% (v/v) on chili pepper have also reduced anthracnose severity and symptom appearance associated to *C. capsici* and *C. gloeosporioides* infections under pot conditions. The ethyl acetate extract of the endophytic fungus *Epicoccum dendrobii* has eventually suppressed *C. gleosporioides* development on detached leaves of Chinese fir.

The present study demonstrated the antifungal activity of cell-free cultural filtrates and organic extracts of microbial agents (*Penicillium* sp. and *Gliocladium* spp.) tested against the in vitro and the in vivo growth of *C. coccodes*. The most effective bio-agents in controlling *C. coccodes* pathogen mycelial growth and anthracnose severity on tomato fruits were CH6 of *Penicillium* sp. and Gc1 of *G. catenulatum*. Further experiments are needed for the identification of the nature of chemical compounds involved in the recorded antifungal activity by the high-performance liquid chromatography (HPLC-MAS) for the corresponding solvent extract.

8.6 Biocontrol

A deeper understanding of the relationships between plants and pathogens along with the environmental factors prevalent in a particular area needs to be understood prior to biocontrol implementation, particularly under widespread disease conditions (Pandit et al., 2022). In plant pathology, biocontrol is defined as the interaction of numerous environmental elements with the goal of reducing the negative impacts of harmful species while promoting the growth of beneficial crops, helpful insects, and microbes. Biological control is dependent on numerous agonistic and antagonistic interconnections between plants and microbes living in the rhizosphere and phyllosphere and their application to minimize disease and subdue pests. Organisms from the rhizosphere can be harnessed from the surrounding environment (the black box approach) or can be introduced into the field from external sources (the silver bullet approach). It is beneficial to apply a consortium of microbes with collaborative properties rather than relying on a single organism since microbial consortia make up a stable rhizosphere that offers more effective control against pathogens. Apart from microbial applications, the utilization of other plant products like extracts, biofertilizers, and biopesticides, natural enemies of pests and pathogens, and gene products also aid in carrying out biological control.

i. **Microbial Biocontrol:** The rhizosphere is the soil area that surrounds the roots and is composed of microbes capable of repressing plant pathogens. It, therefore, aids in providing natural protection to the plants against a variety of organisms either directly by synthesizing metabolites antagonistic towards the pathogens or indirectly by suppressing pathogen growth and improving the host's defense mechanisms (Pandit et al., 2022). Antibiosis caused by the release of antibiotics, organic compounds, toxins, and various hydrolytic enzymes like beta-xylosidase, chitinase, pectin methylesterase, β-1,3-glucanase, etc. is one of the mechanisms employed by the rhizosphere microbial population to carry out the destruction of the pathogen including disintegration of the glycosidic linkages in its cell wall. Plant growth-promoting rhizobacteria (PGPR) residing in the rhizosphere also perform biocontrol by reducing the incidence of plant disease thereby assisting in plant growth. The PGPR also promote antibiosis, competition, production of metabolites that induce systemic acquired resistance (SAR) and induction of systemic resistance (ISR), parasitism, production of hydrolytic enzymes such as cellulase, glucanase, chitinase, and protease that break down the cell wall along with a number of antibiotics like oomycin A, 2,4-diacetyl phloroglucinol (DAPG), pyoluteorin etc against the pathogens.

For example genus, *Serratia* belonging to Enterobacteriaceae is a PGPR that produces secondary metabolites having attractive biocontrol properties (Soenens and Imperial, 2019).

ii. **Fungal Biocontrol:** Apart from their ability to improve nutrient uptake and nitrogen use in plants, fungi also have biocontrol capabilities. They can aid in the fight against pests like nematodes and microbial pathogens that infect various parts of the plant such as roots, foliage, and fruits (Pandit et al., 2022). They offer protection against diseases with the help of processes like mycoparasitism, competition for resources with pathogens, antibiosis, conferring ISR to the host plant, and mycovirus mediated cross-protection or MMCP. Some of the well-known fungal biocontrol agents include the *Trichoderma* species, ectomycorrhizas, arbuscular mycorrhizas (AMF), yeasts, and endophytes. Even the nonvirulent strains of certain pathogens can utilize hypovirulence-associated mycoviruses in order to function as biocontrol fungi. With improved biotechnological and genetic advances, it is not only possible to introduce beneficial fungal genes into the genomes of the host plants but also to interrupt or overexpress these genes in order to improve biocontrol ability. A comprehensive list of fungal biological control agents that were used against fungal plant pathogens according to modern taxonomic concepts, and clarifies their phylogenetic relationships was provided. Furthermore, they clarify that this is important in view of the wrong names are frequently used in the literature of biocontrol. They list details of some 300 fungal antagonists belonging to 13 classes and 113 genera together with the target pathogens and corresponding plant diseases. According to them, *Trichoderma* is identified as the genus with greatest potential of biocontrol and it comprises 25 species as biocontrol agents that have been used against a number of plant fungal diseases. In addition, nine more genera were recognized by them as significant in this regard that comprise five or more known antagonistic species, namely, *Alternaria*, *Aspergillus*, *Candida*, *Fusarium*, *Penicillium*, *Pichia*, *Pythium*, *Talaromyces*, and *Verticillium*. Majority of the plant growth-promoting fungi (PGPF), viz., *Trichoderma*, *Penicillium*, *Aspergillus* and *Fusarium* spp. are reported for their abilities to stimulate the plant immune responses upon enemy attack and are considered as one of the safest modes for induced systemic resistance (ISR) and growth promotion in crop plants (Divya et al., 2021). In addition, PGPFs are also known for being beneficial to plants in reducing the impacts of various fungi, bacteria, viruses and nematodes by eliciting ISR. *Trichoderma* species are soil-borne filamentous fungi known for its utility in many plant health benefit applications. Its strains deploy a complex mechanism in pathogen control that includes colonizing the soil and root of the host, inhabiting a physical space and evading the multiplication of the phytopathogens while concomitantly producing cell wall-degrading enzymes, antimicrobial metabolites to kill the pathogens, inducing plant defense mechanisms, promoting plant development and improving plant tolerance to biotic and abiotic stressors.

iii. Plant virus and Biocontrol: Qu et al., (2021) have elucidated the effects of a single-stranded DNA virus, *Sclerotinia sclerotiorum* hypovirulence-associated DNA virus 1 (SsHADV-1) that infects the fungus *Sclerotinia sclerotiarum*, a disease causative agent of many crops (Pandit et al., 2022). Qu et al., (2021) have further elucidated the altered expression of phenotype related genes upon SsHADV-1 infection by using digital RNA sequencing.

Some common and well characterized useful viruses that harbour the plants are those that are known to enhance the beauty of ornamental plants. Tulip breaking virus was the first of this lot. However, many other prized ornamentals owe their value to some extent to the viruses that infect them. Other examples of beneficial plant viruses include several acute viruses, such as Brome mosaic virus, family Bromoviridae, Cucumber mosaic virus, family Bromoviridae, Tobacco rattle virus, family Virgaviridae, and Tobacco mosaic virus, family Virgaviridae, which confer tolerance to drought and freezing temperatures in various crops, and persistent viruses, such as White clover cryptic virus (family Partitiviridae), which can suppress nodulation in legumes during proper supply of nitrogen. Further elaborated by stating that the mild symptom causing plant virus strains have been used for cross-protection against more severe strains and this attribute is utilized in pathogen-derived transgenic resistance strategies. A good number of literature reveal that plant infected with viruses do not show any apparent ill effects on their hosts in the beginning. However, their persistence as displayed by the virus families Chrysoviridae, Endornaviridae, Partitiviridae and Totiviridae that are a group of the most common viruses found in wild plants, has got significance from scientific perspectives pertaining to biocontrol of plant diseases. These viruses are considered as having very long relationships with their plant hosts. Persistent viruses are also common in crops, including peppers, rice, beans, carrots, figs, radish, white clover, melons, barley and avocados (Roossinck et al., 2020).

Plant viruses are controlled either by host resistance, e.g., Plum Pox Virus (PPV) by activation of members of a cluster of meprin and TRAF-C homology domain (MATHd)-containing genes that were designated as possible PPV resistance genes, mild strain cross protection, e.g., Pepino Mosaic Virus (PepMV)-based cross protection in the crops or by biocontrols of their insect vectors, e.g., parasitoids of mealybugs that vector GLRaV-3. The latter approach has been tested for vine mealybug, *Planococcus ficus* that feeds through a membrane feeding system on GLRaV-3, which was blocked with some blocking molecules in a test study for such molecules tapping the feeding membrane system of the parasitoid vector. Pechinger et al. (2019) have provided a detailed list of protective virus isolates and their respective challenging isolates tested for their mild strain cross-protection capabilities. In case of host resistance, host R genes typically induce race-specific resistance in response to the Avr genes of pathogens. During plant–virus interactions occurring in a single cell, an R gene triggered HR response is vital that kills infected cells and restricts the viral invasion and this phenomenon is associated with several molecular events, such as the activation and expression of salicylic (SA), jasmonic acid (JA), mitogen-activated protein kinase signaling, calcium ion influx, callose deposition at the plasmodesmata, membrane permeability modification, pathogenesis-related (PR) protein expression, and immediate accumulation of reactive oxygen species and nitric oxide.

iv. **Arbuscular Mycorrhizal Fungi (AMF) Biocontrol:** A number of studies lay emphasis on the biocontrol abilities of Arbuscular mycorrhizal fungi (AMF) as they have been shown to reduce the incidence of fungal diseases and nematode attacks on host plants by 30 to 42% and 44–57%, respectively (Pandit et al., 2022). The biocontrol properties of AMF are broad-spectrum and more pronounced against fungal root pathogens in comparison to the shoot ones. AMF offers defense against a number of fungal pathogens belonging to the genera *Colletotrichum*, *Alternaria*, *Erysiphe*, *Gaeumannomyces*, *Macrophomina*, *Botrytis*, *Rhizoctonia*, *Fusarium*, *Cylindrocladium*, *Sclerotium*, and *Verticillium*. On the other hand, they do not offer much protection against a large number of bacterial and viral pathogens but some bacteria like *Pseudomonas syringae pv. glycinia* that causes bacterial blight on soybean can be checked by AMF. In the case of viral pathogens, the presence of mycorrhizal fungi seems to increase the damage caused by viral infections as seen with Tomato spotted wilt virus (TSWV), Potato virus Y, Citrus tristeza virus and Citrus leaf rugose virus and Tobacco mosaic virus. Therefore the role of AMF against viral pathogens is largely unclear and mostly points towards a supportive influence resulting in intensified disease rigor (Maffei et al., 2014). Moreover, reduced colonization and spore formation is shown by the AMF when the host plant is infected with a viral pathogen like the yellow mosaic virus.

v. **Biocontrol Yeast**: Yeasts such as *Aureobasidium pullulans*, *Cryptococcus albidus*, *Candida oleophila*, *Saccharomyces cerevisiae*, and *Metschnikowia fructicola* are currently being employed as biocontrol agents as they are effective adversaries of various plant pathogens (Pandit et al., 2022). Yeasts are a category of unicellular fungi that grow in most environments, have simple culture needs and few if any biosafety concerns. They apply competition, volatiles, enzymes and toxins, mycoparasitism, and initiation of immune response mechanisms for plant protection. Due to these properties, they can be exploited as biocontrol effectors but a paucity of studies on their role limits their full utilization. Yeasts are known to exert their biocontrol activity through Phage based competition, enzyme secretion, toxin production, volatiles, mycoparasitism, induction of resistance activity. Ferraz et al. (2019) have extensively listed the success cases of using yeasts to antagonize the spoilage of fruits by filamentous fungi.

Thambugala et al. (2020) have prepared an exhaustive list of some commercialized fungal biocontrol agents for plant fungal diseases and their specifications, which is suggested for further reading. Some important yeast species, such as *Candida oleophila*, *Aureobasidium pullulans*, *Metschnikowia fructicola*, and many others are reported to have been registered as biocontrol agents and have been suggested to having potential for being utilized as commercial biocontrol agents. Di Canito et al. (2021) suggests that Saccharomyces and non-Saccharomyces yeasts as potential antagonists against phytopathogenic fungi of the genera *Penicillium* and *Aspergillus* and the species *Botrytis cinerea* on table grapes, wine grapes, and raisins. They suggest that several non-conventional species are largely unexplored till date in both basic research and for their possible utilization in commercialization. They further say that this group constitutes a huge, untapped reservoir of yeasts having potential for biotechnological innovations constituting selection of species and strains with new metabolic traits, such as the secretion of proteins, adhesiveness, antimicrobial properties, etc. that are required for yeasts to manifest their applications as biocontrol agents. Application of yeasts in prevention of infections represents a new strategic frontier for maintaining the post-harvest quality of table and wine grapes. The genomes of several non-conventional yeast species have been completely sequenced and their number is growing continuously Wend land. Thus, expected novel methods for the genetic analysis and their further modifications in yeasts, as well as their genomic and post-genomic analysis before and after such modifications, will represent a platform for understanding the molecular mechanisms underlying both the simple and complex biological features that are supposed to be useful for the development of new and eco-compatible applications.

vi. **Phage-based Biocontrol:** Phages have been in use as biocontrol agents against bacterial pathogens for a long time. The earliest study demonstrating their biocontrol ability was done by Mallmann and Hemstreet in 1924 in which they isolated *Xanthomonas campestris* pv. *campestris* from plant tissues suffering from the cabbage-rot disease (Pandit et al., 2022). Future studies showed that phages could inhibit soft-rot caused by *Pectobacterium carotovorum* subsp. *carotovorum* in carrots, a bacterial spot of tomato by *X. campestris* pv. *vesicatoria* and *Pectobacterium atrosepticum* in potato slices. More recent explorations into phage biocontrol usage have focussed on improving their durability under field conditions. Exploring the use of phage cocktails and systemic acquired resistance activator in disease management against *X. citri subsp. citri* and *Xanthomonas axonopodis* pv. *citrumelo* that causes citrus bacterial canker and citrus bacterial spot respectively showed positive results in field trials (Ibrahim et al., 2017). On the other hand, some studies showed a better disease management response in laboratory-based bioassays rather than in field trials, like in the case of phage treatment against *Pseudomonas syringae pv. porri* that causes bacterial blight of leek. However various economically significant bacterial pathogens like *Xanthomonas* spp. and *Pseudomonas syringae* can be effectively controlled by phages. Peptidoglycan hydrolases, lysins from phages Atu_ph02 and Atu_ph03 are capable of blocking cell division in *Agrobacterium tumefaciens* (causes crown gall disease) resulting in its lysis. Other lysins from CMP1 and CN77 phages have also shown lytic capacity against *Clavibacter michiganensis* subsp. michiganensis, that causes bacterial wilt and canker of tomato. The incorporation of phage lysins into transgenic crops can aid in their easy application and overcome production issues. Mostly, the application of phages and phage lysins in plant disease management is a progressive step and has shown positive outcomes in a number of instances. Focus now needs to be on developing better delivery methods and guaranteeing a longer shelf life for the phage and its enzymes on the host plant (Dy et al., 2018).

vii. **Natural compounds against plant diseases:** Bioactive natural compounds can be of plant or animal origin and are capable of controlling plant diseases thereby promoting plant growth (Pandit et al., 2022). A number of bioactive molecules belonging to phenolic, terpenoid, or alkaloid categories such as chitin, laminarin, allicin, terpenes, chitosan, naringin, and carrageenans have been identified for use as biopesticides in organic cultivation. Allicin, acquired from garlic exhibits antibacterial and antifungal properties under field conditions, garlic juice inhibits the growth of a number of bacteria belonging to the genus *Pseudomonas*, *Agrobacterium*, *Xanthomonas* and *Erwinia* and fungi *Cercospora arachidicola*, *Botrytis cinerea*, *Rhizoctonia solani*, *Alternaria alternata*, *Fusarium moniliforme*, *Colletotrichum coccodes*. Naringin (40,5,7-trihydroxyflavanone-7-β-D-α-l-rhamnosyl(1-2)-β-D-glucoside) is another potent bioactive molecule found in seeds and pulp of grapefruit that displays effectiveness against fusariosis, alternariosis, and gray mold infections in soybean, ornamental plants, and vegetables such as potato. Tea tree oil (*Melaleuca alternifolia* L.) contains terpenes like terpinen-4-ol, gamma-terpinene, 1,8-cineole and exhibits strong antimicrobial properties against a variety of bacteria and fungi. It is particularly effective against *Bremia lactucae* and downy mildew that attack lettuce (Yu et al., 2015).

At times the use of bioactive compounds like garlic pulp is more beneficial than synthetic compounds like azoxystrobin as seen in the case of sweet pepper plants. Chitin which is the second most abundant polysaccharide in nature and a component of the fungal cell wall and exoskeleton of crustaceans and insects shows bioactivity against a number of bacterial, viral, and fungal pathogens. It is known to have a strong antifungal influence against soil-borne pathogenic fungi that infect soybeanand is a fungal microbe-associated molecular pattern (MAMP) molecule that is able to activate immune responses in the host plant]. It can be isolated using enzymatic reactions and chitosan distillation. Bioactive compounds, therefore, show a variety of modes of action not only to limit pathogen growth and multiplication but also inactivation of the host defense response. They usually act via binding to the membrane receptors on plants and produce a signal that is capable of initiating an immune response.

viii. **Algal and Cyanobacterial Biocontrol:** Apart from being an abundant source of vitamins, saccharides, enzymes, amino acids, phytohormones and elements like molybdenum, boron, manganese, iron, iodine, and zinc, algae and cyanobacteria extracts are a rich source of bioactive elicitors with antifungal, antiviral and antibacterial properties (Pandit et al., 2022). These extracts are usually applied in agriculture to improve productivity and plant vitality. Use of extracts from the algae *Sargassum filipendula*, *Ulva lactuca*, *Caulerpa sertularioides*, *Padina gymnospora* and *Sargassum liebmannii* ease symptoms of fungal infection on tomato produced by *Alternaria solani* and *Xanthomonas campestris* pv. *vesicatoria*. Studies on tomato seedlings infected by *Macrophomina phaseolina* showed improvement after the application of *Kappaphycus alvarezii*. The algal action was propagated through improved levels of phytohormones (salicylic acid, indole-3-acetic acid and abscisic acid), transcription of PR-1b1, PR-3, and PR-4 genes, and the cytokinin zeatin. The activity of polyphenol oxidase and peroxidase enzymes important in plant defense in tomatoes was also shown to improve when extracts from *Cystoseira myriophylloides*, *Laminaria digitata* and *Fucus spiralis* were utilized against *Verticillium dahliae* wilt.

Cyanobacteria have been applied against plant pathogens both at the levels of soil and leaves. Employing *Nostoc entophytum* and *Nostoc muscorum* in the soil against *Rhizoctonia solani* greatly enhanced seedling endurance along with improving root and shoot dry weight and plant length. In tomato, application of *Nostoc linckia* in soil against *Fusarium oxysporum* f. sp. *Lycopersici* decreased wilt while an improved state of similarly infected seedlings of tomato was observed with *Nostoc commune*. Usage of cyanobacteria, *Anabaena* sp. on zucchini cotyledons infected with powdery mildew (*Podosphaera xanthii*) resulted in enhanced enzymatic activity of peroxidases, endochitinase, chitin 1,4-β-chitotriosidase, β-N-acetylhexosaminidase, and β-1,3-glucanase. Similar enzymatic activation was observed by Prasanna et al., (2015) upon employing a biofilm composed by *Anabaena* sp. on maize roots and shoots. Cyanobacteria, like algae are also capable of high polysaccharide production in response to various categories of plant pathogens but there is a lack of data which limits their use as biocontrol agents (Righini et al., 2019).

8.6.1 Emerging Biocontrol Strategies

i. **Microbial Volatilome and its role in the biological control**: One of the most resilient and encouraging solutions in biocontrol approaches is the employment of microorganisms as biological control agents (BCAs) (Pandit et al., 2022). Among the several microbiological strategies used by BCAs, the production of volatile organic compounds (VOCs) is a method that is helpful in situations where the straightforward association between the pathogen and its competitor is not possible. All living forms synthesize VOCs and these can be exploited for usage in biocontrol of plant pathogens like bacteria, oomycetes, and fungi. VOCs are a sustainable preference for synthetic fungicides due to their ease of application, low residue deposition in the environment and on crops, and their biocontrol efficacy (Tilocca et al., 2020). According to Tahir and colleagues, VOCs produced by *Bacillus* species are known to function at a number of levels against the tobacco wilt agent *Ralstonia solanacearum*. In vitro studies indicate that *Bacillus* volatile compounds reduced Ralstonia growth and viability and caused significant problems in cell integrity and motility in addition to considerable alterations in *Ralstonia* genes expression that controls disease progression. Furthermore, tobacco plants treated with *Bacillus* emissions and purified detected VOCs elevated transcription levels in critical defense-related genes such NPR1 and EDS1, leading to inhibition of systemic resistance. It's possible that bacterial volatiles has a role in *Bacillus* reported biocontrol properties both directly and indirectly, and that bacterial VOC bouquets function as multifactorial, sequential, or simultaneous signals on pathogens and hosts.

ii. **Microbiome-based solutions for plant protection:** New findings demonstrate a remarkable microbial diversity among all plants, as well as unique phytopathogen antagonistic bacteria. Mosses, which are the world's oldest land plants, exhibit a unique microbial diversity, and their ecology allows them to contain a large number of enemies (Pandit et al., 2022). Apart from mosses, medicinal and endemic plants are also likely sources of rare biodiversity and enemies. A characteristic acquired by them due to their unique metabolism that alters the architecture of the plant microbiome. We expect endophytes, particularly seed endophytes, to serve as sources for novel biocontrol agents as a result of new discoveries. Until now, bacteria and fungi have been used mostly for biocontrol. Archaea have just recently been recognized as part of the plant microbiome; their effects on plants and potential for biocontrol are unknown. Microbial invasion can affect the network of microorganisms that are linked with plants. These network models of soil and plant microbiomes can be interpreted for biocontrol and present new prospects for disease management. While single organisms were commonly utilized in the past, their effects were often uneven, microbiome-based biocontrol techniques are now possible (Berg et al., 2017).

    In the future, microbial consortia and biocontrol agents can be employed to improve biodiversity associated with crops via microbiome engineering so as to achieve definitive microbiome outcomes as desired. Crop-specific biological consortia can be assembled from a pool of selected biocontrol agents in this setting. Taking a holistic approach and incorporating microbiome-based solutions allows for targeted and predictive biocontrol measures. Furthermore, integrated breeding and biocontrol measures are essential to sustain ecosystem variety and health. These systemic techniques are necessary to prevent further biodiversity losses and promote sustainable agriculture operations.

iii. **Phage Cocktail:** Phage cocktails are a feasible option for controlling a variety of plant diseases; however, further study and solutions to technical obstacles are needed to achieve successful biocontrol. Thorough knowledge of interactions between plant, phage, and pathogen is required since the habitat of each plant system is unique and complex (Pandit et al., 2022). This can only be achieved by conducting more extensive field experiments, as in vitro and in vivo tests under laboratory conditions do not accurately reflect the real circumstances in the field. More advanced protective formulations are needed to ensure the survival of phage mixtures during long-term storage under ambient conditions. The use of already existing phages from the phyllosphere can provide better protection against the phytopathogens in that environment. To improve phage persistence in the phyllosphere, light-absorbing compounds and/or protective formulations could be added to phages that have evolved to resist UV-induced damage. Synthetic phage cocktails with customized host ranges can also be created using genetically engineered phages. More research is needed in order to obtain well-characterized phages with defined and configurable host ranges. Finally, due to the great diversity of phytobacteria, a single universal phage cocktail for all diseases is not viable. Designing tests that can identify the disease-causing bacteria and its antagonistic phage can greatly aid in its control. To date, no such simple and economical option is either available or implemented.

iv. **Genetically Modified biocontrol agents:** To improve the efficacy of BCAs, techniques for genetic engineering of all organisms can be used. *Rhizoctonia solani* infection in beans can be brought under control by transferring a gene coding for the enzyme chitinase from *Serratia* to a *Pseudomonas* endophyte (Pandit et al., 2022). While transferring a gene for glucanase to *Trichoderma* produced resistance to pathogens such as *Rhizoctonia, Rhizopus*, and *Pythium*. Cloning of 2,4-diacetylphloroglucinol (2,4-DAPG) biosynthetic locus phlACBDE from strain CPF-10 into a mini-Tn5 transposon by Zhou et al. (2009) into a mini-Tn5 transposon and its insertion into the chromosome of *Pseudomonas fluorescens* P32 improved resistance of wheat to *Gaeumannomyces graminis var. tritici* and tomato to *Ralstonia solanacearum* bacterial wilt. Regardless of the findings of this research, these newly produced BCAs are subject to the same restrictions that apply to organisms that have been genetically changed using recombinant DNA technology. Clermont et al. (2011 employed genome shuffling to create superior *Streptomyces melanosporofaciens* EF76 biocontrol strains. Four strains with improved antagonistic activity against the potato diseases *Streptomyces scabies* and *Phytophthora infestans* were isolated after two rounds of genome shuffling. Biological control ability can also be improved by employing chemical mutagenesis. Examples include the use of nitrosoguanidine mutagenesis in *Pseudomonas aurantiaca* B-162 to produce a strain with better phenazine synthesis leading to improved biocontrol activity and *Trichoderma harzianum* strains that exhibited enhanced biocontrol ability after UV mutagenesis. The addition of the required mutation can at times produce altered gene expression in non-targeted genes resulting in undesired effects. These constraints can be overcome using more recently established genome editing approaches. We can insert mutations into specific regions in the genome with high precision and efficiency using techniques like Crispr/Cas. Another benefit is that mutations can be induced in numerous genes at the same time, which will aid in determining the role of different genes in biocontrol. The gene editing approach could also help in commercialization of BCA's through ease of regulatory clearances.

v. **Microbiome Engineering:** Many researchers suggest the microbiome to be representative of a "second genome", but some prefer the term "holobiont" to describe the variety of microbes linked with plant and animal hosts (Pandit et al., 2022). Microbiome engineering has the potential to have a big impact on agriculture. As a result, the creation of an altered microbiome with the desired properties is required. Many recent investigations have revealed that certain endophytic strains can alter the structure and species richness of plant tissues. Very few studies have investigated the internal microbiome of plants for subsequent generations post introduction of a specific strain(s). Moreover, little research has been carried out on the importance of manipulated microbiomes from disease-suppressive soils on control of phytopathogens. From a practical standpoint, it would be immensely beneficial to establish microbiomes that are durable and stress-tolerant thereby capable of increasing agricultural output. Finally, plant microbiome bioengineering is an intriguing option for improving a plant's biological capabilities, an approach that, while still in its infancy, has the potential to be of immense agricultural value (del Carmen et al., 2018).

vi. **Mycoviruses as biocontrol agents:** Recently, mycoviruses having capability to infect fungal pathogens are known to have the potential to be used as biological control agents against plant diseases. The mycoviruses were recognized to induce hypovirulence (reduced virulence) in their hosts and this notion elicited great interest in characterization of viruses from phytopathogenic fungi as being utilized as biocontrol agents (Pandit et al., 2022). Following this information, scores of mycoviruses started to be worked upon and Garcia-Pedrajas et al. (2019) further reports that majority of viruses from filamentous fungi possess either double-stranded RNA (dsRNA) genomes or positive sense (+) single-stranded RNA (ssRNA) genomes with dsRNA replicative intermediates, and can possess a capsid forming true virions or be sometimes capsidless. Surprisingly, numerous plant pathogenic fungi are found to harbor mycoviruses that reduce the virulence of their fungal host. Lacking extracellular transmission route to other isolate, mycoviruses are transmitted primarily through hyphal anastomosis or via conidia in vertical transmission while giving rise to progenies also. Transmission efficiencies are dependent on both the fungal host and the infecting virus and hence it was possible to utilize artificial transfection methods to infect a variety of fungi, thus expanding their possible use to the control of pathogens other than those where they were identified. Hypovirulence-associated mycoviruses are those mycoviruses that reduce the pathogenicity of their inhabiting fungal hosts. However, it is difficult to transmit these mycoviruses easily between vegetatively incompatible groups and hence it has been difficult to develop commercial mycovirus biocontrol strategies for these phytopathogenic fungi.

8.6.2 How to improve the efficacy of microbial biocontrol agents?: Besides the specific indication given above for each group of biocontrol agents there are few ways to increase the efficacy of these tools. In the first instance wrong or bad uses/application of such products should be avoided (Ilaria Pertot et al., 2020). Most of the time, failure is due to the scarce knowledge on the way or conditions of application which results in a wrong use. Secondly we should distinguish failures caused by the application of low quality bio-inoculum. Often, inefficacy is due to the fact that fertilizers or compost containing biocontrol agents are used instead of registered plant protection products. This bad practice generated a lot of confusion around biocontrol agents.

The specific way to increase efficacy of these products includes:

- Use in combination with agronomic practices in an IPM strategy: in this case the biocontrol agents are applied not just as substitution of a chemical soil fumigant but in combination of agronomic practices, for example after solarisation, anaerobic soil disinfestation or biofumigation.
- Verify the efficacy of the strain against the specific disease and the right conditions of application;
- Verify the compatibility between the strain and the crop or cultivar, in particular for rhizosphere colonizers;
- Give sufficient time to the biocontrol agent to reduce the inoculum of the disease;
- Apply under the right conditions;
- Calibrate the quantity per unit of soil to a level that can guarantee efficacy (often, the quantity per hectare is limited by the price of the product);
- Develop carriers or specific substrates to increase their efficacy;
- Improve production systems to reduce the cost of biocontrol agents;
- Develop multi-year strategies to control pathogens with biocontrol agents, especially for those supposed to reduce soil-borne inoculum of pathogens; Apply only when the inoculum is not too high;
- Identify more effective strains (most of the commercial strains have been isolated more than 20 years ago, most of the time without a targeted selection and wide comparison among strains).

8.6.3 The 'Grey zone' and Research Priorities: There are several commercial products that are sold as fertilizers or biostimulants which contains biocontrol agents. These products usually do not claim a direct fungicide efficacy however they play with the concept that they may improve plant health and contain the most common species used in biocontrol (Ilaria Pertot et al., 2020). This 'grey zone' of product creates confusion not only for farmers, but also for some experts. These products do not have to declare the CFUs neither guarantee the viability of the microorganisms. Most of the time they fail to give any efficacy and thus contribute to give a bad image of the microbial biocontrol agents. Although for many strains the distinction between plant protection products and biofertilisers is rather difficult (they possess both activities) a more strict regulation is needed to guarantee the quality of microorganisms contained in the biofertilisers and to avoid misuse of biofertilisers (as plant protection products). Today there are only few genera, species and strains of BCAs [*Coniothyrium minitans* (1 strain), *Gliocladium catenulatum* (1 strain), *Pseudomonas chlororaphis* and sp. (2 strains), *Streptomyces griseovirides,S. lydicus* and *Trichoderma asperellum, T. atroviride* and *T. harzianum* – several strains) that are registered against some soil borne diseases on certain crops.

Regarding nematodes, *Bacillus firmus*, and *Purpureocillium lilacinum* are the major BCAs approved for use against nematodes, and also, not in any crop. Even if we take the whole list some of these products will be never registered in some countries, so, finally the availability in different European countries of microbial biocontrol products to be used against soil-borne diseases and nematodes is practically very limited for the growers. In addition in order to improve microbiological control it would be useful to associate several strains of the same genus-specie, or several genus-specie and it would be useful also to associate several modes of action. But from the point of view of regulation, that would mean to register any single strain with unsustainable costs for companies (8-10 or more successful efficacy trials per crop/disease in each zone). Due to this contrain, biological control of soil-borne plant diseases in Europe is still very limited; as stated above there are only a few products registered in Europe and with possibilities to be available to the growers.

Research Priorities: The microbial biocontrol agents can target directly the problem and solve it when other agronomic practices are insufficient or cannot be applied (Ilaria Pertot et al., 2020). For example crops rotation, which is an excellent agronomic solution, is useless in case of polyphagous pathogens that can grow on several crops or when farms are highly specialised as in horticulture or viticulture (vineyards can be planted on specific areas, i.e. ADC, DOC, terroirs, etc.). Another example is when technologies as solarisation, biofumigation or anaerobic soil disinfestation cannot be applied because of one or more compulsory factors are missing (enough temperature, water, time,....). The identified needs from practice can be summarised as follows.

- Isolate a new generation of biocontrol agents (several among the registered ones have been isolated more than 20 years ago, when this field of research was at its infancy) with higher efficacy, good technological characteristics (high productivity in fermenters), good shelf-life and possibility to store at room temperature, high compatibility with other means of control (i.e. survival during solarisation, biofumigation conditions, with low dosages of pesticides, etc.).
- Identification of the best protocols to apply biocontrol agent against soil-borne disease (timing, quantities, integration with other means of control, etc.);
- Implementation of IPM strategies which include the use of microbial biocontrol agents in integration with other means of control;
- Development of decision support systems to guide the right application of the available biocontrol agents;
- Enlarge the knowledge on the efficacy of the microbial biocontrol agents against other soilborne pathogens than the targeted in the label and possible use with carriers which can increase survival in soil, in particular to prove environmental safety;
- Optimize and reduce the cost of production by improving the technologies fermentation or use of low cost substrates;

- Support and push industries to register the microbial products that fit with 'low risk substances' against pathogens especially for the use in minor crops (either speed-up the process of registration of low risk substances or subsidies to farmers choosing low risk substances).
- Improving microbiological control by integrating several strains of the same genusspecie, or several genus-specie

Ideas and priorities that could be interesting for operational groups: The following ideas can be interesting for operational groups (Ilaria Pertot et al., 2020).

- Development of an integrated strategy (protocols integrating microbial biocontrol agents, choice of crops/varieties and agronomic practices) to reduce the problems related to replanting perennial crops (soil-borne pathogens including replanting disease syndrome), when crop rotation is not a feasible option.
- Development of protocols to reduce the inoculum of soil-borne disease in horticulture, which include targeted use of biocontrol agents.
- Use of specific biocontrol agents in combination/sequence with anaerobic soil disinfestation, biofumigation or solarisation in order to improve the efficacy
- Combination of compost and other organic substrates or carriers to increase the efficacy of biocontrol agents
- Selection of compatible varieties to specific rhizosphere competent biocontrol agents in order to improve the resistance induction mechanism in hydroponic or soilless substrates
- Use of biocontrol agents to recycle/reuse substrates in greenhouse production (i.e. to 'clean' peat after the growing cycle and reuse it)

Improve the cost / benefits: One of the major limiting factors in the wide use of microbial biocontrol agents is their cost compared to the chemical fumigants. The high cost is due to the cost of production (fermentation), the logistics (they comply should be kept at low temperatures and their shelf life is very short 0.5-2 years) and the cost of registration (Ilaria Pertot et al., 2020). Possible solutions can be found in

- Improving the production systems and the logistics
- Make them more attractive to companies for example by speeding-up the entrance to the market for those which fit with the 'low risk substance' by speeding up the registration process.
- Subsidies for growers to make them more familiar with these products

The effects on the whole soil-health: Commonly these biocontrol agents have no effect or only minor and transient effect on the soil microflora and microfauna, because many of them are highly specific against pathogens, their persistence is limited, they are biodegradable and do not bioaccumulate (Ilaria Pertot et al., 2020). Therefore if they are implemented in a good agronomic strategy they can be a sustainable tool to reduce the soil inoculum potential and improve soil and plant health.

8.6.4 Biotechnological aspects in biocontrol of plant pathogens: Most sustainable and environmentally acceptable control may be achieved using biocontrol agents due to the effort to reduce the use of agrochemicals and their residues in the environment and in food. These BCA controlling plant diseases through various mechanism viz., hyperparasitism, predation, antibiosis, cross protection, competition for site and nutrient and induced resistance (Khokhar et al., 2012). Identifying, understanding and utilizing microorganisms or microbial products to control of plant diseases and to enhance crop production are integral parts of sustainable agriculture. Biological control has the potential to control crop diseases while causing no or minimal detrimental environmental impact. In this review, different aspects of biological control of fungal plant diseases including modes of action, application strategies, modern tools of microbial genetics or molecular techniques, which can use to improve the biocontrol activity, metabolites and products that could have important environmental benefits and outlooks, have been discussed.

Some research criteria will advance our understanding of biological control and the conditions under which it can be most fruitfully applied. Ecological factors play very important roles in the performance and activity of biocontrol-active microorganisms. Application strategies still there are some areas which should be investigated and developed for the enhancement of the effectiveness of biocontrol microorganisms. Introducing new strains and mechanisms of fungal plant pathogens are very diverse and their pathogenicity is different on host plants, it is therefore very important to look for new and novel biocontrol microorganisms with different mechanisms.

Application of Molecular Genetic to improve biocontrol agents

i. **Genetical improvement of bioagents:** Genetically improved antagonistic hyperparasitic microorganisms tend to increase their effectiveness as biological control agents. Effective performance requires either improvement of the environment to favour the biocontrol agent or genetic improvement of the agent. Genetic manipulation will be done to enhance antifungal metabolites productivity of biocontrol agents, to improve antagonistic potential of biocontrol agents, to control a broad spectrum of phytopathogens and to develop biocontrol agents tolerance of some stress conditions. Thus it can be achieved by chemical and physical mutation, sexual hybrids, homokaryons, and genetic manipulation e.g, directed mutagenesis, protoplast fusion, recombination, transformation or isolation of useful genes from biocontrol fungi without functional sexual stages.

ii. **Mutation as tool for the improvement:** Many authors were applied the mutagensis by either the physical or the chemical mutagens to generate new biotypes with improve the potentialities of biocontrol agents and / or antifungal metabolites producers. Induced mutation is one of the most commonly used routine to restrain the genetic construction of microorganisms (Khokhar et al., 2012).

iii. **Use of Protoplast Fusion for Genetic Improvement or Manipulation:** Protoplast fusion is a quick and easy method for combining the advantageous properties of distinct promising strains. Protoplast fusion techniques were used to combine genetic traits desirable for improving biocontrol activity by *T. harzianum* and increased amounts of specific proteins. Fusion of protoplast derived from two efficient biocontrol strains of *T. harzianum* resulted in the recovery of a progeny strain with greatly improved biocontrol ability. After that, directed transfer of fungal genetic sequences encoding for factors important in biocontrol will be possible as soon as such sequences are available.

iv **Directed Mutagenesis**: Many authors were applied the mutagensis by either the physical mutagens (i.e. gamma irradiation or UV-irradiation) or the chemical mutagens to generate new biotypes with improve the potentialities of biocontrol agents and / or antifungal metabolites producers (Khokhar et al., 2012). Mutation technique was used by UV light and improved the production of important *P. fluorescens* of antibiotics, phenazine, pyrrolnitrin and phloroglucinol as well as siderophore pigment production against some tomato damping-off pathogens (*Fusarium solani, Fusarium oxysporum f.sp. lycopersici and Rhizoctonia solani*). These mutant strains increased the antibiosis and the fluorescens on King's medium comparing to the wild-type. A mitogen-activated protein kinase encoding gene, tvk1, from *Trichoderma virens* was cloned, and its role during the mycoparasitism, conidiation, and biocontrol was examined in tvk1 null mutants. These mutants showed, a clear increase in the level of the expression of mycoparasitism-related genes during direct confrontation with the plant pathogen *Rhizoctonia solani.* The null mutants displayed an increased protein secretion phenotype as measured by the production of lytic enzymes in culture supernatant compared to the wild-type. Consistently, biocontrol assays demonstrated that the null mutants were considerably more effective in disease control than the wildtype strain or a chemical fungicide. In addition, tvk1 gene disruptant strains sporulated abundantly in submerged cultures, a condition that is not conducive to sporulation in the wild type. These data suggest that Tvk1 acts as a negative modulator during host sensing and sporulation in *T. virens.*

v. **Genetic modification and Transformation:** A genetic modification approach was used to further enhance the biocontrol ability of biocontrol agents. Genetically modified (GM) was used in strains, *P. fluorescens* F113Rif (pCU8.3) and *P. fluorescens* F113Rif (pCUP9), were developed for enhanced phenazine-1-carboxylate (Phl) production and assessed for biocontrol efficacy and impact on sugar beet in microcosm experiments. *Pseudomonas aeruginosa* PNA1, isolated from the rhizosphere of chickpeas in India, mutant (FM13) deficient in phenazine production were obtained following transposon mutagenesis of PNA1. Anthranilate, an intermediate in the tryptophan biosynthesis pathway, suppressed mycelial growth of *Pythium* spp. in culture and damping-off of *P. vulgaris* and lettuces. It is concluded that anthranilate, excreted by FM13 as a consequence of the trpC mutation, may have contributed to the suppression of Pythium damping-off by the mutant.

8.6.5 Challenges and Opportunities in biocontrol: The growing concerns about the overuse of synthetic chemical pesticides and their residues, increased significance of insect pests and pathogens due to increased food demand, the withdrawal of several chemical pesticides, including soil fumigants, the appearance of new invasive species, and pesticide-resistant strains of pests, climate change, and specialized monoculture are all factors have contributed to the expansion of the biological control domain of plant protection under the sustainable agriculture goal (Jaiswal et al., 2022). However, bio-control agents (BCAs) have advantages over traditional crop protection (CPs) methods but are not yet ready to take their place. In many cases, the adaptability of BCA in a non-native environment is poor. Further, their efficacy against multiple pathogens/insect pests is also low. As a result, it hasn't been widely used.

Another major challenge is the lack of adequate characterization of bio-agents coupled with the poor marketing strategy of bioagent-producing firms; for example, many PGPRs/biofertilizers projected as a biocontrol agent; low efficacy in non-native soils/environment work on adaptability, and its contains is lacking appropriate research on the efficacy, growth promoting activities; lacking response of bio-agents at the physiological and molecular level, and poor of characterization and product formulation of bio-agents. The mild disease/pest suppression by these bio-agents may probably be due to their growth-promoting effect on plants. Therefore, this product confuses end users because they expect anti-pest activity from these products. As a result, they are often ineffective, which helps diminish microbial biocontrol agents' reputation. Even though many strains blur the distinction between plant protection products and biopesticide/biofertilizers, strict regulation is required to ensure the efficacy of biopesticide microorganisms and prevent their misuse as plant protection products. Presently, only a few genera, species, and strains of BCAs (*Coniothyrium minitans*, *Gliocladium catenulatum*, *Pseudomonas chlororaphis* and spp., *Streptomyces griseovirides* and *Streptomyces lydicus*, and *Trichoderma asperellum*, *T. atroviride*, and *T. harzianum*) are registered against some soil-borne pathogens. Similarly, *Bacillus firmus* and *Purpureocillium lilacinum* are the only BCAs approved for use against nematodes.

Experts have encountered great difficulties in developing various BCA products, in addition to the cost and scalability challenges of BCA. Many alternative solutions, such as those based on fermentation or pheromones, are prohibitively expensive to manufacture, providing customers with a little financial incentive to switch away from using known BCA-containing products. As a result, several companies are looking into novel ways to reduce production costs. Due to the low barriers to entry and high market attractiveness in this domain, hundreds of companies, ranging from major CP firms to many mid-tier firms, engage in the BCA and bio-stimulants industries. Many new businesses emerge due to the influx of venture capital, but they frequently lack the funds to register their company, develop their products, and enter the market (Shoham et al., 2020). To register a single strain for commercial usage, firms would have to conduct extensive, statistically significant efficacy trials for each crop/disease in each zone. This limitation has resulted in a dearth of products for the biological control of insect pests and diseases in Europe and Asia. As was previously stated, only a handful of these products have been approved for use by European growers.

Opportunities: Biological control is a cost-effective, eco-friendly, and long-term solution for crop protection against biotic stresses. Progressive farmers increasingly use the conservation and management of endangered species of biocontrol microorganisms, among other biologicals, to combat plant diseases (Jaiswal et al., 2022). The most successful approach to biological management for conservation objectives is to concentrate on the most critical aspects of natural enemy ecology. There are two strategies to boost natural enemy effectiveness, changing the habitat so that natural enemies benefit at the expense of pests or decreasing the detrimental effects of pesticides on natural enemies. Furthermore, the significance of biological control conservation in developing countries has been emphasized. Numerous microorganisms have been shown to be effective against soil-borne diseases and nematodes over the last 50 years. Among these are the active ingredients in at least one biopesticide that is already on the market. Even though several of these strains were developed a few years ago, none have achieved widespread commercial success due to competition from synthetic chemical fumigants, which are often more cost-effective, easier to apply, have a wider spectrum of activity, and are highly effective. Since the ban on methyl bromide and other chemicals, there has been a revived interest in microbial biocontrol agents against soil-borne diseases. These agents operate best in conjunction with other agronomic practices or resistant/tolerant plant varieties. The mechanism of action of microbial biocontrol agents against plant pathogens includes direct antibiosis, hyper-parasitism, resistance induction, and competition for space and nutrients.

In addition, researchers are investigating the role of non-pathogenic beneficial rhizobacteria in increasing plant resistance to pathogens, a process known as induced systemic resistance (ISR). Plant pathogen infection can result in systemic acquired resistance (SAR). Some microorganisms (such as *Bacillus* spp., *Pseudomonas* spp., *Acinetobacter calcoaceticus*; *Azotobacter* spp., *Azospirillum* spp., *Mesorhizobium*, *Bradyrhizobium*, *Burkholderia*) act as bio-stimulators by producing indole-acetic acid, nitrogen fixation, P-solubilizing, siderophore, HCN production, 1-aminocyclopropane-1-carboxylate (ACC) deaminase, degrading organic matter to improving the plant growths & yields, controlling disease & pest and maintaining soil health's. Furthermore, soil and plant microbiomes can act as inoculants, aid in nutrient absorption, biocontrol products, help protect plants from pests and diseases, or both. Some soil amendments may be required to ensure beneficial microbes' survival. Perhaps "probiotics" can be identified to maintain plant microbiomes healthy (Harutyunyan et al., 2022).

Research priorities in exploring the biocontrol agents: Basic biological research, particularly in taxonomy, ecology, and behavior, has tremendously aided procedures employed in the exploration, selection, and risk evaluation of biological control agents (Jaiswal et al., 2022). However, some questions remain unanswered in the field of biocontrol, such as the lacking efficacy in profiling plant-associated microbial bio-controlling agents, the lack of an overall understanding of a pathogen's biology, and the epidemiology of the resulting disease, which hinder the development of disease and pest management strategies. Therefore, scientists and researchers must keep the following goals in mind as they investigate emerging issues in the field in order to establish an eco-friendly bio-control strategy.

- Exploration of a new generation of biocontrol agents with higher efficacy, high productivity in fermenters, long shelf life and the ability to be stored at room temperature, and high compatibility with other control methods
- Standardizing of identification of BCA protocol against soil-borne disease & pest
- Population genetics research presents opportunities to better understand how the impact of biological control can be optimized.
- Improving microbiological control by integrating several strains of the same genus specie, or several genus-specie
- Increase our understanding of microbial biocontrol agents' potential against other soilborne pathogens beyond those listed on labels, as well as their potential use with carriers that can increase survival in soil, in order to demonstrate their environmental safety.
- Exploring cutting-edge genomic tools like CRISPR genome editing can reduce fewer desirable traits in biological control agents and insert new desirable characteristics such as insecticide resistance.
- Implementation of IPM strategies which include the use of microbial biocontrol agents with other management strategies
- Optimize and reduce the cost of production of BCA by improving the technologies of fermentation or use of low-cost carrier substrates for BCA
- To protect food and ornamental crops from pathogens, it is important to encourage and assist businesses in registering microbial products that meet the criteria of 'low-risk substances' either by expediting the registration of low-risk substances or by providing subsidies to farmers who choose low-risk substances).

Bibliography

Aamir, M., Rai, K.K., Zehra, A., Dubey, M.K., Kumar, S., Shukla, V., Upadhyay, R.S.,

Adrien Anckaert, Anthony Arguelles Arias, Grégory Hoff, Maryline Calonne-Salmon, Stéphane Declerck and Marc Ongena, 2022. The use of *Bacillus* spp. as bacterial biocontrol agents to control plant diseases. Chapter taken from: Köhl, J. and Ravensberg, W. (ed.), Microbial bioprotectants for plant disease management, Burleigh Dodds Science Publishing, Cambridge, UK, 2022, (ISBN: 978 1 78676 813 1)

Alblooshi, A.A., Purayil, G.P., Saeed, E.E., Ramadan, G.A., Tariq, S., Altaee, A.S., El-Tarabily, K.A. and AbuQamar, S.F., 2022. Biocontrol potential of endophytic actinobacteria against *Fusarium solani*, the causal agent of sudden decline syndrome on date palm in the UAE. *J. Fungi,* **8:** 8

Anonymous, 2020. Biopesticides Registration Action Document *Coniothyrium minitans* strain CON/M/91-08 (PC Code 028836) U.S. Environmental Protection Agency, Office of Pesticide Programs , Biopesticides and Pollution Prevention Division *Coniothyrium minitans* strain CON/M/91-08 (PC Code 028836).

Azmi, N. S. A., Singkaravanit-Ogawa, S., Ikeda, K., Kitakura, S., Inoue, Y., Narusaka, Y., et al., 2018. Inappropriate expression of an NLP effector in *Colletotrichum orbiculare* impairs infection on cucurbitaceae cultivars via plant recognition of the C-Terminal region. *Mol. Plant Microbe Interact.*, **31:** 101–111

Basit, H.A., Angle, J.S., Salem, S. and Gewaily, E.M., 1992. Phage coating of soybean seed reduces nodulation by indigenous soil bradyrhizobia. *Canadian J Microbiol.,* **38:**1264–1269

Bastas, K.K. and Baysal, Ö., 2022. Microbial battling of fire blight disease on pome fruits. Microbial biocontrol: Food security and post harvest management, Vol 2. Springer, pp 211–226

Belen Guijarro, Inmaculada Larena, Carla Casals, Neus Teixidó, Paloma Melgarejo and Antonieta De Cal, 2019. Compatibility interactions between the biocontrol agent *Penicillium frequentans* Pf909 and other existing strategies to brown rot control. *Biol. Contr.*, **129**: 45-54

Belur Satyan Kumudini, Sunita Mahadik, Amrisha Srivastava and Savita Veeranagouda Patil, 2023. Chapter 14- Plant-microbe interactions: different perspectives in promoting plant growth and health. *In: Microbiome Research in Plants and Soil,Rhizobiome*. Javid A. Parray, Nowsheen Shameem, Dilfuza Egamberdieva, R.Z. Sayyed (Eds.,). Academic Press, P 309-328.

Berg, G., Köberl, M., Rybakova, D., Müller, H., Grosch, R. and Smalla, K., 2017. Plant microbial diversity is suggested as the key to future biocontrol and health trends. *FEMS Microbiol. Ecol.*,**93**: 5

Berg, G., Rybakova, D., Fischer, D., Cernava, T., Vergès, M.C.C., Charles, T., Chen, X., Cocolin, L., Eversole, K., Corral, G.H., et al., 2020. Microbiome definition re-visited: Old concepts and new challenges. *Microbiome,* **8**: 103.

Bertinetti, B., Rodriguez, M., Godeas, A. *et al.,* 2010. 1H,1′H-[3,3′]biindolyl from the terrestrial fungus *Gliocladium catenulatum*. *J. Antibiot.,* **63:** 681–683

Campo S., Martín-Cardoso H., Olivé M., Pla E., Catala-Forner M., Martínez-Eixarch M., et al., Rice, 2020; **13**: 42. DOI 10.1186/ s12284-020-00402-7.

Carlos Henrique Barbosa Santos, Luana Alves De Andrade, Edvan Teciano Frezarin, Luziane Ramos Sales and Everlon Cid Cid Rigobelo, 2023. *Purpureocillium lilacinum* for Biocontrol, Bioremediation and Biofertilization. Preprints (www.preprints.org). doi: 10.20944/preprints202305.1926.v1

Castillo, H., Rojas, R. and Villalta, M., 2015. *Gliocladium* sp., agente biocontrolador con aplicaciones prometedoras. Tecnología en Marcha. *Edición Especial Biocontrol.,* 65-73.

Chen, W., Modi, D. and Picot, A., 2023. Soil and Phytomicrobiome for Plant Disease Suppression and Management under Climate Change: A Review. *Plants,* **12**:2736

Chevallereau, A., Pons, B.J., van Houte, S. and Westra, E.R., 2022. Interactions between bacterial and phage communities in natural environments. *Nat. Rev. Microbiol.,* **20**:49–62

Clermont, N., Lerat, S. and Beaulieu, C., 2011. Genome shuffling enhances biocontrol abilities of *Streptomyces* strains against two potato pathogens. *J. Appl. Microbiol.,* **111:**671–682.

Collinge, D.B., Jensen, D.F. Rabiey, M., Sarrocco, S., Shaw, M.W. and Shaw, R.H., 2022. Biological control of plant diseases-what has been achieved and what is the direction? *Plant Pathol.,* **71:**1024–1047

Collins, D.P. and Jacobsen, B.J., 2003. Optimizing a *Bacillus subtilis* isolate for biological control of sugar beet *Cercospora* leaf spot. *Biol. Contr.,* **26:**153–161.

Dar, M.H. and Reshi, Z.A.. 2017. *Russ. Agric Sci.*, **43:** 138–143. DOI 10.3103/ s1068367417020057.

Darya Zargar, Jahanshir amini and Jafar abdollahzadeh, 2019. Biocontrol potential of endophytic *Penicillium* spp. against strawberry anthracnose. *Biol. Contr. Pests & Pl. Dis.*, **8:** 47-58

del Carmen, M., del Carmen Rocha-Granados, M. and Glick, G., 2018. Microbiome engineering to improve biocontrol and plant growth-promoting mechanisms. *Microbiol. Res.,* **208:** 25–31.

del Carmen, M., Rodríguez, H., Evans, H.C., de Abreu, L.M., de Macedo, D.MNdacnou,., M.K., Bekele, K.B. and Barreto, R.W., 2021. New species and records of *Trichoderma* isolated as mycoparasites and endophytes from cultivated and wild coffee in Africa. *Sci. Rep.*, Article 5671, 10.1038/s41598-021-84111-1

Di Canito, A., Mateo-Vargas, M.A. and Mazzieri, M., 2021. The Role of Yeasts as Biocontrol Agents for Pathogenic Fungi on Postharvest Grapes: A Review. *Foods,* **10**:1650

Divya, K.S., Mahadeva Murthy, S. and Sudisha, 2021. Ecological Studies of Fungal Biodiversity in Freshwater and Their Broad-Spectrum Applications. J. *Biocontr. Agents & Sec. Metabolites,* Woodhead Publishing; Cambridge, UK, pp. 631–648.

Dy, R.L., Rigano, L.A. and Fineran, P.C., 2018. Phage-based biocontrol strategies and their application in agriculture and aquaculture. *Biochem. Soc. Trans.,* 46:1605–1613

Fang, J. G. and Tsao, P. H.,1995. Efficacy of *Penicillium funiculosum* as a Biological control agent against *Phytophthora* Root Rots of Azalea and Citrus. *Phytopathology* **85**:871-878.

FAO, Conservation Agriculture. Available online: https://www.fao.org/conservation-agriculture/en/ (accessed on 10 January 2023).

Ferraz, P., Fernanda, C. and Cândida, L., 2019. Potential of Yeasts as Biocontrol Agents of the Phytopathogen Causing Cacao Witches Broom Disease: Is Microbial Warfare a Solution? *Front. Microbiol.,* doi: 10.3389/fmicb.2019.01766.

García-Pedrajas, M.D., Cañizares, M.C., Sarmiento-Villamil, J.L., Jacquat, A.G. and Dambolena, J.S., 2019. Mycoviruses in Biological Control: From Basic Research to Field Implementation. *Phytopathology,* **109**:1828–1839

Goicoechea, N., 2020. *Plants*, **9**: 1468. DOI 10.3390/plants9111468

Halawa, E.M., 2023. Challenges of bacteriophages application in controlling bacterial plant diseases and how to overcome them. *J. Genet. Eng. Biotechnol.,* **21**: 98.

Harutyunyan, N., Kushugulova, A., Hovhannisyan, N. and Pepoyan, A., 2022. One health probiotics as biocontrol agents: one health tomato probiotics. *Plant,* **11**:1334.

Hashem, A., Tabassum, B. and Fathi Abd Allah, E., 2019. *Bacillus subtilis*: A plant-growth promoting rhizobacterium that also impacts biotic stress. *Saudi J. Biol. Sci.,* **26**:1291-1297

Hassine, M., Aydi-Ben-Abdallah, R., Jabnoun-Khireddine, H. *et al.,* 2022. Soil-borne and compost-borne *Penicillium* sp. and *Gliocladium* spp. as potential microbial biocontrol agents for the suppression of anthracnose-induced decay on tomato fruits. *Egypt. J. Biol. Pest Control*, **32:** 20.

Hatfull, G.F., Dedrick, R.M. and Schooley, R.T., 2022. Phage therapy for antibiotic-resistant bacterial infections. *Ann. Rev. Med.*, **73**: 197–211

Hoda Ahmed, Caroline Strub, Frederique Hilaire and Sabine Schorr-Galindo, 2015. First report: *Penicillium adametzioides,* a potential biocontrol agent for ochratoxin-producing fungus in grapes, resulting from natural product pre-harvest treatment. *Food Control*, **51**: 23-30

Huizhang Zhao, Ting Zhou, Jiatao Xie, Jiasen Cheng, Tao Chen, Daohong Jiang and Yanping Fu, 2020. Mycoparasitism illuminated by genome and transcriptome sequencing of *Coniothyrium minitans*, an important biocontrol fungus of the plant pathogen *Sclerotinia sclerotiorum. Microbial Genomics,* **6**: https://doi.org/10.1099/mgen.0.000345

Ibrahim, Y.E., Saleh, A.A. and Al-Saleh, M.A., 2017. Management of Asiatic citrus canker under field conditions in Saudi Arabia using bacteriophages and acibenzolar-S-methyl. *Plant Dis.,* **101**:761–765

Ilaria Pertot, Claude Alabouvet e, Estefanía Hinarejos Esteve and Soraya Franca, 2020. Focus Group Soil-Borne Diseases Mini-paper - The use of microbial biocontrol agents against soil-borne diseases. *eip-agri.* Funded by European Commission.

Jaiswal, D.K., Gawande, S.J., Soumia, P.S. *et al.,* 2022. Biocontrol strategies: an eco-smart tool for integrated pest and diseases management. *BMC Microbiol,* **22:** 324.

Jamil Shafi, Hui Tian and Mingshan Ji, 2017. *Bacillus* species as versatile weapons for plant pathogens: A review. *Biotechnol. & Biotechnological* Equip., **31**:446-459

Kasun, M., Dinushani, A., Alam, J.L., Phillips, S., Kannangara, D. and Itthayakorn, P., 2020. Fungi Vs. Fungi in Biocontrol: An Overview of Fungal Antagonists Applied against Fungal Plant Pathogens. *Front. Cell. Infect. Microbiol.,*

Kateřina Bělonožníková, Veronika Hýsková, Josef Chmelík, Daniel Kavan, Noemi Čeřovská and Helena Ryšlavá, 2022. *Pythium oligandrum* in plant protection and growth promotion: Secretion of hydrolytic enzymes, elicitors and tryptamine as auxin precursor. *Microbiol. Res.,* **258:** 126976

Khan, S., Srivastava, S., Karnwal, A. and Malik, T., 2023. *Streptomyces* as a promising biological control agents for plant pathogens. *Front. Microbiol.,* **14:**1285543.

Khokhar, M.K., Gupta, R. and Sharma, R., 2012. Biological Control of Plant Pathogens using Biotechnological Aspects:- A Review. **1**:277.

Kloepper, J.W., Ryu, C.M. and Zhang, S., 2004. *Phytopathology,* **94**:1259-1266

Larrahondo-Rodríguez, E., Liao, Y.Y. and Huerta, A.I., 2022. Diagnostic guide for bacterial spot of tomato and pepper. *Plant Health Progr.,* **23**:355–361

Le, K.D., Yu, N.H., Park, A.R., Park, D.J., Kim, C.J. and Kim, J.C., 2022. *Streptomyces* sp. AN090126 as a Biocontrol Agent against Bacterial and Fungal Plant Diseases. *Microorganisms,* **10:** 791.

Lee, C., Kim, H. and Ryu, S., 2022. Bacteriophage and endolysin engineering for biocontrol of food pathogens/pathogens in the food: recent advances and future trends. *Crit. Rev. Food Sci. Nutr.,* **54:**1–20

Lerat, S., Lapointe, L., Piché, Y. and Vierheilig, H., 2003. *Can. J. Bot*., **81**: 886–889.

Liang, L.J., Jeewon, R., Dhandevi, P., Durairajan, S.S.K., Li, H., Lin, F.C. and Wang, H.K., 2021. A Novel Species of *Penicillium* With Inhibitory Effects against *Pyricularia oryzae* and Fungal Pathogens Inducing Citrus Diseases. *Front Cell Infect Microbiol.,* **10**:604504

López-Quintero, C.A., Atanasova, L., Franco-Molano, A.E., Gams, W., Komon-Zelazowska, M., Theelen, B., Müller, W.H., Boekhout, T., Druzhinina, I., 2013. DNA barcoding survey of *Trichoderma* diversity in soil and litter of the Colombian lowland Amazonian rainforest reveals *Trichoderma strigosellum* sp. *nov.* and other species. *Antonie Van. Leeuwenhoek,* **104:** 657-674

Łoś, J., Zielińska, S., Krajewska, A., Michalina, Z., Małachowska, A., Kwaśnicka, K. and Łoś, M.,2021. Temperate phages, prophages, and lysogeny. *Bacteriophages Biol. Technol. Ther.,* **8**:119–150

Maffei, G., Miozzi, L., Fiorilli, V., Novero, M., Lanfranco, L. and Accotto, G.P., 2014. The arbuscular mycorrhizal symbiosis attenuates symptom severity and reduces virus concentration in tomato infected by Tomato yellow leaf curl Sardinia virus (TYLCSV). *Mycorrhiza,* **24:**179–186.

Majdinasab, M., Daneshi, M. and Marty, J.L., 2021. Recent developments in nonenzymatic (bio) sensors for detection of pesticide residues: Focusing on antibody, aptamer and molecularly imprinted polymer. *Talanta*, **232**:122397

Maria Villarino, Antonieta De Cal, Paloma Melgarejo, Inmaculada Larena and Eduardo A. Espeso, 2016. The development of genetic and molecular markers to register and commercialize *Penicillium rubens* (formerly *Penicillium oxalicum*) strain 212 as a biocontrol agent. *Micro. Biotech.,* **9**: 89-99

Masip, G., Sabalza, M., Pérez-Massot, E., Banakar, R., Cebrian, D., Twyman, R.M. and Christou, P., 2013. Paradoxical EU agricultural policies on genetically engineered crops. *Tre. Plant Sci.,* **18:** 312-324

Mateus Torres Nazari, Vera Analise Schommer, Julia Catiane Arenhart Braun, Lara Franco dos Santos, Samuel Teixeira Lopes, Viviane Simon, Bruna Strieder Machado, Valdecir Ferrari, Luciane Maria Colla, and Jeferson Steffanello Piccin, 2023. Using *Streptomyces* spp. as plant growth promoters and biocontrol agents. *Rhizosphere*, **27:** 100741

Mcquilken, M., Gemmell and Lahdenperä, 2008. *Gliocladium catenulatum* as a Potential Biological Control Agent of Damping□off in Bedding Plants. *J.Phytopathol.*, **149:** 171 - 178.

Meile, S., Du, J., Dunne, M., Kilcher, S. and Loessner, M.J., 2022. Engineering therapeutic phages for enhanced antibacterial efcacy. *Curr. Opin. Virol.*, **52:**182–191

Milica Mihajlović, Emil Rekanović, Jovana Hrustić, Mila Grahovac and Brankica Tanović, 2017. Methods for management of soil borne plant pathogens. *Pestic. Phytomed. (Belgrade),* **32:** 9–24

Monica Höfte, 2022. The use of Pseudomonas spp. as bacterial biocontrol agents to control plant diseases. Ghent University, Belgium. Chapter taken from: Köhl, J. and Ravensberg, W. (ed.), Microbial bioprotectants for plant disease management, Burleigh Dodds Science Publishing, Cambridge, UK, 2022,

Morales Moreira, Z.P., Chen, M.Y., Yanez Ortuno, D.L. and Haney, C.H., 2023. Engineering plant microbiomes by integrating eco-evolutionary principles into current strategies. *Curr. Opin. Plant Biol.,* **71:** 102316

Muzio, F. M., Agaras, B. C., Masi, M., Tuzi, A., Evidente, A. and Valverde, C., 2020. 7-hydroxytropolone is the main metabolite responsible for the fungal antagonism of Pseudomonas donghuensis strain SVBP6. *Environ. Microbiol.,* **22:** 2550–2563.

Nair, B., Abraham, D., Saranya, K. and Mallikarjunaswamy, G.E., 2020. Bio-control potentiality of *Penicillium multicolor* Grig.-Man. and Porad., against important root pathogens. *Curr.Res.in Environ. & Appl.Mycol.,***10:** 514–523,

Nakao, S., Watanabe, H., Yano, T., Yamaoka, Y. and Ishii, H., 2021. Control efficacy of the systemic acquired resistance (SAR) inducer acibenzolar-S-methyl against *Venturia nashicola* in Japanese pear orchards. *J. Gen. Plant Pathol.,* **87**:307–315

Narender Kumar, Sumit Kumar, Rajendra Singh and Mohan Lal, 2019. Integrated disease management is a holistic approach in modern agriculture. *J. Pharmacognosy & Phytochem.*, **SP4:** 98-101

Nawaz, A., Zafar, S., Shahzadi, M. *et al.,* 2023. Bacteriophages: an overview of the control strategies against phytopathogens. *Egypt J. Biol. Pest Control,* **33**: 108

Neshev, G., 2008. Major soil-borne phytopathogens on tomato and cucumber in Bulgaria, and methods for their management. *In: Alternatives to replace methyl bromide for soil-borne pest control in East and Central Europe*. Labrada, R. (Ed.), P 1-14. FAO, UNEP

Nisa, R.T., Bhat, K.A., Basu, U., Rashid, R., Wang, X., Jan, S.K., Viswanath, H.S., Soufan, W., Mushtaq, M and Rahman, M.H., et al., 2021. Identification and Characterization of Triple Action Bioagents (TAB) and Their Potency against Fusarium Wilt of Lentil. *Horticulturae,* **7**:587.

Olumayowa Mary Olowe, Lidia Nicola, Michael Dare Asemoloye, Akinlolu Olalekan Akanmu and Olubukola Oluranti Babalola, 2022. *Trichoderma*: Potential bio-resource for the management of tomato root rot diseases in Africa. *Microbiol. Res.*, **257:** 126978,

Omoboye, O. O., Oni, F. E., Batool, H., Yimer, H. Z., De Mot, R. and Höfte, M.,2019. *Pseudomonas* cyclic lipopeptides suppress the rice blast fungus *Magnaporthe oryzae* by induced resistance and direct antagonism. *Front. Plant Sci.,* **10:** 901

Pandit, M.A., Kumar, J., Gulati, S., Bhandari, N., Mehta, P., Katyal, R., Rawat, C.D., Mishra, V. and Kaur, J., 2022. Major Biological Control Strategies for Plant Pathogens. *Pathogens,* **11:** 273.

Pechinger, K., Chooi, K.M. MacDiarmid, R.M., Harper, S.J. and Ziebell, H., 2019. A New Era for Mild Strain Cross-Protection. *Viruses,* **11**:670.

Pocurull, M. *et al.,* 2020. Commercial formulates of *Trichoderma* induce systemic plant resistance to *Meloidogyne incognita* in tomato and the effect is additive to that of the Mi-1.2 resistance gene. *Front. Microbiol.,* p. 10 https://doi.org/10.3389/fmicb.2019.03042

Poveda, J., 2021. Biological control of *Fusarium oxysporum* f. sp. *ciceri* and *Ascochyta rabiei* infecting protected geographical indication Fuentesaúco-Chickpea by *Trichoderma* species. *Eur. J. Plant Pathol.,* **160:** 825–840

Prasad, P., Varshney, D. and Adholeya Alok, 2022. Whole genome annotation and comparative genomic analyses of bio-control fungus *Purpureocillium lilacinum*. Deakin University. Journal contribution. https://hdl.handle.net/10779/DRO/DU:20632071.v1

Prasanna, R., Bidyarani, N., Babu, S., Hossain, F., Shivay, Y.S. and Nain, L., 2015. Cyanobacterial inoculation elicits plant defense response and enhanced Zn mobilization in maize hybrids. *Cogent. Food Agric.,* **1:**998507

Qu, Z., Fu, Y., Lin, Y., Zhao, Z., Zhang, X., Cheng, J., Xie, J., Chen, T., Li, B. and Jiang, D., 2021. Transcriptional Responses of *Sclerotinia sclerotiorum* to the Infection by SsHADV-1. *J. Fungi,* **7:** 493

Ram, R.M., Rajput, R.S. and Vaishnav, A., 2020. Management of *Sclerotium rolfsii* Induced Diseases in Crops by *Trichoderma* Species. *Rhizosphere Microbes*, *Springer,* 593-617

Righini, H., Baraldi, E., García Fernández, Y., Martel Quintana, A. and Roberti, R., 2019. Different antifungal activity of *Anabaena* sp., *Ecklonia* sp., and *Jania* sp. against *Botrytis cinerea*. *Mar. Drugs,* **17**:299

Roossinck, M.J., 2020. Persistent plant viruses: Molecular hitchhikers or epigenetic elements? *In*: *Viruses: Essential Agents of Life.* Witzany G., (Ed.). Springer; Dordrecht, The Netherlands, pp. 177–186.

Ruangwong, O.U., Pornsuriya, C., Pitija, K. and Sunpapao, A., 2021. Biocontrol mechanisms of *Trichoderma koningiopsis* PSU3-2 against postharvest anthracnose of chili pepper. *J. Fungi*, **7**:276

Ruwanthika Kalamulla, Samantha C. Karunarathna, Suhail Asad, Saowaluck Tibpromma, Steven L. Stephenson, Itthayakorn Promputtha and Neelamanie Yapa.Biocontrol, 2022. Potential of Arbuscular Mycorrhizal Fungi on Soil-borne Pathogens in Rice and Other Crops. *Chiang Mai J. Sci.*, **49:** 730-741.

Ryan, P. R., Delhaize, E. and Jones, D. L., 2001. Function and mechanism of organic anionexudation from plant roots. *Ann. Rev. Plant Physiol. Plant Mol. Biol.,* **52**:527-560

Saleh, M.M. and El-Akshar, Y.S., 2020. Integration Between Arbuscular Mycorrhizal Fungi, Bacterial and Fungal Bioagents for Controlling Rice Brown Spot Disease. *Egypt. J. Phytopathol.*, **48:** 81–93 .

Schouteden, N., De Waele D., Panis, B. and Vos, C.M., 2015. Arbuscular Mycorrhizal Fungi for the Biocontrol of Plant-Parasitic Nematodes: A Review of the Mechanisms Involved. *Front. Microbiol.*, **6:** 1280.

Shoham, J., 2020. The rise of biological products in the crop protection and plant nutrition markets. *Outlooks Pest Manag.,* **31**:129–31.

Sikandar, A., Zhang, M. Y., Zhu, X. F., Wang, Y. Y., Ahmed, M., Iqbal, M. F., Javeed, A., Xuan, Y. H., Fan, H. Y., Liu, X. Y., Chen, L. J. and Duan, Y. X., 2019. Efficacy of *Penicillium chrysogenum* strain snef1216 against root-knot nematodes (*Meloidogyne incognita*) in cucumber (*Cucumis sativus* L.) under greenhouse conditions. *Appl. Ecol. & Environ. Res.,* **17**:12451-12464

Sivagnanapazham, K., Karthikeyan, M., Raguchander, T., Swarna Priya, R. and Kamalakannan, A., 2022. Biological control of cabbage head rot (*Sclerotinia sclerotiorum*) by *Coniothyrium minitans*. *Egyptian J. Biol. Pest Contr.*, **32**:132.

Soenens, A., 2019. Imperial J. Biocontrol capabilities of the genus Serratia. *Phytochem. Rev.,* **19**: 577–587.

Sood, M., Kapoor, D., Kumar, V., Sheteiwy, M.S., Ramakrishnan, M., Landi, M. and Sharma, A., 2020. *Trichoderma*: The "secrets" of a multitalented biocontrol agent. *Plants,* **9**: 762

Sreevidya, M. and Gopalakrishnan, S., 2016. *Penicillium citrinum VFI-51 as biocontrol agent to control charcoal rot of sorghum (Sorghum bicolor (L.) Moench). African J. Microbiol. Res.,* **10**: 669-674

Stanley, N.R., Britton, R.A., Grossman, A.D. and Lazazzera, B.A., 2003. Identification of catabolite repression as a physiological regulator of biofilm formation by *Bacillus subtilis* by use of DNA microarrays. *J. Bacteriol.,* **185**:1951–1957.

Subirats, J., Sànchez-Melsió, A., Borrego, C.M., Balcázar, J.L. and Simonet, P., 2016. Metagenomic Analysis Reveals That Bacteriophages Are Reservoirs of Antibiotic Resistance Genes. *Int. J. Antimicrob. Agents,* **48:**163–167.

Tian, L., Chang, C., Ma, L., Nasir, F., Zhang, J., Li, W., et al., 2019. Comparative study of the mycorrhizal root transcriptomes of wild and cultivated rice in response to the pathogen *Magnaporthe oryzae. Rice*, **12**: 35.

Tilocca, B., Cao, A. and Migheli, Q., 2020. Scent of a killer: Microbial volatilome and its role in the biological control of plant pathogens. *Front. Microbiol.,* **11**:41.

Topolovec-Pintaric, S., 2019. *Trichoderma:* Invisible partner for visible impact on agriculture. https://doi.org/10.5772/intechopen.83363.

Umrao, P.D., Kumar, V. and Kaistha, S.D., 2021. Biocontrol potential of bacteriophage ϕsp1 against bacterial wilt-causing *Ralstonia solanacearum* in Solanaceae crops. *Egypt. J. Biol. Pest Control*, **31**:1–12

Vallance, J., Le Floch, G., Déniel, F., Barbier, G., Lévesque, C.A. and Rey, P., 2009. Influence of *Pythium oligandrum* biocontrol on fungal and oomycete population dynamics in the rhizosphere. *Appl. Environ. Microbiol.*, **75**: 4790-800.

Villalpando-Aguilar, J.L., Matos-Pech, G., López-Rosas, I., Castelán-Sánchez, H.G. and Alatorre-Cobos, F., 2022. Phage Therapy for Crops: Concepts, Experimental and Bioinformatics Approaches to Direct Its Application. *Int. J. Mol. Sci.,* **24**:325.

Wang, C., Wang, X., Jin, Z., Müller, C., Pugh, T.A., Chen, A., Wang, T., Huang, L., Zhang, Y., Li, L.X. and Piao, S., 2022. Occurrence of crop pests and diseases has largely increased in China since 1970. *Nat. Food,* **3**:57–65

Whipps, J. M., Lewis, K. And Cooke, R. C., 1988. Mycoparasitism and plant disease control. *Fungi in Biological Control Systems*, 161-187.

Woo, S.L., Donzelli, B., Scala, F., Mach, R., Harman, G.E., Kubicek, C.P., Sorbo, G.D., Lorito, M., 1999. Disruption f the *ech42* (endochitinase encoding) gene affects biocontrol activity in *Trichoderma harzianum* P1. *Am. Phytopathol. Soc.,* **12**: 219-229

Xu, X.M., Jeffries, P., Pautasso, M. and Jeger, M. J., 2011. Combined use of biocontrol agents to manage plant diseases in theory and practice. *Phytopathology,* **101**:1024-1031.

Yang, K., Chen, C., Wang, Y., Li, J., Dong, X., Cheng, Y., Zhang, H., Zhai, Y., Ai, G., Song, Q., Wang, B., Liu, W., Yin, Z., Peng, H., Shen, D., Fang, S., Dou, D. and Jing, M., 2022. Nep1-Like Proteins from the Biocontrol agent *Pythium oligandrum* enhance Plant Disease Resistance Independent of Cell Death and Reactive Oxygen Species. *Front. Plant Sci.,* **13:**830636.

Yu, D., Wang, J., Shao, X., Xu, F. and Wang, H., 2015. Antifungal modes of action of tea tree oil and its two characteristic components against *Botrytis cinerea. J. Appl. Microbiol.,* **119**:1253–1262.

Zerihun Tsegaye, et al., 2018. Concept, Principle and Application of Biological Control and their Role in Sustainable Plant Diseases Management Strategies. *Intel. J.Res. Studies in Biosci., (IJRSB*), **6** 18-34

Zhou, H.Y., Wei, H.L., Liu, X.L., Wang, Y., Zhang, L.Q. and Tang, W.H., 2009. Improving biocontrol activity of *Pseudomonas fluorescens* through chromosomal integration of 2,4-diacetylphloroglucinol biosynthesis genes. *Chin. Sci. Bull.,* **50**:775–781.

a significant change in the specificity of phage to different strains in a bacterial species.

